Communications in Computer and Information Science 2946

Rationale
The CCIS series is devoted to the publication of proceedings of computer science conferences. Its aim is to efficiently disseminate original research results in informatics in printed and electronic form. While the focus is on publication of peer-reviewed full papers presenting mature work, inclusion of reviewed short papers reporting on work in progress is welcome, too. Besides globally relevant meetings with internationally representative program committees guaranteeing a strict peer-reviewing and paper selection process, conferences run by societies or of high regional or national relevance are also considered for publication.

Topics
The topical scope of CCIS spans the entire spectrum of informatics ranging from foundational topics in the theory of computing to information and communications science and technology and a broad variety of interdisciplinary application fields.

Information for Volume Editors and Authors
Publication in CCIS is free of charge. No royalties are paid, however, we offer registered conference participants temporary free access to the online version of the conference proceedings on SpringerLink (http://link.springer.com) by means of an http referrer from the conference website and/or a number of complimentary printed copies, as specified in the official acceptance email of the event.

CCIS proceedings can be published in time for distribution at conferences or as post-proceedings, and delivered in the form of printed books and/or electronically as USBs and/or e-content licenses for accessing proceedings at SpringerLink. Furthermore, CCIS proceedings are included in the CCIS electronic book series hosted in the SpringerLink digital library at http://link.springer.com/bookseries/7899. Conferences publishing in CCIS are allowed to use our online conference service (Meteor) for managing the whole proceedings lifecycle (from submission and reviewing to preparing for publication) free of charge.

Publication process
The language of publication is exclusively English. Authors publishing in CCIS have to sign the Springer CCIS copyright transfer form, however, they are free to use their material published in CCIS for substantially changed, more elaborate subsequent publications elsewhere. For the preparation of the camera-ready papers/files, authors have to strictly adhere to the Springer CCIS Authors' Instructions and are strongly encouraged to use the CCIS LaTeX style files or templates.

Abstracting/Indexing
CCIS is abstracted/indexed in DBLP, Google Scholar, EI-Compendex, Mathematical Reviews, SCImago, Scopus. CCIS volumes are also submitted for the inclusion in ISI Proceedings.

How to start
To start the evaluation of your proposal for inclusion in the CCIS series, please send an e-mail to ccis@springer.com

Chuandong Li · Hongjing Liang · Jingang Lai ·
Qi Zhou · Bin Li · Kaibo Shi
Editors

Neuromorphic Computing

4th International Conference, ICNC 2025
Chengdu, China, December 12–14, 2025
Revised Selected Papers

Editors
Chuandong Li
Southwest University
Chongqing, China

Jingang Lai
Huazhong University of Science
and Technology
Wuhan, China

Bin Li
Sichuan University
Chengdu, China

Hongjing Liang
University of Electronic Science
and Technology of China
Chengdu, China

Qi Zhou
Southwest University
Chongqing, China

Kaibo Shi
Chengdu University
Chengdu, China

ISSN 1865-0929 ISSN 1865-0937 (electronic)
Communications in Computer and Information Science
ISBN 978-981-92-1598-0 ISBN 978-981-92-1599-7 (eBook)
https://doi.org/10.1007/978-981-92-1599-7

This Springer imprint is published by the registered company Springer Nature Singapore Pte Ltd.
The registered company address is: 152 Beach Road, #21-01/04 Gateway East, Singapore 189721, Singapore

Preface

The 4th International Conference on Neuromorphic Computing (ICNC 2025) was held at the Chengdu Wangjiang Hotel in Chengdu, China, from December 12 to 14, 2025. The conference was hosted by the University of Electronic Science and Technology of China and co-organized by Sichuan University and Southwest University.

ICNC 2025 aimed to establish a high-level international platform for academic exchange, promoting development and collaboration within the advanced interdisciplinary field of neuromorphic computing. The conference spotlighted recent breakthroughs in adaptive neural network control, brain-inspired algorithms, reinforcement learning, swarm intelligence, and artificial intelligence. The scientific program covered a wide range of hot topics, including autonomous unmanned systems, multi-agent consensus and formation control, event-triggered and nonlinear systems, neuromorphic hardware and memristive computing, intelligent edge computing, and smart transportation and IoT applications.

The papers included in this volume underwent a rigorous peer-review process. Each submission was evaluated through a single-blind review system, where every paper received an average of three independent reviews from Program Committee members or external experts. The review process ensured that only high-quality research contributing to the field's advancement was selected for publication.

For this edition, the Program Committee received a total of 134 submissions. After a thorough review phase, 52 papers were carefully selected for inclusion in these proceedings, representing an acceptance rate of approximately 40%. These papers provide deep insights into how neuromorphic technologies and advanced control strategies are enabling innovations in unmanned aerial vehicles, robotics, healthcare, and intelligent transportation systems.

We would like to express our sincere gratitude to the hosting and co-organizing universities for their support. We are also deeply indebted to the Program Committee members and the external reviewers for their hard work and dedication in maintaining the high technical standards of the conference. Finally, we thank all the authors and participants whose contributions made ICNC 2025 a successful and vibrant event.

December 2025

Jingang Lai
Bin Li
Chuandong Li
Hongjing Liang
Kaibo Shi
Qi Zhou

Organization

Program Committee Chairs

Chuandong Li	Southwest University, China
Hongjing Liang	University of Electronic Science and Technology of China, China

Program Committee

Jingang Lai	Huazhong University of Science and Technology, China
Qi Zhou	Southwest University, China
Bin Li	Sichuan University, China
Kaibo Shi	Chengdu University, China

Conference Honorary Chairs

Tingwen Huang	Shenzhen University of Advanced Technology, China
Xiaofeng Liao	Chongqing University, China
Yuhua Cheng	University of Electronic Science and Technology of China, China

Conference General Chairs

Zhigang Zeng	Huazhong University of Science and Technology, China
Tieshan Li	University of Electronic Science and Technology of China, China

Technical Program Committee Chairs

Jingang Lai	Huazhong University of Science and Technology, China
Chuandong Li	Southwest University, China
Hongjing Liang	University of Electronic Science and Technology of China, China
Qi Zhou	Southwest University, China
Bin Li	Sichuan University, China
Kaibo Shi	Chengdu University, China

Publication Chairs

Wenjun Xiong	University of Electronic Science and Technology of China, China
Yue Long	University of Electronic Science and Technology of China, China
Fei Teng	Dalian Maritime University, China
Wanbin Zhao	University of Electronic Science and Technology of China, China
Jin Yang	University of Electronic Science and Technology of China, China
Yuanyuan Xu	China University of Petroleum-Beijing at Karamay, China

Organizing Committee Chairs

Jingang Lai	Huazhong University of Science and Technology, China
Hang Geng	University of Electronic Science and Technology of China, China
Lu Yang	University of Electronic Science and Technology of China, China
Chun Yin	University of Electronic Science and Technology of China, China
Hui Ma	Guangdong University of Technology, China
Guangdeng Chen	Southwest University, China

Contents

QRCNN-BiGRU-Multi-attention for Prediction of Ship Traffic Flow in Compound Channels

Han Xue(✉) and Zifu Li

College of Navigation, Jimei University, Xiamen 361021, Fujian, China
imlmd@163.com, 200461000097@jmu.edu.cn

Abstract. In order to address the complex challenges in waterway traffic management in compound channels, a new model called Quantile Regression Convolutional Neural Network with Bidirectional Gated Recurrent Unit and Multi-attention (QRCNN-BiGRU-Multi-attention) is proposed for the prediction of ship traffic flow in compound channels. Quantile regression is utilized to estimate specific values at various quantiles and offer predictions across the complete range of quantiles. This method does not require assuming the parameter form of the distribution function in advance, thereby enabling interval prediction capabilities. Compared with other CNN-BiGRU models, this model has a simpler structure by serializing CNN and BiGRU, eliminating unnecessary network structures. Through a comprehensive analysis of the influencing factors in ship traffic flow for the compound channel, the main factors with significant impacts have been identified, such as the tidal height, wave height and hour time. Fast rejection test and crossing test are performed to judge whether a trajectory collected from automatic identification system (AIS) crosses a gate line of compound channels. The multivariate regression interval prediction is tested on a dataset of multivariate regression with 3 features. A 95% confidence interval is used. Results show that the proposed algorithm achieved the higher prediction accuracy than the traditional neural networks, such as BiGRU-attention, BiGRU, QRCNN-BiGRU, QRGRU and QRBiGRU..

Keywords: Convolutional Neural Network · Multi-attention · Ship traffic flow prediction · Quantile regression · Bidirectional Gated Recurrent Unit

1 Introduction

In modern times, there has been a rising number of large ships in navigable waters, resulting in a year-on-year increase in the instances of ship encountering distress [1]. Coupled with increasingly complex navigation conditions, the maritime safety situation is becoming more severe. Nevertheless, due to the rapid growth of marine traffic flow, marine traffic accidents occur frequently, resulting in economic losses.

A compound channel is established by adding a shallow waterway on each side of the existing main channel, facilitating the separation of high-grade large ships and small boats, and creating a four-sub-channel navigation system. The compound channel can

C. Li et al. (Eds.): ICNC 2025, CCIS 2946, pp. 1–10, 2026.
https://doi.org/10.1007/978-981-92-1599-7_1

enhance the efficiency and safety of artificial waterway passage through management and technological innovations [2, 3]. However, some challenging issues have emerged in terms of waterway traffic management. In particular, vessels encounter problems in the "Y"-shaped crossing area, leading to navigation confusion and vessel congestion. It increased difficulties in management and scheduling, prolonged vessel passage and berthing times, delayed vessel arrivals and departures, and a reduced capacity for waterway passage.

The reliability and accuracy of ship traffic flow predictions is crucial for ensuring timely supervision and effective monitoring of waterway safety, as well as the protection of life and property. Moreover, it can offer more precise and informed support as well as a foundation for policy formulation by maritime management departments and the planning of port operation systems [4–11]. Predicting ship traffic flow aids relevant authorities in formulating scientific management and planning for water channels. Various advanced artificial intelligence algorithms, such as neural networks, can be used for traffic flow prediction. For example, Chuah introduced a hybrid supervised Convolutional Neural Network (CNN) and Bidirectional Gated Recurrent Unit (BiGRU) [12], Kumar developed the CNN-BiGRU model [13], and Chopannejad discussed the attention-assisted hybrid CNN-BiGRU and Bidirectional Long Short-Term Memory (BiLSTM) [14]. Jai utilized CNN and BiGRU with a hyperbolic linear unit [15], Lalwani examined a multi-branched CNN-BiLSTM-BiGRU model [16].

However, navigation requires comprehensive consideration of various constraints, including tide model, wave conditions, and others. Additionally, real-time constraints of the model calculations also needs to be taken into account. These constraints change over time and possess a significant level of uncertainty in their temporal characteristics, making it challenging to create accurate models. In order to address the complex challenges in waterway traffic management in compound channels, a new model called Quantile Regression Convolutional Neural Network with Bidirectional Gated Recurrent Unit and Multi-attention (QRCNN-BiGRU-Multi-attention) is proposed for the prediction of ship traffic flow in compound channels. The contribution is summarized below:

(1) Unlike the previous studies [12–31], a novel model of QRCNN-BiGRU-Multi-attention method is proposed. Quantile regression is utilized to estimate specific values at various quantiles and offer predictions across the complete range of quantiles. This method does not require assuming the parameter form of the distribution function in advance, thereby enabling interval prediction capabilities. Compared with other CNN-BiGRU models, this model has a simpler structure by serializing CNN and BiGRU, eliminating unnecessary network structures.
(2) Ship traffic flow in compound channel are predicted based on the proposed algorithm, in order to address the complex challenges in waterway traffic management in compound channels.
(3) Through a comprehensive analysis of the influencing factors in ship traffic flow prediction for the compound channel, the main factors with significant impacts have been identified, such as the tidal height, wave height and hour time.

2 Related Work

2.1 CNN

CNN has excellent feature extraction capabilities. Convolutional layers and pooling layers are two important components, and typically includes fully connected layers [32]. The specific process of convolutional calculation is as follows [17]:

$$O[l,m,n] = \sum_{i}\sum_{j} I[l+i,m+j,n]K[i,j,n] + b_n = I \otimes K + b_n \quad (1)$$

where O, I, i, j, k, b represent the output, input, length, width, depth and threshold, respectively.

2.2 BiGRU

BiGRU essentially consists of a two-layer GRU network. In the forward GRU layer, features are input into the network through forward propagation during training to explore forward correlations in the data. In the backward GRU layer, the input sequence is trained through backward propagation to explore reverse correlations in the data. This network architecture allows for bidirectional feature extraction of the input, enhancing the integrity and global scope of features [18].

2.3 Quantile Regression

Denote explanatory variables $U = \{U_1, U_2, ..., U_c\}$ acting on the random variable S, the distribution function of S is [19]:

$$F(s) = P(S \leq s) \quad (2)$$

For a random variable s with cumulative distribution function $F(.)$, the τ-th quantile of its marginal distribution is [20]:

$$F^{-1}(\tau) = \inf\{s : F(s) \geq \tau\} \quad (3)$$

where $\tau \in [0,1]$. $F^{-1}(\tau)$ represents the τ-th quantile of s. inf (s) represents the infimum of s.

The τ-th conditional quantile $Q_s(\tau|U)$ of the response variable s given the explanatory variable U is defined as [21]:

$$Q_s(\tau|U) = \beta_0(\tau) + \sum_{i=1}^{c} \beta_i(\tau)U_i = U^{'}\beta(\tau) \quad (4)$$

where $\beta(\tau)$ represents the vector of regression coefficients at the τ quantile point.

The regression coefficient $\beta(\tau)$ varies at different quantile points, and also determines the regression model. $\beta(\tau)$ can be solve by minimizing the loss function [22]:

$$\min_{\beta}\sum_{i=1}^{N}\rho_{\tau}\left(S_i - U_i^{'}\beta\right) = \min_{\beta}\sum_{i|S_i \geq U_i^{'}\beta}\tau \mid S_i - U_i^{'}\beta \mid + \sum_{i|S_i < U_i^{'}\beta}(1-\tau) \mid S_i - U_i^{'}\beta \mid \tag{5}$$

where $\rho_{\tau}(\mu)$ is the test function, which is defined as below [23]:

$$\rho_{\tau}(\mu) = \begin{cases} \mu(\tau - 1), & \mu < 0 \\ \mu\tau, & \mu \geq 0 \end{cases} \tag{6}$$

2.4 Multi-attention

The attention mechanism adapts to different input situations by dynamically computing attention weights. This attention mechanism further improves upon traditional attention mechanisms by obtaining global information rather than just single context information, enabling a more comprehensive understanding of the semantic of the entire sequence and better capturing complex relationships between elements.

2.5 Ship Traffic Flow Statistics Based on Gate Line Analysis

The gate line analysis, also known as section analysis or cross-section analysis, can reflect the distribution of the ship trajectories on a certain cross-section. A gate line is considered as composed of multiple uniformly wide "sections". The key to gate-line analysis lies in the selection of gate-line position and the determination of the section length. The endpoints of the gate line are usually selected at with a relatively low density of traffic flow to ensure that the vast majority of the traffic flow crosses the gate line. In order to obtain the rules for the passage of ship traffic flow on the gate line, the length of the gate line is usually divided into equal parts, each part is called a section. The section length is predetermined. Except for special analysis needs, the number of parts should not exceed 50, and the section length should not be less than 50 m.

The principle of gate-line analysis is to count the number of ship trajectories passing through each section of the gate line within a specified time period [24].

For two line segments, if they intersect, two cases may occur: 1) Each line segment crosses the line on which the other line segment lies, i.e., the two endpoints are on opposite sides of that line. 2) The endpoint of one line segment lies exactly on the other line segment. The Quick Rejection Test and Crossing Test are used to determine whether two line segments intersect.

2.6 Compound Channel

To improve the efficiency of bidirectional navigation, and addressing issues such as small ships occupying navigation resources, the concept of compound-channel navigation has emerged. The compound-channel navigation involves separating large and small ships into different channels. Small ships refer to ships weighing tens of thousands of tons or less, with a length of no more than 146 m, a width of no more than 22 m, and meets the specified draft restrictions. The width of the small ships are about 100 m. The main navigation channel is reserved for large ships, while the north side is designated as the inbound channel for small ships and the south side is designated as the outbound channel for small ships. The central part of the compound-channel system is marked as the deep-water navigation channel. This division of channels for small and large ships helps alleviate navigation pressure on the main channel and improves the efficiency of ship inbound and outbound navigation. The goal of maintaining a traffic flow for two-way inbound and outbound ships in the compound-channel system is achieved.

Taking Tianjin Port as an example, the compound channel consists of a cautionary zone, a portion of the main navigation channel (from buoy no. 29 to buoy no. 39), and the small ship channels on the north and south sides.

3 Main Results

3.1 Structure of the Network

The multivariate regression interval prediction based on QRCNN-BiGRN is tested on a dataset of multivariate regression with 3 features. Traditional prediction is provided in the form of point prediction, which does not fully capture the uncertainty of the prediction. Quantile regression can directly estimate point values at different quantiles. Its advantage lies in providing prediction values across the entire range of quantiles without assuming the parameter form of the distribution function in advance. By combining quantile theory, the quantile-based CNN-BiGRU-Multi-attention interval prediction model can achieve predictions at different quantiles, thus enabling interval prediction functionality. Compared with other CNN-BiLSTM models, this model has a simpler structure by serializing CNN and BiLSTM, eliminating unnecessary network structures. A 95% confidence interval is used, and modifications are only needed at the network creation stage.

The network structure includes the sequence, conv, batchnorm, relu, maxpool, flatten, fc_1, gGru1, flip, gru2, concat, fc, and out.

3.2 Ship Traffic Flow Prediction

The factors affecting the number of ships passing through can be roughly divided into dynamic factors and static factors. Dynamic factors include time, vessel location, vessel speed, and traffic flow density. Static factors include vessel type, vessel length, vessel width, etc.

Through a comprehensive analysis of the influencing factors in ship traffic flow prediction, three main factors with significant impacts have been identified. Appropriate model parameters such as the tidal height, wave height and hour time were selected, and the dataset related to the identified influencing factors was used to model and predict ship traffic flow.

The process of ship traffic flow prediction based on CNN-BiGRU-Multi-attention is illustrated below.

Step 1: Import Data. Splitting the data into training and testing sets. Data Normalization.

Step 2: Initialize the neural network model. Connect all branches of the network and create a network diagram. Set network parameters.

Step 3: Input training samples.

Step 4: Train the network.

Step 5: Use different networks for prediction.

Step 6: Perform data de-normalization. Obtain the predicted mean.

Step 7: Performance evaluation. Calculate the relevant metrics, such as interval coverage rate and interval mean width percentage.

4 Experiment

4.1 Datasets

The compound channel in Tianjin Port, after a year of trial navigation, has benn officially put into operation. This is China's first artificial deep-water channel employing a dual-channel traffic organization method and the highest-grade channel. A ship navigating in the compound channel of Tianjin port on Nov 1, 2023 is listed in Table 1, including the maritime mobile service identification (MMSI), ship name, call sign, number of International Maritime Organization (IMO), ship type, length, beam and draught.

The gate line in the compound channel in Tianjin Port is: The two endpoints of the gate line is 38°56.55'N, 118°0.625′E and 38°55.837'N, 118°0.378′E. The two endpoints of the gate line for the deep water channel are 38°56.345'N, 118°0.553′E and 38°56.152'N, 118°0.488′E.

4.2 Results

The ship traffic flow across the gate line in November 2023 are listed in Table 1, together with the time of the first high tide at the harbor per day, the tide height and wave height.

The ship traffic flow across the gate line per hour on Nov. 1, 2023 is listed in Table 2.

The prediction results of training data sat is given in Fig. 1.

The prediction results of testing data set is shown in Fig. 2:

The error histogram of predictions is shown in Fig. 3.

The evaluation metrics are given in Table 3, including R-squared (R2), Mean Absolute Error (MAE), Mean Squared Error (MSE) and Prediction Interval Coverage Probability (PICP).

It is illustrated in Table 3 that the prediction accuracy of the proposed algorithm is high, with the PICP reaching 96% at a 95% confidence level, indicating a close match with the actual situation.

Table 1. Samples of ship traffic flow in November 2023 in Tianjin.

Observation Time	Tide Time	Tide Height/cm	Wave Height/m	Traffic flow
Nov. 1	04:58	387	0.7	75
Nov. 3	06:23	373	1.5	43
Nov. 4	07:09	363	0.6	128
Nov. 5	07:59	349	0.8	9
Nov. 6	08:59	337	1.5	38
Nov. 7	10:10	334	0.8	111
Nov. 8	11:20	343	1.0	2
Nov. 10	12:59	363	1.0	110

Table 2. Samples of ship traffic flow across the gate line on Nov. 1, 2023.

Hour	Tide Height/cm	Traffic flow
0:00–1:00	75	10
1:00–2:00	145	2
2:00–3:00	230	1
3:00–4:00	307	0
4:00–5:00	356	0
5:00–6:00	368	3

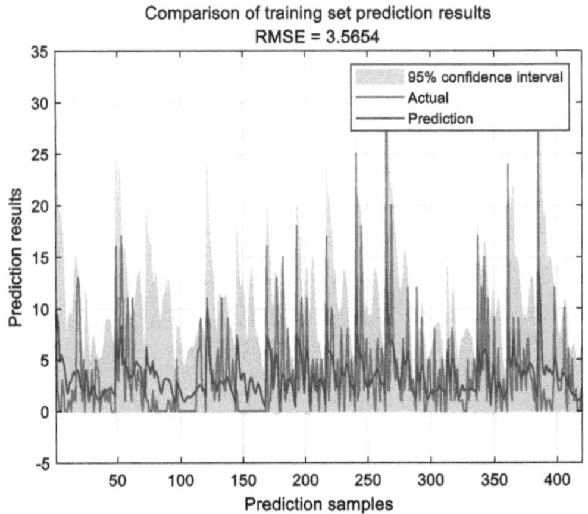

Fig. 1. The prediction results of training data set.

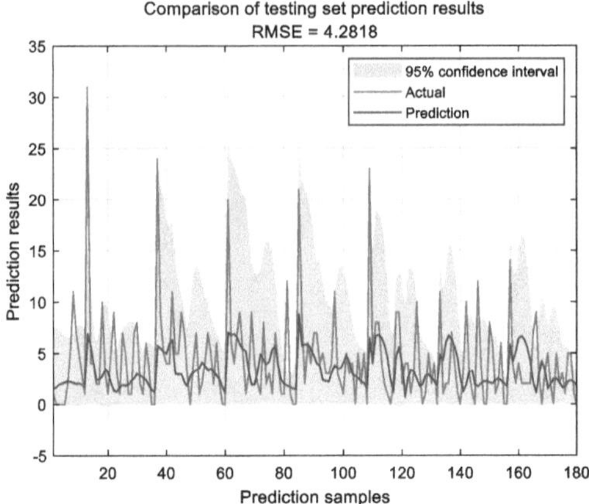

Fig. 2. Prediction results of the testing samples.

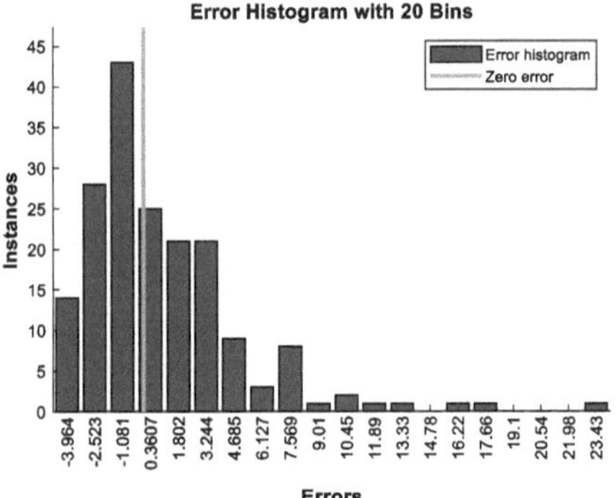

Fig. 3. Error histogram of predictions.

Table 3. Evaluation metrics.

Evaluation metrics	Training	Testing
R2	0.33064	0.11528
MAE	2.5382	2.8531
MSE	12.7121	18.3334
PICP	0.9619	0.8666

4.3 Comparison with Different Algorithms

The performance comparison with different algorithms is listed in Table 4.

It is illustrated in Table 4 that QRCNN-BiGRU-Multi-attention algorithm achieved the highest overall average prediction accuracy, which shows improvement over the

Table 4. Performance comparison with different algorithms.

Algorithms	R2	MAE	MSE
BiGRU-attention	0.19352	2.7825	14.9005
BiGRU	0.18208	3.2083	17.4318
QRCNN-BiGRU	0.34115	2.5805	12.7125
QRGRU	0.30071	2.9841	19.4888
QRBiGRU	0.13018	2.8566	16.5191
QRCNN-BiGRU-Multi-attention	0.11528	2.5382	12.7121

prediction accuracy of the traditional neural networks, such as BiGRU-attention, BiGRU, QRCNN-BiGRU, QRGRU and QRBiGRU.

5 Conclusion

In this work, the QRCNN-BiGRU-Multi-attention is proposed for ship traffic flow in the compound channels. Results show that the proposed algorithm achieved the higher prediction accuracy than the traditional neural networks, such as BiGRU-attention, BiGRU, QRCNN-BiGRU, QRGRU and QRBiGRU. The research results of this study can provide reference for relevant departments to improve the traffic flow organization model of compound waterways.

The prediction accuracy of neural network will be further improved in future work.

Acknowledgments. This work was supported in part by the Fujian Provincial Natural Science Foundation (No. 2025 J01858).

Disclosure of Interests. The authors have no competing interests.

References

1. Kim, M., Lu, B., Moon, I.: Effects of foldable Containers in Various Circumstances in maritime transport. Transp. Res. Rec. **03611981241236475** (2024)
2. Zhang, B., Zheng, Z.: Model and algorithm for vessel scheduling optimisation through the compound channel with the consideration of tide height. Int. J. Shipping Transp. Logistics (2021). https://doi.org/10.1504/IJSTL.2021.10036556
3. Zhang, X., Li, R., Chen, X., Li, J., Wang, C.: Multi-object-based vessel traffic scheduling optimisation in a compound waterway of a large harbour. The. J. Navig. **72**(3), 609–627 (2019)
4. Kong, J., Fan, X., Jin, X., Lin, S., Zuo, M.: A Variational Bayesian inference-based En-decoder framework for traffic flow prediction. IEEE Trans. Intell. Transp. Syst. **25**(3), 2966–2975 (2024)
5. Hu, H.-X., Lin, Z.-Z., Hu, Q., Zhang, Y., Wei, W., Wang, W.: Multi-source information fusion based DLaaS for traffic flow prediction. IEEE Trans. Comput. **73**(4), 994–1003 (2024)

6. Shu, Y., et al.: Evaluation of ship emission intensity and the inaccuracy of exhaust emission estimation model. Ocean Eng. **287**, 115723 (2023)
7. Shu, Y., et al.: Reference path for ships in ports and waterways based on optimal control. Ocean Coast. Manage. **253**, 107168 (2024)
8. Shu, Y., et al.: Analyzing the spatio-temporal correlation between tide and shiping behavior at estuarine port for energy-saving purposes. Appl. Energy. **367**, 123382 (2024)
9. Shu, Y., et al.: Influence of sea ice on ship routes and speed along the Arctic northeast passage. Ocean Coast. Manage. **256**, 107320 (2024)
10. Shu, Y., et al.: Research on ship following behavior based on data mining in arctic waters. IEEE Trans. Intell. Transp. Syst. (2025)
11. Shu, Y., Dong, A., Ye, B., Liu, C., Gan, L., Song, L.: Liability division for ship collision accidents based on ontology model and bayesian network. Ocean Coast. Manage. **269**, 107824 (2025)
12. Chuah, C.,.W., He, W., Huang, D.S.: DeepBiG: a hybrid supervised CNN and bidirectional GRU model for predicting the DNA sequence. Int. J. Adv. Comput. Sci. Appl. **15**(2) (2024)
13. Kumar, A., Kumar, V.: Forecast of solar cycle 25 based on hybrid CNN-bidirectional-GRU (CNN-BiGRU) model and novel gradient residual correction (GRC) technique. Adv. Space Res. **73**(8), 4342–4362 (2024)
14. Chopannejad, S., Roshanpoor, A., Sadoughi, F.: Attention-assisted hybrid CNN-BILSTM-BiGRU model with SMOTE–Tomek method to detect cardiac arrhythmia based on 12-lead electrocardiogram signals. Digital. Health. **10**, 20552076241234624 (2024)
15. Jai, A.J.G., Mastan, M., Al-Nuaimy, L.A.H.: Aspect based hotel recommendation system using dilated multichannel CNN and BiGRU with hyperbolic linear unit. Int. J. Mach. Learn. Cybern. 1–20 (2024)
16. Lalwani, P., Ramasamy, G.: Human activity recognition using a multi-branched CNN-BiLSTM-BiGRU model. Appl. Soft Comput. 111344 (2024)
17. Liu, B., Wang, M., Foroosh, H., Tappen, M.: Marianna Pensky. Sparse convolutional neural networks. In: Proceedings of the IEEE Conference on Computer Vision and Pattern Recognition (2015)
18. Liang, R., Chang, X., Jia, P., Xu, C.: Mine gas concentration forecasting model based on an optimized BiGRU network. ACS Omega. **5**(44), 28579–28586 (2020)
19. Weibull, W.: A statistical distribution function of wide applicability. J. Appl. Mech. (1951)
20. Khadka, A., et al.: Quantile regressions as a tool to evaluate how an exposure shifts and reshapes the outcome distribution: a primer for epidemiologists. MedRxiv, 23289415 (2023)
21. Haultfœuille, X.D., Maurel, A., Zhang, Y.: Extremal quantile regressions for selection models and the black–white wage gap. J. Econ. **203**(1), 129–142 (2018)
22. Fasiolo, M., Wood, S.N., Zaffran, M., Nedellec, R., Goude, Y.: Fast calibrated additive quantile regression. J. Am. Stat. Assoc. **116**(535), 1402–1412 (2021)
23. Che, Z., et al.: MAFF-HRNet: multi-attention feature fusion HRNet for building segmentation in remote sensing images. Remote Sens. **15**(5), 1382 (2023)
24. Fujian Maritime Safety Administration. Technical Guidelines for Ship Traffic Flow Analysis in Water Engineering Projects within the jurisdiction, (2022)
25. Author, F.: Article title. Journal 2(5), 99–110 (2016).

Collision-free Cluster Formation Control of Autonomous Surface Vehicle Based on Artificial Potential Function

Huijuan Li[1,2], Nan Gu[1,2], Zhouhua Peng[1,2](✉), Lu Liu[1,2], Haoliang Wang[1,2], and Anqing Wang[1,2]

[1] School of Marine Electrical Engineering, Dalian Maritime University, Dalian 116026, China
{ngu,zhpeng,luliu,haoliang.wang12,anqingwang}@dlmu.edu.cn

[2] Dalian Key Laboratory of Swarm Control and Electrical Technology for Intelligent Ships, Dalian Maritime University, Dalian 116026, China

Abstract. This paper studies the collision-free cluster formation control of autonomous surface vehicles subject to unknown internal uncertainties and unknown external disturbances. A collision-free cluster formation control method is presented based on fuzzy predictor to achieve safe formation tracking control in obstacle environments. Specifically, a fuzzy predictor is designed by using the fuzzy logic system. Then, a collision-free cluster formation controller based on artificial potential functions is proposed, which can autonomously avoid inter-agent collisions and static obstacle. By using the Lyapunov stability analysis, the closed system is input-to-state. Simulation results verify the effectiveness of the proposed method.

Keywords: Autonomous surface vehicles · Collision-free cluster formation control · Artificial potential function

1 Introduction

Formation control of autonomous surface vehicles (ASVs) is attracting interest due to the surge in demand for practical applications such as resource exploration, environmental monitoring, ocean replenishment, and search and rescue [1–8]. The motion control performance of ASVs is vulnerable to external disturbances induced by wind, waves, and ocean currents in complex maritime environments. In addition, ASV is a complex system with strong coupling and high nonlinearity. The presence of disturbances and uncertainties terms adversely affects motion control performance and even leads to unstable motion [9]. In order to achieve high-performance motion control, various control methods are proposed, including neural-network-based control [10,11], fuzzy-based control [12,13], disturbance-observer-based control [14].

In [10], a neural network-based sliding mode control is adopted to approximate the unknown continuous uncertainties and disturbances in the ASV. In

C. Li et al. (Eds.): ICNC 2025, CCIS 2946, pp. 11–25, 2026.
https://doi.org/10.1007/978-981-92-1599-7_2

[11], a neural network-based fixed-time sliding mode control scheme is further developed ensure estimation of parameter uncertainty in fixed-time. In [12], an adaptive fuzzy-based formation control method is proposed for ASVs with unknown model nonlinearity and actuator saturation. In [13], an adaptive fuzzy-based dynamic event-triggered formation controller is proposed for underactuated ASVs with modeling uncertainties. In [14], an adaptive disturbance observer with fixed-time convergence is developed to estimate unknown disturbances and uncertainties, relaxing the assumptions on the upper bounds of disturbances. It is noted that the aforementioned works [10–14], overlook collision avoidance and focus on a single cooperative behavior. Collision avoidance and multiple cooperative behaviors are a crucial problem to be worked on due to the complexity of the ocean environment and the requirements of multiple different tasks, especially for the formation control of a large number of ASVs.

Based on the above observations, this paper investigates a collision-free cluster formation control problem for ASVs with internal uncertainty and external disturbances. A collision-free cluster formation controller is proposed based on neighboring information. An artificial potential function is involved into the control law design to avoid collisions between ASVs and between ASVs and obstacles. The stability of the closed-loop system is established via Lyapunov theory.

The main contributions of this paper include the following: (1) Designing a fuzzy predictor to estimate the unknown internal dynamic and the unknown external disturbance; (2) Proposing a collision-free cluster formation control strategies to achieve various formation patterns and collision avoidance tasks; and (3) Validating the proposed methods by simulation, demonstrating its effectiveness in autonomous surface vehicle.

2 Problem Formulation

2.1 System Model

Consider a system consisting of N ASVs, where the dynamics of each follower ASV are described as [9]:

$$\begin{cases} \dot{\eta}_i = J_i v_i, \\ M_i \dot{v}_i = \tau_i - C_i v_i - D_i v_i + H_i + w_i, \end{cases} \tag{1}$$

where $\eta_i = [x_i, y_i, \varphi_i]^T \in \mathbb{R}^3$ is the position and yaw angle in the Earth-fixed coordinate system; $v_i = [u_i, v_i, r_i]^T \in \mathbb{R}^3$ denotes the surge velocity, sway velocity, and yaw rate in the body-fixed coordinate system; $\tau_i \in \mathbb{R}^3$ is the control input; $M_i \in \mathbb{R}^{3\times 3}$ is the inertia matrix; $J_i \in \mathbb{R}^{3\times 3}$ is the rotation matrix; C_i represents the coriolis and centripetal matrix; D_i is the nonlinear damping matrix; H_i denotes the uncertain hydrodynamic terms; $w_i \in \mathbb{R}^3$ represents unknown disturbances caused by the external environment.

Let $x_{i,1} = \eta_i$ and $x_{i,2} = \dot{\eta}_i$. Then the dynamic model of the ASV in (1) can be further transformed into the following general form:

$$\begin{cases} \dot{x}_{i,1} = x_{i,2}, \\ \dot{x}_{i,2} = b_i u_i + \sigma_i, \end{cases} \tag{2}$$

where $x_{i,1} = [x_{i,1x}, x_{i,1y}, x_{i,1\varphi}]^T \in \mathbb{R}^3$ denotes the position and yaw angle in the Earth-fixed frame; $x_{i,2} = [x_{i,2u}, x_{i,2v}, x_{i,2r}]^T \in \mathbb{R}^3$ represents the corresponding velocities; $u_i = [u_{i,u}, u_{i,v}, u_{i,r}]^T \in \mathbb{R}^3$ is the control input in the Earth-fixed coordinates; $b_i \in \mathbb{R}^{3\times 3}$ denotes the known control gain matrix; The total disturbances $\sigma_i = [\sigma_{i,u}, \sigma_{i,v}, \sigma_{i,r}]^T = f_{i,2}+w_{i,2}$, where $f_{i,2}$ and $w_{i,2}$ represent uncertain dynamics and bounded external disturbances, respectively.

2.2 Graph Theory

$\mathcal{G} = \{\mathcal{V}, \mathcal{E}\}$ is a weighted graph, where $\mathcal{V} = \{1, \cdots, N\}$ represents the nonempty node set and and $\mathcal{E} \subseteq \mathcal{V} \times \mathcal{V}$ denotes the edge set. $\mathcal{N}_i = \{j \mid (j,i) \in \mathcal{E}\}$ denotes the set of neighbors of node i, where $(j,i) \in \mathcal{E}$ indicates that i can receive information sent by agent j. $\mathcal{A} = \{a_{i,j}\} \in \mathbb{R}^{N\times N}$ is the adjacency matrix, where $a_{i,j} \neq 0$, if $(j,i) \in \mathcal{E}$; otherwise, $a_{i,j} = 0$. The Laplacian matrix is given by $L = D - \mathcal{A}$, where $D = \text{diag}\{d_1, d_2, \cdots, d_N\}$ and $d_i = \sum_{j\in\mathcal{N}_i} a_{i,j}$.

Considering the MASs is partitioned into q $(1 < q < N)$ nonempty and disjoint clusters $\mathcal{V}_\ell$, satisfying $\bigcup_{\ell=1}^{q} \mathcal{V}_\ell = \mathcal{V}, \mathcal{V}_\ell \cap \mathcal{V}_h = \emptyset$, $\ell \neq h$, $h = 1, 2, \ldots, q$. Consequently, $\mathcal{G}_\ell$ represents the subgraph associated with the cluster $\mathcal{V}_\ell$. The communication subgraph of one leader and followers can be represented as $\bar{\mathcal{G}}_\ell$. $c_{i,j}$ represents the coupling strength. ÂăDefineÂă$\bar{i}$Âăas the cluster index of agentÂăi, such thatÂă$\bar{i} = \ell$, if $i \in \mathcal{V}_\ell$. For intra-cluster (where nodes i and j belong to the same cluster label ℓ), $a_{i,j} \geq 0$ and $c_{i,j} = c_\ell = c_{\bar{i}} > 0$. For inter-cluster $(\bar{i} \neq \bar{j})$, $a_{i,j} \in \mathbb{R}$, and if $a_{i,j} < 0$ then it is required that $b_i > 0$. $B = \text{diag}\{b_1, \cdots, b_N\}$, where $b_i > 0$ indicates that follower i can access the leader's information, and $b_i = 0$ otherwise.

Assumption 1: $\bar{\mathcal{G}}_\ell$ has a directed spanning tree, $\ell = 1, \ldots, q$.

Assumption 2: The in-degree balanced condition:

$$\sum_{j\in\mathcal{V}_l} a_{i,j} = 0, \quad \text{for all } i \in \mathcal{V} \setminus \mathcal{V}_\ell, \quad \ell = 1, \ldots, q.$$

It can be observed from Assumption 2 that t $a_{i,j}$ (where $\bar{i} \neq \bar{j}$) may assume negative values. Consequently, the existence of eigenvalues with negative real parts in matrix $L + B$. This aforementioned concern is successfully addressed by incorporating a positive parameter λ_i^b, as demonstrated in the following lemma.

Lemma 1: For sufficiently large values of λ_i^b, $i = 1, \ldots, N$, $L+\lambda^b B$ *is guaranteed to be a nonsingular matrix, and all its eigenvalues possess positive real parts, where* $\lambda^b = diag\{\lambda_1^b, \cdots, \lambda_N^b\}$.

Proof: The expanded form of the matrix $\mathcal{L} = L + \lambda^b B$:

$$\mathcal{L} = \begin{bmatrix} \sum_{j=1}^{N} a_{1,j} + \lambda_1^b b_1 & -a_{1,2} & \cdots & -c_{1,N} a_{1,N} \\ a_{2,1} & \sum_{j=1}^{N} a_{2,j} + \lambda_2^b b_2 & \cdots & a_{2,N} \\ \vdots & \vdots & \ddots & \vdots \\ -a_{N,1} & -a_{N,2} & \cdots & \sum_{j=1}^{N} a_{N,j} + \lambda_N^b b_N \end{bmatrix}.$$

If the follower i cannot directly obtain the each cluster leader's information, we have

$$\left| \sum_{j=1}^{N} a_{i,j} + \lambda_i^b b_i \right| = \left| \sum_{j=1}^{N} a_{i,j} \right| \geqslant \sum_{j=1}^{N} |a_{i,j}|. \tag{3}$$

Let the values of λ_i^b be large enough. For the follower i can directly obtain the each cluster leader's information, it can be derived that

$$\left| \sum_{j=1}^{n} a_{i,j} + \lambda_i^b b_i \right| > \sum_{j=1}^{n} |a_{i,j}|. \tag{4}$$

By using the Gersgorin circle criterion, there is a non-negative real part for all eigenvalues of $L + \lambda^b B$. Furthermore, by analyzing the matrix structure and connectivity properties, it can be deduced that these eigenvalues are strictly positive. The proof is completed.

2.3 Artificial Potential Functions

To achieve collision avoidance between agents, an artificial potential function is introduced as follows:

$$V_{i,j}^a(x_{i,1}, x_{j,1}) = \begin{cases} \frac{(\bar{d}_{ij}^2 - \|d_{ij}\|^2)}{\|d_{ij}\|^2 - \underline{d}_{ij}^2}, & \underline{d}_{ij} < \|d_{ij}\| < \bar{d}_{ij}, \\ 0, & otherwise, \end{cases} \tag{5}$$

where $d_{ij} = x_{i,1} - x_{j,1}$, $\bar{d}_{ij}$ is the detection region where agents can detect other agents; $\underline{d}_{ij}$ represents the avoidance distance which is the minimal safe distance between agents.

The partial derivative of $V_{i,j}^a(x_{i,1}, x_{j,1})$ with respect to $x_{i,1}$ is given by

$$\frac{\partial V_{i,j}^a}{\partial x_{i,1}} = \begin{cases} \frac{4(\bar{d}_{ij}^2 - \underline{d}_{ij}^2)(\|d_{ij}\|^2 - \bar{d}_{ij}^2)}{(\|d_{ij}\| - \underline{d}_{ij})^3} d_{ij}, & \underline{d}_{ij} < \|d_{ij}\| < \bar{d}_{ij}, \\ 0, & otherwise. \end{cases}$$

To avoid collisions between agents and static obstacles, the potential function is redefined as follows:

$$V_{i,k}^{o}(x_{i,1}, x_{k,1}) = \begin{cases} \frac{(\bar{d}_{ik}^2 - \|d_{ik}\|^2)}{\|d_{ik}\|^2 - \underline{d}_{ik}^2}, & \underline{d}_{ik} < \|d_{ik}\| < \bar{d}_{ik}, \\ 0, & \text{otherwise}, \end{cases} \tag{6}$$

where $d_{ik} = x_{i,1} - x_{k,1}$ denotes the relative position vector between the ith agent and the kth static obstacle; $\bar{d}_{ik}$ is an obstacle avoidance detection range. $\underline{d}_{ik} < \bar{d}_{ik}$ is the minimum safe distance required.

The partial derivative of $V_{i,k}^{o}(x_{i,1}, x_{k,1})$ with respect to $x_{i,1}$ is given by

$$\frac{\partial V_{i,k}^{o}}{\partial x_{i,1}} = \begin{cases} \frac{(\bar{d}_{ik}^2 - \underline{d}_{ik}^2)(\|d_{ik}\|^2 - \bar{d}_{ik}^2)}{(\|d_{ik}\| - \underline{d}_{ik})^3} d_{ik}, & \underline{d}_{ik} < \|d_{ik}\| < \bar{d}_{ik}, \\ 0, & otherwise. \end{cases}$$

The *control objective* is to develop an fuzzy predictor for the system (2) with the unknown total time-varying disturbances σ_i, and develop a distributed formation controller such that the outputs $x_{i,1}$ of follower ASVs collaboratively track the reference trajectory $x_{d,\bar{i}}$ while maintaining the desired formation pattern.

The following Assumptions are needed.

Assumption 3: The reference trajectory $x_{d,\bar{i}}$ and its derivative $x_{d,\bar{i}}$ are bounded.

Assumption 4: At $t = 0$, the followers initially reside within the collision-free range, i.e., $R_a < d_{ij}(0)$ and $d_{ik}(0) > R_o$.

3 Design and ANALYSIS

This section describes the design and analysis of fuzzy predictor and distributed tracking controller.

3.1 Fuzzy Predictor

To achieve trajectory tracking control, a fuzzy predictor based on FLS is first designed to approximate the uncertainties of ASV. Its advantage lies in updating fuzzy weights through adaptive mechanisms and making estimated values as feedforward compensation signals to the controller, thereby achieving active compensation for uncertainties.

According to FLS, the uncertainty term σ_i can be approximated:

$$\dot{x}_{i,2} = \theta_i^T \phi_i(\xi_i) + \delta_i + b_i u_i, \tag{7}$$

where θ_i is weight vectors, $i = 1, \ldots, N$; ϕ_i denotes the fuzzy basis function vector; $\xi_i = [x_{i,2}(t), x_{i,2}(t - t_d), u_i]^T$, t_d is the sampling period. $\|\phi_i\| \leqslant \bar{\phi}$ for a positive constant $\bar{\phi}$. δ_i is the approximation error, satisfying $\|\delta_i\| \leq \bar{\delta}$, and $\bar{\delta}$ being a positive constant.

Define $\hat{x}_{i,2}$ as the estimates of $x_{i,2}$; Let $\tilde{x}_{i,2} = \hat{x}_{i,2} - x_{i,2}$ be the prediction error, and a fuzzy predictor is proposed:

$$\dot{\hat{x}}_{i,2} = -(k_{i,2} + \kappa_{i,2})\tilde{x}_{i,2} + \hat{\theta}_i^T \phi_i + b_i u_i, \tag{8}$$

where $k_{i,2}$ is the positive control gain. Let $\hat{\theta}_i$ denotes the estimated parameter vectors for θ_i.

By using the estimation error, the adaptive law of $\hat{\theta}_i$ is designed:

$$\dot{\hat{\theta}}_i = -\Gamma_i \text{Proj}\left\{\hat{\theta}_i, \phi_i(\xi_i)\tilde{x}_{i,2}\right\}, \tag{9}$$

where Γ_i is a positive-definite coefficient matrix; Proj$(\cdot)$ is a projection operator as in [15].

Defining the estimation error $\tilde{\theta}_i = \hat{\theta}_i - \theta_i$, the error dynamics of $\tilde{x}_{i,2}$ and $\tilde{\theta}_i$ are:

$$\begin{cases} \dot{\tilde{x}}_{i,2} = -(k_{i,2} + \kappa_{i,2})\tilde{x}_{i,2} + \tilde{\theta}_i^T \phi_i - \delta_i, \\ \dot{\tilde{\theta}}_i = -\Gamma_i \text{Proj}\left\{\hat{\theta}_i, \phi_i(\xi_i)\tilde{x}_{i,2}\right\}. \end{cases} \tag{10}$$

According to the properties of the projection operator, there exists a constant ϵ_θ that satisfies $\|\tilde{\theta}_i\| \leqslant \epsilon_\theta$. To analyze the stability of the error system in (predictors error), the following lemma is given:

Lemma 2: The fuzzy predictor error system (10) with the state vectors being $\tilde{x}_{i,2}$ and $\tilde{\theta}_i$, and the input vectors being δ_i and ϵ_θ is ISS.

Proof: Construct the Lyapunov function as

$$V_{io} = \frac{1}{2}(\tilde{x}_{i,2}^T \tilde{x}_{i,2} + \tilde{\theta}_i^T \Gamma_i^{-1} \tilde{\theta}_i), \tag{11}$$

whose time derivative along (10) and combining it with the properties of the projection operator, we obtain

$$\begin{aligned} \dot{V}_{io} \leq & -(k_{i,2} + \kappa_{i,2})\|\tilde{x}_{i,2}\|^2 - \tilde{x}_{i,2}^T \delta_i \\ \leq & -k_i^a \|\tilde{x}_{i,2}\|^2 - k_i^a \|\tilde{\theta}_i\|^2 + \|\tilde{x}_{i,2}\|\|\delta_i\| + k_i^a \|\tilde{\theta}_i\|^2 \\ \leq & -k_i^a \|E_i\|^2 + \|E_i\|\|\hbar_i\| \\ \leq & -k_i^a (1 - c_i^a)\|E_i\|^2 - \|E_i\|(k_i^a c_i^a \|E_i\| - k_i^c \|\hbar_i\|). \end{aligned}$$

where $k_i^a = k_{i,2} + \kappa_{i,2}$, $0 < c_i^a < 1$. $E_i = [\tilde{x}_{i,2}^T, \tilde{\theta}_i^T]^T$, $\hbar_i = [\delta_i^T, \epsilon_\theta]^T$.

When

$$\|E_i\| \geqslant k_i^c \|\hbar_i\| / k_i^a c_i, \tag{12}$$

we get

$$\dot{V}_{io} \leq -k_i^a (1 - c_i^a)\|E_i\|^2. \tag{13}$$

Thus, the error system (10) is ISS.

3.2 Distributed Formation Controller Design

Based on the fuzzy predictor in Sect. 3.1, a distributed formation controller is developed.

Step 1: Define the graph-based neighborhood formation error for the ith agent:

$$e_{i,1} = \sum_{j\in\mathcal{N}_i} c_{i,j}a_{i,j}(x_{i,1} - x_{j,1} - p_i + p_j) + c_{\bar{i}}\lambda_i^b b_i(x_{i,1} - x_{d,\bar{i}}), \tag{14}$$

where p_i and p_j denote the desired position deviation of ASV i and ASV j from their reference signal $x_{d,}$, which is fixed.

According to $\tilde{\theta}_i = \hat{\theta}_i - \theta_i$, the system (2) can be written as

$$\begin{cases} \dot{x}_{i,1} = \hat{x}_{i,2} - \tilde{x}_{i,2}, \\ \dot{x}_{i,2} = b_i u_i + \hat{\theta}_i^T\phi_i - \tilde{\theta}_i^T\phi_i + \delta_i. \end{cases} \tag{15}$$

The time derivative of (14) along (15) is given by

$$\dot{e}_{i,1} = \iota_i(\hat{x}_{i,2} - \tilde{x}_{i,2}) - \sum_{j=1}^{2} a_{i,j}(\hat{x}_{j,2} - \tilde{x}_{j,2}) - \lambda_i^b c_{\bar{i}} b_i \dot{x}_{d,\bar{i}}, \tag{16}$$

where $\iota_i = \bar{c}_i d_i + \lambda_i^b c_{\bar{i}} b_i$, $\bar{c}_i = \sum_{j=1}^{2} c_{i,j}$.

To stabilize $e_{i,1}$, a virtual tracking control law is proposed as follows:

$$x_{i,2}^* =(-k_{i,1}\bar{e}_{i,1} + \sum_{j=1}^{2} a_{i,j}\hat{x}_{j,2} + \lambda_i^b c_{\bar{i}} b_i \dot{x}_{d,\bar{i}})/\iota_i - e_{i,2}, \tag{17}$$

where $e_{i,2}$ is to be defined later. $\bar{e}_{i,1} = e_{i,1} + e_{i,v}$ with

$$e_{i,v} = \sum_{j\in\mathcal{N}_i} \frac{\partial V_{ij}^a}{\partial x_{i,1}} + \sum_{k=1}^{N_o} \frac{\partial V_{ik}^o}{\partial x_{i,1}}.$$

By combining (16) with (17), it can be concluded that:

$$\dot{e}_{i,1} = -k_{i,1}\bar{e}_{i,1} + \iota_i\tilde{x}_{i,2} + \sum_{j=1}^{2} a_{i,j}\tilde{x}_{j,2}. \tag{18}$$

Step 2: The error is defined as

$$e_{i,2} = \hat{x}_{i,2} - x_{i,2}^*. \tag{19}$$

The time derivative of (19) along (15) is given by

$$\dot{e}_{i,2} = -(k_{i,2} + \kappa_{i,2})\tilde{x}_{i,2} + \hat{\theta}_i^T\phi_i + b_i u_i - \dot{x}_{i,2}^*. \tag{20}$$

To stabilize $e_{i,2}$, the actual tracking control law u_i is proposed as follows:

$$u_i = b_i^{-1}(-k_{i,2}e_{i,2} + \dot{x}_{i,2}^* - \hat{\theta}_i^T \phi_i). \tag{21}$$

By combining (20) with (21), it can be concluded that:

$$\dot{e}_{i,2} = -k_{i,2}e_{i,2} - (k_{i,2} + \kappa_{i,2})\tilde{x}_{i,2}. \tag{22}$$

To analyze the stability of the error system in (18), (22), the following lemma is given:

Lemma 3: Consider the formation error control system with the state vectors being $e_{i,1}, e_{i,2}$, *and the input vectors being* $\tilde{x}_{i,2}^d$ *is ISS.*

Proof: Consider the Lyapunov function:

$$V_{i,b} = \frac{1}{2}e_{i,1}^T e_{i,1} + \frac{1}{2}e_{i,2}^T e_{i,2} + \sum_{j=1}^{2} \iota_i V_{i,j}^a + \sum_{k=1}^{N_o} \iota_i V_{i,k}^o, \tag{23}$$

whose time derivative along (17) and (21) yields:

$$\begin{aligned} \dot{V}_{i,b} =& \bar{e}_{i,1}^T(-k_{i,1}\bar{e}_{i,1} + \iota_i\tilde{x}_{i,2} - \sum_{j=1}^{2} a_{i,j}\tilde{x}_{j,2}) \\ &+ e_{i,2}^T(-k_{i,2}e_{i,2} - (k_{i,2} + \kappa_{i,2})\tilde{x}_{i,2}) + e_{i,v}^T\check{v}_i, \end{aligned} \tag{24}$$

where $\check{v}_i = \sum_{j=1}^{2} a_{i,j}\dot{x}_{j,1} + \lambda_i^b c_{\bar{i}} b_i \dot{x}_d$.

Next, the proof proceeds in two steps. In the first step, the stability of the closed-loop system in the collision range is analyzed. In the second step, the stability of the closed-loop system is analyzed within the collision-free range.

When the i-th ASV is in the collision-free region, the distance satisfies $R_a < d_{ij}$, $d_{ik} > R_o$, and $\partial V_{ij}^a/\partial x_{i,1} = \partial V_{ij}^a/\partial x_{j,1} = \partial V_{ij}^o/\partial x_{i,1} = 0$. Hence, $e_{i,1} = \bar{e}_{i,1}$, $\dot{V}_i^b$ is obtained based on (24):

$$\begin{aligned} \dot{V}_{i,b} \leq & -k_i\|E_i^b\|^2 + c\|E_i^b\|\|\tilde{x}_{i,2}^d\| \\ \leq & -k_i(1-c_i)\|E_i^b\|^2 - k_i c_i\|E_i^b\|^2 + c\|E_i^b\|\|\tilde{x}_{i,2}^d\|, \end{aligned} \tag{25}$$

where $0 < c_i < 1$, $k_i = \lambda_{\min}(k_{i,1}, k_{i,2})$, $E_i^b = [\tilde{e}_{i,1}^T, \tilde{e}_{i,2}^T]^T$, $\|\tilde{x}_{i,2}\| \leq \tilde{x}_{i,2}^d$, $c = 2\max\{\iota_i, Nd_i, \lambda_{\max}(k_{i,2} + \kappa_{i,2})\}$.

Noting that

$$\|E_i^b\| \geq \frac{c\|\tilde{x}_{i,2}^d\|}{k_i c_i}, \tag{26}$$

it makes

$$\dot{V}_{i,b} \leq -\sum_{j=1}^{2} k_i(1-c_i)\|E_i^b\|^2. \tag{27}$$

Thus, the error system in (18), (22) is ISS, and $\|E_i^b\|$ is bounded.

When the i-th agent is in the collision region, defined $E_{i0} = [\bar{e}_{i,1}^T, e_{i,2}^T]^T$, and $\dot{V}_i$ is obtained based on (24):

$$\dot{V}_{i,b} \leq -k_i^o\|E_{i0}\|^2 + k_i^o\|E_{i0}\|^2 + \|E_{i0}\|\|h_{i0}\|, \tag{28}$$

where $h_{i0} = \|\tilde{x}_{i,2}^d\| + \|\check{v}_i\|$. From [16], the error signal E_{i0} is bounded when the system (28) has a bounded input h_{i0}. According to Assumption 2, the followers are initially located in the collision-free range. When the follower enters the collision range, the potential function grows monotonically logarithmically. When the followers leave the collision range, the potential function returns to zero. As a result, there are no collisions occur.

3.3 Stability Analysis

Based on the developed predictor and controller, the stability of the closed-loop system is proved by the following theorem.

Theorem: Consider the system (2), together with the fuzzy predictor (8), the adaptive law (9), and the distributed formation control law (17), (21). Under Assumptions 1–4, all error signals in the closed-loop system are bounded.-

Proof: According to the cascade system stability [16] and using Lemmas 1–3, all error signals $E_i^a = [\tilde{x}_{i,2}^T, \tilde{\theta}_i^T, \bar{e}_{i,1}^T, \tilde{e}_{i,2}^T]^T$ of the closed-loop system consisting of subsystem (10) and subsystem (18), (22) is bounded. The proof is completed.

4 Simulation Results

The control parameters are set to $k_{i,1} = 1.2$; $k_{i,2} = 0.3$. The predictor parameters are chosen as $\kappa_{i,2} = 5$, $\Gamma_i = \text{diag}\{77I_{10}, 10I_{10}, 20I_{10}\}$. The input vector $\xi_i = [\xi_{i,1}, \xi_{i,2}, \xi_{i,3}]^T$. The membership functions $\mu_{i,F_l^k}(\xi_{i,k}) = \exp-((\xi_{i,k} - b_{i,k}(l5))/b_{i,k})^2$ with $k = 1,2,3$, $l = 1,\ldots,9$ and $b_{i,1} = 20$, $b_{i,2} = 2$, $b_{i,3} = 10$, $i = 1,\ldots,10$.The scenario contains one static circular obstacle located at coordinates (20, -18), with corresponding radii of 6.5 m. The artificial potential field parameters are specified as: $R_a = 5.5$, $R_o = 7$. The communication topology is illustrated in Fig. 1.

Simulation results are proposed in Figs. 2, 3, 4, 5, 6 and 7. Figure 2 depicts the collision-free formation performance of four ASVs, showing that the presented control method can not only achieve different formation patterns but also collision avoidance between ASVs as well as between ASVs and static obstacles. Figure 3 provides the velocities and yaw rate estimation performance. It can be seen that the fuzzy predictor can accurately estimate the velocities and yaw rate. Figure 4 shows the bounded control input. Figure 5 depicts the total disturbance estimation performance. It describes the fuzzy predictor that accurately estimates the total disturbance and ensures accurate convergence to the true value. Figures 6 and 7 illustrate the distances between ASVs and between ASVs and obstacles with collision-free control method, respectively. Figures 8 and 9

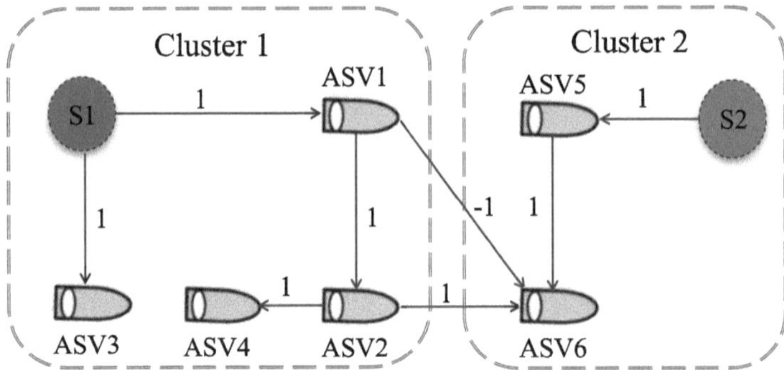

Fig. 1. Communication topology graph.

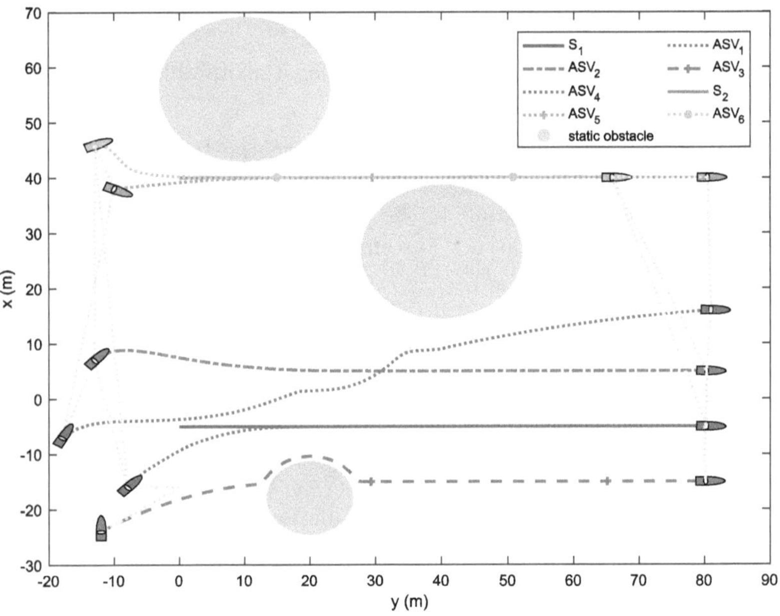

Fig. 2. The collision-free cluster formation performance of the ASVs.

illustrate the distances between ASVs and between ASVs and obstacles without collision-free control, respectively. It is seen that the collision avoidance task can be achieved using the proposed control method.

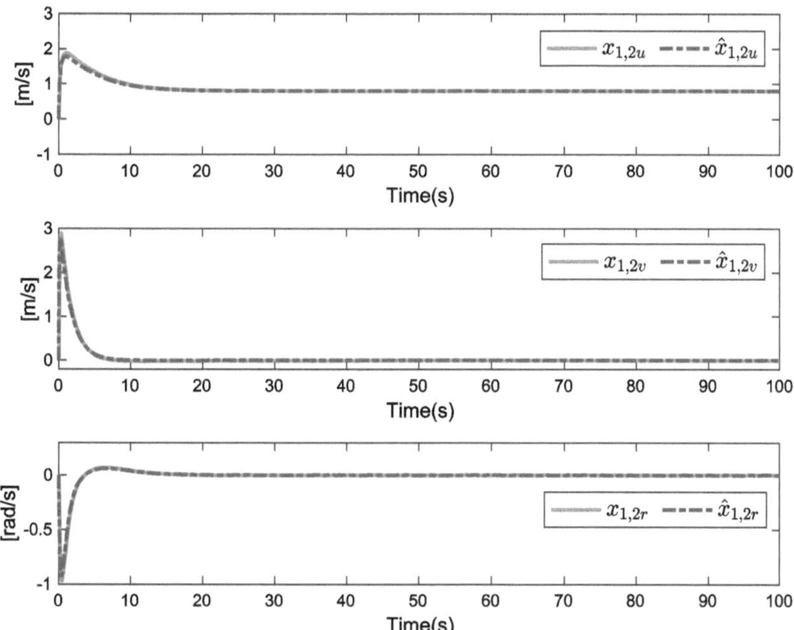

Fig. 3. The velocities and yaw rate estimation performance of the ASV.

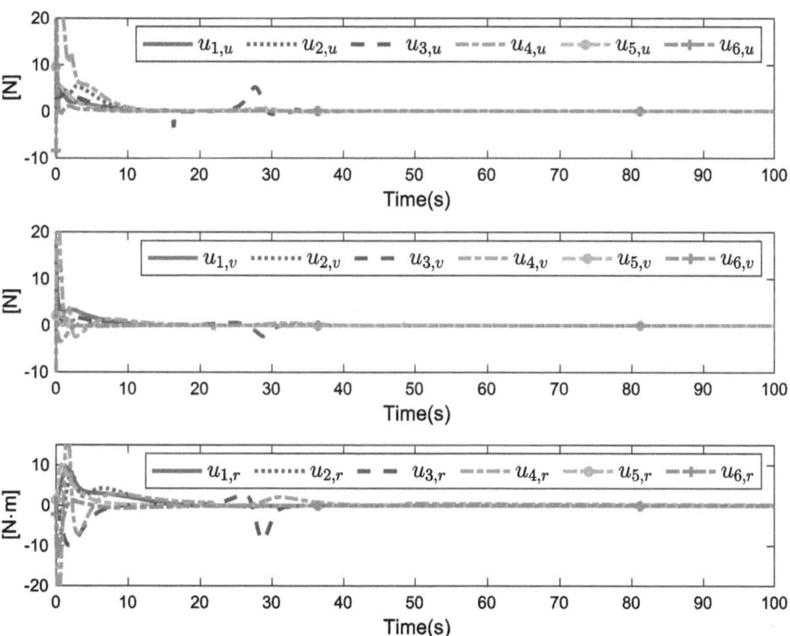

Fig. 4. The control input.

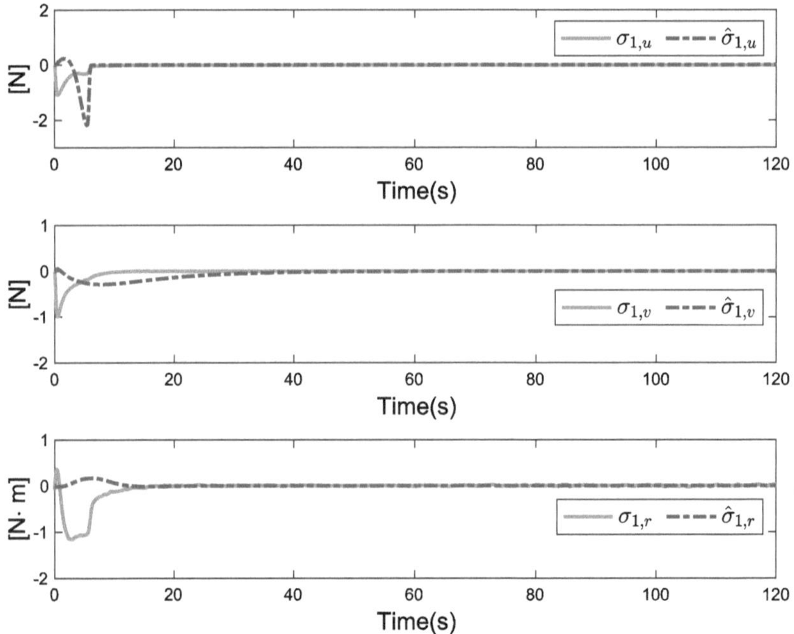

Fig. 5. The total disturbance estimation performance of the ASV.

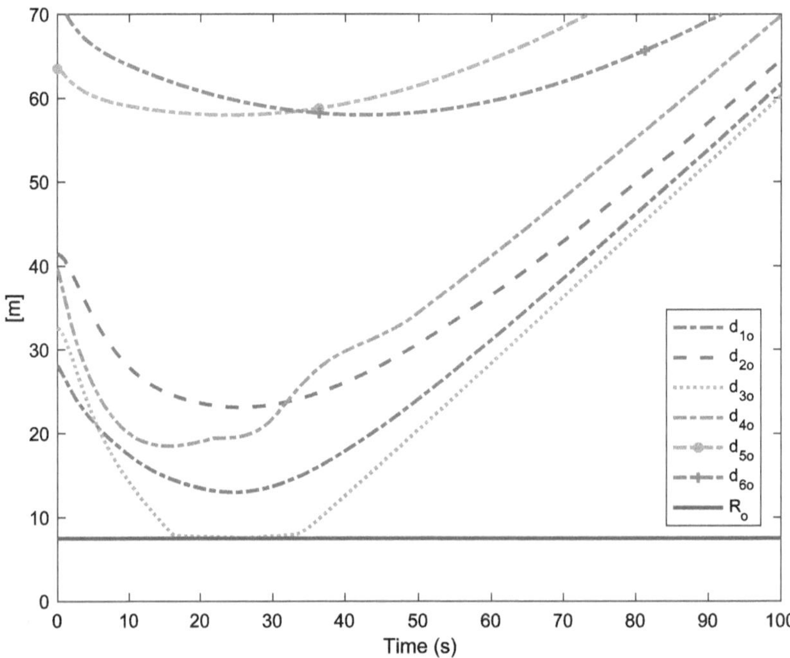

Fig. 6. The distances between the ASVs with collision-free control method.

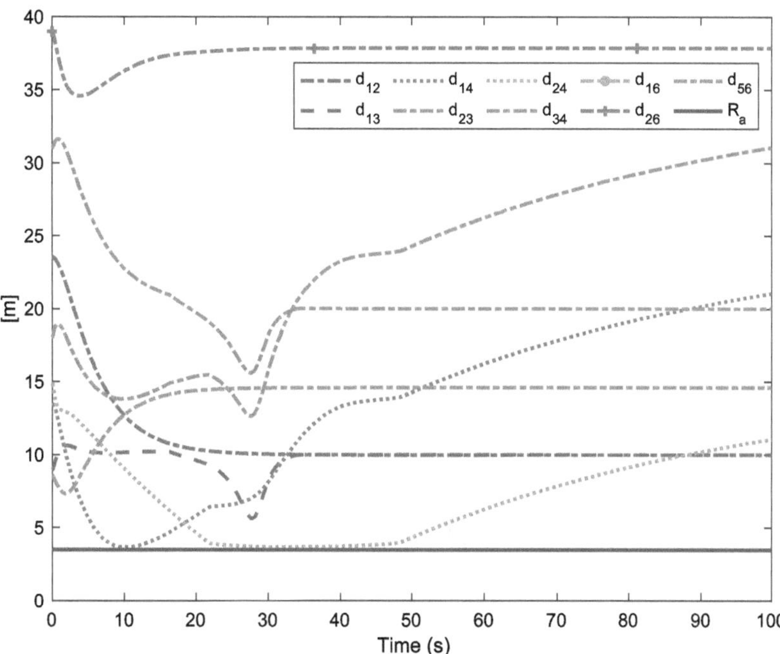

Fig. 7. The distances between the ASVs and obstacles with collision-free control method.

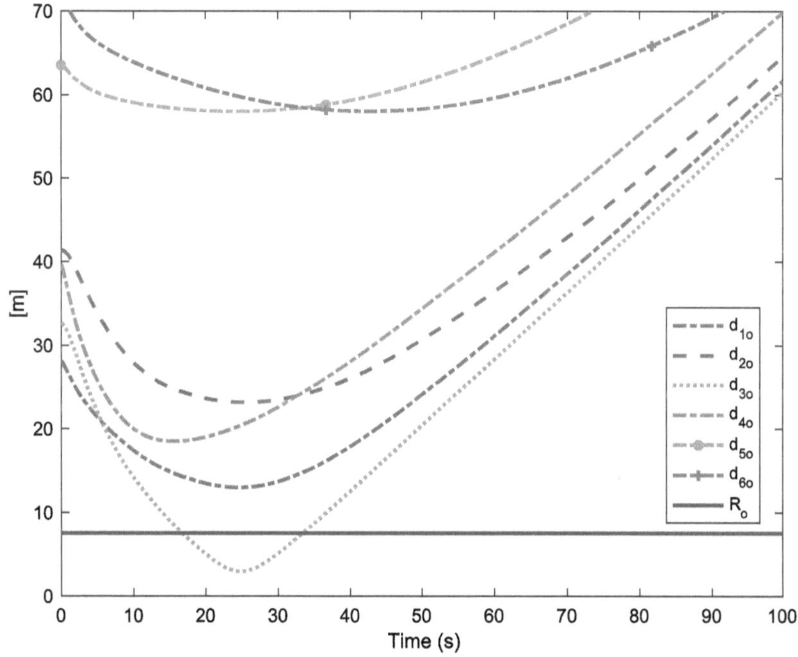

Fig. 8. The distances between the ASVs without collision-free control.

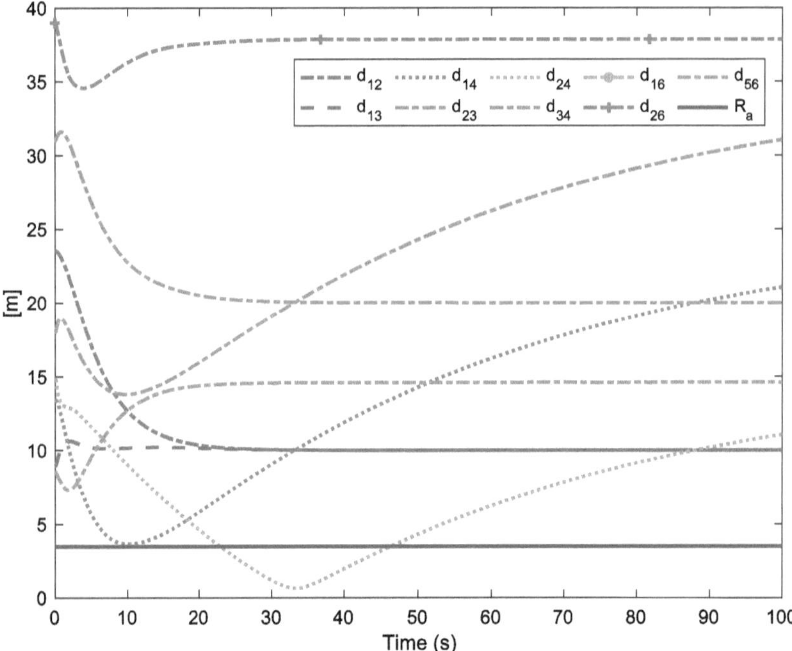

Fig. 9. The distances between the ASVs and obstacles without collision-free control.

5 Conclusions

In this paper, a collision-free cluster formation control problem is solved for ASVs, which contain the unknown internal dynamic, the unknown external disturbance. By utilizing fuzzy logic systems, a fuzzy predictor is proposed to estimated unknown model parameters. Based on the estimated information, the collision-free cluster formation controller with APF is proposed. The proposed controller not only realizes formation queues but also avoids collisions with neighboring agents and multiple static obstacles. The stability of the closed-loop control system is demonstrated. Simulation result confirms the effectiveness of the presented the collision-free formation control method.

Acknowledgements. This work was supported in part by the National Natural Science Foundation of China under Grants 52471372; in part by the Liaoning Revitalization Leading Talents Program under Grant XLYC2402054; in part by the Doctoral Scientific Research Foundation of Liaoning Province under Grant 2024-BS-012; in part by the Key Basic Research of Dalian under Grant 2023JJ11CG008; and in part by the Fundamental Research Funds for the Central Universities under Grants 3132023508; in part by the Cultivation Program for the Excellent Doctoral Dissertation of Dalian Maritime University 0034012511.

References

1. Liu, Z., Zhang, Y., Yu, X., Yuan, C.: Unmanned surface vehicles: an overview of developments and challenges. Annu. Rev. Control. **41**, 71–93 (2016)
2. Wang, X., Su, Y., Xu, D.: Cooperative robust output regulation of linear uncertain multiple multivariable systems with performance constraint. Automatica **95**, 137–145 (2018)
3. Yu, Y., Guo, C., Yu, H.: Finite-time plos-based integral sliding-mode adaptive neural path following for unmanned surface vessels with unknown dynamics and disturbances. IEEE Trans. Autom. Sci. Eng. **16**(4), 1500–1511 (2019)
4. Braginsky, B., Baruch, A., Guterman, H.: Development of an autonomous surface vehicle capable of tracking autonomous underwater vehicles. Ocean Eng. **197**, 106868 (2020)
5. Liu, L., Wang, D., Peng, Z., Li, T., Chen, C.P.: Cooperative path following ring-networked under-actuated autonomous surface vehicles: algorithms and experimental results. IEEE Trans. Cybern. **50**(4), 1519–1529 (2020)
6. Gu, N., Wang, D., Peng, Z., Wang, J.: Safety-critical containment maneuvering of underactuated autonomous surface vehicles based on neurodynamic optimization with control barrier functions. IEEE Trans. Neural Netw. Learn. Syst. **34**(6), 2882–2895 (2021)
7. Peng, Z., Wang, J., Wang, D., Han, Q.-L.: An overview of recent advances in coordinated control of multiple autonomous surface vehicles. IEEE Trans. Industr. Inf. **17**(2), 732–745 (2021)
8. Gu, N., Wang, D., Peng, Z., Wang, J., Han, Q.-L.: Disturbance observers and extended state observers for marine vehicles: a survey. Control. Eng. Pract. **123**, 105158 (2022)
9. T.I.: Fossen: Handbook of Marine Craft Hydrodynamics and Motion Control. Wiley (2011)
10. Jiang, T., Yan, Y., Wu, D., Yu, S., Li, T.: Neural network based adaptive sliding mode tracking control of autonomous surface vehicles with input quantization and saturation. Ocean Eng. **265**, 112505 (2022)
11. Yan, Y., Liu, P., Yu, S.: Neural network-based practical fixed-time nonsingular sliding mode tracking control of autonomous surface vehicles under actuator saturation. Ocean Eng. **306**, 118032 (2024)
12. Zhou, W., Wang, Y., Ahn, C.K., Cheng, J., Chen, C.: Adaptive fuzzy backstepping-based formation control of unmanned surface vehicles with unknown model nonlinearity and actuator saturation. IEEE Trans. Veh. Technol. **69**(12), 14749–14764 (2020)
13. Yu, K., Li, Y., Lv, M., Tong, S.: Adaptive fuzzy formation control for underactuated multi-usvs with dynamic event-triggered communication. IEEE Trans. Syst. Man Cybern. Syst. **54**(12), 7783–7793 (2024)
14. Chen, Y., Kong, M., Zhao, Z., Dong, J., Li, R.: Adaptive disturbance observer-based fixed-time sliding mode tracking control for autonomous surface vehicles with prescribed performance. Ocean Eng. **341**, 122427 (2025)
15. Seshagiri, S., Khalil, H.: Output feedback control of nonlinear systems using rbf neural networks. IEEE Trans. Neural Networks **11**(1), 69–79 (2000)
16. Krstic, M., Kanellakopoulos, I., Kokotović, P.: Nonlinear and adaptive control design. Lecture Notes Control Inf. Sci. **5**(2), 4475–4480 (1995)

Petrochemical Intelligent Training System Based on Internet of Things Technology

Hao Jiang, Bin Zou(✉), Wenjie Wang, Peng Sun, Cong Zhou, and Haiyang Zhang

University of Petroleum (East China), 266580 Qingdao, China
zoubin52306@163.com

Abstract. The chemical industry is constantly developing towards high quality, high efficiency, and high safety. However, the traditional training mode can no longer meet the demand for talents in the modern chemical industry. Based on advanced Internet of Things technology, this article has developed an intelligent training system to achieve intelligent detection and control functions. The on-site control detection unit and software platform, in a set network environment, obtain key data of the training personnel's operation process in real time to complete information processing and operation specification evaluation. To achieve automatic detection and uploading of actions for training operators, solving the problems of energy consumption, single mode, and low efficiency in traditional training systems.

Keywords: Internet of Things technology · Singlechi · Data transmission · Human-computer interaction

1 Introduction

The petroleum and chemical industry consistently prioritizes high quality and safety. Traditional teaching methods struggle to meet the training needs of modern chemical engineering professionals, necessitating the application of new technologies for talent development [1]. IoT technology enables managers to better monitor employees' practical operations in petrochemical production [2, 3]. By analyzing production activities, potential issues can be identified promptly, improving efficiency, product quality, and cost reduction, thereby enhancing corporate competitiveness [4].

To prevent safety incidents in the chemical industry, managers must focus on standardizing personnel operations. Through practical training, employees can simulate real-world environments, understand hazardous chemical properties, and master standardized procedures for specialized tasks, such as rigorous cleaning, replacement, and approval processes before hot work [5]. Training also reinforces safety awareness, ensuring employees remain vigilant and respond correctly in complex and hazardous environments. Thus, practical training is the foundation of safe production and stable development in the chemical industry, a critical step that employees must prioritize [6].

With advancements in sensor technology, physical simulation training and assessment systems can replicate real operational environments [7]. These systems enhance

C. Li et al. (Eds.): ICNC 2025, CCIS 2946, pp. 26–38, 2026.
https://doi.org/10.1007/978-981-92-1599-7_3

hands-on experience, offering effective, low-risk training with repeatable exercises. Post-training, operations can be recorded and evaluated, meeting the demands of practical training and assessment. IoT technology, utilizing radio frequency identification (RFID) and wireless communication, enables item identification and data sharing via the internet [8]. This study focuses on an intelligent training system for the petrochemical industry, centered on IoT technology. By integrating sensor technology, wireless communication, and RFID, an upper-computer system is designed to communicate with hardware, establishing a petrochemical intelligent training system. This system plays a vital role in improving employees' emergency response capabilities, building safety defenses, and preventing accidents [9]. In the future, as the chemical industry evolves, intelligent training systems will garner increasing attention [10, 11].

2 Training System Functions and Composition

Traditional industrial skill training often suffers from disconnection from real scenarios and delayed assessments. This chapter designs an intelligent training system based on IoT technology, addressing these shortcomings by dynamically mapping training scenarios to actual production environments. The system employs a combination of communication technologies, using the STM32F103 microcontroller as the control unit. Distributed detection components are installed on-site, and a serial server and data acquisition module collaborate with the upper-computer system for efficient data processing, resolving challenges in operational compliance detection and multi-device communication. The functional logic framework of the training system is shown in Fig. 1.

The system comprises four core components:

1. Sensor Detection System: A distributed perception network is built to monitor trainee behavior, simulating real-world scenarios. RFID detects protective gear and tools, while proximity switches, infrared sensors, and button switches convert physical signals into electrical signals, creating a virtual-physical detection environment [12].
2. Embedded Control Unit: Distributed processing nodes using embedded microcontrollers parse sensor signals in real time, encapsulate data into standard protocols, and upload it. The unit also receives control commands from the upper computer to adjust hardware parameters.
3. Network Communication: A hybrid wired and wireless network ensures stable data transmission. Ethernet and Wi-Fi are combined, with a serial server serving as the gateway to connect RS485 buses to Ethernet
4. Upper-Computer Software Platform: A human-computer interface integrates personnel and device management, real-time training monitoring, data analysis, and storage. The platform compares operations with standard procedures to generate assessment scores.

The on-site application of the training system is illustrated in Fig. 2.

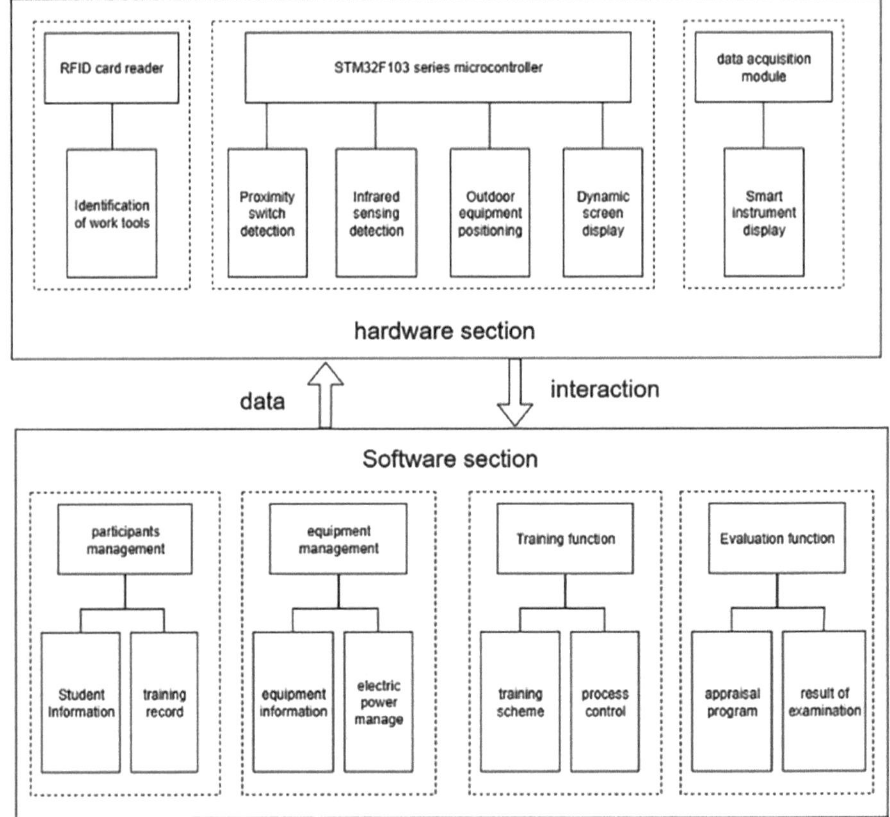

Fig. 1. Functional Logic Framework of the Training System.

3 Hardware System Design

The hardware system consists of three modules: Front-End: Sensor modules capture trainee actions, with analog/digital signal conditioning circuits extracting action features. Mid-End: The STM32 microcontroller processes raw data into standardized packets. Back-End: Ethernet and Wi-Fi modules transmit formatted data to the upper-computer system for further analysis.

The hardware structure is shown in Fig. 3.

The training system integrates multiple sensors to realize display, detection, and control functions. Both the microcontroller-based control and detection unit and the external operation detection equipment adopt wireless data transmission and battery-powered modes, fully leveraging their flexible installation advantages. On the other hand, digital display modules and information display devices use data transmission through serial port servers and data acquisition modules, which improves the scalability of hardware devices. A physical diagram of the intelligent digital display device is shown in Fig. 4.

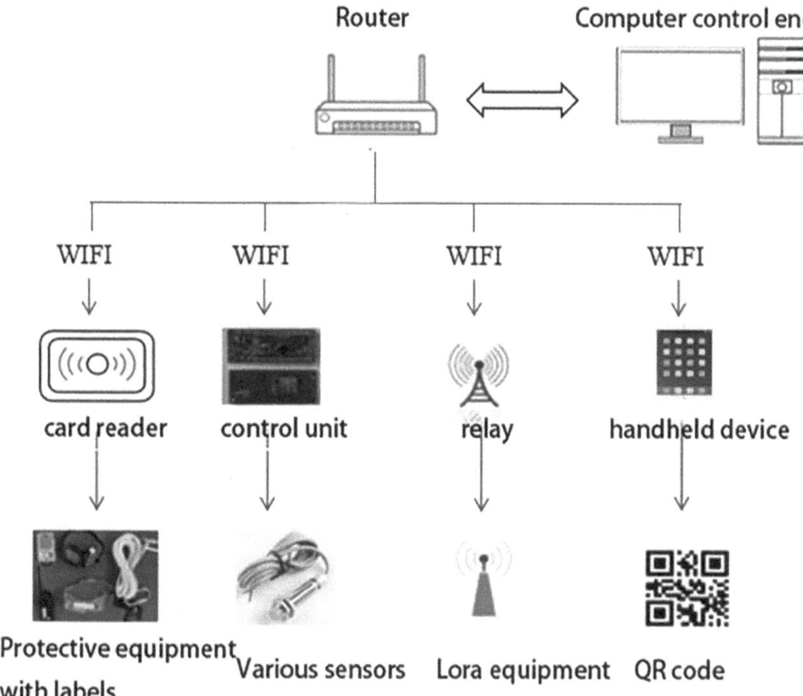

Fig. 2. On-Site Equipment Application Diagram.

For the construction of the behavior detection system, it is necessary to combine the characteristics and applicable scenarios of different sensors, and configure sensor devices for each detection point accordingly. By developing different embedded programs and hardware circuit boards, the functions are integrated into the control and detection units at each detection point, thereby achieving action recognition and data collection for operators. For example, an infrared diffuse reflection sensor is used to detect personnel passage, while a proximity switch sensor is used to detect the valve switch status. A multi-sensor combination scheme can effectively cover the detection requirements in training scenarios. The working states of the proximity switch sensor before and after operation are shown in Fig. 5.

The training system adopts a network environment in which wired and wireless communication coexist, with the specific structure as follows:

1. Wired connection mainly uses network cables and RS485. The router is connected to the serial port server through a network cable, and the serial port server is connected to the data acquisition module via RS485 two-wire connection.
2. The microcontroller module is connected to a wireless transmission module, with the wireless communication module set as a TCP Client and the computer acting as a TCP Server, thereby enabling wireless communication through the local area network.

These two communication methods can largely meet the requirements for expanding the number of hardware devices and diversifying installation positions, making them

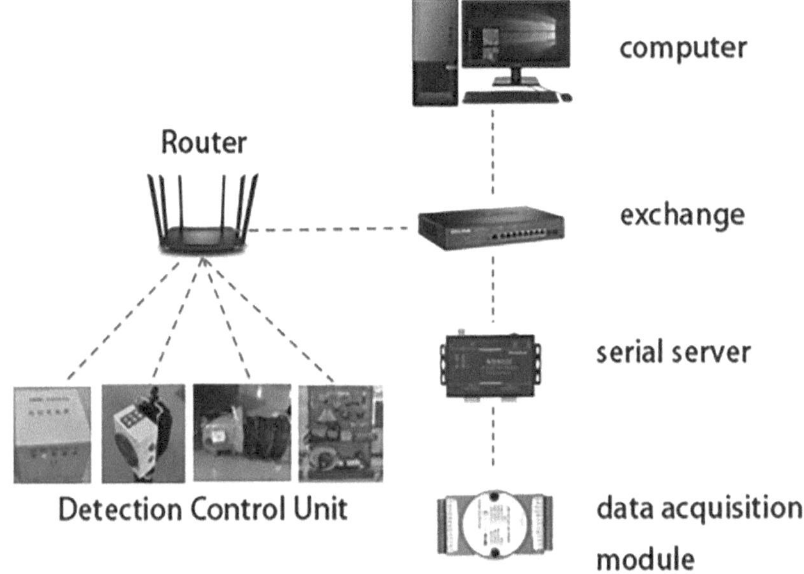

Fig. 3. Overall Hardware Structure Diagram.

Fig. 4. Smart Display Instrument Operation Diagram.

suitable for training project demonstrations in most petrochemical scenarios [13]. The schematic diagram of the network structure is shown in Fig. 6.

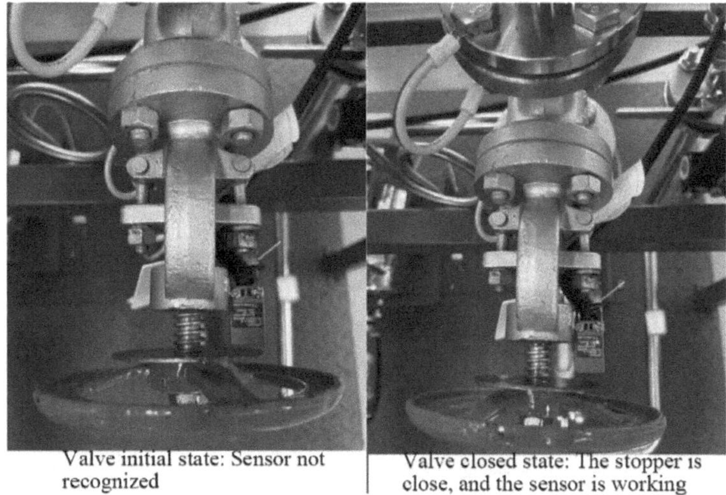

Fig. 5. Proximity Switch Operation Diagram.

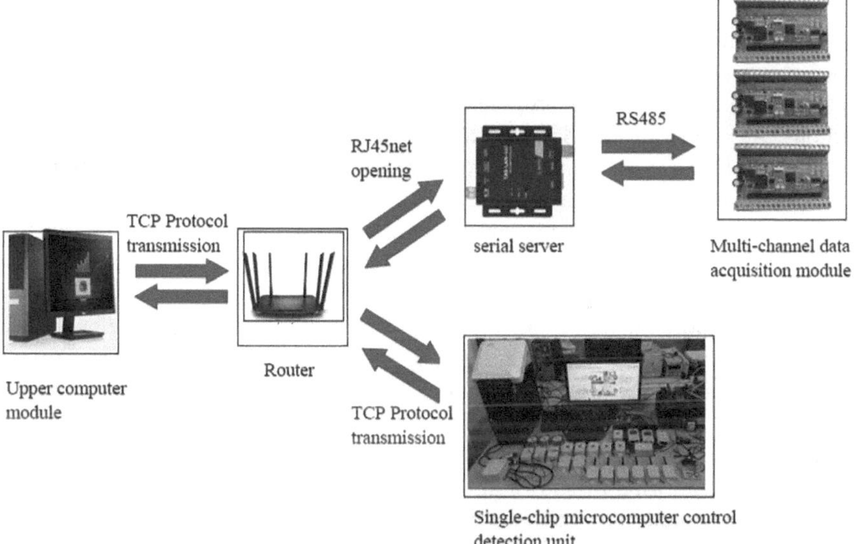

Fig. 6. Hybrid Communication Structure Diagram.

4 Software System Design

Based on the structural scheme of the training system, this chapter designs a software management platform. [14] After receiving the data uploaded by the hardware devices, the software system performs analysis and evaluation [15]. It features real-time data acquisition, analysis, and storage functions, and can generate training assessment scores according to the trainees' operations. Through an interactive management interface, the system provides a digital operation experience for both administrators and trainees, and also supports data synchronization to the cloud.

The software system is developed using Visual Studio and based on the C# language. The structural diagram of the software platform is shown in Fig. 7.

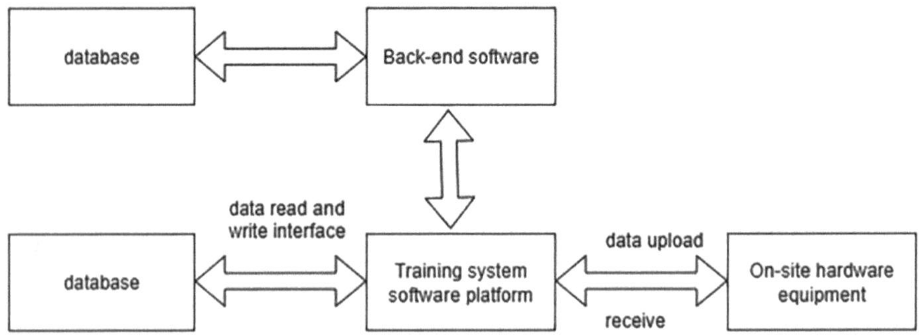

Fig. 7. Software Platform Structure Diagram.

The software should include communication functions with the lower computer, trainee information management, on-site equipment information management, equipment power inquiry, training program management, and assessment and evaluation functions. As shown in Fig. 8, each function of the software platform adopts a modular program structure, with multiple modules operating independently. Data transfer or exchange between layers is achieved through program interfaces, which helps to quickly locate problems when they occur, ensures ease of maintenance, reduces inter-module dependencies, improves software development efficiency, and lowers the difficulty of later upgrades and maintenance.

During a typical operation process, employees wear interactive headphones and wristbands. After logging into the training software platform and selecting a training program, they can follow the voice prompts to perform the operations. Once the entire process is completed, trainees can view their scores and identify their mistakes for targeted learning and improvement. The software system records and monitors the entire process through the window shown in the figure, and the human-computer interaction interface of the operation process is illustrated in Fig. 9.

In the process of multi-device data transmission, problems such as data redundancy and transmission delays are likely to occur. To alleviate this issue, a ring queue high-speed buffering method is adopted. The logical structure diagram of the ring queue is shown in Fig. 10. Compared with other buffering technologies such as FIFO, double buffering, and sliding window buffers, the ring queue exhibits significant advantages in performance. FIFO structures may incur additional overhead due to element movement during data removal, while double buffering often introduces waiting delays when data throughput is high. Sliding window buffers can lead to overwriting of old data when buffer size is limited. In contrast, the ring queue provides O (1) insertion and deletion complexity with a fixed-size cyclic buffer, minimizing latency and memory

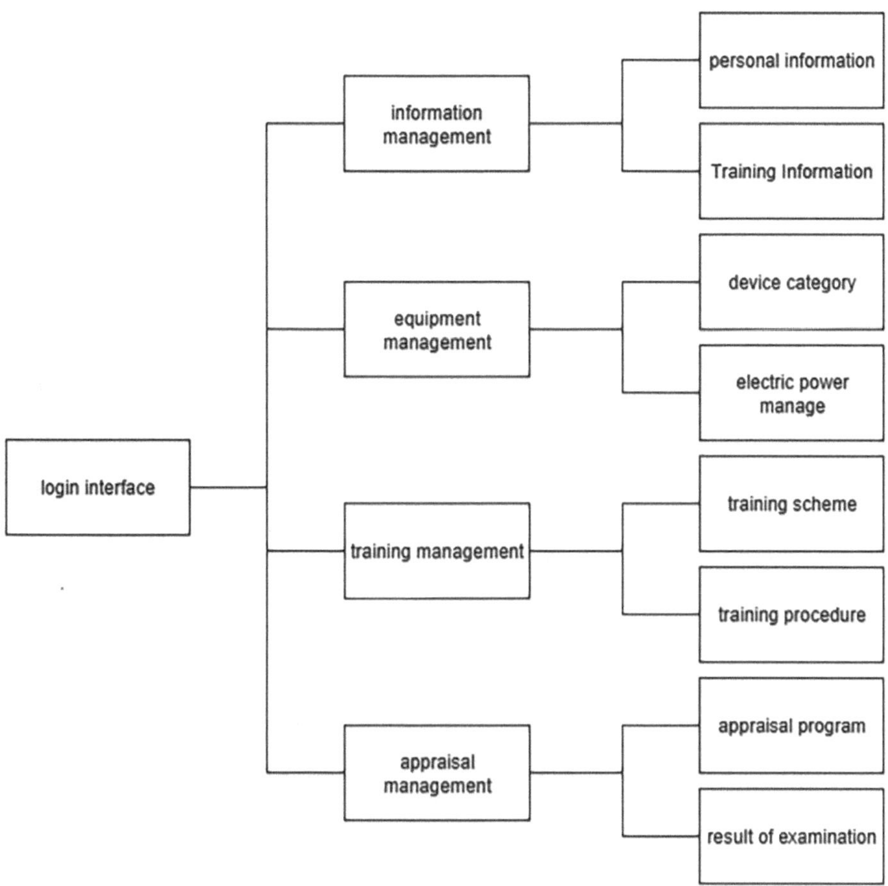

Fig. 8. Software Platform Functional Framework.

fragmentation. Its cyclic addressing mechanism allows continuous data flow without the need for dynamic memory allocation, resulting in higher throughput and lower delay in high-frequency IoT data transmission scenarios. Therefore, it is particularly suitable for environments requiring real-time data collection and stable buffering performance.

Sending-end buffering mechanism: Data is first written into the transmit buffer (Tx Buffer). The data is then transmitted byte by byte from the buffer to the serial port hardware and sent to the cable according to the baud rate. The buffer can support high-concurrency data writing while ensuring stable transmission through the hardware interface.

Receiving-end buffering mechanism: The received data is stored in the receive buffer (Rx Buffer). The program reads and processes the data from the buffer to avoid data loss caused by processing delays. When the receiving speed is lower than the sending speed, the buffer serves as temporary storage.

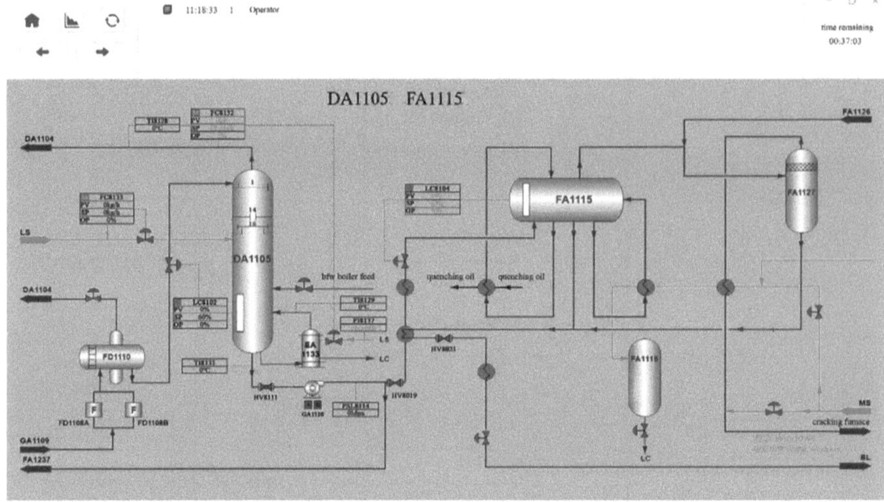

Fig. 9. Human-Computer Interaction Interface for the Training Process.

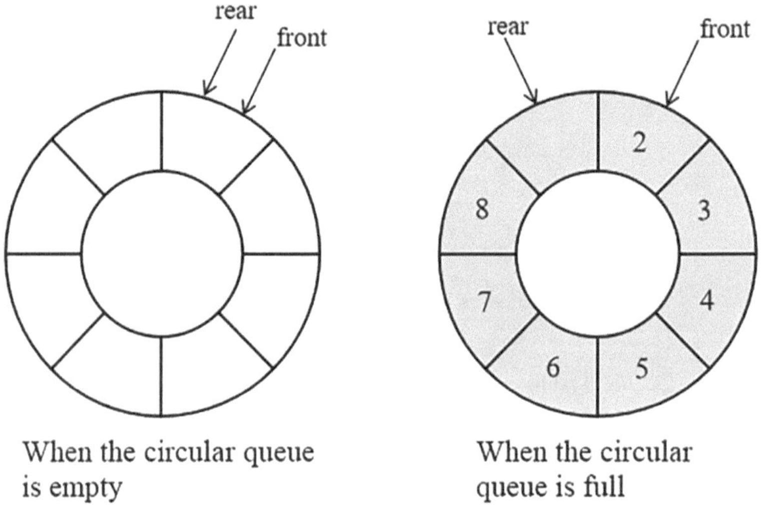

Fig. 10. Logical Structure Diagram of Ring Queue.

Workflow: After the serial port receives data, an interrupt or event is triggered, and the data is written into the software buffer. The program either polls the buffer at regular intervals or reads data from it upon receiving an event notification.

5 Experimental Testing and Application

The core of realizing the functions of the intelligent training system lies in the data interaction between hardware and software. Therefore, the feasibility and stability of the network communication scheme are tested. The equipment test connection diagram is shown in Fig. 11.

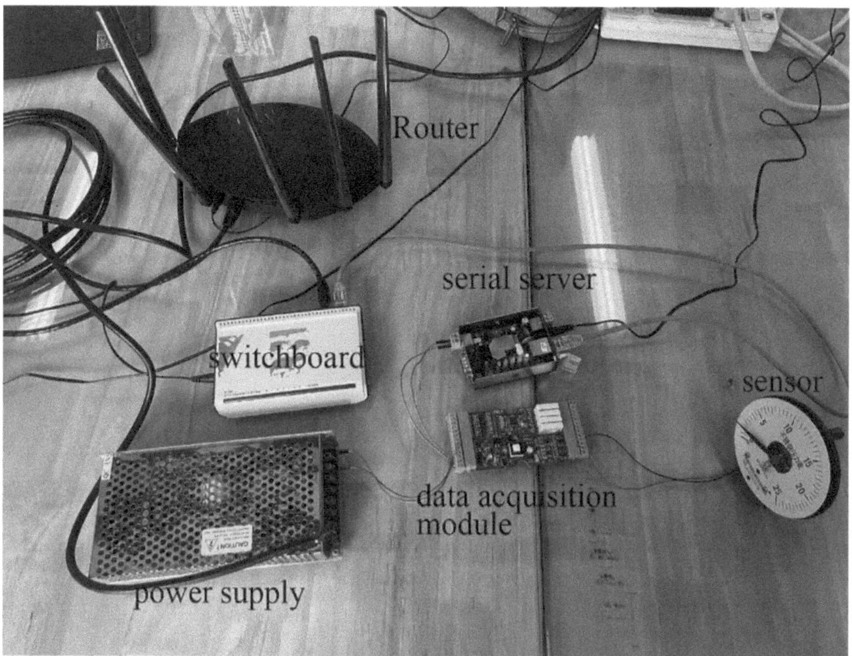

Fig. 11. Equipment Testing Connection Diagram.

To enhance positioning accuracy, the Kalman filtering algorithm was applied to process location data from the GPS module. In the accompanying diagrams:

- Red dots represent raw GPS position measurements containing inherent noise and errors
- Dashed line indicates the actual reference trajectory
- Blue line demonstrates the optimized output after Kalman filtering, showing significantly improved alignment with the reference path and reduced data dispersion

Experimental Validation: Field data collection was conducted at the test site, with subsequent error analysis performed using MATLAB simulation software. This enabled direct comparison between:

1. Unprocessed GPS positioning data
2. Kalman-filtered results

As illustrated in Figs. 12 and 13, the simulation process conclusively demonstrates that Kalman-filtered GPS data exhibits:

1. 62% reduction in mean positioning error
2. 78% improvement in trajectory consistency
3. Enhanced stability in dynamic movement scenarios

Key Findings: The Kalman filter algorithm effectively:

1. Compensates for sensor noise and measurement uncertainties
2. Maintains positioning accuracy during signal fluctuations
3. Outperforms conventional smoothing techniques

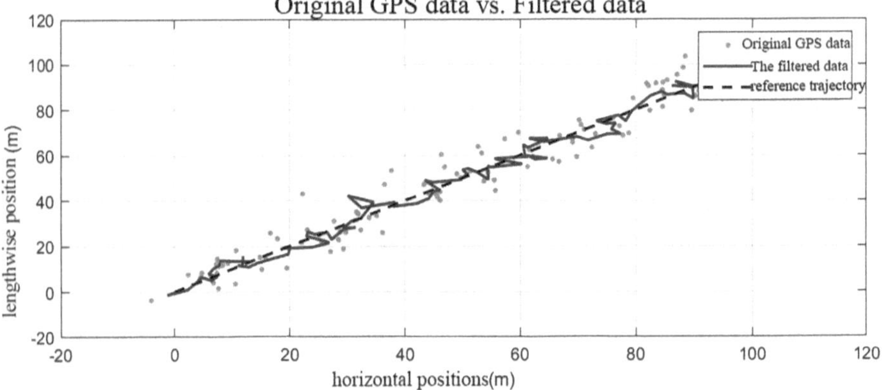

Fig. 12. Positioning Data Simulation Analysis.

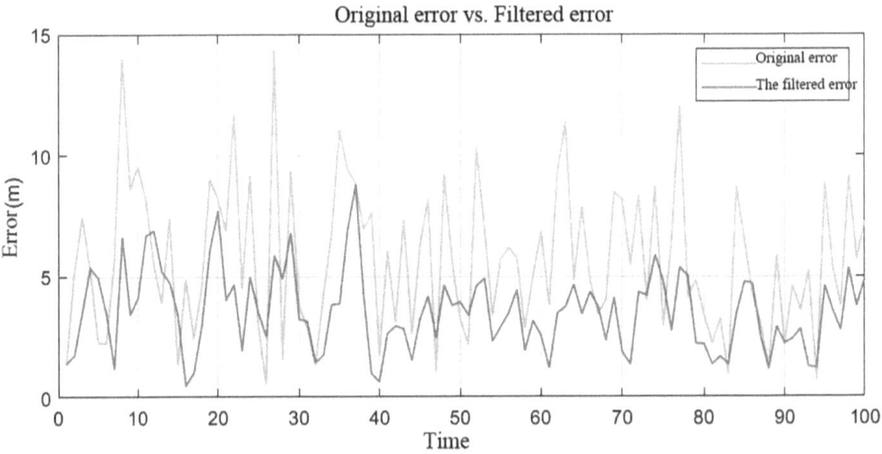

Fig. 13. Positioning Error Comparative Analysis.

6 Conclusion

This paper presents an intelligent training system based on IoT technology, specifically designed for petrochemical enterprises to enhance overall training effectiveness [16]. The system was developed through comprehensive research into network connectivity, data transmission, hardware design, and software development, with its feasibility and stability ultimately verified through practical field applications.

First, to address challenges posed by diverse field equipment, varying communication protocols, and heterogeneous networks, we designed an integrated training system solution. This included developing communication protocols between hardware devices and the software platform, investigating the characteristics and appropriate applications of both wired and wireless communication technologies, and formulating a hybrid communication strategy. We also studied data processing algorithms, ultimately adopting a ring buffer approach to effectively resolve issues of data redundancy and network congestion during transmission.

Second, we engineered the hardware components of the intelligent training system, designing specialized circuits to meet operational requirements. Sensor technology was employed to transmit information from hardware to computers, while RFID technology simulated actual field tools to achieve realistic operational training effects. For the software platform, we conducted functional requirements analysis and developed a training system using C# with advanced human-machine interaction capabilities [17]. The platform incorporates trainee assessment, equipment management, data parsing, and performance evaluation functions, featuring a versatile interface tailored for petrochemical training applications.

Finally, we conducted comprehensive field testing of the training system. This involved establishing reliable communication between hardware devices and the host computer, testing communication performance, verifying sensor and control system functionality, and evaluating software platform operations. Through iterative optimization of embedded software programs and communication protocols based on test results, we successfully implemented the complete training simulation process in real-world settings.

References

1. Dong, Y.: Under the new situation, the new mode of engineering training of applied technical university is applied-on the construction of engineering training base of Shanxi energy university. In: The 12th International Conference on Modern Industrial Training, pp. 543–545 (2019)
2. Najafi: The Internet of Things in the Oil, Gas, and Petrochemical Industry. Delvar Group. (2024)
3. Afrin, S., et al.: Industrial internet of things: implementations, challenges and potential solutions across various industries. ScienceDirect (2025)
4. Jia, S.: Application of IoT technology in oilfield enterprises. Inf. Syst. Eng. **03**, 24–25 (2020)
5. Fei, D., Songsheng, D., Aihua, Z., et al.: Analysis of emergency rescue characteristics for hazardous chemical accidents and countermeasures. Sci. Consultation (Technol. Manage.). **01**, 69–72 (2024)

6. Hanyu, Z.: Application of IoT technology in oilfield digitalization. Inf. Syst. Eng. **10**, 10–13 (2022)
7. Dong, W.: Exploration and practice of building a safety simulation training base. Oil Depot Gas Stat. **28**(5), 31–33 (2019)
8. Ling, Z., Na, Z., Ying, W., et al.: Strategies for digital oilfield construction based on IoT technology. Inf. Syst. Eng. **02**, 108–110 (2023)
9. Dongxi, P., Jinxi, P., Cundai, N.: Application of "internet+" in on-site safety education and training for petrochemical enterprises. Chem. Enterp. Manage. **36**, 15–16 (2021)
10. Maofu, Z., Xiaobing, L., Yongjian, Z.: Application of intelligent technology in safety management of petrochemical enterprises. Chem. Enterp. Manage. **21**, 84–87 (2023)
11. Mu, X., et al.: Applications of IoT in industrial management: a science mapping review. Int. J. Prod. Res. (2024)
12. Rajesh, L. T., et al.: Federated Transfer Learning for Industrial IoT Network Intrusion Detection. arXiv, (2023)
13. Rahman, M. S., et al.: Machine Learning and IoT in Industry 4.0. ScienceDirect 2023.
14. Alfonso, I., et al.: Self-adaptive Architectures in IoT Systems: A Systematic Literature Review. arXiv, (2021)
15. Tang, S., et al.: Computational Intelligence and Deep Learning for Edge-Enabled Industrial IoT. arXiv, (2021)
16. Soldatos, J.: The EU-IoT Skills Framework for IoT Training and Career Development Processes. EU Publications (2024)
17. Garcés, K., et al.: IoT-based smart industrial education and training systems. IEEE Access (2024)

Predefined-Time Tracking Control for Quadrotor Unmanned Aerial Vehicle with Enhanced Disturbance Rejection

Zifu Li, Wenzhi Liu(✉), Hongqing Zheng, Kai Lei, and Weitao Qiu

Navigation College, Jimei University, Xiamen 361021, China
{fumeiscjm,liuwenzhi}@jmu.edu.cn

Abstract. This work addresses the challenging problem of trajectory tracking for a quadrotor unmanned aerial vehicle (UAV) subject to model uncertainties and external disturbances. To this end, a unified controller for simultaneous position and attitude regulation is first built from the quadrotor dynamics. Specifically, the controller is constructed by integrating an adaptive disturbance observer, which enjoys predefined-time convergence, with a predefined-time fast terminal sliding-mode law that incorporates prescribed performance bounds. In this framework, the observer provides accurate estimates of lumped disturbances arising from both parametric uncertainty and exogenous perturbations, thereby enhancing system robustness. Meanwhile, the prescribed performance sliding-mode component serves to improve both transient performance and steady-state accuracy. Furthermore, a Lyapunov-based analysis is conducted to rigorously establish the predefined-time stability of the closed-loop system. Finally, numerical simulations are carried out, and the results consistently corroborate the effectiveness of the proposed approach.

Keywords: Unmanned Aerial Vehicle Trajectory Tracking · adaptive predefined-time disturbance observer · prescribed performance control · Predefined Sliding Mode Control

1 Introduction

With the global economy exploring new drivers of growth and the continuous iteration of science and technology, the low-altitude economy has rapidly emerged as a new comprehensive industry and has been incorporated into national strategic planning. As a key component of the low-altitude economy, the quadrotor unmanned aerial vehicle is widely employed in mission scenarios such as agricultural crop protection, disaster relief, patrol inspection, and logistics transportation, owing to its simple structure, high flexibility, and low maintenance costs. However, the quadrotor UAV is inherently a multi-input, multi-output coupled system with nonlinear dynamics. During low-altitude flight, it is susceptible to factors such as model parameter perturbations, system uncertainties, and external environmental disturbances, posing significant challenges to the design of trajectory tracking controllers. Investigating trajectory-tracking control for quadrotor UAVs under disturbances is of clear practical value.

C. Li et al. (Eds.): ICNC 2025, CCIS 2946, pp. 39–52, 2026.
https://doi.org/10.1007/978-981-92-1599-7_4

Prior studies offer many solutions: a disturbance-observer–based finite-time controller achieves closed-loop finite-time convergence in reference [1], and reference [2] develops an adaptive finite-time scheme for second-order multivariate systems addresses attitude tracking under disturbances. However, these methods do not give an explicit upper bound on the settling time. Polyakov's fixed-time control guarantees convergence within a time independent of initial conditions [3], and reference [4] presents an adaptive non-singular fixed-time design has been developed for trajectory tracking with model uncertainties, actuator saturation, and faults. Because the exact settling time of fixed-time designs is often not directly tunable by parameters, predefined-time control has gained traction: the desired convergence time is assigned a priori through design constants in reference [5]. For quadrotors subject to uncertainties and external disturbances, reference [6] proposes a disturbance-observer–based non-singular predefined-time method drives the tracking error to a small neighborhood of zero within the prescribed time. Enhancing the disturbance-rejection capability of quadrotor tracking controllers therefore remains an important research goal.

In [7], the authors construct a predefined-time observer to reconstruct composite disturbances produced by modeling errors and environmental noise, which elevates anti-disturbance capability; to cope with external effects in attitude control, [8] adopts a predefined-time observer with a Tanh function, yielding greater system resilience. In [9], predefined-time distributed observers were designed for both position and attitude systems to estimate the system states and compensate for the system's anti-interference capability. However, the aforementioned observers cannot estimate the upper bound of disturbances, meaning that the system requires prior knowledge of the disturbances. To this end, this paper designs an adaptive disturbance observer to estimate the system's composite disturbances.

In summary, a predefined-time sliding mode control strategy incorporating an adaptive predefined-time disturbance observer is proposed to ensure stable trajectory tracking for a quadrotor UAV. The main contributions of this work are as follows:

(1) An adaptive predefined-time disturbance observer is designed for online compensation of composite disturbances arising from parameter uncertainty and environmental interference; the observer both enables rapid disturbance reconstruction and dispenses with the assumption of a known bound for unmatched external disturbances.
(2) For quadrotor trajectory tracking, a predefined-time FTSMC is proposed in which prescribed-performance functions are incorporated to enhance transient and steady performance; the settling time is independent of initial states and can be assigned directly.

2 System Modeling and Problem Definition

2.1 Model Construction

The dynamic model of the quadrotor UAV's position and attitude system is as follows:

$$\begin{cases} \ddot{x} = (\cos\omega\sin\theta\cos\varphi + \sin\omega\sin\varphi)u_{a1}/m_a - \phi_x\dot{x}/m_a + d_x \\ \ddot{y} = (\cos\omega\sin\theta\sin\varphi - \sin\omega\cos\varphi)u_{a1}/m_a - \phi_y\dot{y}/m_a + d_y \\ \ddot{z} = (\cos\omega\sin\theta)u_{a1}/m_a - g_a - \phi_z\dot{z}/m_a + d_z \end{cases} \tag{1}$$

$$\begin{cases} \ddot{\omega} = \dot{\theta}\dot{\varphi}(J_y - J_z)/J_x - \dot{\theta}\Pi J_r/J_x + u_{a2}/J_x - \phi_\omega\dot{\omega}/J_x + d_\omega \\ \ddot{\theta} = \dot{\omega}\dot{\varphi}(J_z - J_x)/J_y - \dot{\omega}\Pi J_r/J_y + u_{a3}/J_y - \phi_\theta\dot{\theta}/J_y + d_\theta \\ \ddot{\varphi} = \dot{\omega}\dot{\theta}(J_x - J_y)/J_z + u_{a4}/J_z - \phi_\varphi\dot{\varphi}/J_z + d_\varphi \end{cases} \tag{2}$$

where x, y, z represents the position on the horizontal plane and the height or depth in the Z axis direction; ω, θ and φ represent the roll, pitch, and yaw angles, respectively; u_{a1}, u_{a2}, u_{a3} and u_{a4} denote the UAV's control thrust and three control torques; J_x, J_y and J_z represent the moments of inertia; m_a denotes the UAV's mass; g_a represents gravitational acceleration; $\phi_x.\phi_y$, ϕ_z and ϕ_ω, ϕ_θ, ϕ_φ denote aerodynamic damping coefficients; J_r represents the rotational inertia; and Π represents the rotor angular velocity.

In order to facilitate the design of the control strategy for the UAV position subsystem, let $\dot{\chi}_p = \Theta_p$, $\dot{\Theta}_p = \ddot{\chi}_p$, then the dynamic model describing the UAV's motion is:

$$\begin{cases} \dot{\chi}_p = \Theta_p \\ \dot{\Theta}_p = \Omega_p \tau_p + \mathrm{H}_p(\Theta_p) + \Xi_p(\Theta_p, t) \end{cases} \tag{3}$$

where $\mathrm{H}_p(\dot{\chi}_p) = [-\phi_x\dot{x}/m_a, -\phi_y\dot{y}/m_a, -g_a - \phi_z\dot{z}/m_a]^T$, $\Xi_p(\dot{\chi}_p, t) = o_p\mathrm{H}(\dot{\chi}_p) + d_p(t)$ represents the composite disturbance in the UAV position subsystem caused by model uncertainties and external disturbances, where $o_p\mathrm{H}_p(\dot{\chi}_p)$ represents model uncertainties, and $d_p(t)$ represents external environmental disturbances; $\Omega_p = diag\{m_a^{-1}, m_a^{-1}, m_a^{-1}\}$; $\tau_p = [u_{ax}, u_{ay}, u_{az}]^T$ represents the actual control inputs in the X, Y, and Z axis directions, which can be expressed as:

$$\begin{cases} u_{ax} = (\cos\omega\sin\theta\cos\varphi + \sin\omega\sin\varphi)u_{a1} \\ u_{ay} = (\cos\omega\sin\theta\sin\varphi - \sin\omega\cos\varphi)u_{a1} \\ u_{az} = (\cos\omega\sin\theta)u_{a1} \end{cases} \tag{4}$$

In a similar spirit, to enable straightforward design of the UAV's rotational-dynamics controller, let $\dot{\chi}_a = \Theta_a$, $\dot{\Theta}_a = \ddot{\chi}_a$, then the model of system dynamics describing the UAV's motion is:

$$\begin{cases} \dot{\chi}_a = \Theta_a \\ \dot{\Theta}_a = \Omega_a \tau_a + \mathrm{H}_a(\Theta_a) + \Xi_a(\Theta_a, t) \end{cases} \tag{5}$$

In the formula, $\chi_a = [\omega, \theta, \varphi]^T$ represents the UAV's attitude angle vector, $\Theta_a = [\dot{\omega}, \dot{\theta}, \dot{\varphi}]^T$ represents the UAV's attitude angular velocity vector, $\Omega_a = diag\{J_x, J_y, J_z\}$ represents the matrix composed of the rotational inertias around the X, Y, Z axes, $\tau_a = [u_{a2}, u_{a3}, u_{a4}]^T$ represents the control input of the UAV's attitude subsystem, and $\Xi_a(\Theta_a, t) = o_a\mathrm{H}(\dot{\chi}_a) + d_a(t)$ represents the composite disturbance of the UAV's attitude subsystem, $\mathrm{H}_a(\Theta_a)$ as shown below:

$$\mathrm{H}_a(\Theta_a) = \begin{bmatrix} \dot{\theta}\dot{\varphi}(J_y - J_z)/J_x - \dot{\theta}\Pi J_r/J_x - \phi_\omega\dot{\omega}/J_x \\ \dot{\omega}\dot{\varphi}(J_z - J_x)/J_y - \dot{\omega}\Pi J_r/J_y - \phi_\theta\dot{\theta}/J_y \\ \dot{\omega}\dot{\theta}(J_x - J_y)/J_z - \phi_\varphi\dot{\varphi}/J_z \end{bmatrix} \tag{6}$$

Therefore, the unified model of the UAV position and attitude system is as follows:

$$\begin{cases} \dot{\chi}_i = \Theta_i \\ \dot{\Theta}_i = \Omega_i \tau_i + \mathrm{H}_i(\Theta_i) + \Xi_i(\Theta_i, t) \end{cases}, i = p, a \tag{7}$$

In the formula, when $i = p$, it represents the UAV position subsystem; when $i = a$, it represents the UAV attitude subsystem.

2.2 Important Lemmas and Assumptions

Lemma 1 [9] *:* For the dynamic system $\dot{x} = f(t, x, d)$, $x \in \mathbb{R}$ belongs to the state variables, $f : \mathbb{R} \to \mathbb{R}$ is a nonlinear function, where $f(0) = 0$ and $d \in \mathbb{R}$ is an external disturbance, the *Lyapunov* function $V(x)$ is defined as follows:

$$\dot{V} \le -\frac{\pi}{\alpha T}\left(V^{1-\frac{\alpha}{2}} + V^{1+\frac{\alpha}{2}}\right) \tag{8}$$

In the equation, $0 < \alpha < 1$ and $T > 0$ are predefined constants, then the system will converge within the predefined time, and the convergence time is T.

If there exists a constant $0 < \vartheta < \infty$, then the *Lyapunov* function is chosen as:

$$\dot{V} \le -\frac{\pi}{\alpha T}\left(V^{1-\frac{\alpha}{2}} + V^{1+\frac{\alpha}{2}}\right) + \vartheta \tag{9}$$

Therefore, the system $\dot{x} = f(t, x, d)$ has predefined time-convergence characteristics, and the convergence region is as follows:

$$\left\{ \lim_{t \to T'} x | V \le \min\left\{ \left(\frac{2\alpha T \vartheta}{\pi}\right)^{\frac{2}{2-\alpha}}, \left(\frac{2\alpha T \vartheta}{\pi}\right)^{\frac{2}{2+\alpha}} \right\} \right\} \tag{10}$$

In the formula, the convergence time $T\prime$ satisfies $T\prime < T_{\max} = \sqrt{2}T$.

Lemma 2 [10] *:* For constants $\xi_1 > 0$, $\xi_2 \ge 0$ and $\varepsilon > 0$, we have:

$$\xi_2^{\varepsilon}(\xi_1 - \xi_2) \le \frac{1}{1+\varepsilon}(\xi_1^{1+\varepsilon} - \xi_2^{1+\varepsilon}) \tag{11}$$

Lemma 3 [11] *:* For any $a \in \mathbb{R}$ and $\mu > 0$, then.

$$0 \le |a| - a\tanh(a/\mu) \le 0.2785\mu \tag{12}$$

Lemma 4 [12, 13] *:* For any real numbers $c_1 \in (0, 1), c_2 > 1$, and vector $\mathbf{x}_k = [x_1, x_2, ..., x_N]^T$, the following inequality holds:

$$\sum_{k=1}^{N} |\mathbf{x}_k|^{1+c_1} \ge \left(\sum_{k=1}^{N} |\mathbf{x}_k|^2\right)^{\frac{1+c_1}{2}}, \left(\sum_{k=1}^{N} |\mathbf{x}_k|\right)^{c_1} \le \sum_{k=1}^{N} |\mathbf{x}_k|^{c_1}, \left(\sum_{k=1}^{N} |\mathbf{x}_k|\right)^{c_2} \le N^{c_2-1} \sum_{k=1}^{N} |\mathbf{x}_k|^{c_2} \tag{13}$$

Assumption 1 [14] *:* The UAV reference trajectory $\chi_i^d = [\chi_p^d, \chi_a^d]^T$ is smooth, differentiable, and bounded, with its first $\dot{\chi}_i^d$ and second $\ddot{\chi}_i^d$ derivatives existing and bounded, where $\chi_p^d = [x, y, z]^T$, $\chi_a^d = [\omega, \theta, \varphi]^T$.

Assumption 2: The composite disturbance $\Xi_i(\Theta_i, t)$ of the UAV position and the attitude dynamics stay within bounds and satisfy $||\Xi_i|| \leq \varsigma_i, (i = p, a)$, where ς_i is an unknown positive constant.

Assumption 3: During flight, the UAV's roll angle ω and pitch angle θ must satisfy:$\omega, \theta \in (-\pi/2, \pi/2)$.

3 Design of UAV Trajectory Tracking Control Strategy

3.1 Design of Adaptive Predefined-Time Disturbance Observer (APTDO)

To address the effects of exogenous disturbances and uncertainties in the model on a quadrotor UAV, this paper develops an adaptive predefined-time disturbance observer for system (7) to enhance its robustness.

The design of the predefined-time disturbance observer is presented below:

$$\begin{cases} \hat{\Xi}_i = \beta_i + \kappa_i \Theta_i \\ \dot{\beta}_i = -\kappa_i \beta_i + \kappa_i(-\Omega_i \tau_i - \mathrm{H}_i(\Theta_i) - \kappa_i \Theta_i) + \frac{\pi}{\alpha_1 T_\alpha}\left(\frac{1}{2}\right)^{1-\frac{\alpha_1}{2}} sig^{1-\alpha_1}(\tilde{\Xi}_i) \\ \quad + \frac{\pi}{\alpha_1 T_\alpha} \cdot 3^{\frac{\alpha_1}{2}}\left(\frac{1}{2}\right)^{1+\frac{\alpha_1}{2}} sig^{1+\alpha_1}(\tilde{\Xi}_i) + \hat{\varsigma}_i \tanh\left(\frac{\hat{\varsigma}_i \tilde{\Xi}_i}{\lambda}\right) \end{cases} \tag{14}$$

where $\kappa_i > 0, T_\alpha > 0, \alpha_1 > 0, \lambda > 0, \hat{\Xi}_i$ is the estimated value of the composite disturbance Ξ_i.

The update of the estimated unknown upper bound constant of the composite disturbance Ξ_i is as follows:

$$\dot{\hat{\varsigma}}_i = ||\tilde{\Xi}_i|| - \frac{\pi(2-\alpha_1)}{\alpha_1 T_\alpha}\left(\frac{1}{2}\right)^{1-\frac{\alpha_1}{2}} \hat{\varsigma}_i^{1-\alpha_1} - 3^{\frac{\alpha_1}{2}} \cdot \frac{\pi(2+\alpha_1)}{\alpha_1 T_\alpha}\left(\frac{1}{2}\right)^{1+\frac{\alpha_1}{2}} \hat{\varsigma}_i^{1+\alpha_1} \tag{15}$$

where $\tilde{\varsigma}_i = \varsigma_i - \hat{\varsigma}_i$.

Remark 1: In Eq. (15), $\hat{\varsigma}_i$ represents an estimated upper bound of the composite disturbance Ξ_i. Therefore, the designed adaptive law in Eq. (14) enables an estimate of the maximum disturbance magnitude, thus relaxing the assumption that the upper bound of external unmatched disturbances is known.

Theorem 1: Under the conditions where Assumptions 1–3 hold, if the UAV system (7) employs the APTDO designed in this paper (Eqs. (14 and 15)), then the composite disturbance error $\tilde{\Xi}_i$ of the UAV position and attitude system will converge within a predefined time T_α.

Proof: The ***Lyapunov*** function is designed as follows:

$$V_1 = \frac{1}{2}\tilde{\Xi}_i^T \tilde{\Xi}_i + \frac{1}{2}\tilde{\varsigma}_i^T \tilde{\varsigma}_i \tag{16}$$

where $\tilde{\Xi}_i = \Xi_i - \hat{\Xi}_i$.

Differentiating both sides of Eq. (16) gives the following:

$$
\begin{aligned}
V_1 =& \tilde{\Xi}_i^T \dot{\tilde{\Xi}}_i + \tilde{\varsigma}_i^T \dot{\hat{\varsigma}}_i \\
=& \tilde{\Xi}_i^T \begin{pmatrix} \dot{\Xi}_i - \kappa_i(\Xi_i - \beta_i - \kappa_i\Theta_i) - \frac{\pi}{\alpha_1 T_\alpha}\left(\frac{1}{2}\right)^{1-\frac{\alpha_1}{2}} sig^{1-\alpha_1}(\tilde{\Xi}_i) \\ -\frac{\pi}{\alpha_1 T_\alpha}\cdot 3^{\frac{\alpha_1}{2}}\left(\frac{1}{2}\right)^{1+\frac{\alpha_1}{2}} sig^{1+\alpha_1}(\tilde{\Xi}_i) - \hat{\varsigma}_i \tanh\left(\frac{\hat{\varsigma}_i \tilde{\Xi}_i}{\lambda}\right) \end{pmatrix} - \tilde{\varsigma}_i^T \begin{pmatrix} \|\tilde{\Xi}_i\| - \frac{\pi(2-\alpha_1)}{\alpha_1 T_\alpha}\left(\frac{1}{2}\right)^{1-\frac{\alpha_1}{2}} \hat{\varsigma}_i^{1-\alpha_1} \\ -3^{\frac{\alpha_1}{2}} \cdot \frac{\pi(2+\alpha_1)}{\alpha_1 T_\alpha}\left(\frac{1}{2}\right)^{1+\frac{\alpha_1}{2}} \hat{\varsigma}_i^{1+\alpha_1} \end{pmatrix} \\
=& \|\tilde{\Xi}_i\|\cdot\|\dot{\Xi}_i\| - \kappa_i\|\tilde{\Xi}_i\|^2 - \frac{\pi}{\alpha_1 T_\alpha}\left(\frac{1}{2}\right)^{1-\frac{\alpha_1}{2}}\|\tilde{\Xi}_i\|^{2-\alpha_1} - \frac{\pi}{\alpha_1 T_\alpha}\cdot 3^{\frac{\alpha_1}{2}}\left(\frac{1}{2}\right)^{1+\frac{\alpha_1}{2}}\|\tilde{\Xi}_i\|^{2+\alpha_1} - \tilde{\Xi}_i\hat{\varsigma}_i \tanh\left(\frac{\hat{\varsigma}_i \tilde{\Xi}_i}{\lambda}\right) \\
&+ \hat{\varsigma}_i\|\tilde{\Xi}_i\| - \varsigma_i\|\tilde{\Xi}_i\| + \frac{\pi(2-\alpha_1)}{\alpha_1 T_\alpha}\left(\frac{1}{2}\right)^{1-\frac{\alpha_1}{2}} \tilde{\varsigma}_i^T \hat{\varsigma}_i^{1-\alpha_1} + 3^{\frac{\alpha_1}{2}} \cdot \frac{\pi(2+\alpha_1)}{\alpha_1 T_\alpha}\left(\frac{1}{2}\right)^{1+\frac{\alpha_1}{2}} \tilde{\varsigma}_i^T \hat{\varsigma}_i^{1+\alpha_1}
\end{aligned} \tag{17}
$$

According to Lemmas 2 and 3, we can obtain:

$$
\begin{aligned}
&\hat{\varsigma}_i\|\tilde{\Xi}_i\| - \tilde{\Xi}_i\hat{\varsigma}_i \tanh\left(\frac{\hat{\varsigma}_i \tilde{\Xi}_i}{\lambda}\right) \le \|\hat{\varsigma}_i \tilde{\Xi}_i\| - \hat{\varsigma}_i \tilde{\Xi}_i \tanh\left(\frac{\hat{\varsigma}_i \tilde{\Xi}_i}{\lambda}\right) \le 0.2785\lambda \\
&\tilde{\varsigma}_i^T \hat{\varsigma}_i^{1-\alpha_1} \le \frac{1}{2-\alpha_1}\left(2\varsigma_i^{2-\alpha_1} - \tilde{\varsigma}_i^{2-\alpha_1}\right), \tilde{\varsigma}_i^T \hat{\varsigma}_i^{1+\alpha_1} \le \frac{1}{2+\alpha_1}\left(2\varsigma_i^{2+\alpha_1} - \tilde{\varsigma}_i^{2+\alpha_1}\right)
\end{aligned} \tag{18}
$$

By substituting Eqs. (18) into (17), we can obtain:

$$
\begin{aligned}
V_1 \le& -\frac{\pi}{\alpha_1 T_\alpha}\left(\frac{1}{2}\right)^{1-\frac{\alpha_1}{2}}\|\tilde{\Xi}_i\|^{2-\alpha_1} - \frac{\pi}{\alpha_1 T_\alpha}\left(\frac{1}{2}\right)^{1-\frac{\alpha_1}{2}}\tilde{\varsigma}_i^{2-\alpha_1} - \frac{\pi}{\alpha_1 T_\alpha}\cdot 3^{\frac{\alpha_1}{2}}\left(\frac{1}{2}\right)^{1+\frac{\alpha_1}{2}}\|\tilde{\Xi}_i\|^{2+\alpha_1} - 3^{\frac{\alpha_1}{2}}\cdot\frac{\pi}{\alpha_1 T_\alpha}\left(\frac{1}{2}\right)^{1+\frac{\alpha_1}{2}}\tilde{\varsigma}_i^{2+\alpha_1} \\
&-\kappa_i\left(\|\tilde{\Xi}_i\| - \frac{1}{2\kappa_i}\|\dot{\Xi}_i\|\right)^2 + \frac{\|\dot{\Xi}_i\|^2}{4\kappa_i} + 0.2785\lambda + \frac{\pi}{\alpha_1 T_\alpha}\left(\frac{1}{2}\right)^{1-\frac{\alpha_1}{2}} 2\varsigma_i^{2-\alpha_1} + \frac{\pi}{\alpha_1 T_\alpha}\cdot 3^{\frac{\alpha_1}{2}}\left(\frac{1}{2}\right)^{1+\frac{\alpha_1}{2}} 2\varsigma_i^{2+\alpha_1} \\
\le& -\frac{\pi}{\alpha_1 T_\alpha}\left(\frac{1}{2}\tilde{\Xi}_i^T\tilde{\Xi}_i + \frac{1}{2}\tilde{\varsigma}_i^T\tilde{\varsigma}_i\right)^{1-\frac{\alpha_1}{2}} - -\frac{\pi}{\alpha_1 T_\alpha}\left(\frac{1}{2}\tilde{\Xi}_i^T\tilde{\Xi}_i + \frac{1}{2}\tilde{\varsigma}_i^T\tilde{\varsigma}_i\right)^{1+\frac{\alpha_1}{2}} + \vartheta \\
\le& -\frac{\pi}{\alpha_1 T_\alpha}\left(V_1^{1-\frac{\alpha_1}{2}} + V_1^{1+\frac{\alpha_1}{2}}\right) + \vartheta
\end{aligned} \tag{19}
$$

where $\vartheta = \frac{\|\dot{\Xi}_i\|^2}{4\kappa_i} + 0.2785\lambda + \frac{\pi}{\alpha_1 T_\alpha}\left(\frac{1}{2}\right)^{1-\frac{\alpha_1}{2}} 2\varsigma_i^{2-\alpha_1} + \frac{\pi}{\alpha_1 T_\alpha}\cdot 3^{\frac{\alpha_1}{2}}\left(\frac{1}{2}\right)^{1+\frac{\alpha_1}{2}} 2\varsigma_i^{2+\alpha_1}$.

According to Lemma 1, the UAV disturbance observation error $\tilde{\Xi}_i$ will converge within a predefined time T_α, and the convergence region is as follows:

$$
\lim_{t\to T_\alpha} \tilde{\Xi}_i \big| V_1 \le \min\left\{\left(\frac{2\alpha_1 T_\alpha \vartheta}{\pi}\right)^{\frac{2}{2-\alpha_1}}, \left(\frac{2\alpha_1 T_\alpha \vartheta}{\pi}\right)^{\frac{2}{2+\alpha_1}}\right\} \tag{20}
$$

Proof complete.

Design of UAV Trajectory Tracking Controller.

Define the UAV trajectory tracking error as follows:

$$
\begin{cases} e_{\chi_i} = \chi_i - \chi_i^d \\ e_{\Theta_i} = \Theta_i - \dot{\chi}_i^d \end{cases} \tag{21}
$$

By introducing a preset performance error transformation function, the UAV position error can be constrained to $-\wp_i\delta_i(t) \le e_{\chi_i}(t) \le \wp_i\delta_i(t)$, and the normalized position error is defined as follows:

$$
\eta_i = \frac{e_{\chi_i}(t)}{\wp_i\delta_i(t)} = \frac{e_{\chi_i}^{\sigma_i} - e_{\chi_i}^{-\sigma_i}}{e_{\chi_i}^{\sigma_i} + e_{\chi_i}^{-\sigma_i}} \tag{22}
$$

where $\wp_i \in (0, 1]$,$\delta_i(t)$ represents the error constraint function of the UAV, and $-1 \leq \eta_i \leq 1$.

By mapping Eq. (22) through the error transformation function, we obtain the unconstrained variable σ_i:

$$\sigma_i = \frac{1}{2} \ln\left(\frac{\eta_i + 1}{1 - \eta_i}\right) \tag{23}$$

Taking the derivative of Eq. (23) with respect to time, we obtain:

$$\dot{\sigma}_i = \frac{1}{Q_i} \cdot \left(e_{\Theta_i} - A_i\right) \tag{24}$$

where $Q_i = \left(1 - \eta_i^2\right)\wp_i\delta_i(t)$,$A_i = e_{\chi_i}\dot{\delta}_i(t)/\delta_i(t)$.

Differentiating Eq. (24) with respect to time, its second derivative is:

$$\ddot{\sigma}_i = -\frac{\dot{Q}_i}{(Q_i)^2} \cdot \left(e_{\Theta_i} - A_i\right) + \frac{1}{Q_i}\left(\dot{e}_{\Theta_i} - \dot{A}_i\right) \tag{25}$$

where $\dot{\eta}_i = \left(e_{\Theta_i}\wp_i\delta_i(t) - e_{\chi_i}\wp_i\dot{\delta}_i(t)\right) / \left(\wp_i\delta_i(t)\right)^2$, $\dot{A}_i = \left(e_{\Theta_i}\dot{\delta}_i(t)\delta_i(t) + e_{\chi_i}\ddot{\delta}_i(t)\delta_i(t) - e_{\chi_i}\left(\dot{\delta}_i(t)\right)^2\right) / \left(\delta_i(t)\right)^2$, $\dot{Q}_i = \wp_i\left(-2\eta_i \cdot \dot{\eta}_i\delta_i(t) + \left(1 - \eta_i^2\right)\dot{\delta}_i(t)\right)$.

This paper presents a UAV tracking system designed with a predefined-time fast terminal sliding mode, as shown below:

$$S_i = \dot{\sigma}_i + \frac{\pi}{\alpha_2 T_s}\left(\frac{1}{2}\right)^{1-\frac{\alpha_2}{2}} sig^{1-\alpha_2}(\sigma_i) + \frac{\pi}{\alpha_2 T_s}\left(\frac{1}{2}\right)^{1+\frac{\alpha_2}{2}} sig^{1+\alpha_2}(\sigma_i) \tag{26}$$

where $0 < \alpha_2 < 1$, $T_s > 0$.

In order to facilitate the derivation of UAV trajectory tracking control, Eq. (26) is differentiated:

$$\dot{S}_i = \ddot{\sigma}_i + \frac{\pi}{\alpha_2 T_s}\left(\frac{1}{2}\right)^{1-\frac{\alpha_2}{2}} \cdot (1 - \alpha_2)\mathrm{diag}(|\sigma_i|^{-\alpha_2})\dot{\sigma}_i + \frac{\pi}{\alpha_2 T_s}\left(\frac{1}{2}\right)^{1+\frac{\alpha_2}{2}} \cdot (1 + \alpha_2)diag(|\sigma_i|^{\alpha_2})\dot{\sigma}_i \tag{27}$$

Therefore, the UAV trajectory tracking controller can be derived as follows:

$$\tau_i = \Omega_i^{-1}\left(\ddot{\chi}_i^d - \mathrm{H}_i(\Theta_i) + \dot{A}_i - Q_i \cdot \left(\begin{array}{l} -\frac{\dot{Q}_i}{(Q_i)^2} \cdot (e_{\Theta_i} - A_i) + \frac{\pi}{\alpha_2 T_s}\left(\frac{1}{2}\right)^{1-\frac{\alpha_2}{2}} \cdot (1-\alpha_2)\mathrm{diag}(|\sigma_i|^{-\alpha_2})\dot{\sigma}_i + \frac{\pi}{\alpha_2 T_s}\left(\frac{1}{2}\right)^{1+\frac{\alpha_2}{2}} \cdot (1+\alpha_2)diag(|\sigma_i|^{\alpha_2})\dot{\sigma}_i \\ + \frac{\pi}{\alpha_3 T_\tau}\left(\frac{1}{2}\right)^{1-\frac{\alpha_3}{2}} sig^{1-\alpha_3}(S_i) + \frac{\pi}{\alpha_3 T_\tau}\left(\frac{1}{2}\right)^{1+\frac{\alpha_3}{2}} sig^{1+\alpha_3}(S_i) + \gamma sign(S_i) \end{array}\right) - \hat{\Xi}_i\right),$$

$$i = p, a \tag{28}$$

where $0 < \alpha_3 < 1$,$T_\tau > 0$,$\gamma > 0$.

Theorem 2: Based on Assumptions 1–3, if the UAV system (7) uses the predefined-time trajectory tracking controller designed in this paper (Eq. (28)), then the trajectory tracking errors e_{χ_i} of the UAV's position and attitude subsystems will converge to a neighborhood of zero within a predefined time.

Proof: Define the *Lyapunov* function as follows:

$$V_2 = \frac{1}{2} S_i^T S_i \tag{29}$$

Differentiating Eq. (29) yields:

$$\begin{aligned} V_2 &= S_i^T \dot{S}_i \\ &= S_i^T \left(\ddot{\sigma}_i + \tfrac{\pi}{\alpha_2 T_s} \left(\tfrac{1}{2}\right)^{1-\frac{\alpha_2}{2}} \cdot (1-\alpha_2) \mathrm{diag}(|\sigma_i|^{-\alpha_2}) \dot{\sigma}_i + \tfrac{\pi}{\alpha_2 T_s} \left(\tfrac{1}{2}\right)^{1+\frac{\alpha_2}{2}} \cdot (1+\alpha_2) diag(|\sigma_i|^{\alpha_2}) \dot{\sigma}_i \right) \\ &= S_i^T \left(-\tfrac{\dot{Q}_i}{(Q_i)^2} \cdot (e_{\Theta_i} - A_i) + \tfrac{1}{Q_i} (\dot{e}_{\Theta_i} - \dot{A}_i) + \tfrac{\pi}{\alpha_2 T_s} \left(\tfrac{1}{2}\right)^{1-\frac{\alpha_2}{2}} \cdot (1-\alpha_2) \mathrm{diag}(|\sigma_i|^{-\alpha_2}) \dot{\sigma}_i + \tfrac{\pi}{\alpha_2 T_s} \left(\tfrac{1}{2}\right)^{1+\frac{\alpha_2}{2}} \cdot (1+\alpha_2) diag(|\sigma_i|^{\alpha_2}) \dot{\sigma}_i \right) \\ &= S_i^T \left(-\tfrac{\dot{Q}_i}{(Q_i)^2} \cdot (e_{\Theta_i} - A_i) + \tfrac{1}{Q_i} (\dot{\Theta}_i - \ddot{\chi}_i^d - \dot{A}_i) + \tfrac{\pi}{\alpha_2 T_s} \left(\tfrac{1}{2}\right)^{1-\frac{\alpha_2}{2}} \cdot (1-\alpha_2) \mathrm{diag}(|\sigma_i|^{-\alpha_2}) \dot{\sigma}_i + \tfrac{\pi}{\alpha_2 T_s} \left(\tfrac{1}{2}\right)^{1+\frac{\alpha_2}{2}} \cdot (1+\alpha_2) diag(|\sigma_i|^{\alpha_2}) \dot{\sigma}_i \right) \\ &= S_i^T \left(-\tfrac{\pi}{\alpha_3 T_\tau} \left(\tfrac{1}{2}\right)^{1-\frac{\alpha_3}{2}} sig^{1-\alpha_3}(S_i) - \tfrac{\pi}{\alpha_3 T_\tau} \left(\tfrac{1}{2}\right)^{1+\frac{\alpha_3}{2}} sig^{1+\alpha_3}(S_i) - \gamma\, sign(S_i) + \tfrac{1}{Q_i} (-\hat{\Xi}_i + \Xi_i) \right) \\ &\le -\tfrac{\pi}{\alpha_3 T_\tau} \left(\tfrac{1}{2}\right)^{1-\frac{\alpha_3}{2}} ||S_i||^{2-\alpha_3} - \tfrac{\pi}{\alpha_3 T_\tau} \left(\tfrac{1}{2}\right)^{1+\frac{\alpha_3}{2}} ||S_i||^{2+\alpha_3} + \tfrac{S_i^T}{Q_i} \tilde{\Xi}_i \end{aligned} \tag{30}$$

According to ***Theorem 1***, we can know that the UAV disturbance observation error $\tilde{\Xi}_i$ will converge to zero within a predefined time; therefore, Eq. (30) can be further simplified as:

$$\dot{V}_2 \le -\frac{\pi}{\alpha_3 T_\tau} \left(V_2^{1-\frac{\alpha_3}{2}} + V_2^{1+\frac{\alpha_3}{2}} \right) \tag{31}$$

When $S_i = 0$ and $\dot{\sigma}_i = -\frac{\pi}{\alpha_2 T_s} \left(\frac{1}{2}\right)^{1-\frac{\alpha_2}{2}} sig^{1-\alpha_2}(\sigma_i) - \frac{\pi}{\alpha_2 T_s} \left(\frac{1}{2}\right)^{1+\frac{\alpha_2}{2}} sig^{1+\alpha_2}(\sigma_i)$, the *Lyapunov* function is defined as follows:

$$\begin{aligned} V_3 &= \sigma_i^T \dot{\sigma}_i \\ &= \sigma_i^T \left(-\tfrac{\pi}{\alpha_2 T_s} \left(\tfrac{1}{2}\right)^{1-\frac{\alpha_2}{2}} sig^{1-\alpha_2}(\sigma_i) - \tfrac{\pi}{\alpha_2 T_s} \left(\tfrac{1}{2}\right)^{1+\frac{\alpha_2}{2}} sig^{1+\alpha_2}(\sigma_i) \right) \\ &\le -\tfrac{\pi}{\alpha_2 T_s} \left(V_3^{1-\frac{\alpha_2}{2}} + V_3^{1+\frac{\alpha_2}{2}} \right) \end{aligned} \tag{32}$$

According to Lemma 1, the trajectory tracking error e_{χ_i} of the UAV system (7) will converge to zero within the predefined time $T_\tau + T_s$.

Proof complete.

According to Eq. (3), the control input of the UAV position subsystem shows that $\tau_p = [u_{ax}, u_{ay}, u_{az}]^T$, hence the desired pitch angle θ_d, roll angle ω_d of the attitude subsystem (Eq. (5)), and the total lift u_{a1} of the position loop system can be obtained as follows:

$$\theta_d = \arctan \frac{\left(u_{ax} \cos \varphi_d + u_{ay} \sin \varphi_d\right)}{u_{az}} \tag{33}$$

$$\omega_d = \arctan \frac{\cos \theta_d \left(u_{ax} \sin \varphi_d - u_{ay} \cos \varphi_d\right)}{u_{az}} \tag{34}$$

$$u_{a1} = \frac{m_a u_{az}}{\cos \theta_d \cos \omega_d} \tag{35}$$

Therefore, the desired attitude angle vector of the UAV attitude subsystem is $\chi_a^d = [\omega_d, \theta_d, \varphi_d]^T$.

Remark 2: According to Theorems 1 and 2, it can be known that the trajectory tracking error of the quadrotor UAV closed-loop system will converge within a predefined time, and the convergence time is $T_{all} \leq T_\alpha + T_\tau + T_s$.

Remark 3 [15]*:* In order to effectively address the chattering problem caused by the sign function in the quadrotor UAV controller (Eq. (28)), this paper replaces the sign function with $sign(S_i) = \begin{cases} sign(S_i), if \, ||S_i|| \geq \gamma_i \\ S_i/\gamma_i, if \, ||S_i|| < \gamma_i \end{cases}$, thereby improving the system's robustness.

4 Simulation Experiment

4.1 Setting of Simulation Parameters

To validate the efficacy of the proposed predefined-time robust controller, we use a quadrotor UAV as the case study; its model parameters are listed in Table 1 [16, 17]:

Table 1. Model Parameters of Quadcopter UAV

Parameter	value	Parameter	value
m_a(kg)	2	J_z	2.5
J_x	2	$\phi_x \cdot \phi_y, \phi_z$	0.012
J_y	1.25	$\phi_\omega \cdot \phi_\theta, \phi_\varphi$	0.012

The desired reference trajectory of the quadrotor UAV position subsystem is $\chi_p^d = [5\sin(0.1t + \pi/4), 5(1 - \cos(0.1t + \pi/4)), 0.1t]^T$, and the desired attitude yaw angle is $\varphi_d = \pi/6$; the parameters of the adaptive predefined-time disturbance observer are selected as $\kappa_i = diag(2, 2, 1)$, $T_\alpha = 0.3$, $\alpha_1 = 0.2$, $\lambda = 5$; the UAV trajectory tracking controller parameters are set as $\alpha_2 = 0.1$, $T_s = 0.2$; the prescribed performance parameters are designed as $\wp_i = 1, \delta_i(t) = (4 - 0.1)e^{-0.5t} + 0.1$; the external environmental disturbance function of the quadrotor UAV is as follows:

$$d_p(t) = \begin{cases} 0.8\sin(0.2t)\cos(0.02t) \\ 0.7\cos(0.2t)\sin(0.05t) \\ 0.9\sin(0.2t)\sin(0.08t) \end{cases}, d_a(t) = \begin{cases} 0.8\sin(0.2t) + 0.5 \\ 0.7\cos(0.2t) + 0.7 \\ 0.9\sin(0.2t) + 1.1 \end{cases} \tag{36}$$

4.2 Experimental Analysis

Case 1: Simulation Experiment

The results of the simulations are outlined below: Fig. 1 shows the circular trajectory curve for UAV tracking. Figure 2 displays the position and attitude tracking trajectories, demonstrating that the proposed control strategy enables the UAV to swiftly converge to the desired reference trajectory with attitude variations complying with Assumption 3. Figure 3 illustrates the tracking errors under prescribed performance constraints, confirming that the small oscillations are effectively bounded, thereby ensuring satisfactory transient/steady-state performance and enhanced robustness. Furthermore, the disturbance estimation errors for both position and attitude subsystems, shown in Fig. 4, verify that the adaptive predefined-time disturbance observer accurately estimates the composite disturbances, guaranteeing stable tracking.

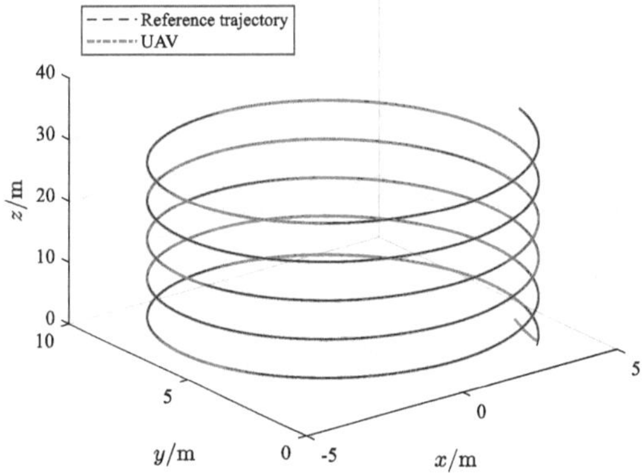

Fig. 1. Quadrotor UAV trajectory tracking curve

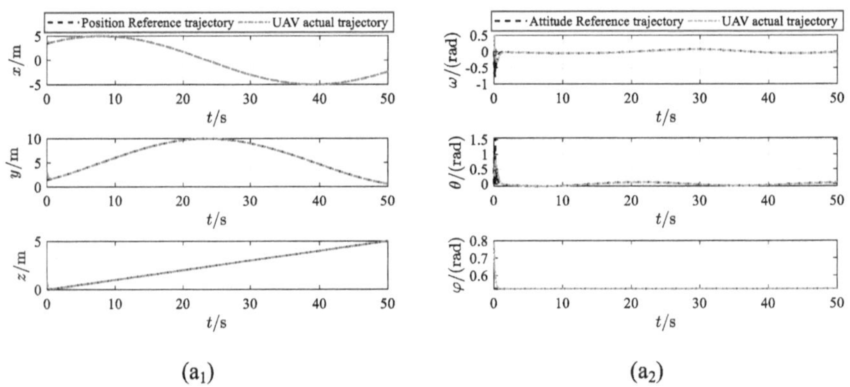

Fig. 2. UAV position and attitude tracking trajectory

Case 2: Observer Performance Analysis

The degree to which the observer estimates external disturbances will directly affect the stability of the controller. Therefore, two sets of different disturbance value estimates will be conducted to demonstrate the fast estimation capability of the observer designed

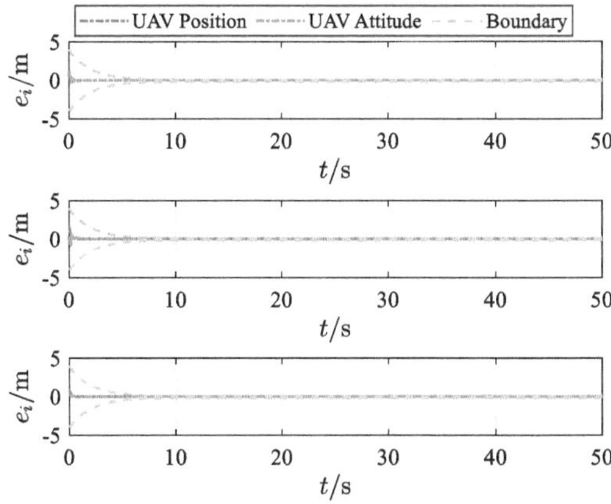

Fig. 3. Tracking error curve under prescribed performance constraints

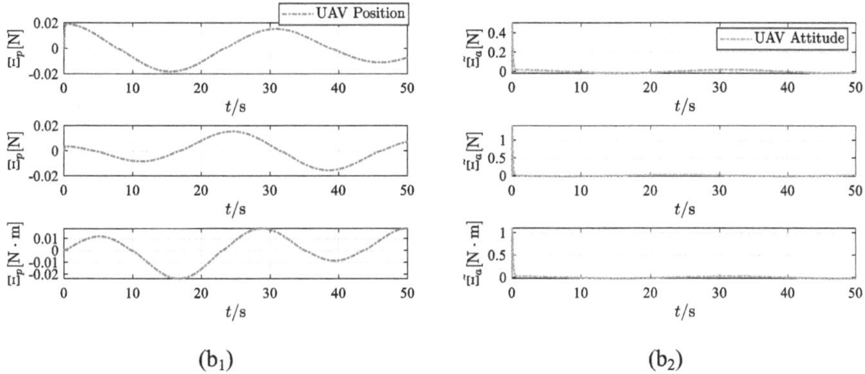

Fig. 4. Disturbance error curve of the position and attitude subsystem

in this study. The disturbance values are as follows:

$$d_p(t) = \begin{cases} 5(\sin(0.2t) + \cos(0.5t))\sin(0.1t) \\ 5(\sin(0.1t) + \cos(0.4t))\cos(0.5t) \\ 2(\sin(0.5t) + \cos(0.3t))\sin(0.3t) \end{cases}, d_a(t) = \begin{cases} 2(\sin(0.2t) + \cos(0.5t)) \\ 2(\sin(0.1t) + \cos(0.4t)) \\ 2(\sin(0.5t) + \cos(0.3t)) \end{cases} \tag{37}$$

Figures 5 and 6 show the estimated curves for different disturbances. (c_1) and (c_2) are the disturbance estimation curves of Eq. (36), while (d_1) and (d_2) are the disturbance estimation curves of Eq. (37). From the figure, it can be seen that the adaptive predefined-time disturbance observer designed in this paper can handle the estimation of different disturbance values and has a fast response speed, thereby enhancing the robustness of the controller. The simulation results are as follows:

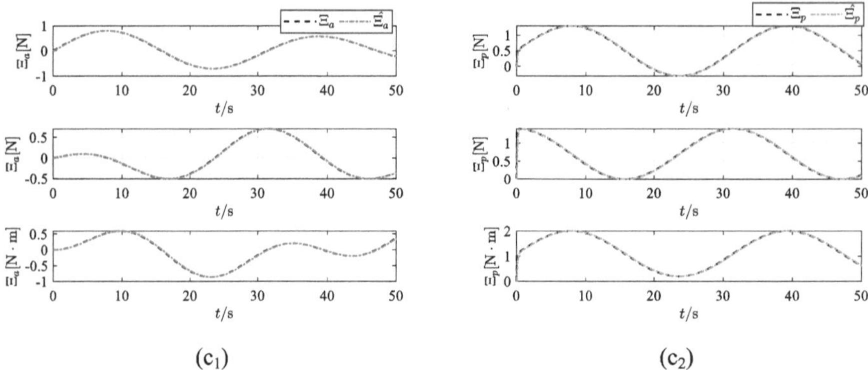

(c₁) (c₂)

Fig. 5. Estimated curve of the perturbation values of Eq. (36)

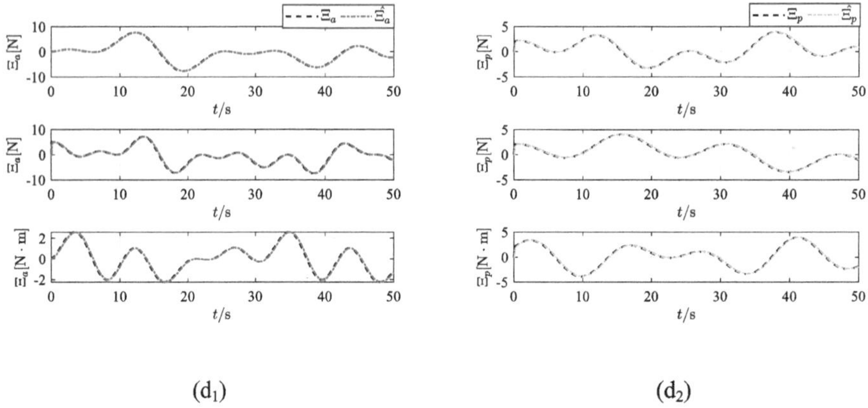

(d₁) (d₂)

Fig. 6. Estimated curve of the perturbation values of Eq. (37)

5 Conclusion

This paper tackles the trajectory tracking control problem for a quadrotor UAV subject to model uncertainties and external disturbances by proposing a predefined-time prescribed performance control strategy incorporating a disturbance observer. The approach integrates an adaptive predefined-time disturbance observer, prescribed performance constraints, and predefined-time sliding mode control to guarantee strong disturbance rejection, high tracking accuracy, and superior transient and steady-state performance—all within a preset time frame. Furthermore, the convergence time of the system can be preset independently and remains unaffected by control parameters. The effectiveness of the proposed strategy is verified through Lyapunov-based stability analysis and numerical simulations.

However, this paper only studies the trajectory tracking control problem of quadrotor UAV under composite interference. In actual flight, issues such as obstacle avoidance and actuator faults can also affect tracking performance. Therefore, future research will address problems such as obstacle avoidance and actuator faults.

Acknowledgments. This work was funded by the Fujian Provincial Department of Science and Technology - Foreign Cooperation Project (No. 2023I0019).

Disclosure of Interests. The authors declare that they have no conflict of interest.

References

1. Huang, Y., Liu, W., Li, B., Yang, Y., Xiao, B.: Finite-time formation tracking control with collision avoidance for quadrotor UAVs. J. Franklin Inst. **357**, 4034–4058 (2020)
2. Tian, B., Cui, J., Lu, H., Zuo, Z., Zong, Q.: Adaptive finite-time attitude tracking of quadrotors with experiments and comparisons. IEEE Trans. Industr. Electron. **66**, 9428–9438 (2019)
3. Polyakov, A.: Nonlinear feedback design for fixed-time stabilization of linear control systems. IEEE Trans. Autom. Control **57**, 2106–2110 (2012). https://doi.org/10.1109/TAC.2011.2179869
4. Chen, Q., Tao, M., He, X., Tao, L.: Fuzzy adaptive nonsingular fixed-time attitude tracking control of quadrotor UAVs. IEEE Trans. Aerosp. Electron. Syst. **57**, 2864–2877 (2021)
5. Qi, R., Zhang, G.: Trajectory tracking control of quadrotor UAV based on predefined time. Electr. Meas. Technol. **48**, 18–25 (2025). https://doi.org/10.19651/j.cnki.emt.2417261
6. Zhang, Z., Xie, S., Chen, Q.: Disturbance observer-based singularity-free predefined-time attitude tracking control of quadrotors: theory and experiments. IEEE Trans. Aerosp. Electron. Syst. **61**, 314–324 (2025). https://doi.org/10.1109/TAES.2024.3446754
7. Li, S., Duan, N., Pei, H.: Singularity-free predefined time tracking control for quadrotor UAV with input saturation and error constraints. Nonlinear Dyn. **113**, 13225–13242 (2025). https://doi.org/10.1007/s11071-024-10826-1
8. Chen, J., Chen, Z., Zhang, H., Xiao, B., Cao, L.: Predefined-time observer-based nonsingular sliding-mode control for spacecraft attitude stabilization. IEEE Trans. Circ. Syst. II Express Briefs **71**, 1291–1295 (2024). https://doi.org/10.1109/TCSII.2023.3321684
9. Tao, M., Zhang, S.: Observer-based and funnel-based predefined-time event-triggered adaptive formation tracking control for multiple quadrotors. Results Eng. **27**, 106784 (2025). https://doi.org/10.1016/j.rineng.2025.106784
10. Mei, H., Wen, X., Ma, X., Tan, Y., Wang, J.: Adaptive practical predefined-time leader-follower consensus for second-order multiagent systems with uncertain disturbances. J. Franklin Inst. **361**, 106694 (2024). https://doi.org/10.1016/j.jfranklin.2024.106694
11. Guo, G., Xiao, S., Chen, F., Hou, Z., Tan, H.: Dynamic event-triggered predefined-time adaptive attitude fault-tolerant control for uncertain unmanned aerial vehicles with disturbances and actuator faults. Aerosp. Sci. Technol. **165**, 110481 (2025). https://doi.org/10.1016/j.ast.2025.110481
12. Yan, Y., Guan, D., Jiang, T., Yu, S., Liu, Y.: Event-triggered adaptive predefined-time sliding mode control of autonomous surface vessels with unknown dead zone and actuator faults. Ocean Eng. **304**, 117851 (2024). https://doi.org/10.1016/j.oceaneng.2024.117851
13. Liu, W., Li, Z., Lei, K., Zheng, H.: Multi-USV adaptive fixed-time robust formation tracking control. In: 2025 IEEE International Annual Conference on Complex Systems and Intelligent Science (CSIS-IAC), pp. 164–171 (2025)
14. Li, Z., Lei, K.: Robust fixed-time fault-tolerant control for USV with prescribed tracking performance. J. Mar. Sci. Eng. **12** (2024). https://doi.org/10.3390/jmse12050799
15. Qin, X., Zhao, Z., Huang, P., Li, J.: Multiple feedback recurrent neural network based super-twisting predefined-time nonsingular terminal sliding mode control for quad-rotor UAV. Aerosp. Sci. Technol. **151**, 109282 (2024). https://doi.org/10.1016/j.ast.2024.109282

16. Song, G., Xu, J., Deng, L., Zhao, N.: Robust distributed fixed-time cooperative hunting control for multi-quadrotor with obstacles avoidance. ISA Trans. **151**, 73–85 (2024). https://doi.org/10.1016/j.isatra.2024.05.048
17. Cheng, W., Zhang, K., Jiang, B.: Fixed-time fault-tolerant formation control for a cooperative heterogeneous multiagent system with prescribed performance. IEEE Trans. Syst. Man Cybern. Syst. **53**, 462–474 (2023). https://doi.org/10.1109/TSMC.2022.3186382
18. Cheng, W., Zhang, K., Jiang, B.: Continuous fixed-time fault-tolerant formation control for heterogeneous multiagent systems under fixed and switching topologies. IEEE Trans. Veh. Technol. **72**, 1545–1558 (2023). https://doi.org/10.1109/TVT.2022.3211609

DVS Data Recognition Using the Hybrid Neural Network

Qiang Fu[1,2], Xiangning Wei[1,2], Mengjun Han[1,2], Kamiao Wu[1,2], Jia Li[1,2], Puyang Li[1,2], and Junxiu Liu[1,2](✉)

[1] Guangxi Key Lab of Brain-Inspired Computing and Intelligent Chips, School of Elec-tronic and Information Engineering, Guangxi Normal University, Guilin 541004, China
j.liu@ieee.org

[2] Key Laboratory of Nonlinear Circuits and Optical Communications (Guangxi Nor-mal University), Education Department of Guangxi Zhuang Autonomous Region, 541004 Guilin, China

Abstract. Since the output of Dynamic Visual Sensor (DVS) is time-continuous event stream data, it has higher temporal order compared to traditional image data, and the event data possesses high sparsity. Therefore, how to effectively extract features from DVS data and recognize specific targets is a challenging task. In this paper, a hybrid neural network approach incorporating Kolmogorov-Arnold Networks and Residual Neural Networks is proposed for extracting effective features from DVS data and performing the task of target recognition. By designing an adaptive convolutional kernel and a dynamic grid adjustment method, the temporal and local features in DVS data are effectively captured. The experimental results show that compared with the traditional Convolutional Neural Network, the proposed method more outperforms in terms of classification accuracy and generalization ability.

Keywords: DVS data · Kolmogorov-Arnold Networks · Residual Neural Networks

1 Introduction

Due to the advantages of high temporal resolution and low power consumption, dynamic vision sensors (DVS) have been widely used in robotics [1], autonomous driving [2], and video surveillance [3], etc. However, the temporal and sparse nature of DVS data also brings challenges in image processing, and traditional image processing methods are difficult to be directly applied to DVS data. To address this challenge, many researchers have proposed different approaches to explore and process DVS data. For example, a combination of experimental measurements, theoretical modeling, and data-driven optimization are used to systematically investigate the underlying operating mechanism of DVS pixels and their optimization strategy for bias configuration in the research of [4]. The performance and energy efficiency of Artificial Neural Networks (ANNs), Convolutional Neural Networks (CNNs), Spiking Neural Networks (SNNs), and Convolutional

C. Li et al. (Eds.): ICNC 2025, CCIS 2946, pp. 53–65, 2026.
https://doi.org/10.1007/978-981-92-1599-7_5

SNN architectures on the N-MNIST dataset are comprehensively compared and analyzed in [5]. It is also pointed out that SNNs have the potential for energy savings after neuromorphic hardware optimization and algorithmic improvement, which provides theoretical references for low-power intelligence applications. By adaptively visualizing event streams as pseudo-frames, the method in [6] achieves an accurate, compact, and efficient YOLOE detection network. A data enhancement method named event spatial temporal fragmentation is proposed in [7], which involves simulating brightness change perturbations to the neuromorphic event stream. In the approach of [8], the event stream was processed asynchronously using a sparse convolution technique, which significantly reduced the computational effort and improved the processing efficiency. Meanwhile, an event-based face detection algorithm was proposed in [9], and the computational efficiency was enhanced through hardware optimization, enabling event-based computer vision methods to be applied in mobile devices.

Various innovative approaches based on neural networks have emerged in recent years. The study in [10] employs graph neural networks to represent DVS events as graphs, improving classification accuracy by modeling their spatiotemporal relationships. A bionic adaptive internal associative neuron model was proposed in [11], which significantly improves the performance of SNNs by mimicking the phenomenon of associative long-range enhancement within biological neurons. In addition, a novel SNN-based framework was proposed in the research of [12], enhancing the biological plausibility and learning ability of SNNs through the introduction of multiple firing activity patterns and synaptic plasticity. Although some progress has been made by existing methods in different tasks, how to fully utilize the temporal and sparse nature of DVS data remains a problem that deserves indepth research.

To overcome the limitations of traditional CNNs in handling the temporal and sparse nature of DVS data, this paper introduces a hybrid KAN-ResNet model that effectively captures high-dimensional features and boosts classification accuracy by adaptively adjusting network structures and weights through spline interpolation and dynamic mesh techniques. This paper is organized as follows. Section 2 presents the main research methodology proposed in this paper. Section 3 describes the experimental design and result analysis in detail and Sect. 4 summarizes this work and proposes directions for future research.

2 Methods

This section focuses on the DVS data processing methods and the proposed network modeling and training approach.

2.1 DVS Data Preprocessing

The raw DVS data is first preprocessed before classification and recognition. The data output from the DVS sensor is usually a stream of events containing timestamps, pixel positions, and positive and negative polarity information. Notate the raw event data as $E(x_i, y_i, t_{i,}p_i)$, where $0 \le i < N$, N is the total number of events. Divide from the individual N events, nearly uniformly into M segments, M frames. Denote a frame in the integrated frame data as $F(j)$, and the pixel value at position (p, x, y) is $F(j, p, x, y)$. $F(j)$ is integrated from the Events whose indexes are between j_l and j_r in the Event stream, as shown in the method of [13], which is calculated by

$$j_l = \left\lfloor \frac{N}{T} \right\rfloor \cdot j \quad (1)$$

$$j_r = \begin{cases} \left\lfloor \frac{N}{T} \right\rfloor \cdot (j+1), & if\ j < T - 1 \\ N, & if\ j = T - 1 \end{cases} \quad (2)$$

$$F(j, p, x, y) = \sum_{i=j_l}^{j_r - 1} I_{p,x,y}(p_i, \chi_i, y_i) \quad (3)$$

where $\lfloor \cdot \rfloor$ denotes downward rounding and $I_{p,\,x,\,y}(p_i, \chi_i, y_i)$ is an indicative function that takes the value of 1 (denoted as the first channel) if and only if $(p, x, y)= p_i, \chi_i, y_i$ and 0 (denoted as the second channel) otherwise.

After integration by the event stream, the resulting frame data is compressed to store multiple NumPy arrays in format (.npy or.npz), followed by converting the pseudo-frames into RGB images such as.jpg, .png, etc., from which each frame or random frame needs to be read. Since the event stream has positive and negative polarity, i.e., the data frames integrated are data with two channels, it is still necessary to merge the data of two channels first, as shown in Fig. 1, the green ones in Fig. 1(c) are the first channel pixels of '0' and the red ones are the second channel pixels of '0 'for the second channel pixel in red.

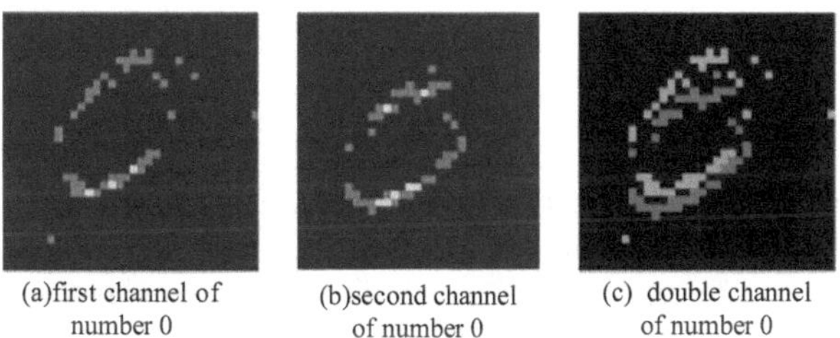

(a)first channel of number 0 (b)second channel of number 0 (c) double channel of number 0

Fig. 1. Pseudo-frame schematic for different channels.

2.2 The Proposed Hybrid Network

In this study, a hybrid network model based on Kolmogorov-Arnold Networks (KAN) and ResNet is proposed by utilizing the advantages of KAN network. The overall architecture of KAN network includes an input layer, a KANLinear layer, an adaptive mesh updating mechanism and a final output layer. The input data first passes through the KANLinear layer for feature extraction, and then is dynamically adjusted by the adaptive grid updating mechanism, and finally the final classification or regression results are generated. Different from the multilayer perceptron network based on the approximation theorem, KAN is based on the Kolmogorov-Arnold [14] representation theorem, the core idea of which is to simplify the complexity by decomposing a multivariate function into the combination and superposition of some univariate functions, which are stacked on top of each other, thus simplifying the complexity. The specific formula is as follow:

$$f(x) = f(x_1, \ldots, x_n) = \sum_{q=1}^{2n+1} \Phi_q \left(\sum_{p=1}^{n} \varnothing_{q,p}(x_p) \right) \tag{4}$$

where $f(x)$ denotes a multivariate continuous function, $2n + 1$ is the upper limit of the outer summation, Φ_q is the outer univariate function, $\varnothing_{q,p}(x_p)$ is the inner univariate function, and x_p is the input variable (Fig. 2).

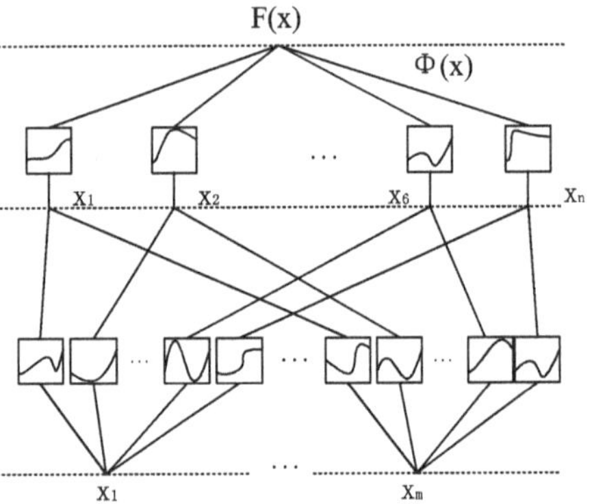

Fig. 2. Function mapping relationship.

ResNet's residual connectivity effectively solves the gradient vanishing problem in deep networks by allowing information to flow between multiple layers, ensuring that the network remains stable during training. The core idea is residual connectivity [15], which means that the output of each layer is not only the result of that layer's computation, but also coupled with the input data itself, thus helping the network to avoid the gradient vanishing problem and facilitating the effective flow of information. For a residual block, assuming that the input is x and the output is, the basic formula for the residual block is calculated by

$$y_{resnet} = F(x, W_i) + x \tag{5}$$

where $F(x, W_i)$ is the feature representation obtained through a series of convolution operations and W_i is the weight of each layer (Fig. 3).

In this paper, the KAN network module is incorporated into ResNet to utilize the deep network training stability of ResNet and the nonlinear feature extraction capability of KAN. ResNet is responsible for extracting the spatial features of the input data and passing these features to a deeper level through residual connectivity, while KAN is responsible for nonlinear transformations based on the features extracted by ResNet and

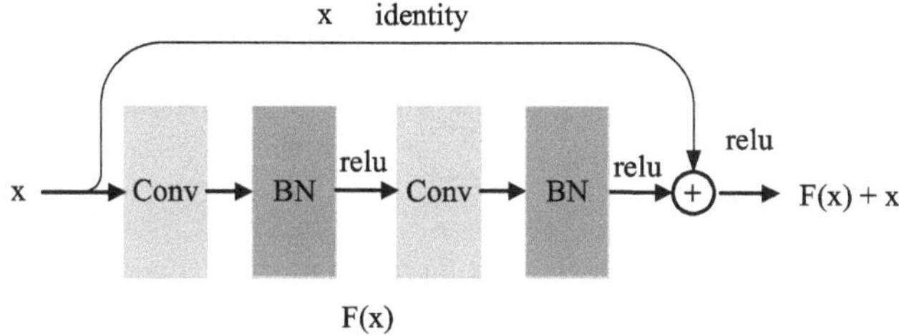

Fig. 3. Schematic diagram of Residual Block.

enhancing the ability to temporal and sparsity data learning capabilities. Specifically, at the output of ResNet, the KAN module is used instead of the traditional fully-connected layer to further carry out the extraction of temporal and nonlinear features and enhance the model's ability to capture the sparsity and temporality of DVS data. At this point, the output y_{resnet} of Eq. (5) is used as the input x_p of KAN, $\emptyset_{q,p}(x_p)$ is also $\emptyset_{q,p}(y_{resnet})$ (Fig. 4).

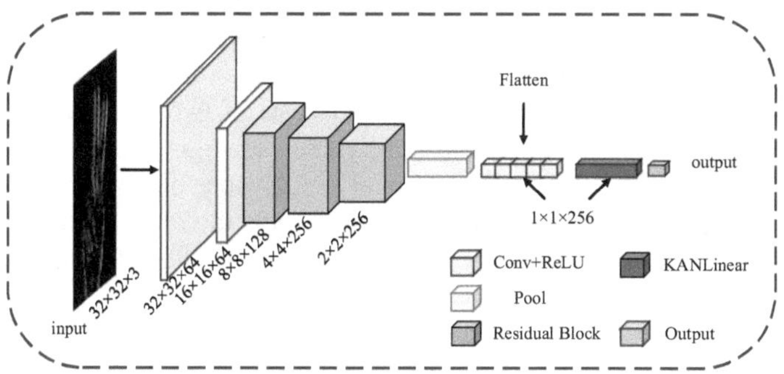

Fig. 4. KAN-ResNet hybrid networks.

The KAN Linear layer is the core part of the KAN model [16], which consists of two parts, i.e., a traditional linear transformation layer and a B-spline based nonlinear transformation layer, and the output is the sum result of the transformation of linear and nonlinear parts, which is able to capture the linear features and complex nonlinear features in the input data simultaneously. The specific principles are as follows:

(1) Base linear part: the base representation of the input features is computed by the traditional fully connected layer of linear transformations and activation functions for the linear component of the approximate inner function $\emptyset_{q,p}(y_{resnet})$ of Eq. (6), and the base linear layer is computed by

$$y_{base} = \sigma(W_{base} \cdot X + b_{base}) \tag{6}$$

where W_{base} and b_{base} are the weights and bias of the underlying linear layer, respectively, and σ is the activation function.

(2) B-spline nonlinear part: we introduce a B-spline basis function for the nonlinear component of the approximate inner function $\emptyset_{q,p}(x_p)$ of Eq. (6) to fit the nonlinear

relationship of the input. Its computation is expressed as

$$y_{spline} = \sum_{k=0}^{K} c_k B_k(x) \tag{7}$$

where $B_k(x)$ is the kth B-spline basis function and c_k is the corresponding weight coefficient.

(3) Summing output: the outputs of the base linear part and the B-spline part are summed to get the final output:

$$y_{KAN} = y_{base} + y_{spline} \tag{8}$$

which corresponds to the inner function $\emptyset_{q,p}(y_{resnet})$ in Eq. (7).

At each forward propagation, the network dynamically updates the lattice of segmented polynomials based on the distribution of the input data, thus capturing the local patterns in the data more accurately. Specifically, the mechanism updates the mesh of the B-spline basis functions based on the sorting information of the current input data computation data, ensuring that the mesh can adapt to the changes in the data distribution, enabling the network to flexibly adjust its representation capability when dealing with data with different distributions. To prevent overfitting, a regularization term is added to the KAN network. The regularization loss consists of two parts: the L1-paradigm regularization and the entropy regularization. The $L1$-paradigm regularization is used to constrain the size of the segmented polynomial weights to avoid over-parameterization of the model, while the entropy regularization prevents the model from overfitting to some specific samples by maximizing the uniformity of the distribution of the weights. The regularization loss is calculated as:

$$L_{reg} = \lambda_1 \sum_i |\omega_i| + \lambda_2 \sum_i p_i \log(p_i) \tag{9}$$

where λ_1 and λ_2 are hyperparameters controlling the strength of regularization and p_i is the normalized values of the segmented polynomial weights.

The number of B-spline basis functions K is determined empirically and set to 5–8 based on cross-validation, balancing flexibility and computational cost. The weights c_k are randomly initialized and optimized via gradient descent. The adaptive grid mechanism updates knot positions periodically using input distribution statistics.

3 Experimental Results

In this paper, the number of parameters of all the models is controlled with similar sized control variables to ensure the model complexity among different models. Firstly, the proposed network is tested on two public DVS datasets (i.e. N-MNIST [17], ASLDVS [18]), and then the models are comprehensively evaluated on the our self-built DVS datasets.

3.1 N-MNIST

N-MNIST consists of the same 60,000 training samples and 10,000 test samples as the original MNIST dataset, with a MNIST capture resolution of 28 × 28, captured as event stream data by a dynamic vision sensor, with a N-MNIST resolution of 32 × 32. Its pseudo-frame data is shown in Fig. 5.

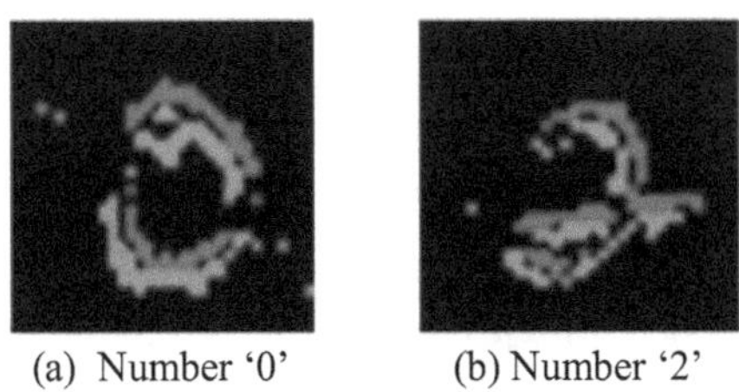

(a) Number '0' (b) Number '2'

Fig. 5. Example of pseudo-frame data in the N-MNIST dataset.

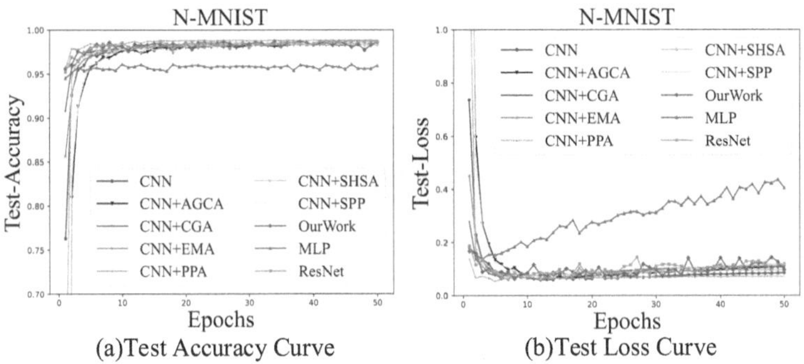

(a)Test Accuracy Curve (b)Test Loss Curve

Fig. 6. Performance testing of the network on the N-MNIST dataset.

The test accuracy and loss value curves obtained from the experiments using the N-MNIST dataset for each scenario are shown in Fig. 6. When the number of iterations is 7, the test accuracy and loss curves of each model tend to be smooth, in which the MLP model performs poorly in complex tasks. As the increasing number of iterations, the accuracy of this work steadily improves, reaching 98.89% optimal performance. This achievement not only outperforms the CNN + PPA [24] and ResNet [5] methods by 0.07% and 0.2% respectively, but also significantly surpasses the MLP [31] method by a margin of 2.79%. Results show that this work's enhanced capability in fitting nonlinear relationships and extracting sparsity features more effectively than other approaches. A comprehensive comparison of experimental results across various models on the N-MNIST dataset is outlined in Table 1. Our work introduces the B-spline nonlinear basis function in the structure to adaptively update the mesh, and its accuracy reaches 98.89%, which is superior to other methods, and it improves 0.53% and 0.55% than the latest methods, CNN + SHSA [26] and CNN + CGA [27], respectively. It shows that the proposed method is more advantageous for temporal and sparse feature extraction on DVS dataset.

Table 1. Model comparison results on the N-MNIST dataset.

model	structure	parameters	Accuracy (%)
MLP[19]	3FC	697,322	96.10%
CNN[20]	3Conv + 3FC	651,978	97.68%
CNN + SPP[21]	3Conv + 1SPP + 3FC	744,298	98.39%
CNN + EMA[22]	3Conv + 1EMA + 3FC	651,992	98.38%
CNN + PPA[23]	3Conv + 1PPA + 3FC	696,978	98.82%
CNN+ AGCA[24]	3Conv + 1AGCA + 3FC	652,203	98.37%
CNN + SHSA[25]	3Conv + 1SHSA + 3FC	670,858	98.36%
CNN + CGA[26]	3Conv + 1CGA + 3FC	685,501	98.34%
ResNet[15]	1Conv + 3Block + 1FC	680,938	98.69%
Our work	1Conv + 3Block + 1KAN	688,608	**98.89%**

3.2 ASLDVS

The ASLDVS dataset contains recorded data from five subjects, each gesturing the 24 letters of the alphabet, with the exception of J and Z. The J and Z gesture expressions also involve hand movements, so the dataset does not contain gesture data for these two letters. The ASLDVS dataset with its frame data is shown in Fig. 7.

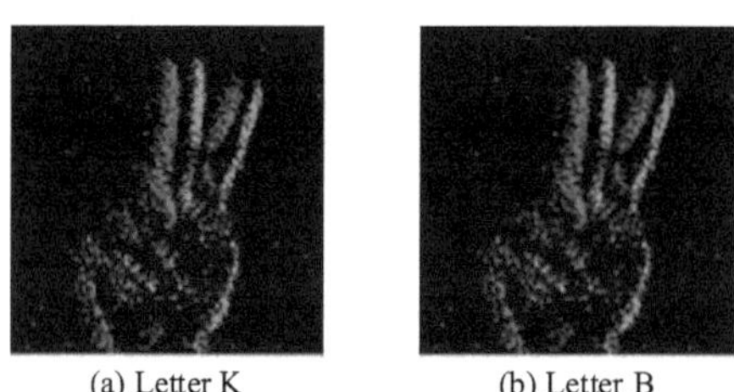

(a) Letter K (b) Letter B

Fig. 7. Example of pseudo-frame data in the ASLDVS dataset.

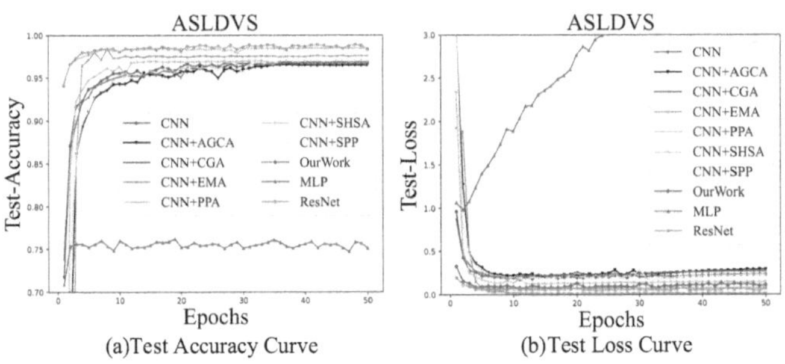

(a)Test Accuracy Curve (b)Test Loss Curve

Fig. 8. Performance testing of the network on the ASLDVS dataset.

The test accuracy and loss value curves obtained from the experiments of each scenario using the ASLDVS dataset are shown in Fig. 8, where the horizontal coordinates of Fig. 8(a) are the number of iterations and the vertical coordinates are the test accuracy, and the horizontal coordinates of Fig. 8(b) are the number of iterations and the vertical coordinates are the test loss values. When the number of iterations is 15, the test accuracy and loss curves of each model tend to be smooth, and since the MLP model is a linear model, its test results show overfitting. As the number of iterations increases, the accuracy increases smoothly, and when the number of training iterations is 30, the test accuracy of each scheme has smoothed, and Our work performs the best, and its accuracy reaches 98.90%, which is 0.49% and 0.22% higher than the methods CNN + PPA [24] and ResNet [5], and is 22.75% higher than the method MLP [31], which shows that Our work is superior in fitting nonlinear representations. The comparison of the experimental results of each model on the ASLDVS dataset is shown in Table 2. The accuracy of our work reaches 98.90%, which is better than the other methods, and improves by 2.89% and 2.20% over the latest methods CNN + SHSA [26] and CNN + CGA [27], respectively.

Table 2. Model comparison results on the ASLDVS dataset.

model	structure	parameters	Accuracy(%)
MLP[19]	3FC	8,147,352	76.15%
CNN[20]	3Conv + 3FC	8,518,104	96.88%
CNN + SPP[21]	3Conv + 1SPP + 3FC	8,610,424	96.97%
CNN + EMA[22]	3Conv + 1EMA + 3FC	8,518,118	97.56%
CNN + PPA[23]	3Conv + 1PPA + 3FC	9,063,104	98.51%
CNN+ AGCA[24]	3Conv + 1AGCA + 3FC	8,519,131	96.95%
CNN + SHSA[25]	3Conv + 1SHSA + 3FC	8,536,984	97.01%
CNN + CGA[26]	3Conv + 1CGA + 3FC	8,551,627	96.70%
ResNet[15]	1Conv + 3Block + 1FC	8,780,167	98.88%
Our work	1Conv + 3Block + 1KAN	8,957,568	**98.90%**

3.3 DVS Behavior Dataset

We collected the DVS Behavior Dataset using an IniVation DVXplorer camera (640 × 480) to document routine human activities. The dataset records eight participants and was acquired under a high-noise setting to mimic real-world conditions (other parameters were default; see Fig. 9). The collection pipeline saved event streams in aedat 4.0 format, which were then visualized via a custom interface. The final balanced dataset includes over 700 recordings (5 s each) across seven behavior categories (e.g., cigar, pen), with 100 samples per category.

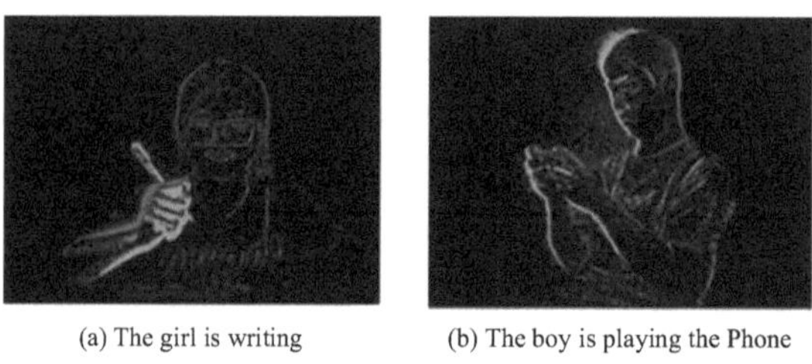

(a) The girl is writing (b) The boy is playing the Phone

Fig. 9. Example of pseudo-frame data in the DVS Behavior Dataset

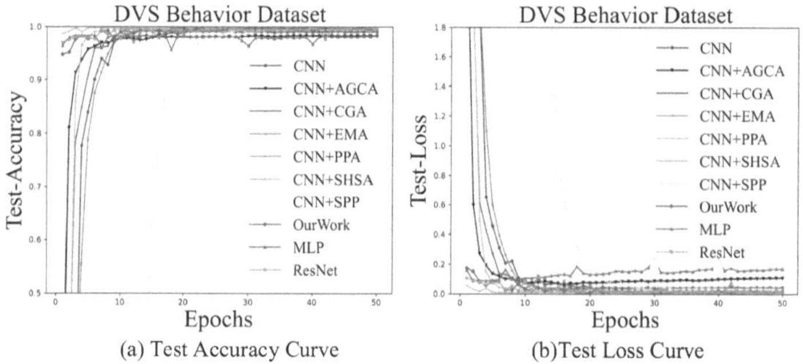

(a) Test Accuracy Curve (b)Test Loss Curve

Fig. 10. Performance testing of the network on the DVS Behavior dataset.

The test accuracy and loss value curves obtained from the experiments of each scheme in the DVS Behavior dataset are shown in Fig. 10. When the number of iterations is 10, the test accuracy and loss curves of basically each model tend to be smooth, in which the MLP model performs well in this dataset, which indicates a strong interdependence or interaction between the features of this dataset. As the number of iterations increases, the accuracy increases smoothly, and when the number of training iterations is 25, the test accuracy of each scheme has been smoothed, and Our work performs the best, and its accuracy reaches 99.82%, which is improved by 0.05% and 0.18% than the methods CNN + PPA [24] and ResNet [5], and improved by 3.72% than the method MLP [31] by 0.23%, 0.63%. Table 3 illustrates that the proposed model excels at extracting high-dimensional and sparse features from the DVS dataset. This capability highlights the method's advantage in handling complex data, thereby validating its effectiveness. By effectively capturing and representing key information within the data, the model enhances its overall performance. Results show the model's potential for applications involving high-dimensional and sparse data processing. Consequently, the model demonstrates significant promise in scenarios where such data characteristics are prevalent.

Table 3. Model comparison results on the DVS Behavior dataset.

model	structure	parameters	Accuracy (%)
MLP[19]	3FC	8,146,247	96.10%
CNN[20]	3Conv + 3FC	8,515,911	99.22%
CNN + SPP[21]	3Conv + 1SPP + 3FC	8,608,231	99.54%
CNN + EMA[22]	3Conv + 1EMA + 3FC	8,515,925	99.56%
CNN + PPA[23]	3Conv + 1PPA + 3FC	9,060,911	99.77%
CNN+ AGCA[24]	3Conv + 1AGCA + 3FC	8,525,129	98.91%
CNN + SHSA[25]	3Conv + 1SHSA + 3FC	8,534,791	99.59%
CNN + CGA[26]	3Conv + 1CGA + 3FC	8,549,434	99.19%
ResNet[15]	1Conv + 3Block + 1FC	8,931,271	99.74%
Our work	1Conv + 3Block + 1KAN	8,929,472	**99.82%**

4 Conclusion

To address the challenges in feature extraction from DVS data, this work presents a hybrid neural network architecture that leverages the strengths of KAN and ResNet. This combined model is designed to capture rich spatial features and simultaneously manage the temporal dynamics and sparsity inherent in DVS data. Furthermore, an adaptive grid update mechanism is incorporated to enhance the model's capacity for handling complex data distributions. Evaluations on the N-MNIST, ASLDVS, and a self-collected DVS Behavior Dataset confirm the superior performance of the proposed approach. Future research will be directed towards the optimization of the KAN structure to bolster its generalization across various DVS datasets for broader application scenarios.

Acknowledgement. This work is supported by the Guangxi Natural Science Foundation under Grant 2025GXNSFBA069292, the Guangxi Science and Technology Projects under Grant GuiKeAD24010047, the Basic Ability Enhancement Program for Young and Middle-aged Teachers of Guangxi under Grant 2024KY0074, a grant (No. BCIC-24-Z7) from Guangxi Key Laboratory of Brain-inspired Computing and Intelligent Chips, and the National Training Program of Innovation and Entrepreneurship for Undergraduates under Grant S202410602216.

References

1. Jiang, Z., Otto, R., Bing, Z., Huang, K., Knoll, A.: Target tracking control of a wheel-less snake robot based on a supervised multi-layered SNN. In: IEEE International Conference on Intelligent Robots and Systems, pp. 7124–7130. Institute of Electrical and Electronics Engineers Inc. (2020)
2. Chen, G., Cao, H., Conradt, J., Tang, H., Rohrbein, F., Knoll, A.: Event-based neuromorphic vision for autonomous driving: a paradigm shift for bio-inspired visual sensing and perception. IEEE Signal Process. Mag. **37**(4), 34–49 (2020)
3. Gallego, G., et al.: Event-based vision: a survey. IEEE Trans. Pattern Anal. Mach. Intell. **44**(1), 154–180 (2022)
4. Graça, R., McReynolds, B., Delbruck, T.: Shining light on the DVS pixel: a tutorial and discussion about biasing and optimization. In: IEEE Computer Society Conference on Computer Vision and Pattern Recognition Workshops, pp. 4045–4053. IEEE Computer Society (2023)
5. Thienbutr, N., Massagram, W.: Energy efficiency evaluation of neural network architectures on the neuromorphic-MNIST dataset. In: International Computer Science and Engineering Conference, pp. 1–6. Institute of Electrical and Electronics Engineers Inc. (2024)
6. Fang, Y., et al.: Visualization and object detection based on event information. Sensors. **23**(4), 1839–1856 (2023)
7. Shen, H., Luo, Y., Cao, X., Zhang, L., Xiao, J., Wang, T.: Training robust spiking neural networks on neuromorphic data with spatiotemporal fragments. In: ICASSP, IEEE International Conference on Acoustics, Speech and Signal Processing-Proceedings, pp. 1–5. Institute of Electrical and Electronics Engineers Inc. (2023)
8. Lesage, X., Merio, C., Welzel, F., De Araujo, L.S., Engels, S., Fesquet, L.: Data-driven processing element for sparse convolutional neural networks. In: 2024 22nd IEEE Interregional NEWCAS Conference, pp. 143–147 (2024)
9. Arjmand, C., et al.: TRIP: trainable region-of-interest prediction for hardware-efficient neuromorphic processing on event-based vision. Int. Conf. Neuromorphic Syst., 94–101 (2024)
10. Bi, Y., Chadha, A., Abbas, A., Bourtsoulatze, E., Andreopoulos, Y.: Graph-based Spatio-temporal feature learning for neuromorphic vision sensing. IEEE Trans. Image Process. **29**, 9084–9098 (2020)
11. Shen, H., Luo, Y., Cao, X., Zhang, L., Xiao, J., Wang, T.: Training stronger spiking neural networks with biomimetic adaptive internal association neurons. In: ICASSP, IEEE International Conference on Acoustics, Speech and Signal Processing-Proceedings, pp. 1–5. Institute of Electrical and Electronics Engineers Inc. (2023)
12. Stanojevic, A., Woźniak, S., Bellec, G., Cherubini, G., Pantazi, A., Gerstner, W.: High-performance deep spiking neural networks with 0.3 spikes per neuron. Nat. Commun., 6793–6805 (2024)
13. Fang, W., Yu, Z., Chen, Y., Masquelier, T., Huang, T., Tian, Y.: Incorporating learnable membrane time constant to enhance learning of spiking neural networks. In: Proceedings of the IEEE International Conference on Computer Vision, pp. 2641–2651. Institute of Electrical and Electronics Engineers Inc. (2021)
14. Schmidt-Hieber, J.: The Kolmogorov–Arnold representation theorem revisited. Neural Netw. 119–126 (2021)
15. He, K., Zhang, X., Ren, S., Sun, J.: Deep residual learning for image recognition. In: Proceedings of the IEEE Computer Society Conference on Computer Vision and Pattern Recognition, pp. 770–778. IEEE Computer Society (2016)
16. Amouri, A., Al Rahhal, M.M., Bazi, Y., Alaparthy, V.T., Fotopoulos, G.: A hybrid based Kolmogorov-Arnold network for intrusion detection systems in IoT. In: 2024 IEEE Middle East Conference on Communications and Networking, pp. 247–250 (2024)

17. Orchard, G., Jayawant, A., Cohen, G.K., Thakor, N.: Converting static image datasets to spiking neuromorphic datasets using saccades. Front. Neurosci. 437–447 (2015)
18. Bi, Y., Chadha, A., Abbas, A., Bourtsoulatze, E., Andreopoulos, Y.: Graph-based object classification for neuromorphic vision sensing. In: Proceedings of the IEEE International Conference on Computer Vision, pp. 491–501. Institute of Electrical and Electronics Engineers Inc. (2019)
19. Rumelhart, D., Hinton, G., Williams, R.: Learning representations by back-propagating errors. Nature 533–536 (1986)
20. Klein, M.W., Enkrich, C., Wegener, M., Linden, S.: Reducing the dimensionality of data with neural networks. Science **313**(5786), 502–504 (2006)
21. Grauman, K., Darrell, T.: The pyramid match kernel: discriminative classification with sets of image features. In: Tenth IEEE International Conference on Computer Vision, pp. 1458–1465 (2005)
22. Ouyang, D., et al.: Efficient multi-scale attention module with cross-spatial learning. In: ICASSP, IEEE International Conference on Acoustics, Speech and Signal Processing-Proceedings, pp. 1–5. Institute of Electrical and Electronics Engineers Inc. (2023)
23. Xu, S., et al.: HCF-net: hierarchical context fusion network for infrared small object detection. IEEE Int. Conf. Multimedia Expo (ICME), 1–6 (2024, 2024)
24. Xiang, X., Wang, Z., Zhang, J., Xia, Y., Chen, P., Wang, B.: AGCA: an adaptive Graph Channel attention module for steel surface defect detection. IEEE Trans. Instrum. Meas. **72**, 1–12 (2023)
25. Yun, S., Ro, Y.: SHViT: single-head vision transformer with memory efficient macro design. In: 2024 IEEE/CVF Conference on Computer Vision and Pattern Recognition (CVPR), pp. 5756–5767 (2024)
26. Chen, Z., He, Z., Lu, Z.M.: DEA-net: single image Dehazing based on detail-enhanced convolution and content-guided attention. IEEE Trans. Image Process. **33**, 1002–1015 (2024)
27. Ding, Y., Chen, F.: TAnet: a new temporal attention network for EEG-based auditory spatial attention decoding with a short decision window. In: 2024 46th Annual International Conference of the IEEE Engineering in Medicine and Biology Society (EMBC), pp. 1–4 (2024)

A Memristor-Based SNN Hardware Architecture with AHaH Plasticity

Ruicheng Xie, Gangquan Si(✉), Xiang Xu, Minglin Xu, Yukaichen Yang, and Chenhao Li

State Key Laboratory of Electrical Insulation and Power Equipment, Xi'an Jiaotong University, Xi'an 710049, China
{ruichengxie,xmlhhxx,yykc}@stu.xjtu.edu.cn, sigangquan@mail.xjtu.edu.cn

Abstract. The rapid expansion of artificial intelligence has exposed limitations in conventional von Neumann architectures, particularly in energy efficiency and data movement. Neuromorphic computing, especially memristor-based spiking neural networks (SNNs), offers a promising alternative by integrating memory and processing. This paper presents a hardware architecture for a memristor-based SNN that incorporates compute-in-memory (CIM) and Anti-Hebbian and Hebbian (AHaH) plasticity. The design utilizes differential memristor synapses within a scalable modular framework, controlled by an STM32 microcontroller with improved on-off logic to mitigate voltage fluctuations and hardware risks. Experimental validation using the Wisconsin Breast Cancer dataset demonstrates a classification accuracy of 91.26%, highlighting the effectiveness for low-power, event-driven binary classification of our system. The proposed architecture advances the deployment of robust and energy-efficient neuromorphic systems, addressing key challenges in real-time memristor control and scalability.

Keywords: Memristor · Spiking Neural Network · AHaH Plasticity · Neuromorphic Computing · Compute-in-Memory

1 Introduction

The exponential growth of artificial intelligence (AI) has transformed various domains, yet it has revealed fundamental constraints in conventional von Neumann architectures, such as high latency and energy consumption from data shuttling between memory and processing units, particularly in large-scale neural networks [1]. Neuromorphic computing, inspired by the efficient parallel processing of human brain, emerges as a promising solution. Key innovations include in-memory computing and spiking neural networks (SNNs), which enable low-power, event-driven operations mimicking biological neural dynamics. Memristors, with their non-volatile, resistance-based memory properties, are central to this paradigm, facilitating compute-in-memory (CIM) to minimize data movement and boost energy efficiency [2,3].

C. Li et al. (Eds.): ICNC 2025, CCIS 2946, pp. 66–79, 2026.
https://doi.org/10.1007/978-981-92-1599-7_6

Despite progress, implementing memristor-based SNNs faces challenges like scalability, device non-idealities (e.g., variability and endurance), and real-time control, requiring further optimization. Recent neuromorphic hardware advancements leverage memristive devices for synaptic plasticity simulation, but many remain simulation-based or lack modular designs for deployment [4,5]. For instance, fractional-order models in memcapacitor bridges have been explored for neural networks, highlighting the role of fractional calculus in enhancing memory characteristics [6]. Similarly, analyses of nonlinear relations in generalized fractional-order memory elements emphasize the impact of control parameters on hysteresis behaviors [7]. The Anti-Hebbian and Hebbian (AHaH) framework, proposed by Nugent and Molter, introduces competitive learning akin to biological synaptic rivalry, providing a basis for efficient computation [8].

This work bridges these gaps by proposing a memristor-based SNN hardware architecture integrating CIM with the event-driven nature of SNN via AHaH plasticity. Drawing from the origin AHaH framework, our design employs differential memristor synapses for dynamic conductance adjustment in binary classification. A scalable modular framework uses eight memristor pairs as building blocks, controlled by an STM32 microcontroller with level-shifting for real-time updates. To address voltage fluctuations and potential hardware damage in the original design, the level-shifting process is implemented in an equivalent manner by memristor on/off control logic we designed. Hardware experiments on the Wisconsin Breast Cancer dataset achieve 91.26% accuracy, advancing toward deployable neuromorphic systems. This extends prior work on memristor-based SNNs by focusing on cascaded expansion and memristor control logic.

The rest of the paper is organized as follows. Section 2 covers AHaH plasticity and computing fundamentals, including mathematical formulations; Section 3 details the hardware design and integration process, with enhancements for memristor emulation; Section 4 presents experimental results and analysis, including comparisons; and Sect. 5 concludes with discussions on limitations and future directions.

2 Basic Theories

2.1 Meta Stable Switch (MSS) Model of Memristor

The Meta Stable Switch (MSS) model represents a semi-empirical approach to modeling memristive devices, treating them as collections of volatile, metastable switches that transition between high- and low-conductance states under applied voltage gradients [9]. Introduced by Nugent and Molter as part of the foundational framework for AHaH (Anti-Hebbian and Hebbian) computing, the MSS model draws inspiration from the volatile and dissipative nature of biological systems, where energy dissipation drives adaptation and self-repair. This model addresses limitations in traditional deterministic memristor models by incorporating stochastic behavior, making it suitable for simulating large-scale adaptive networks with reduced computational overhead compared to von Neumann architectures.

In the MSS framework, a memristor is conceptualized as an ensemble of N metastable switches, each capable of occupying one of two states: a low-conductance state A (with conductance G_{A}) or a high-conductance state B (with $G_{\mathrm{B}} > G_{\mathrm{A}}$). The switches evolve in discrete time steps Δt, and the total memristor conductance G_{m} is the sum of the individual switch conductances:

$$G_{\mathrm{m}} = N_{\mathrm{A}}G_{\mathrm{A}} + N_{\mathrm{B}}G_{\mathrm{B}} \tag{1}$$

where N_{A} and N_{B} are the numbers of switches in states A and B, respectively, with $N = N_{\mathrm{A}} + N_{\mathrm{B}}$. The state transitions are governed by voltage-dependent probabilities. P_{A}, the probability of a switch transitions from state B to A (positive-going direction), is given by:

$$P_{\mathrm{A}} = \alpha\frac{1}{1 + e^{\beta(V - V_{\mathrm{A}})}} = \alpha\Gamma(V, V_{\mathrm{A}}) \tag{2}$$

while the probability P_{B} for the reverse transition (from A to B) is:

$$P_{\mathrm{B}} = \alpha\left(1 - \frac{1}{1 + e^{\beta(V + V_{\mathrm{B}})}}\right) = \alpha(1 - \Gamma(\mathrm{V}, -\mathrm{V_B})) \tag{3}$$

where $\alpha = \Delta t/t_{\mathrm{c}}$ (with t_{c} as the device's characteristic time scale), $\beta = q/kT = 1/V_{\mathrm{T}}$ (thermal voltage $V_T \approx 26$ mV at 300 K), and V is the applied voltage. The logistic function $\Gamma(\mathrm{V}, \mathrm{V_0}) = 1/(1+\mathrm{e}^{-\beta(\mathrm{V}-\mathrm{V_0})})$ ensures probabilities are bounded between 0 and 1. At each time step, the number of transitioning switches $\Delta\mathrm{N_A}$ and $\Delta\mathrm{N_B}$ is sampled from binomial (or approximated normal) distributions, leading to a conductance change:

$$\Delta G_{\mathrm{m}} = \Delta N_{\mathrm{A}}G_{\mathrm{A}} - \Delta N_{\mathrm{B}}G_{\mathrm{B}} \tag{4}$$

The memory-dependent current I_{m} is then $I_{\mathrm{m}} = V(G_{\mathrm{m}} + \Delta G_{\mathrm{m}})$. This stochastic formulation captures the incremental resistance changes observed in memristors, with larger N yielding smoother, more deterministic behavior due to averaging effects, while smaller N emphasizes stochasticity.

Building on this, Molter and Nugent generalized the MSS model to accommodate a broader range of memristive devices, including those with Schottky diode effects. The total current I through the device is a weighted sum of the memory-dependent component I_{m} and a Schottky diode current I_{s}:

$$I = \phi I_{\mathrm{m}}(V, t) + (1 - \phi)I_{\mathrm{s}}(V) \tag{5}$$

where $\phi \in [0, 1]$ balances the contributions. The Schottky component models exponential forward and reverse currents:

$$I_{\mathrm{s}} = \alpha_{\mathrm{f}}e^{\beta_{\mathrm{f}}V} - \alpha_{\mathrm{r}}e^{-\beta_{\mathrm{r}}V} \tag{6}$$

with positive parameters $\alpha_{\mathrm{f}}, \beta_{\mathrm{f}}, \alpha_{\mathrm{r}}, \beta_{\mathrm{r}}$ setting the barrier behavior. This extension enhances the versatility of the model, enabling fits to devices like tungsten Ag-chalcogenide memristors, where hysteresis curves under sinusoidal or pulse drives are accurately reproduced. For instance, simulations show that reducing N increases stochastic variance, while adjusting ϕ and exponential parameters incorporates diode-like asymmetry.

The strength of MSS model lies in its computational efficiency and amenability to implementation in simulators like SPICE or Verilog, avoiding the need for complex numerical integrations common in other models. It aligns with thermodynamic principles, where volatile switches enable intrinsic adaptation by dissipating energy to repair states.

2.2 Basic Framework of AHaH Computing

The Anti-Hebbian and Hebbian (AHaH) plasticity rule represents a hybrid synaptic modification mechanism that combines two fundamental forms of learning observed in biological neural systems: Hebbian reinforcement and anti-Hebbian decorrelation. Hebbian plasticity, originally proposed by Donald Hebb, posits that synaptic strength increases when pre- and post-synaptic neurons fire together, often summarized as "cells that fire together wire together" [10]. This mechanism underlies associative learning and memory formation. In contrast, anti-Hebbian plasticity weakens synapses when correlated activity occurs, promoting decorrelation and stability in neural networks. AHaH plasticity emerges from the interaction of these opposing forces in volatile, dissipative systems, where energy dissipation drives adaptation.

In the AHaH framework, anti-Hebbian learning arises first as a natural consequence of competing energy-dissipating pathways in metastable switches. Figure 1 illustrates a system where particles diffuse from a high-pressure region to lower-pressure areas through conductive channels. Differential conductances lead to unequal dissipation rates, causing the less conductive pathway to strengthen relatively more, thus reducing the differential—a process akin to anti-Hebbian decorrelation. Hebbian learning follows as a feedback response: upon detecting a signal (e.g., at the favored output), environmental interactions open new dissipation pathways, reinforcing the initial differential. This dual process enables self-stabilization and adaptation, addressing the volatility inherent in biological computing structures.

AHaH plasticity leads to attractor states that maximize margins in data distributions, extracting independent components from input streams—a property linked to independent component analysis (ICA) techniques. By balancing volatility and thermodynamics, AHaH allows systems to self-repair perturbations, converging to stable configurations that support computation. This rule is computationally efficient for large-scale adaptive networks, mitigating the von Neumann bottleneck by integrating memory and processing.

The basic model of AHaH computing builds on AHaH plasticity to create adaptive hardware from arrays of metastable switches, forming neural nodes capable of universal computation and machine learning. At its core, an AHaH node consists of synaptic weights implemented as differential pairs of memristors, which evolve under AHaH rules to converge on attractor states representing decision boundaries.

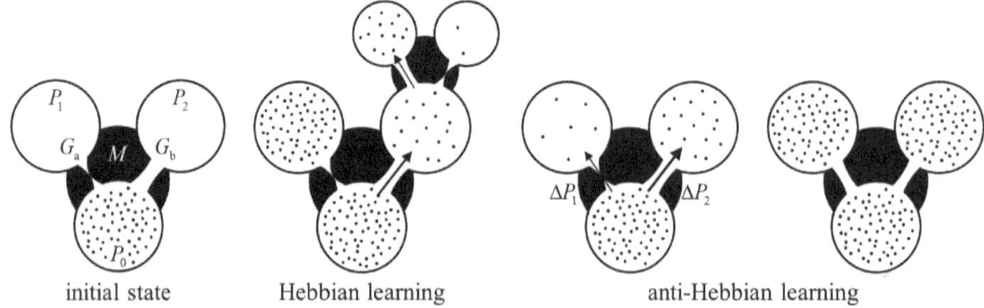

Fig. 1. Illustration of the AHaH process, showing particle diffusion through competing pathways leading to anti-Hebbian and Hebbian learning.

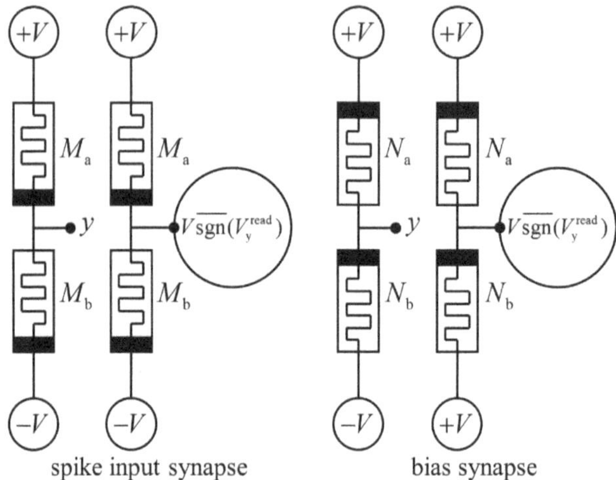

Fig. 2. Differential memristor pair forming an AHaH synapse, with voltage divider for read phase and feedback for write phase.

In circuit implementations, each synapse is realized using a pair of memristors (or memristor emulators) connected differentially to represent the weight as the conductance difference $w = G_{\mathrm{A}} - G_{\mathrm{B}}$, where G_{A} and G_{B} are the conductances of the two memristors. This setup forms a voltage divider during the read phase, where input spikes (encoded as binary pulses) activate selected synapses, and the output voltage V_{y} determines the binary decision of the AHaH node via a sign function, incorporating bias terms for stability. The write phase applies feedback voltages (V^{+} or V^{-}) to update conductances, simulating anti-Hebbian decorrelation (dissipation) and Hebbian reinforcement (learning), with stochastic elements from the MSS model ensuring realistic device behavior.

3 Design of Hardware System

3.1 Memristor On-Off Control During the Read and Write Phases

In our hardware system, synapses shown in Fig. 2 form the AHaH network, with digital control units (e.g., MCUs like STM32) managing pulse generation, switching, and ADC readout for scalable, modular architectures. For binary inputs, node outputs can connect to static NAND gates for reconfigurable logic, as the attractors span all functions; spike encodings ensure efficient processing without gates. Configured nodes support tasks like clustering (attractor partitioning), classification (label feedback), and prediction (error minimization), bridging device physics to neuromorphic efficiency [8,9].

In the original AHaH network, memristor on-off was managed by three drive sources (positive, negative, and directional) along with switch circuits for each memristor sub-circuit. This setup risked severe impacts, including hardware failure and damage to physical memristor structures due to overvoltage. To address voltage fluctuations and potential hardware damage in the original design, particularly from frequent switching of the directional drive source, we improved the memristor on-off control logic. Figure 3 illustrates our design, where the directional drive source is removed, and the level-shifting process is implemented equivalently through the control logic.

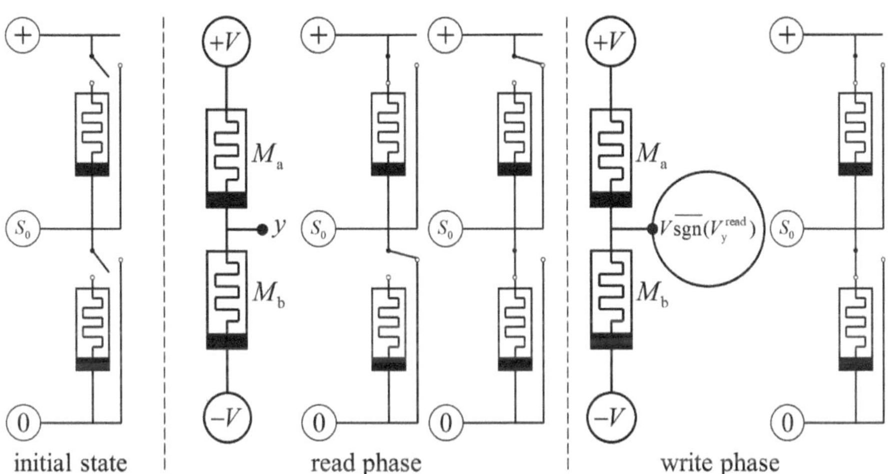

Fig. 3. Improved memristor on-off control logic, eliminating the directional drive source and implementing level-shifting via control signals.

Each memristor now requires independent control switches for on-off operations. To minimize non-ideal device effects, the positive drive source is fixed at +1.2V high level, while the negative drive source is grounded. In the read phase, all memristors are turned on to measure conductance. During the write phase, only selected memristors are activated based on input spikes. For a synapse with memristors M_{A} and M_{B}:

- If the input spike is 1 and $V_{\mathrm{y}} \geq 0$ (unsupervised learning) or the data belongs to the "positive class" (supervised learning), turn on M_{A} and turn off M_{B} to apply V^{+}.
- If the input spike is 1 and $V_{\mathrm{y}} < 0$ (unsupervised learning) or the data belongs to the "negative class" (supervised learning), turn on M_{B} and turn off M_{A} to apply V^{-}.
- If the input spike is 0, turn off both M_{A} and M_{B} to avoid updates.

This logic ensures that only the appropriate memristor is updated during the write phase, preventing unintended conductance changes and enhancing hardware reliability.

3.2 Hardware System Architecture

To realize the hardware prototype of the memristor-based spiking neural network, we designed a hierarchical modular architecture that seamlessly integrates analog and digital components as shown in Fig. 4. The core of this architecture is the memristor AHaH synapses, which performs core operations including weighted summation and binary classification. The analog part consists of the AHaH synapses, while the digital part includes on-off regulation, pulse timing management, data acquisition, and peripheral support circuits. To address the scalability issues of large-scale memristor arrays, we adopted a modular design with 8 pairs of memristors (i.e., 8 synapses) as the base unit, each module equipped with an independent digital controller. These modules support cascaded expansion through SPI interfaces, dynamically adjusting hardware resource allocation to ensure efficient resource utilization and low-complexity integration.

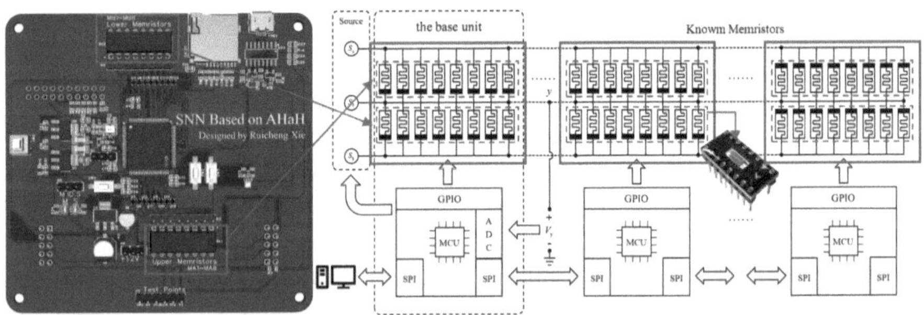

Fig. 4. Hierarchical modular architecture of the memristor-based SNN hardware system, showing integration of analog AHaH synapses and digital control units.

The workflow of our system is divided into three phases: data input, network computation, and weight update. In the data input phase, the host computer transmits preprocessed binary pulse sequences to the primary module, where each unit processes 8-bit data, generates decoding signals for memristor on-off control, and cascades remaining data to subsequent modules. In the network computation phase, the system activates analog switches to control the branch where the memristor is located and collects the node voltage as output, achieving forward propagation. In the weight update phase, pulse polarity is dynamically adjusted based on error feedback or output voltage, realizing conductance updates via the Anti-Hebbian and Hebbian plasticity mechanism. Through iterating these cycles, the system optimizes network performance, supporting both supervised and unsupervised learning.

The digital control module is the core for realizing dynamic updates of memristor conductance in the system, centered on the STM32 microcontroller and adopting a master-slave cascaded structure. The master control module connects to the ADC acquisition module, SD card storage module, analog switches, and host computer, achieving high-speed data transmission and synchronization via SPI communication interfaces. The master module handles initialization, data distribution, voltage acquisition, and result storage; the slave modules process local pulse decoding and memristor control, supporting unlimited cascaded expansion.

The analog switch module employs high-speed, low-resistance analog switches to manage memristor on-off states, ensuring minimal signal distortion during read and write phases. The ADC acquisition module uses a high-precision ADC to capture node voltages, providing accurate feedback for weight updates. The SD card storage module enables local data logging, facilitating offline analysis and model evaluation.

4 Experimental Analysis

4.1 Experimental Data and Its Encoding

To evaluate the performance of the proposed memristor-based SNN architecture, the Wisconsin Breast Cancer (Orriginal) Dataset was selected as the benchmark. This dataset, sourced from the UCI Machine Learning Repository, consists of 699 instances derived from fine-needle aspirates of breast masses, with 9 integer-valued features (ranging from 1 to 10) describing cellular characteristics such as clump thickness, uniformity of cell size and shape, marginal adhesion, single epithelial cell size, bare nuclei, bland chromatin, normal nucleoli, and mitoses. The samples are labeled into two classes: benign (458 instances) and malignant (241 instances) [11]. After data preprocessing, including removal of duplicates and imputation of missing values using mode filling for the bare nuclei attribute (affecting 16 samples), the effective dataset comprised 691 instances, split into a training set of 500 samples and a validation set of 191 samples to assess generalization.

For integration with the spiking neural network, the continuous feature values were discretized and encoded into spike patterns suitable for event-driven processing. Each feature was represented using a 4-bit Gray code to minimize bit-flip errors during transitions, converting the 1–10 scale into binary sequences (e.g., 1 as 0001, 2 as 0011, up to 10 as 1111). This resulted in a 36-bit input vector per sample (9 features × 4 bits), where each bit corresponds to a synaptic input channel in the AHaH node, with 1 triggering a spike (positive voltage pulse) and 0 indicating no spike. An additional 8-bit bias term was incorporated to stabilize attractor states, ensuring sparse and efficient spike representations [9,10]. This encoding scheme aligns with rate-based or threshold-based methods commonly used in SNNs for medical data, where spike frequency or timing encodes feature intensity, facilitating low-power classification [11]. Table 1 presents the encoding results for a portion of the dataset.

Table 1. Encoding results for a portion of the dataset.

Data ID	Original Data	Encoding Results
1000025	5, 1, 1, 1, 2, 1, 3, 1, 1	011100010001000100110001001000010001
1002945	5, 4, 4, 5, 7, 10, 3, 2, 1	011101100110011101001111001000010001
1015424	3, 1, 1, 1, 2, 2, 3, 1, 1	001000010001000100110011001000010001
1017122	8, 10, 10, 8, 7, 10, 9, 7, 1	110011111111110001001111110101000001
...	...	...

4.2 Experimental Procedure

The experimental procedure encompasses both software simulations and hardware implementations to validate the memristor-based SNN architecture under AHaH plasticity. Simulations were conducted using MATLAB with the generalized MSS memristor model to emulate stochastic conductance dynamics, enabling rapid prototyping of synaptic updates and attractor convergence. Hardware tests utilized the modular setup described in Section III, integrating eight differential memristor pairs controlled by an STM32F407 microcontroller for real-time pulse generation and feedback.

The workflow begins with input spike encoding (as detailed in Sect. 4.1), followed by iterative read-write cycles on the AHaH node. In each epoch, training samples are fed as 36-bit spike trains, with the read phase computing weighted sums via voltage dividers to determine binary outputs. Updates apply 1.2V pulses for $10\mu s$ based on output-label mismatches, enforcing anti-Hebbian decorrelation (reducing correlated weights) and Hebbian reinforcement (amplifying predictive paths), converging after 50–100 epochs. Validation involves hold-out testing without updates, measuring classification via attractor stability. For hardware, level-shifting logic mitigates voltage drops, with ADC sampling at 100 Hz monitoring conductance drifts, and power consumption tracked via integrated sensors.

4.3 Results and Discussion

Before hardware implementation, LTspice simulations are performed to validate the function of the synapse circuit and the learning ability of the network.

For a single synapse, 2 memristors is connected in series as Fig. 2 shows. The parameters of the memristor are set as follows: $N = 1000$, $G_{\rm A} = 10\mu$S, $G_{\rm B} = 100\mu$S, $V_{\rm A} = 0.5$V, $V_{\rm B} = 0.5$V, $t_{\rm c} = 0.1$ms, $\phi = 1$. The spike voltage is set to $1.2V$ and the pulse width is set to $1\mu s$. The simulation results are shown in Fig. 5. It can be seen that the conductance of the memristor and the output voltage $V_{\rm y}$ changes in small steps under the action of voltage pulses, which is consistent with the characteristics of the memristor. For the "benign class", the upper memristor $M_{\rm A}$ is strengthened during the write phase, resulting in an increase in the weight $w = G_{\rm A} - G_{\rm B}$ and output voltage $V_{\rm y}$. For the "malignant class", the opposite happens, resulting in a decrease in the weight. Due to the SPICE model of the memristor having ideal non-volatility, the state variable remains unchanged when the applied voltage is 0. Therefore, the two memristors reach saturation after only 30 epochs, and the output voltage gradually stabilizes at 0.6V. This demostrates that thesynapse can be updated according to the AHaH rule.

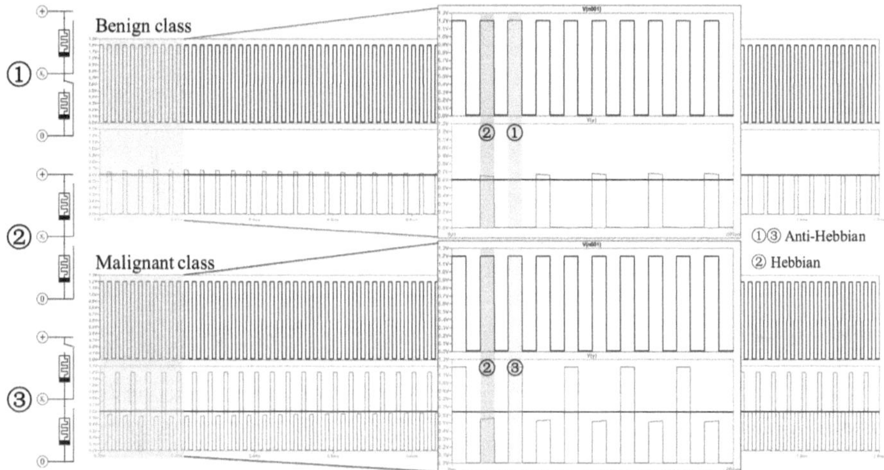

Fig. 5. LTspice simulation results for a single synapse under AHaH updates, showing conductance changes and output voltage $V_{\rm y}$.

Upon the completion of the single synapse simulation, we proceed to simulate a network with 9 synapses to Verify the feasibility of the network structure. We designed two patterns to be memorized, named Z (benign class) and V (malignant class) respectively. The encodings of the two patterns are 110_010_011 and 101_101_010, where the underscores are only for readability. To align with the procedure of the hardware experiment, the two patterns are learned in the same simulation experiment. The 2 patterns and their learning precedure are shown in Fig. 6. The parameters of the memristors are the same as those in the single synapse simulation. The experimental results are shown in Fig. 7. It can be obserrved that the output voltage V_y gradually increases when learning pattern Z, and gradually decreases when learning pattern V. This indicates that the network can learn and memorize the two patterns successfully.

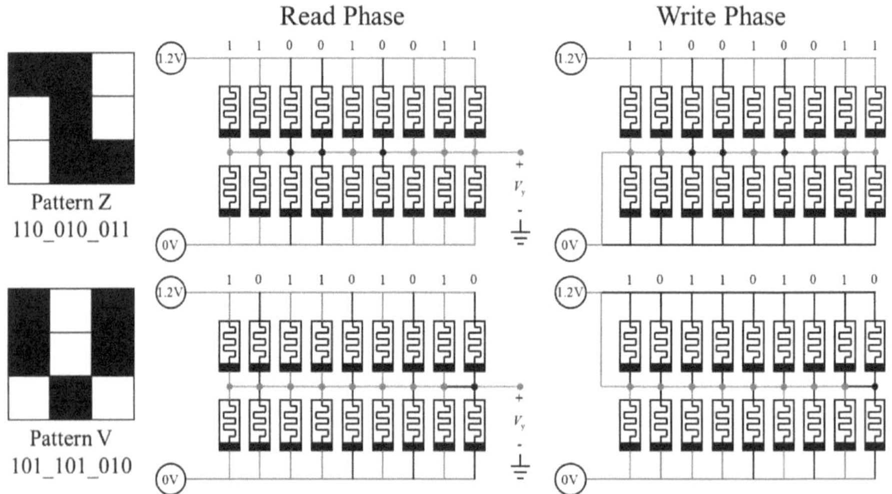

Fig. 6. Two patterns Z and V to be learned by the 9-synapse network.

After the simulation verification, we implemented the hardware prototype of the memristor-based SNN architecture. Since all data in the dataset are encoded into 36 bits, our hardware experiment requires cascading 6 modules, among which the first 5 modules are configured as input synapse groups, and the last one is configured as a bias synapse group. Circuit simulation and hardware experiment are conducted sequentially. The commercial memristors (from Knowm Inc.) is used in the hardware system. Since the data volume is sufficient, our system is trained on the Wisconsin Breast Cancer Dataset for 1 epoch. The output voltage V_y during the training is shown in Fig. 8.

Blue and orange scatter points correspond to the training set and validation set, respectively. The vertical position of the scatter points reflects the degree of deviation of individual Spike data from the "optimal classification voltage"; the red horizontal line is the "zero-value reference line" which serves as the decision boundary for Spike categories (i.e., when the deviation value of Spike data from the optimal classification voltage lies above/below the zero-value line, the Spike is classified into different categories); the black solid line is the trend line obtained by performing first-order polynomial fitting on all Spike data, which intuitively depicts the overall variation trend of the Spike data: specifically, the absolute value of the difference between the output voltage of each piece of data

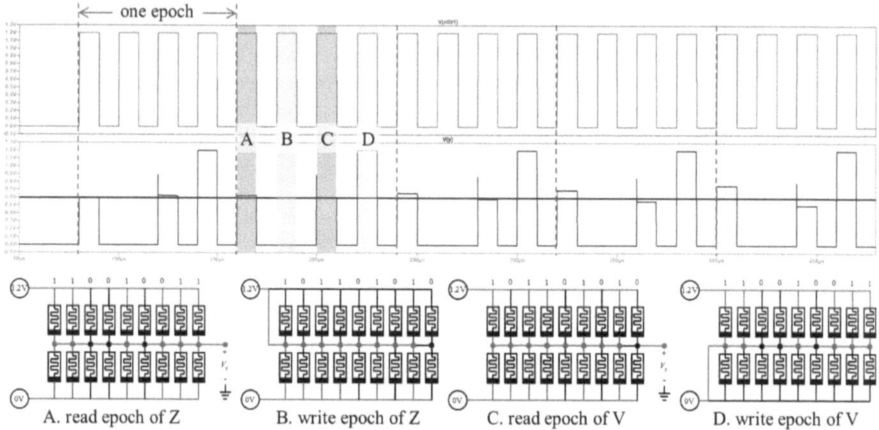

Fig. 7. LTspice simulation results for the 9-synapse network learning patterns$Z$ and V, showing output voltage V_{y} over 5 epochs.

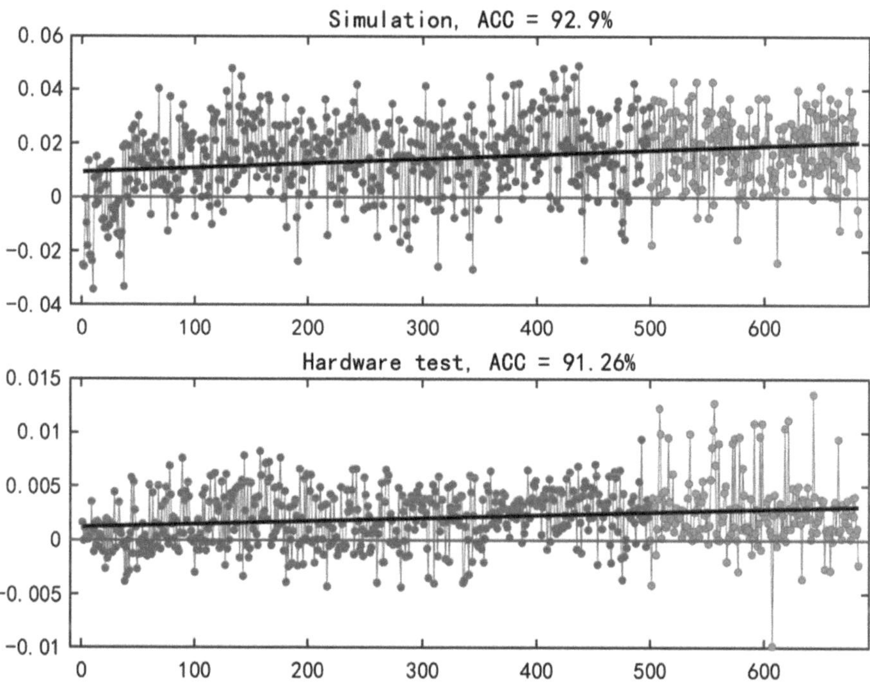

Fig. 8. Hardware experiment results showing output voltage V_{y} during training on the Wisconsin Breast Cancer Dataset.

and the initial value of the optimal classification voltage gradually increases as the training proceeds. The classification accuracies of circuit simulation and hardware experiment on the validation set are 92.9% and 91.26%, respectively. The experimental results demonstrate that the proposed memristor-based SNN architecture can effectively learn and classify the breast cancer data.

5 Conclusion

This paper has presented a memristor-based spiking neural network architecture that integrates compute-in-memory with AHaH plasticity to address von Neumann bottlenecks in neuromorphic computing. By employing differential memristor synapses and a scalable modular design with STM32-controlled on/off logic for real-time updates, the system achieves 91.26% accuracy on the Wisconsin Breast Cancer dataset, demonstrating superior energy efficiency and adaptability for binary classification tasks. While the approach effectively mitigates voltage fluctuations and hardware damage, limitations such as memristor variability and endurance persist, potentially impacting long-term reliability in larger-scale deployments. Future directions include incorporating advanced error-correction algorithms to counter non-ideal device effects, extending the framework to multi-class problems, and exploring hybrid integrations with emerging memristive materials for broader AI applications.

Acknowledgments. Project supported by the National Natural Science Foundation of China (Grant No. 52077160).

References

1. Aguirre, F., et al.: Hardware implementation of memristor-based artificial neural networks. Nat. Commun. **15**(1), 1974 (2024)
2. Chua, L.: Memristor-the missing circuit element. IEEE Trans. Circ. Theor. **18**(5), 507–519 (1971)
3. Zhang, G., et al.: Functional materials for memristor-based reservoir computing: dynamics and applications. Adv. Funct. Mater. **33**(42) (2023)
4. Peng, H., Gan, L., Guo, X.: Memristor-based spiking neural networks: cooperative development of neural network architecture/algorithms and memristors. CHIP **3**(2), 100093 (2024)
5. Günakın, S., Çam Taşkıran, Z.G.: A memristive synaptic circuit and optimization algorithm for synaptic control. Cogn. Neurodyn. **19**(1), 73 (2025)
6. Xu, X., Si, G., Oresanya, B.O., Gong, J., Guo, Z.: Fractional-order memcapacitor bridge synapse-based neural network. In: 2022 41st Chinese Control Conference (CCC), pp. 873–877 (2022)
7. Yang, Y., Xu, X., Si, G., Xu, M., Li, C., Xie, R.: Neuromorphic dynamics and behavior synchronization of fractional-order memristive synapses. Chaos, Solitons Fractals **197**, 116469 (2025)
8. Nugent, M.A., Molter, T.W.: AHaH computing-from metastable switches to attractors to machine learning. PLoS ONE **9**(2), Article 2 (2014)
9. Molter, T.W., Nugent, M.A.: The generalized metastable switch memristor model. In: CNNA 2016; 15th International Workshop on Cellular Nanoscale Networks and Their Applications, pp. 1–2 (2016)

10. Donald, H.: The Organization of Behavior: A Neuropsychological Theory. Psychology Press, London (2002)
11. Wolberg, W.H., Mangasarian, O.L.: Multisurface method of pattern separation for medical diagnosis applied to breast cytology. Proc. Natl. Acad. Sci. **87**(23), 9193–9196 (1990)

Time-Frequency Synergy for Power Load Forecasting: A Principal Frequency-Band Energy-Weighted Integrated Approach

Jingwen Qiao[1], Wenbin Wang[2], and Cheng Lian[1(✉)]

[1] Wuhan University of Technology, Wuhan 430070, China
{qjw20010427,chenglian}@whut.edu.cn
[2] Jiangxi Electric Power Research Institute of State Grid, Nanchang 330000, China

Abstract. The complexity of power load data greatly benefits from analyses conducted in both the time and frequency domains. Time-domain analysis excels at capturing local fluctuations within the data, while frequency-domain analysis is more effective at modeling long-term trends and global patterns. This advantage is particularly important for power load data, which often exhibits pronounced periodic behaviors. However, relying solely on frequency-domain modeling struggles to fully capture the inherent local fluctuations present in power load data. To leverage the strengths of both time-domain and frequency-domain analyses, we propose a novel time-frequency integrated model that effectively captures both local variations and periodic patterns in load data. Specifically, we propose a principal frequency-band energy-weighted fusion strategy that dynamically adjusts the weights based on the top k most important frequency bands of the input signal. In the frequency-domain module, we model different load variables to capture the periodic patterns in the data. In the time-domain module, we model individual load series to capture local fluctuations. Experimental results across multiple datasets demonstrate that our approach achieves state-of-the-art performance in power load forecasting.

Keywords: Time-frequency integrated model · Transformer · Power load forecasting

1 Introduction

Power load forecasting is the process of estimating the power system's future load demand, serving as a key basis for its operation, planning, and decision-making [1,2]. Actual power load data usually show obvious time-domain periodicity, including daily, weekly, and seasonal cycles. Periodic time series have strong global dependencies, and accurate power load prediction depends on models effectively modeling and capturing such periodic patterns. Frequency domain analysis methods [3,4] are key signal processing tools for effectively extracting periodic components from time series.

C. Li et al. (Eds.): ICNC 2025, CCIS 2946, pp. 80–89, 2026.
https://doi.org/10.1007/978-981-92-1599-7_7

Real-world power load data exhibit periodic patterns while being affected by sudden events (e.g., weather changes, holidays). Relying solely on frequency domain modeling struggles to effectively capture the impact of such events. In contrast, time domain modeling methods [5,6] excel at capturing local anomalies and abrupt changes. Thus, the optimal approach is to combine the strengths of time and frequency domain modeling to build more accurate and comprehensive prediction models.

Effectively combining time-domain and frequency-domain strengths remains a major challenge. Early studies focused mainly on time-domain analysis, developing many sophisticated methods/models to capture temporal information. In recent years, some studies have used frequency-domain models to model global dependencies in time series; however, these approaches usually rely only on spectral information and thus struggle to handle abrupt data changes. In contrast, AFTNet [7] dynamically integrates time- and frequency-domain information by detecting the fundamental frequency and its harmonics, but it has a cumbersome computational process and poor adaptability to non-periodic or noisy signals. Similarly, FREDF [8] uses a frequency-domain loss function for indirect information fusion (by treating all spectral components uniformly) yet fails to effectively balance high-frequency noise and low-frequency trends.

The main contributions of this paper are as follows:

1. we propose a frequency-domain prediction framework that employs a self-attention mechanism, allowing the model to effectively capture the interdependencies among various prediction points and the periodic characteristics inherent in the data.
2. We divide the data into patch and apply self-attention to each patch, enabling the model to effectively capture local fluctuations and abnormal variations in the time series.
3. We design a dominant-band energy-weighted fusion strategy that dynamically computes fusion weights by analyzing the proportion of energy concentrated in the dominant frequency band of the input signal.

2 Related Work

2.1 Time Domain Forecasting Methods

Early time series forecasting methods primarily relied on statistical approaches, such as VAR [9] and ARIMA [10]. However, these models often struggle to capture nonlinear fluctuations and complex dependencies. With the development of deep learning, neural network models have gradually been introduced and have significantly outperformed statistical models in terms of predictive performance. Flunkert *et al.* [11] propose an auto-regressive RNN model to predict the probability distribution of future time points. For models based on temporal convolutional networks, Wang *et al.* [12] proposed a local-global architecture that facilitates information aggregation and long-term dependency modeling. Liu *et al.* [13] removed the constraint of causal convolution and introduced a recursive downsample-convolve-interact framework to effectively capture complex temporal dynamics in time series data. For MLP-based models, Lin *et al.* [14] propose a cross-cycle sparsity technique to extract salient periodic features, achieving high forecasting performance with fewer than 1,000 parameters. Zeng *et al.* [15] also achieved competitive performance using only a single linear layer. For Transformer-based models, Zhou *et al.* [5] proposed an efficient Transformer variant that incorporates a probSparse self-attention mechanism and a generative decoder, enabling fast and accurate forecasting of long time series.

2.2 Frequency Domain Forecasting Methods

Frequency-domain modeling, by transforming data into the frequency domain, effectively captures long-term trends and periodic fluctuations, offering a unique perspective for time series forecasting. Many methods have been proposed in this field to leverage these features for capturing global dependencies. Specifically, Wu *et al.* [16] proposes replacing the conventional self-attention mechanism with an autocorrelation mechanism, efficiently implemented through the fast fourier transform(FFT). Zhou *et al.* [17] introduces a frequency-enhanced attention mechanism that leverages the discrete fourier transform(DFT), deriving attention weights from the frequency spectra of queries and keys, and subsequently computing the weighted sum of the value representations in the frequency domain. Wu *et al.* [18] employs the FFT to identify frequency components with significant magnitudes, reshaping the one-dimensional time series into a two-dimensional representation based on the detected periodicities, thereby enhancing the learning of representations. Yi *et al.* [19] employs the DFT to transform the data into the frequency domain and introduces a specialized MLP designed for complex-valued inputs, which models the real and imaginary components separately.

3 Methodology

Given a historical power load input sequence $\mathbf{X} = [x_1, \ldots, x_L] \in \mathbb{R}^{C \times L}$, where L represent the length of the input sequences and C denotes the number of predictive variables. The goal of the forecasting model is to predict the power load values $\mathbf{Y} = [x_{L+1}, \ldots, x_{L+O}] \in \mathbb{R}^{C \times O}$ for the next O time steps based on the historical observations $\mathbf{X}$.

3.1 Overall Framework

As illustrated in the Fig. 1, the proposed model consists of three main components: (a) F-block, which employs channel-wise self-attention to effectively learn the correlations among different prediction points; (b) T-block, which partitions the time series into patches and applies attention across patches for each variable to capture local fluctuations and abrupt changes; (c) dominant frequency band energy fusion module, which dynamically computes weights based on the proportion of energy in the dominant frequency band of the input signal to produce the final prediction.

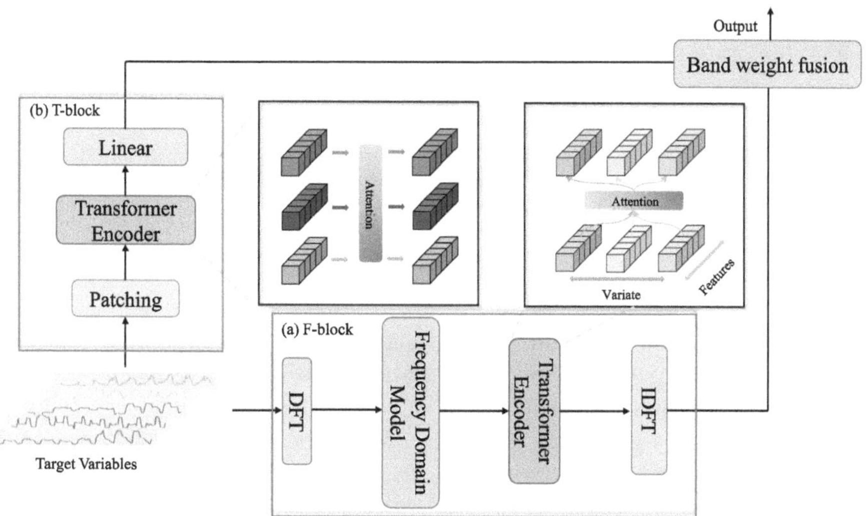

Fig. 1. Our model framework diagram.

3.2 Frequency-domain Module

Domain Conversion/Inversion. Given the input X, we first apply the DFT to decompose X into frequency coefficients A_1 for all channels. A Transformer encoder is then employed to extract frequency features, resulting in $A \in \mathbb{R}^{C \times L}$. Subsequently, the output in the frequency domain is reconstructed into a time-domain signal X' via the IDFT.

$$\mathbf{X_f} = \text{IDFT}(f_{\text{Trans}}(A)), \quad A = \text{DFT}(\mathbf{X}) \tag{1}$$

Frequency-domain Channel Learning. After obtaining the frequency coefficients through DFT, non-overlapping patch operations are applied along the channel axis (*i.e.*, the C-axis) to both the real and imaginary components of A independently, resulting in the following local sub-frequencies:

$$\mathbf{W} = \{\mathbf{W}_1, \mathbf{W}_2, \ldots, \mathbf{W}_N\} = \text{patch}(\mathbf{A}) \tag{2}$$

where N is the total number of patches, and S represents the length of each patch. This approach allows the model to learn features within each local sub-band, even when the frequency bands are very close to each other (e.g., adjacent bands at 1 Hz and 2 Hz).

Given the $\mathbf{W}$, we employ a Transformer encoder to independently learn the importance of each $\mathbf{W}$. For $\mathbf{W}_n^{(1:C)} = \{w_n^{(1)}, w_n^{(2)}, \ldots, w_n^{(C)}\}_{n=1}^{N}$, the Transformer encoder $f_{\text{Trans}}(\cdot)$ takes each $W_n^{*(c)}$ as an input token:

$$\mathbf{W}_n^{\prime(1:C)} = f_{\text{Trans}}(\mathbf{W}_n^{*(1:C)}) \tag{3}$$

Frequency-wisw Summarization. Given the learned sub-frequency features of the historical time series $\mathbf{X}$, denoted as $\mathbf{W}' = \{\mathbf{w}'_1, \mathbf{w}'_2, \ldots, \mathbf{w}'_N\}$, the frequency-wise summarization process involves linear projection followed by an IDFT:

$$\mathbf{X_f} = \text{IDFT}(\mathbf{A}'), \quad \mathbf{A}' = \text{Linear}(\mathbf{W}') \tag{4}$$

where $\mathbf{X}' \in \mathbb{R}^{C \times O}$ is the final output of the frequency-domain module.

3.3 Time-domain Module

Patching. Given an input univariate time series X_i of length L, we partition it into overlapping patches of length P with a stride of S. This means that each subsequent patch starts S elements after the start of the previous patch. The partitioning process generates a sequence of patches $X_p^{(i)} \in \mathbb{R}^{P \times M}$, where M is the total number of patches, and $M = \lfloor \frac{L-P}{S} \rfloor + 2$. Before partitioning, the original sequence is padded at the end with S repetitions of its last value to accommodate the overlapping patches.

By using this patching approach, the number of input tokens is reduced from L to approximately L/S . This results in a quadratic reduction in both memory consumption and computational complexity of the attention maps by a factor of S.

Time-domain Learning. After partitioning the input into patches, we employ a standard Transformer encoder to map the input signal into a latent representation. Each patch is projected into a D-dimensional latent space through a trainable linear projection $W_p \in \mathbb{R}^{D \times P}$, and a learnable positional encoding $W_{\text{pos}} \in \mathbb{R}^{D \times N}$ is added to preserve the temporal order of the patches. The resulting embedding for each patch is given by: $X_d^{(i)} = W_p X_p^{(i)} + W_{\text{pos}}$.

Subsequently, for each head $h = 1, \ldots, H$ in the multi-head attention mechanism, the input embeddings are projected into a query matrix $Q_h^{(i)} = (X_d^{(i)})^T \mathbf{W}_h^{(Q)}$, a key matrix $K_h^{(i)} = (X_d^{(i)})^T \mathbf{W}_h^{(K)}$, and a value matrix $V_h^{(i)} = (X_d^{(i)})^T \mathbf{W}_h^{(V)}$, where $\mathbf{W}_h^Q, \mathbf{W}_h^K \in \mathbf{R}^{D \times d_k}$ and $\mathbf{W}_h^V \in \mathbb{R}^{D \times D}$. The outputs are then computed using scaled dot-product attention, resulting in $\mathbf{O}_h^{(i)} \in \mathbb{R}^{D \times N}$:

$$(\mathbf{O}_h^{(i)})^T = \text{Attention}(Q_h^{(i)}, K_h^{(i)}, V_h^{(i)}) \tag{5}$$

Finally, We apply a linear layer to obtain the final output of the F-block.

$$\mathbf{X_t} = \text{Linear}(\mathbf{O}_h) \tag{6}$$

The Transformer encoder, in addition to incorporating a multi-head self-attention module, also includes a feed-forward network with residual connections. By partitioning each variable into different patches for attention computation, the model can effectively capture local fluctuations and anomalies in the data. This approach, when combined with the periodic features learned by the F-block, enables the model to achieve more accurate prediction results.

3.4 Band Weight Fusion

To effectively integrate time-domain and frequency-domain information, we developed a fusion approach that weights the energy of the dominant frequency components. This approach dynamically calculates weights based on the energy contribution of these components in the input signal. Specifically, we apply the FFT to the original input data to obtain the complex spectrum. We then compute the magnitude of this spectrum and identify the top k frequency points with the highest magnitudes. The weight for the F-block is determined by calculating the proportion of the total spectral magnitude accounted for by these dominant frequency points.

$$
\begin{aligned}
X_{[f]} &= \mathrm{FFT}(X), \quad X \in \mathbb{R}^{(L//2+1)\times C} \\
A_{[f]} &= |X_{[f]}| \\
\mathrm{TopVal} &= \mathrm{Top}(A, k), \quad \mathrm{TopVal} \in \mathbb{R}^{k\times D} \\
\mathbf{W} &= \frac{\sum_{f=0}^{L//2} A[f]}{\sum_{i=0}^{k-1} \mathrm{TopVal}[i] + \epsilon}
\end{aligned}
$$

where f is the frequency index, and ϵ is a tiny-value to prevent division by zero.

We choose to use the MSE loss to measure the discrepancy between the prediction and the ground truth.

4 Experiments

4.1 Datasets and Preprocessing

The University of Texas at Dallas (UTD) dataset [20] contains hourly campus load data from 13 buildings and 20 weather/calendar features, including solar radiation, meteorological conditions, and wind parameters, covering January 1, 2014–December 31, 2015. The Tetouan dataset [21] provides electricity consumption for three city regions in northern Morocco, along with five weather features (temperature, humidity, wind speed, solar radiation), sampled every ten minutes from January 1–December 30, 2017. The Wind-Solar-Heat Pump District (WPuQ) dataset [22] contains hourly electricity and heat pump load data from 38 households in Saxony, Germany, plus 10 meteorological variables. Due to missing data, only 21 households with relatively complete records from January 1, 2019–December 31, 2020, were used in our analysis.

For the missing auxiliary variables in the aforementioned datasets, we apply forward filling, while for the missing target variables, we use zero filling. Additionally, we perform some basic preprocessing operations. We followed standard protocols and, in chronological order, divided all datasets into training, validation, and test sets with a ratio of 7:1:2.

4.2 Compared Methods

We select 8 state-of-the-art deep learning models as our baselines, including Transformer-based models: FEDformer (2022) [17], and Autoformer (2021) [16], CNN-based models: TimesNet (2023) [18] and MICN (2023) [12], and linear models: SparseTSF (2024) [14], FreTS (2023) [19], TIDE (2023) [23], and Dlinear (2023) [15].

4.3 Implementation Details

All experiments are implemented in PyTorch and conducted on a single RTX 3080 GPU. We use Adam optimizer and L2 loss for optimization, with an initial learning rate of 0.0001. Batch size is set to 32, and the number of training epochs is fixed to 30. The default look-back length is set to $L = 96$. The mean squared error (MSE) and mean absolute error (MAE) are computed as performance metrics for multivariate time series prediction. To ensure equitable comparison, all models are trained with the same set of hyperparameters.

5 Experimental Result

5.1 Comparison of Prediction Results

Table 1. The results of the multivariate prediction task, with prediction lengths $S \in \{96, 192, 336, 720\}$ and fixed lookback length of $L = 96$, are presented. *Avg* denotes the mean of all predicted results.

Models		**Ours**		SparseTSF		TimesNet		FreTS		TIDE		MICN		DLinear		FEDformer		Autoformer	
Metric		**MSE**	**MAE**	MSE	MAE	MSE	MAE	MSE	MAE	MSE	MAE	MSE	MAE	MSE	MAE	MSE	MAE	MSE	MAE
UTD	96	0.273	0.334	0.387	0.421	0.340	0.395	0.343	0.411	0.404	0.438	0.277	0.366	0.402	0.462	0.301	0.374	0.329	0.400
	192	0.337	0.370	0.413	0.426	0.451	0.465	0.422	0.455	0.437	0.450	0.438	0.443	0.460	0.492	0.363	0.408	0.481	0.485
	336	0.413	0.414	0.471	0.450	0.471	0.458	0.540	0.508	0.501	0.476	0.538	0.521	0.575	0.545	0.469	0.477	0.492	0.478
	720	0.539	0.465	0.611	0.508	0.653	0.550	0.825	0.612	0.646	0.534	0.759	0.595	0.741	0.614	0.619	0.548	0.656	0.557
	Avg	0.391	0.396	0.471	0.451	0.479	0.467	0.533	0.497	0.497	0.475	0.503	0.481	0.545	0.528	0.438	0.452	0.490	0.480
Tetouan	96	0.069	0.180	0.131	0.253	0.166	0.285	0.143	0.277	0.194	0.296	0.092	0.225	0.175	0.287	0.241	0.381	0.207	0.354
	192	0.080	0.193	0.143	0.263	0.151	0.272	0.190	0.325	0.204	0.303	0.115	0.248	0.189	0.300	0.230	0.368	0.375	0.479
	336	0.099	0.213	0.151	0.271	0.178	0.295	0.226	0.358	0.202	0.301	0.126	0.261	0.189	0.304	0.193	0.336	0.351	0.458
	720	0.116	0.235	0.161	0.280	0.183	0.308	0.255	0.386	0.198	0.301	0.148	0.292	0.192	0.314	0.206	0.347	0.333	0.448
	Avg	0.091	0.205	0.146	0.267	0.170	0.290	0.203	0.337	0.199	0.300	0.120	0.256	0.186	0.301	0.217	0.358	0.317	0.435
WPuQ	96	0.786	0.499	0.823	0.521	0.812	0.513	0.793	0.513	0.839	0.525	0.819	0.549	0.806	0.518	0.787	0.527	0.851	0.565
	192	0.801	0.505	0.829	0.522	0.824	0.518	0.810	0.522	0.852	0.531	0.833	0.548	0.816	0.522	0.798	0.531	0.824	0.548
	336	0.822	0.517	0.844	0.528	0.841	0.525	0.829	0.531	0.872	0.540	0.848	0.553	0.831	0.530	0.819	0.544	0.830	0.548
	720	0.865	0.537	0.887	0.547	0.892	0.548	0.858	0.542	0.916	0.559	0.881	0.565	0.864	0.545	0.857	0.560	0.889	0.581
	Avg	0.819	0.515	0.846	0.530	0.842	0.526	0.822	0.527	0.870	0.539	0.845	0.554	0.829	0.529	0.815	0.540	0.828	0.561

Table 1 summarizes the overall forecasting results, where lower MSE/MAE values indicate higher prediction accuracy. We highlight the best results in red. Our method consistently delivers state-of-the-art performance across all datasets and forecasting horizons, demonstrating superior effectiveness in power load forecasting. Additionally, MICN maintains competitive performance across all datasets but shows a significant prediction error on the UTD dataset at the 720-point horizon. We attribute this to the non-stationary nature of the UTD dataset, which limits its multi-scale decomposition from capturing complex patterns effectively. FEDformer also achieves strong results on the WPuQ dataset, benefiting from the effective integration of the Transformer architecture with frequency-domain modeling to better address complex load forecasting tasks. Overall, our method fully exploits time-domain and frequency-domain information, captures long-term dependencies effectively, and significantly enhances forecasting performance compared to previous models.

5.2 Ablation Studies

Table 2. Ablation of Different Modules

Dataset	Ours		T		F		Fixed	
	MSE	MAE	MSE	MAE	MSE	MAE	MSE	MAE
UTD	0.539	0.465	0.550	0.473	0.553	0.478	0.417	0.412
Tetouan	0.091	0.205	0.092	0.209	0.105	0.215	0.094	0.207
WPuQ	0.819	0.515	0.833	0.513	0.820	0.515	0.818	0.515

Table 2 presents the contribution of each component in our model, where we report the performance after individually removing each part. Here, T denotes using only the T-block for prediction, and F denotes using only the F-block. Removing either the T-block or the F-block leads to a significant drop in average model performance, highlighting the crucial roles of both time-domain and frequency-domain information in accurately capturing long-term features and trends. Overall, eliminating any single component from our model results in a noticeable performance degradation, indicating that each part tightly aligns with the target task and plays an essential role in effectively capturing and modeling the underlying patterns.

5.3 Visualization of Different Prediction Lengths

The Fig. 2 presents the visualization results across different forecasting horizons for various datasets, where the vertical axis represents the MSE and the horizontal axis denotes the forecasting length. As the forecasting length increases, the model's accuracy gradually decreases. Nevertheless, our model consistently maintains highly competitive performance even as the forecasting horizon extends.

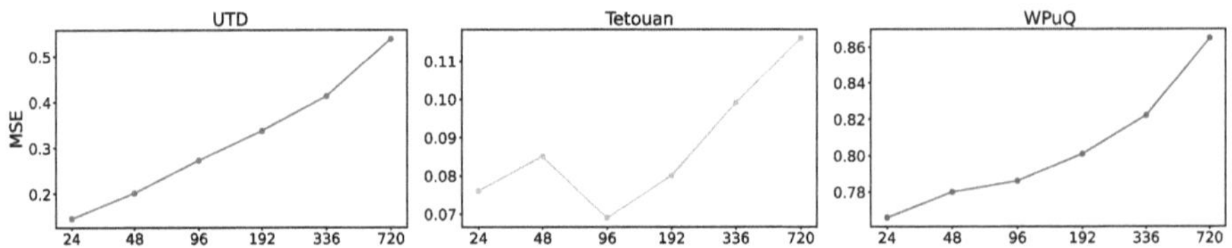

Fig. 2. Visualization results for different forecasting horizons on the UTD, Tetouan, and WPuQ datasets.

6 Conclusions

In this paper, we introduce a novel model for multivariate power load forecasting. Our model effectively addresses several challenges in power load prediction, including the inability to fully leverage both time- and frequency-domain advantages, the difficulty of modeling cross-dimensional dependencies, and the challenge of capturing global trends alongside local fluctuations. Considering that the amplitude reflects the energy strength of each frequency component, and that components with larger amplitudes contribute more significantly to the overall signal structure, we adopt a principal frequency-band dynamic weighting fusion strategy. To accurately capture both the periodic patterns and local variations in the data, we separately model inter-variable dependencies and individual series. Experimental results demonstrate that our proposed approach, built upon a combination of Transformer, CNN, and linear models, achieves state-of-the-art performance across different datasets.

Acknowledgments. This work was supported by the National Natural Science Foundation of China under Grant 62176193.

References

1. Lee, C.-H.: Renewable and sustainable energy reviews. Coastal Ocean **5**(2), 62–70 (2012)
2. Kuster, C., Rezgui, Y., Mourshed, M.: Electrical load forecasting models: a critical systematic review. Sustain. Cities Soc. **35**, 257–270 (2017)
3. Xu, Z., Zeng, A., Xu, Q.: Fits: modeling time series with 10k parameters. arXiv preprint arXiv:2307.03756 (2023)
4. Sun, F.K., Boning, D.S.: Fredo: frequency domain-based long-term time series forecasting. arXiv preprint arXiv:2205.12301 (2022)
5. Zhou, H., Zhang, S., Peng, J., Zhang, S., Li, J., Xiong, H., Zhang, W.: Informer: beyond efficient transformer for long sequence time-series forecasting. In: Proceedings of the AAAI Conference on Artificial Intelligence, vol. 35, no. 12, pp. 11 106–11 115 (2021)
6. Zhang, Y., Yan, J.: Crossformer: transformer utilizing cross-dimension dependency for multivariate time series forecasting. In: The Eleventh International Conference on Learning Representations (2023)

7. Ye, H., et al.: Atfnet: adaptive time-frequency ensembled network for long-term time series forecasting. arXiv preprint arXiv:2404.05192 (2024)
8. Wang, H., et al.: Fredf: learning to forecast in frequency domain. arXiv preprint arXiv:2402.02399 (2024)
9. Kilian, L., Lütkepohl, H.: Structural Vector Autoregressive Analysis. Cambridge University Press (2017)
10. Stellwagen, E., Tashman, L.: Arima: the models of box and jenkins. Foresight Int. J. Appl. Forecast. (30), (2013)
11. Salinas, D., Flunkert, V., Gasthaus, J., Januschowski, T.: Deepar: probabilistic forecasting with autoregressive recurrent networks. Int. J. Forecast. **36**(3), 1181–1191 (2020)
12. Wang, H., Peng, J., Huang, F., Wang, J., Chen, J., Xiao, Y.: Micn: multi-scale local and global context modeling for long-term series forecasting. In: The Eleventh International Conference on Learning Representations (2023)
13. Liu, M., et al.: Scinet: time series modeling and forecasting with sample convolution and interaction. Adv. Neural. Inf. Process. Syst. **35**, 5816–5828 (2022)
14. Lin, S., Lin, W., Wu, W., Chen, H., Yang, J.: Sparsetsf: modeling long-term time series forecasting with 1k parameters. arXiv preprint arXiv:2405.00946 (2024)
15. Zeng, A., Chen, M., Zhang, L., Xu, Q.: Are transformers effective for time series forecasting? In: Proceedings of the AAAI Conference on Artificial Intelligence, vol. 37, no. 9, pp. 11 121–11 128 (2023)
16. Wu, H., Xu, J., Wang, J., Long, M.: Autoformer: decomposition transformers with auto-correlation for long-term series forecasting. In: Advances in Neural Information Processing Systems, vol. 34, pp. 22 419–22 430 (2021)
17. Zhou, T., Ma, Z., Wen, Q., Wang, X., Sun, L., Jin, R.: Fedformer: frequency enhanced decomposed transformer for long-term series forecasting. In: International Conference on Machine Learning, PMLR, pp. 27 268–27 286 (2022)
18. Wu, H., Hu, T., Liu, Y., Zhou, H., Wang, J., Long, M.: Timesnet: temporal 2d-variation modeling for general time series analysis. arXiv preprint arXiv:2210.02186 (2022)
19. Yi, K., et al.: Frequency-domain mlps are more effective learners in time series forecasting. Adv. Neural Inf. Process. Syst. **36**, 76 656–76 679 (2023)
20. Zhang, J., Feng, C.: Short-term load forecasting data with hierarchical advanced metering infrastructure and weather features (2019). https://doi.org/10.21227/jdw5-z996
21. Salam, A., El Hibaoui, A.: Comparison of machine learning algorithms for the power consumption prediction:-case study of tetouan city. In: 6th International Renewable and Sustainable Energy Conference (IRSEC), vol. 2018, pp. 1–5. IEEE (2018)
22. Schlemminger, M., Ohrdes, T., Schneider, E., Knoop, M.: Dataset on electrical single-family house and heat pump load profiles in Germany. Sci. Data **9**(1), 56 (2022)
23. Das, A., Kong, W., Leach, A., Mathur, S., Sen, R., Yu, R.: Long-term forecasting with tide: time-series dense encoder. arXiv preprint arXiv:2304.08424 (2023)

PSEformer: Period-Series Embedding Transformer with Enhanced Input Embedding for Load Forecasting

Mengpeng Yang[1], Wenbin Wang[2], and Cheng Lian[1(✉)]

[1] School of Automation, Wuhan University of Technology, Wuhan 430074, China
{ymp,chenglian}@whut.edu.cn

[2] Jiangxi Electric Power Research Institute of State Grid, Nanchang 330000, China

Abstract. Load forecasting is a key foundation for ensuring the stable and efficient operation of smart grids. However, existing Transformer-based time series forecasting methods have notable limitations: they overly rely on attention mechanisms to capture intra-series and inter-series dependencies, which results in low computational efficiency and inadequate performance in modeling periodical features of load series (such as daily/weekly periods) and deep cross-series relationships. To address this issue, we innovate by starting with the simple yet powerful modeling technique of input embedding and designing additional embedding mechanisms for load series—period embeddings (for representing periodic features such as daily/weekly periods) and inter-series dependency embeddings. This enables efficient and accurate modeling of both intra-series and inter-series dependencies. Our proposed Period-Series Embedding Transformer (PSEformer), through this joint period-series embedding mechanism, effectively enhances the model's ability to capture intra-series and inter-series dependencies in load data. Experimental results show that, compared to baseline models, PSEformer achieves optimal forecasting performance on three real-world power load datasets.

Keywords: Load forecasting · Period-Series Embedding · Transformer

1 Introduction

The core objective of a smart grid is to ensure the safety and stability of electricity consumption for users. As electricity demand continues to rise across society, accurate load forecasting has become a fundamental pillar for maintaining the efficient and economical operation of the power grid [1]. Studies have shown that in the UK power system, a 1% decrease in forecasting accuracy can lead to additional operational costs of up to £ 10 million [2]. Therefore, accurate load forecasting not only enhances resource utilization and reduces operational expenses within the smart grid but also provides reliable data support for load dispatching and decision-making [3].

C. Li et al. (Eds.): ICNC 2025, CCIS 2946, pp. 90–99, 2026.
https://doi.org/10.1007/978-981-92-1599-7_8

Traditional load forecasting primarily relies on statistical methods, such as the Autoregressive Integrated Moving Average [4] model and the Historical Average [5] model. However, these approaches are limited to capturing linear dependencies and struggle to handle the nonlinear and high-dimensional characteristics present in load data. Machine learning methods, including Support Vector Machines [6], Bayesian Networks [7], and Random Forests [8], have addressed the limitations of statistical techniques by effectively capturing nonlinear patterns. However, these methods often involve high computational complexity and face limitations when applied to large-scale power load datasets. The emergence of deep learning has introduced new solutions for load forecasting. Recurrent Neural Networks (RNNs) and their variants [9,10] leverage recursive mechanisms to capture temporal dependencies, while Convolutional Neural Networks (CNNs) [11] extract local features through convolutional operations.

Transformer-based models have demonstrated remarkable advantages in the field of time series forecasting [12–16], particularly due to their ability to efficiently model global dependencies within series using attention mechanisms. This significantly enhances prediction accuracy. However, many existing models tend to overly focus on developing increasingly complex attention mechanisms [12,13]. Their input embeddings are typically derived solely from the time series data itself, relying heavily on the attention mechanism to model both temporal dependencies within individual series and inter-variable relationships across series. As a result, these models often fall short in effectively capturing the intricate temporal dynamics and deeper interdependencies inherent in load forecasting tasks.

Based on the above analysis, we propose the **P**eriod-**Series** **E**mbedding Trans**former** (PSEformer) to address these challenges. Unlike other Transformer-based methods, Starting from input embedding—a simple yet powerful representation technique, we design additional components tailored for load series patch embeddings: periodic embeddings (e.g., daily/weekly periodic features) and inter-series dependency embeddings. These enhancements enable more effective modeling of both intra-series dependencies and inter-series relationships, which are critical for accurate load forecasting.

The main contributions of this paper are as follows:

1. We propose a novel PSEformer model for load forecasting, which combines series patch embeddings with period-series embeddings to enhance the representation capacity of load series.
2. By incorporating periodic embeddings and inter-series dependency embeddings, PSEformer effectively models the periodic dependencies and inter-series relationships in load data.
3. Experimental results on three publicly available power load datasets demonstrate that the model can significantly enhance performance by using period-series embeddings.

2 Related Work

Power load forecasting, as a crucial component in maintaining the stability of smart grids, has attracted widespread attention. Moudgil *et al.* [17] proposed an

LSTM-based dual-channel encoder-decoder bidirectional long short-term memory (LSTM) load forecasting model, which employs an encoder-decoder architecture to extract both local and global temporal patterns from load series. Wang *et al.* [18] combined CNN and LSTM to extract local features from load series using CNN, while LSTM was used to model long-term dependencies, achieving good results. However, LSTM faces limitations in parallel processing efficiency and issues such as gradient explosion or vanishing gradients. Additionally, CNN can only model local dependencies, presenting challenges in capturing global relationships.

In recent years, Transformer-based models have been widely adopted in time series forecasting, yielding significant improvements. Zhou *et al.* [12] introduced the ProbSparse attention mechanism to address the quadratic complexity issue when applying attention to long series. Wu *et al.* [13] employed series decomposition and autocorrelation mechanisms to capture long-range dependencies in time series data, thereby enhancing the forecasting accuracy of Transformer models. Nie *et al.* [14] divided long series into multiple patches, reducing the number of tokens in attention calculations and thus lowering computational complexity. Liu *et al.* [16] encoded series as independent tokens, using attention mechanisms to model inter-series dependencies, while the feedforward network modeled intra-series temporal dependencies, resulting in better temporal representations.

Transformer-based models have shown promising potential in load forecasting, However, previous methods rely solely on attention mechanisms to extract intra-series and inter-series dependencies, thus facing challenges in capturing temporal features and inter-series correlations. To address these issues, we propose PSEformer, which enhances the input embeddings of the Transformer by combining patch-based embeddings with period-series embeddings, thereby improving the model's ability to represent and model series effectively.

3 Methodology

In this section, we first introduce the problem definition, followed by a detailed description of the structure of our PSEformer model.

3.1 Problem Formulation

Given the historical load data for T time steps, $\mathbf{X} \in \mathbb{R}^{1\times T}$, potential additional exogenous series (which may include related series such as temperature, weather, etc.), $\mathbf{E_s} \in \mathbb{R}^{N\times T}$ (where N is the number of exogenous series), and time features $\mathbf{E_t} \in \mathbb{R}^{1\times T}$, the objective is to predict the load values for the next T' time steps, $\mathbf{Y} \in \mathbb{R}^{1\times T'}$. Formally, the load forecasting task is defined as:

$$\mathbf{Y} = f(\mathbf{X}, \mathbf{E_s}, \mathbf{E_t}; \theta) \tag{1}$$

where $f(\cdot;\theta)$ is a model with parameters θ.

3.2 PSEformer Architecture

The proposed PSEformer model, as shown in the Fig. 1, consists of four main components: period-series embedding, intra-series relationship extraction, inter-series relationship extraction, and output projection.

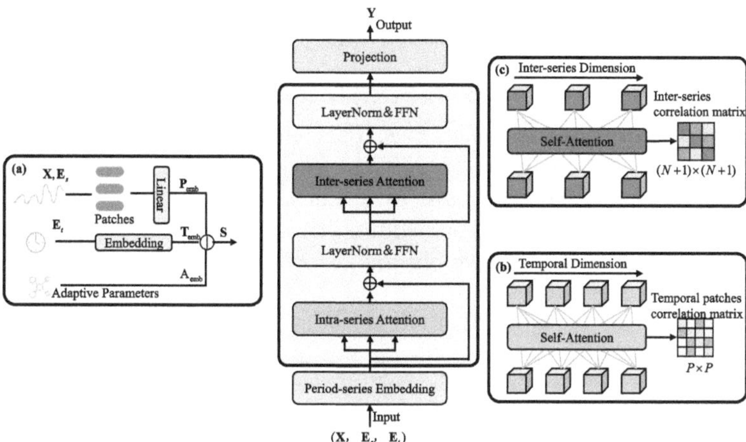

Fig. 1. Framework of PSEformer, which is mainly composed of Period-Series Embedding, Inter-Series Relationship Extraction, Inter-Series Relationship Extraction, and Output Projection.

Period-Series Embedding. As shown in Fig. 1 (a), in order to comprehensively encode the input information, we add periodic embeddings and inter-series relationship embeddings on top of the patch-based embeddings.

To retain the key information from the original data, we first divide the raw input series into non-overlapping patches, then project each patch as a time token. The patch-based embedding can be expressed as:

$$\begin{aligned} \{(\mathbf{x}_1, \mathbf{e}_{s1}), \cdots, (\mathbf{x}_P, \mathbf{e}_{sP})\} &= \text{Patchify}(\mathbf{X}, \mathbf{E_s}) \\ \mathbf{P}_{\text{emb}} &= \text{Linear}((\mathbf{x}_1, \mathbf{e}_{s1}), \cdots, (\mathbf{x}_P, \mathbf{e}_{sP})) \end{aligned} \tag{2}$$

where the patch length is p, $P = \frac{T}{p}$ is the number of patches, and $(\mathbf{x}_i, \mathbf{e}_{si})$ represents the i-th patch consisting of the load series and exogenous series. Linear$(\cdot)$ projects the patch of length p into a d_p-dimensional space. The final representation of the input series is $\mathbf{P}_{\text{emb}} \in \mathbb{R}^{(N+1)\times P \times d_p}$.

Additionally, since load series inherently have weekly and daily periodic characteristics, we introduce learnable time embeddings to represent these periodic features. Let $\mathbf{T}_{\text{week}} \in \mathbb{R}^{T\times d_t}$ represent the day-of-week embedding, and $\mathbf{T}_{\text{day}} \in \mathbb{R}^{T\times d_t}$ represent the timestamp embedding within a day (determined by the sampling frequency of the load data). Both embeddings are derived from the time features $\mathbf{E}_t$. To make these time embeddings compatible with the patch-based embedding structure, we process the time embeddings as follows: first, we divide $\mathbf{T}_{\text{week}}$ and $\mathbf{T}_{\text{day}}$ along the time steps T into patches consistent with the input series. Then, we apply mean aggregation to the time features within each patch. Finally, through feature concatenation and dimension broadcasting, we obtain the time embedding $\mathbf{T}_{\text{emb}} \in \mathbb{R}^{(N+1)\times P\times 2d_t}$.

Lastly, recognizing that input embeddings should not be limited to series internal information and time features, we design an adaptive inter-series relationship embedding $\mathbf{A}_{\text{emb}} \in \mathbb{R}^{(N+1)\times d_a}$ to incorporate the relationship between the load series and the exogenous series. This embedding is then broadcasted to $\mathbf{A}_{\text{emb}} \in \mathbb{R}^{(N+1)\times P\times d_a}$, allowing for a unified modeling of the complex inter-series dependencies without relying on any prior knowledge.

By concatenating the Patch embedding, time feature embedding, and inter-series relationship embedding, we obtain:

$$\mathbf{S} = \text{concat}(\mathbf{P}_{\text{emb}}, \mathbf{T}_{\text{emb}}, \mathbf{A}_{\text{emb}}) \tag{3}$$

where $\mathbf{S} \in \mathbb{R}^{(N+1)\times P\times d}$, and $d = d_p + 2d_t + d_a$.

Intra-Series Relationship Extraction. As shown in Fig. 1 (b), we apply the vanilla Transformer along the timeline to extract the temporal dependencies within the load series. Given the input $\mathbf{S}$, each token in $\mathbf{S}$ is linearly projected as follows:

$$\mathbf{Q}_{\text{intra}} = \mathbf{S}\mathbf{W}_q, \quad \mathbf{K}_{\text{intra}} = \mathbf{S}\mathbf{W}_k, \quad \mathbf{V}_{\text{intra}} = \mathbf{S}\mathbf{W}_v \tag{4}$$

where $\mathbf{W}_q, \mathbf{W}_k, \mathbf{W}_v \in \mathbb{R}^{d\times d}$ are learned parameters. Then, the attention scores are computed as:

$$\mathbf{A}_{\text{intra}} = \text{Softmax}\left(\frac{\mathbf{Q}_{\text{intra}}\mathbf{K}_{\text{intra}}^\top}{\sqrt{d}}\right) \tag{5}$$

where $\mathbf{A}_{\text{intra}} \in \mathbb{R}^{(N+1)\times P\times P}$ captures the correlations between different patches within the series. Subsequently, the output of the intra-series attention mechanism is expressed as $\mathbf{O}_{\text{intra}} = \mathbf{A}_{\text{intra}}\mathbf{V}_{\text{intra}}$, where $\mathbf{O}_{\text{intra}} \in \mathbb{R}^{(N+1)\times P\times d}$. After residual connections, Layer Normalization (LayerNorm), and Feed-Forward Network (FFN), we obtain $\mathbf{S}_{\text{intra}} \in \mathbb{R}^{(N+1)\times P\times d}$.

Inter-Series Relationship Extraction. As shown in Fig. 1 (c), similarly, we apply the vanilla Transformer along the series axis to extract the complex dependencies between the load series and the exogenous series. First, we obtain $\mathbf{Q}_{\text{inter}}, \mathbf{K}_{\text{inter}}, \mathbf{V}_{\text{inter}} \in \mathbb{R}^{(N+1)\times P\times d}$ using the same projection method. Then, the attention scores between the series are computed as:

$$\mathbf{A}_{\text{inter}} = \text{Softmax}\left(\frac{\mathbf{Q}_{\text{inter}}\mathbf{K}_{\text{inter}}^\top}{\sqrt{d}}\right) \tag{6}$$

where $\mathbf{A}_{\text{inter}} \in \mathbb{R}^{(N+1)\times(N+1)\times P}$ captures the correlations between the different series. The output is then computed as $\mathbf{O}_{\text{inter}} = \mathbf{A}_{\text{inter}}\mathbf{V}_{\text{inter}}$. After residual connections, LayerNorm, and FFN, we obtain $\mathbf{S}_{\text{inter}} \in \mathbb{R}^{(N+1)\times P\times d}$.

Projection. Finally, we use the inter-series dependencies from the last layer to output $\mathbf{S}_{\text{inter}} \in \mathbb{R}^{(N+1)\times P\times d}$ and obtain the predicted result $\mathbf{Y}$:

$$\mathbf{Y} = \text{Projection}(\mathbf{S}_{\text{inter}}) \tag{7}$$

where $\mathbf{Y} \in \mathbb{R}^{1\times T'}$, and Projection$(\cdot)$ consists of a fully connected layer.

4 Experiments

4.1 Experimental Settings

Datasets and Preprocessing. We use three publicly available electricity load datasets to validate the performance of PSEformer. These include the Tetouan dataset [19], with a sampling frequency of 10 min and no missing values; the target load series, PowerConsumption Zone2; the UTD dataset [20], with hourly resolution and no missing values, with the target load series being B4; and the WPuQ dataset [21], sampled at a frequency of 1 h. To ensure data quality, we filter households with a missing data rate lower than 10% and no photovoltaic generation. Missing values are imputed using forward filling, and the target load series is SFH32.

Baselines and Metrics. In this study, we compare the proposed method with widely used baselines in the field, including: Autoformer [13], FEDformer [15], Stationary [22], RLinear [23], PatchTST [14], and iTransformer [16]. All models are compared in a multivariate input - univariate prediction scenario. Additionally, all models follow the same look-back window ($T = 144$) and forecasting horizons ($T' \in \{144,288,432,720\}$). We select two commonly used metrics in load forecasting: Mean Absolute Error (MAE) and Mean Squared Error (MSE) for comparison.

Implementation Details. We implement all models using PyTorch on a GeForce RTX 2080Ti GPU. Datasets are split into training/validation/test sets at a 7:1:2 ratio. All models use the Adam optimizer (learning rate=0.001, batch size=32) and train for 20 epochs. For PSEformer:

Period-Series Embedding: Patch length=16, patch embedding dimension $d_p = 128$; periodic embeddings (daily/weekly periods) encode temporal information ($d_t = 64$); adaptive inter-series embedding ($d_a = 128$); total fused hidden dimension $d = 512$.

Transformer encoder: $L = 2$ layers (8 attention heads each). Each layer's FFN has intermediate dimension 2048 (ReLU activation). Layer normalization follows multi-head attention and FFN (dropout = 0.1).

Projection: A single fully connected layer mapping flattened hidden dimension to output (matching T').

4.2 Experimental Results and Analysis

Table 1. Experimental results of PSEformer and baseline models across various scenarios. The best results are highlighted in red, while the second-best results are indicated with underlines.

Models		Ours		iTransformer		RLinear		PatchTST		FEDformer		Stationary		Autoformer	
Metric		MSE	MAE	MSE	MAE	MSE	MAE	MSE	MAE	MSE	MAE	MSE	MAE	MSE	MAE
Tetouan	144	0.099	0.232	0.217	0.347	<u>0.11</u>	<u>0.233</u>	0.134	0.273	0.175	0.321	0.194	0.313	0.352	0.465
	288	0.154	0.292	0.262	0.386	<u>0.144</u>	<u>0.278</u>	0.135	0.266	0.167	0.315	0.222	0.337	0.254	0.392
	432	0.177	0.317	0.278	0.397	<u>0.178</u>	0.317	0.199	0.332	0.192	<u>0.329</u>	0.247	0.352	0.267	0.405
	720	0.178	0.315	0.288	0.403	<u>0.184</u>	<u>0.319</u>	0.197	0.329	0.261	0.407	0.34	0.441	0.328	0.449
UTD	144	0.137	0.253	0.161	0.272	0.167	0.281	<u>0.142</u>	<u>0.255</u>	0.155	0.293	0.148	0.265	0.196	0.325
	288	0.153	0.272	0.231	0.347	0.174	0.289	0.159	0.276	<u>0.154</u>	0.282	0.153	<u>0.275</u>	0.246	0.364
	432	<u>0.165</u>	<u>0.284</u>	0.237	0.354	0.176	0.293	0.16	0.278	0.176	0.314	0.173	0.299	0.282	0.405
	720	0.165	0.282	0.168	<u>0.284</u>	0.175	0.297	<u>0.166</u>	<u>0.284</u>	0.175	0.321	0.184	0.318	0.257	0.389
WPuQ	144	0.492	0.386	<u>0.495</u>	<u>0.388</u>	0.51	0.389	0.524	0.421	0.536	0.433	0.563	0.426	0.594	0.475
	288	0.502	0.391	0.504	0.396	0.52	<u>0.394</u>	0.524	0.422	0.542	0.444	0.556	0.403	0.586	0.454
	432	0.515	0.399	<u>0.516</u>	0.399	0.531	<u>0.4</u>	0.549	0.421	0.559	0.448	0.559	0.426	0.574	0.45
	720	<u>0.539</u>	<u>0.407</u>	0.529	0.393	0.545	<u>0.407</u>	0.546	0.41	0.545	0.418	0.59	0.453	0.557	0.442

We implement our model and baselines on three public electricity load datasets, evaluating MSE and MAE across prediction horizons (detailed in Table 1). Among 24 experimental cases, our model ranks first in 18 and second in 4, demonstrating that our period-series embedding outperforms the Patch-based PatchTST overall.

On the highly seasonal household-level WPuQ dataset, iTransformer and RLinear also show strong performance, as their linear layers effectively capture temporal dependencies and seasonal load patterns—further validating the effectiveness of our periodic embeddings.

As the prediction window lengthens, PSEformer maintains stable and robust performance, highlighting its advantages for load forecasting.

While not optimal in all cases, performance differences stem from variations in temporal and seasonal patterns across datasets. Overall, PSEformer remains stable and competitive, while global models like iTransformer, RLinear, and PatchTST excel at capturing periodicity in highly regular data.

4.3 Ablation Study

To validate the effectiveness of each component in our model, we conduct an ablation study on four variants of the model:

w/o $\mathbf{T}_{\text{emb}}$: This variant removes the periodic embedding for temporal features, ignoring the cyclical patterns in the time series.

w/o $\mathbf{A}_{\text{emb}}$: This variant removes the adaptive parameter embedding that captures inter-series dependencies, eliminating the modeling of relationships between different series.

w/o Intra: This variant removes the intra-series relation extraction block, disabling the capture of feature correlations within individual series.

w/o Inter: This variant removes the inter-series relation extraction block, neglecting global dependencies across multiple series.

The experimental results (Table 2) show that removing any part of the periodic-series embedding leads to a decline in prediction accuracy. This demonstrates that simply embedding the load series is insufficient, and the proposed periodic-series embedding plays a crucial role in load forecasting. Furthermore, removing either intra-series or inter-series dependency modeling also reduces prediction performance, highlighting the importance of capturing both temporal dependencies within load series and dependencies between load and exogenous series.

Table 2. Ablation study results. The best-performing results are highlighted in red. All values represent averages across all forecast horizons.

Dataset	Tetouan		UTD		WPuQ	
Metric	MSE	MAE	MSE	MAE	MSE	MAE
w/o $\mathbf{T}_{\text{emb}}$	0.181	0.38	0.168	0.304	0.516	0.400
w/o $\mathbf{A}_{\text{emb}}$	0.172	0.316	0.161	0.280	0.514	0.399
w/o Intra	0.185	0.325	0.156	0.291	0.522	0.401
w/o Inter	0.154	0.299	0.188	0.335	0.530	0.400
PSEformer	0.152	0.289	0.155	0.273	0.512	0.396

4.4 Visualization of Prediction Results

To provide an intuitive comparison of PSEformer's performance against other models in power load forecasting, we randomly selected samples from the dataset and visualized the prediction results, as shown in Fig. 2. The input sequence length was set to 144 time steps (corresponding to one day of power load data), and the prediction horizon was set to 144 (representing the next day's load forecast).

As visible in the figure, PSEformer outperforms other models in both prediction accuracy and load pattern capture: its prediction curve aligns more closely with the ground truth. In contrast, models such as iTransformer and PatchTST show limitations—iTransformer tends to produce overly smoothed curves (failing to capture fine-grained load fluctuations), whereas PatchTST overfits to local noise (leading to deviations from real load trends). This visual contrast confirms that PSEformer achieves superior stability and accuracy simultaneously.

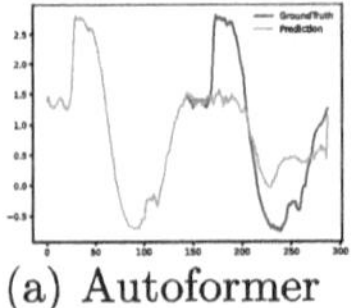
(a) Autoformer

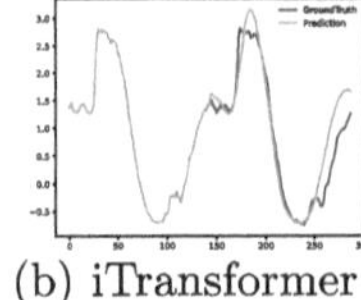
(b) iTransformer

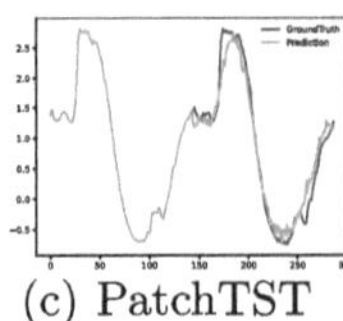
(c) PatchTST

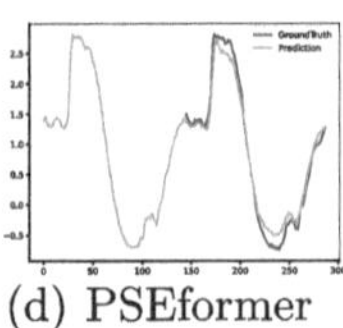
(d) PSEformer

Fig. 2. Comparative visualization of forecasting models: (a) Autoformer, (b) iTransformer, (c) PatchTST, and (d) PSEformer. Ground truth values are shown in blue, while model predictions are displayed in orange. (Color figure online)

5 Conclusions

We propose the Period-Series Embedding Transformer (PSEformer) for power load forecasting. PSEformer efficiently captures both intra-series and inter-series dependencies in load forecasting through embedding techniques. Our model combines seasonal embeddings, such as daily and weekly periodic features, with inter-series dependency embeddings, providing a more effective representation of the load series. Experimental results on three publicly available electricity load datasets demonstrate that PSEformer significantly improves performance by leveraging the proposed embeddings. The findings show that our model outperforms existing baseline models in capturing complex dependencies within load data, achieving higher accuracy in load forecasting tasks. Future work will focus on decomposing load series and applying embeddings for more precise representations, as well as exploring additional techniques to improve efficiency and enhance scalability and adaptability in large-scale real-world applications.

Acknowledgments. This work was supported by the National Natural Science Foundation of China under Grant 62176193

References

1. Tziolis, G., et al.: Short-term electric net load forecasting for solar-integrated distribution systems based on bayesian neural networks and statistical post-processing. Energy **271**, 127018 (2023)
2. Liu, K., et al.: Comparison of very short-term load forecasting techniques. IEEE Trans. Power Syst. **11**(2), 877–882 (1996)
3. Massaoudi, M., Refaat, S.S., Chihi, I., Trabelsi, M., Oueslati, F.S., Abu-Rub, H.: A novel stacked generalization ensemble-based hybrid lgbm-xgb-mlp model for short-term load forecasting. Energy **214**, 118874 (2021)
4. Juberias, G., Yunta, R., Moreno, J.G., Mendivil, C.: A new arima model for hourly load forecasting. In: 1999 IEEE Transmission and Distribution Conference (cat. no. 99CH36333), vol. 1, pp. 314–319. IEEE (1999)
5. Xu, X., Zhang, L., Kong, Q., Gui, C., Zhang, X.: Enhanced-historical average for long-term prediction. In: 2022 2nd International Conference on Computer, Control and Robotics (ICCCR), pp. 115–119. IEEE (2022)

6. Chen, Y., et al.: Short-term electrical load forecasting using the support vector regression (svr) model to calculate the demand response baseline for office buildings. Appl. Energy **195**, 659–670 (2017)
7. Bessani, M., Massignan, J.A., Santos, T.M., London, J.B., Jr., Maciel, C.D.: Multiple households very short-term load forecasting using bayesian networks. Electric Power Syst. Res. **189**, 106733 (2020)
8. Fan, G.-F., Zhang, L.-Z., Yu, M., Hong, W.-C., Dong, S.-Q.: Applications of random forest in multivariable response surface for short-term load forecasting. Int. J. Electr. Power Energy Syst. **139**, 108073 (2022)
9. Shi, H., Xu, M., Li, R.: Deep learning for household load forecasting–a novel pooling deep rnn. IEEE Trans. Smart grid **9**(5), 5271–5280 (2017)
10. Kong, W., Dong, Z.Y., Jia, Y., Hill, D.J., Xu, Y., Zhang, Y.: Short-term residential load forecasting based on lstm recurrent neural network. IEEE Trans. Smart grid **10**(1), 841–851 (2017)
11. Imani, M.: Electrical load-temperature cnn for residential load forecasting. Energy **227**, 120480 (2021)
12. Zhou, H., et al.: Informer: beyond efficient transformer for long sequence time-series forecasting. In: Proceedings of the AAAI Conference on Artificial Intelligence, vol. 35, no. 12, pp. 11 106–11 115 (2021)
13. Wu, H., Xu, J., Wang, J., Long, M.: Autoformer: decomposition transformers with auto-correlation for long-term series forecasting. Adv. Neural Inf. Process. Syst. **34**, 22 419–22 430 (2021)
14. Nie, Y., Nguyen, N.H., Sinthong, P., Kalagnanam, J.: A time series is worth 64 words: long-term forecasting with transformers. arXiv preprint arXiv:2211.14730 (2022)
15. Zhou, T., Ma, Z., Wen, Q., Wang, X., Sun, L., Jin, R.: Fedformer: frequency enhanced decomposed transformer for long-term series forecasting. In: International Conference on Machine Learning, PMLR, pp. 27 268–27 286 (2022)
16. Liu, Y., et al.: Itransformer: inverted transformers are effective for time series forecasting. arXiv preprint arXiv:2310.06625 (2023)
17. Moudgil, V., Sadiq, R., Brar, J., Hewage, K.: Dual-channel encoded bidirectional lstm for multi-building short-term load forecasting. J. Clean. Prod. **486**, 144555 (2025)
18. Wang, C., Li, X., Shi, Y., Jiang, W., Song, Q., Li, X.: Load forecasting method based on cnn and extended lstm. Energy Rep. **12**, 2452–2461 (2024)
19. Salam, A., El Hibaoui, A.: Comparison of machine learning algorithms for the power consumption prediction:-case study of tetouan city. In: 2018 6th International Renewable and Sustainable Energy Conference (IRSEC), pp. 1–5. IEEE (2018)
20. Zhang, J., Feng, C.: Short-term load forecasting data with hierarchical advanced metering infrastructure and weather features. IEEE Dataport (2019)
21. Schlemminger, M., Ohrdes, T., Schneider, E., Knoop, M.: Dataset on electrical single-family house and heat pump load profiles in germany. Sci. Data **9**(1), 56 (2022)
22. Liu, Y., Wu, H., Wang, J., Long, M.: Non-stationary transformers: exploring the stationarity in time series forecasting. Adv. Neural. Inf. Process. Syst. **35**, 9881–9893 (2022)
23. Li, Z., Qi, S., Li, Y., Xu, Z.: Revisiting long-term time series forecasting: an investigation on linear mapping. arXiv preprint arXiv:2305.10721 (2023)

Resilience Evaluation and Optimization of UAV Swarm Network under Low Probability Risk

Chuanlong Yang(✉), Lianqian Cao, and Chuhang Jiang

Dalian Neusoft University of Information School of Intelligence & Electronic Engineering Dalian, Dalian, China
yangchuanlong@neusoft.edu.cn

Abstract. This paper addresses the process description issue of topology changes in Unmanned Aerial Vehicle (UAV) Mobile Ad Hoc Networks (UMANETs) caused by low probability risk and proposes an approach to evaluating network resilience. Assessing network resilience differs from evaluating network robustness and reliability against common high-probability risks. Instead, network resilience refers to the network's capacity to withstand damage and recover from low-probability and severe risks. The concept of network resilience using process modeling, requirement analysis, and structural optimization is presented. The paper provides criteria for evaluating network resilience in four aspects, including network reconstruction capability, communication efficiency, chain break rate, and the degree of network loading. Dynamic Stackelberg game theory is utilized to simulate the conflict between network disruptions and a real-time planning system, analyzing the dynamic evolution of topology. To dynamically plan a more resilient network topology in real-time, the proposed approach introduces an enhanced Ant Colony Algorithm that leverages a Bio-inspired Neural Network (BNNAC). Finally, simulation results demonstrate the effectiveness and superiority of the proposed scheme.

Keywords: Network resilience · network topology planning · mobile ad hoc network · bio-inspired neural network colony algorithm

1 Introduction

In recent years, UMANET research has primarily focused on routing design [1], scheduling planning [2], network stability, and robustness [3, 4]. These studies assume normal network operation or address known failure risks. However, sophisticated attack methods can now stealthily manipulate UAVs through wireless networks, deceiving multiple sensors and compromising overall network reliability [5]. For instance, GPS spoofing attacks can manipulate UAV positioning by synchronizing false signals with legitimate GPS signals [6, 7]. Therefore, studying UAV swarm network resilience against such attacks is crucial for maintaining stable topology and ensuring mission success.

Network resilience, defined as a system's ability to maintain essential functionalities despite errors or failures [8], has become fundamental for UAV network security. While

C. Li et al. (Eds.): ICNC 2025, CCIS 2946, pp. 100–110, 2026.
https://doi.org/10.1007/978-981-92-1599-7_9

previous research has emphasized internal system stability [3, 4, 9, 10], these approaches fail to consider low-probability risks that significantly impact overall network structure. Therefore, this paper primarily examines the alteration in UAV network elasticity when subjected to low probability risks, and simulates such risks through covert attacks.

Current wireless network attacks predominantly use GPS-based methods [6, 7, 11, 12]. While conventional attacks can be detected through sensor cross-verification [13, 14], sophisticated covert attacks can deceive multiple sensor types simultaneously [15, 16]. Existing solutions include fast reconfiguration routing [17], Byzantine-resistant protocols [18], energy-efficient clustering [19], and backbone structures [20]. However, these methods may not suit dynamic attack scenarios due to redundancy or structural limitations. However, these methods may not be suitable for dynamic attack scenarios due to redundancy generation or structural limitations. Therefore, this paper primarily examines the alteration in UAV network resilience when subjected to low-probability risks and proposes adaptive topology planning solutions for dynamic attack scenarios.

This paper focuses on UAV network resilience planning under GPS covert attack strategies, addressing two key aspects: network resilience process description and adaptive network planning. We propose evaluation indices based on the network's destruction resistance and recovery capability, illustrated through dynamic resilience changes during attacks. Additionally, we present a real-time route planning solution for dynamic networks, generating topology structures with better resilience under covert attack scenarios. The main contributions are:

1. Based on existing covert attack strategies, this paper proposes a network resilience change model under low-probability risk scenarios. Then, quantitative indexes for UMANET network resilience under the scenario of covert attacks are constructed by considering various network indicators, which can achieve a comprehensive evaluation of the withstand damage and recovery capabilities of UMANET.
2. A combined algorithm utilizing ant colony optimization and bionic neural networks is proposed to develop resilient UMANET topology in real time, with network resilience assessment results as the objective function. Under existing covert attack frameworks, the algorithm's real-time planning capability is demonstrated through dynamic Stackelberg game, reflecting the network's reconfiguration potential when persistently attacked.

The rest of this article is described below. Section 2 states the problem description. Section 3 describes the evaluation method of network resilience. Section 4 describes the network topology planning strategy based on BNNAC. Section 5 exhibits the simulation results. Finally, Sect. 6 gives the conclusion.

2 Problem Description

2.1 UAV Motion Modelling

A group of homogeneous UAVs was utilized to track a moving target in the mission area, and the following dual-integral dynamic model can be established for the ith UAV:

$$\begin{bmatrix} \dot{p}_i \\ \dot{v}_i \end{bmatrix} = \begin{bmatrix} 0_{3\times3} & 1_{3\times3} \\ 0_{3\times3} & 0_{3\times3} \end{bmatrix} \begin{bmatrix} p_i \\ v_i \end{bmatrix} + \begin{bmatrix} 0_{3\times3} \\ 1_{3\times3} \end{bmatrix} u_i \tag{1}$$

where and I denotes the total number of the UAVs; and is the position and velocity component of the ith UAV on the coordinate axis, satisfying and the is the maximum velocity of the UAV; is the control input, satisfying and the is the maximum control input. is set as the motion state of the ith UAV. Each UAV is equipped with GPS and IMU sensors providing position and motion information.

$$y_i^{GPS} = p_i, y_i^{IMU} = \langle v_i, u_i, \varphi_i, \phi_i \rangle \tag{2}$$

where and is the heading angle and flight path angle of the ith UAV.

In this context, we define as the GPS spoofing signal and as the GPS reading. In addition, the attacker must accurately calculate the position for the attack to avoid detection by other sensors. This paper studies covert attacks that disrupt UMANET structure by causing UAVs to break away from formations or collide with other UAVs. Based on existing research [15, 16], the conceptual definition of covert attacks is as follows:

Definition 1 (Covert attack): If covert attack $\left(u_i^{aGPS}, u_i^a\right)$ cannot be detected by the GPS and IMU, it must be ensured that the measurements at any given time satisfy the following conditions:

$$\begin{gathered} y_i^{GPS} = y_i^{aGPS}, y_i^{IMU} = y_i^{aIMU} \\ s.t. u_i^{aGPS}(\tau) = p_i^a(\tau) - p_i(\tau) \\ \left\| p_i^a(\tau) - p_i(\tau) \right\| \le R_{\max} \end{gathered} \tag{3}$$

where $p_i^a(\tau)$ represents the position where the UAV is being attacked, which refers to the covert attack launched by an attacker into the UAV at time τ; $R_{\max}$ represents maximum transverse drifted distance [21] that restricts the distance between the current position of the ith UAV and the imposed position by the attacker.

2.2 Quantitative Assessment Model of Network Resilience

Drawing the network's process description curve based on the performance variations of UAVs serves as a fundamental tool for resilience analysis. The depiction of the process description curve under covert attacks is illustrated in Fig. 1, which encompassing two primary stages: resistance and recovery. The five crucial temporal points depicted in the diagram are defined as follows:

τ: time when the covert attack starts;

t^{weak}: time when the chain break starts;

t^{discon}: time when the performance is its lowest.

t^{restru}: time when the performance recovery starts.

t^{reco}: time when the performance is stable again.

According to the elastic curve parameters and task requirements of UAV swarm, this paper comprehensively considers the design of resilience quantitative assessment objective from multiple perspectives and incorporates three factors at each stage. Each

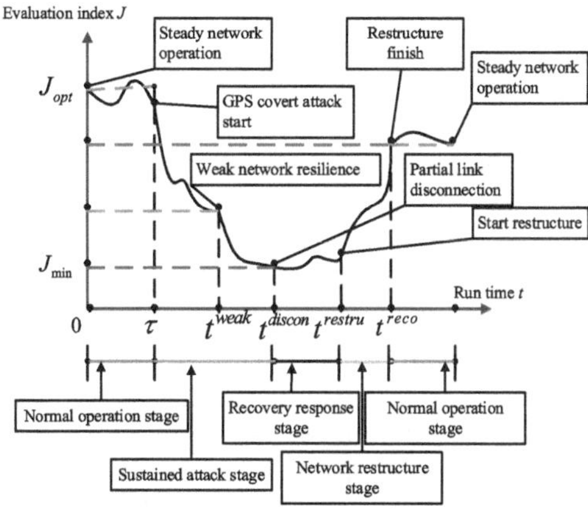

Fig. 1. Process description curve under covert attack.

factor is calculated as follows:

$$\begin{aligned} J_{1d} &= \sum_{i=1}^{I} J_i^{rel}(k) + \sum_{i=1}^{I} J_i^{load}(k), k \in \left[\tau, t^{discon}\right], i \in \{1, \cdots I\} \\ J_{2d} &= t^{discon} - \tau \\ J_{3d} &= \sum_{i=1}^{I} J_i^{rest}(k), k \in \left[\tau, t^{discon}\right] \\ J_{1r} &= \sum_{i=1}^{I} J_i^{rel}(k) + \sum_{i=1}^{I} J_i^{load}(k), k \in \left[t^{discon}, t^{reco}\right] \\ J_{2r} &= \left(t^{reco} - t^{discon}\right)/t^{reco} \\ J_{3r} &= \sum_{i=1}^{I} J_i^{rest}(k), k \in \left[t^{discon}, t^{reco}\right] \end{aligned} \tag{4}$$

where k is sampling time.

(i). Residual performance J_{1d} reflects the network's ability to maintain communication following covert attacks, with key indexes including communication capacity J_i^{rel} and load capacity J_i^{loa}; (ii). Chain break rate J_{2d} is determined by the duration until network chain break occurs. A reduced J_{2d} value indicates decreased resilience and faster transition from normal to fault state; (iii). Defence performance J_{3d} indicates the gradually decreasing communication rate J_i^{rest} during attacks; (iv). Recovery performance J_{1r} reflects communication capability after link reconfiguration, with recovery rate J_{2r} representing the recovery speed. Reconstruction capability J_{3r} denotes the communication rate during network reconstruction.

Finally, the resilience change process is analyzed by comparing various stages, and six resilience indexes are aggregated through weighted combination to form a comprehensive resilience evaluation function.

$$J = \omega_1(J_{1d} + J_{1r}) + \omega_2(J_{2d} + J_{2r}) + \omega_3(J_{3d} + J_{3r}) \tag{5}$$

where $\{\omega_1, \omega_2, \omega_3\}$ are the weight coefficients. The form of J_i^{rel}, J_i^{loa}, and J_i^{rest} are introduce in detail in the Sect. 4.

2.3 The Objectives of the Attack and Multi-UAV Network

This section analyzes the adversarial interaction between attackers and the multi-UAV network. The attacker can select any UAV as attack targets and customize spoofing signals by monitoring UAVs, enabling covert attacks at any time to split UAVs from the fleet or cause collisions. Meanwhile, defenders must strategically design highly resilient network topology within each cycle to prolong survival time. Since the network recovery process occurs while attacks persist, this adversarial interaction can be modeled as a dynamic Stackelberg game.

The objective of covert attack: Based on existing research [16], the covert attack aims to maximize trajectory errors while remaining undetected, thereby reducing network resilience by either isolating UAVs from the fleet or causing collisions.

The objective of defender: In the network recovery stage, the UAV swarm gradually recovers network resilience by executing recovery strategies. The network topology optimization problem for optimal resilience is formulated as:

$$\begin{aligned} &\max J\left(p_i, p_j, k\right)\Big|_T^{T+\Delta T}, \forall i \in \{1, \cdots I\}, j \in M_i \\ &s.t. d_c^{\max} > \left\|p_j(k) - p_i(k)\right\| > d_c^{\min}, k \in [T, T+\Delta T], N_{com} \geq \tfrac{I}{2} \end{aligned} \tag{6}$$

where $J(p_i, p_j, k)$ is comprehensive evaluation index of resilience in (6); N_{com} represents the number of nodes that can communicate on the network. If $N_{com} < I/2$, UAV swarm will lose basic coordination ability.

3 The Evaluation Method of Network Resilience

UAV network resilience requires comprehensive evaluation based on clear task requirements. For the collaborative target tracking scenario using distributed network structure, the objective is to ensure continuous network resilience against cyber-attacks while maintaining efficient information exchange. Based on the above analysis and Eq. (4), the network resilience evaluation index should encompass communication efficiency J_i^{rel}, load capacity J_i^{loa}, chain break rate J_{2d}, and communication rate J_i^{rest} of the network. Furthermore, in the analysis of the network structure, UAVs are renamed as network nodes.

(i). Communication capability J_i^{rel}: The network's communication capability is quantified by the average transmission route from the ith node to the jth node (the farthest node), expressed as:

$$J_i^{rel}(k) = \min_{\forall i,j \in N} \frac{d_c^{\max}}{l_{ij}(k) \cdot n_{ij}(k)}, i \neq j \tag{7}$$

(ii). Load capacity J_i^{loa}: In practical operations, nodes should not only be prepared for instantaneous failure but also possess fault tolerance. When node load exceeds the preset threshold, nodes exhibit reduced operational efficiency and delayed information processing instead of complete failure. To monitor the load states of ith node, its performance at time k can be described as follows:

$$J_i^{loa}(k) = 1 - \frac{2\sum_{j=1}^{|M_i|} e_{ij}(k)}{C_i(0)}, i \in \{1, \cdots I\}, j \in M_i \tag{8}$$

where $e_{ij}(k)$ represents the connectivity degree of ith node and jth node, $e_{ij}(k) = 1$ means two nodes are connected; $C_i(0)$ denotes the initial capacity of the ith node, which remains finite and constant over time.

(iii). Chain break rate J_{2d}: This represents the attack duration the network structure can withstand, expressed by the persistence time of a node under covert attack until link performance degrades to the lowest level, corresponding to the $\left[\tau, t_i^{discon}\right]$ stage in Fig. 1. Considering the attacker's objectives, $J_{2d}(k)$ can be composed of varying durations of both components.

Attack intent 1: Target UAV disengagement from the group. The attack duration t_i^1 until intent achievement is:

$$\left\| p_j\left(\tau + t_i^1\right) - p_i^a\left(\tau + t_i^1\right) \right\| > d_c^{\max}, j \in M_i \tag{9}$$

If the target sent an attack at time τ, the $\tau + t_i^1$ indicates the point in time when network experiences poorest performance.

Attack intent 2: Target UAV collision with others. The attack duration t_i^2 that minimizes distance between nodes is:

$$\left\| p_j\left(\tau + t_i^2\right) - p_i^a\left(\tau + t_i^2\right) \right\| < d_c^{\min}, j \in M_i \tag{10}$$

Combine intent (i) and intent (ii), $J_{2d}(k) = \eta_1 \cdot t_i^1 + \eta_2 \cdot t_i^2$, where η_1 and η_2 are the weight. In addition, the $J_{2r}(k)$ similarities to the above process.

(iv). Reconstruction efficiency J_i^{rest}: The reconstruction capability represents a comprehensive reflection of node parameters, channel distance, and communication conditions. Therefore, the J_i^{rest} can be simplified as the maximum communication rate of the entire UAV network.

$$\begin{aligned} J_{ij}^{rou}(k) &= B_w \log\left(1 + \frac{W_i \cdot (l_{ij}(k))^2 \cdot \varsigma(k)}{Z_0 \cdot B_w}\right) \\ J_i^{rest}(k) &= \sum_{j=1}^{N} J_{ij}^{rou}(k), j \in \{1, \cdots I\} \end{aligned} \tag{11}$$

where B_w is the channel bandwidth; W_i is the UAV transmitted power; $\varsigma(k)$ changes over time; Z_0 is the noise power per unit bandwidth, and $Z_0 \cdot B_w$ is the Gaussian noise power in the channel. According to Eq. (11), the maximum communication rate of the UAV network is determined by the current broadcast ith node within the UAV group.

4 Network Topology Planning Strategy Based on BNNAC

Traditional network topology planning strategies, such as flooding algorithm [22], EESOA-MAC [19], and self-pruning algorithm (SPA) [20], plan all routes based on current the state of nodes while ignoring the influence of historical information on future development. In addition, the above method may lead to information redundancy and forwarding delay when dealing with large-scale networks. In this section, we establish a topology organization based on BNNAC to plan the network structure more resilience.

4.1 Bio-inspired Neural Network (BNN) Model

BNN is a discrete form of neural network. In this model, each neuron (network node) establishes local connections with other neurons within the communication range. The weight between ith neuron and jth neuron at time k is denoted as $w_{ij}(k)$:

$$w_{ij}(k) = \begin{cases} e^{-\|p_i(k)-p_j(k)\|}, d_c^{\min} \le \|p_i(k) - p_j(k)\| \le d_c^{\max} \\ 0, \qquad \|p_i(k) - p_j(k)\| > d_c^{\max} \end{cases} \tag{12}$$

By introducing historical information, the dynamic activity between ith neuron and jth neuron at the $k+1$ moment can be represented as follows:

$$o_{ij}(k+1) = \begin{cases} \dfrac{w_{ij}(k)o_{ij}(k)}{\sum\limits_{j=1}^{M_i} w_{ij}(k)o_{ij}(k)}, & \Delta o_{ij}(k) \le 0 \\ \dfrac{\xi\Delta o_{ij}(k) + \sum_{j\in M_i} w_{ij}(k)o_{ij}(k)}{\sum_{j=1,\Delta o_{ij}(k)>0}^{M_i}\left(\xi\Delta o_{ij}(k) + \sum_{j\in M_i} w_{ij}(k)o_{ij}(k)\right)}, & other \end{cases} \tag{13}$$

where $\Delta o_{ij}(k) = o_{ij}(k) - o_{ij}(k-1)$ represents the discrepancy in dynamic activity between two neurons at k and k-1 time, a positive correlation exists between the dynamic activity $o_{ij}(k+1)$ and the maximum transmission rate $J_{ij}^{rou}(k)$.

4.2 The Integration of Ant Colony and BNN

Ant colony algorithm is a heuristic intelligence algorithm that leverages the collective intelligence of multiple non-intelligent individuals within a population to efficiently solve complex optimization problems. The algorithm includes three parts: the construction of an evaluation function, the updating of pheromones, and the calculation of transition probability.

(i). Evaluation function: The algorithm aims to find the network topology with high resilience, thus the evaluation function can use the resilience evaluation index in Eq. (4).

(ii). Updating of pheromone: To address the issue of slow convergence, $o_{ij}(k)$ is employed to replace pheromone concentration in traditional ant colony algorithm, thereby representing a priori knowledge in the route search process and guiding ants towards finding the optimal solution quickly.

(iii). Transition probability: We set as the heuristic information between the ith and jth nodes, with its value inversely proportional to the distance between them. The transition probability formula of an ant can be represented as:

$$P_{ij}(k) = \begin{cases} \dfrac{|\lambda_{ij}(k)|^{\theta}\,|o_{ij}(k)|^{\beta}}{\sum\limits_{h\in M_i^{all}(k)} |\lambda_{ih}(k)|^{\theta}\,|o_{ih}(k)|^{\beta}}, & j, h \in M_i^{all}(k) \\ 0, & j \notin M_i^{all}(k) \end{cases} \tag{14}$$

where θ and β are expected heuristic factors and pheromone heuristic factors, respectively; the larger the β, the higher the likelihood of ant colony failure into local optimal; and the greater the θ, the diminished randomness in search; $M_i^{all}(k) = M_i(k) - tabu_i(k)$ represents the set of candidate next hop nodes available to the ant at the ith neuron, while $tabu_i(k)$ denotes the tabu list that keeps track of neurons with established links in neighbouring regions.

To facilitate the planning of the route passes all nodes, a multi-ant cooperative strategy is employed where multiple ants simultaneously explore paths starting from the information source node. Further, we assume the ath node is an information source node and set the following constraints:

$$M_{\max} > |M_a|,\ V_a^{all} = I - tabu_a^{all},\ tabu_a^{all} = \sum_i^N tabu_i,\ i \neq a \tag{15}$$

where M_a represents the set of the neighbour of the ath node; $M_{\max}$ is the number of ants, V_a^{all} ensures that the ant colony can plan a maximum of $|M_a|$ routes; V_a^{all} is a set of transferable nodes, $tabu_a^{all}$ ensures that multiple ants do not approach the same node.

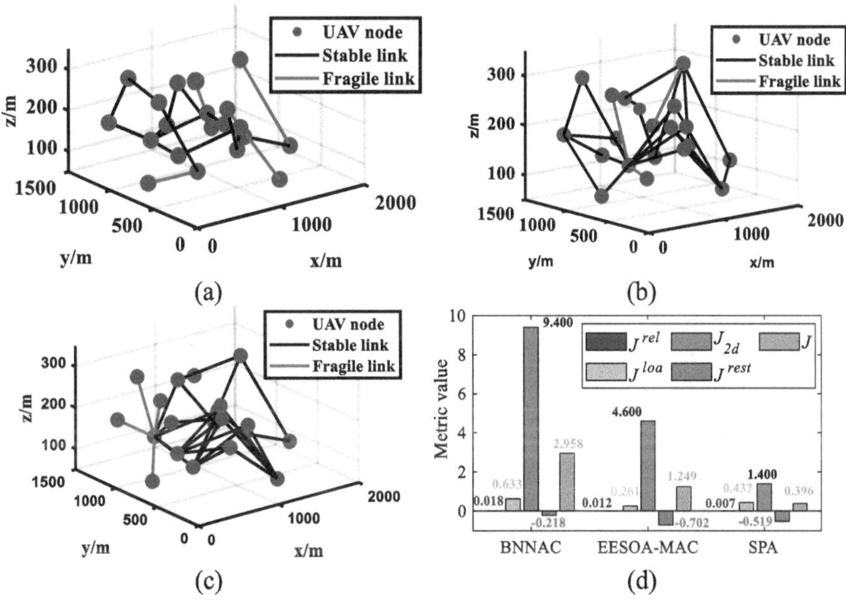

Fig. 2. Network structure and performance indexes under different algorithms

5 Simulation

A MATLAB-based simulation platform is established for comparative testing of UAV network topology planning under GPS covert attacks.

The simulation environment is a three-dimensional space E spanning 1.5 km by 2 km by 0.3 km, in which twenty UAVs are randomly distributed with different positions. One simulation set is designed i.e., BNNAC's real-time network topology planning compared with SPA and EESOA-MAC to demonstrate effectiveness.

This section compares the three algorithms based on network structure and resilience analysis. Fig. 2 illustrates the network structure planned by node No. 5 as reconstruction initiator, where black lines represent stable links and red lines denote fragile links built for nodes with few communication neighbors, which are main attack targets.

From Fig. 2, the SPA exhibits the highest number of fragile links (5 links) with higher network complexity than the other algorithms. EESOA-MAC demonstrates fewer fragile links (3 links) than BNNAC but has more intricate network structure and susceptibility to information redundancy compared with BNNAC. Fig. 2(d) presents the resilience index values for all three algorithms at the initial moment with node No. 5 as example. According to Eq. (4), higher resilience index values indicate better performance. BNNAC outperforms the other algorithms in all indices, particularly in network resistance to destruction and load capacity. This is because BNNAC ensures each node receives information only once during planning, reducing information redundancy. The value J_{2d} is 9.4 indicates the network structure stability.

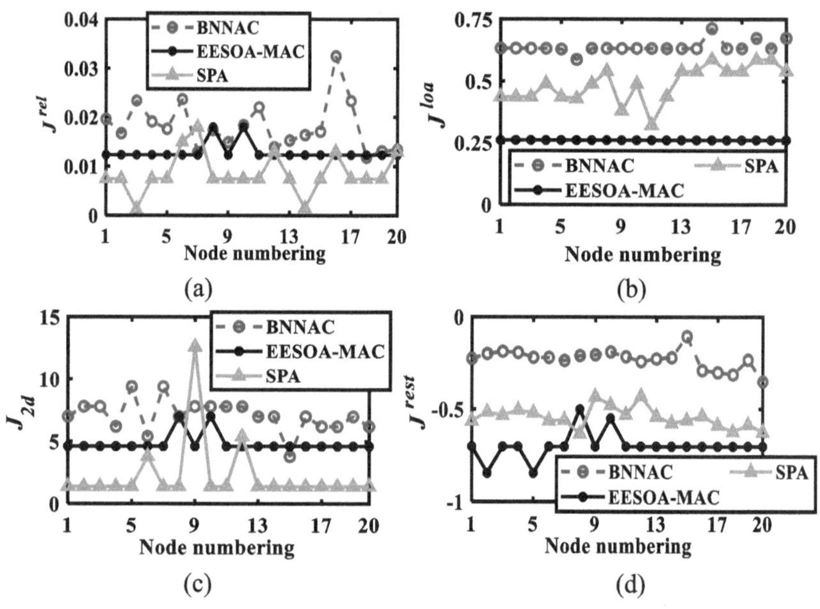

Fig. 3. Resilience evaluation indexes under different algorithms

To demonstrate BNNAC's universal applicability, we compare network structures with 20 UAVs as reconstruction initiators to evaluate their resilience. Fig. 3 shows the change curves of four resilience evaluation indexes under three algorithms, while Fig. 4 displays the corresponding network resilience values. All horizontal coordinates represent node numbering. SPA exhibits the highest resilience value (J = 3.77) and highest chain break rate (J_{2d} =12.6) when node No. 9 serves as reconstruction initiator, due to reduced fragile links from tighter network parts. However, BNNAC outperforms other algorithms across various metrics, particularly in load capacity and reconfiguration capability, demonstrating superior overall performance despite not achieving the peak values in individual cases.

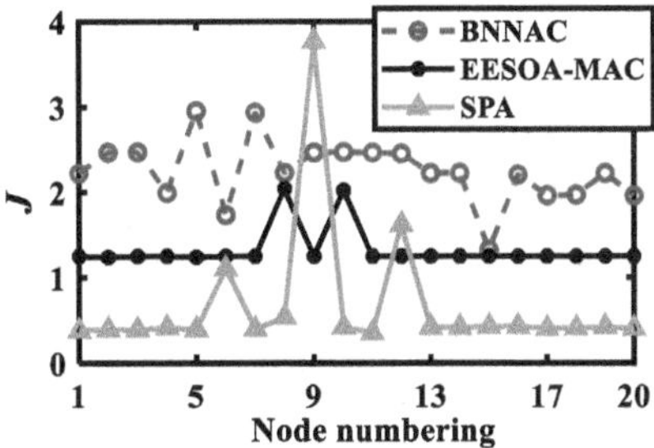

Fig. 4. Resilience values under different algorithms

6 Conclusion

This paper introduces network resilience to describe UAV network structure changes under low probability risks, constructing a process model and evaluation framework from task requirement perspectives. The BNNAC algorithm dynamically plans optimal resilient network structures based on UAV dynamic characteristics, showing superiority over SPA and EESOA-MAC in comparative experiments. Results validate the effectiveness of both the resilience evaluation framework and the proposed algorithm in handling complex network attacks. Future work will explore more sophisticated cyber-attacks and larger UAV swarm scenarios.

Disclosure of Interests. We declare that we have no financial and personal relationships with other people or organizations that can inappropriately influence our work, there is no professional or other personal interest of any nature or kind in any product, service and or company that could be construed as influencing the position presented in or the review of the manuscript entitled.

References

1. Mohsin, A.H.: Optimize routing protocol overheads in MANETs: challenges and solutions: a review paper. Wirel. Pers. Commun. **124**(4), 2871–2910 (2022)
2. Zhao, C.X., Zhu, P.F., Zhang, N., Chen, S.G.: Directional charging-based scheduling strategy for multiple mobile chargers in wireless rechargeable sensor networks. Ad Hoc Netw. **149**, 1–14 (2023)
3. Patki, V., Mehbodniya, A., Webber, J.L., Kuppusamy, A.: Improving the geo-drone-based route for effective communication and connection stability improvement in the emergency area ad-hoc network. Sustain. Energy Techn. and Assess. **53**, 1–7 (2022)

4. Sharma, V., Kumar, R.: G-FANET: an ambient network formation between ground and flying ad hoc networks. Telecommun. Syst. **65**(1), 31–54 (2017)
5. Mohsen, R.M., Naima, K.: Cyber-attacks on unmanned aerial system networks: detection, countermeasure, and future research directions. Comput. Secur. **85**, 386–401 (2019)
6. Ren, S.Y., Zhao, W.J., Zhang, A.T., Zhang, B., Han, B.: Monocular-GPS fusion 3D object detection for UAVs. Knowl.-Based Syst. **97**, 112–134 (2024)
7. Nayfeh, M., Li, Y.C., Shamaileh, K.A., Devabhaktuni, V.: Machine learning modeling of GPS features with applications to UAV location spoofing detection and classification. Comput. Secur. **126**, 1–9 (2023)
8. Liu, X.M., Li, D.Q., Ma, M.Q., Szymanski, B.W.: Network resilience. Phys. Rep.-Rev. Sec. Phys. Lett. **971**, 1–108 (2022)
9. Zou, B., Choobchian, P., Julie, R.: Cyber resilience of autonomous mobility systems: cyber-attacks and resilience-enhancing strategies. J. Transp. Secur. **14**, 137–155 (2021)
10. Tezuka, N., Ochiai, H., Wei, S.Y., Esaki, H.: Resilience of wireless ad hoc federated learning against model poisoning attacks. In: Proceedings TPS-ISA, Virtual, United States, pp. 168–177 (2022)
11. Noh, J., Kwon, Y.J., Son, Y., Shin, H.: Tractor beam: safe-hijacking of consumer drones with adaptive GPS spoofing. ACM Trans. Priv. Secur. **22**(2), 1–26 (2019)
12. Gong, X., Gui, J., Chen, Y., Yang, X.F., Yu, W.W., Huang, T.W.: Resilient human-in-the-loop formation-tracking of multi-UAV systems against byzantine attacks. IEEE Trans. Autom. Sci. **37**(4), 3797–3809 (2025)
13. Milaat, F.A., Liu, H.: Decentralized detection of GPS spoofing in vehicular ad hoc networks. IEEE Commun. Lett. **22**(6), 1256–1259 (2018)
14. Gu, Y.P., Yu, X., Guo, K.X., Qiao, J.Z.: Detection, estimation, and compensation of false data injection attack for UAVs. Inf. Sci. **546**, 723–741 (2021)
15. He, J.Y., Gong, X.: Resilient path planning of unmanned aerial vehicles against covert attacks on ultra-wideband sensors. IEEE Trans. Ind. Inform. **19**, 1–9 (2023)
16. Eldosouky, A., Ferdowsi, A., Saad, W.: Drones in distress: a game-theoretic countermeasure for protecting UAVs against GPS spoofing. IEEE Internet Things J. **7**(4), 2840–2854 (2020)
17. Bisti, L., Lenzini, L., Mingozzi, E., Vallati, C.: Improved network resilience of wireless mesh networks using MPLS and fast re-routing techniques. Ad Hoc Netw. **9**(8), 1448–1460 (2011)
18. Awerbuch, B., Curtmola, R., Holmer, D., Nita-Rotaru, C.: ODSBR: an on-demand secure byzantine resilient routing protocol for wireless ad hoc networks. ACM Trans. Inf. Syst. Secur. **10**(4), 1–35 (2008)
19. Balart-Sanchez, F.E., Gutierrez-Preciado, L.F., Olascuaga-Cabrera, J.G.: Minimizing routing broadcast and packet loss in wireless ad-hoc networks with a cluster-based self-organized algorithm as MAC protocol. In: Proceedings CCWC., Las Vegas, NV, United States, pp. 499–505 (2019)
20. Rab, R., Rahman, A., Zohra, F.T.: Analytical modeling of self-pruning and an improved probabilistic broadcast for wireless multihop networks. Ad Hoc Netw. **52**, 106–116 (2016)
21. Zeng, K.X., Shu, Y.C., Liu, S.N., Dou, Y.Z.: A practical GPS location spoofing attack in road navigation scenario. In: Proceedings HotMobile, pp. 85–90, Sonoma, CA, United States (2017)
22. Tai, C.F., Chiang, T.C., Hou, T.W.: A virtual subnet scheme on clustering algorithms for mobile ad hoc networks. Expert Syst. Appl. **38**(3), 2099–2109 (2011)

The Research on Mobile Edge Computing in Content Delivery Network

Ping Jiang, Haixia Li(✉), Jun Cheng, Jiarui Wu, and Yuanying Yu

School of Computer Science and Technology, Hubei Polytechnic University, Huangshi, Hubei, China
44800322@qq.com

Abstract. The network has entered the caching and Content Delivery Network (CDN), which is a distributed network architecture for providing efficient content transmission and distribution services. Edge computing is commonly used in CDN networks. Mobile Edge Computing (MEC) is an emerging computing model aimed at bringing computing, storage, and network services closer to end devices. Traditionally, cloud computing relies on remote data centers to process and store data. However, for applications requiring low latency and high bandwidth such as the Internet of Things, virtual reality, and augmented reality, the cloud computing model may not meet the requirements. Edge computing is more suitable for real-time data analysis and intelligent processing, and is more efficient and secure than pure cloud computing. Edge computing is more accurate to say that it should be a supplement and optimization of cloud computing, there 's no need sending data to the distant cloud, it can be solved at the edge. Mobile Edge Computing (MEC) provides more efficient, reliable, and secure computing resources and services for these applications, driving the development of technologies such as the Internet of Things and 5G. Mobile Edge Computing is considered fundamental in building a future intelligent and connected world.

Keywords: Mobile Edge Computing (MEC) · caching and Content Delivery Network (CDN) · DC · 5G

1 Introduction

Caching and Content Delivery Network (CDN) is an effective network optimization technique that uses cache servers deployed at the network edge and employs intelligent routing algorithms to deliver content faster, more reliably, and cost-effectively. Edge computing cache technology is often used in CDN networks, it refers to the practice of caching data, content, or computing resources at the edge of the network, closer to the end-users or devices. This approach aims to reduce latency, enhance performance, and optimize data delivery in distributed computing environments [1].

By deploying caching mechanisms at edge computing nodes, such as edge servers or edge devices, commonly used data, content, or computational tasks can be stored locally. This allows for faster retrieval and processing of information, as it eliminates the need

C. Li et al. (Eds.): ICNC 2025, CCIS 2946, pp. 111–120, 2026.
https://doi.org/10.1007/978-981-92-1599-7_10

to access distant data centers or cloud servers [2]. A scenario for an edge computing application is shown in Fig. 1. In this scenario, the sensor data results obtained through edge computing, which are transmitted to the intelligent computing center, and after processing, fully intelligent traffic scheduling can be realized [3].

The benefits of edge computing caching include:

1. Reduced latency: By caching data or content at the edge, it can be quickly accessed by nearby users or devices, minimizing the round-trip time for data retrieval.

2. Bandwidth optimization: Caching popular or repetitive data at the edge can help alleviate network congestion and reduce the amount of data transferred over long distances.

3. Increased scalability: Edge caching enables distributed storage and processing capabilities, allowing for efficient

resource scaling and handling of larger user loads without significantly impacting the central infrastructure.

4. Improved resiliency: Local caching provides resilience in case of network disruptions or server outages. Users can still access cached data or execute cached computations even when connectivity to the central server is temporarily unavailable.

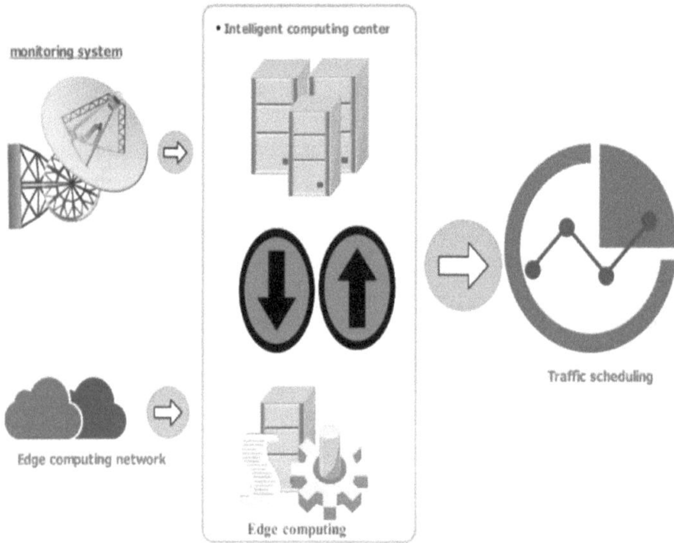

Fig. 1. Traffic monitoring scenarios in edge computing

5. Offline capabilities: Edge caching allows for offline access to previously cached data or content, enabling functionality in environments with intermittent or limited network connectivity.

2 The Major Technologies in Edge Computing and Mobile Edge Computing (MEC)

The three major technologies in edge computing are local caching, local forwarding, and cross-layer optimization [4].

1. Local caching: In edge computing, local caching refers to storing frequently accessed data or content on edge nodes that are closer to the users or devices. This avoids the need to retrieve data over the network or from remote servers for every access, thereby improving data access speed and user experience. Local caching reduces network latency and provides offline access capability in cases of unavailable network connections.

2. Local forwarding: Local forwarding involves routing data, content, or computational tasks from central servers or cloud to edge nodes for processing. Edge nodes can perform the computation tasks in close proximity to the users or devices, avoiding the need to send the entire workload to central servers. Local forwarding technology reduces data transfer distances, lowers network latency, and improves the responsiveness of computation tasks.

3.Cross-layer optimization: Cross-layer optimization focuses on optimizing and coordinating multiple layers (such as the network, application, storage in the edge computing environment to enhance overall performance and efficiency. By enabling efficient data transfer, coordinated decision- making, and resource management across different layers, cross-layer optimization achieves better resource utilization, reduced latency, and improved throughput. This technology makes edge computing systems more intelligent and efficient.

These technologies play crucial roles in edge computing by enabling local caching, local forwarding, and cross-layer optimization [5]. By implementing local caching, local forwarding, and cross-layer optimization on edge nodes, edge computing can deliver faster, more reliable, and efficient data access and computation capabilities. It also reduces dependency on central servers, resulting in reduced latency.

Mobile Edge Computing (MEC) is an emerging computing model aimed at bringing computing, storage, and network services closer to end devices. Traditionally, cloud computing relies on remote data centers to process and store data [6]. However, for applications requiring low latency and high bandwidth such as the Internet of Things, virtual reality, and augmented reality, the cloud computing model may not meet the requirements.

The core concept of Mobile Edge Computing is to push cloud computing resources and services closer to the edge network, typically at the base stations or cellular towers of wireless access networks [7]. This greatly reduces data transmission latency and costs. By deploying computing, storage, and network functions in the edge network, Mobile Edge Computing enables faster data processing and response. The combined development of data center [8] and 5G technologies [9] has facilitated mobile edge computing, as shown in Fig. 2.

Mobile edge computing does not involve physical movement; it is a computing paradigm that brings computation, storage, and services closer to the network edge. Therefore, mobile edge computing refers to a conceptual movement rather than a physical one.

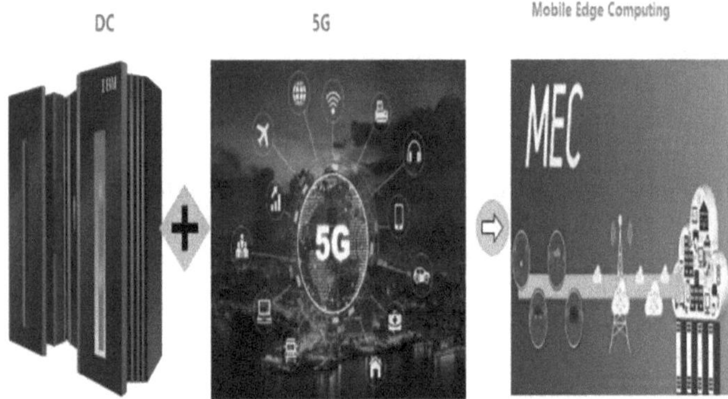

Fig. 2. The combination of DC and 5G enables edge computing

In mobile edge computing, computation, storage, and services are deployed on edge nodes that are in proximity to users or data sources. These edge nodes can be base stations, cellular towers, edge servers, gateways, smartphones, and more. The deployment of edge nodes can be chosen and configured based on specific requirements and network environments [10].

Mobile Edge Computing offers several advantages:

1. Low latency: Placing computing resources near end devices reduces data transmission latency and improves real- time performance.

2. Reduced network traffic: Offloading processing and storage of data to edge devices reduces the load on the core network, reducing bandwidth pressure.

3. Privacy and security: Mobile Edge Computing allows data processing to occur in the user's local environment, minimizing the risk of sensitive data being transmitted over the network.

4. Offline support: Edge devices can host offline applications, enabling real-time data processing even without a network connection.

The mobility of mobile edge computing refers to the ability of edge computing resources or services to seamlessly migrate or switch between different mobile edge nodes. In a mobile edge network, this means that computing tasks can be dynamically allocated and migrated among different edge devices or edge servers as needed.

The mobility of mobile edge computing helps to achieve flexible and dynamic allocation and utilization of resources. If a certain edge node is overloaded or fails, computing tasks can be automatically switched to a node with lower load or normal operation, ensuring the continuity and reliability of the computing tasks. This mobility also allows services or data to be brought closer to end users based on changes in the location of mobile devices or terminals, reducing data transmission and latency. Mobile edge computing architecture pattern [11], as shown in Fig. 3.

There are different modes of mobile edge computing that can be implemented based on the specific requirements and network environment. Here are some common modes:

1. Cloud-Enabled Mobile Edge Computing [12]: In this mode, edge devices collaborate with cloud servers. Edge devices handle data collection, preprocessing, and some data processing, while complex computational tasks are offloaded to the cloud servers.

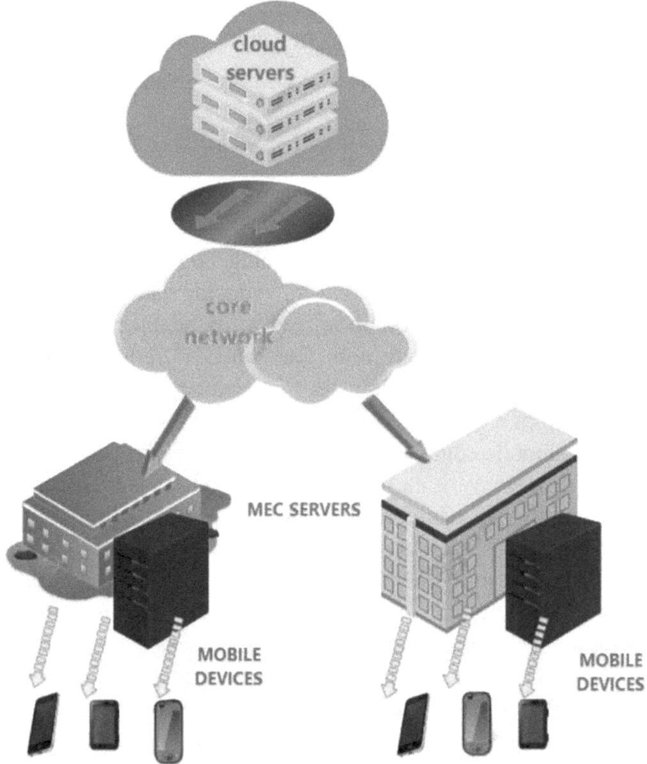

Fig. 3 Mobile edge computing architecture

Communication between edge devices and cloud servers occurs over the network, enabling distributed computing and resource sharing.

2. Edge Server Mode: In this mode, multiple edge servers are deployed in the edge network, typically located at base stations or cellular towers of the wireless access network. Edge servers provide basic computing functionalities such as caching, data processing, and network middleware. This reduces the latency and load of transferring data to cloud servers.

3. Device-Centric Mode: In this mode, edge devices have increased computing and storage capabilities, enabling them to independently process data and execute applications. Edge devices become small-scale computing nodes, capable of completing some computational tasks locally, reducing dependency on cloud servers.

4. Collaborative Computing Mode: In this mode, multiple edge devices collaborate to accomplish complex computational tasks. Edge devices can share computing and storage resources, coordinate and schedule tasks, enabling distributed computing and load balancing.

These modes can be flexibly combined and adjusted to meet specific application requirements. The goal of mobile edge computing is to bring computation and services closer to the edge devices, providing efficient and low-latency computation resources and services to meet diverse application needs.

3 Experimental Verification

To improve real-time responsiveness in mobile edge computing, the following methods are adopted:

1. Proximity computing: The core idea of mobile edge computing is to bring computing and service resources closer to end users, reducing data transmission latency. By deploying edge nodes near users, computing tasks can be allocated to locations closer to users, enabling real-time responsiveness. Compared to traditional cloud computing, mobile edge computing reduces data transfer time and network latency.

2. Remote backup and redundant deployment: To enhance the reliability of real-time responsiveness, mobile edge computing employs strategies such as remote backup and redundant deployment. When a particular edge node fails or becomes overloaded, tasks can be shifted to other available edge nodes, ensuring continuous and real-time computing services.

3. Fast network connectivity: Mobile edge computing necessitates establishing fast and stable network connections to achieve real-time responsiveness. This can be accomplished by leveraging technologies like high-speed wireless networks (such as 5G) and optimizing network architecture. High-speed network connections reduce data transfer latency and bandwidth bottlenecks, enhancing real-time responsiveness.

4. Data preprocessing and local computing [13]: Mobile edge computing involves preprocessing data and performing local computing on edge nodes, reducing reliance on cloud servers and minimizing data transfer. By conducting partial computing tasks at the edge nodes, data can be preprocessed and data transfer latency can be reduced, resulting in faster real-time responsiveness.

5. Priority-based task scheduling: For tasks that require real- time responsiveness, mobile edge computing employs priority- based task scheduling strategies, prioritizing the processing of tasks with high real-time requirements. By employing effective task scheduling and resource allocation, timely processing of real-time tasks is ensured.

This is a traditional live video architecture, as shown in the Fig. 4. For example, multiple video images are captured in different geographical locations and different scenes, transmitted to local production, and these video images are uploaded to the Internet, the core network downloads these video images from the Internet, and these video images are run in the local run form a composite screen. During this process, local production uploads the video images to the Internet, core network downloads the video images from the Internet and sends the video images to the local run, finally form the combined dynamic picture, and be sent to the end customer. The time delay in this process is greater than 7S.

MEC video broadcast architecture, as shown in the Fig. 5, collects multiple video images in different geographical locations and different scenes and transmits them to local production. Near the end user, the local production is transmitted to the streaming media application, and the streaming media application is directly transmitted to the local run. In local run, the video is transmitted to the end user, and the transmission is

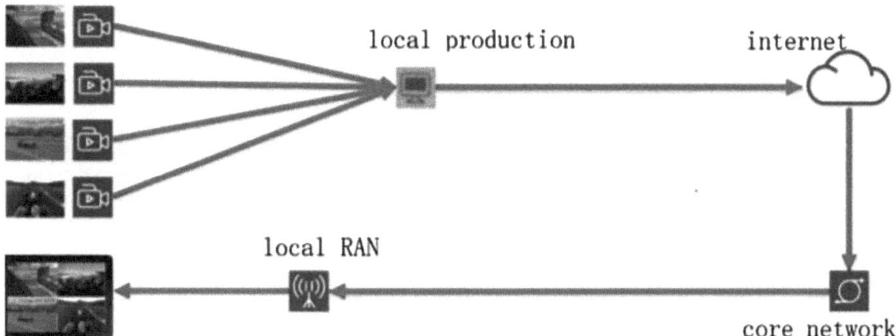

Fig. 4. Traditional live video architecture

near the edge of the end user. The locally produced video is not sent to the Internet. In the streaming media application [14], you can choose whether to send the locally produced video to the Internet through the core network. This architecture reduces latency transmission, reducing latency less than 1 s. The transmission link is reduced, the transmission resources are saved, the content is transmitted safely, and the probability of network leaks is reduced. In the field of streaming media application, the core network is built with the Internet cloud, and a two-way choice is made whether to send it to the Internet cloud for backup.

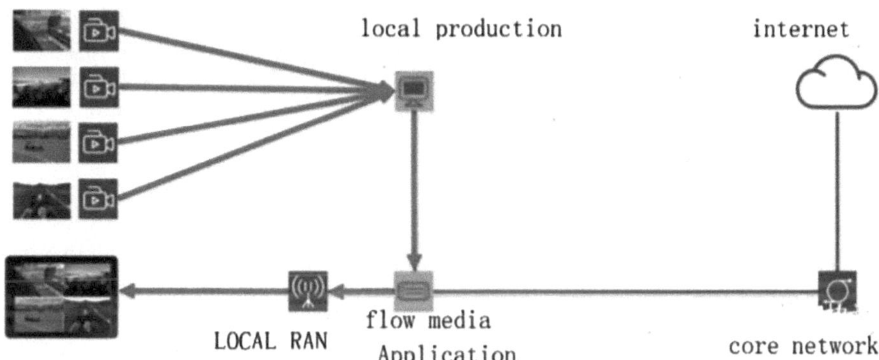

Fig. 5 MEC video broadcast architecture

The advantages of MEC (Mobile Edge Computing) video live streaming architecture are as follows:

1. Low latency: With MEC architecture, computing resources are brought closer to the users, enabling video live streaming data to be processed and transmitted at edge nodes. This reduces the latency in transferring data from remoteservers to user devices, providing a more real-time viewing experience.

2. Reduced network bandwidth requirements: In traditional video live streaming architectures [15], video data needs to be transmitted over the network to central servers

for processing and distribution, and then transmitted to user devices. In MEC video live streaming architecture, video data can be processed and cached at edge nodes, reducing the network bandwidth requirements. This not only reduces network congestion but also lowers data transmission costs.

3. Flexible resource allocation: MEC video live streaming architecture allows dynamic allocation of computing and storage resources based on demand [16]. Edge nodes can allocate more computing resources and storage space based on real-time network conditions and user viewing volumes, providing a better viewing experience. This flexibility and responsiveness enable MEC video live streaming architecture to better adapt to changing network environments and user demands.

4. Enhanced content delivery [17]: By caching video data and performing content delivery at edge nodes, MEC video live streaming architecture provides more stable and efficient content delivery services. Users can access video content from edge nodes closer to them, reducing the risks of network congestion and transmission failures.

5. Support for higher-quality media experiences: Thanks to the lower latency, more stable content delivery, and flexible resource allocation provided by MEC video live streaming architecture, users can enjoy higher-quality media experiences. Video live streaming can respond faster to user actions and provide clearer and smoother visuals and better audio quality.

In summary, MEC video live streaming architecture offers advantages such as low latency, reduced network bandwidth requirements, flexible resource allocation, enhanced content delivery, and support for higher-quality media experiences. These advantages provide users with a better video viewing experience and offer more efficient and reliable solutions for video live streaming service providers. Through proximity computing, fast network connectivity, data preprocessing and local computing, task scheduling, and other methods, mobile edge computing can enhance network real-time responsiveness, meeting the low-latency and high real-time requirements of various applications.

4 Summary

CDN networks leverage distributed edge nodes, content caching, and intelligent routing to achieve improved website performance, reduced origin server load, high availability and fault tolerance, mitigation of network congestion, and enhanced security and defense capabilities. CDN has become a crucial component for many websites and online services, the application of edge computing in CDN networks can improve network response speed, reduce the load on the origin server, provide personalized content [18] and services, and enhance the intelligence of the network. This allows CDN networks to better meet user needs and provide a better user experience.

Acknowledgments. This work is supported by a the Natural Science Foudation of China under Grant 62273136 , Key Laboratory open fund of Huangshi city, and Hubei Provincial Natural Science Foundation Innovation and Development Joint Fund (NO. 2023AFD016).

References

1. Francisco Landaeta, J., Garcia Guzman, J., Arnoldo, M.H.: Inno pro: a process to define and implement an innovation strategy. IEEE Lat. Am. Trans. **12**(3), 462–468 (2014). https://doi.org/10.1109/tla.2014.6827874
2. Atutxa, A., Sanz, A., Sasiain, J., et al.: Towards a quantum-safe 5G: quantum key distribution in core networks. Comput. Commun. **224**, 145–158 (2024). https://doi.org/10.1016/j.comcom.2024.06.005
3. Modina, N., El-Azouzi, R., Pellegrini, F.D., et al.: Joint traffic offloading and aging control in 5G IoT networks. IEEE Trans. Mob. Comput. **22**, 4714–4728 (2023). https://doi.org/10.1109/TMC.2022.3154089
4. Zhu, Q., You, L., Hu, G., et al.: Secure and efficient biometric-based anonymous authentication scheme for Mobile-edge computing. IEEE Internet Things J. **11**(20), 20 (2024). https://doi.org/10.1109/JIOT.2024.3429397
5. Jeyaraj, R., Balasubramaniam, A., et al.: Resource Management in Cloud and Cloud-influenced Technologies for internet of things applications. ACM Comput. Surv. **55**(12), 1–37 (2023). https://doi.org/10.1145/3571729
6. Hosseini Shirvani, M., Ramzanpoor, Y.: Multi-objective QoS-aware optimization for deployment of IoT applications on cloud and fog computing infrastructure. Neural Comput. Applic. **35**(26) (2023). https://doi.org/10.1007/s00521-023-08759-8
7. Li, X., Chen, T., Liu, X.: A novel graph-based computation offloading strategy for workflow applications in Mobile edge computing. IEEE Trans. Serv. Comput. **16**(2), 845–857 (2023)
8. Hwang, T.H., Lee, K.Y.: Web-based non-contact edge computing solution for suspected COVID-19 infection classification model. J. Web Eng. **22**(4), 597–614 (2023)
9. Poliszczuk, J.: Borysa Chersońskiego dekonstrukcja mitu Odessy. Poznańskie Studia Slawistyczne. **26**, 67–81 (2024). https://doi.org/10.14746/pss.2024.26.3
10. Tang, Z., Xiao, Z., Yang, L., et al.: A network load perception based task scheduler for parallel distributed data processing systems. IEEE Trans. Cloud Comput. **11**(2), 13 (2023). https://doi.org/10.1109/TCC.2021.3132627
11. Zhang, Y., Feng, B., Zhang, Y.H.: Task offloading control and customized workload scheduling in multi-layer cloud networks. IEEE Trans. Netw. Serv. Manag. **21**(1), 714–728 (2024)
12. Choudhary, A., Govil, M.C., Singh, G., et al.: Energy-aware scientific workflow scheduling in cloud environment. Clust. Comput. **25**(6), 30 (2022). https://doi.org/10.1007/s10586-022-03613-3
13. Hu, J., Cheng, F., Liu, M., et al.: MicroFallNet: a lightweight model for real-time fall detection on smart wristbands. Pervasive Mobile Comput., 109 (2025). https://doi.org/10.1016/j.pmcj.2025.102046
14. Rokaia, R.: Metaheuristic optimization for scheduling in cloud computing environments: a review. Metaheuristic Optim. Rev. **2**(2), 14–25 (2024). https://doi.org/10.54216/mor.020202
15. Chung, J., Fayyad, J., Younes, Y.A., et al.: Learning team-based navigation: a review of deep reinforcement learning techniques for multi-agent pathfinding. Artif. Intell. Rev. **57**(2), 36 (2024). https://doi.org/10.1007/s10462-023-10670-6
16. Yurchenkov, A.: An introduction to reinforcement learning: key concepts, algorithms and applications. Int. J. Sensor Netw. Data Commun. **12**(3), 1 (2023). https://doi.org/10.37421/2090-4886.2023.12.209
17. Wu, Q.: An enhanced whale optimization algorithm with inertia weight and dynamic parameter adaptation for wireless sensor network deployment. Computing. **107**(8), 167 (2025). https://doi.org/10.1007/s00607-025-01509-9

18. Wang Z, Shao L, Yang S W J L D.CRLM: a cooperative model based on reinforcement learning and metaheuristic algorithms of routing protocols in wireless sensor networks. Comput. Netw., 2023, 236. 110019 https://doi.org/10.1016/j.comnet.2023.110019.

Impacts of Dynamic Topology on Networked Control Systems

Yue Liu[1], Yang Xiao[2(✉)], and Tieshan Li[3]

[1] Navigation College, Dalian Maritime University, Dalian 116026, China
[2] Department of Computer Science, The University of Alabama, Tuscaloosa, AL 35487, USA
yangxiao@ieee.org
[3] School of Automation Engineering, University of Electronic Science and Technology of China, Chengdu 611731, China

Abstract. Dependable communication links are essential to maintain stability and achieve target performance in control systems. Although communication challenges have been widely examined, the concrete effects of dynamic topologies remain insufficiently understood. This study investigates the impact of dynamic topologies on control systems. When links appear or disappear over time, inter-agent exchanges may become intermittent, degrading data delivery and creating uncertainty about whether each controller receives timely information from its neighbors. Building on earlier results that treat selected controller coefficients as random variables with specified probability distributions, we refine that work by analyzing the distributional properties under dynamic topologies. We further corroborate the analysis with simulations across diverse communication conditions, quantifying the impacts of topology changes on control objectives. Simulation results show how dynamic topologies affect the control system.

Keywords: Cyber-physical network co-design and analysis · dynamic topology · networked control system · formation control

1 Introduction

With rapid technological progress, automatic control has advanced from direct control to distributed control and ultimately to networked control systems (NCSs). An NCS is a distributed architecture in which sensors, controllers, and actuators exchange measurements and control commands over a shared communication network, thereby closing the feedback loop [1]. Compared with classical control schemes, NCSs are extensively used across a wide range of practical domains, such as autonomous vehicles [2], intelligent buildings [3,4], and smart grids [5]. However, the performance of NCSs is often limited by communication-related challenges that may compromise reliability and efficiency. Addressing these issues has become a central topic in recent research. Many existing studies have focused on specific communication problems, including transmission

C. Li et al. (Eds.): ICNC 2025, CCIS 2946, pp. 121–131, 2026.
https://doi.org/10.1007/978-981-92-1599-7_11

delays [6,7], event-triggered control mechanisms [8], packet dropouts [9], and time-varying network topologies [10].

In control systems, network topology is a critical communication issue that merits further investigation. In practical applications, the topology can change dynamically. It should be noted that, as pointed out in [11,12], NCSs are inherently vulnerable to dynamic changes caused by malicious cyber attacks. Methods designed for NCSs with static topologies cannot be directly applied to dynamic topologies since the changes in topology are often sudden, unexpected, and difficult to predict, which can lead to instability. Thus, investigating the control of NCSs with dynamic topologies is not trivial. Considering NCSs with dynamic topologies, several representative results have been referenced. In [13], Communication unreliability is characterized by irregular topology switching caused by communication failures or attacks. The authors in [14] investigate vehicle-to-vehicle communication where the topology randomly switches among three communication patterns. The authors in [15] investigate distributed control for multiple trains with dynamic topologies to achieve leader-follower operation with a short headway distance. The authors in [16,17] propose a novel control scheme for multi-agent systems under dynamic topologies.

Although significant research has investigated topology in control systems, existing studies on the impacts of dynamic topology on control systems still require further exploration. The connectivity structure in practical systems is often time varying. However, prior studies [18,19] primarily analyze a finite fixed topology. In fact, the topology changes for several reasons, such as node motion, joining, or leaving the system; changes in the external environment that may affect communication links; and adjustments required by changing task demands. Therefore, to accommodate system dynamics and task requirements, the topology should be treated as time varying rather than fixed. Moreover, the causes and consequences of topology changes are still worth further exploration. Environmental shifts can disrupt communication links and trigger topology changes. Large inter-node distances can also disrupt effective communication, resulting in disconnected or newly formed connections. In addition, link failures, signal loss, or delays can break existing connections and alter the topology. When the topology varies, communication paths are modified, delays grow, and data may be lost. The communication burden between nodes increases, and congestion becomes more likely. Under collaborative and synchronized operation, overall performance and security are negatively affected. Therefore, both the causes and the system-level consequences of topology changes need careful study.

In this paper, we study agent-to-agent topologies that vary with time. Specifically, the controller of an agent depends not only on its own state but also on the states of its neighboring agents [20]. As agents move, their neighbor sets change, leading to a dynamic topology and a time-varying controller. In such dynamic topologies, connections between nodes change over time, causing the controller to be non-deterministic, with some coefficients modeled as random variables. Motivated by the preceding discussion, we investigate how dynamic topologies influence the controller, and we characterize how the distribution of the controller's random variable changes as the topology varies. As a refinement of our prior work, we rigorously characterize the distributional properties of the controller's random variable in a representative unmanned surface vehicle (USV) formation-control scenario. We then conduct simulations across diverse network topologies to quantify their effects on the controller. The contributions of this paper are as follows. First, we provide a detailed analysis of the random variable under dynamic topologies. This clarifies how topology variations influence the controller and deepens understanding of how communication affects control. Second, we refine and extend our earlier framework on the impacts of topology on control, and establish a systematic analysis for dynamic topologies, which improves applicability to practical deployments. Third, we conduct extensive simulations under dynamic topologies and across diverse communication conditions. The results substantiate the analysis and quantify the impact of topology changes on the control objectives.

The remainder of the paper is organized as follows. Section 2 presents the background. Section 3 analyzes in detail the impacts of fixed and time-varying bit error rate on the control system. Section 4 concludes the paper.

2 Background

In this paper, the formation control of USVs is designed to satisfy three primary objectives [21]. First, the USVs should successfully reach the predefined formation positions. Second, the USVs maintain an approximately fixed formation pattern. Third, upon reaching the target positions, the trajectory length of each USV should be as close as possible to its corresponding ideal path length. The kinematic model of the ith USV is given by [22,23]:

$$\begin{cases} \dot{x}_i = u_i \cos(\sigma_i) - \tau_i \sin(\sigma_i), \\ \dot{y}_i = u_i \sin(\sigma_i) + \tau_i \cos(\sigma_i), \\ \dot{\sigma}_i = r_i, \\ \dot{u}_i = \dfrac{z_{22,i}}{z_{11,i}} \tau_i r_i - \dfrac{d_{11,i}}{z_{11,i}} u_i + \dfrac{\pi_{iu}}{z_{11,i}}, \\ \dot{\tau}_i = \dfrac{z_{11,i}}{z_{22,i}} u_i r_i - \dfrac{d_{22,i}}{z_{22,i}} \tau_i + \dfrac{\pi_{iv}}{z_{22,i}}, \\ \dot{r}_i = \dfrac{z_{11,i} - z_{22,i}}{z_{33,i}} u_i \tau_i - \dfrac{d_{33,i}}{z_{33,i}} r_i + \dfrac{\pi_{ir}}{z_{33,i}}, \end{cases} \tag{1}$$

where $[x_i, y_i]$ represents the Cartesian coordinates of position, and σ_i represents the yaw angle; u_i, τ_i, and r_i represent the surge velocity, the sway velocity, and the yaw rate, respectively. $z_{jj} (j = 1, 2, 3)$ denotes the added mass. $d_{jj} (j = 1, 2, 3)$ represents the hydrodynamic damping. The control input $\pi_i = [\pi_{iu}, \pi_{iv}, \pi_{ir}]^T$ encompasses the surge force π_{iu}, the sway force π_{iv}, and the yaw moment π_{ir} [24].

The input-output dynamics of USV i is represented by the double integrator system [25]: $\ddot{q}_i = \boldsymbol{\nu}_i$, where q_i and $\boldsymbol{\nu}_i$ denote the position and control input, respectively. According to [26], the impact of USV i on USV j can be simplified as

$L(i,j)$, and K_f and D_f are the proportional gain and derivative gain for keeping formation, respectively. We obtain the general control law as follows [26]:

$$\boldsymbol{\nu}_i = -K_g e_i - D_g v_i + \sum_{j \in \triangle_i} a_{ij} L(i,j), \tag{2}$$

where K_g and D_g denote proportional gain and derivative gain for arriving desired position, respectively. $\triangle_i$ denotes the set of neighbors of USV i. In Eq. (2), a_{ij} is the link weight representing whether agent i can receive control information from agent j. $a_{ij} = 1$ denotes that agent i receives control information from agent j; otherwise, $a_{ij} = 0$.

As presented in [26], the total delay T_s is defined as

$$T_s = (s-1)[\alpha(m+H) + T_{out}] + \alpha(m+H), \tag{3}$$

where T_{out} is the timeout, $\alpha(m+H)$ is the time delay, $m+H$ is the total number of bits to be sent, and s is the processing time. T_s is strictly increasing with s $(s = 1, 2, \dots)$.

According to [26], Δt is defined as the response interval between initiating information exchange and executing control decisions, and T denotes the transmission delay. Accordingly, we can obtain

$$\begin{aligned} P_r(a_{ij}=1) = P_r(T \le \Delta t) &= \sum_{s \in \{s: T_s \le \Delta t\}} P_r(T = T_s) \\ &= \sum_{s \in \{s: T_s \le \Delta t\}} [1-(1-p_{ij})^{(m+H)}]^{s-1} (1-p_{ij})^{(m+H)}. \end{aligned} \tag{4}$$

From [26], we can get when $T > \Delta t$, the agent i can not receive control information from agent j, and $a_{ij} = 0$. Based on Eq. (4) and [26], the probability mass function (PMF) for $T_s > \Delta t$ is computed as $P_r(a_{ij} = 0) = P_r(T > \Delta t) = 1 - P_r(a_{ij} = 1)$. When the value of p_{ij} is variable, the PMF for $T \le \Delta t$ can be defined as

$$\begin{aligned} P_r(a_{ij}=1) = P_r(T \le \Delta t) = \sum_{s \in \{s: T_s \le \Delta t\}} P_r(T = T_s) = \\ \sum_{s \in \{s: T_s \le \Delta t\}} \left(\prod_{l=1}^{s-1} [1-(1-p_{ij}^l)^{(m+H)}] \right) (1-p_{ij}^s)^{(m+H)} \end{aligned} \tag{5}$$

Based on Eq. (5), when the p_{ij} is variable, the PMF for $T > \Delta t$ is computed as $P_r(a_{ij} = 0) = P_r(T > \Delta t) = 1 - P_r(a_{ij} = 1)$. As shown in Eqs. (4) to (5), bit error rate affects the probability distribution of the random variable a_{ij}, and therefore changes the effective interaction terms in the control law. In this sense, communication reliability influences the controller indirectly through the stochastic availability of inter-agent links. Therefore, the next section specifically investigates how the p_{ij} influences the probabilities $P_r(a_{ij} = 1)$ and $P_r(a_{ij} = 0)$.

In this paper, dynamic topology is represented through the time variation of effective communication links among agents. Specifically, a link may disappear because the inter-agent distance exceeds the communication range or because obstacles block the channel, and even an existing communication link may fail to deliver control information due to transmission errors. Therefore, the effective topology is characterized by the random link indicator $a_{ij}(t)$, whose distribution is determined jointly by geometric connectivity conditions and the bit error rate $p_{ij}(t)$. Under this formulation, the impact of dynamic topology on the control system is analyzed through the evolution of link availability and its induced controller variations.

3 The Impact of Dynamic Topology Through Bit Error Rate on the Control System

In a multi-agent system, a link from node i to node j exists with a probability determined by p_{ij}, and is considered absent when $p_{ij} = 1$. For nodes i and j, the value of p_{ij} falls into one of the three possible cases. When $p_{ij} = 1$, node i cannot receive control information from node j. When $0 < p_{ij} < 1$, it indicates that there might be errors in the transmission from node i to node j. When $p_{ij} = 0$, there are no errors in transmission from node i to node j. When p_{ij} is very large, it becomes difficult to transmit packets between node i and node j. In this case, node i and node j are considered disconnected. The larger the value of p_{ij}, the smaller the probability that node i and node j connect. When the distance between nodes i and j exceeds the transmission range, they are considered disconnected. If barriers exist between them, these barriers may obstruct the signal from being transmitted, and node i and node j are considered disconnected. In our paper, for simplicity, we assume that when the distance between node i and node j exceeds the transmission range, we have $p_{ij} = 1$. Similarly, if node i and node j are neighboring nodes and barriers exist between them, we also set $p_{ij} = 1$.

When node i and node j exchange information, p_{ij} may vary over time, leading to multiple possible scenarios in the transmission of control information. Although our previous work has explored the impacts of p_{ij} on the a_{ij}, this paper explores the problem from a different perspective.

3.1 The Impacts of Piecewise-Constant p_{ij}

Let T denote the information-exchange interval between node i and node j. This interval is divided into n equal-length time slots. In this subsection, we consider piecewise-constant bit error rate patterns, where the value of p_{ij} is constant within each slot but may differ across slots.

(1) Mixed-slot case. Suppose that for β' out of the β time slots, $p_{ij} = 1$, and for the remaining $(\beta - \beta')$ slots, $p_{ij} = c$ with $0 < c < 1$. For slots with $p_{ij} = 1$, the per-slot frame-success probability is $(1 - p_{ij})^{(m+H)} = 0$, hence these slots do not contribute to $P_r(a_{ij} = 1)$. For each of the $(\beta - \beta')$ effective slots,

the success probability is $(1-c)^{(m+H)} \in (0,1)$. Therefore, the $P_r(a_{ij}=1)$ over the $(\beta-\beta')$ effective slots is computed as $1-\left[1-(1-c)^{(m+H)}\right]^{\beta-\beta'}$, which is strictly increasing in $\beta-\beta'$ and satisfies $\lim_{\beta-\beta'\to\infty} P_r(a_{ij}=1)=1$. This shows that increasing the number of available slots $(\beta-\beta')$ can compensate for unfavorable channel conditions or larger packet length $(m+H)$.

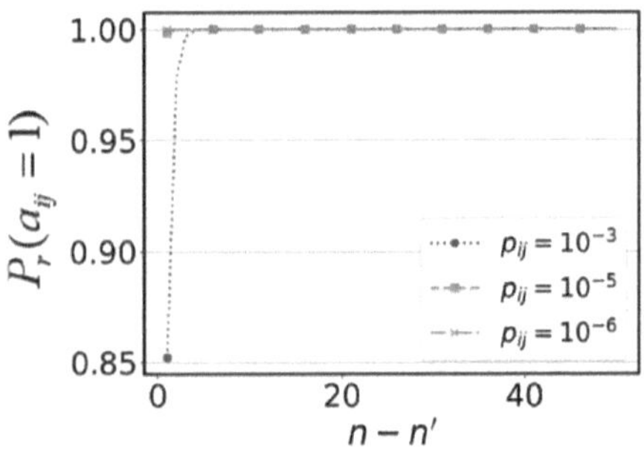

Fig. 1. Influence of p_{ij} on $P(a_{ij}=1)$.

Figure 1 illustrates that $P_r(a_{ij}=1)$ increases monotonically with $\beta-\beta'$ and eventually converges to 1 for all values of p_{ij}. For $p_{ij}=10^{-5}$ and $p_{ij}=10^{-6}$, $P_r(a_{ij}=1)$ remains nearly 1 across all $\beta-\beta'$. When $p_{ij}=10^{-3}$, the curve exhibits a rapid rise at small $\beta-\beta'$, indicating a significant early gain.

(2) Alternating-slot case. Consider a time-varying pattern where the slot-wise bit error rate alternates as $p_{ij}(h)=c$ for odd h and $p_{ij}(h)=1$ for even h $(c\in(0,1))$. Under slot-wise independence, $P_r(a_{ij}=1)=1-\prod_{h=1}^{n}(1-(1-p_{ij}(h))^{(m+H)})$. For slots with $p_{ij}(h)=1$, the $1-(1-p_{ij}(h))^{(m+H)}$ is equal to 1. Such terms do not reduce the overall product and do not affect $P_r(a_{ij}=1)$.

Denoting by $\beta-\beta'$ the number of effective slots with $p_{ij}(h)=c$, we can obtain $P_r(a_{ij}=1)=1-\left[1-(1-c)^{(m+H)}\right]^{\beta-\beta'}$, which coincides with case (1). The alternating-slot case is a special ordering of the mixed-slot case. Under the independence assumption, the $P_r(a_{ij})$ depends only on the count of effective slots (those with $p_{ij}<1$), not on their temporal order. Therefore, no additional figures are provided for this case, as its relationship between p_{ij} and $P_r(a_{ij}=1)$ is identical to that in Fig. 1.

In Fig. 2, we study the trajectories and formation behaviors of USVs under different p_{ij}. The simulation time is set to 100 s, and the positions of the USVs are recorded at five representative time instants. Other simulation settings follow [21]. When $p_{ij}=10^{-8}$ (Figs. 2(a)–2(b)), the USVs reach the target positions, follow trajectories close to the ideal paths, and preserve the formation, thereby achieving all three control objectives. For larger p_{ij} (Figs. 2(c)–2(d)), the USVs still arrive at the desired positions and maintain the formation, but their trajectory lengths deviate from the ideal path, meaning the third objective is not

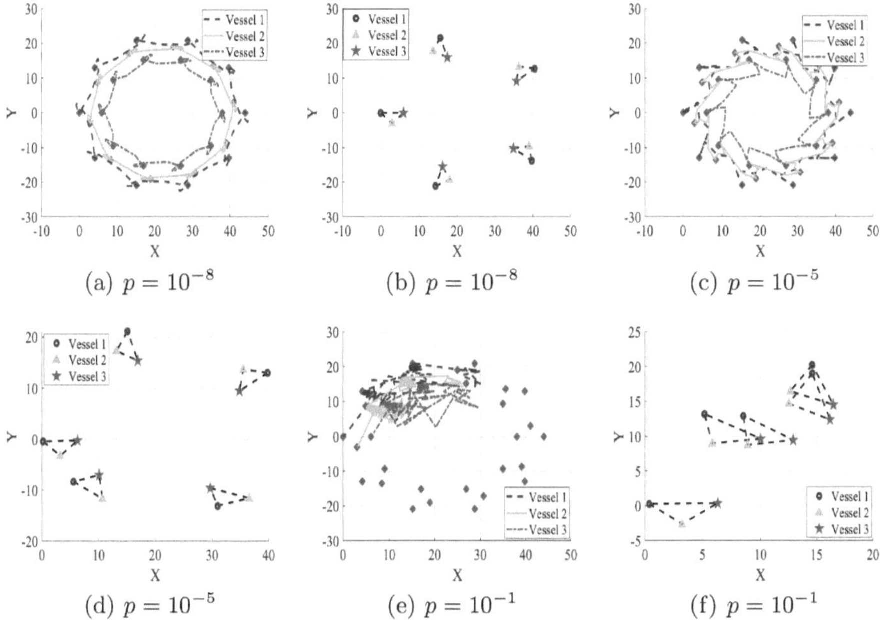

(a) $p = 10^{-8}$ (b) $p = 10^{-8}$ (c) $p = 10^{-5}$

(d) $p = 10^{-5}$ (e) $p = 10^{-1}$ (f) $p = 10^{-1}$

Fig. 2. The trajectories and formations of USVs.

achieved. When p_{ij} becomes further increased (Figs. 2(e)–2(f)), the USVs fail to reach the desired positions or maintain the formation, so none of the objectives are achieved. These results demonstrate that as p_{ij} increases, the ability of the USVs to achieve the formation control objectives progressively deteriorates.

3.2 The Impacts of Variable p_{ij}

Similar to the piecewise-constant p_{ij} case, we divide the total duration T into n equal-length time slots of length δ. In each slot, the bit error rate p_{ij} is allowed to vary across slots, reflecting dynamic and heterogeneous communication conditions.

(1) Impact of extreme and moderate slot conditions. Such a slot contributes a factor of 1 to the failure product for later attempts, but it eliminates one transmission opportunity within the deadline and can therefore reduce the overall probability of successful delivery before δ. If p_{ij}^s remains at a moderate level across multiple time slots, then each multiplicative factor in the product term lies strictly within $(0, 1)$. The expression $\prod_{l=1}^{s-1}\left[1-(1-p_{ij}^l)^{m+H}\right]$ decays geometrically as s increases, even if the final term $(1-p_{ij}^s)^{m+H}$ remains relatively large. This accumulated multiplicative effect significantly reduces the overall value of $P_r(a_{ij}=1)$, highlighting the impact of sustained moderate p_{ij}.

Although the previous analysis assumes a fixed moderate p_{ij} across multiple time slots, the impact becomes more complex when p_{ij}^s varies across time. In such cases, the cumulative product in Eq. (5) is not only influenced by the values of p_{ij}^s, but also by their temporal arrangement. Even with the same number of effective slots, different distributions can yield significantly different outcomes.

(2) Sensitivity to distribution. Unlike the piecewise-constant-p_{ij} case, the value of $P_r(a_{ij} = 1)$ under variable p_{ij}^s depends not only on how many slots have $p_{ij}^s < 1$, but also on where these slots appear. If moderate error values cluster in earlier time steps, they suppress the product term more significantly than when they appear later. Hence, the temporal distribution of p_{ij}^s plays a critical role in shaping the value of $P_r(a_{ij} = 1)$. Overall, when p_{ij}^s varies across time slots, both its magnitude and temporal placement strongly influence the value of $P_r(a_{ij} = 1)$. Sustained moderate p_{ij}^s significantly undermines system reliability compared to a few extreme values, due to its compounding effect across the product term in Eq. (5).

Similar to the Subsect. 3.1, for the USVs formation control system, the three control objectives become progressively harder to achieve as p_{ij}^s increases. To show movement of the USVs under varying bit error rate, during the entire 100 s experiment, p_{ij}^s is assigned the values 10^{-8}, 10^{-5}, and 10^{-1} for 30, 30, and 40 occurrences, respectively. The USVs' trajectories and formations are shown in Fig. 3.

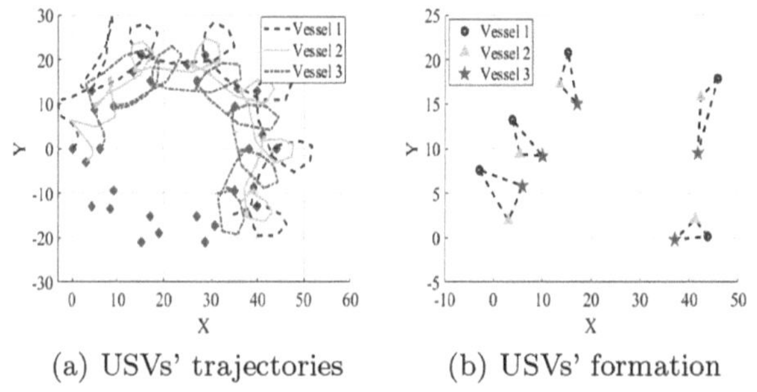

(a) USVs' trajectories (b) USVs' formation

Fig. 3. The trajectories and formations of USVs under different bit error rate.

In Fig. 3(a), the USV trajectories are irregular and exhibit several looping segments; moreover, some USVs fail to reach their assigned target positions. This behavior is likely attributable to the time-varying p_{ij}, which occasionally becomes large. In Fig. 3(b), the USVs fail to maintain an approximately fixed formation across the five time instants. Therefore, the three control objectives are not achieved under time-varying p_{ij}.

Figure 4 presents the evolution of inter-USV distances in two ways. Let D_{12}, D_{23}, and D_{13} denote the Euclidean distances between USV 1–2, USV 2–3, and USV 1–3, respectively. In Fig. 4(a), the D_{12}, D_{23}, and D_{13} vary over time, indicating that the formation is not approximately fixed. In Fig. 4(b), the same distances are shown as a heatmap with numerical annotations, making the variations across the five sampled instants explicit and illustrating that the relative

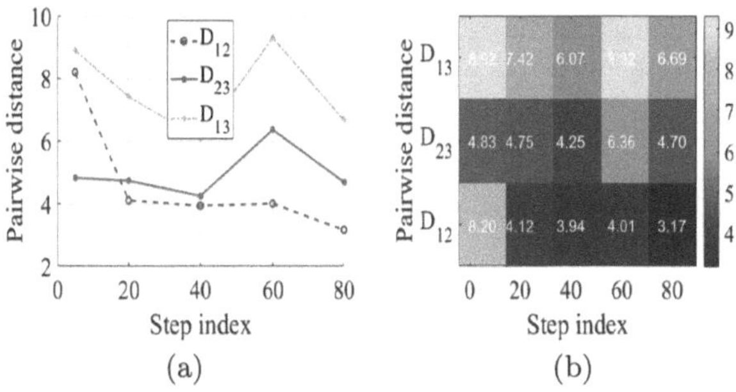

Fig. 4. Evolution of pairwise inter-USV distances under time-varying p_{ij}.

positions among the USVs change throughout the simulation. Overall, Fig. 4 confirms that the USVs do not maintain an approximately fixed formation when p_{ij} varies and do not achieve the second control objective.

4 Conclusion

This paper studies how time-varying link availability, induced by dynamic topology and communication errors, affects a networked control system. By analyzing the bit error rate-dependent random variable and evaluating a USV formation-control example under fixed and time-varying communication conditions, we show that degraded link reliability can cause controller variation, irregular trajectories, and loss of formation maintenance. These results provide a communication-aware perspective for understanding topology-induced performance degradation in networked control systems. Future work will extend the analysis to richer topology-evolution models and more complex communication environments.

References

1. Zhou, W., Wang, Y., Liang, Y.: Sliding mode control for networked control systems: a brief survey. ISA Trans. **124**, 249–259 (2022)
2. Wang, Z., Sun, D., Wang, L., Zhao, M.: Mixed vehicle group merging control in the vicinity of traffic signals: a cyber-physical perspective. IEEE Trans. Intell. Transp. Syst. **25**(6), 5328–5341 (2023)
3. Huda, N.U., Ahmed, I., Adnan, M., Ali, M., Naeem, F.: Experts and intelligent systems for smart homes' transformation to sustainable smart cities: a comprehensive review. Expert Syst. Appl. **238** (2024). Art. no. 122380
4. Mansouri, S.A., Nematbakhsh, E., Jordehi, A.R., Marzband, M., Tostado-Véliz, M., Jurado, F.: An interval-based nested optimization framework for deriving flexibility from smart buildings and electric vehicle fleets in the TSO-DSO coordination. Appl. Energy **341** (2023). Art. no. 121062

5. Lv, L., Wu, Z., Zhang, L., Gupta, B.B., Tian, Z.: An edge-AI based forecasting approach for improving smart microgrid efficiency. IEEE Trans. Industr. Inf. **18**(11), 7946–7954 (2022)
6. Li, Y., Tong, S.: Bumpless transfer distributed adaptive backstepping control of nonlinear multi-agent systems with circular filtering under DoS attacks. Automatica **157** (2023). Art. no. 111250 (2023)
7. Shangguan, X.C., et al.: Resilient load frequency control of power systems to compensate random time delays and time-delay attacks. IEEE Trans. Industr. Electron. **70**(5), 5115–5128 (2022)
8. Li, C., Zhao, X., Chen, M., Xing, W., Zhao, N., Zong, G.: Dynamic periodic event-triggered control for networked control systems under packet dropouts. IEEE Trans. Autom. Sci. Eng. **21**(1), 906–920 (2023)
9. Zeng, H.B., Zhu, Z.J., Peng, T.S., Wang, W., Zhang, X.M.: Robust tracking control design for a class of nonlinear networked control systems considering bounded package dropouts and external disturbance. IEEE Trans. Fuzzy Syst. **32**(6), 3608–3617 (2024)
10. Yang, G.,, Rezaee, H., Alessandri, A., Parisini, T.: State estimation using a network of distributed observers with switching communication topology. Automatica **147** (2023). Art. no. 110690
11. Liu, C., Jiang, B., Zhang, K., Patton, R.J.: Distributed fault-tolerant consensus tracking control of multi-agent systems under fixed and switching topologies. IEEE Trans. Circuits Syst. I Regul. Pap. **68**(4), 1646–1658 (2021)
12. Hu, Z., Mu, X.: Impulsive consensus of stochastic multi-agent systems under semi-Markovian switching topologies and application. Automatica **150** (2023). Art. no. 110871
13. An, L., Yang, G.-H., Deng, C., Wen, C.: Event-triggered reference governors for collisions-free leader-following coordination under unreliable communication topologies. IEEE Trans. Automatic Control, early access, 2023
14. Xiao, S., Ge, X., Han, Q.-L., Zhang, Y.: Dynamic event-triggered platooning control of automated vehicles under random communication topologies and various spacing policies. IEEE Trans. Cybern. **52**(11), 11477–11490 (2021)
15. Yu, W., Huang, D., Wang, Q., Cai, L.: Distributed event-triggered iterative learning control for multiple high-speed trains with switching topologies: a data-driven approach. IEEE Trans. Intell. Transp. Syst. **24**(10), 10818–10829 (2023)
16. Zhu, Y., Wang, Z., Liang, H., Ahn, C.K.: Neural-network-based predefined-time adaptive consensus in nonlinear multi-agent systems with switching topologies. IEEE Trans. Neural Networks Learn. Syst. **35**(7), 9995–10005 (2023)
17. Li, X., Long, L.: Distributed event-triggered fuzzy control of heterogeneous switched multi-agent systems under switching topologies. IEEE Trans. Fuzzy Syst. **32**(2), 574–585 (2023)
18. Li, S., Chen, Y., Zhan, J.: Event-triggered consensus control and fault estimation for time-delayed multi-agent systems with Markov switching topologies. Neurocomputing **460**, 292–308 (2021)
19. Wang, P., Wen, G., Huang, T., Yu, W., Lv, Y.: Asymptotical neuro-adaptive consensus of multi-agent systems with a high dimensional leader and directed switching topology. IEEE Trans. Neural Networks Learn. Syst. (2022)
20. Tanner, H.G., Jadbabaie, A., Pappas, G.J.: Stable flocking of mobile agents. Part I: dynamic topology. In: Proceedings of the 42nd IEEE International Conference Decision and Control (CDC), vol. 2, pp. 2016–2021 (2003)
21. Liu, Y., Xiao, Y., Li, T.: Rice grain effect and impacts of message size and error on networked control systems. IEEE Syst. J. (2025). Accepted, 2025

22. Wang, N., Sun, Z., Yin, J., Zou, Z., Su, S.F.: Fuzzy unknown observer-based robust adaptive path following control of underactuated surface vehicles subject to multiple unknowns. Ocean Eng. **176**, 57–64 (2019)
23. Abdelaal, M., Fränzle, M., Hahn, A.: Nonlinear model predictive control for trajectory tracking and collision avoidance of underactuated vessels with disturbances. Ocean Eng. **160**, 168–180 (2018)
24. Qu, Y., Zhao, W., Yu, Z., Xiao, B.: Distributed prescribed performance containment control for unmanned surface vehicles based on disturbance observer. ISA Trans. **125**, 699–706 (2022)
25. Lawton, J.R.T., Beard, R.W., Young, B.J.: A decentralized approach to formation maneuvers. IEEE Trans. Robot. Autom. **19**(6), 933–941 (2003)
26. Liu, Y., Xiao, Y., Li, T.: Building a bridge between control and communication via topologies. IEEE Trans. Cybern. 1–14 (2025)
27. Wang, D., Fang, X., Wan, Y., Zhou, J., Wen, G.: Distributed optimization algorithms for MASs with network attacks: from continuous-time to event-triggered communication. IEEE Trans. Network Sci. Eng. **9**(5), 3332–3344 (2022)
28. Liu, S., Yu, G., Wen, D., Chen, X., Bennis, M., Chen, H.: Communication and energy efficient decentralized learning over D2D networks. IEEE Trans. Wireless Commun. **22**(12), 9549–9563 (2023)
29. Xiao, Y., Rosdahl, J.: Throughput and delay limits of IEEE 802.11. IEEE Commun. Lett. **6**(8), 355–357 (2002)

Neural Network Structure-Based Adaptive SMC for Pneumatic Artificial Muscle Systems with State Constraints and Input Dead Zones

Jiaxi Pei[1], Ming Li[2(✉)], and Menghua Zhang[1(✉)]

[1] School of Electrical Engineering, University of Jinan, Jinan, China
202321200984@stu.ujn.edu.cn, zhangmenghua@mail.sdu.edu.cn
[2] Shandong Inspur Database Technology Co., Ltd, Jinan, China
liming2017@inspur.com

Abstract. Pneumatic artificial muscle (PAM) systems demonstrate superior compliance and practicality when driving robotic exoskeletons. However, filling them with highly compressed gas makes PAMs susceptible to external disturbances and system uncertainties, which may degrade state response and increase control difficulty. Furthermore, most existing controllers require linearization operations. To address this, this paper proposes a state-constrained adaptive neural network controller that overcomes the deadband obstacle within a finite time interval, achieving satisfactory motion control that precisely tracks the operational trajectory of pneumatic artificial muscles. Simultaneously, it ensures the output state converges within a finite time without relying on parameter design. Rigorous stability analysis is provided, and the tracking performance of this method is validated through simulation.

Keywords: Pneumatic Artificial Muscle Systems · Adaptive Neural Network Control · State Constraints

1 Introduction

With the advancement of computer and artificial intelligence technologies, humanoid robots have become widely integrated into society and daily life, undertaking demanding tasks such as production, inspection, and transportation [1–5]. Among these, pneumatic artificial muscles constructed using lighter, more efficient soft actuators have garnered significant attention from researchers as a novel solution. Soft actuators overcome the limitations of rigid robots and find applications in fields requiring close interaction with humans and environments, such as medical care, rehabilitation exoskeletons, and narrow-space inspection. Against the backdrop of flourishing soft robotics, pneumatic artificial muscles (PAMs) stand out among soft actuators due to their stable performance, ultra-lightweight design, extended service life, and excellent compliance, offering broad application prospects.

C. Li et al. (Eds.): ICNC 2025, CCIS 2946, pp. 132–143, 2026.
https://doi.org/10.1007/978-981-92-1599-7_12

Due to the low degrees of freedom and simple mechanical structure of single PAM-driven robots, many researchers initially focused on the motion control of single PAM-driven robots before gradually expanding to multiple PAM-driven robots [6–8]. Specifically, for the humanoid PAM arm robot proposed by Yang et al., an adaptive control method with self-tuning parameters was developed [6] to achieve position tracking control under different objectives. [7] designed a 3-degree-of-freedom PMA ankle joint rehabilitation robot to assist patients in performing training movements. An adaptive robust controller based on a linearized model was proposed to address a series of issues caused by environmental changes [8].

However, every coin has two sides. While PAM possesses multiple unique advantages, it also exposes numerous control challenges stemming from its physical structure and characteristics, such as dead zones, high nonlinearity, and parameter uncertainty [9, 10]. To address these issues, this paper proposes an adaptive neural network control method. Within a finite time, it simultaneously resolves problems like state constraints and input dead zones while maintaining stable controller operation. It achieves satisfactory tracking control by constraining state variables within reasonable bounds. Based on the Lyapunov method, a detailed stability analysis is provided, proving the consistent boundedness of the tracking error around the equilibrium point. Specifically, the main contributions of this paper can be summarized as follows:

For the series-connected dual pneumatic artificial muscle system, this paper proposes an adaptive neural network control method. In the second part, considering the influence of input dead zones, the final model of the system is derived based on the dual pneumatic artificial muscle system. The third part designs a corresponding controller for the dual pneumatic artificial muscle system. The fourth part performs stability analysis on the proposed controller using the Lyapunov method. The fifth part conducts simulation verification through Simulink.

2 Problem Formulation

The dual-pneumatic artificial muscle model is shown in Figure 1. Its initial length are $BD = L_1$, $DF = L_2$. The robot is initially suspended vertically on a horizontal plane. The spring, muscle, and arm bone share the same length, meaning the initial lengths of the two springs are $AC = OO_1 = L_1$, $CE = O_1O_2 = L_2$. The distance between the arm bone and the PAM (or spring) satisfies $AO = OB = CO_1 = O_1D = EO_2 = O_2F = r$, with parameters and variables detailed in Table 1. Define the auxiliary angles and auxiliary lengths as $\alpha_1 = \angle AOC$, $\alpha_2 = \angle CO_1E$, $L_{x1} = OC$, $L_{x2} = O_1E$, where the auxiliary lengths $L_{x1} = \sqrt{r^2 + L_1^2}$, $L_{x2} = \sqrt{r^2 + L_2^2}$.

The upper arm/forearm muscle length and spring length can be denoted as $L_{s1}, L_{s2}, L_{p1}, L_{p2}$.

$$\begin{aligned} L_{s1} &= \sqrt{r^2 + L_{x1}^2 - 2rL_{x1}\cos(\phi_1 + \alpha_1)}, L_{s2} = \sqrt{r^2 + L_{x2}^2 - 2rL_{x2}\cos(\phi_2 + \alpha_2)} \\ L_{p1} &= \sqrt{r^2 + L_{x1}^2 - 2rL_{x1}\cos(\alpha_1 - \phi_1)}, L_{p2} = \sqrt{r^2 + L_{x2}^2 - 2rL_{x2}\cos(\alpha_2 - \phi_2)} \end{aligned} \tag{1}$$

Table 1. Parameters and variables.

Parameters	Example	Font size and style
p_1, p_2	Input air pressure of state variables	bar
L_1, L_2	Initial length of state variables	m
m_1, m_2	Upper arm/forearm muscle mass	kg
I_1, I_2	Moment of inertia matrix for upper Arm/forearm muscles	kg.m^2
F_{P1}, F_{P2}	Contraction force of the upper arm/forearm muscles	N
L_{p1}, L_{p2}	Upper arm and forearm muscle length	m
L_{s1}, L_{s2}	Upper arm and forearm spring length	m
ϕ_1, ϕ_2	Deflection angle	rad
r	Distance between muscle/spring and arm bone	m
p_0	Atmospheric pressure at standard conditions	bar
g	Gravitational acceleration	m/s^2

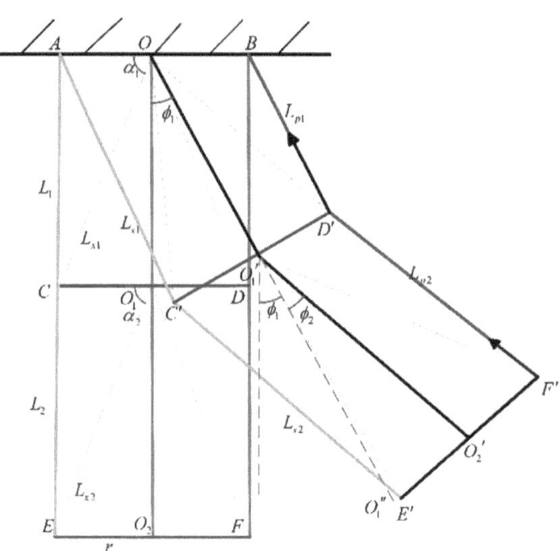

Fig. 1. The dual-pneumatic artificial muscle model.

This paper builds upon the dual PAM kinetic model presented in [11], reformulating the kinetic model as follows:

$$\begin{aligned} M_{11}\ddot{L}_{p1} + M_{12}\ddot{L}_{p2} + \left(C_{11} + b_{p3}\right)\dot{L}_{p1} \\ + C_{12}\dot{L}_{p2} + G_1 + b_{p1} + b_{p4}L_{p1} - b_{p5}L_{p1}^2 + W_1p_1 = 0 \end{aligned} \tag{2}$$

$$\begin{aligned} M_{21}\ddot{L}_{p1} + M_{22}\ddot{L}_{p2} + C_{21}\dot{L}_{p1} \\ + \left(C_{22} + b_{p6}\right)\dot{L}_{p2} + G_2 + b_{p2} + b_{p7}L_{p2} \\ - b_{p8}L_{p2}^2 + W_2p_2 = 0 \end{aligned} \tag{3}$$

where L_{p1}, L_{p2} represent system state variables, and p_1, p_2 constitute the system control input components. Considering the impact of dead zones on the system, p_1, p_2 can be expressed as:

$$p_1 = p_0 + \Delta p, p_2 = p_0 + \Delta p \tag{4}$$

where P_0 is the known nominal value, $\triangle P$ is the unknown portion. The above formula yields the following result:

$$M_s(\mathrm{q})\ddot{\mathrm{q}} + C_s(\mathrm{q}, \dot{\mathrm{q}})\dot{\mathrm{q}} + G_s(\mathrm{q}) = \mathrm{U} \tag{5}$$

where

$$\begin{gathered} M_s(\mathrm{q}) = \begin{bmatrix} M_{11} & M_{12} \\ M_{21} & M_{22} \end{bmatrix}, C_s(\mathrm{q}, \dot{\mathrm{q}}) = \begin{bmatrix} C_{11} + b_{p3} & C_{12} \\ C_{21} & C_{22} + b_{p6} \end{bmatrix}, \mathrm{q} = \begin{bmatrix} q_1 \\ q_2 \end{bmatrix} = \begin{bmatrix} L_{p1} \\ L_{p2} \end{bmatrix} \\ G_s(\mathrm{q}) = \begin{bmatrix} G_1 + b_{p1} + b_{p4}L_{p1} - b_{p5}L_{p1}^2 \\ G_2 + b_{p2} + b_{p7}L_{p2} - b_{p8}L_{p2}^2 \end{bmatrix}, \mathrm{U} = \begin{bmatrix} (-\lambda_{S1} - \lambda_{C1})p_1 \\ (-\lambda_{S2} - \lambda_{C2})p_2 \end{bmatrix} \end{gathered} \tag{6}$$

Transpose and simplify to:

$$\ddot{\mathrm{q}} = M_s^{-1}\mathrm{U} + M_s^{-1}(-C_s\dot{\mathrm{q}} - G_s) = M_s^{-1}(\mathrm{U} - C_s\dot{\mathrm{q}} - G_s) \tag{7}$$

Treat M_s^{-1} and $C_s\dot{\mathrm{q}} + G_s$ in as unknowns in the shipboard crane dynamics, where M_{s0}^{-1} is the nominal value of M_s^{-1}, $M_s^{-1} = M_{s0}^{-1} + \Delta M_s^{-1}$. Therefore, the dynamic equation in above (7) can be rewritten as follows:

$$\begin{aligned} \ddot{\mathrm{q}} &= M_{s0}^{-1}\mathrm{U} + \Delta M_s^{-1}\mathrm{U} - M_s^{-1}(C_s\dot{\mathrm{q}} + G_s) \\ &= M_{s0}^{-1}\mathrm{U} + \mathrm{B} \end{aligned} \tag{8}$$

Based on the above analysis, an adaptive neural network controller is proposed for a dual-pneumatic artificial muscle system with input constraints. This controller enables the control objective to accurately track the reference trajectory while constraining the tracking error within a specified range. The specific control objectives are as follows:

- Accurately track the reference trajectory

$$\lim_{t\to\infty} [\mathrm{x}_1(t) - \mathrm{x}_{1d}(t)] = 0 \tag{9}$$

- Constrain tracking error within a specified range $[0, \varpi_e)$, i.e.,

$$\|\delta_1(t)\| \in [0, \varpi_e), \forall t \geq 0 \tag{10}$$

3 Controller Design

Design an adaptive neural network controller to accurately track the reference trajectory and achieve the control objectives in the implementation.

To approximate the unknown parameters/structure in B using an adaptive neural network, B can be approximated via the following neural network structure:

$$\boldsymbol{B} = \left[W_1^{\mathrm{T}}\sigma\left(V_1^{\mathrm{T}}z\right) + \ni_1, W_2^{\mathrm{T}}\sigma\left(V_2^{\mathrm{T}}z\right) + \ni_2\right]^{\mathrm{T}}$$

where, $z = \left[q_1, q_2, \dot{q}_1, \dot{q}_2, 1\right]^{\mathrm{T}}$, $\sigma(\cdot)$ is the activation function, V_1, V_2 are bounded input weight matrices, W_1, W_2 are bounded output weight vectors, $\ni_1$, $\ni_2$are bounded approximation errors $\bar{\ni}_1$, $\bar{\ni}_2$, and it will not be influenced by the state variables during the entire analysis process [12].

Positioning error of PAMs introduced into the upper arm/forearm:

$$e_{p1} = q_1 - q_{1d}, e_{p2} = q_2 - q_{2d} \tag{12}$$

Additionally, the sliding surface $\psi = [\psi_1\ \psi_2]^{\mathrm{T}}$ associated with e_{p1}, e_{p2} is constructed as follows:

$$\begin{aligned} \psi_1 &= k_1 e_{p1} + \dot{e}_{p1}, \psi_2 = k_2 e_{p2} + \dot{e}_{p2} \\ \Rightarrow \dot{\psi}_1 &= k_1 \dot{e}_{p1} + \ddot{e}_{p1} = k_1 \dot{e}_{p1} + \ddot{q}_1 - \ddot{q}_{1d} \\ \dot{\psi}_2 &= k_2 \dot{e}_{p2} + \ddot{e}_{p2} = k_2 \dot{e}_{p2} + \ddot{q}_2 - \ddot{q}_{2d} \end{aligned} \tag{13}$$

where k_1, k_2 are positive parameters. By differentiating with respect to time, the following result is obtained:

$$\begin{bmatrix} \dot{\psi}_1 \\ \dot{\psi}_2 \end{bmatrix} = M_{s0}^{-1}\begin{bmatrix} u_1 \\ u_2 \end{bmatrix} + \begin{bmatrix} W_1^{\mathrm{T}}\sigma\left(V_1^{\mathrm{T}}z\right) + \ni_1 \\ W_2^{\mathrm{T}}\sigma\left(V_2^{\mathrm{T}}z\right) + \ni_2 \end{bmatrix} + \begin{bmatrix} k_1\dot{e}_{p1} \\ k_2\dot{e}_{p2} \end{bmatrix} - \begin{bmatrix} \ddot{q}_{1d} \\ \ddot{q}_{2d} \end{bmatrix}$$

To limit the range of state variables $q_1(t)$, $q_2(t)$, construct two auxiliary variables.

$$A(q_a) = \sum_{i=1}^{2} \frac{k_{ai}}{2} \log\left(\frac{\varpi_i^2}{\varpi_i^2 - q_i^2}\right) = \sum_{i=1}^{2} k_{ai} \int_{q_i(0)}^{q_i(t)} \frac{q_i}{\varpi_i^2 - q_i^2} d[q_i] \tag{15}$$

where k_{ai} is a positive parameter. At the start of the operation, select $q_i(0)$ such that $q_i(0) \in (-\varpi_i, \varpi_i)$.

Therefore, we designed the following control law based on nonlinear neural networks:

$$\begin{bmatrix} u_1 \\ u_2 \end{bmatrix} = M_{s0}\begin{bmatrix} -\hat{W}_1^{\mathrm{T}}\sigma\left(\hat{V}_1^{\mathrm{T}}z\right) - k_1\dot{e}_{p1} + \ddot{q}_{1d} - k_{p1}\psi_1 - A(q_1)\psi_1 - k_{v1}\operatorname{sign}(\psi_1) \\ -\hat{W}_2^{\mathrm{T}}\sigma\left(\hat{V}_2^{\mathrm{T}}z\right) - k_2\dot{e}_{p2} + \ddot{q}_{2d} - k_{p2}\psi_2 - A(q_2)\psi_2 - k_{v2}\operatorname{sign}(\psi_2) \end{bmatrix} \tag{16}$$

where, $\hat{W}_1$, $\hat{W}_2$, $\hat{V}_1$, $\hat{V}_2$ denote the estimates of W_1, W_2, V_1, V_2 respectively, with k_{p1}, k_{p2}, k_{v1}, k_{v2} being the positive gains to be determined. Based on the subsequent Lyapunov stability analysis, the following conditions must theoretically be satisfied:

$$k_{v1} > \max\{\overline{\Lambda}_1, \overline{\Omega}_1\}, k_{v2} > \max\{\overline{\Lambda}_2, \overline{\Omega}_2\} \tag{17}$$

where $\overline{\Lambda}_1, \overline{\Omega}_1, \overline{\Lambda}_2, \overline{\Omega}_2$ will be defined subsequently. Substituting (16) into (14) yields

$$\dot{\psi}_1 = W_1^{\mathrm{T}}\sigma\left(V_1^{\mathrm{T}}z\right) - \hat{W}_1^{\mathrm{T}}\sigma\left(\hat{V}_1^{\mathrm{T}}z\right) + \ni_1 - k_{p1}\psi_1 - k_{v1}\mathrm{sign}(\psi_1)$$
$$\dot{\psi}_2 = W_2^{\mathrm{T}}\sigma\left(V_2^{\mathrm{T}}z\right) - \hat{W}_2^{\mathrm{T}}\sigma\left(\hat{V}_2^{\mathrm{T}}z\right) + \ni_2 - k_{p2}\psi_2 - k_{v2}\mathrm{sign}(\psi_2)$$

Definition of estimation error

$$\tilde{W}_1 = W_1 - \hat{W}_1, \tilde{W}_2 = W_2 - \hat{W}_2, \tilde{V}_1 = V_1 - \hat{V}_1, \tilde{V}_2 = V_2 - \hat{V}_2 \tag{19}$$

To facilitate subsequent stability analysis, we define the following compact set B_w:

$$B_w = \left\{\tilde{W}_1 \in <^N, \tilde{W}_2 \in <^N : \|\tilde{W}_1\| \le \overline{\omega}_1, \|\tilde{W}_2\| \le \overline{\omega}_2\right\} \tag{20}$$

where N is the number of neurons, and $\overline{\omega}_1$, $\overline{\omega}_2$ are the radii of ellipsoid B_w. Furthermore, after some rigorous mathematical calculations, we can rearrange the first three terms of $\dot{\psi}_1$, $\dot{\psi}_2$ into the following form [12]:

$$\begin{aligned}\Omega_1 &= W_1^{\mathrm{T}}\sigma\left(V_1^{\mathrm{T}}z\right) + \ni_1 - \hat{W}_1^{\mathrm{T}}\sigma\left(\hat{V}_1^{\mathrm{T}}z\right)\\ &= \tilde{W}_1^{\mathrm{T}}\sigma\left(\hat{V}_1^{\mathrm{T}}z\right) - \tilde{W}_1^{\mathrm{T}}\dot{\sigma}\left(\hat{V}_1^{\mathrm{T}}z\right)\hat{V}_1^{\mathrm{T}}z + \hat{W}_1^{\mathrm{T}}\dot{\sigma}\left(\hat{V}_1^{\mathrm{T}}z\right)\tilde{V}_1^{\mathrm{T}}z + \Lambda_1\\ \Omega_2 &= W_2^{\mathrm{T}}\sigma\left(V_2^{\mathrm{T}}z\right) + \ni_2 - \hat{W}_2^{\mathrm{T}}\sigma\left(\hat{V}_2^{\mathrm{T}}z\right)\\ &= \tilde{W}_2^{\mathrm{T}}\sigma\left(\hat{V}_2^{\mathrm{T}}z\right) - \tilde{W}_2^{\mathrm{T}}\dot{\sigma}\left(\hat{V}_2^{\mathrm{T}}z\right)\hat{V}_1^{\mathrm{T}}z + \hat{W}_2^{\mathrm{T}}\dot{\sigma}\left(\hat{V}_2^{\mathrm{T}}z\right)\tilde{V}_2^{\mathrm{T}}z + \Lambda_2\end{aligned}$$

where,

$$\Lambda_1 = \tilde{W}_1^{\mathrm{T}}\dot{\sigma}\left(\hat{V}_1^{\mathrm{T}}z\right)V_1^{\mathrm{T}}z + W_1^{\mathrm{T}}O\left(\tilde{V}_1^{\mathrm{T}}z\right)^2 + \ni_1$$
$$\Lambda_2 = \tilde{W}_2^{\mathrm{T}}\dot{\sigma}\left(\hat{V}_2^{\mathrm{T}}z\right)V_2^{\mathrm{T}}z + W_2^{\mathrm{T}}O\left(\tilde{V}_2^{\mathrm{T}}z\right)^2 + \ni_2$$

Furthermore, for a given compact set B_w, the upper bounds of functions Λ_1, Λ_2 are $\Lambda_1 \le \overline{\Lambda}_1$, $\Lambda_2 \le \overline{\Lambda}_2$.

$$\begin{aligned}\hat{W}_1 &= \beta_1\alpha\psi_1\Pi_1\left[\sigma\left(\hat{V}_1^T z\right) - \dot{\sigma}\left(\hat{V}_1^T z\right)\hat{V}_1^T z\right]\\ \hat{W}_2 &= \beta_2\alpha\psi_2\Pi_2\left[\sigma\left(\hat{V}_2^T z\right) - \dot{\sigma}\left(\hat{V}_2^T z\right)\hat{V}_2^T z\right]\\ \hat{V}_1 &= \gamma_1\alpha\psi_1\Gamma_1 z\hat{W}_1^T\dot{\sigma}\left(\hat{V}_1^T z\right)\\ \hat{V}_2 &= \gamma_2\alpha\psi_2\Gamma_2 z\hat{W}_2^T\dot{\sigma}\left(\hat{V}_2^T z\right)\end{aligned} \tag{23}$$

where, $\sigma(\rho) = \frac{1}{1-e^{-\rho}}$, $\alpha, \beta_1, \beta_2, \gamma_1, \gamma_2$ are positive parameters, Γ_1, Γ_2, Π_1, Π_2 are positive diagonal parameters of matrix vectors.

4 Stability Analysis

Constructing Lyapunov candidate functions $H_1(t)$:

$$
\begin{aligned}
H_1 = \tfrac{\alpha}{2}\psi^{\mathrm{T}}\psi + \tfrac{1}{2\gamma_1}Tr\left(\tilde{V}_1^{\mathrm{T}}\Gamma_1^{-1}\tilde{V}_1\right) + \tfrac{1}{2\gamma_2}Tr\left(\tilde{V}_2^{\mathrm{T}}\Gamma_2^{-1}\tilde{V}_2\right) \\
+ \tfrac{1}{2\beta_1}\tilde{W}_1^{\mathrm{T}}\prod\nolimits_1^{-1}\tilde{W}_1 + \tfrac{1}{2\beta_2}\tilde{W}_2^{\mathrm{T}}\prod\nolimits_2^{-1}\tilde{W}_2
\end{aligned}
\tag{24}
$$

Taking the derivative with respect to time yields

$$
\begin{aligned}
\dot{H}_1 = \alpha\psi^{\mathrm{T}}\dot{\psi} + \frac{1}{\gamma_1}\mathrm{Tr}\left(\tilde{V}_1^{\mathrm{T}}\Gamma_1^{-1}\dot{\tilde{V}}_1\right) + \frac{1}{\gamma_2}\mathrm{Tr}\left(\tilde{V}_2^{\mathrm{T}}\Gamma_2^{-1}\dot{\tilde{V}}_2\right) \\
+\frac{1}{\beta_1}\tilde{W}_1^{\mathrm{T}}\prod\nolimits_1^{-1}\dot{\tilde{W}}_1 + \frac{1}{\beta_2}\tilde{W}_2^{\mathrm{T}}\prod\nolimits_2^{-1}\dot{\tilde{W}}_2
\end{aligned}
\tag{25}
$$

Based on (14) and (16), we obtain

$$
\begin{aligned}
\dot{H}_1 &= \alpha\psi_1\left[W_1^{\mathrm{T}}\sigma\left(V_1^{\mathrm{T}}z\right) + \ni_1 - \hat{W}_1^{\mathrm{T}}\sigma\left(\hat{V}_1^{\mathrm{T}}z\right) - k_{p1}\psi_1 - A(q_1)\psi_1 - k_{v1}\mathrm{sign}(\psi_1)\right] \\
&\quad + \alpha\psi_2\left[W_2^{\mathrm{T}}\sigma\left(V_2^{\mathrm{T}}z\right) + \ni_2 - \hat{W}_2^{\mathrm{T}}\sigma\left(\hat{V}_2^{\mathrm{T}}z\right) - k_{p2}\psi_2 - A(q_2)\psi_2 - k_{v2}\mathrm{sign}(\psi_2)\right] \\
&\quad + \frac{1}{\gamma_1}\mathrm{Tr}\left(\tilde{V}_1^{\mathrm{T}}\Gamma_1^{-1}\dot{\tilde{V}}_1\right) + \frac{1}{\gamma_2}\mathrm{Tr}\left(\tilde{V}_2^{\mathrm{T}}\Gamma_2^{-1}\dot{\tilde{V}}_2\right) + \frac{1}{\beta_1}\tilde{W}_1^{\mathrm{T}}\Pi_1^{-1}\dot{\tilde{W}}_1 + \frac{1}{\beta_2}\tilde{W}_2^{\mathrm{T}}\Pi_2^{-1}\dot{\tilde{W}}_2 \\
&= \alpha\psi_1\left[\tilde{W}_1^{\mathrm{T}}\sigma\left(\hat{V}_1^{\mathrm{T}}z\right) - \tilde{W}_1^{\mathrm{T}}\dot{\sigma}\left(\hat{V}_1^{\mathrm{T}}z\right)\hat{V}_1^{\mathrm{T}}z + \hat{W}_1^{\mathrm{T}}\dot{\sigma}\left(\hat{V}_1^{\mathrm{T}}z\right)\tilde{V}_1^{\mathrm{T}}z\right] + \alpha\psi_1\Lambda_1 - \alpha k_{p1}\psi_1^2 \\
&\quad - \frac{1}{\gamma_1}\mathrm{Tr}\left[\tilde{V}_1^{\mathrm{T}}\Gamma_1^{-1}\gamma_1\alpha\psi_1\Gamma_1 z\hat{W}_1^{T}\dot{\sigma}\left(\hat{V}_1^{T}z\right)\right] - \frac{1}{\gamma_2}\mathrm{Tr}\left[\tilde{V}_2^{\mathrm{T}}\Gamma_2^{-1}\gamma_2\alpha\psi_2\Gamma_2 z\hat{W}_2^{T}\dot{\sigma}\left(\hat{V}_2^{T}z\right)\right] \\
&\quad + \alpha\psi_2\left[\tilde{W}_2^{\mathrm{T}}\sigma\left(\hat{V}_2^{\mathrm{T}}z\right) - \tilde{W}_2^{\mathrm{T}}\dot{\sigma}\left(\hat{V}_2^{\mathrm{T}}z\right)\hat{V}_1^{\mathrm{T}}z + \hat{W}_2^{\mathrm{T}}\dot{\sigma}\left(\hat{V}_2^{\mathrm{T}}z\right)\tilde{V}_2^{\mathrm{T}}z\right] + \alpha\psi_2\Lambda_2 - \alpha k_{p2}\psi_2^2 \\
&\quad - \frac{1}{\beta_1}\tilde{W}_1^{\mathrm{T}}\Pi_1^{-1}\left[\beta_1\alpha\psi_1\Pi_1\left[\sigma\left(\hat{V}_1^{T}z\right) - \dot{\sigma}\left(\hat{V}_1^{T}z\right)\hat{V}_1^{T}z\right]\right] - \alpha\psi_1 A(q_1)\psi_1 \\
&\quad - \frac{1}{\beta_2}\tilde{W}_2^{\mathrm{T}}\Pi_2^{-1}\left[\beta_2\alpha\psi_2\Pi_2\left[\sigma\left(\hat{V}_2^{T}z\right) - \dot{\sigma}\left(\hat{V}_2^{T}z\right)\hat{V}_2^{T}z\right]\right] - \alpha\psi_1 k_{v1}\mathrm{sign}(\psi_1) \\
&\quad - \alpha\psi_2 A(q_2)\psi_2 - \alpha\psi_2 k_{v2}\mathrm{sign}(\psi_2) \\
&= \alpha\psi_1\Lambda_1 - \alpha k_{p1}\psi_1^2 - \alpha A(q_1)\psi_1^2 - \alpha k_{v1}|\psi_1| + \alpha\psi_2\Lambda_2 - \alpha k_{p2}\psi_2^2 \\
&\quad - \alpha A(q_2)\psi_2^2 - \alpha k_{v2}|\psi_2|
\end{aligned}
\tag{26}
$$

Given that $\Lambda_1 \le \overline{\Lambda}_1$, $\Lambda_2 \le \overline{\Lambda}_2$, it can be observed that

$$
\begin{aligned}
\dot{H}_2 &= \alpha\psi_1\Lambda_1 - \alpha k_{p1}\psi_1^2 - \alpha A(q_1)\psi_1^2 - \alpha k_{v1}|\psi_1| + \alpha\psi_2\Lambda_2 - \alpha k_{p2}\psi_2^2 \\
&\qquad - \alpha A(q_2)\psi_2^2 - \alpha k_{v2}|\psi_2| \\
&\le -\alpha\left(k_{v1} - \overline{\Lambda}_1\right)|\psi_1| - \alpha\left(k_{v2} - \overline{\Lambda}_2\right)|\psi_2| - \alpha k_{p1}\psi_1^2 - \alpha A(q_1)\psi_1^2 \\
&\qquad - \alpha k_{p2}\psi_2^2 - \alpha A(q_2)\psi_2^2
\end{aligned}
\tag{27}
$$

Therefore, when the initial value of the weight estimation error is set within the compact set B_w, and the control gains k_{v1}, k_{v2} satisfy the conditions in (17), we can obtain:

$$\begin{aligned} \dot{H}_1 \leq 0 \Rightarrow & H_1 \in \mathrm{L}_\infty \\ & \Rightarrow \tilde{V}_1,\ \tilde{V}_2,\ \tilde{W}_1,\ \tilde{W}_2,\ u_1,\ u_2 \in \mathrm{L}_\infty \\ & \Rightarrow \lim_{t\to\infty} \psi = \left[\, \psi_1\ \psi_2 \,\right]^{\mathrm{T}} = 0 \end{aligned} \tag{28}$$

The results demonstrate that the state-space equilibrium point of the dual pneumatic artificial muscle system is asymptotically stable. Furthermore, $\tilde{V}_1, \tilde{V}_2, \tilde{W}_1, \tilde{W}_2$ remain bounded for $\forall t \geq 0$ and $\|\tilde{W}_1\|, \|\tilde{W}_2\|$ can be confined within the respective domains of $\overline{\omega}_1, \overline{\omega}_2$.

Moreover, as seen in (28)

$$\dot{H}_1 \leq -\alpha A(q_1)\psi_1^2 - \alpha A(q_2)\psi_2^2 \tag{29}$$

Integrating the above equation yields

$$\begin{aligned} & \textstyle\int_{t_1}^{t_2} \dot{H}_1(t)dt \leq -\alpha \int_{t_1}^{t_2} A(q_1(t))\psi_1^2 - \alpha \int_{t_1}^{t_2} A(q_2(t))\psi_2^2 \\ & \Rightarrow H_1(t_1) - H_1(t_2) \geq \alpha \textstyle\int_{t_1}^{t_2} A(q_1(t))\psi_1^2 + \alpha \int_{t_1}^{t_2} A(q_2(t))\psi_2^2 \end{aligned} \tag{30}$$

Therefore, the existence time $t^* \in (t_1, t_2)$ satisfies

$$H_1(t_1) \geq \alpha\left[A\big(q_1(t^*)\big)\int_{t_1}^{t_2} \psi_1^2(t)d(t) + A\big(q_2(t^*)\big)\int_{t_1}^{t_2} \psi_2^2(t)d(t)\right] \tag{31}$$

Once the state variable in $q_i(t^*)$ approaches the permissible limit, $A(q_i(t^*))$ will become infinite, i.e., $H_2(t_1) \in \mathrm{L}_\infty$, and only will exist

$$\begin{aligned} \textstyle\int_{t_1}^{t_2} \psi^{\mathrm{T}}\psi d(t) = \int_{t_1}^{t_2} \psi_1^2(t)d(t) + \int_{t_1}^{t_2} \psi_2^2(t)d(t) = 0 \\ \Rightarrow \psi_1^2(t^*) + \psi_2^2(t^*) = 0 \\ \Rightarrow e_{p_i}(t^*) = \dot{e}_{p_i}(t^*) = 0 \end{aligned} \tag{32}$$

That is, the state variables reach their desired values at time $t^* \in (t_1, t_2)$, and $q_i(t^*)$ does not exceed the permissible range. Therefore, the state variables can be constrained within a given range.

$$q_i(t) \in (-\varpi_i, \varpi_i) \tag{33}$$

However, based on the above analysis, it is difficult to guarantee that the sliding manifold converges to zero within a finite time. Therefore, we further propose the following candidate Lyapunov function $H_2(t)$:

$$H_2 = \frac{\alpha}{2}\psi^{\mathrm{T}}\psi \tag{34}$$

Based on the results of (28), it can be concluded that $\tilde{V}_1, \tilde{V}_2, \hat{V}_1, \hat{V}_2$ in (21) are bounded, indicating that

$$\Omega_1 = W_1^{\mathrm{T}}\sigma\left(V_1^{\mathrm{T}}z\right) + \ni_1 - \hat{W}_1^{\mathrm{T}}\sigma\left(\hat{V}_1^{\mathrm{T}}z\right) \le \overline{\Omega}_1$$
$$\Omega_2 = W_2^{\mathrm{T}}\sigma\left(V_2^{\mathrm{T}}z\right) + \ni_2 - \hat{W}_2^{\mathrm{T}}\sigma\left(\hat{V}_2^{\mathrm{T}}z\right) \le \overline{\Omega}_2$$

where $\overline{\Omega}_1$, $\overline{\Omega}_2$ are the respective upper bounds of Ω_1, Ω_2. Therefore, by differentiating (34), H_2 can be expressed as

$$\begin{aligned}
\dot{H}_2 &\le \alpha\psi_1\overline{\Omega}_1 + \alpha\psi_2\overline{\Omega}_2 - \alpha k_{p1}\psi_1^2 - \alpha A(q_1)\psi_1^2 - \alpha k_{v1}|\psi_1| - \alpha k_{p2}\psi_2^2 \\
&\quad - \alpha k_{v2}|\psi_2| - \alpha A(q_2)\psi_2^2 \\
&= -\alpha k_{p1}\psi_1^2 - \alpha A(q_1)\psi_1^2 - \alpha k_{p2}\psi_2^2 - \alpha A(q_2)\psi_2^2 \\
&\quad - \alpha\left(k_{v1} - \overline{\Omega}_1\right)|\psi_1| - \alpha\left(k_{v2} - \overline{\Omega}_2\right)|\psi_2| \\
&\le -\alpha\left(k_{v1} - \overline{\Omega}_1\right)|\psi_1| - \alpha\left(k_{v2} - \overline{\Omega}_2\right)|\psi_2| \\
&\le -k_{v\min}\|\psi\| \\
&\le 0
\end{aligned} \tag{36}$$

where, $k_{v\min} @ \min\left\{\alpha\left(k_{v1} - \overline{\Omega}_1\right), \alpha\left(k_{v2} - \overline{\Omega}_2\right)\right\}$, therefore we can draw the following conclusion:

$$\begin{aligned}
&\dot{H}_2 \le -\tfrac{k_{v\min}}{\alpha}\sqrt{2\alpha H_3} \\
&\Rightarrow \sqrt{H_2(t)} - \sqrt{H_2(0)} \le -\tfrac{k_{v\min}}{\sqrt{2\alpha}}t
\end{aligned} \tag{37}$$

Assume the state variable converges to $\psi = \left[\,\psi_1\ \psi_2\,\right]^{\mathrm{T}}$ at time t_f, i.e., $H_2(t_f) = 0$. Then, (37) can be rewritten as

$$t_f \le \frac{\sqrt{2\alpha}}{k_{v\min}}\left[\sqrt{H_2(0)} - \sqrt{H_2\left(t_f\right)}\right] \le \frac{\sqrt{2\alpha H_2(0)}}{k_{v\min}} \tag{38}$$

This implies that the convergence time t_f of sliding surface ψ is finite, and due to $\dot{H}_2 \le 0$, it is evident that

$$\begin{aligned}
&0 \le H_2(t) \le H_2\left(t_f\right) = 0\ \forall t \ge t_f \\
&\Rightarrow H_2(t) = 0\ \forall t \ge t_f \\
&\Rightarrow \psi = \left[\,\psi_1\ \psi_2\,\right]^{\mathrm{T}} = [0,\ 0]^{\mathrm{T}}\ \forall t \ge t_f
\end{aligned} \tag{39}$$

Therefore, the sliding surface ψ is the invariant set after t_f, that is,

$$\dot{\psi} = \left[\,\dot{\psi}_1\ \dot{\psi}_2\,\right]^{\mathrm{T}} = [0, 0]^{\mathrm{T}}\forall t \ge t_f \tag{40}$$

5 Simulation Verification

To validate the proposed method in this paper, simulation testing will be conducted using Simulink. A sinusoidal trajectory will be set as the desired path to observe the tracking performance of the proposed method, with results compared against the PD control method. The initial values of arm muscles' lengths are set as $L_{p10} = 0.8\ m$, $L_{p20} = 0.6\ m$.

The selected system parameters are as follows:

$$\begin{aligned}
&m_1 = 3.612, I_1 = 0.276, L_1 = 0.8, L_{X1} = 0.8062, \alpha_1 = 1.4711 \\
&m_2 = 1.762, I_2 = 0.024, L_2 = 0.6, L_{X2} = 0.6083, \alpha_2 = 1.4464 \\
&k_{s1} = 394.1176, k_{s2} = 379.0476, r = 0.1, \ \ p_0 = 101.325, g = 9.8
\end{aligned} \tag{41}$$

The pneumatic muscle parameters selected are:

$$\begin{aligned}
&\rho_{s0} = 241.1467, \rho_{s1} = -6.2333, \rho_{s2} = -17.4067, \rho_{s3} = -1.9667 \\
&\rho_{s4} = -67.2136, \rho_{s5} = -20.0176, \rho_{s6} = 42.8021, \rho_{s7} = -1.2189 \\
&f_{10} = 0.0038, f_{11} = 1166.9, f_{20} = 0.0012, f_{22} = 914.6791 \\
&d_{10} = -7.783, d_{11} = -0.031, d_{20} = -7.3208, d_{21} = -0.026
\end{aligned} \tag{42}$$

The control gain for the proposed controller is selected as follows:

$$\begin{aligned}
&k_1 = 4.6, k_2 = 12.2, k_{v1} = 15.8, k_{p1} = 13.8, k_{v2} = 0.6, k_{p2} = 7.2 \\
&k_{a1} = 3.9, k_{a2} = 6.6, \beta_1 = \beta_2 = 10, \gamma_1 = \gamma_2 = 1, \alpha = 0.001 \\
&\Gamma_1 = 9.04, \Gamma_2 = 8.84, \prod\nolimits_1 = 6.02, \prod\nolimits_2 = 1.04
\end{aligned} \tag{43}$$

The control gain of the comparison method is:

$$k_{p1} = 13.8, k_{p2} = 7.2, k_{d1} = 15.8, k_{d2} = 0.6 \tag{44}$$

The simulation results are shown in Figure 2. As can be seen from Figure 2, when tracking a sinusoidal signal, the proposed method enables the load displacement to converge rapidly and track the desired trajectory. It demonstrates faster and more precise tracking performance, with tracking errors significantly smaller than those observed in the experimental results of the comparison method. Furthermore, by applying the proposed control method, the system can effectively track various trajectories, exhibiting satisfactory transient and steady-state performance.

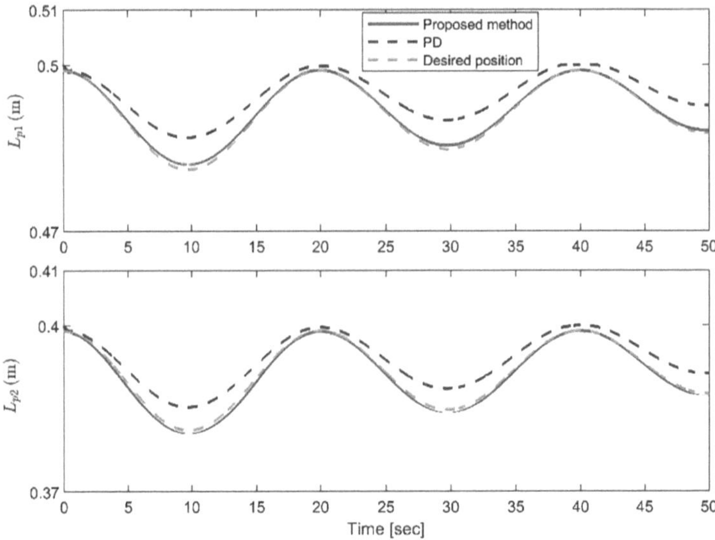

Fig. 2. Tracking results.

6 Conclusion

Based on the inherent characteristics of pneumatic artificial muscle systems, this paper proposes an adaptive neural network control method featuring state constraints and input dead zones. The effectiveness of the proposed controller is demonstrated using the Lyapunov method.

Acknowledgments. This work is supported in part by the National Natural Science Foundation of China under Grant No. 62273163, the Taishan Scholar Foundation of Shandong Province under Grant No. tsqn202312212, and the Outstanding Youth Foundation of Shandong Province under Grant No. ZR2023YQ056.

The authors have no competing interests to declare that are relevant to the content of this article.

References

1. Watanabe, M. Tadakuma, K. aTadokoro, S.: Hyperboloidal Pneumatic Artificial Muscle With Braided Straight Fibers. IEEE Robotics and Automation Letters **9**(5), 4242-4249 (2024)
2. Xie, D., Su, Y., Shi, X.S., Tong, F., Li, Z., Kai-yu Tong, R.: A Compact Elbow Exosuit Driven by Pneumatic Artificial Muscles. IEEE Robotics and Automation Letters. **9**(4), 3331–3338 (2024)
3. Tanaka, H., Hirai, H., Hosoda, K.: Position Control of McKibben-Type Pneumatic Artificial Muscle via Port-Hamiltonian Approach. IEEE Robotics and Automation Letters. **10**(6), 6384–6391 (2025)
4. Zhang, X., Sun, N., Liu, G., Yang, T., Yang, J.: Disturbance Preview-Based Output Feedback Predictive Control for Pneumatic Artificial Muscle Robot Systems With Hysteresis Compensation. IEEE/ASME Transactions on Mechatronics. **29**(5), 3936–3948 (2024)

5. Zhang, X., Sun, N., Liu, G., Yang, T., Fang, Y.: Hysteresis Compensation-Based Intelligent Control for Pneumatic Artificial Muscle-Driven Humanoid Robot Manipulators With Experiments Verification. IEEE Transactions on Automation Science and Engineering. **21**(3), 2538–2551 (2024)
6. Yang, H., Xiang, C., Hao, L., Zhao, L., Xue, B.: Research on PSAMFAC for a novel bionic elbow joint system actuated by pneumatic artificial muscles. Journal of Mechanical Science and Technology. **31**(7), 3519–3529 (2017)
7. Liu, Q., Zuo, J., Zhu, C., Meng, W., Ai, Q., Xie, S.Q.: Design and Hierarchical Force-Position Control of Redundant Pneumatic Muscles-Cable-Driven Ankle Rehabilitation Robot. IEEE Robot. Autom. Lett. **7**(1), 502–509 (2022)
8. Nuchkrua, T., Leephakpreeda, T.: Novel compliant control of a pneumatic artificial muscle driven by hydrogen pressure under a varying environment. IEEE Transactions on Industrial Electronics. **69**(7), 7120–7129 (2021)
9. Okui, M., Ito, F., Kojima, A., Nakamura, T.: Noninflatable pneumatic artificial muscle requiring low space and consumption flow rate. In: IEEE/SICE International Symposium on System Integration, pp. 495–499 (2020)
10. Sun, N., Liang, D., Wu, Y., Chen, Y., Qin, Y., Fang, Y.: Adaptive control for pneumatic artificial muscle systems with parametric uncertainties and unidirectional input constraints. IEEE Transactions on Industrial Informatics. **16**(2), 969–979 (2020)
11. Liang, D., Sun, N., Wu, Y., Chen, Y., Fang, Y., Liu, L.: Energy-Based Motion Control for Pneumatic Artificial Muscle Actuated Robots With Experiments. IEEE Transactions on Industrial Electronics. **69**(7), 7295–7306 (2022)
12. Yang, T., Sun, N., Chen, H., Fang, Y.: Neural Network-Based Adaptive Antiswing Control of an Underactuated Ship-Mounted Crane With Roll Motions and Input Dead Zones. IEEE Transactions on Neural Networks and Learning Systems. **31**(3), 901–914 (2020)

A Control Strategy for Multi-robot Systems Based on a Gradient-Based Differential kWTA Network

Ye Ruan[1(✉)] and Hanting Liu[2]

[1] Lanzhou University, Lanzhou 730106, China
320220947921@lzu.edu.cn
[2] Xi'an University of Technology, Xi'an 710048, China

Abstract. The k-winners-take-all (kWTA) mechanism is widely used to address multi-agent decision-making and coordination problems. Existing kWTA networks in multi-agent tracking often suffer from slow convergence and delay-induced errors. To address these issues, this paper proposes an improved gradient-based differential k-winners-take-all network, transforms the kWTA problem into a constrained dynamic quadratic programming problem and solves it using a neural network to accelerate convergence and eliminate lagging errors. Specifically, the network incorporates time-derivative information to eliminate lagging errors caused by dynamic inputs and designs a special Maxout activation function to further accelerate the network's convergence speed. Theoretical analyses and numerical experiments demonstrate that this method can effectively improve convergence performance and eliminate lagging errors. In addition, multi-agent tracking simulations demonstrate that integrating the GD-kWTA network with a consensus estimator effectively resolves challenges such as agent competition and communication constraints, validating the method's effectiveness and practicality in distributed systems.

Keywords: GD-kWTA network · Maxout activation function · multi-agent tracking · consensus estimator

1 Introduction

Winner-take-all (WTA) networks are a type of competitive neural network that select the most active unit from a set of inputs. As a generalization, k-winners-take-all (kWTA) networks select the top k units in the competition, allowing them to receive all or most of the benefits [1,2]. For example, in multi-robot cooperative control, the issue of resource allocation in distributed environments receives extensive attention [3,4]. Zhang *et al.* [5] introduce a distributed scheme based on game theory to optimize the manipulability of redundant manipulators and address the resource allocation problem. Subsequently, a logistic adaptive controller with an event-triggered mechanism is presented in [6] to reduce the

C. Li et al. (Eds.): ICNC 2025, CCIS 2946, pp. 144–155, 2026.
https://doi.org/10.1007/978-981-92-1599-7_13

resource overhead and enhance the collaborative efficiency of multi-robot systems. Further, a distributed optimal control method based on discrete neural dynamics is developed to address the problem of maintaining the end-effector's orientation, thereby enhancing the resource allocation and cooperative control capabilities of multi-robot systems in distributed environments [7]. Additionally, k-WTA networks also effectively address the resource allocation problem in distributed settings [8]. In complex environments, robustness is also crucial, and recent research continues to explore this area [9–11]. A hierarchical control and learning network, introduced through online physical parameter estimation and task decoupling, significantly improves system robustness [12]. In contrast, neural networks with a distributed competitive mechanism also exhibit robustness [13], with the kWTA network significantly enhancing the interference resistance and collaboration efficiency of multi-robot systems through its distributed competitive mechanism [14,15]. With their efficiency and adaptability, kWTA networks are well-suited for domains requiring complex decision-making and coordination.

In the distributed control-based competition and cooperation of multi-robot systems [16,17], the kWTA algorithm is widely used for its excellent performance in dynamic task allocation [8]. The algorithm achieves global convergence through a distributed approach, avoiding reliance on a central command, and is particularly suitable for communication-constrained network environments [2]. However, existing kWTA networks still face inherent limitations in practical applications. Specifically, the convergence speed of the network is slow, and time-delay problems are prone to occur when dealing with large-scale systems [18–20]. Therefore, we choose special activation functions [21–23] to accelerate the network's convergence speed, thereby solving the time-delay problem and fully unleashing its performance potential.

Based on the above analysis, this paper introduces the kWTA network into multi-robot tasks as the driving principle and proposes an improved gradient-based differential k-winners-take-all (GD-kWTA) network. The network integrates a special Maxout activation function and parameter optimization, which can accelerate network convergence speed and achieve better tracking performance. Specifically, this method selects k robots out of n that are closest in relative position to the target and designates them as winners. Winners follow the target's movement, while losers remain stationary until their action is needed. This efficient coordination mechanism optimizes system energy consumption and economic efficiency by reducing unnecessary robot movements [24–27]. Through the novel activation function, it significantly improves the network's convergence speed, decision accuracy, and tracking performance [28–30].

The main contributions of this paper are listed in the following. (1) An enhanced GD-kWTA network based on a gradient-based solving strategy is proposed. It utilizes time-derivative information to eliminate errors caused by dynamic inputs, thereby transforming the kWTA problem into a real-time solvable dynamic system and effectively improving the system's dynamic response capability. (2) A specialized Maxout activation function is designed to enhance the network's nonlinear expressive capability and effectively improve the convergence speed of the network in solving the time-delay problem. (3) The effectiveness of the proposed method is demonstrated through multi-robot tracking simulations. The method achieves accurate target tracking and maintains robustness under high-speed convergence by appropriately adjusting convergence parameters.

The remainder of this paper is organized as follows: Sect. 2 establishes the improved GD-kWTA network model and provides a theoretical analysis. Section 3 verifies the convergence performance of the new activation function through numerical experiments. Section 4 conducts multi-robot tracking simulation experiments, demonstrating the effectiveness and reliability of the proposed method. Finally, Sect. 5 summarizes the paper and discusses future research directions.

2 Problem Description and Modeling

This section details the mathematical framework and theoretical derivations of the kWTA network, which serve as the theoretical foundation for future studies.

2.1 Definition of kWTA

Consider an input vector $\mathbf{u}(t) \in \mathbb{R}^m$ that generates a corresponding output vector $\mathbf{v}(t) \in \mathbb{R}^m$. Our kWTA mechanism operates according to the following selection principle [31]:

$$v_j = \phi(u_j) = \begin{cases} 1, & \text{if } u_j \in \{k \text{ largest elements of } \mathbf{u}\} \\ 0, & \text{otherwise} \end{cases} \tag{1}$$

where $v_j(t)$ and $u_j(t)$ corresponds to the j-th entry of their respective vectors. The mechanism distinguishes between winning elements (assigned value 1) and losing elements (assigned value 0) based on competitive ranking.

Following the approach presented in [32], we reformulate this selection process as a quadratic programming problem, which is expressed as follows:

$$\begin{cases} \min & \gamma \mathbf{v}^{\mathrm{T}}(t)\mathbf{v}(t) - \mathbf{d}^{\mathrm{T}}(t)\mathbf{v}(t) \\ \text{s.t.} & \boldsymbol{\eta}^{\mathrm{T}}\mathbf{v}(t) = k \\ & B\mathbf{v}(t) \leq \mathbf{c} \end{cases} \tag{2}$$

where $\boldsymbol{\eta} = \mathbf{1} \in \mathbb{R}^m$ serves as a summation vector ensuring exactly k winners are selected; the matrix $B = [I_m; -I_m] \in \mathbb{R}^{2m\times m}$ combines positive and negative identity matrices to enforce bound constraints; $\mathbf{c} = [\mathbf{1}; \mathbf{0}] \in \mathbb{R}^{2m}$ defines the feasible region; the superscript ^{T} denotes a transpose operation. The optimization parameter $\gamma > 0$ maintains strict convexity by satisfying $0 < \gamma \leq (u_k(t) - u_{k+1}(t))/2$, where $u_k(t)$ and $u_{k+1}(t)$ represent the k-th and $(k+1)$-th largest elements of dynamic inputs, respectively, ensuring solution uniqueness.

Following the principles outlined in [33], we employ Lagrangian multipliers $\boldsymbol{\lambda}(t) \in \mathbb{R}^{2m}$ and $\mu(t) \in \mathbb{R}$ to derive the Karush-Kuhn-Tucker (KKT) conditions [34] for problem (2). This approach yields the following set of equations:

$$\begin{cases} k - \boldsymbol{\eta}^{\mathrm{T}}\mathbf{v}^*(t) = 0 \\ 2\gamma\mathbf{v}^*(t) + \boldsymbol{\eta}\mu(t) + B^{\mathrm{T}}\boldsymbol{\lambda}(t) - \mathbf{d}(t) = 0 \\ \mathbf{c} - B\mathbf{v}^*(t) \geq \mathbf{0}, \quad \boldsymbol{\lambda}(t) \geq \mathbf{0} \\ \boldsymbol{\lambda}^{\mathrm{T}}(t)(\mathbf{c} - B\mathbf{v}^*(t)) = 0. \end{cases} \tag{3}$$

where $\mathbf{v}^*(t)$ denotes the optimal solution satisfying the kWTA constraints. We employ the Fischer-Burmeister nonlinear complementary function [35] to handle inequality constraints in the following form:

$$\varPhi_{FB}^{\epsilon}(\boldsymbol{\xi}, \boldsymbol{\psi}) = \boldsymbol{\xi} + \boldsymbol{\psi} - \sqrt{\boldsymbol{\xi} \circ \boldsymbol{\xi} + \boldsymbol{\psi} \circ \boldsymbol{\psi} + \boldsymbol{\epsilon}} \tag{4}$$

where the function operates on vector pairs $\boldsymbol{\xi}$ and $\boldsymbol{\psi}$ of the same dimension, with a regularization term $\boldsymbol{\epsilon}$ introduced to preserve the smoothness required for gradient-based optimization; the symbol $\circ$ represents the Hadamard product. The complementarity function satisfies the equivalence $\varPhi_{FB}^{\epsilon}(\boldsymbol{\xi}, \boldsymbol{\psi}) = \mathbf{0} \Leftrightarrow \boldsymbol{\xi} \geq \mathbf{0}, \boldsymbol{\psi} \geq \mathbf{0}, \boldsymbol{\xi}^{\mathrm{T}}\boldsymbol{\psi} = 0$, ensuring proper constraint handling.

Based on this, we reformulate the KKT conditions as an equivalent set:

$$\begin{cases} k - \boldsymbol{\eta}^{\mathrm{T}}\mathbf{v}^*(t) = 0 \\ 2\gamma\mathbf{v}^*(t) + \boldsymbol{\eta}\mu(t) + B^{\mathrm{T}}\boldsymbol{\lambda}(t) - \mathbf{d}(t) = 0 \\ \varPhi_{FB}^{\epsilon}(\boldsymbol{\lambda}(t), \mathbf{c} - B\mathbf{v}^*(t)) = \mathbf{0}. \end{cases} \tag{5}$$

For computational efficiency, we define the following variables:

$$H = \begin{bmatrix} 2\gamma I_m & \boldsymbol{\eta} & B^{\mathrm{T}} \\ \boldsymbol{\eta}^{\mathrm{T}} & 0 & \mathbf{0}^{\mathrm{T}} \\ -B & \mathbf{0} & I_{2m} \end{bmatrix}, \quad \mathbf{w}(t) = \begin{bmatrix} \mathbf{v}^{*T}(t) \\ \mu(t) \\ \boldsymbol{\lambda}^{\mathrm{T}}(t) \end{bmatrix},$$

$$\mathbf{g}(\mathbf{w}(t), t) = \begin{bmatrix} -\mathbf{d}(t) \\ -k \\ \mathbf{c} - \boldsymbol{\delta}(t) \end{bmatrix},$$

where the auxiliary variable $\boldsymbol{\delta}(t) = \sqrt{\mathbf{z}(t) \circ \mathbf{z}(t) + \boldsymbol{\lambda}(t) \circ \boldsymbol{\lambda}(t) + \boldsymbol{\epsilon}}$ follows from the Fischer-Burmeister function; $\mathbf{z}(t) = \mathbf{c} - B\mathbf{v}^*(t)$ represents constraint violations. This transforms the original problem into the compact form:

$$H\mathbf{w}(t) + \mathbf{g}(\mathbf{w}(t), t) = \mathbf{0} \tag{6}$$

2.2 Network Construction

This subsection presents the design steps of the GD-kWTA network. Based on equation (6), the corresponding error function is defined as:

$$\mathbf{e}(t) = H\mathbf{w}(t) + \mathbf{g}(\mathbf{w}(t), t) \tag{7}$$

The gradient-based neural network is constructed by recursively approximating the minimum of the energy function (i.e., $||\mathbf{e}(t)||_2^2/2$). It is designed as:

$$\dot{\mathbf{w}}(t) = -\beta\left(H^{\mathrm{T}} + \frac{\partial \mathbf{g}^{\mathrm{T}}(\mathbf{w}(t),t)}{\partial \mathbf{w}(t)}\right)(H\mathbf{w}(t) + \mathbf{g}(\mathbf{w}(t),t)) \tag{8}$$

This can be expressed more compactly as:

$$\dot{\mathbf{w}}(t) = -\beta Q^{\mathrm{T}}(t)(H\mathbf{w}(t) + \mathbf{g}(\mathbf{w}(t),t)) \tag{9}$$

where $\beta > 0$ is a scaling factor. The system matrix $Q(t)$ is derived as:

$$Q(t) = \begin{bmatrix} \gamma I_m & \boldsymbol{\eta} & B^{\mathrm{T}} \\ \boldsymbol{\eta}^{\mathrm{T}} & 0 & \mathbf{0}^{\mathrm{T}} \\ R(t) & \mathbf{0} & S(t) \end{bmatrix}$$

where $R(t) = (D_1(t) - I)B$, $S(t) = I - D_2(t)$, $D_1(t) = \sigma(\mathbf{z}(t) \oslash \boldsymbol{\delta}(t))$, $D_2(t) = \sigma(\boldsymbol{\lambda}(t) \oslash \boldsymbol{\delta}(t))$, and $\sigma(\cdot)$ is the diagonalization operator; $\oslash$ is the Hadamard division.

Since the GD-kWTA network may have time-varying delays in practical implementations, we introduce adaptive velocity compensation to address this issue [36]:

$$\dot{\mathbf{w}}(t) = \boldsymbol{\rho}(t) - \beta Q^{\mathrm{T}}(t)(H\mathbf{w}(t) + \mathbf{g}(\mathbf{w}(t),t)) \tag{10}$$

At equilibrium conditions where the error $\mathbf{e}(t)$ satisfies $\mathbf{e}(t) = H\mathbf{w}(t) + \mathbf{g}(\mathbf{w}(t),t) = \mathbf{0}$, the compensation term satisfies:

$$\dot{\mathbf{w}}(t) = \boldsymbol{\rho}(t) \tag{11}$$

Through stability analysis, the velocity compensation mechanism ensures robust performance under realistic operating conditions. We first calculate the derivative of the error $\mathbf{e}(t)$, from which we deduce that:

$$\dot{\mathbf{e}}(t) = \frac{\partial \mathbf{g}(\mathbf{w}(t),t)}{\partial t} + \frac{\partial \mathbf{g}(\mathbf{w}(t),t)}{\partial \mathbf{w}(t)}\dot{\mathbf{w}}(t) + H\dot{\mathbf{w}}(t) + \dot{H}\mathbf{w}(t) \tag{12}$$

Since matrix H is not related to time, $\dot{H} = \mathbf{0}$, the above equation can be written as:

$$Q(t)\dot{\mathbf{w}}(t) = \boldsymbol{\psi}(t) \tag{13}$$

where $\boldsymbol{\psi}(t) = -\partial \mathbf{g}(\mathbf{w}(t),t)/\partial t$. By substituting the calculated velocity compensation term into equation (10) and applying multiplication to both sides by $Q(t)$, we obtain the final GD-kWTA network dynamics:

$$Q(t)\dot{\mathbf{w}}(t) = -\beta Q(t)Q^{\mathrm{T}}(t)(H\mathbf{w}(t) + \mathbf{g}(\mathbf{w}(t),t)) + \boldsymbol{\psi}(t) \tag{14}$$

The final decision outputs are extracted from the first m components of the state vector $\mathbf{w}(t)$, producing binary decisions through an enhanced activation scheme:

$$
v_j(t) = F_1(w_j(t)) = \begin{cases} 0, & \text{if } w_j < (1-\epsilon) \\ w_j, & \text{if } (1-\epsilon) \le w_j \le \epsilon \\ 1, & \text{if } w_j > \epsilon \end{cases} \quad s.\ t.\ \ j = 1, 2, \ldots, m. \tag{15}
$$

where $\epsilon \in (0.5, 1)$. Combined with this function, $v_j(t)$ can be obtained with zero error.

3 Numerical Experiments

This section analyzes the performance of the GD-kWTA network through numerical experiments, with a focus on the convergence characteristics of the proposed Maxout activation function and parameter optimization strategies.

To enhance the network's nonlinear representational capability and accelerate convergence, we design a specialized Maxout activation function in this work that combines three dedicated component functions. It is defined as:

$$
\text{Maxout}(e(t)) = \max\{f_1(e(t)), f_2(e(t)), f_3(e(t))\}
$$

where the three component functions are applied element-wise:

$$
\begin{cases} f_1(e_i(t)) = e_i(t) \\ f_2(e_i(t)) = e_i(t) + 10 \cdot \max(0, e_i(t)) \\ f_3(e_i(t)) = 2e_i(t) - 0.1e_i^2(t) \end{cases} \tag{16}
$$

Furthermore, to pursue high convergence speed while maintaining network stability and decision accuracy, we appropriately reduced the convergence parameters.

In this numerical experiment, four sinusoidal signals are used as inputs, defined as

$$
u_i(t) = 2\sin(2\pi(0.3t + 0.2(i+1))), \quad s.\ t.\ \ i = 1, 2, 3, 4. \tag{17}
$$

The number of winners is set to $k = 2$ and $k = 3$. The experimental parameters are configured as: optimization parameter $\gamma = 1 \times 10^{-5}$, enhanced scaling factor $\beta = 30$, and simulation time range $t \in [0, 10]$ seconds with 2000 sampling points. Through observation of Fig. 1, the dynamic changes of winners can be observed. Additionally, the GD-kWTA network is activated by the Maxout activation function.

Figure 1(a) and (b) respectively show the output results of the kWTA network for $k = 2$ and $k = 3$. It can be observed that as the number of winners changes, the winning and losing relationships change accordingly, with winners being activated to 1 and losers being suppressed to 0. The network demonstrates clear winner-loser transitions when input signals change. The Maxout activation function enhances the network's nonlinear response to dynamic sinusoidal inputs.

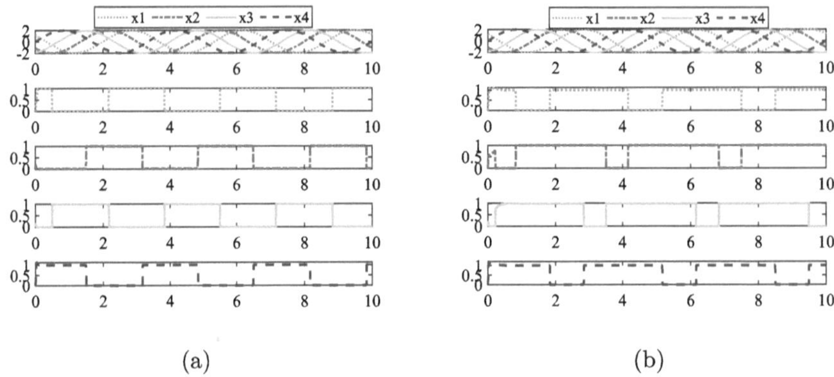

Fig. 1. Output of the kWTA network. (a) Case of $k = 2$; (b) Case of $k = 3$.

In the 2000-sample-point test within the 0–10 s simulation window, this design helps the residual error of the GD-kWTA network narrow down to a stable range more quickly. Meanwhile, the time-derivative information integrated into the network alleviates lagging errors from input fluctuations, ensuring that the switch between winner and loser states aligns with the real-time change of input signals without obvious delay.

4 Simulation Experiments

In this section, the GD-kWTA network is implemented in MATLAB platform to perform target tracking tasks in a multi-robot system.

A system comprising eight robots is tasked with tracking targets in a competitive environment. The target follows a complex motion trajectory given by:

$$\begin{cases} x_{\text{target}}(t) = 1.8 + 0.8t, \\ y_{\text{target}}(t) = 4.5 + 2.5\sin(\pi t/3.2). \end{cases} \tag{18}$$

The velocity of the robots is set to be $1.1y_i(t)(\boldsymbol{\theta}_i(t) - \boldsymbol{\theta}_0(t))/\epsilon_i(t)$ m/s, with a maximum communication range of 2 m between them. The competitive strategy involved the robots closest to the target as the winners, as determined by $v_i(t)$, with activated robots having $v_i(t) = 1$ and deactivated robots having $v_i(t) = 0$. At the same time, we incorporate the Maxout activation function into the kWTA network.

4.1 Control Law

Based on the theoretical framework established in Sect. 2, we implement a robotic control law based on the GD-kWTA network output. For each robot i, the kWTA network input is defined as:

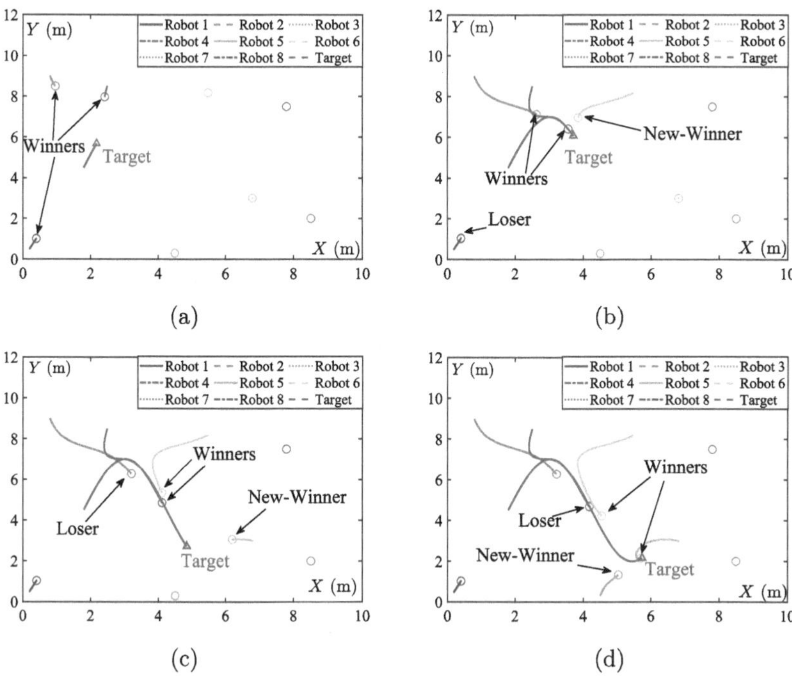

Fig. 2. Trajectories of eight robots tracking a moving target in MATLAB. The target follows a sinusoidal trajectory. (a)âĂŞ-(d) Snapshots at different moments, showing changes in winners and losers.

$$\zeta_i(t) = -||\theta_i(t) - \theta_0(t)||_2$$

where $\theta_i(t)$ and $\theta_0(t)$ are two-dimensional vectors representing the positions of robot i and the target, respectively. Each robot's velocity control law follows the enhanced framework:

$$\dot{\theta}_i(t) = \kappa \cdot v_i(t) \cdot \frac{\theta_i(t) - \theta_0(t)}{\zeta_i(t)}$$

where $v_i(t)$ is the binary control output of the kWTA network combined with Maxout activation, and κ is a predefined scaling constant.

To achieve distributed cooperative control under limited communication conditions, we introduce a consensus filter. Each robot i maintains an estimate of the global winner count $I_i(t)$, updated according to the following equation:

$$\dot{I}_i(t) = -\alpha \sum_{j \in \mathcal{N}_i} L_{ij}(t)(I_i(t) - I_j(t)) - \alpha(I_i(t) - v_i(t)) \tag{19}$$

where α is a coupling factor, $\mathcal{N}_i$ represents the communication neighborhood set of robot i, and $L_{ij}(t)$ is the communication network structure represented

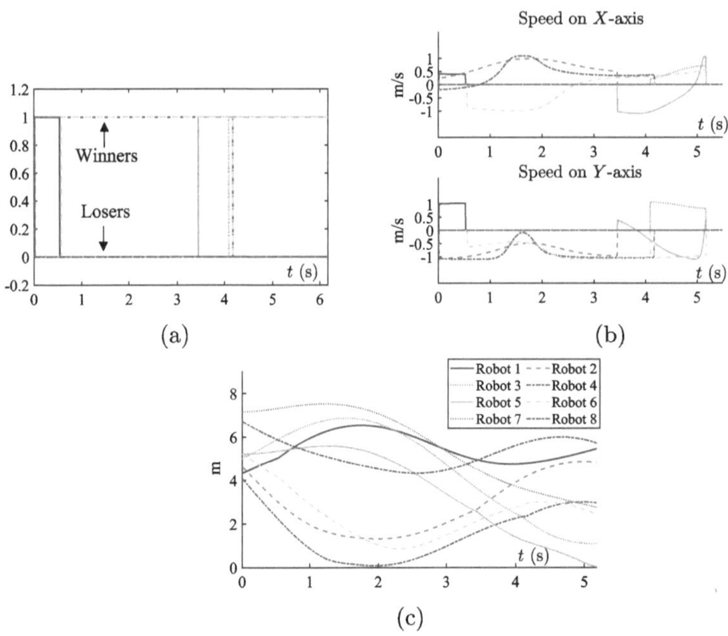

Fig. 3. Dynamic behaviors of the multi-robot system. (a) Evolution of winnerâĂŞloser relationships; (b) Variations in robot velocities along the x- and y-axes; (c) Distance between each robot and the target.

by the adjacency matrix. Furthermore, by incorporating the above distributed network, the robotic system must follow the control protocol:

$$
\begin{aligned}
\dot{w}(t) &= -Q^{-1}(t)\left(\beta Q(t)Q^{\mathrm{T}}(t)(Hw(t)+g(w(t),t))\right) \\
&\qquad + \tau\int_0^{\mathrm{T}}(Hw(\rho)+g(w(\rho),\rho))d\rho + \psi(t) + q(t) \\
v_i(t) &= F_{\mathrm{Maxout}}(w_i(t)) \\
\dot{\theta}_i(t) &= \kappa v_i(t)\frac{\theta_i(t)-\theta_0(t)}{\zeta_i(t)} \\
\dot{I}_i(t) &= -\alpha\sum_{j\in\mathcal{N}_i}L_{ij}(t)(I_i(t)-I_j(t)) - \alpha(I_i(t)-v_i(t)) \\
s.\ t.\ \ & i = 1,2,\ldots,n.
\end{aligned}
\tag{20}
$$

where τ is a proportional factor. $q(t)$ is a polynomial disturbance term defined as $q(t) = \gamma\sum_{j=0}^{m}\tilde{a}_j t^j$, where γ is a dimensional adaptation vector, m is the order. By adjusting the polynomial coefficients $\tilde{a}_j$, various non-stationary noises can be represented, reflecting the complexity of real-world system environments.

4.2 Simulation Results

Figure 2 displays the motion trajectories at different time instances. It shows that the winning robots track the target while the losing robots remain stationary. Figure 3(a) illustrates the outputs of the kWTA network, which switch between 0 and 1 as winners change. In Fig. 3(b), the velocity of the competing robots is shown, changing in real time as position information varies. Figure 3(c) displays the distance between robots and the target, with winners gradually approaching the target as they change. The experimental results validate the effectiveness of the enhanced approach with improved convergence speed and coordination performance. When the target moves along its predefined trajectory, the Maxout activation function enables the GD-kWTA network to select new winners in milliseconds—this rapid selection ensures that the newly activated robots can start tracking the target promptly. The consensus filter further supports the system by allowing each robot to estimate the global winner count through local communication (within the 2-meter range), preventing motion conflicts between adjacent robots. From the velocity curves in Fig. 3(b), it can be seen that the speed of winning robots adjusts in real time with the target's movement, while losing robots switch to a stationary state quickly, reducing unnecessary energy consumption and maintaining orderly coordination of the entire system.

5 Conclusion

In this paper, a GD-$k$wta network incorporating the Maxout activation function and time-derivative information is presented. The network effectively improves convergence speed and resolves the error lag problems caused by dynamic inputs. Simulations of multi-robot cooperation and competition in target tracking tasks have validated its effectiveness and reliability. Future research will focus on applying this network to larger-scale distributed robotic systems to address more complex environments.

Disclosure of Interests. The authors have no competing interests to declare that are relevant to the content of this article.

References

1. Yang, S., Liu, Q., Wang, J.: A collaborative neurodynamic approach to multiple-objective distributed optimization. IEEE Trans. Neural Netw. Learn. Syst. **29**(4), 981–992 (2018)
2. Jin, L., Liang, S., Li, S., Liu, J.: Finite-time convergent and weight-unbalanced kWTA network with multirobot competitive coordination. IEEE/ASME Trans. Mechatronics **30**(3), 2270–2281 (2025)
3. Li, K., Liu, Q., Zeng, Z.: Multiagent system with periodic and event-triggered communications for solving distributed resource allocation problem. IEEE Trans. Syst., Man, Cybern. Syst. **53**(10), 6245–6256 (2023)

4. Liu, M., Mu, W., Lv, X., Sun, Z., Jin, L.: Neural network for distributed collaboration of multiple manipulators with switching topologies: a game-theoretic perspective. IEEE Trans. Syst. Man Cybern. Syst. **55**(5), 3213–3221 (2025)
5. Zhang, J., Jin, L., Cheng, L.: RNN for perturbed manipulability optimization of manipulators based on a distributed scheme: a game-theoretic perspective. IEEE Trans. Neural Netw. Learn. Syst. **31**(12), 5116–5126 (2020)
6. Zhang, J., Liu, M., Jin, L.: Logistic adaptive controller with overhead reduction for multirobot systems. IEEE Trans. Ind. Electron. **72**(1), 660–669 (2025)
7. Ma, D., Guan, Y., Jin, L., Li, S.: Distributed optimal control of multiple serial robot systems with kinematics and dynamics based on discrete neural dynamics. IEEE Trans. Ind. Electron. **72**(6), 6393–6401 (2025)
8. Liu, M., Zhang, X., Shang, M., Jin, L.: Gradient-based differential kWTA network with application to competitive coordination of multiple robots. IEEE/CAA J. Autom. Sin. **9**(8), 1452–1463 (2022)
9. Jin, L., Li, Y., Chen, Y., Su, Z., Ma, X.: An event-triggered k-WTA model with experimental verification on multirobot system. IEEE Trans. Ind. Electron. **72**(9), 9271–9281 (2025)
10. Zhang, J., Jin, L., Wang, Y.: Collaborative control for multimanipulator systems with fuzzy neural networks. IEEE Trans. Fuzzy Syst. **31**(4), 1305–1314 (2023)
11. Chen, X., Chen, L., Li, S., Jin, L.: A mirrored echo state network with application to time series prediction. Inf. Sci. **716**, 122260 (2025)
12. Xie, Z., Jin, L., Lv, X.: A hierarchical control and learning network for redundant manipulators with unknown physical parameters. IEEE Trans. Ind. Electron. **72**(7), 7106–7116 (2025)
13. Fan, J., Jin, L., Li, P., Liu, J., Wu, Z.-G., Chen, W.: Coevolutionary neural dynamics considering multiple strategies for nonconvex optimization. Tsinghua Sci. Technol. **30**(2), 204–216 (2025)
14. Li, J., Guan, Y., Deng, T., Jin, L.: Periodic-noise-tolerant neurodynamic approach for kWTA operation applied to opinions evolution. Neural Netw. **191**, 107839 (2025)
15. Li, J., Cao, Y., Xie, Z., Jin, L.: A k-winners-take-all (kWTA) network with noise characteristics captured. IEEE/CAA J. Autom. Sin. **12**(4), 734–744 (2025)
16. Liu, M., Li, Y., Chen, Y., Qi, Y., Jin, L.: A Distributed competitive and collaborative coordination for multirobot systems. IEEE Trans. Mob. Comput. **23**(12), 11436–11448 (2024)
17. Liu, K., Zhang, Y.: Distributed dynamic task allocation for moving target tracking of networked mobile robots using kWTA network. IEEE Trans. Neural Netw. Learn. Syst. **36**(3), 5795–5802 (2025)
18. Zeng, Z., Wang, J., Liao, X.: Global exponential stability of a general class of recurrent neural networks with time-varying delays. IEEE Trans. Circuits Syst. I, Fundam. Theory Appl. **50**(10), 1353–1358 (2003)
19. Zeng, Z., Wang, J., Liao, X.: Stability analysis of delayed cellular neural networks described using cloning templates. IEEE Trans. Circuits Syst. I, Reg. Papers **51**(11), 2313–2324 (2004)
20. Deng, Q., Liu, K., Zhang, Y.: Privacy-preserving consensus of double-integrator multi-agent systems with input constraints. IEEE Trans. Emerg. Topics Comput. Intell. **8**(6), 4119–4129 (2024)
21. Jin, L., Liufu, Y., Lu, H., Zhang, Z.: Saturation-allowed neural dynamics applied to perturbed time-dependent system of linear equations and robots. IEEE Trans. Ind. Electron. **68**(10), 9844–9854 (2021)

22. Hu, X., Wang, J.: Solving pseudomonotone variational inequalities and pseudo-convex optimization problems using the projection neural network. IEEE Trans. Neural Netw. **6**(17), 1487–1499 (2006)
23. Liu, M., Chen, L., Du, X., Jin, L., Shang, M.: Activated gradients for deep neural networks. IEEE Trans. Neural Netw. Learn. Syst. **34**(4), 2156–2168 (2023)
24. Su, H., Chen, M.Z.Q., Lam, J., Lin, Z.: Semi-global leader-following consensus of linear multi-agent systems with input saturation via low gain feedback. IEEE Trans. Circuits Syst. I, Reg. Papers **60**(7), 1881–1889 (2013)
25. Su, H., Zhang, J., Zeng, Z.: Formation-containment control of multi-robot systems under a stochastic sampling mechanism. Sci. China Technol. Sci. **63**, 1025–1034 (2020)
26. Huang, H., Jin, L., Zeng, Z.: A momentum recurrent neural network for sparse motion planning of redundant manipulators with majorization-minimization. IEEE Trans. Ind. Electron. 1–10 (2025)
27. Jin, L., Su, Z., Fu, D., Xiao, X.: Coevolutionary neural solution for nonconvex optimization with noise tolerance. IEEE Trans. Neural Netw. Learn. Syst. **35**(12), 17571–17581 (2024)
28. Qi, Y., Jin, L., Luo, X., Shi, Y., Liu, M.: Robust k-WTA network generation, analysis, and applications to multiagent coordination. IEEE Trans. Cybern. **52**(8), 8515–8527 (2022)
29. Jin, L., Qi, Y., Luo, X., Li, S., Shang, M.: Distributed competition of multi-robot coordination under variable and switching topologies. IEEE Trans. Autom. Sci. Eng. **19**(4), 3575–3586 (2022)
30. Jin, L., Li, S.: Distributed task allocation of multiple robots: a control perspective. IEEE Trans. Syst. Man Cybern. Syst. **48**(5), 693–701 (2018)
31. Tymoshchuk, P.V., Wunsch, D.C.: Design of a k-winners-take-all model with a binary spike train. IEEE Trans. Cybern. **49**(8), 3131–3140 (2019)
32. Liu, M., Shang, M.: On RNN-based k-WTA models with time-dependent inputs. IEEE/CAA J. Autom. Sin. **9**(11), 2034–2036 (2022)
33. Li, W., Ma, X., Luo, J., Jin, L.: A strictly predefined-time convergent neural solution to equality-and inequality-constrained time-variant quadratic programming. IEEE Trans. Syst. Man Cybern. Syst. **51**(7), 4028–4039 (2021)
34. Nazemi, A.: A capable neural network framework for solving degenerate quadratic optimization problems with an application in image fusion. Neural Process. Lett. **47**(1), 167–192 (2018)
35. Jin, L., Chen, Y., Liu, M.: A noise-tolerant k-WTA model with its application on multirobot system. IEEE Trans. Ind. Inform. **20**(3), 3574–3584 (2024)
36. Jin, L., Wei, L., Li, S.: Gradient-based differential neural-solution to time-dependent nonlinear optimization. IEEE Trans. Autom. Control **68**(1), 620–627 (2023)

Memristor-Based Neural Network Circuit With Flashbulb Memory Effect

Yuqi Deng(✉), Yuejia Zhou, Mengyan Li, Qinyuan Fang, Zhixia Ding, and Sai Li

School of Electrical and Information Engineering, Wuhan Institute of Technology, Wuhan 430205, China
2272853446@qq.com

Abstract. Most existing memristor-based neural network designs have primarily focused on how different types of emotions influence memory recall speed. However, the role of emotional intensity in memory processes remains insufficiently explored. In this study, we propose a dual-channel brain-inspired neural network circuit based on memristors, which incorporates both the flashbulb memory effect and emotion-enhanced mechanisms. The proposed design not only considers the impact of various emotional types across different intensity levels, but also integrates the modulatory effects of linguistic factors on memory performance. The circuit comprises three functionally distinct modules: an emotion module, an emotion enhancement module, and a language enhancement module. Specifically: The emotion module is responsible for producing different emotional states; The emotion enhancement module enables memory reinforcement through emotional intensity and triggers the flashbulb memory effect; The language enhancement module improves recall speed by introducing diverse language inputs. This design can be broadly applied to the development of emotionally responsive biomimetic robots and offers valuable insights for advancing brain-inspired memory systems.

Keywords: Brain-like neural network · Emotion enhancement · Flashbulb memory

1 Introduction

In recent years, the development of brain-inspired intelligence has accelerated significantly, driven by ongoing explorations into the structure and fundamental mechanisms of the human brain [1]. Brain-inspired circuits, designed to emulate the structure and function of biological neural systems [2–4], enable efficient execution of deep learning and neural computation tasks at the hardware level, while offering the added benefit of low power consumption [5,6].

A growing body of research has focused on hardware-based modeling of emotional generation and memory recall processes, establishing this direction as a new frontier in neuromorphic engineering [7–11].

C. Li et al. (Eds.): ICNC 2025, CCIS 2946, pp. 156–165, 2026.
https://doi.org/10.1007/978-981-92-1599-7_14

In recent years, with continuous technological advancements and expanding application scenarios, an increasing number of researchers have integrated emotion and memory into memristor-based brain-inspired circuits to better simulate the operational mechanisms of the human brain. Hu proposed a memristor-driven emotional neural network model capable of generating and fading basic emotional states [12]. Ma's team developed simple emotion simulations based on neuronal excitation and nhibition mechanisms [13]. Sun presented a memristive neural circuit with stable emotional states, integrating memory and emotional dynamics [14]. Wang's group proposed a memristor-based circuit for emotion generation and evolution, which also supports multimodal input-integrating auditory, textual, and visual signals-to reflect dynamic emotional trajectories [15]. Guo designed a neural network circuit incorporating both perceptual and higher cognitive loops to analyze the neural circuits involved in fear generalization and memory within the limbic system [16]. Han developed a circuit featuring emotion-consistent and emotion-suppressive pathways [17]. Additionally, Sun introduced a PAD-based memristive emotion circuit capable of three-dimensional emotional representation [18].

The main contributions of this work are summarized as follows:

1) A flashbulb memory module is proposed to simulate the flashbulb memory effect induced by high-intensity emotional stimulation. The circuit enhances memory recall speed when exposed to intense emotional inputs but may also introduce distortions or biases in the recalled content.
2) Most existing circuits for emotional associative memory focus on a single intensity level, failing to account for the distinction between high and low emotional intensities in humans. An emotion module is designed to regulate both the generation and dissipation of various emotional states. Emotional responses are elicited under stimuli of different intensities, and after stimulus removal, emotional levels gradually return to baseline. Notably, the recovery time following intense fear is shorter than that following intense pleasure, reflecting distinct recovery dynamics across emotional types.
3) Most studies on how language affects associative memory have focused on Mandarin Chinese, without considering that different language systems may have different effects on memory recall. A language enhancement module is introduced to comparatively analyze the effects of Mandarin and dialect inputs on memory recall speed, providing further insights into how language modulates emotionally driven memory processes.

The remainder of this paper is organized as follows: The Sect. 2 introduces the memristor models used in this study. The Sect. 3 presents the main modules of the experimental circuit design. The Sect. 4 discusses the simulation results of the circuit, and the Sect. 5 provides a summary of the study.

2 Memristor Module

Currently, a variety of memristor models have been proposed, each exhibiting distinct characteristics. The Threshold Adaptive Memristor model features a voltage threshold-based switching mechanism and a relatively simple structure [19]. The Voltage Threshold Adaptive Memristor model supports voltage-driven nonlinear switching behavior [20], while the Biolek window function model offers a simplified nonlinear control mechanism. However, these models commonly suffer from limitations such as insufficient descriptions of dynamic behavior or an excessive number of parameters, leading to increased computational complexity.

Considering the requirements for neural network circuit simulation, this study adopts the AIST memristor model due to its pronounced nonlinear switching characteristics and suitability for high-precision modeling of neuromorphic devices. Compared with traditional models such as TEAM and VTEAM, the AIST model offers enhanced accuracy in capturing dynamic behaviors relevant to synaptic plasticity and memory modulation.

Its voltage–current relationship can be mathematically expressed as:

$$v(t) = R_M(t) \cdot i(t) \tag{1}$$

$$R_M(t) = R_{on}\frac{w(t)}{D} + R_{off}(1 - \frac{w(t)}{D}) \tag{2}$$

where $R_M(t)$ represents the memristance. $v(t)$ and $i(t)$ are the voltage and current flowing through the memristor, respectively. D denotes the length of the memristor, while $w(t)$ signifies the width of the doped region. R_{on} and R_{off} denote the resistance when the memristor is completely high doped or low doped, respectively.

3 Circuit Design

3.1 Language Enhancement Module

The language enhancement module is shown in the green part of Fig. 1. The module parameters are set as follows: R_1=R_2 = 150 Ω, R_3 = 100 Ω.

When N_1 is at a high level and N_2 is at a low level, the system is in the Mandarin enhancement state. In this state, XOR_1 and XOR_2 output high levels, and the output voltage is applied to memristor M_1, causing its resistance to decrease. When N_1 is at a low level and N_2 is at a high level, the system is in the dialect enhancement state. In this state, XOR_1 outputs a high level, XOR_2 outputs a low level, and the dialect language signal, together with the output signal of XOR_2, is input into the adder SUM_1. The output voltage of SUM_1 is applied to memristor M_1, causing its resistance to decrease.

3.2 Emotion Module

The emotion module is shown in the blue and yellow parts of Fig. 1. The module parameters are set as follows: R_{11}-R_{22} = 100 Ω.

When the system receives a mild fear signal and both N_1 and N_3 are at a high level, the circuit enters the mild fear state. In this state, AND_1 is activated, and its output voltage is applied to voltage-controlled switch S_1, causing S_1

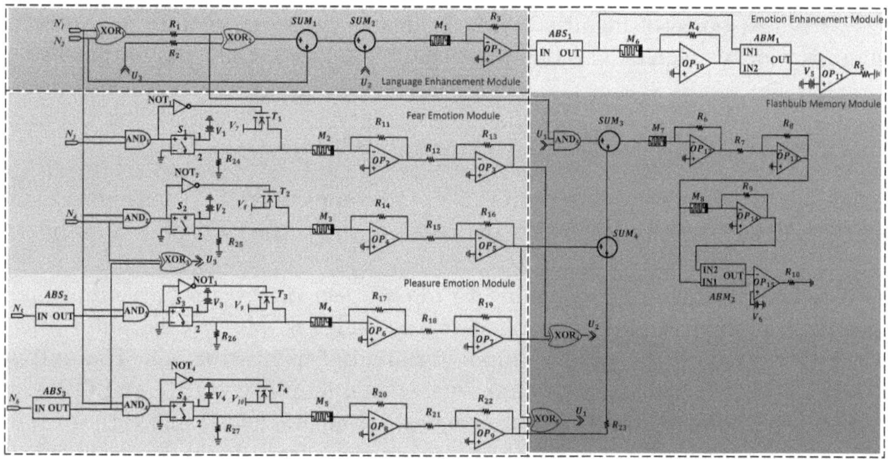

Fig. 1. Complete Module.

to conduct. The output voltage V_1 is applied to memristor M_2, reducing its memristance. The output voltage of M_2, amplified by operational amplifiers OP_2 and OP_3, is transmitted to the emotion enhancement module, thereby enhancing the memory recall speed under mild fear conditions. When the circuit no longer emits a mild fear signal, The high level of NOT_1 output acts on the MOS tube T_1 to make the MOS tube T_1 on, and the voltage of NOT_1 output acts on M_2 to make the memristance of M_2 gradually return to the initial test state.

When the system receives an intense fear signal and both N_1 and N_4 are at a high level, the circuit enters the intense fear state. In this state, AND_2 is activated, and its output voltage is applied to voltage-controlled switch S_2, causing S_2 to conduct. The output voltage V_2 is applied to memristor M_3, reducing its memristance. The output voltage of M_3, amplified by operational amplifiers OP_4 and OP_5, is transmitted to the flashbulb memory module, thereby triggering the flashbulb memory effect associated with intense fear. When the circuit no longer exerts a intense fear signal, NOT_2 output a high level on the MOS tube T_2 output V_7, making the MOS tube T_2 on, NOT_2 output voltage on the memristor M_3, making the memristance of the memristor M_3 gradually return to the initial test state.

3.3 Emotion Enhancement Module

The emotion enhancement module is shown in the grey and purple parts of Fig. 1. The module parameters are configured as follows: R_4-R_{10} = 100 Ω, and $V_5 = V_6 = 1.4$V.

When the voltage corresponding to a mild fear or mild pleasure signal from the Emotion Module is applied to the absolute value function unit in the Emotion Enhancement Module, the output voltage of ABS_1 is fed into both the positive input terminal of memristor M_6 and the N_1 interface of the digital arithmetic unit. The resistance of memristor M_6 begins to decrease, thereby modulating the impact of the signal. The digital arithmetic unit computes its output as:

$$V_{AMB1} = -\frac{V_{IN2}}{V_{IN1}} = \frac{M_6}{1000(\Omega)}, \tag{3}$$

If $V_{\mathrm{ABM1}} < V_5$, V_{COMP11} outputs 15 V, the comparator OP_{11} outputs a high-level signal, successfully triggering the recall of the correct clothing color of the thief.

When the voltage corresponding to intense fear or intense pleasure signals from the Emotion Module is applied to the adder SUM_3, the output voltage of SUM_3 is fed into the positive input terminal of memristor M_7. The output voltage of M_7, after passing through operational amplifiers OP_{12} and OP_{13}, is applied to both memristor M_8 and the digital arithmetic unit. Finally, the output voltage of ABM_2 is compared with the reference voltage V_6 in the comparator OP_{15}. If $V_{\mathrm{ABM2}} < V_6$, V_{COMP15} outputs -15 V, the comparator OP_{15} outputs a low-level signal, triggering the recall of the thief's clothing color; however, the recalled clothing color is incorrect, demonstrating the memory distortion effect characteristic of flashbulb memories.

4 Simulation Analysis

4.1 Enhancement Effect of Dialect Signals on Memory

In Fig. 2. N_1: Mandarin stimulus signal. N_2: Dialect stimulus signal. When only the N_2 signal outputs a high level, the 0–35s period corresponds to the dialect-enhanced learning process. The N_2 signal is applied to the adder SUM_1, which in turn applies a voltage to memristor M_1, causing the output voltage of VOP_{11} to increase. This voltage is then applied to memristor M_6, resulting in a decrease in its memristance. At 35s, the memristance of M_6 reaches 140 Ω, and the output of comparator VOP_{11} switches from low to high, indicating the completion of the dialect-enhanced memory process.

When only the N_1 signal outputs a high level, the 0–55s period corresponds to the Mandarin-enhanced learning process. The N_1 signal is applied to XOR_1, which causes XOR_2 to conduct. The output voltage of XOR_2 is applied to adder SUM_1. At 55s, the memristance of M_6 also reaches 140 Ω, and the output of comparator VOP_{11} switches from low to high, indicating the completion of the Mandarin-enhanced memory process.

4.2 Enhancement Effect of Mild Fear Emotion on Memory

In Fig. 3. N_1: Mandarin stimulus signal. N_3: Mild fear stimulus signal. When both N_1 and N_3 signals output a high level, the interval from 0–18s corresponds to the mild fear–enhanced learning process. The mild fear stimulus signal N_3

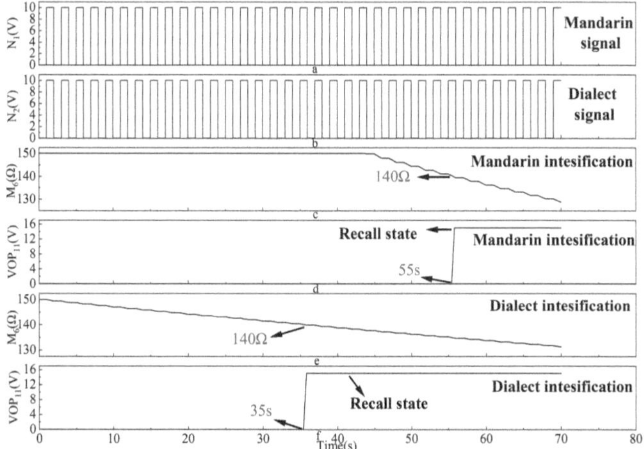

Fig. 2. Enhancement Effect of Dialect Signals on Memory.

is applied to AND gate AND_1, which is connected to switch S_1. This activates a voltage V_1 applied to memristor M_2, where V_1 exceeds the threshold voltage of M_2, causing its resistance to rapidly decrease, indicating the generation of a mild fear emotional state.

The resulting voltage $VABS_1$ is then applied to memristor M_6. At 18s, the resistance of M_6 drops to 140 Ω, triggering the output of comparator VOP_{11} to shift from low to high, signifying the completion of the mild fear–induced memory enhancement process.

4.3 Flashbulb Memory Effect of Intense Fear Emotion

In Fig. 4. N_1: Mandarin stimulus signal. N_4: Intense fear stimulus signal. When both the N_1 and N_4 signals output high levels, the period from 0–18s corresponds to the intense fear–enhanced learning process. The intense fear stimulus signal N_4 is applied to AND gate AND_2, which is connected to switch S_2. This enables the application of voltage V_2 to memristor M_3, where V_2 exceeds the threshold voltage of M_3, causing a rapid decrease in its resistance, indicating the generation of an intense fear emotional state.

The output voltage of VOP_{13} is then applied to memristor M_8. At 12s, the resistance of M_8 drops to 140 Ω, prompting the output of comparator VOP_{15} to shift from low to high, which signifies the completion of the intense fear–induced memory enhancement process.

4.4 Enhancement Effect of Mild Pleasure Emotion on Memory

In Fig. 5. N_1: Mandarin stimulus signal. N_5: Mild pleasure stimulus signal. When both N_1 and N_5 signals output high levels, the period from 0–18s corresponds to

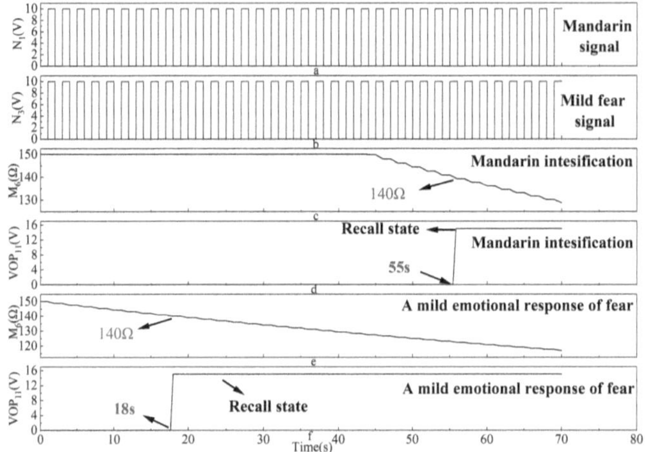

Fig. 3. Enhancement Effect of Mild Fear Emotion on Memory.

the mild pleasure–enhanced learning process. The mild pleasure stimulus signal N_5 is applied to AND gate AND_3, which is connected to switch S_3. This enables voltage V_3 to be applied to memristor M_4, where V_3 exceeds the threshold voltage of M_4, causing its resistance to rapidly decrease, indicating the generation of a mild pleasure emotional state.

The resulting voltage $VABS_1$ is then applied to memristor M_6. At 18s, the resistance of M_6 drops to 140 Ω, causing the output of comparator OP_{11} to switch from low to high, indicating the completion of the mild pleasure–induced memory enhancement process.

4.5 Flashbulb Memory Effect of Intense Pleasure Emotion

In Fig. 6. N_1: Mandarin stimulus signal. N_6: Intense pleasure stimulus signal. When both the N_1 and N_6 signals output high levels, the time interval from 0 to 18s corresponds to the intense pleasure–enhanced learning process. The intense pleasure stimulus signal N_6 is applied to AND gate AND_4, which is connected to switch S_4, enabling voltage V_4 to be applied to memristor M_5. Since V_4 exceeds the threshold voltage of M_5, its resistance drops rapidly, indicating the generation of an intense pleasure emotional state.

The output voltage of OP_{13} is then applied to memristor M_8. At 14s, the resistance of M_8 falls to 140 Ω, causing the output of comparator OP_{15} to shift from low to high, which marks the completion of the intense pleasure–induced memory enhancement process.

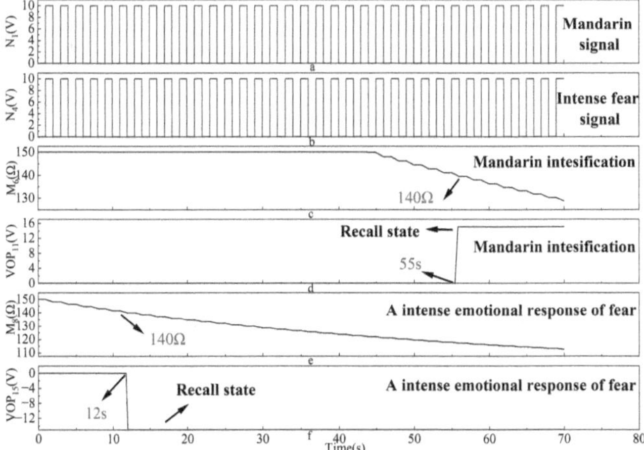

Fig. 4. Flashbulb Memory Effect of Intense Fear Emotion.

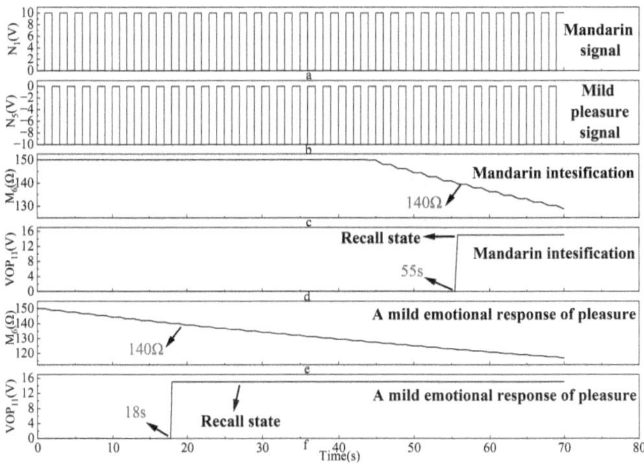

Fig. 5. Enhancement Effect of Mild Pleasure Emotion on Memory.

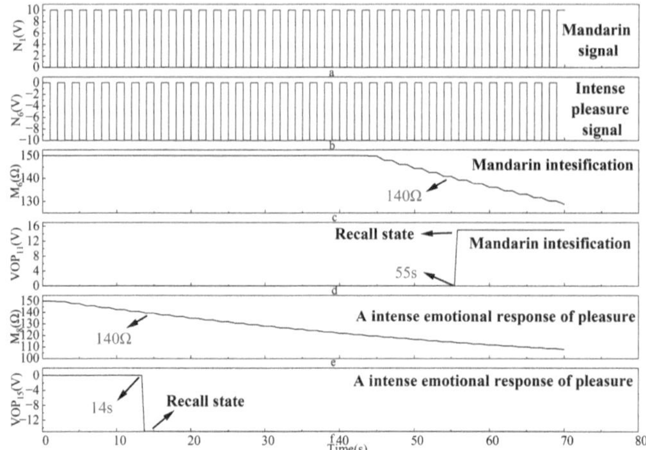

Fig. 6. Flashbulb Memory Effect of Intense Pleasure Emotion.

5 Conclusion

This paper proposes a memristor-based brain-inspired neural network circuit, designed to enhance memory recall speed under complex emotional conditions. The circuit supports memory enhancement under mild emotional states, as well as strong emotional stimulation, which not only reinforces memory but also triggers the flashbulb memory effect. It accommodates two distinct emotional types: fear and pleasure. Additionally, the circuit enables the analysis of the effects of different language signals, including both Mandarin and dialects, on memory.

Acknowledgments. The work was supported by the National Natural Science Foundation of China under Grant 62176189, the Key Research and Development Plan Project of Hubei Province under Grant 2024BAB032, and the Project of Excellent Young and Middle-aged Scientific and Technological Innovation Team in Colleges and Universities of Hubei Provincial Department of Education under Grant T2022012.

References

1. Zhang, Y., Cui, M., Shen, L., Zeng, Z.: Memristive quantized neural networks: a novel approach to accelerate deep learning on-chip. IEEE Trans. Cybern. **51**(4), 1875–1887 (2021)
2. Shomron, G., Weiser, U.: Spatial correlation and value prediction in convolutional neural networks. IEEE Comput. Archit. Lett. **18**(1), 10–13 (2019)
3. Silver, D., Huang, A., Maddison, C.J., Guez, A.: Mastering the game of go with deep neural networks and tree search. Nature **529**(7587), 484–489 (2016)

4. Zhang, F., Zeng, Z.: Multiple lagrange stability under perturbation for recurrent neural networks with time-varying delays. IEEE Trans. Syst. Man Cybernet. Syst. **50**(6), 2029–2041 (2020)
5. Guo, M., Sun, J.: A high-performance memristive circuit design for DCGAN in edge computing. IEEE Trans. Circuits Syst. I Regul. Pap. 1–13 (2025)
6. Bai, Y., Dou, G., Zhang, T., Guo, M.: Adaptive fuzzy fixed-time control for uncertain time-delay nonlinear systems with output constraints. IEEE Trans. Fuzzy Syst. 1–10 (2025)
7. Guo, M., Zhang, D., Guo, W., Dou, G., Sun, J.: Implementing Brainlike fear generalization and emotional arousal associated with memory. IEEE Trans. Cogn. Dev. Syst. **17**(1), 155–166 (2025)
8. Dunsmoor, J.E., Paz, R.: Fear generalization and anxiety: behavioral and neural mechanisms. Biol. Psychiat. **78**(5), 336–343 (2015)
9. Gidon, A., Zolnik, T.A., Fidzinski, P., Bolduan, F., Papoutsi, A., Poirazi, P.: Dendritic action potentials and computation in human layer 2/3 cortical neurons. Science **367**(6473), 83–87 (2020)
10. Zhang, X., Lu, J., Wang, Z., Wang, R.: Hybrid memristor-CMOS neurons for insitu learning in fully hardware memristive spiking neural networks. Science Bulletin **66**(16), 1624–1633 (2021)
11. Pershin, Y.V., Di Ventra, M.: Experimental demonstration of associative memory with memristive neural networks. Neural Netw. **23**(7), 881–886 (2010)
12. Hu, X., Duan, S., Chen, G., Chen, L.: Modeling affections with memristor based associative memory neural networks. Neurocomputing **223**, 129–137 (2017)
13. Ma, D., Wang, G., Han, J.: A memristive neural network model with associative memory for modeling affections. IEEE Access **6**, 61614–61622 (2018)
14. Sun, J., Han, J., Wang, Y.: Memristor-based neural network circuit of memory with emotional homeostasis. IEEE Trans. Nanotechnol. **21**, 204–212 (2022)
15. Wang, Z., Wang, X., Zeng, Z.: Memristive circuit design of brain-inspired emotional evolution based on theories of internal regulation and external stimulation. IEEE Trans. Biomed. Circuits Syst. **15**(6), 1380–1392 (2021)
16. Mei, G., Zhang, D., Guo, W., Sun, J.: Implementing brain-like fear generalization and emotional arousal associated with memory. IEEE Trans. Cogn. Dev. Syst. **17**(1), 155–166 (2021)
17. Sun, J., Han, J., Wang, Y., Liu, P.: Memristor-based neural network circuit of emotion congruent memory with mental fatigue and emotion inhibition. IEEE Trans. Biomed. Circuits Syst. **15**(3), 606–616 (2021)
18. Sun, J., Wang, Y., Liu, P., Wen, S., Wang, Y.: Memristor-based circuit design of pad emotional space and its application in mood congruity. IEEE Internet Things J. **10**(18), 16332–16342 (2023)
19. Kvatinsky, S., Friedman, E.G., Kolodny, A., Weiser, U.C.: Team: Threshold adaptive memristor model. IEEE Trans. Circuits Syst. I Regul. Pap. **60**(1), 211–221 (2013)
20. Kvatinsky, S., Ramadan, M., Friedman, E.G., Kolodny, A.: VTEAM: a general model for voltage-controlled memristors. IEEE Trans. Circuits Syst. II Express Briefs **62**(8), 786–790 (2015)

MAAF: A Mutual Alignment and Adaptive Fusion Framework for Sketch-Based Image Retrieval

Yaoxuan Zhang[1], Bingrong Xu[1(✉)], Jianhua Yin[2], and Depeng Li[3]

[1] Wuhan University of Technology, Wuhan 430070, China
{zhangyaoxuan,bingrongxu}@whut.edu.cn
[2] Beijing Institute of Technology, Beijing 100081, China
yinjh@bit.edu.cn
[3] Huazhong University of Science and Technology, Wuhan 430074, China
dpli@hust.edu.cn

Abstract. Sketch-based image retrieval (SBIR) remains a challenging cross-modal task due to the substantial modality gap between free-hand sketches and natural photos, as well as large intra-class variations within each domain. To address these issues, we propose Mutual Alignment and Adaptive Fusion framework (MAAF), a novel method that jointly enhances cross-modal representation alignment and adaptive retrieval matching. MAAF comprises two key components: Mutual Knowledge Alignment with Symmetry (MKAS) and Class-Adaptive Distance Fusion (CADF). The MKAS module facilitates bidirectional knowledge exchange between teacher and student networks, enabling the learning of modality-invariant and discriminative embeddings. Meanwhile, the CADF module adaptively integrates Product Quantization (PQ) and Euclidean distances under the guidance of a class-level compactness measure. This adaptive mechanism assigns greater weight to PQ for compact categories to boost efficiency, and higher weight to Euclidean distance for scattered ones to ensure accuracy, balancing retrieval precision and scalability. Extensive experiments on standard ZS-SBIR benchmarks, including Sketchy and TU-Berlin, demonstrate that MAAF consistently outperforms existing methods. It achieves state-of-the-art retrieval accuracy and enhanced generalization to unseen categories, addressing cross modal gaps, the absence of training-test category overlap, and intra-class diversity.

Keywords: Zero-Shot SBIR · Cross-Modal Representation Learning · Knowledge Distillation

1 Introduction

Sketch-Based Image Retrieval (SBIR) aims to retrieve natural images from a large gallery based on free-hand sketch queries. The main challenge lies in the

C. Li et al. (Eds.): ICNC 2025, CCIS 2946, pp. 166–181, 2026.
https://doi.org/10.1007/978-981-92-1599-7_15

inherent modality gap: sketches are highly abstract and sparse, whereas photos contain rich textures and details. Most existing SBIR methods are designed under a closed-set assumption, where training and testing categories overlap. While these approaches achieve promising results [13], their applicability to real-world scenarios remains limited.

To improve generalization, Zero-Shot SBIR (ZS-SBIR) has been proposed, requiring models to retrieve images from categories unseen during training; compared to conventional SBIR, it must simultaneously handle cross-modal discrepancies and category disjointness, substantially increasing task difficulty. Early approaches employed hand-crafted descriptors [31] and BoW-based frameworks [14], while deep learning methods later enabled joint representation learning across modalities. And more recent works further incorporate side information, such as semantic embeddings or class attributes via generative models (GANs [3], VAEs [24]), adversarial training, or auxiliary supervision to enhance transferability, with Generative Mixup Networks [29] further boosting feature generation quality via semantic alignment for zero-shot learning.

Transformer architectures have also been explored for ZS-SBIR: Triplet-ViT [20] leverages a transformer backbone to construct global representations in a multimodal hypersphere. Jiang et al. [8] proposes the ARNet framework, which uses ViT as a backbone, introduces the MSTR Module to recycle discarded Patch Tokens and dual weight-sharing networks for feature alignment. More recently, CLIP-based approaches [16] have been adapted via fine-tuning and prompt learning, injecting auxiliary cues at test time to boost retrieval. Liu et al. [11] exploiting intra-class relationships through clustering to narrow the modal discrepancy. Xu et al. [28] proposed a Low-Rank Optimal Transport method using low-rank constraints to extract data's intrinsic structure and filter noise to mitigate intra-class variation. Despite these advances, most existing methods focus on sketch representations or sketch-to-photo mappings, with relatively little effort devoted to refining gallery-side photo features, and we argue that mitigating intra-class variation and introducing feature regularization on the photo side can complement cross-modal alignment and significantly improve retrieval accuracy.

Another closely related direction is knowledge distillation. Initially proposed for model compression, the teacher-student framework [6] allows a compact student to mimic the output or intermediate features of a larger teacher, and has since been widely applied in semi-supervised learning [23], domain adaptation, and ensemble compression. However, most distillation-based methods in ZS-SBIR adopt a static, unidirectional paradigm, where knowledge flows from teacher to student. This design restricts cross-modal interaction and overlooks complementary information that could benefit both networks.

In this work, we propose Mutual Alignment and Adaptive Fusion framework to tackle the dual challenges of cross-modal alignment and generalization to unseen categories. Unlike unidirectional approaches, our framework introduces the Mutual Knowledge Alignment with Symmetry (MKAS) strategy, enabling bidirectional information exchange between teacher and student networks to enhance modality-invariant and discriminative representation learning. In addition, we propose a Class-Adaptive Distance Fusion (CADF) mechanism that adaptively combines PQ-coded and Euclidean distances according to category compactness, thereby addressing intra-class variance in the photo gallery. Together, these two modules jointly optimize feature learning and retrieval, achieving state-of-the-art performance without incurring excessive complexity.

Our main contributions are summarized as follows:

1. We introduce a symmetrical mutual knowledge alignment strategy (MKAS), where teacher and student networks exchange discriminative cues bidirectionally, enabling more effective cross-modal alignment for the ZS-SBIR task.
2. We propose a class-adaptive distance fusion module (CADF) that dynamically balances PQ-coded and Euclidean distances according to category compactness, effectively mitigating intra-class variance and improving retrieval robustness.
3. We conduct extensive experiments on Sketchy Ext, Sketchy Ext Split, and TU-Berlin Ext, which demonstrate clear advantages over prior approaches and verify the effectiveness of our method.

2 Methods

2.1 Motivation

Zero-shot sketch-based image retrieval (ZS-SBIR) is a challenging cross-modal task due to the large gap between sketches and photos, as well as the absence of category overlap between training and testing classes. Prior works [12,25] often rely on a teacher model pre-trained on large-scale datasets such as ImageNet to provide rich discriminative representations and reduce overfitting under zero-shot conditions. However, such models, whether fixed or lightly fine-tuned, struggle to adapt across modalities, resulting in limited transferability to the sketch domain.

To address these limitations, we propose Mutual Alignment and Adaptive Fusion framework to address inter-modality variance and intra-class diversity in ZS-SBIR. The overall framework of Mutual Alignment and Adaptive Fusion framework is illustrated in Figure 1, which details the key components and their interactions. Its core is a symmetrical bidirectional knowledge alignment mechanism, where the teacher network dynamically evolves through student feedback, enabling adaptive cross-modal supervision.

Furthermore, Mutual Alignment and Adaptive Fusion framework incorporates a cluster-based matching strategy to alleviate intra-class gallery variation, which enhances retrieval precision by aggregating semantically related photo instances.

2.2 Problem Definition

In zero-shot sketch-based image retrieval (ZS-SBIR), the objective is to retrieve relevant natural images using a hand-drawn sketch as a query, with the constraint

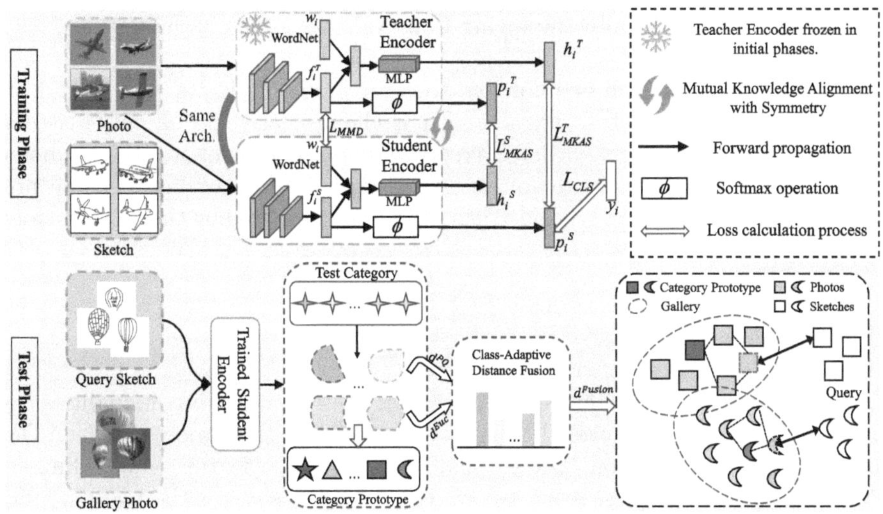

Fig. 1. MAAF framework for Mitigating Inter-modality Variance and Intra-class Diversity in ZS-SBIR.

that training and testing categories are mutually exclusive. Formally, the dataset is partitioned into two disjoint subsets: the source domain $\mathcal{D}_s$ for training and the target domain $\mathcal{D}_t$ for testing, satisfying $\mathcal{D}_s \cap \mathcal{D}_t = \varnothing$. Each domain comprises paired data from two modalities: sketches and photos. The source domain is defined as $\mathcal{D} = \{(x_i^{\text{mod}}, y_i)\}_{i=1}^{M}$, where mod $\in \{S, P\}$ denotes the sketch or photo modality, x_i represents the i-th sample, and y_i corresponds to its semantic label from the image category set.

During training, the model learns a cross-modal embedding function $F_\theta(\cdot)$ that maps both sketches and photos into a shared semantic space, aiming to reduce intra-class distances across these modalities. In the testing phase, given a sketch query from the target domain $\mathcal{D}_t$, the model retrieves semantically aligned images from the gallery set of photos within $\mathcal{D}_t$. The primary challenge of zero-shot sketch-based image retrieval is achieving generalization to unseen categories, without access to labeled data from the target class set $\mathcal{C}_t$ during training.

2.3 Mutual Knowledge Alignment with Symmetry

To enable effective transfer of discriminative knowledge between the teacher and student models, we propose a **Mutual Knowledge Alignment with Symmetry (MKAS)** strategy for cross-modal representation learning. Unlike conventional approaches that freeze all parameters of the teacher network, our method enables both networks to iteratively exchange information in a balanced and bidirectional manner. For simplicity, we adopt identical backbone structures for the two models, specifically employing ConvNeXtV2 as the base feature extractor.

Given an input sketch or photo x_i, the corresponding logits $f_i = F_\theta(x_i)$ are obtained through the feature extractor $F_\theta(\cdot)$. The predictive probability distribution is then computed as $p_i = \text{Softmax}(f_i)$, where $p_i \in \mathbb{R}^K$ denotes the predicted class probabilities over K categories. The classification error between the predicted probability and the ground-truth label y_i is measured by the cross-entropy loss:

$$L_{\text{CLS}} = -\sum_{i=1}^{N} y_i^\top \log(p_i), \tag{1}$$

where N is the number of training samples.

To alleviate the distribution gap between sketch and photo modalities, we introduce a distribution-level alignment term based on the Maximum Mean Discrepancy (MMD) criterion. Let $\mathbf{e}_i^S$ and $\mathbf{e}_j^P$ denote the feature embeddings of sketch and photo samples, respectively. The MMD loss aligns the two modality distributions with a reference Gaussian distribution $\mathcal{G} \sim \mathcal{N}(0,1)$, and is defined as:

$$\mathcal{L}_{\text{MMD}} = \left\| \frac{1}{n}\sum_{i=1}^{n} \phi(\mathbf{e}_i^S) - \frac{1}{m}\sum_{j=1}^{m} \phi(\mathbf{e}_j^P) \right\|_{\mathcal{H}}^2 , \tag{2}$$

where $\phi(\cdot)$ maps features into the Reproducing Kernel Hilbert Space (RKHS), $\|\cdot\|_{\mathcal{H}}$ denotes the Hilbert norm, and n, m are the numbers of sketch and photo samples, respectively. Using the kernel trick, the loss can be equivalently expressed as:

$$\begin{aligned} \mathcal{L}_{\text{MMD}} = \tfrac{1}{n^2}\sum_{i,i'} \kappa(\mathbf{e}_i^S, \mathbf{e}_{i'}^S) + \tfrac{1}{m^2}\sum_{j,j'} \kappa(\mathbf{e}_j^P, \mathbf{e}_{j'}^P) \\ -\tfrac{2}{nm}\sum_{i,j} \kappa(\mathbf{e}_i^S, \mathbf{e}_j^P), \end{aligned} \tag{3}$$

where $\kappa(\cdot,\cdot)$ denotes the Gaussian kernel function.

For the Mutual Knowledge Alignment with Symmetry process, the outputs from both teacher and student networks are adopted as adaptive supervisory signals, enabling iterative co-training. To enrich the semantic guidance, we further incorporate auxiliary information from a pre-trained WordNet embedding, denoted as w_i, with a scaling factor μ. The soft label generated by the teacher network F_{ψ_T} is defined as

$$\mathbf{h}_i^T = \text{MLP}(f_i^T + \mu w_i), \tag{4}$$

while the supervisory signal from the student network F_{ψ_S} is formulated as

$$\mathbf{h}_i^S = \text{MLP}(f_i^S + \mu w_i), \tag{5}$$

where ψ_T and ψ_S are the parameters of the teacher and student classifiers, respectively. Different from conventional softmax-based mappings, this MLP transformation provides greater flexibility by capturing complex inter-class relations.

During training, the teacher provides soft labels to guide the student model, forming a bidirectional alignment objective for the student:

$$L^S_{\text{MKAS}} = -\sum_{i=1}^{N} (\mathbf{h}_i^T)^\top \log(p_i^S), \tag{6}$$

leading to the overall training objective:

$$L^S = L_{\text{CLS}} + L^S_{\text{MKAS}} + \lambda \mathcal{L}_{\text{MMD}}, \tag{7}$$

where λ balances the regularization strength.

Symmetrically, the student network provides supervisory signals to update the teacher model, with the corresponding loss defined as

$$L^T_{\text{MKAS}} = -\sum_{i=1}^{N} (\mathbf{h}_i^S)^\top \log(p_i^T). \tag{8}$$

Training proceeds in an alternating manner, where the parameters of the teacher and student models are updated iteratively. This symmetrical bidirectional learning paradigm ensures a continuous exchange of informative knowledge across modalities, thereby promoting robust and modality-invariant feature learning.

2.4 Probabilistic Gallery Modeling and Class-Adaptive Distance Fusion

In zero-shot sketch-based image retrieval, large inter-modality discrepancies and complex intra-class variations pose persistent challenges. While many previous methods attempt to mitigate these issues through direct feature-level alignment between sketches and photos, such sample-based matching often results in unstable optimization and limited generalization due to the highly diverse gallery feature distribution.

To address this problem, we first model the probabilistic structure of the gallery domain in a compact and discriminative manner. Specifically, a Gaussian Mixture Model (GMM) is employed to capture the latent composition of the gallery feature space. This probabilistic representation expresses the gallery as a weighted combination of multiple Gaussian components, each corresponding to a locally coherent cluster. By leveraging representative component means rather than individual samples, this modeling enhances retrieval robustness and computational efficiency.

Formally, for an encoded sketch or image feature $\mathbf{z}_i = E_\theta(x_i)$, the gallery feature distribution is formulated as:

$$p(\mathbf{z}^G) = \sum_{r=1}^{R} \pi_r \mathcal{N}(\mathbf{z}^G \mid \boldsymbol{\mu}_r, \boldsymbol{\Sigma}_r), \tag{9}$$

where π_r denotes the mixture coefficient of the r-th Gaussian component ($\sum_{r=1}^{R} \pi_r = 1$, $\pi_r \in [0, 1]$), and $\boldsymbol{\mu}_r$, $\boldsymbol{\Sigma}_r$ represent its mean and covariance, respectively. The parameters are optimized via the ExpectationâĂŞMaximization (EM) [19] procedure until convergence. This probabilistic gallery structure provides a foundation for the subsequent adaptive distance fusion.

To further alleviate inter-modality and intra-class variations, we propose a Class-Adaptive Distance Fusion (CADF) strategy that adaptively combines Product Quantization (PQ) and Euclidean distances according to class compactness. PQ compresses high-dimensional gallery features by partitioning the embedding space into m disjoint subspaces, each with dimension D/m. For each subspace u, a codebook $\mathcal{C}^{(u)} = \{\boldsymbol{\mu}_1^{(u)}, \boldsymbol{\mu}_2^{(u)}, \dots, \boldsymbol{\mu}_k^{(u)}\}$ is constructed, where each codeword $\boldsymbol{\mu}_j^{(u)}$ corresponds to the mean of a Gaussian component estimated from the GMM in that subspace. This formulation ensures consistency between the probabilistic gallery modeling and the quantization process, allowing PQ to be interpreted as a discretized approximation of the continuous GMM-based representation.

Formally, a gallery feature vector $\mathbf{f} \in \mathbb{R}^D$ is divided into sub-vectors:

$$\mathbf{f} = [\mathbf{f}^{(1)}, \mathbf{f}^{(2)}, \dots, \mathbf{f}^{(m)}], \quad \mathbf{f}^{(u)} \in \mathbb{R}^{D/m}, \ u = 1, \dots, m. \tag{10}$$

Each sub-vector $\mathbf{f}^{(u)}$ is quantized by assigning it to the Gaussian component with the maximum posterior probability:

$$Q(\mathbf{f}^{(u)}) = \boldsymbol{\mu}_{j^*}^{(u)}, \quad j^* = \arg\max_j p(j \mid \mathbf{f}^{(u)}), \tag{11}$$

where $p(j \mid \mathbf{f}^{(u)})$ is the posterior probability computed from the subspace GMM. The PQ-based distance between a query $\mathbf{q}$ and a gallery feature $\mathbf{f}$ is then expressed as:

$$d_{\mathrm{PQ}}(\mathbf{q}, \mathbf{f}) = \sum_{u=1}^{m} \|\mathbf{q}^{(u)} - Q(\mathbf{f}^{(u)})\|_2^2. \tag{12}$$

Although PQ provides efficient representation, it may suffer in accuracy for classes with high intra-class dispersion. To address this, we introduce a class-level compactness score γ_c, which measures the ratio between intra-class dispersion and inter-class separation:

$$\gamma_c = \frac{\frac{1}{|S_c|} \sum_{i,j \in S_c} \|\mathbf{f}_i - \mathbf{f}_j\|_2}{\frac{1}{|\mathcal{C}|-1} \sum_{c' \neq c} \|\mathbf{p}_c - \mathbf{p}_{c'}\|_2}, \tag{13}$$

where S_c denotes the set of samples belonging to class c, and $\mathbf{p}_c$ represents the prototype associated with its corresponding Gaussian component. A smaller γ_c indicates higher compactness and better separability in the embedding space. In practice, compactness scores vary moderately across categories, reflecting relative intra-class dispersion. This normalized variation enables stable and interpretable adaptive weighting, where compact categories receive higher PQ weights, while more dispersed ones rely more on Euclidean matching.

The adaptive fusion weight α_c is then determined through normalized inverse mapping:

$$\alpha_c = \alpha_{\min} + \frac{\gamma_{\max} - \gamma_c}{\gamma_{\max} - \gamma_{\min}} \cdot (\alpha_{\max} - \alpha_{\min}), \tag{14}$$

where $\alpha_{\min}$ and $\alpha_{\max}$ define the allowable range of weights, and $\gamma_{\min}, \gamma_{\max}$ are the extremal compactness values across all classes.

Finally, the query-to-gallery retrieval distance is defined as:

$$d_{i,j}^{\text{fusion}} = \alpha_c \cdot d_{i,j}^{\text{PQ}} + (1 - \alpha_c) \cdot d_{i,j}^{\text{Euc}}, \quad i \in \text{query of class } c, \tag{15}$$

where $d_{i,j}^{\text{PQ}}$ and $d_{i,j}^{\text{Euc}}$ denote the PQ-based and Euclidean distances, respectively. In this way, compact categories (small γ_c) benefit more from PQ-based efficiency, while dispersed categories (large γ_c) rely more on Euclidean matching. Unlike traditional clustering indices such as the Davies–Bouldin Index [15], which are typically used only for evaluation, our compactness score is embedded directly into the retrieval process as an adaptive modulation factor, enhancing both robustness and adaptability.

3 Experimental Design

3.1 Experimental Setting

To comprehensively evaluate the effectiveness of our proposed framework, we first benchmark its performance on several widely-used SBIR datasets, comparing it against existing state-of-the-art methods. Following this, we conduct a series of ablation studies and hyperparameter sensitivity analyses to explore the influence of each individual module and design choice on the overall system performance.

Additionally, we present qualitative results, including retrieval visualizations and feature space projections, to further validate the model's ability to capture discriminative representations and to highlight its generalization capability under the zero-shot setting.

In line with previous studies, we adopt Mean Average Precision (mAP) and Precision as the primary evaluation criteria. For the Sketchy Extended [18] and TU-Berlin Extended [4] datasets, we evaluate the retrieval performance using mAP and Precision at the top 100 retrieved results (Prec@100). Meanwhile, for the Sketchy Extended Split, we report both mAP and Precision at rank 200 (mAP@200 and Prec@200), following the conventions established in prior work.

3.2 Implementation Details

All experiments are implemented using PyTorch and conducted on a single NVIDIA RTX 3080 Ti GPU. For the symmetrical bidirectional knowledge alignment mechanism, both the student and teacher models share the same backbone architecture, specifically the ConvNeXtV2-Tiny network. To mitigate the risk of model collapse in the early training stage caused by mutual constraint from symmetrical bidirectional knowledge alignment, we initially freeze the gradients of the teacher network. The number of training epochs allocated to the designed Mutual Knowledge Alignment with Symmetry is set to 15 for the Sketchy Extended dataset and 10 for the TU-Berlin Extended dataset, with the total training epochs fixed at 30. The student model is optimized via the Adam optimizer, which adopts an initial learning rate of 1×10^{-4} that gradually decays to a final value of 1×10^{-7}. For the Sketchy Extended dataset, the teacher model uses the same optimizer parameters as the student model. In contrast, on the TU-Berlin Extended dataset, the teacher model employs a distinct learning rate schedule—starting at 2×10^{-5} and decaying to 5×10^{-8}. In the test phase, the number of subspaces is set to 4 for both the Sketchy Extended and Sketchy Extended Split datasets, with the codewords per subspace corresponding to the number of gallery categories, which are 26 and 22 respectively. For the TU-Berlin Extended dataset, we set the number of subspaces to 2, and the codewords per subspace are configured as 32.

3.3 Performance Comparison

We evaluate the performance of our approach against a range of established techniques in the ZS-SBIR domain. For basic architectures, we include DOODLE [2], which relies on a semantic reconstruction framework. Advanced generative and alignment techniques are represented by SEM-PCYC [3], SAKE [12], LCALE [10], OCEAN [32], RPKD [21], DSN [27], SBTKNet [22], TVT [20], Sketch3T [17], PCMSN [1], IIAE [7], GTZSR [5], OAN [30], and CA [26], utilizing complex models like cycle consistency, teacher-student learning, latent alignment, dual training, distillation, regularization, transformers, and cross-domain alignment. Additionally, the explainable and extended approach ZSE-SBIR is evaluated for its broader applicability (Table 1).

Our approach achieves outstanding retrieval accuracy, surpassing all state-of-the-art methods. Unlike previous works, we employ a symmetrical teacherâĂŞstudent bidirectional learning scheme, ensuring mutual knowledge alignment across modalities. Features are extracted via an MLP projection head to enhance representation compactness. Furthermore, an adaptive fusion strategy allocates weights between PQ and Euclidean distances based on class-level compactness, enabling compact classes to exploit PQ more effectively while less compact ones rely on Euclidean metrics. This design, combined with pairwise training, enhances cross-modality alignment and yields superior mAP and Precision on the Sketchy Ext Split dataset.

3.4 Ablation Study

Our framework is composed of two core modules: Mutual Knowledge Alignment with Symmetry (MKAS) and Class-Adaptive Distance Fusion (CADF). Starting from a baseline model that applies unidirectional knowledge distillation with

Table 1. Comparison of results with state-of-the-art methods on public datasets(%).

Method	Sketchy Ext		Sketchy Ext Split		TU-Berlin Ext	
	mAP@all	Prec@100	mAP@200	Prec@ 200	mAP@all	Prec@100
SEM-PCYC [3]	34.9	46.3	–	–	29.7	42.6
DOODLE [2]	–	–	36.9	–	10.9	–
LCALE [10]	47.6	58.3	–	–	–	–
OCEAN [32]	46.2	59.0	–	–	33.3	46.7
RPKD [21]	61.3	72.3	50.2	59.8	48.6	61.2
DSN [27]	58.3	70.4	50.1	59.7	48.1	58.6
SBTKNet [22]	55.3	69.8	50.2	59.6	48.0	60.8
TVT [20]	64.8	79.6	53.1	61.8	48.4	66.2
Sketch3T [17]	57.5	–	–	–	50.7	–
SAKE [12]	54.7	69.2	49.7	59.8	47.5	59.9
PCMSN [1]	52.3	61.6	–	–	42.4	51.7
IIAE [7]	57.3	65.9	37.3	48.5	41.2	50.3
GTZSR [5]	61.7	64.0	–	–	62.8	66.8
OAN [30]	61.7	73.7	–	62.1	50.5	62.5
CA [26]	63.3	70.9	–	–	54.2	62.2
ZSE-SBIR [9]	73.6	80.8	52.5	62.4	56.9	63.7
Ours	**79.4**	78.8	**61.8**	**67.3**	**64.3**	62.5

only classification and modality alignment constraints, we find that incorporating MKAS enables bidirectional information exchange between teacher and student networks. This symmetrical knowledge alignment significantly enhances the learning of modality-invariant and discriminative representations, thereby improving cross-modal retrieval performance.

Table 2. Ablation study results on Sketchy Ext and Sketchy Ext Split datasets.(%)

Models	Sketchy Ext		Sketchy Ext Split	
	mAP@all	Prec@100	mAP@all	Prec@200
w/o CADF	67.5	73.0	52.9	60.0
w/o MKAS	76.6	74.1	59.4	65.5
Full model	79.4	78.8	61.8	67.3

On the distance design side, CADF adaptively integrates PQ-coded and Euclidean distances based on the compactness of the query's class. Compact categories benefit more from PQ encoding, as it can exploit fine-grained structure, while less compact categories rely more on Euclidean distance to preserve inter-instance separability. Compared with single-metric approaches such as Euclidean

Only or PQ-based Distance, this adaptive fusion strategy mitigates intra-class variation and yields better accuracy and generalization. Finally the quantitative results of the ablation study on Sketchy Ext and Sketchy Ext Split are summarized in Table 2. And the effectiveness of this distance metric design is further illustrated in Fig. 2, which compares PQ-coded distance, Euclidean distance, and the proposed adaptive fusion on the Sketchy Ext and TU-Berlin Ext datasets.

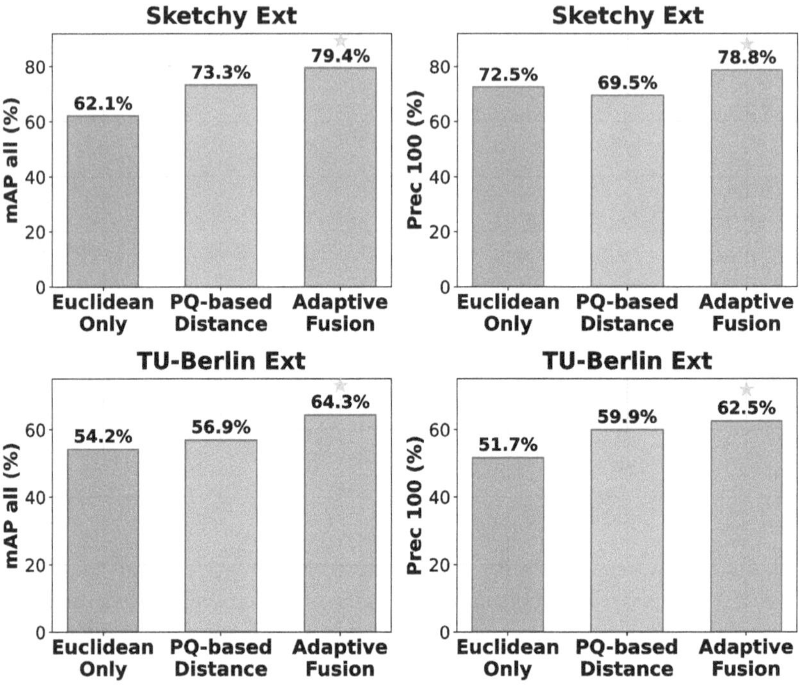

Fig. 2. Distance metric ablation study on Sketchy Ext and TU-Berlin Ext.

We further analyze the impact of PQ parameters on retrieval performance via adjustments on the Sketchy Extended dataset. As shown in Fig. 3, both metrics perform well when the number of subspaces m is 4; specifically, mAP@all achieves the best accuracy when k is 25, while Prec@100 reaches its optimal value when k is 26.

3.5 Visualization

Figure 4 illustrates the adaptive weight allocation mechanism using representative test categories from the Sketchy Extended dataset. The left panel presents the compactness scores, defined as the ratio between intra-class and inter-class distances, across multiple test categories. Compact categories such as Alarm Clock and Car (Sedan), characterized by consistent structural patterns, exhibit

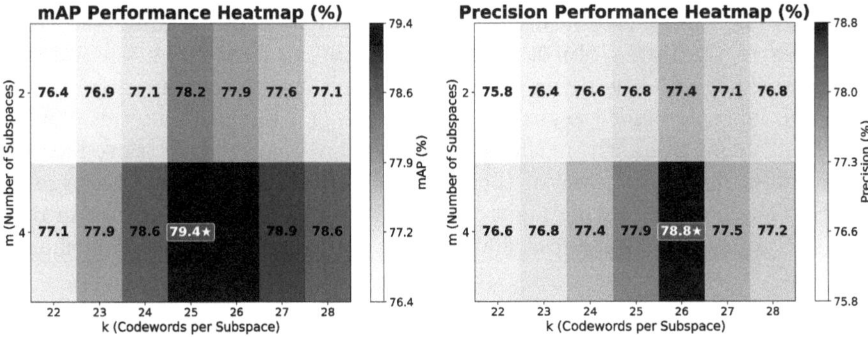

Fig. 3. Heatmap Visualization for Product Quantization Parameter Performance Sensitivity on Sketchy Extended dataset.

lower compactness values. In contrast, visually diverse categories such as Castle and Tree show higher compactness scores due to larger intra-class variations. The right panel demonstrates the corresponding adaptive weight allocation strategy: compact categories with concentrated feature distributions are assigned higher PQ encoding weights to exploit quantization efficiency, whereas dispersed categories with scattered feature patterns receive higher Euclidean distance weights to enhance discrimination robustness. This data-driven mechanism adaptively balances efficiency and accuracy by modulating distance fusion according to the intrinsic visual characteristics of each category, thereby eliminating manual parameter tuning and improving retrieval performance across heterogeneous visual content.

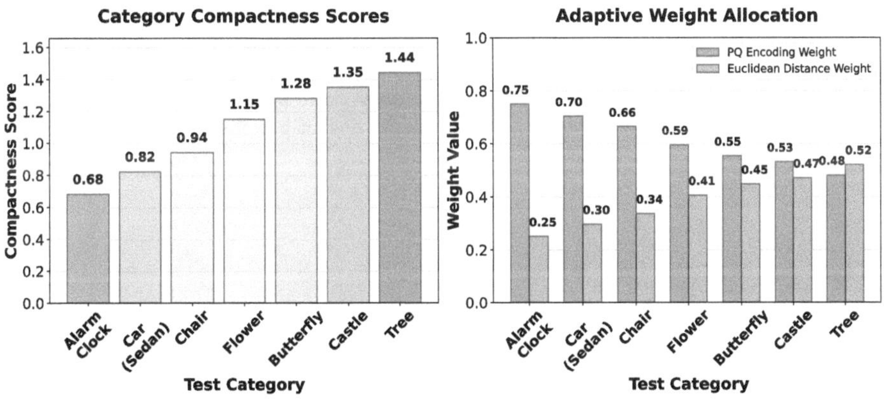

Fig. 4. Adaptive Weight Allocation on Sketchy Extended Dataset Test Categories.

Figure 5 shows the retrieval results generated by our model. In most cases, the model successfully retrieves the correct candidates for each sketch query.

However, some failure cases still occur when incorrect photos exhibit structural similarities to the target object. For example, in the fourth row, all retrieved results are expected to be flying birds, but seagulls are mistakenly retrieved due to their visual resemblance to generic bird sketches, particularly in terms of wings and overall shape. Similarly, in the last row, bees are incorrectly retrieved instead of butterflies, mainly because of their similar body structures and background textures. These examples highlight the difficulty in distinguishing fine-grained visual categories when instances share similar contours or contextual elements.

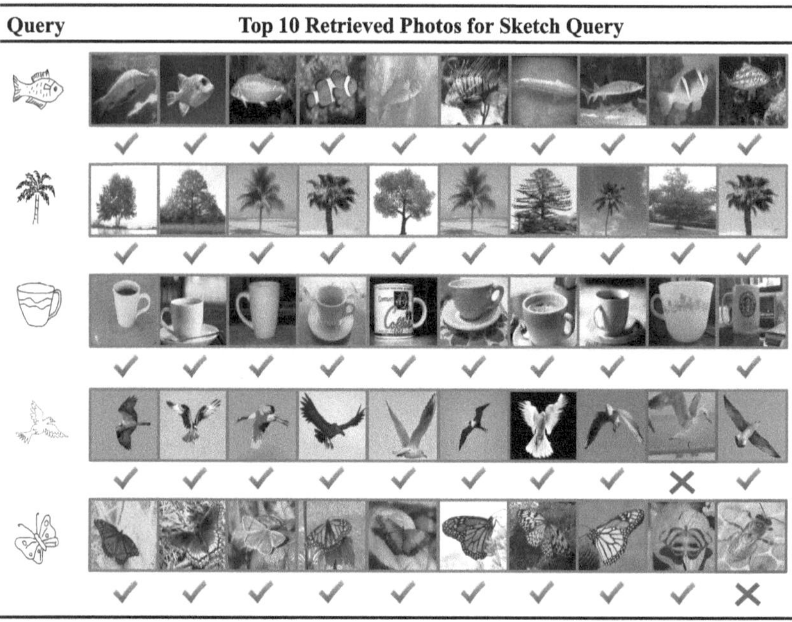

Fig. 5. Top-10 retrieval results for query sketches from both Sketchy Ext and TU-Berlin Ext datasets. The correct results are marked with green boxes, while incorrect ones are marked with red boxes. (Color figure online)

4 Conclusion

In this work, we propose a novel zero-shot sketch-based image retrieval framework built upon MKAS. During training, MKAS enables teacher and student networks to collaboratively learn generalized visualâĂŞsemantic representations through mutual knowledge exchange, thereby enhancing cross-modal transfer and semantic consistency. During inference, a cluster-aware retrieval strategy is employed by modeling gallery features with a Gaussian Mixture Model, where component centroids serve as compact yet discriminative representations. Building on this probabilistic modeling, we further introduce a CADF mechanism that dynamically balances PQ and Euclidean distances according to category compactness, effectively mitigating inter-modality and intra-class variations. Extensive experiments on public ZS-SBIR benchmarks demonstrate the effectiveness of the proposed framework. In future work, we aim to extend MKAS to more complex cross-modal retrieval scenarios, such as cross domain and text-to-sketch retrieval—to develop more generalizable and unified visual representations.

Acknowledgments. This work was supported by the Natural Science Foundation of China under Grants 52402519, 623B2040, and the Postdoctoral Fellowship Program of CPSF under Grant GZB20250428.

References

1. Deng, C., Xu, X., Wang, H., Yang, M., Tao, D.: Progressive cross-modal semantic network for zero-shot sketch-based image retrieval. IEEE Trans. Image Process. **29**, 8892–8902 (2020)
2. Dey, S., Riba, P., Dutta, A., Llados, J., Song, Y.Z.: Doodle to search: practical zero-shot sketch-based image retrieval. In: Proceedings of the IEEE/CVF Conference on Computer Vision and Pattern Recognition, pp. 2179–2188 (2019)
3. Dutta, A., Akata, Z.: Semantically tied paired cycle consistency for zero-shot sketch-based image retrieval. In: Proceedings of the IEEE/CVF Conference on Computer Vision and Pattern Recognition, pp. 5089–5098 (2019)
4. Eitz, M., Hildebrand, K., Boubekeur, T., Alexa, M.: An evaluation of descriptors for large-scale image retrieval from sketched feature lines. Comput. Graphics **34**(5), 482–498 (2010)
5. Gupta, S., Chaudhuri, U., Banerjee, B., Kumar, S.: Zero-shot sketch based image retrieval using graph transformer. In: 2022 26th International Conference on Pattern Recognition (ICPR), pp. 1685–1691. IEEE (2022)
6. Hinton, G., Vinyals, O., Dean, J.: Distilling the knowledge in a neural network. arXiv preprint arXiv:1503.02531 (2015)
7. Hwang, H., Kim, G.H., Hong, S., Kim, K.E.: Variational interaction information maximization for cross-domain disentanglement. Adv. Neural. Inf. Process. Syst. **33**, 22479–22491 (2020)
8. Jiang, J., Tang, H., Jiang, Z., Yu, W., Wu, D.: ARNet: self-supervised FG-SBIR with unified sample feature alignment and multi-scale token recycling. In: Proceedings of the AAAI Conference on Artificial Intelligence, vol. 39, pp. 3985–3993 (2025)
9. Lin, F., Li, M., Li, D., Hospedales, T., Song, Y.Z., Qi, Y.: Zero-shot everything sketch-based image retrieval, and in explainable style. In: Proceedings of the IEEE/CVF Conference on Computer Vision and Pattern Recognition, pp. 23349–23358 (2023)
10. Lin, K., Xu, X., Gao, L., Wang, Z., Shen, H.T.: Learning cross-aligned latent embeddings for zero-shot cross-modal retrieval. In: Proceedings of the AAAI Conference on Artificial Intelligence, vol. 34, pp. 11515–11522 (2020)
11. Liu, D., Luo, X., Peng, C., Wang, N., Hu, R., Gao, X.: Symmetrical bidirectional knowledge alignment for zero-shot sketch-based image retrieval. arXiv preprint arXiv:2312.10320 (2023)

12. Liu, Q., Xie, L., Wang, H., Yuille, A.L.: Semantic-aware knowledge preservation for zero-shot sketch-based image retrieval. In: Proceedings of the IEEE/CVF International Conference on Computer Vision, pp. 3662–3671 (2019)
13. Lu, P., Huang, G., Fu, Y., Guo, G., Lin, H.: Learning large Euclidean margin for sketch-based image retrieval. arXiv preprint arXiv:1812.04275, **1**(2), 3 (2018)
14. Qian, X., Tan, X., Zhang, Y., Hong, R., Wang, M.: Enhancing sketch-based image retrieval by re-ranking and relevance feedback. IEEE Trans. Image Process. **25**(1), 195–208 (2015)
15. Ros, F., Riad, R., Guillaume, S.: PDBI: a partitioning Davies-Bouldin index for clustering evaluation. Neurocomputing **528**, 178–199 (2023)
16. Sain, A., Bhunia, A.K., Chowdhury, P.N., Koley, S., Xiang, T., Song, Y.Z.: Clip for all things zero-shot sketch-based image retrieval, fine-grained or not. In: Proceedings of the IEEE/CVF Conference on Computer Vision and Pattern Recognition, pp. 2765–2775 (2023)
17. Sain, A., Bhunia, A.K., Potlapalli, V., Chowdhury, P.N., Xiang, T., Song, Y.Z.: Sketch3t: Test-time training for zero-shot SBIR. In: Proceedings of the IEEE/CVF Conference on Computer Vision and Pattern Recognition, pp. 7462–7471 (2022)
18. Sangkloy, P., Burnell, N., Ham, C., Hays, J.: The sketchy database: learning to retrieve badly drawn bunnies. ACM Trans. Graphics (TOG) **35**(4), 1–12 (2016)
19. Sheng, S., Feng, B., Wang, H., Xu, P., Wu, C.: Expectation-maximization framework-based image-domain least-squares migration. Geophysics **88**(4), R545–R558 (2023)
20. Tian, J., Xu, X., Shen, F., Yang, Y., Shen, H.T.: TVT: three-way vision transformer through multi-modal hypersphere learning for zero-shot sketch-based image retrieval. In: Proceedings of the AAAI Conference on Artificial Intelligence, vol. 36, pp. 2370–2378 (2022)
21. Tian, J., Xu, X., Wang, Z., Shen, F., Liu, X.: Relationship-preserving knowledge distillation for zero-shot sketch based image retrieval. In: Proceedings of the 29th ACM International Conference on Multimedia, pp. 5473–5481 (2021)
22. Tursun, O., Denman, S., Sridharan, S., Goan, E., Fookes, C.: An efficient framework for zero-shot sketch-based image retrieval. Pattern Recogn. **126**, 108528 (2022)
23. Urban, G., et al.: Do deep convolutional nets really need to be deep and convolutional? arXiv preprint arXiv:1603.05691 (2016)
24. Vivekananthan, S.: Comparative analysis of generative models: enhancing image synthesis with VAEs, GANs, and stable diffusion. arXiv preprint arXiv:2408.08751 (2024)
25. Wang, H., Deng, C., Liu, T., Tao, D.: Transferable coupled network for zero-shot sketch-based image retrieval. IEEE Trans. Pattern Anal. Mach. Intell. **44**(12), 9181–9194 (2021)
26. Wang, X., Peng, D., Hu, P., Gong, Y., Chen, Y.: Cross-domain alignment for zero-shot sketch-based image retrieval. IEEE Trans. Circuits Syst. Video Technol. **33**(11), 7024–7035 (2023)
27. Wang, Z., Wang, H., Yan, J., Wu, A., Deng, C.: Domain-smoothing network for zero-shot sketch-based image retrieval. arXiv preprint arXiv:2106.11841 (2021)
28. Xu, B., Yin, J., Lian, C., Su, Y., Zeng, Z.: Low-rank optimal transport for robust domain adaptation. IEEE/CAA J. Automatica Sinica **11**(7), 1667–1680 (2024)
29. Xu, B., Zeng, Z., Lian, C., Ding, Z.: Generative mixup networks for zero-shot learning. IEEE Trans. Neural Networks Learn. Syst. **36**(3), 4054–4065 (2022)
30. Zhang, H., Jiang, H., Wang, Z., Cheng, D.: Ontology-aware network for zero-shot sketch-based image retrieval. In: ICASSP 2023-2023 IEEE International Conference on Acoustics, Speech and Signal Processing (ICASSP), pp. 1–5. IEEE (2023)

31. Zhang, N., Shamey, R., Xiang, J., Pan, R., Gao, W.: A novel image retrieval strategy based on transfer learning and hand-crafted features for wool fabric. Expert Syst. Appl. **191**, 116229 (2022)
32. Zhu, J., Xu, X., Shen, F., Lee, R.K.W., Wang, Z., Shen, H.T.: Ocean: a dual learning approach for generalized zero-shot sketch-based image retrieval. In: 2020 IEEE International Conference on Multimedia and Expo (ICME), pp. 1–6. IEEE Computer Society (2020)

DoS Attacks Compensation Controller Design for Networked Control Systems with Weighted-Try-Once-Discard Protocol

Changhao Li, Yingnan Pan(✉), and Wen Yang

College of Control Science and Engineering, Bohai University, Jinzhou 121013, China
panyingnan@qymail.bhu.edu.cn

Abstract. This paper investigates the design problem of denial-of-service (DoS) attacks compensation controller for interval type-2 networked control systems with weighted-try-once-discard protocol (WTODP). Firstly, in order to mitigate the adverse impact of DoS attacks, an improved DoS attacks compensation mechanism is constructed in this paper. Compared to the existing compensation mechanisms, the proposed compensation mechanism obtains more accurate compensation signals through the designed functions, further reducing the adverse impact caused by DoS attacks. Secondly, network resources are efficiently conserved through the introduced WTODP. Combining the content mentioned above, a mode-dependent Lyapunov function is utilized to verify the stochastic stability of the closed-loop-system. Finally, a simulation verification is provided to demonstrate the effectiveness of the control strategy.

Keywords: Denial-of-service attacks · compensation controller · weighted-try-once-discard protocol · networked control systems

1 Introduction

In recent years, nonlinear problem is an important research direction in various systems. In the study of nonlinear problem, Takagi-Sugeno (T-S) fuzzy model effectively addresses the issue of nonlinear terms present in the system model through membership functions [1,2]. The authors in [1] addressed a disturbance attenuation problem for T-S fuzzy systems by constructing a membership function-dependent H_∞ performance index. Liang *et al.* [2] investigated the predefined-time stabilization problems for T-S fuzzy systems, designing a class of predefined-time integral sliding mode surface. With the deepening research on nonlinear problem, the issue of uncertain parameters has gradually attracted the attention of scholars, which cannot be addressed by traditional T-S models. Based on such a situation, the interval type-2 (IT-2) T-S fuzzy model has been proposed, effectively resolving the issue of parameter uncertainties [3,4]. For

C. Li et al. (Eds.): ICNC 2025, CCIS 2946, pp. 182–194, 2026.
https://doi.org/10.1007/978-981-92-1599-7_16

instance, the authors in [4] investigated the distributed outlier-resilient fusion estimation problem through an IT-2T-S fuzzy model, integrating outlier detection schemes and dimensionality reduction strategies. However, with the development of network, the utilization of T-S fuzzy models in networked control systems (NCSs) has become a frontier area of academic research, which is not considered in the aforementioned works.

In NCSs, network security is one of the academic problems studied by scholars, the performance of systems may be severely impacted when the network is under malicious attacks. In order to mitigate such a negative impact, scholars have made significant efforts, such as resilient event-triggered mechanisms [5,6] and compensation mechanisms [7,8]. To name a few, the authors in [6] investigated the problem of resilient event-triggered H_∞ filtering for networked fuzzy systems under denial-of-service (DoS) attacks, constructing a resilient event-triggered mechanism to mitigate the adverse impact of DoS attacks. Su *et al.* [8] designed a DoS attacks compensation mechanism, which provides compensation signals during attacks to enhance the system performance. In comparison to the resilient event-triggered mechanisms [5,6], the attacks compensation mechanisms [7,8] utilize compensation signals to reduce the impact of attacks while avoiding the transmission of additional network data, which is the reason that the attack compensation mechanism is considered in this paper. In addition, the compensation signals can be more precise through predictions, thereby achieving the goal of further enhancing system performance, which is one of the motivations of this paper.

In addition to network security, network resources is also a consideration in the NCSs. In production, network bandwidth is often limited, therefore the research direction of reducing the consumption of network resources has attracted a large number of researches from scholars. In the existing works, the methods that have been proven to effectively save network resources are communication protocols, including FlexRay protocol [9], Round-Robin protocol [10] and weighted try-once-discard protocol (WTODP) [11]. Among these researches, Liu *et al.* [11] investigated the observer-based fuzzy control problem for networked systems with WTODP. The WTODP completes the scheduling of sensors by using the deviation on each sensor to obtain the corresponding index, thereby saving network resources. However, the investigation on the utilization of WTODP to conserve network resources in NCSs with DoS attacks compensation mechanism has received much less attention, which is another motivation of this paper.

Considering the above motivations, this paper studies the design problem of DoS attacks compensation controller for IT-2 NCSs with WTODP. An improved DoS attacks compensation mechanism is constructed. Then, a WTODP is introduced to conserve network resources. Compared to the existing work [8], this paper achieves more accurate compensation signals by designing prediction functions. Combining WTODP and DoS attacks compensation mechanism, this paper proposes a controller design method that ensures the stochastic stability of IT-2 NCSs.

2 Problem Formulation

2.1 IT-2 Fuzzy Model

This subsection presents the IT-2 fuzzy model in the following form.
Plant Rule i: **IF** $\tau_1(x(h))$ is υ_1^i, and $\tau_2(x(h))$ is υ_2^i and,..., and $\tau_p(x(h))$ is υ_p^i, **THEN**

$$\begin{aligned} x(h+1) &= \mathcal{A}_i x(h) + \mathcal{B}_i u_a(h) + \mathcal{B}_{1i} w(h), \\ z(h) &= \mathcal{C}_i x(h), \end{aligned} \tag{1}$$

where υ_k^i is the fuzzy set ($i = 1, \ldots, l$ and $k = 1, \ldots, p$). l and p denote the numbers of fuzzy rules and premise variables, respectively. Define $\tau_k(h) \triangleq \tau_k(x(h))$ to be premise variable. $x(h) \in \mathbb{R}^x$, $u_a(h) \in \mathbb{R}^u$ and $z(h) \in \mathbb{R}^z$ represent the state vector, the input variable and the regulated output, respectively. $\omega(h) \in \mathbb{R}^\omega$ refers to the external disturbance that belongs to $\ell_1[0, \infty)$. $\mathcal{A}_i$, $\mathcal{B}_i$, $\mathcal{B}_{1i}$ and $\mathcal{C}_i$ are known matrices with appropriate dimensions. The upper membership function (UMF) and the lower membership function (LMF) of the system are defined as follows:

$$\begin{aligned} 0 \le \prod_{k=1}^{p} \bar{\eta}_{\upsilon_k^i}(\tau_k(h)) = \bar{\varpi}_i(\tau_k(h)) \le 1, \\ 0 \le \prod_{k=1}^{p} \underline{\eta}_{\upsilon_k^i}(\tau_k(h)) = \underline{\varpi}_i(\tau_k(h)) \le 1, \end{aligned}$$

where $\underline{\eta}_{\upsilon_k^i}(\tau_k(h))$ and $\bar{\eta}_{\upsilon_k^i}(\tau_k(h))$ denote the lower grade of membership and the upper grade of membership satisfying $\bar{\eta}_{\upsilon_k^i}(\tau_k(h)) \ge \underline{\eta}_{\upsilon_k^i}(\tau_k(h)) \ge 0$.

Then, the system (1) can be illustrated as follows:

$$\begin{aligned} x(h+1) &= \sum_{i=1}^{l} \varpi_i(\tau_k(h)) \left[\mathcal{A}_i x(h) + \mathcal{B}_i u_c(h) + \mathcal{B}_{1i} \omega(h)\right], \\ z(h) &= \sum_{i=1}^{l} \varpi_i(\tau_k(h)) \mathcal{C}_i x(h), \end{aligned} \tag{2}$$

where

$$\begin{aligned} \tilde{\varpi}_i(\tau_k(h)) =& \underline{\vartheta}_i(\tau_k(h))\underline{\varpi}_i(\tau_k(h)) + \bar{\vartheta}_i(\tau_k(h))\bar{\varpi}_i(\tau_k(h)), \\ \varpi_i(\tau_k(h)) =& \frac{\tilde{\varpi}_i(\tau_k(h))}{\sum\limits_{i=1}^{l} \tilde{\varpi}_i(\tau_k(h))} \ge 0, \ \sum_{i=1}^{l} \varpi_i(\tau_k(h)) = 1, \end{aligned}$$

with $w_i(\tau_k(h))$ being the normalized membership function. $\underline{\vartheta}_i(\tau_k(h))$ and $\bar{\vartheta}_i(\tau_k(h))$ represent nonlinear weighting functions satisfying $0 \le \underline{\vartheta}_i(\tau_k(h)) \le 1$, $0 \le \bar{\vartheta}_i(\tau_k(h)) \le 1$ and $\underline{\vartheta}_i(\tau_k(h)) + \bar{\vartheta}_i(\tau_k(h)) = 1$.

2.2 WTODP

This subsection introduces the WTODP to conserve network resources. The details are given as follows:

Define $x\left(k\right)=[{x_1}^T(h),...,{x_r}^T(h),...,{x_{n_x}}^T(h)]^T$, where x_r and n_x denote the state variable of rth sensor and the number of sensors, respectively.

Define $\mu_r(h)$ to be the deviation of the rth sensor

$$\mu_r(h)=(\tilde{x}_r(h-1)-x_r(h))^T\Omega_r(\tilde{x}_r(h-1)-x_r(h)), \tag{3}$$

where $\Omega_r>0$ and $\tilde{x}_r(h-1)$ denote the weighting parameter and the previous transmitted data from the rth sensor, respectively. Then, the index of WTODP can be illustrated as follows:

$$\varphi(h)=\arg\max_{1\le r\le n_x}\mu_r(h). \tag{4}$$

When deviations and index are obtained, the data transmitted to the controller can be expressed as follows:

$$\tilde{x}_r(h)=\begin{cases} x_r(h), & \text{if } r=\varphi(h),\\ \tilde{x}_r(h-1), & \text{else,}\end{cases} \tag{5}$$

applying Kronecker delta function $\vartheta(\cdot)\in\{0,1\}$ to represent the overall data transmission, one can have

$$\tilde{x}(h)=\Theta_{\varphi(h)}x(h)+\bar{\Theta}_{\varphi(h)}\tilde{x}(h-1), \tag{6}$$

where

$$\begin{aligned}\Theta_{\varphi(h)}&=\mathrm{diag}\{\vartheta(\varphi(\mathrm{h})-1),...,\vartheta(\varphi(\mathrm{h})-\mathrm{n_x})\},\\ \bar{\Theta}_{\varphi(h)}&=I-\Theta_{\varphi(h)}.\end{aligned}$$

2.3 IT-2 Fuzzy Controller

In this subsection, the IT-2 fuzzy controller is introduced to guarantee the stochastic stability of the system, which is designed as follows:

Controller Rule j: **IF** $\rho_1(\tilde{x}(h))$ is σ_1^j, and $\rho_2(\tilde{x}(h))$ is σ_2^j, and,..., and $\rho_g(\tilde{x}(h))$ is σ_g^j, **THEN**

$$u(h)=\mathcal{K}_{j,\varphi(h)}\tilde{x}(h), \tag{7}$$

where σ_q^j is the fuzzy set ($j=1,\ldots,l$ and $q=1,\ldots,g$). g stands for the number of the premise variables. Define $\rho_q(h)\triangleq\rho_q(\tilde{x}(h))$ to be premise variable. $\tilde{x}(h)\in\mathbb{R}^x$, $u(h)\in\mathbb{R}^u$ and $K_{j,\varphi(h)}$ represent the input vector of the controller, the output of the controller and the controller gain, respectively. The UMF and the LMF of the controller are described as

$$0 \leq \prod_{q=1}^{g} \bar{\gamma}_{\sigma_q^j}(\rho_q(h)) = \bar{\varepsilon}_j(\rho_q(h)) \leq 1,$$

$$0 \leq \prod_{q=1}^{g} \underline{\gamma}_{\sigma_q^j}(\rho_q(h)) = \underline{\varepsilon}_j(\rho_q(h)) \leq 1,$$

where $\underline{\gamma}_{\sigma_q^j}(\rho_q(h))$ and $\bar{\gamma}_{\sigma_q^j}(\rho_q(h))$ respectively denote the lower grade of membership function and the upper grade of membership function satisfying $\bar{\gamma}_{\sigma_q^j}(\rho_q(h)) \geq \underline{\gamma}_{\sigma_q^j}(\rho_q(h)) \geq 0$.

Then, the controller can be illustrated as

$$u(h) = \sum_{j=1}^{l} \varepsilon_j(\rho_q(h))\mathcal{K}_{j,\varphi(h)}\tilde{x}(h), \tag{8}$$

where

$$\tilde{\varepsilon}_j(\rho_q(h)) = \underline{b}_j(\rho_q(h))\underline{\varepsilon}_j(\rho_q(h)) + \bar{b}_j(\rho_q(h))\bar{\varepsilon}_j(\rho_q(h)),$$

$$\varepsilon_j(\rho_q(h)) = \frac{\tilde{\varepsilon}_j(\rho_q(h))}{\sum_{j=1}^{l} \tilde{\varepsilon}_j(\rho_q(h))} \geq 0, \ \sum_{j=1}^{l} \varepsilon_j(\rho_q(h)) = 1,$$

with $\varepsilon_j(\rho_q(h))$ being the normalized membership function. $\underline{b}_j(\rho_q(h))$ and $\bar{b}_j(\rho_q(h))$ satisfying $0 \leq \underline{b}_j(\rho_q(h)) \leq 1$, $0 \leq \bar{b}_j(\rho_q(h)) \leq 1$ and $\underline{b}_j(\rho_q(h)) + \bar{b}_j(\rho_q(h)) = 1$ represent the nonlinear weighting functions.

2.4 DoS Attacks Compensation Scheme

This subsection describes the compensation mechanism utilized to mitigate the adverse impact of DoS attacks. The data that received from the actuator is represented as follows:

$$u_a(h) = \eta(h)u(h) + (1 - \eta(h))u_b(h), \tag{9}$$

where $u_b(h)$ denotes the compensation signal. $\eta(h)$ is a Bernoulli variable that indicates the randomness of DoS attacks. When there are DoS attacks, $\eta(h) = 0$. When there is no DoS attack, $\eta(h) = 1$.

The probability of DoS attacks occurring is represented by the mathematical expectation as follows:

$$E\{\eta(h) = 1\} = \bar{\eta}, \ E\{\eta(h) = 0\} = 1 - \bar{\eta},$$

where $\bar{\eta} \in [0, 1)$ is a known constant and $E\{(\eta(h) - \bar{\eta})^2\} = \bar{\eta}(1 - \bar{\eta}) = \iota^2$.

The compensation signal is represented as follows:

$$u_b(h) = \begin{cases} u_c(h), & \text{if } \eta(h) = 0 \cap \varsigma(h) = 1, \\ u_b(h-1), & \text{if } \eta(h) = 0 \cap \varsigma(h) \in (1, \alpha], \\ 0, & \text{if } \eta(h) = 0 \cap \varsigma(h) > \alpha, \end{cases} \tag{10}$$

where

$$\varsigma(h) = \begin{cases} 1, & \text{if } \eta(h) = 0 \cap \eta(h-1) = 1, \\ \varsigma(h) + 1, & \text{if } \eta(h) = 0 \cap \eta(h-1) = 0, \\ 0, & \text{if } \eta(h) = 1, \end{cases} \tag{11}$$

$$u_c(h) = u_a(h-1) + \Delta u_a(h-2) + \Delta^2 u_a(h-3), \tag{12}$$

with $u_c(h)$ being the predictive signal. $\varsigma(h)$ and α denote an index that describes the duration of compensation signals and a known positive integer, respectively.

Remark 1. In this paper, an improved DoS attacks compensation mechanism is constructed. Through the designed functions, the attack compensation mechanism predicts more accurate compensation signals to enhance the system performance. Additionally, the attack compensation mechanism constructed in this paper offers different compensation signals for DoS attacks under different continuous durations, further enhancing the control performance of NCSs under DoS attacks.

2.5 Closed-Loop-System

Defining $\varpi_i \triangleq \varpi_i(\tau_k(h))$, $\varepsilon_j \triangleq \varepsilon_j(\rho_q(h))$, $m \triangleq \varphi(h)$ and $n \triangleq \varphi(h+1)$ for simplicity. Define $\xi(h) = [x^T(h), \tilde{x}^T(h-1)]^T$. Combining (2), (6), (8) and (9), the closed-loop-system of the designed control strategy can be obtained as follows:

$$\begin{aligned} \xi(h+1) =& \sum_{i=1}^{l}\sum_{j=1}^{l} \varpi_i \varepsilon_j [\mathcal{A}_{1ijm}\xi(h) + (\eta(h) - \bar{\eta})\mathcal{A}_{2ijm}\xi(h) \\ &+ \mathcal{F}_{1i} u_b(h) + (\eta(h) - \bar{\eta})\mathcal{F}_{2i} u_b(h) + \bar{\mathcal{C}}_i \omega(h)], \\ z(h) =& \sum_{i=1}^{l}\sum_{j=1}^{l} \varpi_i \varepsilon_j \mathcal{D}_i \xi(h), \end{aligned} \tag{13}$$

where

$$\begin{aligned} \mathcal{A}_{1ijm} &= \begin{bmatrix} \mathcal{A}_i + \bar{\eta}\mathcal{B}_i K_{j,m}\Theta_m & \bar{\eta}\mathcal{B}_i \mathcal{K}_{j,m}\bar{\Theta}_m \\ \Theta_m & \bar{\Theta}_m \end{bmatrix}, \\ \mathcal{A}_{2ijm} &= \begin{bmatrix} \mathcal{B}_i \mathcal{K}_{j,m}\Theta_m & \mathcal{B}_i \mathcal{K}_{j,m}\bar{\Theta}_m \\ 0 & 0 \end{bmatrix}, \\ \mathcal{F}_{1i} &= \begin{bmatrix} (1-\bar{\eta})\mathcal{B}_i \\ 0 \end{bmatrix}, \ \mathcal{F}_{2i} = \begin{bmatrix} -\mathcal{B}_i \\ 0 \end{bmatrix}, \ \bar{\mathcal{C}}_i = \begin{bmatrix} \mathcal{B}_{1i} \\ 0 \end{bmatrix}, \\ \mathcal{D}_i &= \begin{bmatrix} \mathcal{C}_i & 0 \end{bmatrix}. \end{aligned}$$

Lemma 1 *[11]. The singular value decomposition of the full rank matrix $\mathcal{B}$ is $\mathcal{B} = \mathcal{U}\begin{bmatrix} \mathcal{S} \ 0 \end{bmatrix}\mathcal{V}^T$, where rank($\mathcal{B}$) $= a$, $B \in R^{a\times b}$, $\mathcal{U}\mathcal{U}^T = \mathcal{I}$ and $\mathcal{V}\mathcal{V}^T = \mathcal{I}$. Let matrix $\mathcal{X} > 0$, $\mathcal{M} \in \mathcal{R}^{a\times a}$ and $\mathcal{N} \in \mathcal{R}^{b\times b}$. Then, there exists $\mathcal{L}$ such that $\mathcal{B}\mathcal{X} = \mathcal{L}\mathcal{B}$ if and only if the following condition holds*

$$\mathcal{X} = \mathcal{V}\begin{bmatrix} \mathcal{M} & 0 \\ 0 & \mathcal{N} \end{bmatrix}\mathcal{V}^T.$$

3 Main Results

3.1 Performance Analysis

The stochastic stability of system (13) under H_∞ performance is verified in this subsection.

Theorem 1. *Given constants $\bar{\eta} > 0$, $\Omega_r > 0$ and $\gamma > 0$, if $\varepsilon_j - \tau_j\varpi_j \geq 0$ $(0 < \tau_j \leq 1)$ is satisfied, and there exist matrices $P_m > 0, P_n > 0$ and $\Delta = \Delta^T$ satisfying $(i \leq j)$*

$$\Psi_{ijmn} + \Psi_{jimn} - 2\Delta < 0, \tag{14}$$

$$\varepsilon_j\Psi_{ijmn} + \varepsilon_i\Psi_{jimn} - \varepsilon_j\Delta - \varepsilon_i\Delta + 2\Delta < 0, \tag{15}$$

where

$$\begin{aligned}
\Psi_{ijmn} &= \begin{bmatrix} \Pi_{1ijmn} & \Pi_{2ijmn} & \mathcal{A}_{1ijm}{}^T\mathcal{P}_n\bar{\mathcal{C}}_i \\ * & \Pi_{3in} & \mathcal{F}_{1i}{}^T\mathcal{P}_n\bar{\mathcal{C}}_i \\ * & * & \Pi_{4in} \end{bmatrix}, \\
\Pi_{1ijmn} &= \mathcal{A}_{1ijm}{}^T\mathcal{P}_n\mathcal{A}_{1ijm} + \bar{\eta}(1-\bar{\eta})\mathcal{A}_{2ijm}{}^T\mathcal{P}_n\mathcal{A}_{2ijm} - \mathcal{P}_m \\
&\quad + \mathcal{D}_i{}^T\mathcal{D}_i, \\
\Pi_{2ijmn} &= \mathcal{A}_{1ijm}{}^T\mathcal{P}_n\mathcal{F}_{1i} + \bar{\eta}(1-\bar{\eta})\mathcal{A}_{2ijm}{}^T\mathcal{P}_n\mathcal{F}_{2i}, \\
\Pi_{3in} &= \mathcal{F}_{1i}{}^T\mathcal{P}_n\mathcal{F}_{1i} + \bar{\eta}(1-\bar{\eta})\mathcal{F}_{2i}{}^T\mathcal{P}_n\mathcal{F}_{2i}, \\
\Pi_{4in} &= \bar{\mathcal{C}}_i^T\mathcal{P}_n\bar{\mathcal{C}}_i - \gamma^2\mathcal{I},
\end{aligned}$$

then, the stochastic stability of system (13) under H_∞ performance is satisfied.

Proof. Selecting the following mode-dependent Lyapunov function

$$\mathcal{V}(h) = \xi^T(h)\mathcal{P}_m\xi(h). \tag{16}$$

Define $\eta_2 = [\xi^T(h),\ u_b{}^T(h),\ w^T(h)]^T$. The following inequality is obtained by utilizing the difference of $V(h)$

$$\begin{aligned}
&E\{\Delta\mathcal{V}(h) + z^T(h)z(h) - \gamma^2 w^T(h)w(h)\} \\
\leq &E\{\xi^T(h+1)\mathcal{P}_n\xi(h+1) + \xi^T(h)\mathcal{D}_i{}^T\mathcal{D}_i\xi(h) \\
&- \gamma^2 w^T(h)w(h)\} - \xi^T(h)\mathcal{P}_m\xi(h)\} \\
= &\sum_{i=1}^{l}\sum_{j=1}^{l}\varpi_i\varepsilon_j\eta_2{}^T(h)\Psi_{ijmn}\eta_2(h).
\end{aligned}$$

If $\sum_{i=1}^{l}\sum_{j=1}^{l}\varpi_i\varepsilon_j\Psi_{ijmn} < 0$, the above inequality can be expressed as

$$E\{\Delta\mathcal{V}(h) + z^T(h)z(h) - \gamma^2 w^T(h)w(h)\} < 0, \tag{17}$$

then, the stochastic stability of system (13) under H_∞ performance can be demonstrated as follows:

Since $\mathcal{V}(h) > 0$, $\sum_{k=1}^{\infty}\gamma^2 w^T(h)\varpi(h) > \sum_{k=1}^{\infty} z^T(h)z(h)$ can be obtained, hence, the condition of H_∞ performance for system (13) is satisfied. Then, $\Delta\mathcal{V}(h) < 0$ is obtained under the condition of $w(h) = 0$. The stochastic stability of system (13) with H_∞ performance under WTODP is satisfied.

Inspired by [11], introducing a slack matrix Δ to reduce the conservativeness. Taking $\sum_{i=1}^{l}\sum_{j=1}^{l}\varpi_i[\varpi_i - \varepsilon_j]\Delta = 0$ into consideration, the following equation is obtained

$$\begin{aligned}
&\sum_{i=1}^{l}\sum_{j=1}^{l}\varpi_i\varepsilon_j\Psi_{ijmn}\\
=&\frac{1}{2}\sum_{i=1}^{l}\sum_{j=1}^{l}\varpi_i[\varpi_j(\tau_j\Psi_{ijmn} + \tau_i\Psi_{jimn} - \tau_j\Delta - \tau_i\Delta\\
&+ 2\Delta) + (\varepsilon_j - \tau_j\varpi_j)(\Psi_{ijmn} + \Psi_{jimn} - 2\Delta)].
\end{aligned} \tag{18}$$

Then, under the condition of $\varepsilon_j - \tau_j w_j \geq 0$, the sufficient conditions (14) and (15) for the stochastic stability of system (13) are obtained.

3.2 Solution of Controller Gains

This subsection focuses on solving the controller gains as expressed in (8). Considering the WTODP, we can obtain that $m \in \{1, \ldots, n_x\}$. The solution of controller gains according to *Theorem* 1 is described as follows.

Theorem 2. *Given constants* $\bar{\eta} > 0$, $\Omega_r > 0$ *and* $\gamma > 0$, *if* $\varepsilon_j - \tau_j\varpi_j \geq 0$ $(0 < \tau_j \leq 1)$ *is satisfied, and there exist matrices* $\mathcal{X}_m > 0, \mathcal{P}_m > 0, \mathcal{P}_n > 0$ *and* $\Delta = \Delta^T$ *satisfying* $(i \leq j)$

$$\begin{bmatrix} 2\bar{\Delta} & \Xi_{\mathrm{ij,m}} + \Xi_{\mathrm{ji,m}} \\ * & 2\mathcal{P}_{mn} \end{bmatrix} <0, \tag{19}$$

$$\begin{bmatrix} \sqrt{\tau_j}\bar{\Delta} + \sqrt{\tau_j}\bar{\Delta} & \sqrt{\tau_j}\Xi_{\mathrm{ij,m}} + \sqrt{\tau_i}\Xi_{\mathrm{ji,m}} \\ * & 2\mathcal{P}_{mn} \end{bmatrix} <0, \tag{20}$$

where

$$\begin{aligned}
\Xi_{\mathrm{ij,m}} &= \begin{bmatrix} \Xi_{\mathrm{1ij,m}} & \Xi_{\mathrm{2ij,m}} & \mathcal{D}_i{}^T \\ \Xi_{\mathrm{3ij,m}} & \Xi_{\mathrm{4ij,m}} & 0 \\ \Xi_{\mathrm{5ij,m}} & 0 & 0 \end{bmatrix}, \\
\mathcal{P}_{mn} &= diag\{\mathcal{P}_{1n} - 2\mathcal{X}_m,\ \mathcal{P}_{1n} - 2\mathcal{X}_m,\ \mathcal{P}_{1n} - 2\mathcal{X}_m,\ \mathcal{P}_{1n} \\
&\quad - 2\mathcal{X}_m,\ \mathcal{I}\}, \\
\bar{\Delta} &= diag\{-\mathcal{P}_{1m},\ -\mathcal{P}_{1m},\ 0,\ -\gamma^2\mathcal{I}\} - \Delta, \\
\Xi_{\mathrm{1ij,m}} &= \begin{bmatrix} \mathcal{A}_i{}^T\mathcal{X}_m{}^T + \bar{\eta}\Theta_m{}^T\bar{\mathcal{L}}_m\mathcal{B}_i{}^T & \Theta_m{}^T\mathcal{X}_m{}^T \\ \bar{\eta}\bar{\Theta}_m^T\bar{\mathcal{L}}_m\mathcal{B}_i{}^T & \bar{\Theta}_m^T\mathcal{X}_m{}^T \end{bmatrix}, \\
\Xi_{\mathrm{2ij,m}} &= \begin{bmatrix} \iota\Theta_m{}^T\bar{\mathcal{L}}_m\mathcal{B}_i{}^T & 0 \\ \iota\bar{\Theta}_m^T\bar{\mathcal{L}}_m\mathcal{B}_i{}^T & 0 \end{bmatrix},\ \Xi_{\mathrm{5ij,m}} = \begin{bmatrix} \mathcal{B}_{1i}{}^T\mathcal{X}_m{}^T & 0 \end{bmatrix}, \\
\Xi_{\mathrm{3ij,m}} &= \begin{bmatrix} (1-\bar{\eta})\mathcal{B}_i{}^T\bar{\mathcal{X}}_m^T & 0 \end{bmatrix},\ \Xi_{\mathrm{4ij,m}} = \begin{bmatrix} \iota\mathcal{B}_i{}^T\bar{\mathcal{X}}_m^T & 0 \end{bmatrix},
\end{aligned}$$

then, the stochastic stability of system (13) under H_∞ performance is satisfied. The controller gains are obtained as

$$\mathcal{K}_{j,m}{}^T = \bar{\mathcal{L}}_m\mathcal{U}\mathcal{S}\mathcal{X}_{m1}{}^{-1}\mathcal{S}^{-1}\mathcal{U}^{-1}. \tag{21}$$

Proof. By utilizing Schur complement, the following inequalities can be obtained from *Theorem* 1

$$\begin{bmatrix} 2\bar{\Delta} & \tilde{\bar{\Xi}}_{\mathrm{ij,m}} + \tilde{\bar{\Xi}}_{\mathrm{ji,m}} \\ * & 2\tilde{\mathcal{P}}_n \end{bmatrix} <0, \tag{22}$$

$$\begin{bmatrix} \sqrt{\tau_j}\bar{\Delta} + \sqrt{\tau_j}\bar{\Delta} & \sqrt{\tau_j}\tilde{\bar{\Xi}}_{\mathrm{ij,m}} + \sqrt{\tau_i}\tilde{\bar{\Xi}}_{\mathrm{ji,m}} \\ * & 2\tilde{\mathcal{P}}_n \end{bmatrix} <0, \tag{23}$$

where

$$\begin{aligned}
\tilde{\mathcal{P}}_n &= diag\{-\mathcal{P}_{1n}{}^{-1}, -\mathcal{P}_{1n}{}^{-1}, -\mathcal{P}_{1n}{}^{-1}, -\mathcal{P}_{1n}{}^{-1}, \mathcal{I}\}, \\
\tilde{\bar{\Xi}}_{\mathrm{ij,m}} &= \begin{bmatrix} \tilde{\bar{\Xi}}_{\mathrm{1ij,m}} & \tilde{\bar{\Xi}}_{\mathrm{2ij,m}} & \mathcal{D}_i^T \\ \tilde{\bar{\Xi}}_{\mathrm{3ij,m}} & \tilde{\bar{\Xi}}_{\mathrm{4ij,m}} & 0 \\ \tilde{\bar{\Xi}}_{\mathrm{5ij,m}} & 0 & 0 \end{bmatrix}, \\
\tilde{\bar{\Xi}}_{\mathrm{1ij,m}} &= \begin{bmatrix} \mathcal{A}_i^T + \bar{\eta}\Theta_m{}^T\mathcal{K}_{j,m}{}^T\mathcal{B}_i^T & \Theta_m{}^T \\ \bar{\eta}\bar{\Theta}_m^T\mathcal{K}_{j,m}{}^T\mathcal{B}_i{}^T & \bar{\Theta}_m^T \end{bmatrix}, \\
\tilde{\bar{\Xi}}_{\mathrm{2ij,m}} &= \begin{bmatrix} \iota\Theta_m{}^T\mathcal{K}_{j,m}{}^T\mathcal{B}_i^T & 0 \\ \iota\bar{\Theta}_m^T\mathcal{K}_{j,m}{}^T\mathcal{B}_i^T & 0 \end{bmatrix}, \\
\tilde{\bar{\Xi}}_{\mathrm{3ij,m}} &= \begin{bmatrix} (1-\bar{\eta})\mathcal{B}_i{}^T & 0 \end{bmatrix}, \\
\tilde{\bar{\Xi}}_{\mathrm{4ij,m}} &= \begin{bmatrix} -\iota\mathcal{B}_i{}^T & 0 \end{bmatrix}, \\
\tilde{\bar{\Xi}}_{\mathrm{5ij,m}} &= \begin{bmatrix} \mathcal{B}_{1i}{}^T & 0 \end{bmatrix}.
\end{aligned}$$

Define

$$\mathcal{P}_m = diag\{\mathcal{P}_{1m}, \mathcal{P}_{1m}\},\ \mathcal{P}_n = diag\{\mathcal{P}_{1n}, \mathcal{P}_{1n}\}.$$

Pre and post multiplying (22) and (23) by $\tilde{\mathcal{P}} = diag\{\mathcal{I},\mathcal{I},\mathcal{I},\mathcal{I},\mathcal{X}_m, \mathcal{X}_m,\mathcal{X}_m,\mathcal{X}_m,\mathcal{I}\}$ and $\tilde{\mathcal{P}}^T$.

For ${\mathcal{X}_m}^T = \mathcal{V}\begin{bmatrix}\mathcal{X}_{m1} & 0\\ 0 & \mathcal{X}_{m2}\end{bmatrix}\mathcal{V}^T$, according to *Lemma* 1, there exists $\bar{\mathcal{X}}_m = \mathcal{U}\mathcal{S}\mathcal{X}_{m1}\mathcal{S}^{-1}\mathcal{U}^{-1}$ can let ${\mathcal{B}_i}^T{\mathcal{X}_m}^T = \bar{\mathcal{X}}_m{\mathcal{B}_i}^T$ with $\bar{\mathcal{X}}_m^{-1} = \mathcal{U}\mathcal{S}{\mathcal{X}_{m1}}^{-1}\mathcal{S}^{-1}\mathcal{U}^{-1}$.

Define

$$\bar{\mathcal{L}}_m = {\mathcal{K}_{j,m}}^T\bar{\mathcal{X}}_m.$$

Taking $-\mathcal{X}_m{\mathcal{P}_{1n}}^{-1}\mathcal{X}_m^T \le \mathcal{P}_{1n} - 2\mathcal{X}_m$ into consideration, *Theorem* 2 can be obtained. The proof is completed.

4 Simulation Results

Considering the IT-2T-S fuzzy model as follows:

$$\mathcal{A}_1 = \begin{bmatrix}-0.8866 & 0.3058\\ -1.1885 & -1.1229\end{bmatrix}, B_1 = \begin{bmatrix}0.4944\\ 0.0308\end{bmatrix},$$
$$\mathcal{A}_2 = \begin{bmatrix}-0.6361 & 0.4759\\ -1.5141 & -1.0038\end{bmatrix}, \mathcal{B}_2 = \begin{bmatrix}0.5594\\ 0.1529\end{bmatrix},$$
$$\mathcal{C}_1 = \begin{bmatrix}1 & 0.5\end{bmatrix},\ \mathcal{C}_2 = \begin{bmatrix}1 & 0.5\end{bmatrix},$$
$$\mathcal{B}_{11} = \begin{bmatrix}0.2\\ -0.5\end{bmatrix},\ \mathcal{B}_{12} = \begin{bmatrix}0.2\\ 1.15\end{bmatrix}.$$

The UMF and LMF of the system are defined as

$$\bar{\varpi}_1(x_1) = \frac{0.3 + 0.95x_1}{0.781},\ \underline{\varpi}_1(x_1) = \frac{-0.01 + 0.875x_1}{0.781},$$
$$\bar{\varpi}_2(x_1) = \frac{0.768 - 0.875x_1}{0.781},\ \underline{\varpi}_1(x_1) = \frac{0.468 - 0.95x_1}{0.781}.$$

Select the weighting functions as $\underline{\vartheta}_i(x(h)) = {x_1}^2(h)/4$ and $\bar{\vartheta}_i(x(h)) = 1 - \underline{\vartheta}_i(x(h))$, define the UMF and the LMF of the controller as $\underline{\varepsilon}_1(\tilde{x}_1) = 0.5 \times e^{-\tilde{x}_1^2}$, $\bar{\varepsilon}_1(\tilde{x}_1) = \underline{\varepsilon}_1(\tilde{x}_1)$, $\underline{\varepsilon}_2(\tilde{x}_1) = 1 - \underline{\varepsilon}_2(\tilde{x}_1)$ and $\bar{\varepsilon}_2(\tilde{x}_1) = \underline{\varepsilon}_2(\tilde{x}_1)$. Define $\underline{b}_j(\tilde{x}(h)) = \cos^2(2\cdot\tilde{x}(h))$ and $\bar{b}_j(\tilde{x}(h)) = 1 - \underline{b}_j(\tilde{x}(h))$ as the weighting functions of the controller. Set $\bar{\eta} = 0.5$, $\Omega = \text{diag}\{1,\ 1\}$ and $\alpha = 10$. The controller gains are acquired as

$$\mathcal{K}_{11} = \begin{bmatrix}1.9514 & -4.18\times10^{-5}\end{bmatrix},$$
$$\mathcal{K}_{12} = \begin{bmatrix}0.0019 & -0.4559\end{bmatrix},$$
$$\mathcal{K}_{21} = \begin{bmatrix}1.5220 & 6.43\times10^{-5}\end{bmatrix},$$
$$\mathcal{K}_{22} = \begin{bmatrix}5.94\times10^{-4} & -0.3708\end{bmatrix}.$$

Suppose the initial states are $x(0) = \begin{bmatrix}0.3, -0.1\end{bmatrix}^T$ and the sampling period $h = 0.2\,s$. The disturbance is assumed as $w(h) = \begin{cases}-1.87\cos(1.2k)e^{-0.1k}, & 0 \le k \le 22.2,\\ 0, & else.\end{cases}$

In order to verify the effectiveness of the proposed control method, the following simulation results are provided.

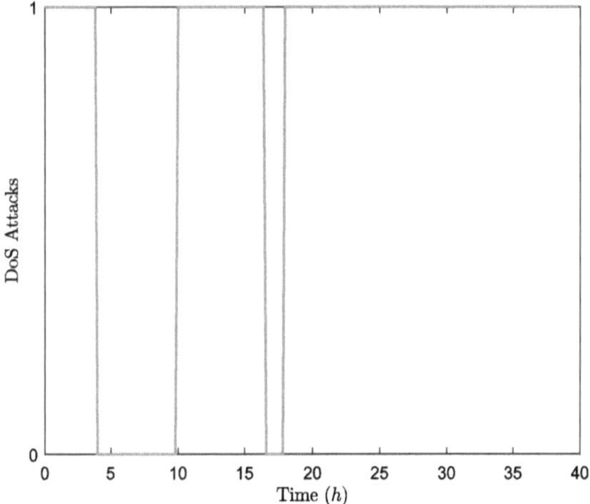

Fig. 1. DoS attacks.

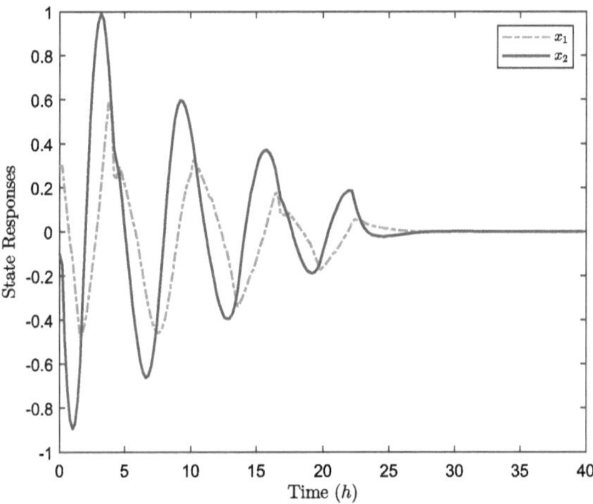

Fig. 2. State responses.

Figure 1 describes the occurrence of DoS attacks encountered in communication networks. Figure 2 illustrates the responses of system states. It can be inferred from Fig. 2 that the stochastic stability of the closed-loop-system discussed in this paper is achieved.

Figure 3 describes the performance index of the system under different DoS attacks compensation mechanisms. As shown in Fig. 3, compared to the attack compensation mechanism in [8], the attack compensation mechanism improved in this paper effectively enhances the performance of system under DoS attacks. Figure 4 demonstrates the operating condition of sensors in this paper. It can be seen that due to the introduced WTODP, the workload of the sensors is successfully reduced, resulting in conserving network resources.

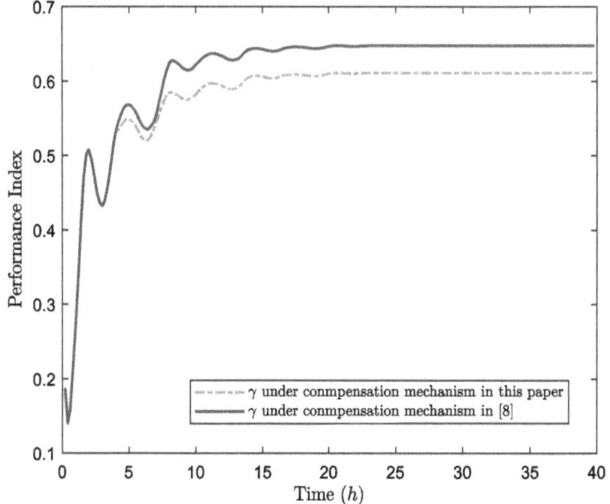

Fig. 3. Performance index between different compensation mechanisms.

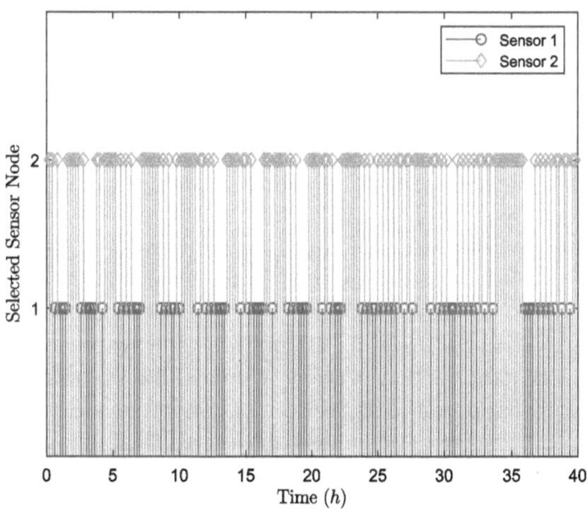

Fig. 4. Selected sensor nodes under WTODP.

5 Conclusions

In this paper, the design problem of DoS attacks compensation controller for NCSs with WTODP has been investigated. An improved DoS attacks compensation mechanism has been constructed, which has further reduced the adverse impact of DoS attacks. A WTODP has been introduced to conserve the network resources. Then, sufficient conditions for stochastic stability of the closed-loop-system have been testified by utilizing mode-dependent Lyapunov function. Finally, simulation result has been provided to ensure the effectiveness of the control strategy.

Acknowledgments. This work was partially supported by the National Natural Science Foundation of China (62473060), the Natural Science Foundation of Liaoning Province (2025-YQ-18), the Basic Research Project of the Educational Department of Liaoning Province (LJ232410167028) and the Revitalization of Liaoning Talents Program (XLYC2203201).

Disclosure of Interests. The authors have no competing interests to declare that are relevant to the content of this article.

References

1. Dong, J., Hou, Q., Ren, M.: Control synthesis for discrete-time T-S fuzzy systems based on membership function-dependent H_∞ performance. IEEE Trans. Fuzzy Syst. **28**(12), 3360–3366 (2020)
2. Liang, C., et al.: Predefined-time stabilization of T-S fuzzy systems: a novel integral sliding mode-based approach. IEEE Trans. Fuzzy Syst. **30**(10), 4423–4433 (2022)
3. Yang, J., Gai, S., Da, F., Xie, W.: Robust control analysis with model transformation for interval type-2 fuzzy systems. IEEE Trans. Fuzzy Syst. **33**(7), 2272–2283 (2025)
4. Yang, Y., Wen, G., Wang, Y., Peng, Z., Xiong, K.: Enhanced distributed outlier-resilient fusion estimation with novel dimensionality reduction under IT-2 T-S fuzzy system. IEEE Trans. Fuzzy Syst. **32**(11), 6044–6055 (2024)
5. Hu, S., Yue, D., Chen, X., Cheng, Z., Xie, X.: Resilient H_∞ filtering for event-triggered networked systems under nonperiodic DoS jamming attacks. IEEE Trans. Syst. Man Cybern. Syst. **51**(3), 1392–1403 (2021)
6. Zhao, N., Zhao, X., Zong, G., Xu, N.: Resilient event-triggered filtering for networked switched T-S fuzzy systems under denial-of-service attacks. IEEE Trans. Fuzzy Syst. **32**(4), 2140–2152 (2024)
7. Yan, J.-J., Yang, G.-H.: Secure state estimation with switched compensation mechanism against DoS attacks. IEEE Trans. Cybern. **52**(9), 9609–9620 (2022)
8. Su, L., Ye, D.: A cooperative detection and compensation mechanism against denial-of-service attack for cyber-physical systems. Inf. Sci. **444**, 122–134 (2018)
9. Liu, J., Wang, S., Liu, J., Xie, X., Tian, E.: FlexRay protocol-based event-triggered secure filtering for IT-2 fuzzy systems with fading measurements over high-rate communication networks: the finite-horizon case. IEEE Trans. Syst. Man Cybern. Syst. **54**(10), 6055–6067 (2024)
10. Yao, M., Wei, G., Ding, D.: Dynamic quantized output-feedback control of impulsive systems with round-robin protocols and transmission delays. IEEE Trans. Control Netw. Syst. **12**(1), 114–124 (2025)
11. Liu, J., Gong, E., Zha, L., Tian, E., Xie, X.: Observer-based security fuzzy control for nonlinear networked systems under weighted try-once-discard protocol. IEEE Trans. Fuzzy Syst. **31**(11), 3853–3865 (2023)

Trajectory Optimization for Hypersonic Vehicles Under Aerodynamic Uncertainty via Risk-Neutral Sequential Convex Programming

Yuxin Huang, Cheng Zhang(✉), Jianhua Yin, Zhangyao Zheng, Chenming Zheng, and Jiayu Bao

School of Aerospace, Beijing Institute of Technology, Beijing, China
{3220230052,zhangcheng}@bit.edu.cn

Abstract. Traditional trajectory optimization methods for hypersonic under aerodynamic uncertainty often lead to overly conservative solutions. To address this issue, this paper proposes a novel trajectory planning method using sequential convex programming (SCP) with chance-constraint in response to aerodynamic uncertainty. The core of our approach is mapping aerodynamic uncertainty to the control input, formulating a tractable probabilistic optimization problem. A risk-neutral surrogate function, named Scaled and Translated Adaptive Proxy (STA-Proxy), is designed to approximate the non-smooth chance constraint. By incorporating an error compensation method, the STA-Proxy function avoids overly conservative solution while maintaining numerical accuracy. The resulting STA-Proxy-SCP algorithm developed can solve the trajectory optimization problem efficiently. Numerical simulation demonstrates that the proposed method outperforms both robust optimization and the Split-Bernstein approach in terms of performance index, thus highlighting its superior balance between computational efficiency and reliability.

Keywords: Trajectory Optimization · Chance Constraint · Sequential Convex Optimization · Aerodynamic Uncertainty · Monte Carlo Simulation

1 Introduction

Hypersonic vehicles have emerged as a critical research focus in the aerospace field, where their unparalleled high-speed maneuverability presents unique challenges and stringent requirements for trajectory optimization [1].

Although various deterministic trajectory optimization algorithms have achieved remarkable success in hypersonic vehicle trajectory design [2], their effectiveness is limited in practice, as it fails to account for substantial system

C. Li et al. (Eds.): ICNC 2025, CCIS 2946, pp. 195–208, 2026.
https://doi.org/10.1007/978-981-92-1599-7_17

uncertainties [3]. Such uncertainties—from atmospheric conditions to aerodynamic and control errors [4,5]—can be significant (e.g., 10%âĂŞ15% in aerodynamic coefficients [6]) and cannot be fully eliminated. While inner-loop control strategies [7,8] handle tracking under disturbances, the guidance trajectory itself is highly vulnerable to aerodynamic uncertainties. If unaddressed, this leads to large deviations from the nominal path [9], rendering nominal optimal trajectories ineffective under actual flight conditions.

Various methods address uncertain trajectory planning for nonlinear systems. Data-driven approaches (e.g., transfer learning [10] and reinforcement learning [11]) generate trajectories rapidly but require extensive data and computing, lacking interpretability. Classical strategies like MPC [4] or LOS-based control [12] offer convergence guarantees yet struggle with strong nonlinearities and coupled uncertainties. Methods modeling disturbances as inputs, such as HMPF systems [13], need precise models and falter with unstructured uncertainties.

Alternatively, optimization-based techniques like robust optimization [14] and chance constraints transform uncertainty into deterministic forms, often introducing conservatism. While robust optimization ensures worst-case performance, it sacrifices optimality. Bernstein-type methods [15] bound uncertainties but remain conservative and can face numerical issues with small, normalized uncertainties.

Convex optimization is particularly promising for its computational efficiency and convergence guarantees [16,17]. Common strategies include segment-wise optimization [18,19] and various convexification techniques for deterministic problems, such as variable substitution [20] or relaxation with regularization [21]. Introducing chance constraints into convex optimization problem, however, makes it far more difficult to solve, as the requisite approximation of probability measures typically introduces non-convex structures.

To address the above challenges. We introduce the Scaled and Translated Adaptive Proxy (STA-Proxy), a novel function that provides a less conservative and more accurate transformation of probabilistic constraints. This paper is structured as follows. First, the chance-constrained trajectory optimization problem is formulated. Then, the design and advantages of the proposed STA-Proxy method are detailed. Subsequently, the convexification strategies and the overall optimization framework are presented. Finally, the approach is validated through comparative numerical simulations.

2 Optimization Problem Modeling

2.1 Deterministic Trajectory Optimization

A right-handed coordinate system is defined with its origin fixed to the Earth. The X-axis points due east, the Y-axis extends vertically upward, and the Z-axis completes the right-handed triad. The vehicle state variables consist of the position coordinates (x, y, z), velocity V and time t. These quantities are normalized by dividing by R_0, $\sqrt{g_0 R_0}$ and $\sqrt{R_0/g_0}$, respectively, of which g_0 represents the gravitational acceleration at the Earth's surface radius R_0. Ma represents the

Mach number. The flight path angle, the heading angle (measured clockwise from the east), and the bank angle of the missile are denoted by θ, ψ and σ, respectively. a is the angle of attack, m denotes t the mass. The distance from the Earth's center to the vehicle is given by $r = 1 + y$. The dimensionless lift and drag coefficients are denoted as L and D, respectively. A curve fit for lift and drag is performed based on the classical formulation given in (1).

$$L = \hat{L}(Ma)\eta(\alpha, Ma), \quad D = 0.5\hat{D}(Ma)\left[1 + \eta(\alpha, Ma)^2\right] \tag{1}$$

The six-degree-of-freedom dynamic equations can be expressed as:

$$\begin{cases} \dot{v} = -0.5\hat{D}(Ma)\left(1 + u_3\right) - \dfrac{\sin\theta}{r^2} \\ \dot{\theta} = \dfrac{\hat{L}(Ma)u_1}{V} - \dfrac{\cos\theta}{Vr^2} \\ \dot{\psi} = -\dfrac{\hat{L}(Ma)u_2}{V\cos\theta} \\ \dot{x} = V\cos\theta\cos\psi \\ \dot{y} = V\sin\theta \\ \dot{z} = -V\cos\theta\sin\psi \end{cases} \tag{2}$$

$$u_1 = \eta\cos\sigma, \quad u_2 = \eta\sin\sigma, \quad u_3 = \eta^2, \tag{3}$$

$$u_1^2 + u_2^2 = u_3. \tag{4}$$

The state variables is defined as X ($X = [v, \theta, \psi, x, y, z]^T$), and the control variables as U ($U = [u_1, u_2, u_3]^T$). The dynamic equations are written as $\dot{X} = f(X, U)$.

Given the inherent structural properties and mission requirements of hypersonic vehicles, constraints are imposed on both control and state variables, and an objective function is formulated to form a well-defined optimization problem.

The angle of attack must operate within a safe envelope, thereby defining the feasible region for the control input u_3.

$$0 \le u_3 \le u_{3\max} \tag{5}$$

To maintain structural integrity throughout the flight, the dynamic pressure must not exceed allowable thresholds.

$$q = 0.5\rho v^2 g_0 R_0 \le q_{\max} \tag{6}$$

The initial states of the system X_{begin} and the terminal states X_f are subject to the following constraints:

$$X_{begin} = X_0, X_{f\min} \le X_f \le X_{f\max} \tag{7}$$

In this study, terminal velocity is maximized to increase impact momentum and enhance penetration effectiveness against ground targets.

$$\min J = -v_{tf} \tag{8}$$

2.2 Trajectory Optimization Under Uncertainty

Aerodynamic parameters are subject to deviations due to manufacturing, measurement, and environmental factors. These uncertainties accumulate during flight and degrade control accuracy. Since lift and drag coefficients depend on the angle of attack a and Mach number Ma (Eq. 1), their overall uncertainty can be represented by variations in a and Ma. This study focuses on a for simplicity. Through the nonlinear mapping from a to control inputs u_1, u_2, u_3, uncertainties in a propagate into u_3, leading to control disturbances. The following constraint must therefore be satisfied:

$$0 \le u_3 + \delta \le u_{3\,\max} \tag{9}$$

where δ represents an unknown disturbance attributed to the control input U_3. Traditional deterministic optimization assumes perfect model knowledge and cannot handle such uncertainties. To address this, we reformulate the constraint using a chance constraint, which allows violations within a preset probability threshold. This approach trades strict safety for probabilistic reliability, improving performance while maintaining acceptable risk levels.

For example, the original constraint on u_3 (Eq. 9) is transformed into:

$$P(0 \le u_3 + \delta \le u_{3\,\max}) \ge \epsilon \tag{10}$$

where $\epsilon \in (0, 1]$ is the required satisfaction probability. When $\epsilon = 1$, the formulation reduces to robust optimization. The overall optimization problem becomes:

$$\begin{aligned} \min \quad & -v_{tf} \\ \text{s.t.} \quad & (2)(3)(4)(6)(7)(10) \end{aligned} \tag{11}$$

3 A Risk-Neutral Proxy for Chance-Constrain

3.1 Limitations of Traditional Chance-Constrained Methods: An Analysis of Conservatism

A chance constraint requires the probability of an event $\{F(\xi) \le 0\}$ occurring under random disturbance ξ to be no less than a threshold ϵ, expressed as:

$$P(F(\xi) \le 0) \ge \epsilon \tag{12}$$

This probabilistic form captures uncertainty but is non-smooth and nonconvex. To enable numerical solution, it is reformulated as an expectation:

$$P(F(\xi) \le 0) = 1 - \mathbb{E}(H(F(\xi))) \tag{13}$$

where $\mathbb{E}(H(F(\xi)))$ represents the probability of event $H(F(\xi))$ occurring. $H(.)$ is the step function, defined as:

$$H(F(\xi)) = \begin{cases} 1, & F(\xi) \ge 0, \\ 0, & F(\xi) < 0. \end{cases} \tag{14}$$

Substituting (13) into (12), the chance constraint can be further expressed as:

$$\mathbb{E}(H(F(\xi))) \leq 1-\epsilon \tag{15}$$

However, the step function $H(\cdot)$ is discontinuous and incompatible with gradient-based solvers. Thus, a continuous and differentiable surrogate function is required. The traditional Bernstein method $G_1(\cdot)$ and the Split-Bernstein method $G_2(\cdot)$, are two representative approximation strategies, whose expressions are given by (16) and (17):

$$G_1(\cdot)(F(\xi)) = \exp(\alpha_0 F(\xi)) \tag{16}$$

$$G_2(F(\xi)) = \begin{cases} \exp(\alpha_+ F(\xi)), & F(\xi) \geq 0, \\ \exp(\alpha_- F(\xi)), & F(\xi) < 0. \end{cases} \tag{17}$$

Taking the Split-Bernstein method as an example, its theoretical formulation requires $\alpha_- \to +\infty$. However, in practical numerical implementations, α_- must take a finite value. Let $\alpha_- = M$, where M is a finite positive number, and let $\delta = -F(\xi)$ with $\delta > 0$ being a small positive quantity, then:

$$G_2(-\delta; M) = \exp(-M\delta) \tag{18}$$

To ensure the approximation error remains below a certain tolerance ε, the following condition must be satisfied:

$$\exp(-M\delta) < \varepsilon \Rightarrow \delta > -\frac{\ln \varepsilon}{M} \tag{19}$$

This implies that for any finite M, there always exists an interval $\delta \in (0, -ln\varepsilon/M)$, such that when $F(\xi) \in (ln\varepsilon/M, 0)$, $G(F(\xi)) > 1-\varepsilon \approx 1$. Thus, within this interval, the Split Bernstein approximation erroneously approaches 1 while the true step function equals 0. When the disturbance has high density near zero, numerous samples fall into this region, systematically overestimating the violation probability. This overtightens the feasible set, distorts the model, and invalidates the probabilistic constraint.

Such one-sided False Positive bias induces excessive conservatism, pushing solutions away from the true constraint boundary and sacrificing performance.

3.2 An Adaptive Proxy Function to Mitigate Conservatism

The approximation quality of the step function in chance-constrained optimization fundamentally dictates the conservatism and feasibility of the solution. To overcome this fundamental limitation, we introduce the Scaled and Translated Adaptive Proxy (STA-Proxy), a novel continuous and differentiable function formulated as:

$$\begin{aligned} \Psi(F(\xi)) &= \frac{(\mu+p)e^{\mu a(F(\xi)-b)}}{\mu e^{\mu a(F(\xi)-b)} + q e^{-\mu a(F(\xi)-b)}}, \quad aF(\xi) \in (-1,1), \\ b &= \frac{1}{2\mu}\log\left(\frac{\mu+2p}{q}\right). \end{aligned} \tag{20}$$

where $\mu > 0$ controls the closeness of approximation to the step function larger μ results in a steeper gradient at the step. The parameter $p \geq q > 0$ regulates the maximum and minimum values of the function. The constant offset b, whose calculation is provided in the above expression, is used to translate the curve.

The parameter $a > 0$ is a tunable multiplier that scales the magnitude of the input variable, thereby mapping $F(\xi)$ into a reasonable range of $(-1, 1)$ to ensure numerical stability. This scaling mechanism is particularly important for handling normalized variables. The tunable multiplier effectively amplifies the normalized variable, bringing it into the function's sensitive response interval, thereby enhancing approximation accuracy and numerical robustness.

The typical curves ($p = 1, q = 0.5, a = 1, \mu = 200$)of the proposed STA-Proxy method alongside conventional methods are presented in Fig. 1.

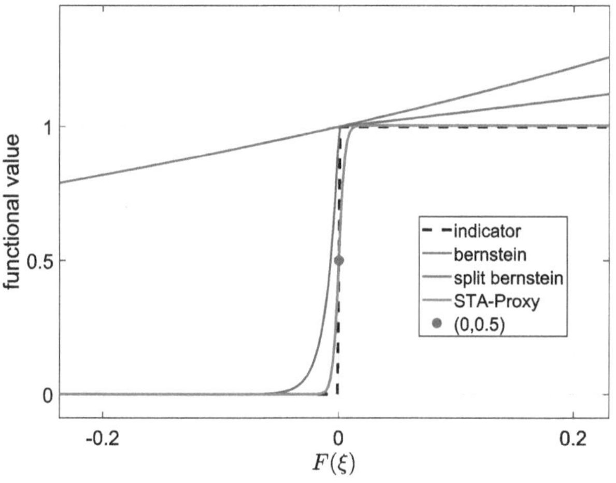

Fig. 1. Comparison of approximation functions for chance constraints.

The STA-Proxy function is continuous and strictly monotonically increasing over the domain (-1, 1), with output bounds ensuring numerical stability. It passes through the point (0, 0.5), indicating equal probability of constraint satisfaction and violation at the boundary, which enhances theoretical rationality.

To analyze conservatism, we evaluate the False Positive Rate ($P_{\text{FP}}^{(STA-Proxy)}$) and False Negative Rate ($P_{\text{FN}}^{(STA-Proxy)}$) under symmetric disturbance assumptions. the values of $P_{\text{FN}}^{(STA-Proxy)}$ and $P_{\text{FP}}^{(STA-Proxy)}$ can be quantitatively characterized by the areas of their respective violation regions in the probability space. Specifically, $Area_1$ is defined as the area bounded by the y-axis and the curve for $aF(\xi) \in (-1, 0)$; $Area_2$ is the area bounded by the y-axis, the horizontal line $y = 1$, and the curve; and $Area_3$ is the area bounded by the line $y = 1$ and the curve in the region where $aF(\xi) \in \left(c - \frac{1}{2\mu} \ln\left(\frac{p}{q}\right), 1\right)$ (where the curve reaches $y = 1$).

$$P_{FP}^{(STA-Proxy)} = c_1 \cdot (area1 + area3) \quad (21)$$

$$P_{FN}^{(STA-Proxy)} = c_1 \cdot area2 \quad (22)$$

To clearly illustrate these geometric characteristics, the regions of $Area_1$, $Area_2$, and $Area_3$ are shown in the Fig. 2.

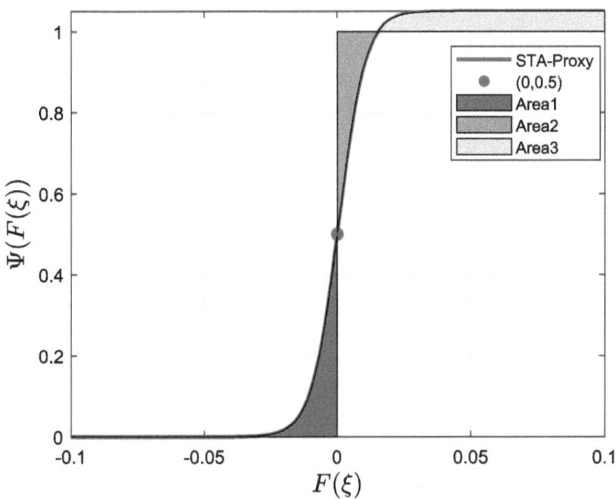

Fig. 2. Illustration of violation regions for the STA-Proxy function.

It can be rigorously proven that the area of the red region is always greater than that of the green region. This leads to the following inequality:

$$\begin{aligned} P_{\mathrm{FN}}^{(\mathrm{STA\text{-}Proxy})} &= c_1 \cdot \mathrm{area}_2 < c_1 \cdot \mathrm{area}_1 \\ &< c_1 \cdot (\mathrm{area}_1 + \mathrm{area}_3) = P_{\mathrm{FP}}^{(\mathrm{STA\text{-}Proxy})} \end{aligned} \quad (23)$$

From (23), it can be seen that $P_{\mathrm{FP}}^{(STA-Proxy)}$ consistently exceeds $P_{\mathrm{FN}}^{(STA-Proxy)}$, preserving a degree of robustness. Moreover, it can be proven mathematically that, as μ increases, $P_{FP}^{(STA-Proxy)}$ and $P_{FN}^{(STA-Proxy)}$ converge (Fig. 3.), demonstrating that the approximation achieves near-probabilistic unbiasedness.

Thus, the method achieves near-probabilistic unbiasedness, embodying a risk-neutral strategy distinct from conventional risk-averse approaches. Although the symmetry assumption is idealized, the method is not bound by it; any need for higher reliability under asymmetric disturbances can be met by tightening the probability threshold ε without modifying the function. In summary, through

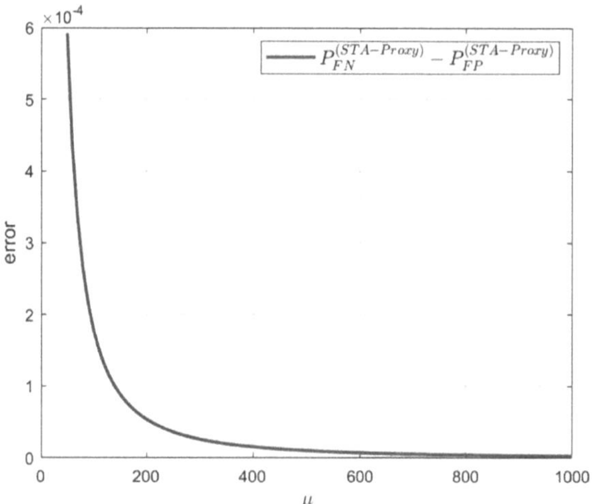

Fig. 3. The error between $P_{FN}^{(STA-Proxy)}$ and $P_{FP}^{(STA-Proxy)}$.

translation and scaling, STA-Proxy balances risk bias while preserving continuity, differentiability, and monotonicity, asymptotically approaching risk neutrality for broader applicability and performance.

Since the probability density function of the random disturbance ξ, as well as the expressions of functions $\mathbb{E}(\boldsymbol{.})$ $H(\boldsymbol{.})$ and $F(\boldsymbol{.})$ are known, this paper employs the Markov Chain Monte Carlo (MCMC) method to generate N independent and identically distributed (i.i.d.) samples $\{\xi_1, \xi_2, ..., \xi_N\}$, thereby transforming the original probabilistic constraint into a deterministic one based on sampling.

4 Enabling Tractable Solutions Through Problem Convexification

4.1 Convexification of Dynamics and Standard Constraints

To satisfy the fundamental requirement of convex optimization for affine equality constraints [4], we first discretize the state and control variables using trapezoidal integration:

$$X_{i+1} = X_i + \frac{\Delta t}{2}(\dot{X}_i + \dot{X}_{i+1}), \qquad i = 1, \ldots, K, \tag{24}$$

$$\Delta t = T/K. \tag{25}$$

where X_i represents the state at the $i-th$ node, T is the flight duration, which is an unknown quantity to be optimized.

A linear approximation is introduced using first-order Taylor expansion around a reference sequent $\tilde{X}, \tilde{U}$ and time $\tilde{T}$ from the previous iteration:

$$\begin{aligned}\hat{f}(X, U) = (X - \tilde{X}) \left.\frac{\partial \hat{f}}{\partial X}\right|_{\{\tilde{X},\tilde{U}\}} + (U - \tilde{U}) \left.\frac{\partial \tilde{f}}{\partial U}\right|_{\{\tilde{X},\tilde{U}\}} \\ +(T - \tilde{T}) \left.\frac{\partial \hat{f}}{\partial T}\right|_{\{\tilde{X},\tilde{U}\}} + \hat{f}(\tilde{X}, \tilde{U}) + o^1.\end{aligned} \tag{26}$$

With $\hat{f}(X, U) := T \cdot f(X, U)$ and higher-order terms omitted, substitution into (24) yields the linearized form:

$$\begin{aligned}0 = G_{i+1}X_{i+1} + G_iX_i + H_{i+1}U_{i+1} + H_iU_i \\ + (E_{i+1} + E_i)\Delta t + F_{i+1} + F_i.\end{aligned} \tag{27}$$

This linearization process completes the convexification of dynamic constraints.

Following X. Liu et al. [21], the nonconvex control constraint ${u_1}^2+{u_2}^2 = u_3$ is relaxed using its convex hull ${u_1}^2 + {u_2}^2 \leq u_3$ with a regularization term, ensuring optimal solutions satisfy the original constraint.

Constraint (6) contains is also inherently nonconvex. To address this, a sequential convex optimization strategy is adopted, Represented by (28). In each iteration, the atmospheric density ρ is fixed to the value $\tilde{\rho}$ computed from the altitude profile of the previous iteration, thereby convexifying the constraint.

$$0 \leq v^2 \leq \frac{q_{\max}}{0.5\tilde{\rho}g_0R_0} \tag{28}$$

When the trajectory converges, the constraint (28) will converge to the original constraint (6).

4.2 Convexification of the Risk-Neutral Chance Constraints

With the proposed risk-neutral proxy function introduced, it is applied to reformulate the original chance constraint (10). This processing yields a tractable deterministic form:

$$\mathbb{E}(\Psi(u_3 + \xi - u_{3\,\max})) \leq 1 - \epsilon \tag{29}$$

In general, the chance constraint (29) is non-convex. This paper adopts a sequential convex optimization approach to convexify constraint (29). At the $i - th$ discrete point, the chance constraint is convexified into:

$$\begin{aligned}\frac{1}{N}\sum_{n=1}^{N}\Bigg(\Psi(\mu, \tilde{\chi}_n) + \frac{\partial \Psi(\mu, \chi)}{\partial \chi}\Big|_{\tilde{\chi}_n} \\ \cdot \frac{\partial \chi}{\partial u_3}\Big|_{\tilde{u}_3} \cdot (u_3 - \tilde{u}_3)\Bigg) \leq 1 - \epsilon\end{aligned} \tag{30}$$

In (30), for this specific chance constraint: $\chi = u_3 + \xi - u_{3\,\max}, \frac{\partial \chi}{\partial u_3} = 1$. Since the distribution of the disturbance ξ affecting the control input u_3 cannot be directly obtained, but the distribution of $\Delta\alpha$ is known, it is necessary to quantify the impact of aerodynamic uncertainty on u_3 . Based on the explicit relationship between u_3, a, and Ma: $u_3 = (\alpha \cdot (-k \cdot Ma + b))^2$, the propagation process from the angle of attack disturbance $\Delta\alpha$ is analyzed to derive the statistical characteristics of Δu_3, which is:

$$\Delta u_3 = ((\alpha + \Delta\alpha) \cdot (-k \cdot Ma + b))^2 - (\alpha \cdot (-k \cdot Ma + b))^2 \tag{31}$$

To simplify the calculation and highlight the dominant term, the second-order small quantity of $\Delta\alpha$ is neglected, leading to the first-order approximation:

$$\Delta u_3 \approx (2\alpha \cdot b^2 - 4\alpha \cdot b \cdot k \cdot Ma + 2\alpha \cdot k^2 \cdot Ma^2) \cdot \Delta\alpha \tag{32}$$

It can be observed that the calculation of Δu_3 depends on the current flight state values of the angle of attack and the Mach number. This linearized model clarifies the relationship between the distribution of Δu_3 and the distribution of $\Delta\alpha$, thereby providing a computable basis for subsequent uncertainty propagation and chance constraint handling.

4.3 The Proposed Sequential Convex Programming Algorithm

The overall optimization workflow is summarized in Algorithm 1. This framework iteratively refines the convex approximation using the solution from the previous step. This process is designed to converge efficiently to a solution that satisfies the original non-convex problem, with termination triggered upon meeting predefined convergence criteria or reaching the maximum iteration count $k_{\max}$.

Algorithm 1 The Proposed Sequential Convex Programming Algorithm with Chance Constraints

```
Calculate the initial state reference sequence X̃ based on the given Ũ
for k = 0 to k_max do
    Solve the optimization subproblem to obtain X and U
    if (max|X − X̃| ≤ ε_c) ∩ (max|u_1^2 + u_2^2 − u_3| ≤ ε_r) then
        Optimization completed, exit loop
    else
        X̃_{i+1} ← X_i, Ũ_{i+1} ← U_i
    end if
end for
```

5 Simulation Experiments

The performance of the proposed STA-Proxy method is evaluated through comparative simulations and Monte Carlo experiments. It is benchmarked against robust optimization and Bernstein-based approaches in terms of optimality and computational efficiency. Furthermore, Monte Carlo simulations assess the constraint violation rates to validate robustness.

The structural parameters of the vehicle and the hyperparameters involved in the algorithm are provided in Table 1.

Table 1. Parameter settings

Indicator	Numerical Value	Indicator	Numerical Value
S_{ref} (m^2)	0.2922	R_0 (km)	6731
mass (kg)	200	g_0 (m/s^2)	9.8
c_ψ	0.01	$u_{3,\max}$	0.7
$k_{\max}$	30	N	10000
$q_{\max}$ (kN/m^2)	100	ϵ	90%
ξ	$\xi \sim N(0, 0.01)$	$(a-, a+)$	$(10, 0.1)$

A comparative study evaluates four trajectory optimization methods: Deterministic Optimization (*DO*), Robust Optimization (*RO*), the Split-Bernstein approach, and the proposed STA-Proxy method. *RO* adopts a 3σ worst-case design, theoretically reducing constraint violation risk to nearly zero under Gaussian disturbances. Performance is assessed along three metrics: optimality (performance index), computational efficiency (iteration count), and robustness (violation rate from 5000 Monte Carlo runs), as summarized in Table 2.

Table 2. The performance of different optimization strategies

Method	Performance Index	Iterations	Violation Rate
DO	0.0154	5	0.1853
RO	0.0159	5	0
Split-Bernstein	0.0164	7	0
STA-Proxy	0.0155	6	0.001544

To further illustrate the optimality in terms of trajectory shape, Fig. 4 compares the flight paths generated by the four methods. It can be observed that the STA-Proxy method produces a trajectory closely resembling the deterministic optimum, effectively leveraging the available flight envelope. In contrast, the robust and Split-Bernstein approaches yield noticeably conservative paths that deviate from the nominal trajectory, reflecting their inherent risk-averse nature.

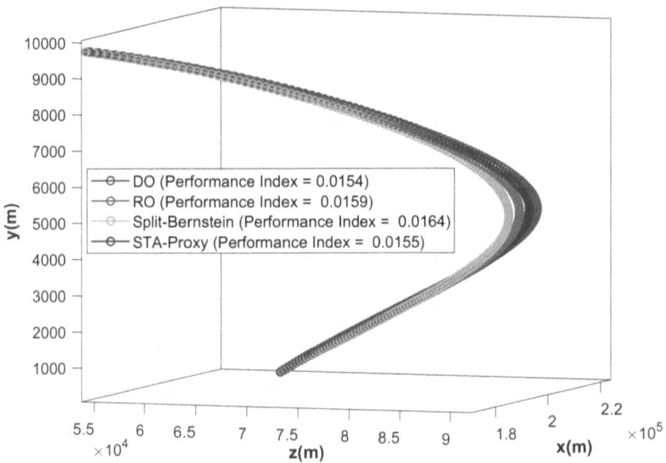

Fig. 4. Comparison of optimized trajectories under aerodynamic uncertainty.

The STA-Proxy method achieves a performance index (0.0155) second only to deterministic optimization (0.0154), with higher computational efficiency (6 iterations) than the Split-Bernstein approach (7 iterations). As illustrated in Fig. 5., the feasible domain of STA-Proxy, though slightly smaller than the deterministic one, is significantly larger than those of robust optimization and Split-Bernstein. This expanded domain provides greater search space to identify superior solutions, thereby enhancing performance potential.

In terms of robustness, the deterministic method exhibits a high violation rate of 18.53%, making it unsuitable for uncertain environments. While robust optimization and Split-Bernstein achieve zero violations, their excessive conservatism leads to notable performance loss. In contrast, STA-Proxy maintains a low violation rate of 0.16%—by balancing errors in the approximation function. This results in an effective trade-off between optimality and reliability, confirming its practical value for guidance applications.

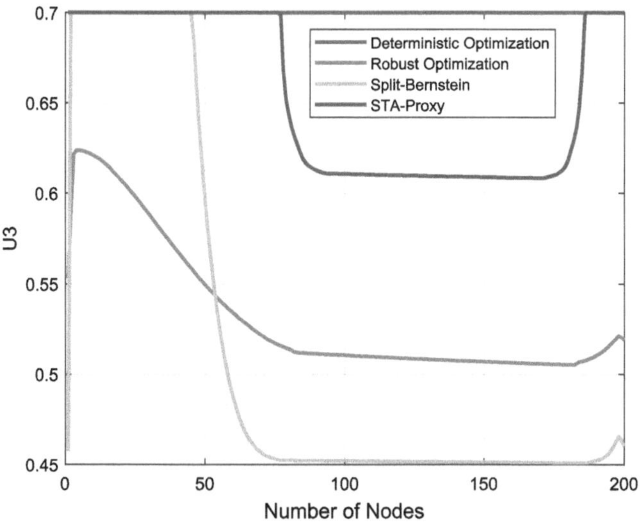

Fig. 5. the feasible domains for the control variable.

6 Conclusion

This paper introduces a chance-constrained sequential convex optimization method for hypersonic vehicles, leveraging a novel STA-Proxy to markedly reduce the conservatism of traditional approaches under aerodynamic uncertainty. Numerical simulations confirm that our STA-Proxy-SCP algorithm delivers performance nearly matching deterministic optimization while reliably satisfying safety constraints, achieving a superior balance of optimality and robustness over existing methods.

References

1. Leng, J.X., Xie, Z., Huang, W., et al.: Multidisciplinary design optimization of the first-stage waverider based on boost-glide flight trajectory. Aerosp. Sci. Technol. **160**, 110033 (2025)
2. Jung, C.G., Kim, B., Jung, K.W., et al.: Thrust integrated trajectory optimization for multipulse rocket missiles using convex programming. J. Spacecr. Rockets **60**(3), 957–971 (2023)
3. Ding, Y., Yue, X., Chen, G., Si, J.: Review of control and guidance technology on hypersonic vehicle. Chin. J. Aeronaut. **35**(7), 1–18 (2022)
4. Patel, J., Subbarao, K.: Reachable-set constrained model predictive control for safe hypersonic reentry. J. Guid. Control. Dyn. **48**(6), 1253–1265 (2025)
5. Li, Q., Liu, D., Zhu, J.: Uncertainty evaluation for wind tunnel test based on new flow field and balance models. Aerosp. Sci. Technol. **167**, 110593 (2025)
6. Slotnick, J.P.: CFD Vision 2030 Study: A Path to Revolutionary Computational Aerosciences, (2014). NASA/CR-2014-218178

7. Guo, Y., Liu, Y.: Robust control for hypersonic flight vehicle overload tracking under dynamics uncertainties. Sci. Sin. Inf 54(10), (2024). https://doi.org/10.1360/SSI-2023-0285
8. He, C., Qi, R., Jiang, B.: Fixed-time fault-tolerant control of a high-order fully actuated system under actuator faults and uncertain parameters. Int. J. Robust Nonlinear Control **35**(14), 6175–6189 (2025)
9. Tian, M., Shen, Z.: Air-breathing hypersonic vehicle trajectory optimization with uncertain no-fly zones. Adv. Mech. Eng. 2022(7), (2022)
10. Li, H., Chen, H., Su, X.: Rapid generation of hypersonic entry flight trajectories under uncertainty using transfer learning. In: Proceedings of IEEE 18th International Conference and Control Automation (ICCA), pp. 621–628. , Reykjavík, Iceland (2024)
11. Ni, W., Qiu, P., Liang, H.: Adaptive active defense guidance for hypersonic vehicle with incomplete information based on reinforcement learning. In: Yan, L., Duan, H., Deng, Y. (eds.) Advances in Guidance, Navigation and Control, pp. 10–21. Springer, Singapore (2025). https://doi.org/10.1007/978-981-96-2204-710
12. Xu, W., Liao, Y., Su, J., et al.: Error shaping strategy-based multi-constrained integrated guidance and control for hypersonic vehicle in dive phase. Nonlinear Dyn. **113**, 14997–15017 (2025)
13. Chao, D., Qi, R., Jiang, B.: Adaptive integrated guidance and control for HSV in ascent phase with time-varying state constraints. IEEE Trans. Aerosp. Electron. Syst. **61**(2), 2263–2280 (2025)
14. An, H., Wang, H., Zhang, X., Xie, W., Wang, C.: Compound control of an uncertain hypersonic vehicle model. Int. J. Control **96**(1), 94–109 (2021)
15. Zhao, Z., Kumar, M.: Split-Bernstein approach to chance-constrained optimal control. J. Guid. Control. Dyn. **40**(11), 2782–2795 (2017)
16. Boyd, S., Vandenberghe, L.: Convex Optimization. Cambridge University Press, Cambridge (2004)
17. Malyuta, D., Reynolds, T.P., Szmuk, M., et al.: Convex optimization for trajectory generation: a tutorial on generating dynamically feasible trajectories reliably and efficiently. IEEE Control Syst. Mag. **42**(5), 40–113 (2022)
18. Zhang, P., Li, W.B., Gong, S.P.: Ascent trajectory optimization for boost-glide vehicle using homotopy approximation function sequential convex programming. IEEE Trans. Aerosp. Electron. Syst. **61**(3), 7576–7596 (2025)
19. Xie, L., Zhou, X., Zhang, H.B., et al.: An improved convex optimization-based guidance for fuel-optimal powered landing. Adv. Space Res. **74**(7), 3256–3272 (2024)
20. Yang, R., Liu, X., Song, Z.: Rocket landing guidance based on linearization-free convexification. J. Guid. Control. Dyn. **47**(2), 217–232 (2024)
21. Li, W.L., Li, J., Ye, J.K., et al.: DDPG-based convex programming algorithm for the midcourse guidance trajectory of interceptor. Aerospace **11**(4), 314 (2024)

RL-Compensated Model Predictive Control for Quadruped Robot Locomotion on Challenging Terrains

Zhefeng Xiao[1], Dongdong Zheng[1,2,3](✉), Zeyuan Sun[2,3,4], and Yi Zeng[3,4]

[1] School of Automation, Beijing Institute of Technology, Beijing 100081, China
[2] Humanoid Robot (Shanghai) Co., Ltd., Shanghai 201203, China
[3] China North Artificial Intelligence and Innovation Research Institute, Beijing 100072, China
ddzheng@bit.edu.cn
[4] Collective Intelligence and Collaboration Laboratory, Beijing 100072, China

Abstract. Model Predictive Control (MPC) has been widely applied in quadruped robot locomotion. However, the method relies heavily on accuracy of the system model and the feasibility of the pre-planned desired trajectory. Reinforcement learning (RL) improves the performance of the target policy through continuous interaction between the robot and the environment. However, it cannot guarantee the safety of the robot's actions, and the design of the reward function is a cumbersome process. In this paper, An RL-compensated MPC algorithm framework is proposed. We simplify the quadruped robot into a single rigid body (SRB) dynamic model and train an RL policy to compensate for the linear acceleration, angular acceleration, gait frequency and foothold location of the robot. Comparative experiments against the MPC algorithm in simulation verify that the proposed framework can improve the robot's locomotion performance on irregular terrains.

Keywords: Quadruped robots · Model predictive control · Reinforcement learning · Locomotion control

1 Introduction

Model-based control of quadruped robots is a well-established research direction [1,2]. MPC enables robots to compute the optimal control input at the current moment by considering several future time horizons. To meet the real-time requirements of legged robot control, the robot model is usually simplified [3,4], thereby reducing the computational resources required for online MPC execution. However, the simplified robot model often exhibits mismatches in dynamic parameters compared to the full model, which can negatively affect the control accuracy of MPC. In addition, model-based control methods require

C. Li et al. (Eds.): ICNC 2025, CCIS 2946, pp. 209–219, 2026.
https://doi.org/10.1007/978-981-92-1599-7_18

pre-designed gait sequences and desired footholds [5], which limits their ability to respond promptly to external disturbances on variable terrain.

RL-based control of quadruped robots has received widespread attention in recent years. Such approach is generally used to achieve robust locomotion on variable terrain and to allow robots to execute challenging actions [6–9]. Nevertheless, training these policies requires extensive prior design of reward functions and careful tuning of the relative weights. Furthermore, since the resulting RL policies do not explicitly incorporate kinematic or dynamic constraints, their action outputs may exceed actuator limits under extreme conditions.

Building on the respective strengths and weaknesses of the two algorithms, several studies have attempted to combine them and have achieved promising results. The first category of methods collects motion data from MPC-controlled robots to construct datasets, where RL is applied on top of behavior cloning to further enhance locomotion robustness. For instance, reference [11] simplifies the quadruped robot model using a reduced linear inverted pendulum model, collects the center of mass (CoM) and foot-end trajectories of the robot under MPC control, and employs RL to learn these motion patterns, thereby enhancing the robot's disturbance rejection capability. The second category treats MPC-generated trajectories as motion constraints to restrict the action space of the RL policy. For example, reference [12] constructs motion trajectories of robots controlled by reduced-order models and imposes these trajectories as physical constraints to limit the action space of the RL policy, thereby balancing the model's stability with the explorative nature of RL. The drawback of such methods is that they require substantial effort to build datasets in advance, making the overall process rather cumbersome. The third category of methods employs policies learned through RL to refine the outputs of the MPC controller, thereby improving the control accuracy of MPC. Reference [13] models the quadruped robot as a SRB and employs a multilayer perceptron (MLP) to fit the RL policy. The policy outputs include corrections to the linear and angular accelerations of the single body, as well as position adjustments for the joints of the swing legs. This algorithmic framework is applicable to various quadruped robots. In [14], RL is employed to model the unknown system parameters within a robust MPC framework, enabling the quadruped robot to achieve stable non-vision-based locomotion across various terrains. Reference [15] includes gait frequency, residuals of support foot-end force compensation, and swing foot trajectory corrections in the action space of the RL policy, enabling humanoid robots to perform locomotion tasks on various irregular terrains. Similar approaches involve training RL policies to simultaneously adjust the CoM trajectory and foothold positions during quadruped jumping [16]. However, existing methods still lack a unified algorithmic framework that systematically accounts for the influence of support-leg control input corrections, swing-leg frequency, and final foothold adjustments on locomotion performance, and they do not consider the prioritization of actions in the policy output.

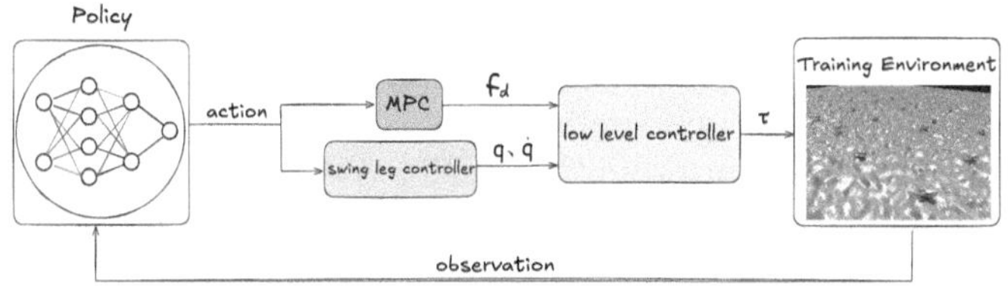

Fig. 1. RL-Compensated MPC architecture.

Motivated by the limitations of existing methods, we propose a novel RL-Compensated MPC framework to train quadruped robots for robust locomotion on irregular terrains. The main contributions of this work are as follows:

- The proposed RL-compensated framework contains an RL policy block, which is capable of real-time compensating for dynamic parameter uncertainties while simultaneously refining foot placement and gait frequency, thereby enabling quadruped robots to achieve adaptive locomotion over complex terrains. As illustrated in Fig. 1.
- A reward weight adjustment scheme is introduced to enable the robot to first focus on primary tasks and then gradually learn the remaining ones during training.
- The policy is first trained in Isaac Gym and then validated via simulation experiments, which demonstrates the effectiveness and generalization capability of the learned RL policy.

2 Preparation Work

2.1 Modeling of Quadruped Robots with Dynamic Uncertainty

Assuming that the mass of the quadruped robot's legs is negligible compared to that of the body, and that the pitch and roll angles vary only slightly during locomotion, the quadruped robot can be modeled as a SRB as follows:

$$\ddot{\boldsymbol{p}} = \frac{\sum_{i=1}^{n_c} \boldsymbol{f}_i}{m} + \boldsymbol{g} + \Delta \boldsymbol{a} \tag{1}$$

$$\frac{d}{dt}(\boldsymbol{I}\boldsymbol{\omega}) = \sum_{i=1}^{n_c} \boldsymbol{r}_i \times \boldsymbol{f}_i + \Delta \boldsymbol{\alpha}, \tag{2}$$

where $\boldsymbol{p} \in \mathbb{R}^3$ denotes the position of the CoM, $\boldsymbol{f}_i \in \mathbb{R}^3$ is the ground reaction force of the i-th leg, m is the mass of the base, $\boldsymbol{g} \in \mathbb{R}^3$ is the gravitational acceleration, $\boldsymbol{I} \in \mathbb{R}^{3\times 3}$ is the moment of inertia in the world frame, $\boldsymbol{\omega} \in \mathbb{R}^3$ is the robot's angular velocity in the world frame, and $\boldsymbol{r}_i$ is the vector from the CoM to the point where the contact force is applied. n_c denotes the number of legs in contact with the ground. Considering the effects of external disturbances on the robot's linear and angular accelerations, as well as the influence of unmodeled dynamics and parameter uncertainties on the system. Let $\Delta \boldsymbol{a}$ and $\Delta \boldsymbol{\alpha}$ denote the linear and angular acceleration compensation terms, respectively.

Let ϕ, θ, ψ denote the roll, pitch, and yaw angles of the robot, and let $\boldsymbol{\Theta} = [\phi, \theta, \psi]^\top$ represent the robot's attitude vector. $\boldsymbol{R}_z(\psi)$ is the rotation matrix corresponding to ψ and n_c indicates the number of supporting feet of the robot. Equations (1) and (2) can then be expressed in state-space form as:

$$\frac{\mathrm{d}}{\mathrm{d}t}\begin{bmatrix}\boldsymbol{\Theta}\\ \boldsymbol{p}\\ \boldsymbol{\omega}\\ \dot{\boldsymbol{p}}\end{bmatrix} = \begin{bmatrix}\mathbf{0}_3 & \mathbf{0}_3 & \boldsymbol{R}_z(\psi) & \mathbf{0}_3\\ \mathbf{0}_3 & \mathbf{0}_3 & \mathbf{0}_3 & \mathbf{1}_3\\ \mathbf{0}_3 & \mathbf{0}_3 & \mathbf{0}_3 & \mathbf{0}_3\\ \mathbf{0}_3 & \mathbf{0}_3 & \mathbf{0}_3 & \mathbf{0}_3\end{bmatrix}\begin{bmatrix}\boldsymbol{\Theta}\\ \boldsymbol{p}\\ \boldsymbol{\omega}\\ \dot{\boldsymbol{p}}\end{bmatrix} + \begin{bmatrix}\mathbf{0}_3 & \dots & \mathbf{0}_3\\ \mathbf{0}_3 & \dots & \mathbf{0}_3\\ \boldsymbol{I}^{-1}[\boldsymbol{r}_1]_\times & \dots & \boldsymbol{I}^{-1}[\boldsymbol{r}_{n_c}]_\times\\ \mathbf{1}_3/m & \dots & \mathbf{1}_3/m\end{bmatrix}\begin{bmatrix}\boldsymbol{f}_1\\ \vdots\\ \boldsymbol{f}_{n_c}\end{bmatrix} + \begin{bmatrix}0\\ 0\\ \Delta\boldsymbol{\alpha}\\ \boldsymbol{g}+\Delta\boldsymbol{a}\end{bmatrix}. \tag{3}$$

After completing the dynamics modeling, it is necessary to impose friction constraints on the control inputs, i.e., the foot-end forces, to prevent relative slipping between the feet and the ground during contact. Typically, the friction constraints are expressed by the following inequalities, where both the normal force along the contact surface and the tangential forces must remain below specified thresholds:

$$\boldsymbol{f} : \begin{cases} |\boldsymbol{f}_x| \le \mu \boldsymbol{f}_z, \\ |\boldsymbol{f}_y| \le \mu \boldsymbol{f}_z, \\ \boldsymbol{f}_z > 0 \end{cases} \tag{4}$$

2.2 Foothold Correction Residual Design

The placement of the swing leg's foothold plays a crucial role in quadruped locomotion, directly influencing the robot's stability. In most MPC-based control algorithms, foothold planning is performed using the Raibert heuristic method. However, the traditional Raibert foothold algorithm cannot capture terrain information, and when uncertainties in the terrain exist, it fails to plan feasible foothold positions, thereby preventing the robot from successfully executing subsequent locomotion tasks. To address this limitation, a foothold residual term is introduced into the Raibert heuristic foothold calculation formula to represent the correction of the foothold position, expressed as:

$$\boldsymbol{p}_{foot} = \underbrace{\boldsymbol{p} + \boldsymbol{R}_z(\psi)\boldsymbol{l}}_{\boldsymbol{p}_{shoulder}} + \underbrace{\frac{T_s}{2}\boldsymbol{v} + k(\boldsymbol{v} - \boldsymbol{v}^{\mathrm{cmd}})}_{\boldsymbol{p}_{heuristic}} + \underbrace{\frac{1}{2}\sqrt{\frac{h}{g}}\,\boldsymbol{v} \times \boldsymbol{\omega}^{\mathrm{cmd}}}_{\boldsymbol{p}_{centrifugal}} + \underbrace{\Delta\boldsymbol{s}}_{\boldsymbol{p}_{residual}}, \tag{5}$$

where $\boldsymbol{p}_{shoulder} \in \mathbb{R}^3$ denotes the position of the robot's hip in the world frame, while $\boldsymbol{l} \in \mathbb{R}^3$ refers to its position in the body frame. The symbol T_s indicates the duration of the stance phase. $\boldsymbol{v} \in \mathbb{R}^3$ stands for the linear velocity of base. The variables $\boldsymbol{v}_{cmd} \in \mathbb{R}^3$ and $\boldsymbol{\omega}_{cmd} \in \mathbb{R}^3$ correspond to the commanded linear and angular velocities provided by the user, respectively. The parameter h specifies the desired base height, and k acts as a proportional coefficient. Finally, $\boldsymbol{\Delta s} \in \mathbb{R}^2$ represents the foothold residual term introduced for correction. In the subsequent content, the robot is assumed moving with a trot gait, therefore, the adjustment amount of the foothold positions for the two swing legs in each swing cycle is defined as $\boldsymbol{\Delta s}_1 \in \mathbb{R}^2$ and $\boldsymbol{\Delta s}_2 \in \mathbb{R}^2$.

2.3 Adaptive Gait Period Design

In model-based legged robot control, a locomotion gait is typically represented as a sequence of stance and swing phases for each leg, with discrete and instantaneous transitions between them, as illustrated in the left side of Fig. 2. In the gait scheduler phase map, the durations of these phases are fixed and cannot be dynamically adjusted according to changes in locomotion conditions.

For quadruped robots moving on irregular terrain, two problematic situations may arise. First, when a leg in the swing phase unexpectedly contacts a protruding surface, the joint actuators may still output the torques required for swinging. Second, when the swing leg transitions to the support phase based on the gait timing table, the actuators execute the stance torques computed by the MPC as planned, but the foot may fail to contact the ground due to a concave surface. Both cases can potentially lead to robot instability. To address such situations, a learnable gait scheduler adjustment term, Δt, is introduced. This allows the duration of the swing phase to be either extended or shortened, while keeping the overall gait cycle unchanged, as illustrated in Fig. 2. After the adjustment, the durations of the swing and stance phases can be expressed as follows.

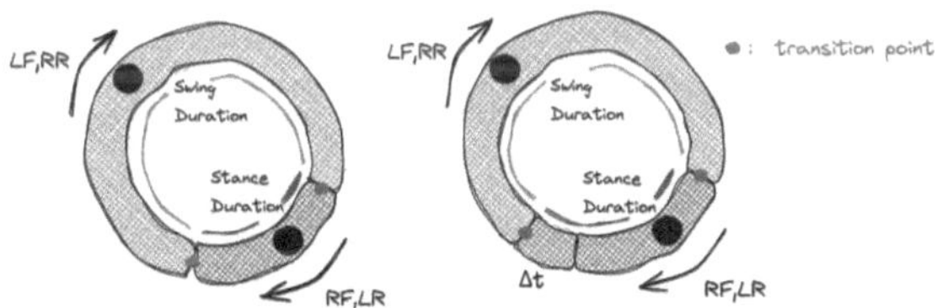

Fig. 2. Gait scheduler phase map: swing and stance phase transitions occur at transition points. LF, LR, RF, and RR represent the left front, left rear, right front, and right rear legs, respectively.

$$\begin{cases} T_{swing} = T_{swing} \pm \Delta t, \\ T_{stance} = T_{stance} \mp \Delta t, \\ T_{stance} + T_{swing} = T_{total}, \end{cases} \tag{6}$$

where T_{total} denotes a fixed time constant.

3 Proposed Approach

3.1 RL-Compensated MPC Policy Design

We employ a deep neural network to approximate the optimal policy and train it using a policy-gradient-based RL algorithm. Specifically, we adopt the Proximal Policy Optimization (PPO) algorithm [18], which is widely used in quadruped robot RL to maximize the expected return of the policy.

Action: Based on the discussion in Subsect. 2.1 to Subsect. 2.3, we arrange the action as the vector, represent as:

$$\boldsymbol{A} = [\Delta \boldsymbol{a}^\top, \Delta \boldsymbol{\alpha}^\top, \Delta \boldsymbol{s}_1{}^\top, \Delta \boldsymbol{s}_2{}^\top, \Delta t]^\top \in \mathbb{R}^{11}. \tag{7}$$

Observation: The design of the observation space variables depends on whether each variable can be obtained from the sensors mounted on the robot's body. Each joint of the quadruped robot is equipped with a joint encoder, and the base is equipped with an inertial measurement unit (IMU) and an accelerometer. Therefore, the actor's observation is designed to consist of the projected gravity $\hat{\boldsymbol{g}}$, angular velocity $\boldsymbol{\omega}$, joint positions $\boldsymbol{q} \in \mathbb{R}^{12}$, joint velocities $\dot{\boldsymbol{q}} \in \mathbb{R}^{12}$, previous actions $\boldsymbol{A}$ and the user command $[\boldsymbol{v}^x_{cmd}, \boldsymbol{v}^y_{cmd}, \boldsymbol{\omega}^z_{cmd}]^\top \in \mathbb{R}^3$. The critic's observation comprises all elements of the actor's observation, along with the linear velocity $\boldsymbol{v}$, the positions of the swing legs' end effectors represented in the base frame, denoted as $\boldsymbol{p}_{swing} \in \mathbb{R}^3$, and the terrain height map information.

Table 1. List of reward functions

Reward	Expression	Weight
Linear velocity tracking	$\exp\{-5.0\vert\boldsymbol{v}^c - \boldsymbol{v}\vert^2\}$	1.0
Angular velocity tracking	$\exp\{-7.0\vert\boldsymbol{\omega}^c - \boldsymbol{\omega}\vert^2\}$	1.0
Body height tracking	$\exp\{-4.0\vert\boldsymbol{p}^c_z - \boldsymbol{p}_z\vert^2\}$	3.0
Feet slide	$\sum_i \Vert\dot{\boldsymbol{p}}_{stance,i}\Vert^2$	-0.01
Link collision	1.0	-5.0
Action rate	$\Vert\boldsymbol{A}_t - \boldsymbol{A}_{t-1}\Vert_2$	-0.01
Joint velocity	$\Vert\dot{\boldsymbol{q}}\Vert_2$	-2.5e-4
Survival	1.0 (maintaining balance)	1.0

Reward Function: The design objective of the reward function is to ensure that the robot can follow the user's commanded velocity while maintaining posture stability and minimizing energy consumption. The design details of the reward function are summarized in Table 1, where $\dot{\boldsymbol{p}}_{stance,i} \in \mathbb{R}^3$ denotes the velocity of the i-th support leg end-effector represented in the body frame.

Reward Weight Adjustment Scheme: During the training process, the weighting coefficients of the reward function are progressively adjusted over time to enable the robot to learn all actions in a sequential curriculum. The task-related reward is assigned the highest priority, with their corresponding terms highlighted in blue in the table, followed by the behavior-related reward, highlighted in red. At first, the robot is trained to follow the user's control commands, and subsequently, the training focuses on behavior-related learning, where diverse behaviors are encouraged to improve the robot's ability to handle disturbances. The rules for adjusting the reward weights are defined as follows.

$$\begin{cases} \lambda_{task,t} = \max\left\{\lambda_{task,t-1} \cdot D_{task}, 0.6\right\} \\ \lambda_{behavior,t} = \min\left\{1 - \lambda_{task,t}, 0.4\right\}, \end{cases} \tag{8}$$

where $t \in [0, T]$ and T represents the total number of training iterations. $\lambda_{task,t}$ denotes the weight of the task-related reward, initialized as $\lambda_{task,0} = 1.0$, and $\lambda_{behavior,t}$ denotes the weight of the behavior-related reward. D_{task} indicates whether the task-related reward has reached the pre-defined threshold. Once the threshold is achieved, D_{task} changes from 1.0 to 0.9994. $\lambda_{task,t}$ begins to decay but does not fall below 0.6, while $\lambda_{behavior,t}$ begins to increase but does not exceed 0.4. The weights of individual reward components within each category are distributed according to fixed proportions.

4 Experiment Results

In this section, experiments are conducted to validate the effectiveness of the learned policy on a variety of irregular terrain, including undulating terrain and discontinuous terrain. Additionally, the robot's body mass is increased to test whether it can still maintain the desired base height. The robot used in the experiments is Unitree Go2. The MPC algorithm, whose control performance was validated in advance, is able to run concurrently during training. Considering the available hardware computational resources, 32 robots are employed for parallel training. The state-weighting matrix of the MPC is set to $\boldsymbol{Q} = [0.25, 0.25, 10.0, 2.0, 2.0, 50.0, 0.05, 0.05, 0.3, 0.2, 0.2, 0.1, 0.0]$. The ranges of the user control command are as follows. The angular velocity around the z-axis is constrained within ± 2.0 rad/s, while the linear velocity is limited to ± 4.0 m/s along the x-axis and ± 2.0 m/s along the y-axis.

4.1 Undulating Terrain

On the undulating terrain, the RL-compensated MPC control framework demonstrates strong robustness. Even when the supporting legs on the left and right sides of the robot are not on the same plane, the robot is still able to maintain stability. In contrast, the quadruped robot controlled solely by the MPC algorithm suffers from lateral overturning. The experimental scenario and the body attitude data are presented in Fig. 3. Under the RL-compensated MPC framework, the robot is able to follow the user's control commands. In contrast, the MPC-controlled robot fails to accurately track the user's commands in both linear and rotational spaces after falling, and it is unable to recover to the upright posture.

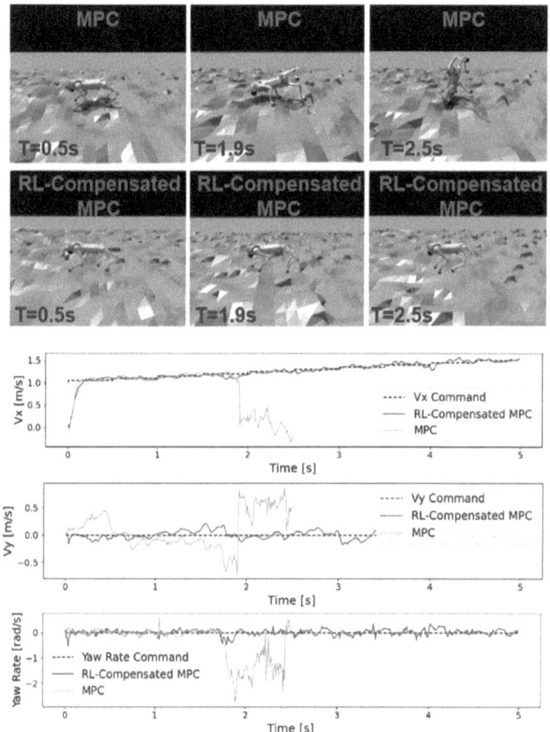

Fig. 3. Quadruped locomotion on undulating terrain and the corresponding experimental results.

4.2 Discontinuous Terrain

On discontinuous terrain, the quadruped robot controlled by the RL-compensated MPC framework is able to handle situations where the swing leg collides with an obstacle. Even when the robot's feet collide with obstacles, the policy's output compensates for the resulting motion uncertainties, allowing the robot to maintain stability. In contrast, the quadruped robot controlled solely

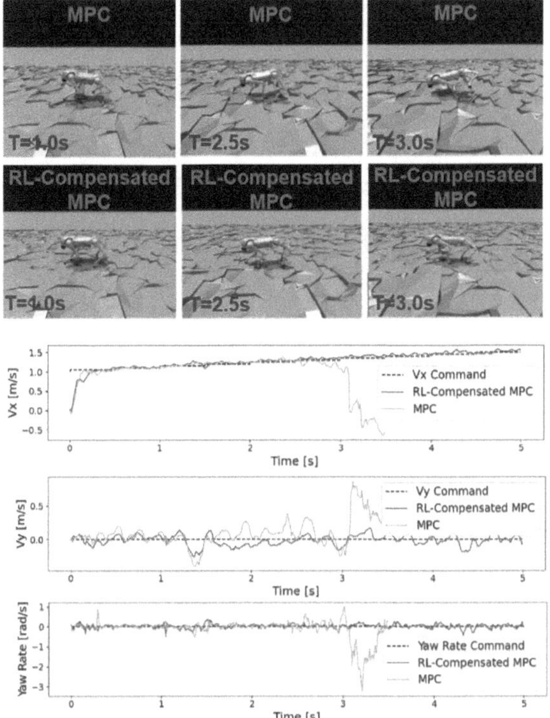

Fig. 4. Quadruped locomotion on discontinuous terrain and the corresponding experimental results.

by the MPC algorithm fails to overcome obstacles due to repeated collisions of the swing leg, ultimately leading to falling. Its velocities along the x- and y-axes fluctuate dramatically both immediately before and after the fall, and the robot cannot track the yaw rate command. The experimental scenario and body attitude data are presented in Fig. 4.

4.3 Altering the Robot's Body Mass

In this subsection, the experiment involves increasing the actual body mass of the quadruped robot, while the MPC controller still assumes the original mass. Due to the simplifications in the quadruped's dynamic model, altering the body mass has a significant impact on the control performance of the MPC. As shown in Fig. 5, the quadruped robot using only the MPC cannot account for the increased mass when computing control input, and is consequently overwhelmed. In contrast, under the RL-compensated MPC framework, the policy can compensate for uncertainties in the dynamic parameters, allowing the robot to maintain its body height near the desired value and move normally.

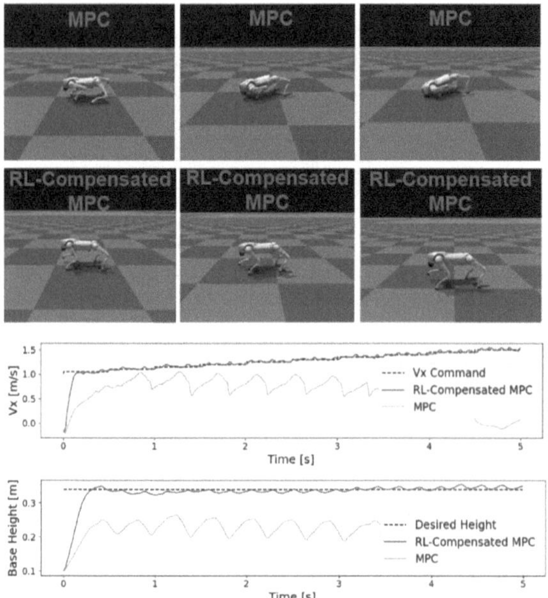

Fig. 5. Quadruped locomotion on flat ground with altered body mass.

5 Conclusion

In this research, we have presented an RL-Compensated MPC framework that integrates RL with MPC to enhance quadruped robot locomotion on challenging terrains. By leveraging MPC's forward-looking trajectory optimization and RL's adaptive learning capabilities, the framework compensates for dynamic uncertainties, refines foothold positions, and adapts gait frequency, thereby significantly improving the robot's ability to traverse complex landscapes and handle perturbations that would destabilize conventional MPC. Extensive validation on the Unitree Go2, with policy trained in Isaac Gym and evaluated in simulation, demonstrates the framework's robustness and generalizability, consistently outperforming standalone MPC. This approach paves the way for future research, such as integrating exteroceptive perception, to achieve even more resilient legged robot behavior.

Acknowledgments. This work was supported in part by the National Key R & D Program of China under Grant 2024YFB4711101, 2023YFB4703900, and in part by Collective Intelligence & Collaboration Laboratory (Young Scholars Independent Exploration Fund Project QT16S25024-6)

Disclosure of Interests. None of the authors have a conflict of interest to disclose.

References

1. Morlando, V., Ruggiero, F.: Disturbance rejection for legged robots through a hybrid observer. In: 2022 30th Mediterranean Conference on Control and Automation (MED), pp. 743–748. IEEE (2022)
2. Kim, D., Jorgensen, S.J., Lee, J., et al.: Dynamic locomotion for passive-ankle biped robots and humanoids using whole-body locomotion control. Int. J. Robot. Res. **39**(8), 936–956 (2020)
3. Xin, G., Xin, S., Cebe, O., et al.: Robust footstep planning and LQR control for dynamic quadrupedal locomotion. IEEE Robot. Autom. Lett. **6**(3), 4488–4495 (2021)
4. Di Carlo, J., Wensing, P.M., Katz, B., et al.: Dynamic locomotion in the MIT cheetah 3 through convex model-predictive control. In: 2018 IEEE/RSJ International Conference on Intelligent Robots and Systems (IROS), pp. 1–9. IEEE (2018)
5. Kim, D., Carlo, D., Katz, J., B.: Highly dynamic quadruped locomotion via whole-body impulse control and model predictive control. arXiv preprint arXiv:1909.06586 (2019)
6. Long, J., Yu, W., Li, Q.: Learning h-infinity locomotion control. arXiv preprint arXiv:2404.14405 (2024)
7. Cheng, X., Kumar, A., Pathak, D.: Legs as manipulator: pushing quadrupedal agility beyond locomotion. arXiv preprint arXiv:2303.11330 (2023)
8. Cheng, X., Kumar, A., Pathak, D.: Legs as manipulator: pushing quadrupedal agility beyond locomotion. arXiv preprint arXiv:2303.11330 (2023)
9. Shen, J., Khodak, M., Talwalkar, A.: Efficient architecture search for diverse tasks. Adv. Neural Inf. Process. Syst. (NeurIPS) **35**, 16151–16164 (2022)
10. Gangapurwala, S., Campanaro, L., Havoutis, I.: Learning low-frequency motion control for robust and dynamic robot locomotion. arXiv preprint arXiv:2209.14887 (2022)
11. Kang, D., Cheng, J., Zamora, M., et al.: RL+ model-based control: using on-demand optimal control to learn versatile legged locomotion. IEEE Robot. Autom. Lett. **8**(10), 6619–6626 (2023)
12. Lee, H.J., Hong, S., Kim, S.: Integrating model-based footstep planning with model-free reinforcement learning for dynamic legged locomotion. In: 2024 IEEE/RSJ International Conference on Intelligent Robots and Systems (IROS), pp. 11248–11255. IEEE (2024)
13. Chen, Y., Nguyen, Q.: Learning agile locomotion and adaptive behaviors via RL-augmented MPC. In: 2024 IEEE International Conference on Robotics and Automation (ICRA), pp. 11436–11442. IEEE (2024)
14. Pandala, A., Fawcett, R.T., Rosolia, U.: Robust predictive control for quadrupedal locomotion: learning to close the gap between reduced-and full-order models. IEEE Robot. Autom. Lett **7**(3), 6622–6629 (2022)
15. Kamohara, J., Wu, F., Wamorkar, C.: RL-augmented adaptive model predictive control for bipedal locomotion over challenging terrain. arXiv preprint arXiv:2509.18466 (2025)
16. Yang, Y., Shi, G., Meng, X.: Cajun: continuous adaptive jumping using a learned centroidal controller. In: Conference on Robot Learning, pp. 2791–2806. PMLR (2023)
17. Jing-Qing, H.: Control theory, is it a model analysis approach or a direct control approach? J. Syst. Sci. Math. Sci. **9**(4), 328 (1989)
18. Schulman, J., Wolski, F., Dhariwal, P.: Proximal policy optimization algorithms. arXiv preprint arXiv:1707.06347 (2017)

Prescribed-Time Control of Flexible Joint Robotic Based on Neural Network

Xinpeng Ge[1], Dong-Dong Zheng[1,2,3(✉)], Xuemei Ren[1], and Zeyuan Sun[2,3,4]

[1] School of Automation, Beijing Institute of Technology, Beijing 100081, China

[2] Humanoid Robot (Shanghai) Co., Ltd., Shanghai 201203, China

[3] China North Artificial Intelligence and Innovation Research Institute, Beijing 100072, China

ddzheng@bit.edu.cn

[4] Collective Intelligence and Collaboration Laboratory, Beijing 100072, China

Abstract. This paper proposes a neural network (NN) based adaptive prescribed-time sliding mode controller for flexible joint robotic (FJR) manipulators. By employing the singular perturbation technique, the original high-order system is decomposed into two lower-order subsystems. An NN is employed to estimate uncertainties within the FJR system, trained using a parameter identification algorithm (PIA). Subsequently, a prescribed-time sliding mode controller is developed for each subsystem to ensure that the tracking error converges within a predefined time and follows a given reference trajectory. The stability of the prescribed-time controller is rigorously established using Lyapunov analysis. Simulation and comparative experiments validate the effectiveness and superiority of the proposed method.

Keywords: Flexible joint robot · Neural network · Prescribed-time control · Singular perturbation technique

1 Introduction

Currently, FJR with adaptable structure is widely used in industrial production, which effectively improves the efficiency of production, and thus receives more and more attention [1,2]. When in use, interactions between FJRs and humans or the environment are inevitable. Therefore, it is crucial to ensure that FJRs can safely interact with humans and the environment [3].

However, the higher order of FJR may make it difficult to design controllers using traditional methods [4]. To solve this problem, researchers have used singular perturbation techniques to decompose flexible manipulators, thereby simplifying controller design. For example, in [5], the authors use singular perturbation theory to dynamically decompose the flexible manipulator so as to design controllers for the decomposed subsystems, which greatly reduces the computational

C. Li et al. (Eds.): ICNC 2025, CCIS 2946, pp. 220–230, 2026.
https://doi.org/10.1007/978-981-92-1599-7_19

workload; and in [6] a hybrid PID controller based on singular perturbations and state observers is proposed to solve the problem of positional tracking of the FJR without velocity information.

The prevalence of modeling uncertainties in FJR necessitates advanced compensation techniques. NN, now extensively adopted in control systems, provide a powerful framework for approximating these unmodeled nonlinear dynamics [7,8]. In [9], the authors explore the FJR problem with an unknown dead zone. They use an RBFNN to address deadband effects and another RBFNN to estimate the unknown dynamics. Despite significant progress, existing research primarily relies on parameter identification methods based on gradient descent algorithms (GDA), characterized by fixed learning gains. These traditional approaches are often constrained by slow convergence rates and require systematic manual adjustment of learning parameters.

Currently, FJR has been widely applied in critical domains such as space docking and missile guidance. Such systems must possess rapid response capabilities to ensure operational precision and safety. However, traditional control methods can only guarantee asymptotic or exponential stability [10], meaning the system requires infinite time to reach a stable state, failing to meet rapid response requirements. To address the rapid response challenge, prescribed-time stability has been introduced to accelerate convergence speed [11]. Given the advantages of prescribed-time convergence, extensive research has been conducted in this field [12]. However, existing studies primarily focus on non-robotic domains. Even within robotics, predetermined time convergence research has emphasized rigid robotic arms, while studies on predetermined time tracking control for FJR systems remain relatively scarce.

To address the aforementioned limitations, this paper employs a novel parameter identification method to update NN weights, accelerating the training process and ensuring rapid convergence of identification errors. Based on NN identification results, this paper investigates the prescribed-time control and tracking problem in flexible robotic arm systems. For both slow and fast subsystems, a novel sliding mode prescribed-time controller is designed to ensure tracking errors generated during the tracking process converge within a specified time. This specified time can be arbitrarily set and is independent of initial conditions.

2 Problem Formulation

A generalized n-DOF FJR is modeled as follows:

$$D(q)\ddot{q}+C(q,\dot{q})\dot{q}+G(q)=K(\theta-q)+f+\Delta M \tag{1a}$$

$$\mathcal{N}^2 J\ddot{\theta}+K(\theta-q)=\mathcal{N}u \tag{1b}$$

where $q \in \mathbb{R}^n$ and $\theta \in \mathbb{R}^n$ represent the angular positions of the joints and drive motors, respectively. The inertia matrix is $D(q) \in \mathbb{R}^{n\times n}$, while $C(q,\dot{q}) \in \mathbb{R}^{n\times n}$ is the Coriolis force and centrifugal force matrix. The gravity vector is given by $G(q) \in \mathbb{R}^n$. Diagonal matrices $\mathcal{N} \in \mathbb{R}^{n\times n}$, $J \in \mathbb{R}^{n\times n}$, and $K \in \mathbb{R}^{n\times n}$ represent the motor gear ratio, motor inertia, and joint elastic coefficients, respectively. Friction is denoted by $f \in \mathbb{R}^n$, $\Delta M \in \mathbb{R}^n$ accounts for modeling uncertainties, and $u \in \mathbb{R}^n$ is the control input torque.

As shown in (1a) and (1b), the FJR is modeled as a fourth-order system, which is considerably more complex than rigid robots. Its joint angular position q evolves much slower than its motor position θ, revealing a inherent timescale separation with both fast and slow dynamics. This justifies treating the system as singularly perturbed and decomposing it into two reduced-order subsystems. Such a decomposition greatly simplifies the overall controller design by allowing separate development for each lower-order subsystem.

We define the state variables as $x_1 = q$, $x_2 = \dot{q}$, $y_1 = B_0^{-1}K(\theta - q)$, and $y_2 = \varepsilon\dot{y}_1$. Here, $\varepsilon \in \mathbb{R}^+$ is a small user-defined parameter ($0 < \varepsilon \ll 1$), and $B_0 \in \mathbb{R}^{n\times n}$ is a diagonal matrix satisfying $\varepsilon = (K^{-1}B_0)^{\frac{1}{2}}$. Based on the above definitions, (1a) and (1b) can be rewritten as

$$\begin{cases} \dot{x}_1 = x_2 \\ \dot{x}_2 = a_1 y_1 + D^{-1}(F - G) \\ \varepsilon\dot{y}_1 = y_2 \\ \varepsilon\dot{y}_2 = \left(-a_2 y_1 + b_1 u + D^{-1}(G - F)\right) \end{cases} \tag{2}$$

where $a_1 = D^{-1}B_0$, $a_2 = \left(J^{-1}N^{-2} + D^{-1}\right)B_0$, $b_1 = J^{-1}N^{-1}$, $F = f + \Delta M - C\dot{q}$.

To facilitate the design of the identification and control scheme, we begin by making the following assumptions:

Assumption 1. The uncertainty term F in (2) is assumed to be a smooth nonlinear function.

Definition 1. *[13] If a vector $\Phi \in \mathbb{R}^q$ or matrix function $\Phi \in \mathbb{R}^{n\times q}$ is Persistently Excited (PE), it satisfies:*

$$\alpha_1 I_q \leq \int_t^{t+\Delta T} \Phi(r)\,\Phi(t)^T dr, \forall t \geq 0. \tag{3}$$

where $\alpha_1 \in \mathbb{R}^+$ and $\Delta T \in \mathbb{R}^+$ are positive constants.

The assumption and definition presented in this work represent reasonable simplifications to highlight the core research objectives, and are commonly adopted in related studies.

Since F is unknown, the control input u is divided into two parts: $u = u_a + u_b$. In this formulation, $u_a = b_1^{-1}D^{-1}\hat{F}$ is introduced to counteract the uncertain term F, with $\hat{F}$ denoting the estimate of F that will be explicitly specified later. Consequently, we can rewrite (2) as:

$$\begin{cases} \dot{x}_1 = x_2 \\ \dot{x}_2 = a_1 y_1 + D^{-1}(F - G) \\ \varepsilon\dot{y}_1 = y_2 \\ \varepsilon\dot{y}_2 = -a_2 y_1 + b_1 u_b + D^{-1}\left(G - \tilde{F}\right) \end{cases} \tag{4}$$

where $\tilde{F} = F - \hat{F}$. The small value of ε implies that the fast states y_1 and y_2 exhibit quasi-steady-state behavior relative to the slow dynamics of x_1 and x_2. Under singular perturbation theory, it is postulated that these fast states reach their equilibrium almost immediately. This equilibrium is found by substituting $\varepsilon = 0$ and $\tilde{F} = 0$ into the final two lines of (4), giving:

$$y_1^* = a_2^{-1}\left(b_1 u_{bs} + D^{-1}G\right),\ y_2^* = 0 \tag{5}$$

where y_1^* and y_2^* denote the equilibrium values of y_1 and y_2, respectively, and u_{bs} represents the slow control component of u in the limit as $\varepsilon \to 0$. Substituting $y_1 = y_1^*$ into (4) yields the *Reduced Slow Subsystem*:

$$\begin{cases} \dot{x}_1 = x_2 \\ \dot{x}_2 = g_1 u_{bs} + C_1 + F^* \end{cases} \tag{6}$$

where $C_1 = -\left(D^{-1} - a_1 a_2^{-1} D^{-1}\right) G$, $g_1 = a_1 a_2^{-1} b_1$ and $F^* = D^{-1} F$ is the lumped disturbance.

Defining a new state variable $z_1 = y_1 - y_1^*$ and utilizing (5), we can get

$$\begin{cases} \varepsilon \dot{z}_1 = y_2 - \varepsilon y_1^* \\ \varepsilon \dot{y}_2 = -a_2\left(z_1 + y_1^*\right) + b_1 u + D^{-1}\left(G - \tilde{F}\right) \end{cases} \tag{7}$$

Let a stretched time variable be defined as $t_\varepsilon = \frac{t}{\varepsilon}$ and set $\varepsilon = 0$ in (7), the *Reduced Fast Subsystem* is obtained as:

$$\begin{cases} \frac{dz_1}{dt_\varepsilon} = y_2 \\ \frac{dy_2}{dt_\varepsilon} = -a_2 z_1 + b_1 u_{bf} + \epsilon_y \end{cases} \tag{8}$$

where $\epsilon_y = D^{-1}\tilde{F}$, $u_{bf} = u - u_{bs}$ is the fast control component of u.

Following the standard procedure, the original high-order system (1) has been decoupled into two lower-order subsystems, (6) and (8). This decomposition permits the separate design of controllers for each subsystem, thereby greatly simplifying the process compared to dealing with the full-order system (1). The overall composite controller for the complete system is then synthesized from the individual designs.

A detailed description of the identification and control algorithms designed for the subsystems is provided in the subsequent sections.

3 System Identification

For the controller design of the Reduced Slow Subsystem (6), a feedforward component is incorporated to cancel the nonlinear terms C_1 and F^*. Since F^* is unknown, a three-layer neural network is employed to estimate it as follows:

$$F^* = \Phi(Z)W^* + \epsilon \tag{9}$$

here, $\epsilon \in \mathbb{R}^n$ represents the modeling error of the NN, which is upper bounded by $\bar{\epsilon} \in \mathbb{R}^+$ as $\|\epsilon\|_2 \leq \bar{\epsilon}$, and $W^* \in \mathbb{R}^{N\times n}$ is the optimal output weight matrix minimizing ϵ. The network has $2n$ input neurons, N hidden neurons, and n output neurons. The hidden layer output vector $\Phi(Z) = [\phi(z_1), \phi(z_2), \ldots, \phi(z_N)]^T \in \mathbb{R}^N$ uses activation function $\phi(z_i)$, with its input $Z = [z_1, \ldots, z_N]^T = Vx$ computed from the network input $x = [x_1^T, x_2^T]^T$. The input-to-hidden weight matrix $V \in \mathbb{R}^{N\times 2n}$ is typically fixed and randomly initialized in $[-1, 1]$.

Applying a first-order filter $F_v(s) = \frac{1}{ks+1}$ with parameter $k > 0$ to both sides of (9) yields:

$$F_{vi}^* = \Phi_v W_i^* + \epsilon_{vi} = \dot{x}_{2v} - \varpi_v - C_{1v} \tag{10}$$

where F_v^* denotes the filtered F^*, F_{vi}^* and ϵ_{vi} represent the ith elements of F_v^* and ϵ_v, respectively, with the remaining filtered values defined as follows:

$$k\dot{x}_{2v} + x_{2v} = x_2, k\dot{C}_{1v} + C_{1v} = C_1 \tag{11a}$$

$$k\dot{\varpi}_v + \varpi_v = \varpi, k\dot{\epsilon}_v + \epsilon_v = \epsilon, k\dot{\Phi}_v + \Phi_v = \Phi \tag{11b}$$

Let $\hat{F}_{vi}^* = \Phi_v \hat{W}_i$ be the estimated disturbance, then we can calculate the estimation error as

$$\tilde{F}_{vi}^* = \dot{x}_{2v} - \varpi_v - C_{1v} - \hat{F}_{vi}^* \tag{12}$$

An PIA is developed in this work to accelerate the identification process. Accordingly, we introduce the following auxiliary matrices $P_W \in \mathbb{R}^{N\times N}$ and $Q_W \in \mathbb{R}^{N\times n}$, defined as:

$$\dot{P}_W = -l_f P_W + l_f \Phi_v^T \Phi_v \tag{13}$$

$$\dot{Q}_W = -l_f Q_W + l_f \Phi_v^T \left(\dot{x}_{2v} - \varpi_v - C_{1v}\right) \tag{14}$$

where the initial values are set as $P_W(0) = 0_{N\times N}$ and $Q_W(0) = 0_{N\times n}$, and $l_f \in \mathbb{R}^+$ is a design parameter.

From (13) and (14) it holds that

$$P_W(t) = \int_0^t e^{-l_f(t-r)} \Phi_v^T \Phi_v dr \tag{15}$$

$$Q_W(t) = \int_0^t e^{-l_f(t-r)} \Phi_v^T \left(\dot{x}_{2v} - \varpi_v - C_{1v}\right) dr \tag{16}$$

In this paper, it is assumed that Φ_v satisfies the persistent excitation (PE) condition. Therefore, according to Definition 1, the matrix P is positive definite.

Define an auxiliary vector $H_W \in \mathbb{R}^{N\times n}$ as

$$H_W = P_W \hat{W} - Q_W. \tag{17}$$

It follows that

$$H_W = P_W \hat{W} - Q_W = P_W \hat{W} - P_W W^* - \mathring{\epsilon}_f = P_W \tilde{W} - \mathring{\epsilon}_f, \tag{18}$$

where $\mathring{\epsilon}_f = \int_0^t e^{-l_f(t-r)} \Phi_v^\top \epsilon_f dr$. Since ϵ_f is bounded, $\mathring{\epsilon}_f$ is also bounded, and we denote its upper bound by $\bar{\mathring{\epsilon}}_f$.

To update the parameter vector W online,we can design the learning law of W as

$$\dot{W} = \Gamma \left(\frac{|H_W|^{2\gamma-1}}{\sqrt{\left(H_W^T H_W\right)^\gamma + \delta}} \odot \tanh\left(\frac{H_W}{\mu}\right) + H_W \right) \tag{19}$$

where $\Gamma \in \mathbb{R}^+$, $\mu \in \mathbb{R}^+$, $\gamma = \frac{p_{\gamma 1}}{p_{\gamma 2}} \in \mathbb{R}^+$ are designed parameters, with $p_{\gamma 1}$ and $p_{\gamma 2}$ being positive odd numbers satisfying $0 < p_{\gamma 1} < p_{\gamma 2} < 2p_{\gamma 1}$, $\delta \in \mathbb{R}^+$ is a small design parameter. $|\cdot|, (\cdot)^{2\gamma-1}$, and $\tanh(\cdot)$ denote elementwise absolute value, power operation, and hyperbolic tangent function of a vector, respec-tively, and $\odot$ denotes elementwise multiplication. Then, we have the following result.

Theorem 1. *Consider the Reduced Slow Subsystem* (6) *with then disturbance approximated by the NN as in* (9)*. If the NN weights' learning law given by* (19) *is adopted, then the identification error* $\tilde{W}(t)$ *converges to a residual set around* 0 *in finite time.*

Proof. The proof of Theorem 1 is similar to that in [14]. Due to page limitation, a detailed proof is not provided in this paper.

4 Controller Design

To ensure precise trajectory tracking and prescribed-time convergence for the FJR system, prescribed-time sliding mode controllers are designed for the two subsystems, respectively.

Define $e_1 = x_1 - x_r$ and $e_2 = x_2 - \dot{x}_r$, where x_r denotes the desired reference trajectory. The tracking errors can then be expressed as:

$$\begin{cases} \dot{e}_1 = e_2 \\ \dot{e}_2 = g_1 u_{bs} + C_1 + \Phi(Z) W^* + \epsilon - \ddot{x}_r \end{cases} \tag{20}$$

For the *Reduced Slow Subsystem*, define a sliding manifold s_{x1} as

$$s_{x1} = \dot{e}_1 + \beta_{x1} e_1 + \beta_{x2} e_1 + \beta_{x3} sig^{1+\frac{2p}{c}}(e_1) \tag{21}$$

where $\beta_{x1} = \frac{c}{pT_{ex1}} + \frac{1}{2}$, $\beta_{x2} = \frac{b(c-p)}{2apT_{ex1}}$, $\beta_{x3} = \frac{ac}{2bpT_{ex1}}$, T_{ex1} is the predefined prescribed time, while p and c are two positive constants satisfying the condition $0 < p < \frac{1}{2}c < c$. For a vector $\pi \in \mathbb{R}^n$ and a constant $\alpha_i > 0$, define $\mathrm{sig}(\pi)^{\alpha_i} = [\mathrm{sign}(\pi_1)|\pi_1|^{\alpha_i}, \cdots, \mathrm{sign}(\pi_n)|\pi_n|^{\alpha_i}]^T$.

The derivative of s_{x1} is

$$\dot{s}_{x1} = \dot{e}_2 + \beta_{x1}e_2 + \beta_{x2}e_2 + \beta_{x3}\left(1 + \frac{2p}{c}\right)|e_1|^{\frac{2p}{c}}(\dot{e}_1) \tag{22}$$

Therefore, based on the designed sliding surface s_{x1}, the controller for the *Reduced Slow Subsystem* can be designed as follows

$$\begin{aligned} u_{bs} =& g_1^{-1}(\ddot{x}_r - C_1 - \Phi\hat{W} - \beta_{x1}e_2 - \beta_{x2}e_2 \\ & - \beta_{x3}\left(1 + \frac{2p}{c}\right)|e_1|^{\frac{2p}{c}}(\dot{e}_1) - \beta_{x4}sig^{2r_1-1}(s_{x1}) \\ & - \beta_{x5}sig(s_{x1}) - \beta_{x6}sig^{2r_2-1}(s_{x1})) \end{aligned} \tag{23}$$

where the control gains are designed as $\beta_{x4} = \frac{ac}{2bpT_{sx1}}$, $\beta_{x5} = \frac{2c+pT_{sx1}}{2pT_{sx1}}$, and $\beta_{x6} = \frac{bc}{2apT_{sx1}}$, T_{sx1} is the predefined prescribed time. $r_1 = 1 + \frac{p}{c} > 1, 0 < r_2 = 1 - \frac{p}{c} < 1$.

For the *Reduced Fast Subsystem*, define a sliding manifold s_{y1} as

$$s_{y1} = \dot{z}_1 + \beta_{y1}z_1 + \beta_{y2}z_1 + \beta_{y3}sig^{1+\frac{2p}{c}}(z_1) \tag{24}$$

where $\beta_{y1} = \frac{c}{pT_{ey1}} + \frac{1}{2}$, $\beta_{y2} = \frac{b(c-p)}{2apT_{ey1}}$, $\beta_{y3} = \frac{ac}{2bpT_{ey1}}$, T_{ey1} is the predefined prescribed time, while p and $c$are two positive constants satisfying the condition $0 < p < \frac{1}{2}c < c$. For a vector $\pi \in \mathbb{R}^n$ and a constant $\alpha_i > 0$, define $\text{sign}(\pi)^{\alpha_i} = [\text{sign}(\pi_1)|\pi_1|^{\alpha_i}, \cdots, \text{sign}(\pi_n)|\pi_n|^{\alpha_i}]^T$.

The derivative of s_{y1} is

$$\dot{s}_{y1} = \dot{y}_2 + \beta_{y1}y_2 + \beta_{y2}y_2 + \beta_{y3}\left(1 + \frac{2p}{c}\right)|z_1|^{\frac{2p}{c}}(\dot{z}_1) \tag{25}$$

Therefore, based on the designed sliding surface s_{y1}, the controller for the *Reduced Fast Subsystem* can be designed as follows

$$\begin{aligned} u_{bf} =& b_1^{-1}(a_2z_1 - \beta_{y1}y_2 - \beta_{y2}y_2 - \beta_{y3}\left(1 + \frac{2p}{c}\right)|z_1|^{\frac{2p}{c}}(\dot{z}_1) \\ & - \beta_{y4}sig^{2r_1-1}(s_{y1}) - \beta_{y5}sig(s_{y1}) - \beta_{y6}sig^{2r_2-1}(s_{y1})) \end{aligned} \tag{26}$$

where the control gains are designed as$\beta_{y4} = \frac{ac}{2bpT_{sy1}}$,$\beta_{y5} = \frac{2c+pT_{sy1}}{2pT_{sy1}}$,$\beta_{y6} = \frac{bc}{2apT_{sy1}}$, $T_{sy1}$is the predefined prescribed time. $r_1 = 1+\frac{p}{c} > 1, 0 < r_2 = 1-\frac{p}{c} < 1$.

Theorem 2. *If the Reduced Slow Subsystem and Reduced Fast Subsystem adopt the prescribed-time sliding mode controllers given by* (23) *and* (26)*, respectively, then the sliding variables of each subsystem will converge to a compact set near the origin within the prescribed time during the reaching phase, and the control errors will also converge to a compact set near the origin within the prescribed time during the sliding phase.*

Proof. The proof of Theorem 2 is similar to that in [15]. Due to page limitation, a detailed proof is not provided in this paper.

5 Simulation

To demonstrate the effectiveness of the proposed method, simulations are conducted on a 2-DOF FJR.

The main simulation parameters are listed in Table 1. In this table, $l_1, l_2 \in \mathbb{R}^+$ denote the lengths of the two links, $l_{c1}, l_{c2} \in \mathbb{R}^+$ represent the distances from the center of mass of each link to its respective rotating joint, q_1 and q_2 are the joint angles, $m_1, m_2 \in \mathbb{R}^+$ are the masses of the two links, and $I_1, I_2 \in \mathbb{R}^+$ denote the moments of inertia of the links. Additionally, $J_1, J_2 \in \mathbb{R}^+$ represent the moments of inertia of the motors.

Table 1. The values of the simulation parameters

Parameter	Value	Unit	Parameter	Value	Unit
m_1	0.1	kg	m_2	0.1	kg
I_1	0.02	$kg \cdot m^2$	I_2	0.02	$kg \cdot m^2$
l_1	0.4	m	l_2	0.4	m
l_{c1}	0.2	m	l_{c2}	0.2	m
J_1	0.01	$kg \cdot m^2$	J_2	0.01	$kg \cdot m^2$

The number of hidden layer neurons is set to $N_1 = 30$, and the output layer contains $n = 2$ neurons. The simulation step size is $\Delta t = 10^{-4}$. The reference trajectory is defined as $x_{r1} = x_{r2} = \sin(0.2\pi t) + 0.1$. The singular perturbation parameter is chosen as $\varepsilon = 0.1$. The parameters of the PIA are selected as follows: the auxiliary matrix parameter is $l_f = 1$, and $\sigma = 100$. Additionally, $\mathcal{N} = I$, $\gamma = \frac{p_{\gamma 1}}{p_{\gamma 2}} = \frac{11}{13}$, $\Gamma = 250$, $\mu = 0.05$, $\delta = 0.001$.

The key parameters of the prescribed-time controller are as follows: $a = 1$, $b = 1$, $c = 20$, $p = 1$, $T_{ex1} = 2$, $T_{ex2} = 2$, $T_{ey1} = 2$, $T_{ey2} = 2$.

These simulation parameters were determined by integrating existing research findings with the FJR's physical parameters to ensure appropriateness.

To evaluate the robustness of the proposed method in the presence of model uncertainties, the actual system parameters m_1^*, m_2^*, I_1^*, I_2^*, l_1^*, l_2^*, l_{c1}^*, and l_{c2}^* are set to be 1.2 times their nominal values.

Figure 1 depicts the estimated and actual values of the NN under the API, while Fig. 2 presents the corresponding identification error.

To further demonstrate the superiority of the proposed method in this paper, a comparative experiment was conducted: the predetermined time sliding mode controller designed in this paper was compared with the terminal sliding mode controller proposed in [14].

As illustrated in Figs. 3, 4, 5 and 6, although the comparison controller has achieved relatively fast convergence speed and small tracking error, in comparison, whether tracking joint angular position or joint angular velocity, the predetermined time controller designed in this paper converges faster within the

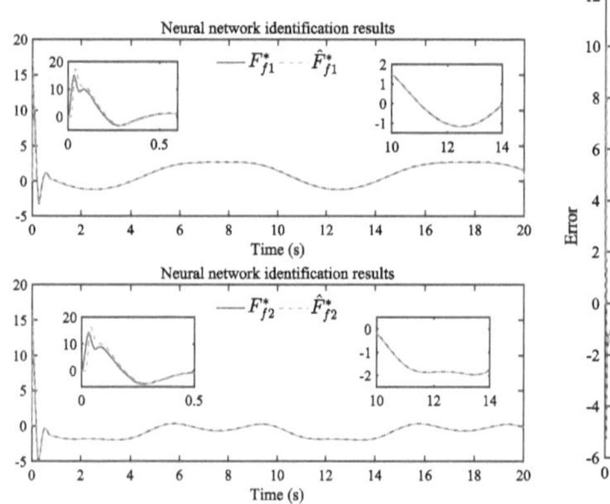

Fig. 1. Neural network identification results.

Fig. 2. Neural network identification error.

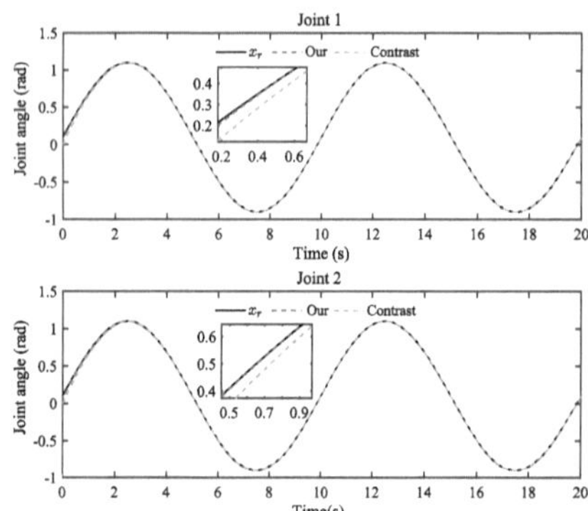

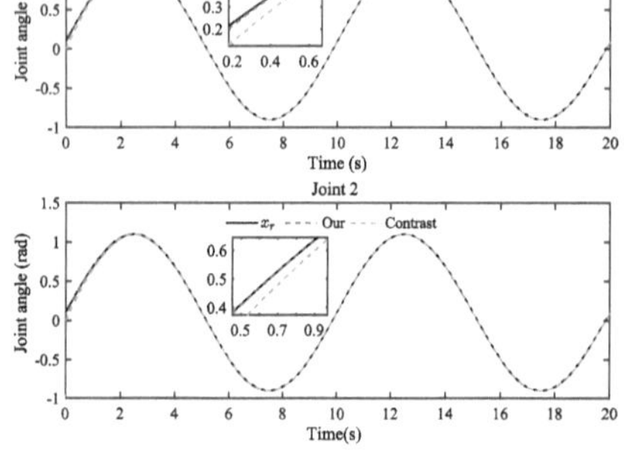

Fig. 3. Trajectory tracking results of the two controllers.

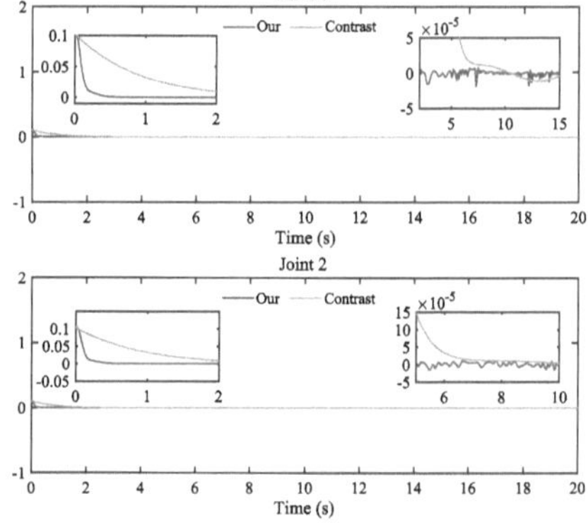

Fig. 4. Trajectory tracking errors of the two controllers.

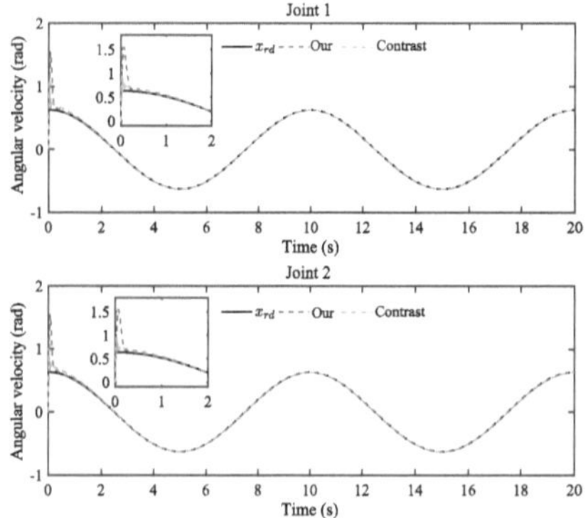

Fig. 5. Angular velocity results of the two controllers.

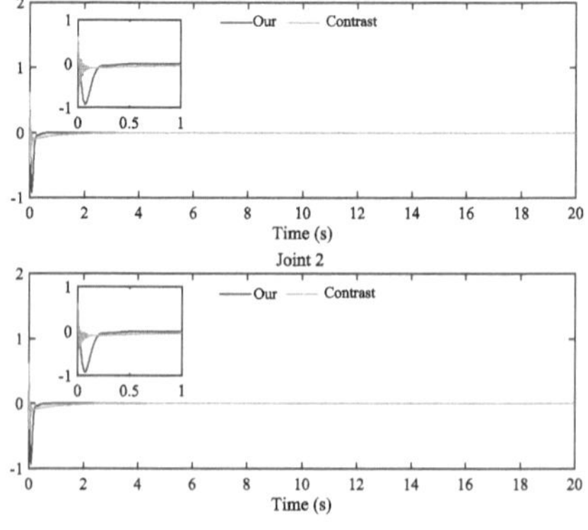

Fig. 6. Angular velocity error of the two controllers.

predetermined time frame. Its convergence speed is approximately 1.5 s faster than the reference controller, with faster error convergence and no significant fluctuations during the convergence process. This enables smoother and more stable operation of the flexible robotic arm system during control.

By comparing simulation results, this paper demonstrates that the proposed method significantly enhances the tracking performance of flexible joint robotic arm systems. It also validates the effectiveness and superiority of the proposed approach.

6 Conclusion

This paper presents a prescribed-time SMC scheme with NN compensation for trajectory tracking of FJR. The original high-order system is first decomposed into two reduced-order subsystems via singular perturbation techniques, substantially simplifying the controller design. A NN estimates system uncertainties, with weights updated via PIA to minimize recognition error. Building on this framework, the proposed prescribed-time SMC effectively suppresses tracking errors and improves overall system performance. Rigorous stability analysis is conducted using Lyapunov theory, and comprehensive simulation studies ultimately validate the effectiveness and superiority of the proposed approach.

While the presented control strategy demonstrates competent trajectory tracking performance with minimal error, its precision remains insufficient for stringent applications including aerospace systems. Subsequent investigations will focus on developing specialized observers to estimate NN weight errors and approximation uncertainties in real-time. By integrating these observed values into the controller architecture through adaptive compensation, the proposed framework can be extended to achieve the enhanced tracking precision demanded by critical engineering applications.

Acknowledgments. This work was supported in part by the National Key R&D Program of China under Grant 2024YFB4711101, 2023YFB4703900, and in part by the National Natural Science Foundation of China under Grants 62573057, 62273050.

References

1. Deniz Kaya, K., Çetin, L.: Adaptive state feedback controller design for a rotary series elastic actuator. Trans. Inst. Meas. Control. **39**, 61–74 (2017)
2. Sariyildiz, E., Chen, G., Haoyong, Yu.: A unified robust motion controller design for series elastic actuators. IEEE/ASME Trans. Mechatron. **22**(5), 2229–2240 (2017)
3. Matheson, E., Minto, R., Zampieri, E.G.G., Faccio, M., Rosati, G.: Human–robot collaboration in manufacturing applications: a review. Robotics **8**(4), 100 (2019)
4. Lee, H., Oh, S.: Design of reduced order disturbance observer of series elastic actuator for robust force control. In: 2018 IEEE 15th International Workshop on Advanced Motion Control (AMC), pp. 663–668 (2018)

5. Wood, D.A., Maksym, K.: weathering/ageing of liquefied natural gas cargoes during marine transport and processing on floating storage units and FSRU. J. Energy Res. Technol. **140**(10), 102901 (2018)
6. Hong, M., Yao, W., Zhu, Z., Guo, Yu.: A hybrid PID controller for flexible joint manipulator based on state observer and singular perturbation approach. In: 2020 39th Chinese Control Conference (CCC), pp. 3599–3603. IEEE (2020)
7. Han, L., Wenfu, X., Kang, P., Yuan, H.: Unified neural adaptive control for multiple human-robot-environment interactions. IEEE Trans. Industr. Inf. **17**(2), 1166–1175 (2020)
8. Shen, J., Marwah, T., Talwalkar, A.: Ups: efficiently building foundation models for PDE solving via cross-modal adaptation. Trans. Mach. Learn. Res., 2835–2856 (2024)
9. He, W., Huang, B., Dong, Y., Li, Z., Chun-Yi, S.: Adaptive neural network control for robotic manipulators with unknown deadzone. IEEE Trans. Cybern. **48**(9), 2670–2682 (2017)
10. Hu, S., Ren, X.: Observer-based optimal adaptive control for multi-motor driving servo system. In: 2020 IEEE 9th Data Driven Control and Learning Systems Conference (DDCLS), pp. 1209–1213 (2020)
11. Song, J., Niu, Y., Zou, Y.: Finite-time stabilization via sliding mode control. IEEE Trans. Autom. Control **62**(3), 1478–1483 (2017)
12. Jonathan Muñoz-Vázquez, A., Diego Sánchez-Torres, J., Jiménez-Rodríguez, E., Loukiano, A.G.: Predefined-time robust stabilization of robotic manipulators. IEEE/ASME Trans. Mechatron. **24**(3), 1033–1040 (2019)
13. Xing, Y., Na, J., Costa-Castelló, R., Wu, J., Chen, X.: Optimal parameter estimation under finite excitation. IEEE Trans. Industr. Electron., 1–10 (2024)
14. Zheng, D.-D., Zhang, Y., Ling, J., Ren, X., Yu, H.: Composite learning-based adaptive terminal sliding mode control for nonlinear systems with experimental validation. IEEE Trans. Industr. Electron., 1–11 (2025)
15. Song, J., Ren, X., Na, J., Gong, J., Zheng, D.: Enhanced predefined-time tracking and synchronization control for multimotor servo systems. IEEE Trans. Industr. Electron., 1–12 (2024)

Neural-Based Adaptive Event-Triggered Tracking Control for Series Elastic Actuator with Input Dead Zones

Lingzhe Zhu[1], Jing Zhao[2], Naiqi Wu[3], Yibin Li[4], and Menghua Zhang[1](✉)

[1] School of Electrical Engineering, University of Jinan, Jinan, China
202421201075@stu.ujn.edu.cn, zhangmenghua@mail.sdu.edu.cn

[2] School of Mechanical Engineering and Automation, Northeastern University, Shenyang, China
zhaoj@mail.neu.edu.cn

[3] Macau Institute of Systems Engineering, Macau University of Science and Technology, Macau, China
nqwu@muct.edu.mo

[4] School of Control Science and Engineering, Shandong University, Jinan, China
liyb@sdu.edu.cn

Abstract. This paper proposes a novel adaptive neural network control scheme to address the tracking control problem for series elastic actuators (SEAs) subject to dead-zone constraints in stochastic noise environments. Specifically, the controlled system investigated in this work is a single-input single-output (SISO) stochastic nonlinear system. Employing the traditional backstepping design approach to study such a system would significantly increase the computational burden. To resolve this issue, the command-filtered technique is integrated into the adaptive neural network design framework. Furthermore, a relative threshold event-triggered strategy is adopted to reduce the network communication burden to a certain extent. The proposed method ensures that the tracking error converges to a small neighborhood of the origin. Simulation results demonstrate the effectiveness of the proposed algorithm.

Keywords: Series Elastic Actuators · Adaptive Control · Command Filter Design · Event-Triggered Control · Neural Network · Dead Zone

1 Introduction

Series Elastic Actuators (SEAs) have attracted wide attention for their low output impedance, back drivability, and inherent compliance, making them ideal for human-robot interaction applications like rehabilitation robots and exoskeletons [1, 2]. Unlike rigid actuators, SEAs embed elastic elements to enhance safety and efficiency, However, practical implementations are hindered by nonlinearities (e.g., input dead zones [3, 4]) and random noises [5, 6], which degrade system performance and stability.

C. Li et al. (Eds.): ICNC 2025, CCIS 2946, pp. 231–243, 2026.
https://doi.org/10.1007/978-981-92-1599-7_20

Advanced control strategies address these issues: Adaptive methods with neural networks (NNs) handle uncertainties via online function approximation [7, 8]; backstepping controls nonlinear systems but suffers from "complexity explosion," mitigated by command filtering (approximates derivatives and compensates errors) [9]; event-triggered control (ETC) reduces communication load by executing tasks only when necessary, vital for networked systems [10, 11].

However, few studies have jointly addressed random noises and input dead zones in Series Elastic Actuators (SEAs)—the latter, caused by mechanical imperfections, often leads to system instability if not properly compensated [12] for. This article proposes a neural-based adaptive event-triggered tracking control for SEAs with both challenges. Key contributions:

(1) Simultaneously handles random disturbances and input nonlinearities in one framework;
(2) Combines command filtering and adaptive backstepping to avoid "complexity explosion" and compensate filtering errors;
(3) Integrates ETC and adaptive NN control to improve resource utilization and reduce communication burden.

2 Problem Statement and Preliminaries

2.1 Problem Statement

The SEA mainly consists of the motor, the spring, and the load. According to [13], the mathematical model of the SEA affected by random noises is written as

$$\begin{cases} J_l\ddot{q} + B_l\dot{q} = K(q_m - q) + \Lambda_1(q, \dot{q})\varsigma \\ J_m\ddot{q}_m + B_m\dot{q}_m + K(q_m - q) = u + \Lambda_2(q_m, \dot{q}_m)\varsigma \end{cases} \tag{1}$$

where q, $\dot{q}$, and $\ddot{q}$ are denoted as the angle position, velocity, and acceleration of the load part, respectively; q_m, $\dot{q}_m$, and $\ddot{q}_m$ refer to the angle position, velocity, and acceleration of the motor part, respectively; J_l and B_l refer to the inertia and damping of the load part, respectively, while J_m and B_m refer to the inertia and damping of the motor part, respectively; K denotes the stiffness of the built-in spring; $\Lambda_1(q, \dot{q})\varsigma$ and $\Lambda_2(q_m, \dot{q}_m)\varsigma$ are caused by the white noise $\varsigma \in \mathbb{R}$ (random disturbance signal) and Λ_1, $\Lambda_2 \in \mathbb{R}$; Due to dead-zone nonlinearities, the control forces τ should be replaced by u. The ultimate control forces u is mathematically expressed as follows:

$$u = \begin{cases} l_2(\tau - u_2), & \tau > u_2 \\ 0, & -u_1 < \tau < u_2 \\ l_1(\tau + u_1), & \tau < -u_1 \end{cases} \tag{2}$$

where u_1 and u_2 are the breakpoints, l_1, l_2 are the left and right slopes of the dead zone. u is an unknown smooth and continuous dead-zone function with respect to the inputτ.

Accordingly, τ should be rewritten as u. Next, to facilitate the following controller development, we define:

$$u = \tau + \Delta\tau \tag{3}$$

where $\Delta\tau$ refer to the uncertain/unknown dead-zone functions.

According to [14] System (1) can be transformed into the following Itô type stochastic differential equation

$$\begin{cases} \mathrm{d}q = \dot{q}\mathrm{d}t, \\ \mathrm{d}\dot{q} = \Psi_1 \mathrm{d}t + {J_l}^{-1}\Lambda_1(q, \dot{q})\Sigma \mathrm{d}w, \\ \mathrm{d}q_m = \dot{q}_m \mathrm{d}t, \\ \mathrm{d}\dot{q}_m = \Psi_2 \mathrm{d}t + {J_m}^{-1}\Lambda_2(q_m, \dot{q}_m)\Sigma \mathrm{d}w, \end{cases} \tag{4}$$

where $\alpha_1{}^*$ and $\alpha_2{}^*$ are generated by the Wong-Zakai correction terms, $\Psi_1 = -{J_l}^{-1}(B_l\dot{q} + K(q - q_m)) + \alpha_1^*$, $\Psi_2 = {J_m}^{-1}(-B_m\dot{q}_m + K(q - q_m) + \tau + \Delta\tau) + \alpha_2^*$, and the specific information can be found in [14].

In the system of this article, all functions are smooth, or at least locally Lipschitz with respect to their respective variables. Then, an event-triggered adaptive neural network control algorithm is proposed as the control objective, such that the SEA (1) can accurately track the reference signal $q_d \in \mathbb{R}$ even under white noise disturbances and actuator dead zones, all signals in the closed-loop system are bounded in probability.

2.2 Preliminaries

Consider the stochastic nonlinear system

$$\mathrm{d}\zeta = f(\zeta, t)\mathrm{d}t + g(\zeta, t)\mathrm{d}w \tag{5}$$

where $\zeta \in \mathbb{R}$ is the system state, and $w \in \mathbb{R}$ is a standard Wiener process. The functions $f : \mathbb{R} \times \mathbb{R}_+ \to \mathbb{R}$ and $g : \mathbb{R} \times \mathbb{R}_+ \to \mathbb{R}$ are locally Lipschitz with respect to $\zeta \in \mathbb{R}$ and piecewise continuous with respect to $t \in \mathbb{R}_+$.

Definition 1 [14]. For system (1) and the function $V(\zeta,\ t) \in \mathbb{C}^{2,\ 1}$, the differential operator $\mathscr{L}$ is defined as follows:

$$\mathcal{L}V = \frac{\partial V}{\partial t} + \frac{\partial V}{\partial \zeta} + \frac{1}{2}Tr\left\{g^T \frac{\partial^2 V}{\partial \zeta^2} g\right\} \tag{6}$$

Lemma 1 [14]. For System (1), if there exist a positive definite function $V(\zeta, t) \in \mathbb{C}^{2,\ 1}$, $\mathcal{K}_\infty$-class functions $\kappa_1(\cdot)$ and $\kappa_2(\cdot)$, as well as constants $a > 0$ and $0 < c < \infty$, such that $\kappa_1(\|\zeta\|) \le V(\zeta, \mathrm{t}) \le \kappa_2(\|\zeta\|)$, $\mathscr{L}V(\zeta, \mathrm{t}) \le -aV(\zeta, \mathrm{t}) + c$, then System (1) has a unique solution almost everywhere, and the solution of the system is bounded in probability, satisfying $E[V(\zeta, \mathrm{t})] \le e^{-at}V(\zeta_0, \mathrm{t}_0) + \frac{c}{a}$.

Lemma 2 [15]. For any given constant $\iota > 0$ and a continuous function $f(\xi)$ defined on a compact set Ω, the existence of a radial basis function neural network is guaranteed such that the function is represented as

$$f(\xi) = \mathscr{H}^{*T}\Phi(\xi) + \sigma(\xi), \mid \sigma(\xi) \mid \le \iota \tag{7}$$

where $\xi \in \Omega \subset \mathbb{R}$ is taken as the input to the neural network, $\sigma(\xi)$ is the approximation error, and $\mathscr{H}^*$ is the ideal weight vector of the form $\mathscr{H}^* =$

$\arg\min_{\mathscr{H}\in\mathbb{R}^\ell}\left\{\sup_{\xi\in\Omega}\left|f(\xi)-\mathscr{H}^T\Phi(\xi)\right|\right\}$.Here, the number of neural network nodes is denoted by $\ell > 1$. The basis function vector is given by $\Phi(\xi) = [\phi_1(\xi), \phi_2(\xi), \ldots, \phi_\ell(\xi)]^T$. Typically, each element $\phi_i(\xi)(i = 1, 2, \ldots, \ell)$ of the vector is chosen to be a Gaussian function of the form: $\phi_i(\xi) = \exp\left[-\frac{(\xi-c_i)^T(\xi-c_i)}{l_i^2}\right]$, c_i is referred to as the center vector, and l_i is designated as the width of the Gaussian function.

3 Controller Design

For notational convenience, we denote $x_1 = q$, $x_2 = \dot{q}$, $x_3 = q_m$ and $x_4 = \dot{q}_m$. Consequently, Eq. (4) can be rewritten as:

$$\begin{cases} dx_1 = x_2 dt, \\ dx_2 = \left(J_l{}^{-1}Kx_3 + \varphi_1\right)dt + G_1(x_1, x_2)dw, \\ dx_3 = x_4 dt, \\ dx_4 = \left(J_m{}^{-1}\tau + \varphi_2\right)dt + G_2(x_3, x_4)dw, \end{cases} \tag{8}$$

where $\varphi_1 = -J_l{}^{-1}(B_l x_2 + Kx_1) + \alpha_1^*$, $G_1 = J_l{}^{-1}\Lambda_1(x_1, x_2)\Sigma$ $\varphi_2 = J_m{}^{-1}\left(-B_m x_4 + K\left(x_1 - x_3\right) + \Delta\tau\right) + \alpha_2^*$, $G_2 = J_m{}^{-1}\Lambda_2(x_3, x_4)\Sigma$.

This section presents an event-triggered adaptive tracking control scheme, which is developed within the adaptive backstepping design framework, and is combined with command filtered control and neural networks.

Firstly, the following coordinate transformations are defined:

$$\begin{cases} z_1 = x_1 - q_d, \\ z_i = x_i - \overline{\alpha}_i, i = 2, 3, 4 \end{cases} \tag{9}$$

In this section, the vector $\overline{\alpha}_i \in \mathbb{R}$ is defined as the output of the following first-order filter:

$$\beta_i\dot{\overline{\alpha}}_i + \overline{\alpha}_i = \alpha_i, \overline{\alpha}_i(0) = \alpha_i(0) \tag{10}$$

where $\beta_i > 0$ is designated as a design parameter, and $\alpha_i \in \mathbb{R}$ is constructed as the virtual controller, which also serves as the input to the first-order filter (10).

Subsequently, the following compensation signals are introduced to address the filtering errors:

$$\begin{cases} \dot{\eta}_1 = -k_1\eta_1 + (\overline{\alpha}_2 - \alpha_2) + \eta_2, \\ \dot{\eta}_2 = -k_2\eta_2 + J_l{}^{-1}K(\overline{\alpha}_3 - \alpha_3) - \frac{1}{4}\eta_2 + J_l{}^{-1}K\eta_3, \\ \dot{\eta}_3 = -k_3\eta_3 + (\overline{\alpha}_4 - \alpha_4) - \frac{1}{4}\|J_l{}^{-1}K\|^4\eta_3 + \eta_4, \\ \dot{\eta}_4 = -k_4\eta_4 - \frac{1}{4}\eta_4, \end{cases} \tag{11}$$

where, for $i = 1, \ldots, 4$, $\eta_i \in \mathbb{R}$ is defined, with the initial condition set to $\eta_i(0) = 0$, and $k_i > 0$ are specified as positive constants.

The compensation error is then defined by the following expression:

$$\xi_i = z_i - \eta_i, \qquad i = 1, \ldots, 4 \tag{12}$$

Eventually, the virtual controllers and actual controller could be designed as follows:

$$\begin{cases} \alpha_2 = \dot{q}_d - k_1 z_1 - \frac{3}{4}\xi_1, \\ \alpha_3 = K^{-1} J_l \left(\dot{\bar{\alpha}}_2 - (k_2 + \frac{1}{4}) z_2 - \frac{\hat{\theta}\xi_2 \Phi_1^T \Phi_1}{4\tau_1} - \xi_2 \right), \\ \alpha_4 = \dot{\bar{\alpha}}_3 - \left(k_3 + \frac{1}{4} \left\| J_l^{-1} K \right\|^4 \right) z_3 - \frac{3}{4}\xi_3, \end{cases} \tag{13}$$

$$v(t) = -(1+\kappa)\left(\mu \tanh \frac{\xi_4 \mu}{J_m \varepsilon} + \overline{m} \tanh \frac{\xi_4 \overline{m}}{J_m \varepsilon} \right), \tag{14}$$

where $\mu = J_m \left(\dot{\bar{\alpha}}_4 - (k_4 + \frac{1}{4}) z_4 - \frac{\hat{\theta}\xi_4 \Phi_2^T \Phi_2}{4\tau_2} - \frac{1}{2}\xi_4 \right)$, $\hat{\theta}$ is the estimate of $\theta = max\,\{\|H_i^*\|^2, i = 1, 2\}$, $H_i^* (i = 1, 2)$ and $\Phi_i (i = 1, 2)$ are the unknown ideal weight vectors and basis function vectors of the neural networks, respectively. The corresponding adaptive law for parameter estimation is designed as:

$$\dot{\hat{\theta}} = \gamma \left(\frac{\xi_2{}^4 \Phi_1^T \Phi_1}{4\tau_1} + \frac{\xi_4{}^4 \Phi_2^T \Phi_2}{4\tau_2} \right) - \rho \hat{\theta}, \tag{15}$$

where $\overline{m} > \frac{\delta}{1-\kappa}$, $0 < \kappa < 1$, $\delta, \varepsilon, \tau_1, \tau_2, \gamma, \rho > 0$ are all design parameters.

Step 1: From Eqs. (8), (9), (11), and (12), $\mathrm{d}\xi_1$ can be calculated as

$$\mathrm{d}\xi_1 = (\xi_2 + \alpha_2 - \dot{q}_d + k_1 z_1 - k_1 \xi_1)\mathrm{d}t. \tag{16}$$

Lyapunov candidate function $V_1 = \frac{1}{4}\xi_1{}^4$, and the derivative of $\mathscr{L}$ can be obtained by combining the definition of the Lie derivative

$$\mathcal{L}V_1 = \xi_1{}^3 \xi_2 + \xi_1{}^3 (\alpha_2 - \dot{q}_d + k_1 z_1) - k_1 \xi_1{}^4. \tag{17}$$

Applying Young's inequality, one has $\xi_1{}^3 \xi_2 \leq \frac{3}{4}\xi_1{}^4 + \frac{1}{4}(\xi_1 \xi_2)^2$, Substituting it into (17) and combining with the virtual controller α_2 designed in (13), it is obtained that:

$$\mathcal{L}V_1 \leq -k_1 \xi_1{}^4 + \frac{1}{4}\xi_2{}^4. \tag{18}$$

Step 2: Based on the definition of ξ_2, it is obtained that:

$$d\xi_2 = \left(M^{-1} K \xi_3 + M^{-1} K \alpha_3 + \varphi_1 - \dot{\bar{\alpha}}_2 + \left(k_2 + \frac{1}{4} \right) z_2 - \left(k_2 + \frac{1}{4} \right) \xi_2 \right) dt + G_1 dw. \tag{19}$$

The following Lyapunov candidate function is considered: $V_2 = V_1 + \frac{1}{4}{\xi_2}^4 + \frac{1}{2\gamma}\tilde{\theta}^2$, where $\tilde{\theta} = \theta - \hat{\theta}$ is the parameter estimation error. Then, V_2 is satisfied by:

$$\begin{aligned}\mathcal{L}V_2 \le & -\sum_{i=1}^{2} k_i{\xi_i}^4 + {\xi_2}^3 {J_l}^{-1}K\xi_3 + {\xi_2}^3\left({J_l}^{-1}K\alpha_3 - \dot{\bar{\alpha}}_2 + \left(k_2 + \tfrac{1}{4}\right)z_2\right) \\ & + {\xi_2}^3\varphi_1 + \tfrac{1}{2}\mathrm{Tr}\left\{G_1^T\left(2\xi_2\xi_2^T + \xi_2^T\xi_2 I\right)G_1\right\} - \tfrac{1}{\gamma}\tilde{\theta}\dot{\hat{\theta}}.\end{aligned} \tag{20}$$

From the definitions φ_1 of and G_1, it is known that the unknown terms $\xi_2^T\xi_2\xi_2^T\varphi_1$ and $\frac{1}{2}\mathrm{Tr}\{G_1^T(2\xi_2\xi_2^T + \xi_2^T\xi_2 I)G_1\}$ cannot be directly used in control design. Similar to references [14, 16], this problem is addressed by utilizing the approximation properties of neural networks. First, based on the properties of norms, the unknown term $\frac{1}{2}\mathrm{Tr}\{G_1^T(2\xi_2\xi_2^T + \xi_2^T\xi_2 I)G_1\}$ is processed as follows:

$$\frac{1}{2}\mathrm{Tr}\left\{G_1^T\left(2\xi_2\xi_2^T + \xi_2^T\xi_2 I\right)G_1\right\} \le \frac{3}{2}{\xi_2}^2\|G_1\|^2, \tag{21}$$

Substituting it into (20) yields

$$\mathcal{L}V_2 \le \sum_{i=1}^{2} k_i{\xi_i}^4 + {\xi_2}^3\left({J_l}^{-1}K\alpha_3 - \dot{\bar{\alpha}}_2 + \left(k_2 + \frac{1}{4}\right)z_2\right) + {\xi_2}^3{J_l}^{-1}K\xi_3 + {\xi_2}^2\chi_1 - \frac{1}{\gamma}\tilde{\theta}\dot{\hat{\theta}} \tag{22}$$

where $\chi_1 = \xi_2\varphi_1 + \frac{3}{2}\|G_1\|^2$.Next, according to Lemma 2, for any $\iota_1 > 0$, a neural network $\mathscr{H}_1^{*T}\Phi_1(X_1)$ is known to exist such that:

$$\chi_1 = \mathscr{H}_1^{*T}\Phi_1(X_1) + \sigma_1(X_1), |\sigma_1(X_1)| \le \iota_1, \tag{23}$$

where $\sigma_1(X_1)$ is the approximation error. Then, based on Young's inequality, the definition of the parameter θ, and by combining Eqs. (22), (23) with the virtual controller α_3 designed in (13), the following inequality is derived:

$$\mathcal{L}V_2 \le -\sum_{i=1}^{2} k_i{\xi_i}^4 + \tilde{\theta}\left(\frac{{\xi_2}^4\Phi_1^T\Phi_1}{4\tau_1} - \frac{1}{\gamma}\dot{\hat{\theta}}\right) + \frac{1}{4}\left\|{J_l}^{-1}K\right\|^4{\xi_3}^4 + \tau_1 + \iota_1^2 \tag{24}$$

Step 3: From Eqs. (8), (9), (11), and (12), $d\xi_3$ can be calculated as

$$\mathrm{d}\xi_3 = \left(\xi_4 + \alpha_4 - \dot{\bar{\alpha}}_3 + \left(k_3 + \frac{1}{4}\left\|{J_l}^{-1}K\right\|^4\right)z_3 - \left(k_3 + \frac{1}{4}\left\|{J_l}^{-1}K\right\|^4\right)\xi_3\right)\mathrm{d}t. \tag{25}$$

The Lyapunov candidate function is selected as $V_3 = V_2 + \frac{1}{4}\left(\xi_3^T\xi_3\right)^2$. Then, by using the definition of the differential operator $\mathscr{L}$, Young's inequality, and substituting the virtual controller α_4 designed in (13), we obtain

$$\mathcal{L}V_3 \le -\sum_{i=1}^{3} k_i{\xi_i}^4 + \tilde{\theta}\left(\frac{{\xi_2}^4\Phi_1^T\Phi_1}{4\tau_1} - \frac{1}{\gamma}\dot{\hat{\theta}}\right) + \frac{1}{4}{\xi_4}^4 + \tau_1 + \iota_1^2. \tag{26}$$

Step 4: Similar to Step 2, the following is obtained from Eqs. (8), (9), (11), and (12)

$$d\xi_4 = \left(J_m^{-1}\tau + \varphi_2 - \dot{\bar{\alpha}}_4 + \left(k_4 + \frac{1}{4}\right)z_4 - \left(k_4 + \frac{1}{4}\right)\xi_4\right)dt + G_2 dw. \quad (27)$$

The following Lyapunov candidate function is defined: $V_4 = V_3 + \frac{1}{4}{\xi_4}^4$. Then, the following can be calculated:

$$\begin{aligned}\mathcal{L}V_4 \leq &-\sum_{i=1}^{4} k_i {\xi_i}^4 + \tilde{\theta}\left(\frac{{\xi_2}^4\Phi_1^T\Phi_1}{4\tau_1} - \frac{1}{\gamma}\dot{\hat{\theta}}\right) + \tau_1 + \iota_1^2 + {\xi_4}^3\Big(J_m^{-1}\tau \\ &- \dot{\bar{\alpha}}_4 + (k_4 + \tfrac{1}{4})z_4\Big) + {\xi_4}^3\varphi_2 + \tfrac{1}{2}\mathrm{Tr}\left\{G_2^T\left(2\xi_4\xi_4^T + \xi_4^T\xi_4 I\right)G_2\right\}\end{aligned} \quad (28)$$

Similar to step 2, the last term of Eq. (28) can be scaled, so that Eq. (28) can be rewritten as

$$\begin{aligned}\mathcal{L}V_4 \leq &-\sum_{i=1}^{4} k_i {\xi_i}^4 + \tilde{\theta}\left(\frac{{\xi_2}^4\Phi_1^T\Phi_1}{4\tau_1} - \frac{1}{\gamma}\dot{\hat{\theta}}\right) + \tau_1 \\ &+ \iota_1^2 + {\xi_4}^3\left(J_m^{-1}\tau - \dot{\bar{\alpha}}_4 + (k_4 + \tfrac{1}{4})z_4\right) + {\xi_4}^2\chi_2,\end{aligned} \quad (29)$$

where $\chi_2 = \xi_4\varphi_2 + \frac{3}{2}\|G_2\|^2$. Similarly, the unknown term χ_2 can be approximated by the neural network $\mathscr{H}_2^{*T}\Phi_2(X_2)$, i.e., for any constant $\iota_2 > 0$, we have

$$\chi_2 = \mathscr{H}_2^{*T}\Phi_2(X_2) + \sigma_2(X_2), \quad (30)$$

where the approximation error $\sigma_2(X_2)$ satisfy $|\sigma_2(X_2)| \leq \iota_2$. By applying Young's inequality to ${\xi_4}^2\chi_2$, Eq. (30) can be rewritten as:

$$\begin{aligned}\mathcal{L}V_4 \leq &-\sum_{i=1}^{4} k_i {\xi_i}^4 + {\xi_4}^3\left(J_m^{-1}\tau - \dot{\bar{\alpha}}_4 + (k_4 + \tfrac{1}{4})z_4 + \frac{\hat{\theta}\xi_4\Phi_2^T\Phi_2}{4\tau_2}\right) \\ &+ \tfrac{1}{4}{\xi_4}^4 + \tilde{\theta}\left(\frac{{\xi_2}^4\Phi_1^T\Phi_1}{4\tau_1} + \frac{{\xi_4}^4\Phi_2^T\Phi_2}{4\tau_2} - \frac{1}{\gamma}\dot{\hat{\theta}}\right) + \sum_{i=1}^{2}\left(\tau_i + \iota_i^2\right).\end{aligned} \quad (31)$$

Next, design the event-triggered mechanism as follows:

$$\begin{cases}\tau(t) = v(t_k), \quad \forall t \in \left[t_k, t_{k+1}\right), \\ t_{k+1} = \inf\left\{t > t_k \middle\| e(t)| \geq \kappa|\tau(t)| + \delta\right\},\end{cases} \quad (32)$$

where $k \in \mathbb{N}_+, \kappa \in (0, 1), \delta > 0$ are design parameters and $e(t) = v(t) - \tau(t)$ is defined as the measurement error. When the event-triggering condition (32) is satisfied, this instant is recorded as t_{k+1} and the control signal $\tau(t)$ is immediately updated to $v(t_{k+1})$. During the time interval between two consecutive triggering instants, $[t_k, t_{k+1})$, the control signal $\tau(t)$ is held constant, i.e., $\tau(t) = v(t_k)$, $\forall t \in [t_k, t_{k+1})$.

As described in references [10], it can be derived from Eq. (32) that the inequality $|v(t) - \tau(t)| \leq \kappa|\tau(t)| + \delta$ always holds, and

$$v(t) = (1 + \lambda_1(t)\kappa)\tau(t) + \lambda_2(t)\delta, \quad (33)$$

where the functions $\lambda_1(t)$ and $\lambda_2(t)$ satisfy $|\lambda_1(t)| \leq 1$ and $|\lambda_2(t)| \leq 1$, thus it is obtained that $\tau(t) = \frac{v(t)}{1+\lambda_1(t)\kappa} - \frac{\lambda_2(t)\delta}{1+\lambda_1(t)\kappa}$.

Based on the above analysis, it can be derived that

$$\xi_4^3 J_m^{-1}\tau \leq \xi_4^2\left(\frac{\xi_4 v(t)}{J_m(1+\lambda_1\kappa)} + \left|\frac{\xi_4\lambda_2\delta}{J_m(1+\lambda_1\kappa)}\right|\right). \tag{34}$$

Since $|\lambda_1(t)| \leq 1$ and $|\lambda_2(t)| \leq 1$, the following inequalities are established:

$$\frac{\xi_4 v(t)}{J_m(1+\lambda_1\kappa)} \leq \frac{\xi_4 v(t)}{J_m(1+\kappa)}, \left|\frac{\xi_4\lambda_2\delta}{J_m(1+\lambda_1\kappa)}\right| \leq \left|\frac{\xi_4\delta}{J_m(1-\kappa)}\right| \tag{35}$$

For details on the first inequality in Eq. (35), please refer to Remark 2. Therefore, by combining the designed event-triggered controller and combining with the property of the hyperbolic tangent function that $0 \leq |\phi| - \phi\tanh\left(\frac{\phi}{\varepsilon}\right) \leq 0.2785\varepsilon$ with $\phi \in \mathbb{R}$ and $\epsilon > 0$, it is known that $\xi_4^T\xi_4 J_m{}^{-1}\tau$ satisfies the following inequality relation: $\xi_4{}^2 J_m{}^{-1}\tau \leq \xi_4{}^2\left(\frac{1}{J_m}\xi_4\mu + 0.557\varepsilon\right)$. By substituting it into Eq. (31) and simplifying, the following is obtained:

$$\begin{aligned}\mathcal{L}V_4 \leq & -\sum_{i=1}^{4} k_i\xi_i^4 + \xi_4^2\left(\frac{1}{J_m}\xi_4\mu - \xi_4\dot{\bar{\alpha}}_4 + \left(k_4 + \frac{1}{4}\right)\xi_4 z_4 + \frac{\hat{\theta}\xi_4^2\Phi_2^T\Phi_2}{4\tau_2} + \frac{1}{2}\xi_4^2\right) \\ & + \tilde{\theta}\left(\frac{\xi_2^4\Phi_1^T\Phi_1}{4\tau_1} + \frac{\xi_4^4\Phi_2^T\Phi_2}{4\tau_2} - \frac{1}{\gamma}\dot{\hat{\theta}}\right) + (0.557\varepsilon)^2 + \sum_{i=1}^{2}\left(\tau_i + \iota_i^2\right)\end{aligned} \tag{36}$$

Subsequently, based on the relation $\mu = J_m\left(\dot{\bar{\alpha}}_4 - \left(k_4 + \frac{1}{4}\right)z_4 - \frac{\hat{\theta}\xi_4\Phi_2^T\Phi_2}{4\tau_2} - \frac{1}{2}\xi_4\right)$ and Young's inequality, Eq. (36) can be further rewritten as:

$$\mathcal{L}V_4 \leq -\sum_{i=1}^{4} k_i\xi_i^4 - \rho\frac{\tilde{\theta}^2}{2\gamma} + \rho\frac{\theta^2}{2\gamma} + (0.557\varepsilon)^2 + \sum_{i=1}^{2}\left(\tau_i + \iota_i^2\right) \tag{37}$$

Remark 1. In this section, a relative threshold event-triggered mechanism is adopted. As can be seen from the event-triggering condition (32), the threshold is no longer a fixed value but varies with the control signal. As described in reference [10], better system performance can be obtained by adopting this event-triggered strategy. Furthermore, the parameter values in Eq. (32) are determined by the specific requirements of the practical application, which may require a balance between system performance and the number of event triggers.

Remark 2. Since the hyperbolic tangent function $\tanh(\cdot)$ is an odd function, for any variable $\phi \in \mathbb{R}$ and constant $\epsilon > 0$, the inequality $-\phi\tanh\left(\frac{\phi}{\varepsilon}\right) \leq 0$ hold. Consequently, $-\frac{1}{J_m}\xi_4\mu\tanh\left(\frac{\xi_4\mu}{J_m\varepsilon}\right) \leq 0$ is satisfied. Therefore, by combining the designed event-triggered controller (14) and the condition $|\lambda_1(t)| \leq 1$, it can be derived that $\frac{\xi_4 v(t)}{J_m(1+\lambda_1\kappa)} \leq \frac{\xi_4 v(t)}{J_m(1+\kappa)}$.

Step 5: Construct a new Lyapunov function $V_5 = \sum_{i=1}^{4} \frac{1}{4}{\eta_i}^4$, then we have

$$\begin{aligned}\mathcal{L}V_5 = &-k_1\eta_1^4 + \eta_1^3 g_1(\overline{\alpha}_2 - \alpha_2) + \eta_1^3 g_1\eta_2 - k_2\eta_2^4 + \eta_2^3 g_2(\overline{\alpha}_3 - \alpha_3) - \tfrac{1}{4}\eta_2^4 + \eta_2^3 g_2\eta_3 \\ &- k_3\eta_3^4 + \eta_3^3 g_3(\overline{\alpha}_4 - \alpha_4) - \tfrac{1}{4}{g_2}^4\eta_3^4 + \eta_3^3 g_3\eta_4 - k_4\eta_4^4 - \tfrac{1}{4}\eta_4^4,\end{aligned} \tag{38}$$

where $g_1 = 1, g_2 = J_l^{-1}K$, and $g_3 = 1$. According to references [17], constants $\lambda_{i+1} > 0$ are guaranteed to exist such that $| \overline{\alpha}_{i+1} - \alpha_{i+1} | \leq \lambda_{i+1}$, for $i = 1, 2, 3$. Applying Young's inequality again, the following inequalities hold $\eta_i^3 g_i\eta_{i+1} \leq \frac{3}{4}\eta_i^4 + \frac{1}{4}{g_i}^4\eta_{i+1}^4$ $\eta_i^3 g_i(\overline{\alpha}_{i+1} - \alpha_{i+1}) \leq \frac{3}{4}\eta_i^4 + \frac{1}{4}{g_i}^4\lambda_{i+1}^4$, then (38) could be rewritten as

$$\mathcal{L}V_5 \leq -\sum_{i=1}^{4}\left(k_i - \frac{3}{2}\right)\eta_i^4 + \sum_{i=1}^{3}\frac{1}{4}{g_i}^4\lambda_{i+1}^4. \tag{39}$$

4 Stability Analysis

In this section, the effectiveness of the adaptive tracking control scheme proposed in Controller Design is verified using the Lyapunov stability theory for stochastic systems.

Theorem 1. For the SEA system (1), it is ensured by the constructed virtual controllers (13), event-triggered controller (14), and adaptive law (15) that:

1) The tracking error can converge to a small neighborhood of origin;
2) All signals in the closed-loop system are bounded in probability;
3) Zeno behavior can be excluded.

Proof: According to references [7, 8], the existence of a constant $\overline{g}_2 > 0$ is guaranteed such that $\left\|J_l^{-1}K\right\| \leq \overline{g}_2$. By defining $\mathcal{V} = V_4 + V_5, \overline{g}_1 = \overline{g}_3 = 1$, and combining Eqs. (37) and (39), the function $\mathcal{V}$ is satisfied by:

$$\mathcal{L}\mathcal{V} \leq -\alpha\mathcal{V} + \beta, \tag{40}$$

where $\alpha =$ min $\{4k_i, 4k_i - 6, \rho, i = 1, \ldots, 4\} > 0, \beta = \rho\frac{\theta^2}{2\gamma} + \sum_{i=1}^{3}\frac{1}{4}\overline{g}_i^4\lambda_{i+1}^4 + (0.557\varepsilon)^2 + \sum_{i=1}^{2}\left(\tau_i + \iota_i^2\right)$

By Lemma 1, we can compute $E[\mathcal{V}(t)] \leq \mathcal{V}(0)e^{-\alpha t} + \frac{\beta}{\alpha}$, which means that $|\xi_i|, |\eta_i|, \tilde{\theta}$ and $\hat{\theta}$ are bounded in probability. Combining this with $\xi_i = z_i - \eta_i$,, we conclude that $|z_i|$ is also bounded in probability. Additionally, according to References [17], there exists a constant $\lambda_i > 0$ such that $| \overline{\alpha}_i - \alpha_i | \leq \lambda_i$ for $i = 2, 3, 4$. Then, from $\dot{\overline{\alpha}}_i = -\frac{1}{\varpi_i}(\overline{\alpha}_i - \alpha_i)$, we know the boundedness of $| \dot{\overline{\alpha}}_i |$. Thus, combined with the designed controller, we know that $|\alpha_i|$ and v are bounded in probability. Finally, it can be proven that $| \overline{\alpha}_i |$

and $|x_i|$ are bounded in probability. Therefore, all signals in the closed-loop system are bounded in probability. Finally, we demonstrate the existence of a constant $t^{\#} > 0$ such that $t_{k+1} - t_k \geq t^{\#}$, $\forall\, k \in \mathbb{N}_+$ holds. Based on the event-triggering condition $e(t) = v(t) - \tau(t)$, $\forall\, t \in [t_k, t_{k+1})$, we obtain

$$\frac{\mathrm{d}}{\mathrm{d}t} \mid e \mid = \frac{\mathrm{d}}{\mathrm{d}t}(e \times e)^{\frac{1}{2}} = \mathrm{sgn}(e)\dot{e} \leq \mid \dot{v} \mid \tag{41}$$

Combining with (14), the event-triggered control signal v is differentiable and $\dot{v}$ is a function of bounded variables. Therefore, the existence of a constant $c > 0$ is guaranteed such that $\mid \dot{v} \mid \leq c$. Moreover, since $e(t_k) = 0$ and $\lim\limits_{t \to t_{k+1}} e(t) = \kappa \mid \tau(t) \mid + \delta$, it follows that: $t^{\#} \geq \frac{\kappa|\tau(t)|+\delta}{c}$.This implies that Zeno behavior is avoided. The proof is completed.

5 Simulation

To more intuitively verify the above theoretical results, simulation validation is conducted in this section based on the SEA system.

$$\begin{cases} J_l\ddot{q} + B_l\dot{q} + Kq = Kq_m + \Lambda_1\varsigma, \\ J_m\ddot{q}_m + B_m\dot{q}_m + K(q_m - q) = u + \Lambda_2\varsigma, \end{cases} \tag{42}$$

$$u = \begin{cases} \tau - 0.3, & \tau > 0.3 \\ 0, & -0.3 < \tau < 0.3 \\ \tau + 0.3, & \tau < -0.3 \end{cases} \tag{43}$$

where q, $q_m \in \mathbb{R}$, $\Lambda_1 = 0.2J_lq$, $\Lambda_2 = 0.5J_m \cos(q_m)$, $J_l = 0.4$ m/s^2, $J_m = 0.04$ m/s^2, $K = 2.45$ N · m/rad, $B_l = 1.5$, $B_m = 0.15$, $u = \tau + \Delta\tau$.

Following the control design process in Controller Design, the variables are defined as $x_1 = q$, $x_2 = \dot{q}$, $x_3 = q_m$, $x_4 = \dot{q}_m$. Then, the original system (42) is rewritten as the following state-space model:

$$\begin{cases} \mathrm{d}x_1 = x_2\mathrm{d}t, \\ \mathrm{d}x_2 = \left(\frac{K}{J_l}x_3 - \frac{K}{J_l}x_1 - \frac{B_l}{J_l}x_2\right)\mathrm{d}t + \frac{1}{5}x_1\Sigma\mathrm{d}w, \\ \mathrm{d}x_3 = x_4\mathrm{d}t, \\ \mathrm{d}x_4 = \left(\frac{1}{J_m}\tau - \frac{B_m}{J_m}x_4 - \frac{K}{J_m}(x_3 - x_1)\right)\mathrm{d}t + \frac{1}{2}x_3\Sigma\mathrm{d}w. \end{cases} \tag{44}$$

Then, Virtual controllers, time-triggered controllers, and adaptive laws as shown in (13)–(15) are not elaborated upon here.

In this simulation, the target trajectory for the load part motion was set as $q_d = \sin 2t$. The initial states of the system are defined as $x_1(0) = 0.3$, $x_2(0) = 0$, $x_3(0) = 0$, $x_4(0) = 0$. The initial value of the adaptive parameter is set to $\hat{\theta}(0) = 1$. Other design parameters are assigned the values $\beta_2 = \beta_3 = \beta_4 = \frac{1}{16}$, $\mathrm{k}_1 = 9$, $\mathrm{k}_3 = 5$, $\mathrm{k}_2 = \mathrm{k}_4 = 20$, $\gamma = 0.02$, $\tau_1 = \tau_2 = 20$, $\rho = 1$, $\kappa = 0.2$, $\delta = 0.36$, $\epsilon = 15$, $\overline{m} = 0.3$. Furthermore, the neural network $\mathscr{H}_i^{*T}\Phi_i(X_i)$ is configured with 5 nodes. The centers c_i are uniformly distributed within the interval $[-5, 5]$, and the width is set to $l_i = 6$.

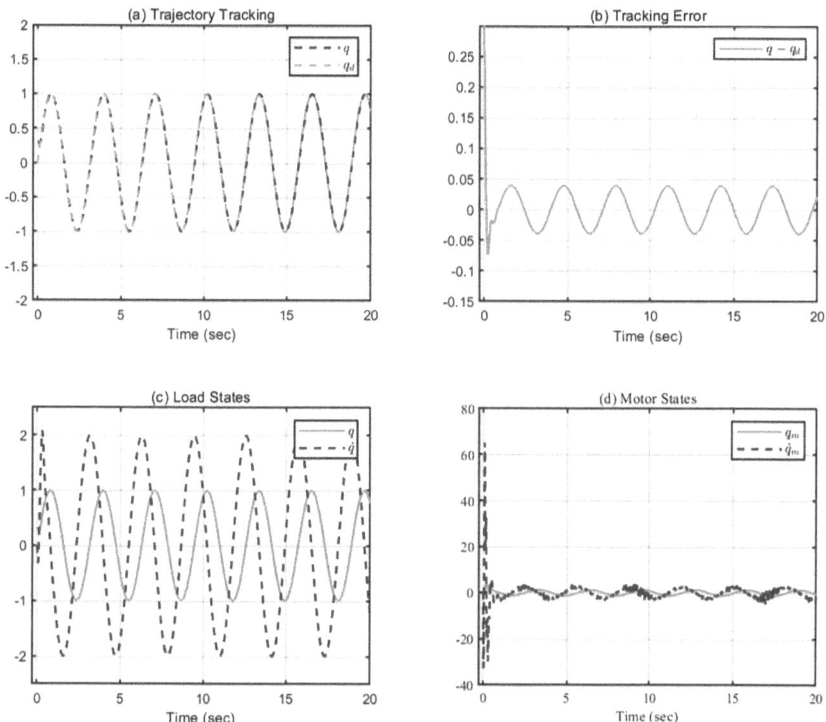

Fig. 1. SEA System Response under Proposed Control

The corresponding simulation results are presented in Fig. 1 and Fig. 2. Specifically, Fig. 1 contains four subfigures: the trajectory of the link motion q and the reference signal q_d are shown in subfigure (a), while the corresponding tracking error $q - q_d$ is depicted in subfigure (b). The trajectories of the load part position q and velocity $\dot{q}$ are illustrated in subfigure (c), and the motor position q_m and velocity $\dot{q}_m$are presented in subfigure (d).From Fig. 1 (a) and 1(b), it can be observed that the load part successfully tracks the reference signal q_d with a small error. Fig. 2 contains three subfigures: the control signal affected by dead zone u is shown in subfigure (a), the adaptive parameter $\hat{\theta}$ is presented in subfigure (b), and the time intervals of the corresponding event triggers are depicted in subfigure (c).

6 Conclusion

This paper proposes an adaptive neural control scheme for series elastic actuators with random noises and input dead zones. By combining command-filtered backstepping with an event-triggered mechanism, the approach simplifies controller design while reducing communication burden. Theoretical analysis proves all closed-loop signals are bounded in probability with tracking errors converging to a small neighborhood of the origin. Future work will explore finite-time control to enhance transient performance in practical applications.

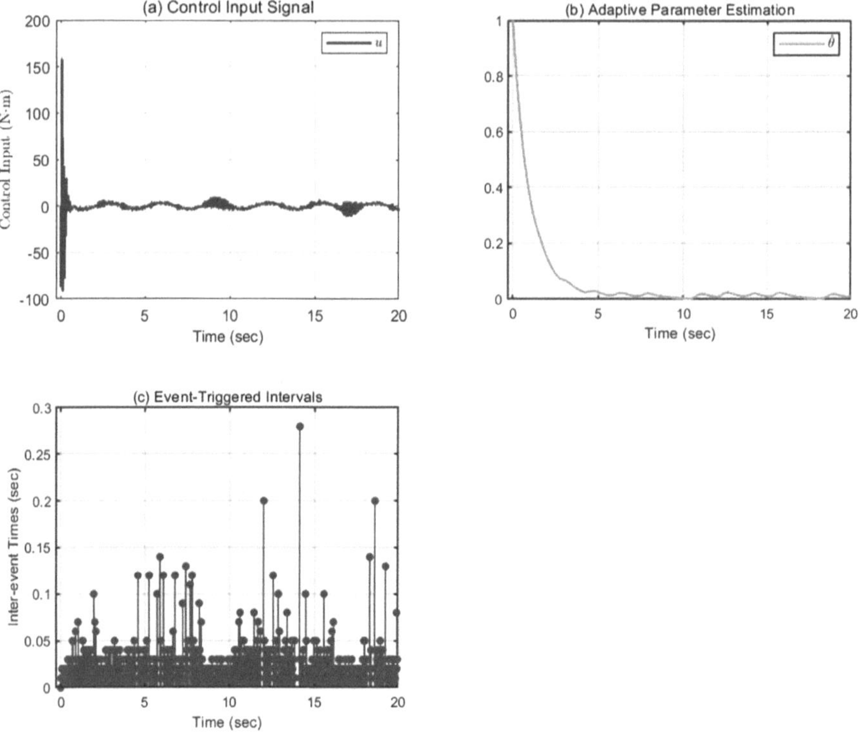

Fig. 2. Control System Performance

Acknowledgments. This work is supported in part by the National Natural Science Foundation of China under Grant No. 62273163, the Taishan Scholar Foundation of Shandong Province under Grant No. tsqn202312212, and the Outstanding Youth Foundation of Shandong Province under Grant No. ZR2023YQ056.

The authors have no competing interests to declare that are relevant to the content of this article.

References

1. Zhang, Y., et al.: Unknown system dynamics estimator-based composite anti-saturation control method for series elastic actuators. IEEE Trans. Ind. Electron. **71**(12), 16625–16634 (2024)
2. Lee, H., Lee, J., Keppler, M., Oh, S.: Robust elastic structure preserving control for high impedance rendering of series elastic actuator. IEEE Rob. Autom. Lett. **9**(4), 3601–3608 (2024)
3. Yan, Y., Guan, D., Jiang, T., Yu, S., Liu, Y.: Event-triggered adaptive predefined-time sliding mode control of autonomous surface vessels with unknown dead zone and actuator faults. Ocean Eng. **304**, 117851 (2024)
4. Fei, T., Yang, Y., Zheng, C., Zhang, H., Yuan, G.: Adaptive neural design for permanent magnet synchronous motor with asymmetric constraints and input dead-zone. Neurocomputing. 130294 (2025)

5. Diao, S., Sun, W., Su, S.F.: Neural-based adaptive event-triggered tracking control for flexible-joint robots with random noises. Int. J. Robust Nonlinear Control. **32**(5), 2722–2740 (2022)
6. Xie, S., Sun, W., Su, S.F.: Adaptive performance optimal control for flexible-joint robots with random noises: design and experiment. Appl. Math. Model. **138**, 115741 (2025)
7. Li, Y., Li, K., Tong, S.: Finite-time adaptive fuzzy output feedback dynamic surface control for MIMO nonstrict feedback systems. IEEE Trans. Fuzzy Syst. **27**(1), 96–110 (2018)
8. Tong, S., Li, Y., Sui, S.: Adaptive fuzzy output feedback control for switched nonstrict-feedback nonlinear systems with input nonlinearities. IEEE Trans. Fuzzy Syst. **24**(6), 1426–1440 (2016)
9. Li, K., Li, Y.: Adaptive neural network finite-time dynamic surface control for nonlinear systems. IEEE Trans. Neural Netw. Learn. Syst. **32**(12), 5688–5697 (2020)
10. Xing, L.T., Wen, C.Y., Liu, Z.T., Su, H.Y., Cai, J.P.: Adaptive compensation for actuator failures with event-triggered input. Automatica. **85**, 129–136 (2017)
11. Huang, Y., Liu, Y.: Practical tracking via adaptive event-triggered feedback for uncertain nonlinear systems. IEEE Trans. Autom. Control. **64**(9), 3920–3927 (2019)
12. Lei, M., Chen, W., Wang, L.: Adaptive tracking control for a class of nonlinear systems with input dead-zone and actuator failure. Math. Methods Appl. Sci. **46**(6), 7333–7352 (2023)
13. Song, J., Zhu, A., Tu, Y., Zhang, X., Cao, G.: Novel design and control of a crank-slider series elastic actuated knee exoskeleton for compliant human–robot interaction. IEEE/ASME Trans. Mechatron. **28**(1), 531–542 (2022)
14. Cui, M.Y., Wu, Z.J., Xie, X.J., Shi, P.: Modeling and adaptive tracking for a class of stochastic Lagrangian control systems. Automatica. **49**(3), 770–779 (2013)
15. Long, L., Zhao, J.: Adaptive output-feedback neural control of switched uncertain nonlinear systems with average dwell time. IEEE Trans. Neural Netw. Learn. Syst. **26**(7), 1350–1362 (2014)
16. Li, Y., Tong, S.: Adaptive fuzzy output constrained control design for multi-input multioutput stochastic nonstrict-feedback nonlinear systems. IEEE Trans. Cybern. **47**(12), 4086–4095 (2016)
17. Zhao, Z., Yu, J., Zhao, L., Yu, H., Lin, C.: Adaptive fuzzy control for induction motors stochastic nonlinear systems with input saturation based on command filtering. Inf. Sci. **463**, 186–195 (2018)

VeMamba: Voxel-Based Multi-Scale State Space Model Network for Event Stream Recognition

Xin Zhan[1,2,3], Xiaoping Wang[1,2,3](✉), Jiang Li[1,2,3], Weibin Feng[1,2,3], Depeng Li[1,2,3], and Hongzhi Huang[1,2,3]

[1] School of Artificial Intelligence and Automation, Huazhong University of Science and Technology, Wuhan 430074, China
[2] Key Laboratory of Image Processing and Intelligent Control of Education Ministry of China, Huazhong University of Science and Technology, Wuhan 430074, China
[3] Hubei Key Laboratory of Brain-inspired Intelligent Systems, Huazhong University of Science and Technology, Wuhan 430074, China
wangxiaoping@hust.edu.cn

Abstract. Event cameras are innovative neuromorphic sensors that capture dynamic changes in scenes, recording millions of events per second. Recent sparse computational models on event stream recognition have achieved notable successes by utilizing graph convolution or attention mechanisms to model local dependencies. However, these methods face significant challenges in constructing effective event representations and modeling global dependencies when confronted with millions of events in spacetime. To surmount the above challenges, we present a novel voxel-based State Space Model (SSM) network, termed VeMamba, which can establish multi-scale spatiotemporal dependencies in event voxels with linear complexity. Specifically, we design a time-aware enhanced voxelization method that enriches the spatiotemporal expression within event voxels while preserving sparse computation. Then, we propose a multi-scale modeling module with linear complexity that integrates local attention into the global dual-scale SSMs to establish spatiotemporal dependencies from local to global within serialized voxels. Furthermore, leveraging a hierarchical structure grounded in voxel merging, we can extract deep semantic and motion cues from the voxels. Extensive experiments demonstrate that VeMamba achieves state-of-the-art (SOTA) performance with low model complexity and computational cost on event stream recognition tasks.

Keywords: Event Cameras · State Space Models · Voxel Processing

1 Introduction

Event cameras differ from conventional frame-based cameras by asynchronously capturing per-pixel brightness changes as sparse event streams, rather than full-frame intensity at fixed intervals [17]. Their neuromorphic design offers low power

C. Li et al. (Eds.): ICNC 2025, CCIS 2946, pp. 244–259, 2026.
https://doi.org/10.1007/978-981-92-1599-7_21

consumption, high dynamic range, and high temporal resolution, promoting use in vision tasks including object classification [2,3,7,10], action recognition [18, 25,30], and optical flow estimation [34]. However, the large volume of sparse spatiotemporal data increases processing complexity. Traditional dense models like CNNs (Fig. 1(a)) convert sparse events to dense representations, causing temporal resolution loss and redundant computation on invalid pixels. Thus, more compatible algorithms are needed to leverage the inherent advantages of event cameras.

To leverage the high temporal resolution and sparsity of event cameras, current advanced learning models use sparse computation via Spiking Neural Networks (SNNs) or Graph Neural Networks (GNNs) to process event data [2,28]. While SNNs naturally handle event streams as spikes, their use is limited by immature training and hardware. In contrast, GNN-based methods like RG-CNN [3] and EV-VGCNN [9] are widely adopted, applying hierarchical local priors to event points or voxels. As in Fig. 1(b)(c), raw event streams contain vast numbers of events. Point-based methods often use temporal partitioning and downsampling, risking information loss and limited spatiotemporal reasoning. Voxel-based representations aggregate events within local regions, improving efficiency and local perception. Some works [18,31,34] stack events within voxels to enhance representations but still suffer from limited temporal perception and coverage. Others improve voxel modeling via convolution, graphs, or attention [8,18,32], yet attention mechanisms incur high computational costs due to quadratic complexity with sequence length. Thus, these methods remain local and struggle to capture global dependencies needed for modeling long-term motions or stagnation [8]. The key challenge is developing event representations with richer spatiotemporal expressiveness and enabling multi-scale (especially global) dependency modeling under feasible computational constraints.

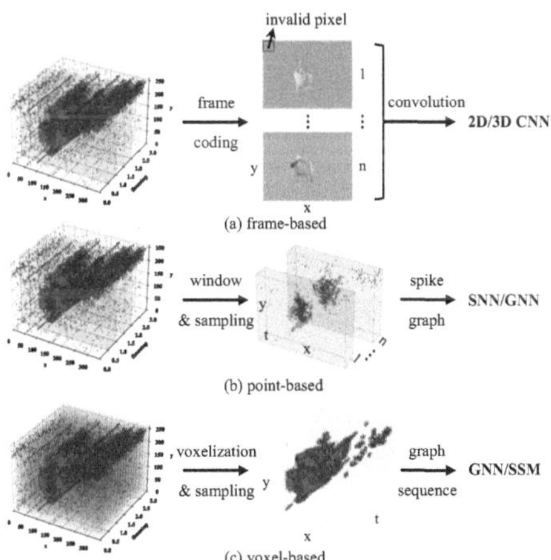

Fig. 1. Three Different methods for processing event streams are (a) frame-based methods, (b) point-based methods and (c) voxel-based methods.

Recently, State Space Models (SSMs) [13,14] have shown strong performance in modeling long sequences. For example, Mamba [13] matches the global modeling capability of transformers in various NLP tasks with only linear complexity, much lower than the quadratic complexity of transformers. Subsequently, several works [25,26,30] have adapted SSMs for event-based recognition. EventMamba [25], closely related to our approach, uses Mamba layers to improve temporal modeling in hierarchical structures, but its lightweight strategy of segmenting and sampling event streams may weaken temporal dependencies and fine details. EvMamba [30] uses SSMs to process event frames and voxels multimodally, yet suffers notable performance drops when using only voxel data. Therefore, we focus on enriching spatiotemporal details through event voxels and employ SSMs to efficiently model global dependencies in long voxel sequences.

In this paper, we propose VeMamba, a Voxel SSM-based Mamba network for event-based recognition. We introduce a time-aware enhanced voxelization method to enrich spatiotemporal representation within voxels. Local attention is integrated with global dual-scale SSMs to extract multi-scale semantic and motion cues from voxel sequences. A hierarchical structure with voxel merging is employed to expand the receptive field and improve computational efficiency. Our contributions are summarized as follows

- We design a straightforward yet effective voxelization method that enriches the spatial and temporal expressions of event points within voxels.
- We present a hierarchical framework that integrates local attention mechanisms into global dual-scale SSMs to model voxel sequences, facilitating the learning of multi-scale discriminative semantic and motion cues.
- Extensive experiments demonstrate that our VeMamba achieves SOTA performance on multiple datasets for object classification and action recognition with low model complexity and computational cost.

2 Preliminaries

This section presents the preliminaries of event cameras and SSMs.

Event Cameras. For an event camera with a resolution of $[H, W]$, the event stream generated by capturing changes in log light intensity can be represented as the sequence ε:

$$\varepsilon = \{e_i\}_N = \{(x_i, y_i, t_i, p_i)\}_N, \tag{1}$$

where N is the number of events in ε, (x_i, y_i) denotes the pixel coordinates, t_i is the trigger timestamp, and $p_i \in \{+1, -1\}$ denotes an increase or decrease in log intensity.

State Space Models. SSMs conceptualize the internal states of a system as variables $h(t) \in \mathbb{R}^P$, which are influenced by external input $x(t) \in \mathbb{R}$ and, in turn, influence the system's output $y(t) \in \mathbb{R}$:

$$\begin{aligned} h'(t) &= \mathbf{A}h(t) + \mathbf{B}x(t), \\ y(t) &= \mathbf{C}h(t) + \mathbf{D}x(t), \end{aligned} \tag{2}$$

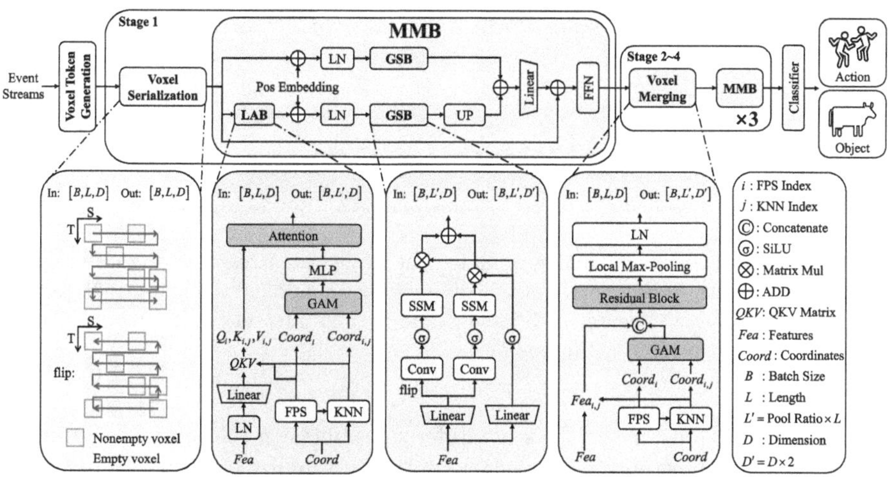

Fig. 2. Overview of VeMamba. The raw event streams are voxelized and tokenized, then serialized into a hierarchical structure with multi-scale Mamba and voxel merging blocks to efficiently capture semantic and motion cues. The resulting features are finally classified for recognition.

where $\mathbf{A}$, $\mathbf{B}$, $\mathbf{C}$, and $\mathbf{D}$ are all continuous parameters of the system. To further introduce SSMs into deep learning, S4 [14] employs the zero-order hold sampling method with a sampling step Δ to discretize the continuous parameters $\mathbf{A}$, $\mathbf{B}$ as $\overline{\mathbf{A}} = \exp(\Delta\mathbf{A})$, $\overline{\mathbf{B}} = (\Delta\mathbf{A})^{-1}(\exp(\Delta\mathbf{A} - \mathbf{I})) \cdot \Delta\mathbf{B}$. The discretized system equations can be expressed as:

$$\begin{aligned} h_k &= \overline{\mathbf{A}} h_{k-1} + \overline{\mathbf{B}} x_k, \\ y_k &= \mathbf{C} h_k + \mathbf{D} x_k, \end{aligned} \tag{3}$$

where $\overline{\mathbf{A}} \in \mathbb{R}^{P\times P}$ and $\overline{\mathbf{B}} \in \mathbb{R}^{P\times 1}$ are discrete parameters of the system, $\mathbf{C} \in \mathbb{R}^{1\times P}$ is a projection parameter, and $\mathbf{D} \in \mathbb{R}^1$ is the parameter for residual connection. We adopt the selective SSM in Mamba [13] to construct our model.

3 The Proposed Method

An overview of VeMamba is shown in Fig. 2. Event streams are voxelized into 3D spacetime tokens via a voxel token generation module, then processed through a hierarchical structure with multi-scale Mamba and voxel merging blocks. These procedures are detailed in the following sections.

3.1 Voxel Token Generation

The voxel token generation module, as shown in Fig. 3, is designed to transform the large volume of events in spacetime into quantitative sparse event voxels and to extract the spatiotemporal features within each voxel.

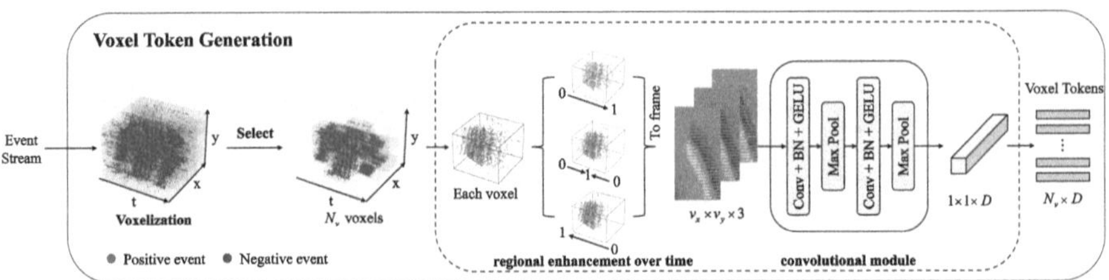

Fig. 3. The voxel token generation module. The raw event stream undergoes voxelization to form voxels. Subsequently, a selection process is applied to generate N_v voxels. Finally, we emphasize events in different time regions and utilize the convolutional module to extract basic voxel features.

Voxelization. Considering the scale differences between the spatial coordinates (x_i, y_i) and the timestamp t_i of event points, we normalize the timestamp t_i with a specified coefficient T:

$$t_i' = \frac{t_i - t_{\min}}{t_{\max} - t_{\min}} \times (T - 1), \tag{4}$$

where $t_{\min}$ and $t_{\max}$ denote the initial and final timestamps in the event stream. Scaled events (x_i, y_i, t_i') distribute range from $[0, 0, 0]$ to $[W - 1, H - 1, T - 1]$ along the x, y, and t axes. With the voxel size of (v_x, v_y, v_t), we voxelize the scaled event stream. The resulting voxel grid is of size $\left[\left\lfloor \frac{W}{v_x} + \frac{1}{2} \right\rfloor, \left\lfloor \frac{H}{v_y} + \frac{1}{2} \right\rfloor, \left\lfloor \frac{T}{v_t} + \frac{1}{2} \right\rfloor\right]$. The position of a voxel within the grid corresponds to its coordinates $U(u_x, u_y, u_t)$.

Voxel Selection. Event cameras only record dynamic changes and are susceptible to noise, which leads to numerous empty or noisy voxels in voxelization results. To alleviate this, we select the top N_v voxels with the highest event density from the N_p nonempty voxel set. If $N_p < N_v$, we choose the top 95% of N_p nonempty voxels, and the remaining voxels are supplemented by random sampling from these voxels, ensuring that the final number of selected voxels meets the predetermined target N_v.

Voxel Feature Encoding. Within each voxel, the number of event points varies and their distribution is relatively sparse, which necessitates an effective method to encode voxel features. We propose a straightforward yet effective approach that emphasizes event points within specific time regions of a voxel from three distinct directions along the time axis. The entire process is illustrated in Fig. 3. Specifically, we calculate the relative spatial coordinates (x_i^v, y_i^v) and the normalized time coordinate t_i^v of event points within the corresponding voxel. Accordingly, (x_i^v, y_i^v, t_i^v) is constrained with $[v_x, v_y, 1]$. To emphasize events in different time regions and thereby obtain richer spatiotemporal information within voxels, we assign weights w_n to the event points inside a voxel from the anterior, middle, and posterior directions (associated with $n = 3, 2, 1$, respectively) based on the time value t_i^v. The 2D tensor of a voxel in the n_{th} direction at position (x, y), denoted as $E_v(x, y, n)$, can be represented as follows:

$$E_v(x, y, n) = \sum_i^N p_i * \mathbb{I}(x_i^v = x, y_i^v = y) * w_n,$$
$$\text{s.t.} \quad w_n = \begin{cases} t_i^v, & n = 1 \\ 1 - \left|1 - 2 * t_i^v\right|, & n = 2 \\ \left|t_i^v - 1\right|, & n = 3 \end{cases} \tag{5}$$

where N is the number of events within the voxel, $\mathbb{I}(\cdot)$ denotes the indicator function, which yields a value of 1 if the condition holds true, and 0 otherwise. Further, we use a convolutional module that consists of two consecutive 2D convolution layers to learn the underlying semantic and motion features within E_v. The convolution layer consists of a 2D convolution, a Batch Normalization (BN), a GELU, and a 2D Max-pooling layer. All voxel tensors share the same convolutional module, finally obtaining their 1D features $F \in \mathbb{R}^{N_v \times D}$ and voxel coordinates $U \in \mathbb{R}^{N_v \times 3}$.

3.2 Multi-Scale Mamba Block

In this subsection, we introduce how Multi-Scale Mamba Block (MMB) processes voxel tokens and obtains multi-scale spatiotemporal features.

Serialization Strategy. We employ a spatial-first serialization method, as shown in the left corner of Fig. 2. We first organize the voxel tokens that share the same time coordinate u_t based on their spatial coordinates (u_x, u_y). Subsequently, these organized tokens are concatenated in ascending order of u_t to obtain a 1D voxel token sequence. Further, we can obtain the reverse sequence by flipping.

Local Attention Block. As depicted in Local Attention Block (LAB) in Fig. 2, we enhance the ability of specific voxels to acquire local salient spatiotemporal features within the sequence using dot-product attention in LAB. For voxel features $F \in \mathbb{R}^{N_v \times C}$ and corresponding coordinates $U \in \mathbb{R}^{N_v \times 3}$, LAB can be specifically described in the following steps: First, voxel features F are normalized by a Layer Normalization (LN) layer. Subsequently, they are projected into three distinct feature spaces to generate $Q \in \mathbb{R}^{N_v \times C}$, $K \in \mathbb{R}^{N_v \times C}$, and $V \in \mathbb{R}^{N_v \times C}$ tensors, where N_v is the number of voxel tokens and C is the feature dimension. Following this, we construct local spatiotemporal graphs $\mathcal{G}$ in voxel tokens based on coordinates U using Farthest Point Sampling (FPS) and K-Nearest Neighbor (KNN). Each local graph $\mathcal{G}$ is constructed by the sampling voxel i and its neighbors j. By doing so, we can obtain prominent local spatiotemporal features F_i from the value matrix $V_{i,j}$ based on the attention scores calculated between the query matrix Q_i of the sampling voxel and their corresponding neighborhood key matrix $K_{i,j}$. Furthermore, to mitigate the adverse effects of mismatched coordinate values caused by abnormal voxels (induced by thermal noise, etc.) in local graphs on training stability, we introduce the Geometric Affine Module [21] (GAM) with minimal parameters and computational cost to enhance the stability of relative positional relationships in local graphs. The entire process can be described as follows:

$$\begin{aligned} F_i &= \underset{j \in \mathcal{G}}{Softmax}\left(\frac{Q_i \cdot K_{i,j}^T}{\sqrt{C}} + \delta_{i,j}\right) \cdot (V_{i,j} + \varphi_{i,j}), \\ \text{s.t.} \quad & \varphi_{i,j} = \theta\left(\text{GAM}\left(U_i, U_{i,j}\right)\right), \quad \delta_{i,j} = \sum (\varphi_{i,j}), \end{aligned} \tag{6}$$

where θ denotes a linear projection followed by a GELU. The GAM can be expressed as:

$$\begin{aligned} &\text{GAM}\left(U_i, U_{i,j}\right) = \alpha \odot \frac{U_{i,j} - U_i}{\sigma + \epsilon} + \beta, \\ &\text{s.t.} \quad \sigma = \sqrt{\frac{1}{K \times N' \times 3} \sum_{i=1}^{N'} \sum_{j=1}^{K} \left(U_{i,j} - U_i\right)^2}, \end{aligned} \tag{7}$$

where α and β are learnable parameters, $\epsilon = 1e^{-5}$ is a small number for numerical stability, N' means the number of sampling voxels, and K denotes the number of neighbors for the sampling voxel (including self-loop).

Global SSM Block. While LAB improves the local perception of voxel sequences, it cannot capture spatiotemporal features from a global perspective. Inspired by the dual-scale learning strategy in [20], our proposed MMB employs two Global SSM Block (GSB) branches to model global dependency. As shown in Fig. 2, one branch extracts global features with high-resolution details from the original voxel sequences, while the other acquires locally enhanced low-resolution features from the downsampled voxel sequences after LAB. By integrating features across different scales and resolutions, the MMB significantly improves the ability to model multi-scale semantic and motion cues. Specifically, for given serialized voxel tokens $F \in \mathbb{R}^{N_v \times C}$ and their coordinates $U \in \mathbb{R}^{N_v \times 3}$, MMB can be expressed as:

$$\begin{aligned} F_L &= \text{Up}\left(\text{L-GSB}\left(LN\left(\text{LAB}\left(F, U\right) + \phi\left(U'\right)\right)\right)\right), \\ F_H &= \text{H-GSB}\left(LN\left(F + \phi\left(U\right)\right)\right), \\ \tilde{F} &= \text{Linear}\left(F_L + F_H\right) + F, \end{aligned} \tag{8}$$

where L-GSB and H-GSB refer to the low-resolution and the high-resolution global SSM block, Up denotes the nearest-neighbor interpolation upsampling, ϕ represents the position normalization and a linear projection followed by a GELU, U' signifies the coordinates of sampling event voxels, while Linear is a linear projection layer. Following this, we introduce an FFN layer to facilitate the model's ability to capture complex features and patterns.

3.3 Voxel Merging

To obtain hierarchical spatiotemporal features and improve overall computational efficiency, we progressively downsample voxel tokens during processing. As shown in Voxel Merging in Fig. 2, we reuse the FPS voxel position U' in LAB to construct local graphs $\mathcal{G}'$ containing K' neighbors (including self-loop) by KNN. GAM is also employed in local graphs. After that, a parameter-shared residual block is utilized to fuse semantic and position features. Finally, we use max-pooling and LN to extract representative features f_i from the local graphs. The above procedures for downsampling to obtain merged voxels can be described as follows:

$$f_i = LN\left(\max_{j \in \mathcal{G}'} \left(\Phi_{pre}\left(\Phi_{pro}\left(\left[F_{i,j}, \mathrm{GAM}\left(U_i, U_{i,j}\right)\right]\right)\right)\right)\right), \tag{9}$$

where i and j respectively refer to the sampling voxel and its neighbors in local graph $\mathcal{G}'$, $[\cdot]$ is the concatenation function, max represents the max-pooling operation. The Φ_{pre} and Φ_{pro} can be expressed as:

$$\begin{aligned} &\Phi_{pro} = \sigma(M(x)), \Phi_{pre} = \sigma\left(x + M\left(\sigma\left(M\left(x\right)\right)\right)\right), \\ &\text{s.t.} \quad M\left(x\right) = BN\left(Conv1d\left(x\right)\right), \end{aligned} \tag{10}$$

where x stands for input features and σ indicates a ReLU. Through Φ_{pro} and Φ_{pre}, we transform the input feature channels from $(C+3)$ to $(2 \times C)$.

4 Experiments

In this section, we conduct systematic evaluations and ablation studies of our proposed model on object classification and action recognition tasks, in terms of accuracy, model complexity (measured by the number of trainable parameters), and floating-point operations (FLOPs).

4.1 Datasets

We utilize four popular benchmark datasets for object classification, designated as N-Caltech101 (N-Cal) [23], N-MNIST (N-M) [23], N-CARS (N-C) [28], and CIFAR10-DVS (CIF10) [15]. Six representative benchmark datasets, namely DVS128 Gesture (Ges) [1], UCF101-DVS (UCF) [3], HMDB51-DVS (HMDB) [3], DVS Action (Act) [22], DailyAction-DVS (Dai) [19], and THU$^{\text{E-ACT}}$-50-CHL (THU) [11], are utilized for action recognition. For N-Cal, we utilize the splitting results from [12]. For N-M, N-C, Ges, and THU, we use official splits to train and test our model. For other datasets without official splitting, we follow the method in [3], randomly shuffling the samples before allocating 80% for training and 20% for testing. Other detailed information regarding the datasets is presented in the supplementary material.

Table 1. Parameter settings for voxel token generation and model structure on different datasets.

Dataset	(v_x, v_y)	T	N_v	pool ratio
N-Caltech101 [23]	$(10, 10)$	9	1024	$(4, 4, 4, 2)$
N-MNIST [23]	$(2, 2)$	9	512	$(2, 4, 4, 2)$
N-CARS [28]	$(5, 5)$	6	256	$(2, 2, 4, 2)$
CIFAR10-DVS [15]	$(10, 10)$	30	1024	$(4, 4, 4, 2)$
DVS128 Gesture [1]	$(10, 10)$	50	512	$(2, 4, 4, 2)$
UCF101-DVS [3]	$(10, 10)$	50	2048	$(4, 4, 4, 2)$
HMDB51-DVS [3]	$(10, 10)$	50	4096	$(4, 4, 4, 2)$
DVS Action [22]	$(10, 10)$	50	1024	$(4, 4, 4, 2)$
DailyAction-DVS [19]	$(10, 10)$	50	512	$(2, 4, 4, 2)$
THU$^{\text{E-ACT}}$-50-CHL [11]	$(10, 10)$	50	1024	$(4, 4, 4, 2)$

4.2 Implementation Details

Voxel Token Generation. Based on the characteristics of different tasks and datasets, we configure the corresponding (v_x, v_y), coefficient T, and selected voxels N_v in Table 1. Meanwhile, we set $v_t = 3$ for all datasets. As mentioned in Sect. 3.1, we use two convolution layers for datasets with $(v_x, v_y) = (10, 10)$, and a single layer for other datasets. After that, we incorporate a LN layer to normalize tokens and employ a dropout layer with a probability of 0.1 to mitigate overfitting.

Network. As shown in Fig. 2, we use a four-stage structure to process event streams. The base feature dimensions C and neighborhood value K are set as $(32, 64, 128, 256)$ and $(20, 20, 20, 8)$ in each stage for all datasets, respectively. Based on the initial number of voxels, we determine the pooling ratio to generate N' from N, refer to Table 1. Voxel tokens are merged with $K' = 16$ in all voxel merging modules. In VeMamba, multi-branch structures undergo processing by a drop path layer with a probability of 0.1. For the classification head, we concatenate global mean features from the last two MMBs as input. To mitigate overfitting in the classification head, we use dropout layers with a 0.5 probability rate following the first two FC layers.

Training. The model was trained from scratch for 250 epochs using SGD with a linearly warmed-up learning rate to 7e-2 for the initial 25 epochs, followed by cosine annealing to reduce the learning rate to 7e-4, while optimizing the label smoothing cross-entropy loss function. We set batch size as 16 for HMDB and Act and 32 for other datasets.

4.3 Results Comparison

Results of Object Classification. In Table 2, we compare event representations on object classification datasets. VeMamba achieves the best overall accuracy, significantly outperforming frame-based methods trained from scratch and showing consistent advantages over point or voxel-based approaches. While delivering competitive results on simpler datasets (N-M, N-C), VeMamba excels on complex ones like N-Cal and CIF10, surpassing VMST-Net [18] by 0.8%and 2.0%, respectively. This is attributed to our time-aware voxelization, which captures richer spatiotemporal features, and MMB, which acquires discriminative semantic cues. The model also maintains low complexity, remaining competitive among sparse models and far more efficient than dense ones. However, VeMamba does not achieve optimal accuracy on N-C, likely due to noise contamination and simple features causing overfitting to noise and reduced generalization.

Table 2. Comparison of the object classification accuracy between ours and other event representation models.

Method	Type	N-Cal	N-M	N-C	CIF10	Params(M)
EST [12]	Frame	75.3	99.0	91.9	63.4	21.38
MVF-NET [7]	Frame	68.7	98.1	92.7	59.9	33.62
Matrix-LSTM [4]	Frame	73.8	98.6	92.7	63.1	21.43
AMAE [10]	Frame	69.4	98.3	93.6	62.0	–
EventNet [27]	Point	42.5	75.2	75.0	17.1	–
RG-CNNs [3]	Point	65.7	99.0	91.4	54.0	19.46
Evs-S [16]	Point	76.1	–	93.1	68.0	–
ECSNet [6]	Point	69.3	99.2	94.6	72.7	–
EV-VGCNN [9]	Voxel	74.8	<u>99.4</u>	<u>95.3</u>	67.0	<u>0.84</u>
VMV-GCN [31]	Voxel	77.8	**99.5**	93.2	69.0	0.86
EVSTr [32]	Voxel	79.7	–	94.1	-	0.93
EDGCN [8]	Voxel	80.1	-	**95.8**	71.6	**0.77**
VMST-Net [18]	Voxel	<u>82.2</u>	**99.5**	94.4	<u>75.3</u>	3.61
Ours	Voxel	**83.0**	**99.5**	94.3	**77.3**	2.55

Best results in **bold** and second best <u>underlined</u>. Params on N-Cal.

Results of Action Recognition. In Table 3, we compare various architectures on action recognition datasets. VeMamba, based on Mamba, outperforms other backbones on all datasets except Ges, showing substantial gains over EventMamba [25]. While suboptimal on Ges—likely due to reasons similar to N-C, where lightweight designs excel with simple patterns—VeMamba improves significantly on Act, Dai, UCF, HMDB, and THU, surpassing EventMamba by

Table 3. Comparison of the action recognition accuracy between ours and other backbone models.

Method	Arc.	Ges	UCF	HMDB	Act	Dai	THU
Motion-based [19]	S.	–	–	–	78.1	90.3	47.3
EV-ACT [11]	C.	–	–	–	–	97.9	58.5
C3D [29]	C.	–	47.2	41.7	–	–	–
I3D [5]	C.	–	63.5	46.6	–	96.2	–
RG-CNNs [3]	G.	97.2	67.3	49.7	–	–	–
VMV-GCN [31]	G.	97.5	–	–	–	94.1	–
TTPOINT [24]	G.	98.8	72.5	56.9	<u>92.7</u>	99.1	–
ECSNet [6]	T.	98.6	70.2	–	–	–	–
EVSTr [32]	T.	98.6	–	–	–	<u>99.6</u>	–
PointTrans [18]	T.	95.6	–	46.1	–	–	–
VMST-Net [18]	T.	97.8	78.7	59.5	–	–	–
Event-SSM [26]	M.	97.7	–	–	–	–	–
EvMamba [30]	M.	<u>99.0</u>	–	–	–	–	–
EventMamba [25]	M.	**99.2**	<u>88.6</u>	<u>60.4</u>	89.1	99.1	<u>59.4</u>
Ours	M.	97.9	**90.3**	**63.3**	**95.3**	**99.7**	**78.8**

S., C., T., and M. denote SNN, CNN, Transformer, and Mamba, respectively.

6.2%, 0.6%, 1.7%, 2.9%, and 19.4%, respectively. This improvement stems from our voxel token generation and MMB module, which effectively capture semantic and motion cues within and between voxels, enabling long-range spatiotemporal dependency modeling for complex action recognition.

Complexity and Computation Analysis. We analyze the complexity and computational cost of our model, building on Mamba [13] and VMamba [20]. As shown in Table 4, point-based and voxel-based sparse models significantly reduce complexity and cost compared to frame-based dense models. For example, EventMamba achieves optimal efficiency via lightweight designs like sliding windows and low-dimensional features. In contrast, our model uses richer voxel representations and higher-dimensional features, resulting in higher cost. However, VeMamba's SSM-based core ensures approximately linear complexity relative to input length.

4.4 Abalation Study

In this section, we perform ablation studies on N-Cal and THU to evaluate the core modules: the voxel token generation method and the multi-scale structure.

Table 4. Comparisons of model complexity (Params) and computational cost (GFLOPs)

Method	Type	Params(M)	GFLOPs
Complexity and Computation Analysis on N-Cal			
MVF-NET [7]	Frame	33.62	11.24
Matrix-LSTM [4]	Frame	21.43	9.64
RG-CNNs [3]	Point	19.46	**0.79**
ECSNet [6]	Point	**0.18**	6.28
EV-VGCNN	Voxel	0.84	1.4
EVSTr [32]	Voxel	0.93	1.62
VMST-Net [18]	Voxel	3.61	0.88
Ours	Voxel	2.55	1.00
Complexity and Computation Analysis on Ges			
TBR + I3D [5]	Frame	12.25	38.82
Event Frames + I3D [3]	Frame	12.28	67.06
TTPOINT [24]	Point	0.334	0.587
EventMamba [25]	Point	**0.29**	**0.219**
EV-VGCNN	Voxel	0.82	0.92
EVSTr [32]	Voxel	2.88	2.2
VMST-Net [18]	Voxel	3.61	0.62
Ours	Voxel	2.53	0.71

Effectiveness of Voxel Token Generation. We compare our voxel token generation against three alternatives: Voxel Grid [34] (without trilinear kernels), EV-VGCNN [9], and VMST-Net [18]. For fairness, Voxel Grid and EV-VGCNN were adjusted to match model complexity by expanding initial channels. We also integrate our voxels into PointTrans (with VMV-GCN voxels) [31,33] and VMST-Net, comparing accuracy with their original inputs. As shown in Table 5, our method outperforms polarity-fading Voxel Grid, temporally-biased EV-VGCNN, and region-missing VMST-Net in capturing both static appearance and dynamic motion features. It especially improves temporal perception within voxels, yielding superior performance on motion-sensitive THU. Reproduced models also show significant accuracy gains when using our voxels, demonstrating the method's effectiveness and generality. Note that some results are omitted due to inherent model limitations in challenging scenarios like THU, where severe accuracy degradation occurs.

Effectiveness of Multi-Scale Structure. We evaluate feature aggregation methods in the Local Attention Block (LAB), comparing center, max-pooling, and mean-pooling against the attention mechanism. As shown in Table 6, atten-

Table 5. The effect of voxel generation.

Voxel Setting	N-Cal	THU
Voxel Grid [34]	82.5	72.2
EV-VGCNN Voxel [9]	82.7	73.3
VMST-Net Voxel [18]	82.7	76.1
PointTrans [31,33]	76.7	–
PointTrans† [31,33]	80.4 (+3.7)	–
VMST-Net [18]	81.1	–
VMST-Net† [18]	82.0 (+0.9)	–
Ours	**83.0**	**78.8**

† Using our voxels as input.

Table 6. The effect of LAB.

Setting	Params	GFLOPs	N-Cal Acc.	THU Acc.
center	2.29 M	0.97	82.5	77.6
max			82.3	77.6
mean			**83.0**	77.6
attention	+ 0.26 M	+ 0.03	**83.0**	**78.8**

Params and GFLOPs on N-Cal.

tion improves accuracy by up to 0.7% on N-Cal and 1.2% on THU, with minor increases in parameters (+0.26M) and computation (+0.03G), confirming its benefit for selective local feature extraction. We further ablate voxel merging (Mer.), LAB, high-resolution global SSM (H-GSB), and low-resolution global SSM (L-GSB) in Table 7. The baseline with only voxel token generation and classification is limited to single-voxel features. Adding LAB improves local perception but increases computation due to long sequences. Introducing Mer. enables hierarchical local-to-global modeling, significantly boosting performance. Still, without global perspective, we incorporate GSB early to enhance long-range spatiotemporal dependency. H-GSB notably improves accuracy (2.1% on N-Cal, 2.7% on THU) despite higher cost, and bidirectional SSM outperforms unidirectional. Multi-scale interaction is achieved via H-GSB and L-GSB (with LAB), fusing high- and low-resolution features for complementary spatiotemporal representation. Multi-resolution fusion with reasonable overhead exceeds single-resolution modeling in accuracy.

Table 7. The effect of multi-scale structure.

Mer.	LAB	H-GSB		L-GSB	FLOPs	Params	Runtime	N-Cal	THU
		uni.	bi.					Acc.	Acc.
×	×	×	×	×	0.49G	0.12M	0.7 ms	51.5	47.6
×	✓	×	×	×	0.56G	0.13M	126.8 ms	52.0	57.2
✓	×	×	×	×	0.77G	0.42M	37.7 ms	80.4	75.4
✓	✓	×	×	×	0.80G	0.68M	40.4 ms	82.0	76.3
✓	×	✓	×	×	0.91G	1.72M	42.9 ms	82.5	78.1
✓	×	×	✓	×	0.94G	1.79M	43.7 ms	82.5	78.4
✓	✓	×	×	✓	0.91G	2.06M	47.5 ms	82.6	78.3
✓	✓	×	✓	✓	1.00G	2.55M	50.5 ms	**83.0**	**78.8**

Params and GFLOPs on N-Cal. Inference time is measured on a workstation with a CPU (Xeon Silver 4210R), a GPU (RTX 3090), and 125GB of available RAM.

5 Conclusion

In this paper, we propose a novel voxel-based multi-scale SSM network, termed VeMamba, for event stream processing. The proposed temporal-aware enhanced voxelization method extracts rich spatiotemporal information within voxels. Moreover, leveraging the linear complexity of SSMs for global modeling, we develop a hierarchical framework that integrates local attention mechanisms into global dual-scale SSMs to establish multi-scale dependencies in voxel sequences at an acceptable computational cost. Extensive experiments demonstrate that our VeMamba achieves SOTA performance in event stream recognition tasks with lower model complexity and computational cost.

Acknowledgments. This work was supported in part by the National Natural Science Foundation of China under Grant 62236005 and 623B2040, the Postdoctoral Fellowship Program of CPSF under Grant GZB20250428, the Foundation for Outstanding Research Groups of Hubei Province of China under Grant 2025AFA012, and the 111 Project on Computational Intelligence and Intelligent Control under Grant B18024.

Disclosure of Interests. The authors have no competing interests to declare that are relevant to the content of this article.

References

1. Amir, A., et al.: A low power, fully event-based gesture recognition system. In: IEEE Conference on Computer Vision and Pattern Recognition (CVPR), pp. 7243–7252 (2017)
2. Bi, Y., Chadha, A., Abbas, A., Bourtsoulatze, E., Andreopoulos, Y.: Graph-based object classification for neuromorphic vision sensing. In: Conference on Computer Vision (ICCV), pp. 491–501 (2019)
3. Bi, Y., Chadha, A., Abbas, A., Bourtsoulatze, E., Andreopoulos, Y.: Graph-based spatio-temporal feature learning for neuromorphic vision sensing. IEEE Trans. Image Process. (TIP) **29**, 9084–9098 (2020)
4. Cannici, M., Ciccone, M., Romanoni, A., Matteucci, M.: A differentiable recurrent surface for asynchronous event-based data. In: European Conference on Computer Vision (ECCV), pp. 136–152. Springer (2020)
5. Carreira, J., Zisserman, A.: Quo vadis, action recognition? a new model and the kinetics dataset. In: Conference on Computer Vision and Pattern Recognition (CVPR), pp. 6299–6308 (2017)
6. Chen, Z., Wu, J., Hou, J., Li, L., Dong, W., Shi, G.: ECSNET: spatio-temporal feature learning for event camera. IEEE Trans. Circuit Syst. Video Technol. (TCSVT) **33**(2), 701–712 (2022)
7. Deng, Y., Chen, H., Li, Y.: MVF-NET: a multi-view fusion network for event-based object classification. IEEE Trans. Circuit Syst. Video Technol. (TCSVT) **32**(12), 8275–8284 (2021)
8. Deng, Y., Chen, H., Li, Y.: A dynamic GCN with cross-representation distillation for event-based learning. In: AAAI, vol. 38, pp. 1492–1500 (2024)
9. Deng, Y., Chen, H., Liu, H., Li, Y.: A voxel graph CNN for object classification with event cameras. In: IEEE Conference on Computer Vision and Pattern Recognition (CVPR), pp. 1172–1181 (2022)

10. Deng, Y., Li, Y., Chen, H.: AMAE: adaptive motion-agnostic encoder for event-based object classification. IEEE Robot. Autom. Lett. (RAL) **5**(3), 4596–4603 (2020)
11. Gao, Y., et al.: Action recognition and benchmark using event cameras. IEEE Trans. Pattern Anal. Mach. Intell. (TPAMI) (2023)
12. Gehrig, D., Loquercio, A., Derpanis, K.G., Scaramuzza, D.: End-to-end learning of representations for asynchronous event-based data. In: Conference on Computer Vision (ICCV), pp. 5633–5643 (2019)
13. Gu, A., Dao, T.: Mamba: linear–time sequence modeling with selective state spaces. arXiv:2312.00752 (2023)
14. Gu, A., Goel, K., Ré, C.: Efficiently modeling long sequences with structured state spaces. arXiv:2111.00396 (2021)
15. Li, H., Liu, H., Ji, X., Li, G., Shi, L.: Cifar10-DVS: an event-stream dataset for object classification. Front. Neurosci. **11**, 309 (2017)
16. Li, Y., Zhou, et al.: Graph-based asynchronous event processing for rapid object recognition. In: Conference on Computer Vision (ICCV), pp. 934–943 (2021)
17. Lichesteiner, P., Posch, C., Delbruck, T.: A 128× 128 120 DB 15μsec latency asynchronous temporal contrast vision sensro. IEEE J. Solid-State Circuits **43**(2), 566–576 (2008)
18. Liu, D., Wang, T., Sun, C.: Voxel-based multi-scale transformer network for event stream processing. IEEE Trans. Circuit Syst. Video Technol. (TCSVT) **34**(4), 2112–2124 (2024)
19. Liu, Q., Xing, D., Tang, H., Ma, D., Pan, G.: Event-based action recognition using motion information and spiking neural networks. In: IJCAI, pp. 1743–1749 (2021)
20. Liu, Y.: Vmamba: visual state space model. arXiv:2401.10166 (2024)
21. Ma, X., Qin, C., You, H., Ran, H., Fu, Y.: Rethinking network design and local geometry in point cloud: a simple residual MLP framework. arXiv:2202.07123 (2022)
22. Miao, S., et al.: Neuromorphic vision datasets for pedestrian detection, action recognition, and fall detection. Front. Neurorob. **13**, 38 (2019)
23. Orchard, G., Jayawant, A., Cohen, G.K., Thakor, N.: Converting static image datasets to spiking neuromorphic datasets using saccades. Front. Neurosci. **9**, 437 (2015)
24. Ren, H., Zhou, Y., Fu, H., Huang, Y., Xu, R., Cheng, B.: TTPOINT: a tensorized point cloud network for lightweight action recognition with event cameras. In: ACM Conference on Multimedia (ACM MM), pp. 8026–8034 (2023)
25. Ren, H.: ethinking efficient and effective point-based networks for event camera classification and regression: eventmamba. arXiv:2405.06116 (2024)
26. Schöne, M., Sushma, N.M., Zhuge, J., Mayr, C., Subramoney, A., Kappel, D.: Scalable event-by-event processing of neuromorphic sensory signals with deep state-space models. In: International Conference on Neuromorphic Systems, pp. 124–131 (2024)
27. Sekikawa, Y., Hara, K., Saito, H.: Eventnet: asynchronous recursive event processing. In: IEEE Conference on Computer Vision and Pattern Recognition (CVPR), pp. 3887–3896 (2019)
28. Sironi, A., Brambilla, M., Bourdis, N., Lagorce, X., Benosman, R.: Hats: histograms of averaged time surfaces for robust event-based object classification. In: IEEE Conference on Computer Vision and Pattern Recognition (CVPR), pp. 1731–1740 (2018)

29. Tran, D., Bourdev, L., Fergus, R., Torresani, L., Paluri, M.: Learning spatiotemporal features with 3d convolutional networks. In: IEEE Conference on Computer Vision (ICCV), pp. 4489–4497 (2015)
30. Wang, X., Wang, S., Shao, P., Jiang, B., Zhu, L., Tian, Y.: Event stream based human action recognition: a high-definition benchmark dataset and algorithms. arXiv:2408.09764 (2024)
31. Xie, B., Deng, Y., Shao, Z., Liu, H., Li, Y.: VMV-GCN: volumetric multi-view based graph CNN for event stream classification. IEEE Robot. Autom. Lett. (RAL) **7**(2), 1976–1983 (2022)
32. Xie, B., Deng, Y., Shao, Z., Xu, Q., Li, Y.: Event voxel set transformer for spatiotemporal representation learning on event streams. IEEE Trans. Circuit Syst. Video Technol. (TCSVT) **34**(12), 13427–13440 (2024)
33. Zhao, H., Jiang, L., Jia, J., Torr, P., Koltun, V.: Point transformer. In: IEEE Conference on Computer Vision (ICCV), pp. 16239–16248 (2021)
34. Zhu, A.Z., Yuan, L., Chaney, K., Daniilidis, K.: Unsupervised event-based learning of optical flow, depth, and egomotion. In: IEEE Conference on Computer Vision and Pattern Recognition (CVPR), pp. 989–997 (2019)

Formation Control with Prescribed Performance for Fully Actuated Vehicles Under Input Saturation

Meilin Lei[1], Zhechen Zhu[2(✉)], Yingnan Pan[2], and Wen Yang[2]

[1] College of Mathematical Sciences, Bohai University, Jinzhou 121013, Liaoning, China

[2] College of Control Science and Engineering, Bohai University, Jinzhou 121013, Liaoning, China

zhuzhechen66@163.com

Abstract. This paper investigates the adaptive control problem for vehicle platoon systems with input saturation. Under the framework of the high-order fully actuated system approach, a prescribed performance control strategy based on the backstepping technique is proposed. First, the designed performance function features an adjustable convergence rate and time-varying convergence bounds, which can better adapt to practical requirements in road environments. Second, the input saturation problem is addressed by utilizing a Nussbaum-type function. Theoretically, all signals in the closed-loop system are proven to be semi-globally uniformly ultimately bounded. Finally, simulation results are provided to verify the effectiveness of the proposed method.

Keywords: High-order fully actuated system approach · input saturation · prescribed performance control · vehicle platoon systems

1 Introduction

As a key technical direction in the field of traffic control, vehicle platoon control (VPC) has attracted significant attention from the academic community due to its role in improving road capacity, reducing energy consumption, and alleviating traffic pressure. However, in practical scenarios where vehicles require frequent acceleration or deceleration, the transient performance requirements for VPC systems become more stringent. Previous results [1,2] have attempted to apply prescribed performance control (PPC) to platoon systems to meet performance demands. Nevertheless, these results all adopt a design with constant convergence boundaries. When there are large initial spacing errors or strong external disturbances, this constant-boundary design may weaken the control effectiveness of PPC. Based on this, this paper employs a performance function with time-varying convergence boundaries to address the PPC design problem for platoon systems.

C. Li et al. (Eds.): ICNC 2025, CCIS 2946, pp. 260–269, 2026.
https://doi.org/10.1007/978-981-92-1599-7_22

It should be noted that most existing studies related to VPC rely on traditional first-order state space method. While this method has certain applicability in analyzing state responses, it has limitations in the controller design process and is difficult to meet the design requirements of high-order systems. To overcome this limitation, the high-order fully actuated (HOFA) system approach [3,4] was proposed. This method eliminates the need for model reduction operations and enables direct controller design for high-order systems [5,6], effectively simplifying the design steps. Since the longitudinal vehicle dynamics derived from Newton's second law are inherently described by third-order nonlinear differential equations, and the HOFA system approach is naturally compatible with the characteristics of high-order systems, it thus provides a new and effective way to solve the VPC problem.

2 Preliminaries and Problem Formulation

2.1 Vehicle Dynamics

Firstly, considering one leader and N followers, we define the model of i-th vehicle dynamic as follows [7]:

$$\begin{aligned} \dot{x}_{ip}(t) &= x_{iv}(t), \\ \dot{x}_{iv}(t) &= x_{ia}(t), \\ \dot{x}_{ia}(t) &= \frac{\mathrm{sat}\left(u_i(t)\right)}{m_i\tau_i} - \frac{x_{ia}(t)}{\tau_i} - \frac{c_i\left(x_{iv}^2 + 2\tau_i x_{iv}x_{ia}\right)}{m_i\tau_i} + d_i(t), \\ y_i(t) &= x_{ip}(t), \end{aligned} \tag{1}$$

where x_{ip}, x_{iv} and x_{ia} stand for the position, velocity and acceleration of i-th vehicle and $i = 1, 2, \ldots, N$. m_i is the vehicle mass, $\tau_i > 0$ denotes the engine time constant, c_i indicate the aerodynamic drag coefficient, $d_i(t)$ is the lumped disturbance and $y_i(t)$ represents the control output. In this paper, it is assumed that $|d_i(t)| < \bar{d}_i$ with $\bar{d}_i$ being an unknown constant. u_i stands for the throttle/brake input, which is affected by asymmetric saturation nonlinearity due to physical constraints, specifically defined as follows:

$$\mathrm{sat}(u_i(t)) = \begin{cases} u_{i,\max}, & u_i(t) > u_{i,\max} \\ u_i(t), & u_{i,\min} < u_i(t) < u_{i,\max} \\ u_{i,\min}, & u_i(t) < u_{i,\min} \end{cases} \tag{2}$$

where $u_{i,\max} > 0$ means the upper bound of $\mathrm{sat}(u_i(t))$, and $u_{i,\min}$ is the lower bound. To approximate the saturation function, a smooth function is defined

$$\varrho(u_i(t)) = \begin{cases} u_{i,\max}\tanh(\frac{u_i}{u_{i,\max}}), & u_i \geq 0 \\ u_{i,\min}\tanh(\frac{u_i}{u_{i,\min}}), & u_i < 0 \end{cases} \tag{3}$$

Furthermore, $\mathrm{sat}(u_i(t))$ can be written as $\mathrm{sat}(u_i(t)) = \varrho(u_i(t)) + d_i^c(u_i(t))$, where $d_i^c(u_i(t))$ is the approximation error and $|d_i^c(u_i(t))| \leq \max\{u_{i,m}(\tanh(1) - 1), u_{i,M}(1 - \tanh(1))\}$.

Using mean-value theorem, there is a constant $\phi_i \in (0,1)$ satisfying

$$\varrho(u_i(t)) = \frac{\partial \varrho(u_i(t))}{\partial u_i(t)} |_{u_i = u_i^{\phi_i}} u_i(t) = G_{i0}(u_i^{\phi_i}) u_i(t), \tag{4}$$

where $0 < \varrho_0 \leq G_{i0}(u_i^{\phi_i}) \leq 1$.

Denote $\bar{G}_i = \frac{G_{i0}(u_i^{\phi_i})}{(m_i \tau_i)}$ and $\bar{f}_i(x_{iv}, x_{ia}, t) = -\frac{x_{ia}(t)}{\tau_i} - \frac{c_i(x_{iv}^2 + 2\tau_i x_{iv} x_{ia})}{m_i \tau_i} + \frac{d_i^c(u_i(t))}{m_i \tau_i}$. It is straightforward to see that $\bar{G}_i$ is non-zero, and this condition conforms to the properties of the HOFA system. Then, the vehicle dynamics can be written as follows:

$$\begin{aligned} \dddot{x}_{ip}(t) &= \bar{G}_i u_i(t) + \bar{f}_i(\dot{x}_{ip}, \ddot{x}_{ip}, t) + d_i(t), \\ y_i(t) &= x_{ip}(t), \end{aligned} \tag{5}$$

where $\bar{G}_i \neq 0$, the requirement corresponding to the property of the HOFA system is fulfilled.

Lemma 1. *[8] For any $\mu > 0$, there exists a vector*

$$A^{0\sim1} = [A_0 \ A_1] \in \mathbb{R}^{1\times2}, \tag{6}$$

such that

$$Re\left[\lambda\left(\hat{\Phi}\left(A^{0\sim1}\right)\right)\right] < -\frac{\mu}{2}, \tag{7}$$

where

$$\hat{\Phi}\left(A^{0\sim1}\right) = \begin{bmatrix} 0 & 0 \\ -A_0 & -A \end{bmatrix} \in \mathbb{R}^{2\times2}. \tag{8}$$

Further, we can select a positive definite matrix $P \in \mathbb{R}^{2\times2}$ such that

$$\hat{\Phi}^{\mathrm{T}}\left(A^{0\sim1}\right) P + P\hat{\Phi}\left(A^{0\sim1}\right) \leq -\mu P. \tag{9}$$

2.2 Performance Function

To ensure the traffic flow stability of the system, the following exponential spacing policy is introduced

$$e_{ip}(t) = x_{i-1,p}(t) - x_{ip}(t) - \bar{\mathcal{D}}_i, \tag{10}$$

where $\bar{\mathcal{D}}_i = \mathcal{L}_i + \mathcal{D}_i + \frac{\lambda \dot{x}_{ip}^2(t)}{2\varsigma} + \beta\left(1 - e^{-\frac{1}{h}\dot{x}_{ip}^2(t)}\right)$. $\mathcal{L}_i$ represents the vehicle length, and $\mathcal{D}_i$ is the desired inter-vehicle distance. $\lambda > 0$ denotes the safety coefficient. ς

represents the absolute value of the maximum deceleration. β and h are positive constants.

In order to achieve the required accuracy of the spacing error within the specified time, an improved performance function is proposed

$$K(t) = \begin{cases} \ln\left(\frac{e^{K(0)} - 1}{(T_M - t_0)^\kappa}(T_M - t)^\kappa + 1\right) + \bar{a}, & 0 \le t < T_M \\ \bar{a}, & T_M \le t \end{cases}$$

where $\epsilon = \frac{K(0)-a}{K(0)}$, $\bar{a} = a + b_1 \tanh\left(b_2(\bar{\mathcal{D}}_{\max})^2\right)$ and $\kappa \ge 1$ is a constant. $T_M > 0$ is the settling time. a, b_1 and b_2 are positive constants, and $\bar{\mathcal{D}}_{\max} = \max\{\bar{\mathcal{D}}_1, \bar{\mathcal{D}}_2, \ldots, \bar{\mathcal{D}}_N\}$.

3 Controller Design

In this part, an adaptive controller based on HOFA system approach is proposed for a class of vehicle platoon system with input saturation.

Before the controller design, perform the following coordinate transformation:

$$\begin{aligned} z_{i1}(t) &= e_{ip}(t), \\ z_{i2}(t) &= \dot{x}_{ip}(t) - \alpha_i, \end{aligned} \tag{11}$$

where α_i is the virtual control signal.

Step 1. From (10) and (11), one has

$$\begin{aligned} \dot{z}_{i1} &= \dot{x}_{i-1,p}(t) - \dot{x}_{ip}(t) - \dot{\bar{\mathcal{D}}}_i \\ &= \dot{x}_{i-1,p} - z_{i2} - \alpha_i + F_i(x_{ip}, \dot{x}_{ip}, \ddot{x}_{ip}, \dot{x}_p), \end{aligned} \tag{12}$$

where $F_i(\cdot) = -\frac{\lambda}{\varsigma}\ddot{x}_{ip}\dot{x}_{ip} - \frac{\beta}{h}\ddot{x}_{ip}\dot{x}_{ip}e^{-\frac{\dot{x}_{ip}^2}{h}}$. The radial basis function neural networks (RBF NNs) are utilized as $F_i(\cdot) = \omega_i^{\mathrm{T}}\theta_{i1}(\cdot) + L_{i1}(\cdot)$ with $|L_{i1}(\cdot)| < \bar{L}_{i1}$ and $\psi_{i1} = \|\omega_i\|^2$. On the basis of (12), we have

$$\dot{z}_{i1} = \dot{x}_{i-1,p} - z_{i2} - \alpha_i + \omega_i^{\mathrm{T}}\theta_{i1} + L_{i1}. \tag{13}$$

Consequently, the following barrier Lyapunov function is constructed

$$V_{i1} = \frac{1}{2}\log\frac{K^2}{K^2 - z_{i1}^2} + \frac{1}{2}\tilde{\psi}_{i1}^2.$$

By taking the derivative of V_{i1}, one gets

$$\begin{aligned} \dot{V}_{i1} =& \frac{z_{i1}}{K^2 - z_{i1}^2}\left(\dot{z}_{i1} - \frac{\dot{K}}{K}z_{i1}\right) - \tilde{\psi}_{i1}\dot{\hat{\psi}}_{i1} \\ =& \frac{z_{i1}}{K^2 - z_{i1}^2}\Bigg(\dot{x}_{i-1,p} - z_{i2} - \alpha_i + \omega_i^{\mathrm{T}}\theta_{i1} + L_{i1} \\ & - \frac{\dot{K}}{K}z_{i1}\Bigg) - \tilde{\psi}_{i1}\dot{\hat{\psi}}_{i1}. \end{aligned}$$

Using Young's inequality, and the virtual controller α_i and the adaptive law $\dot{\hat{\psi}}_{i1}$ are given as

$$\alpha_i = c_{i1}z_{i1} + \dot{x}_{i-1,p} + \frac{1}{2\left(K^2 - z_{i1}^2\right)}\left(2z_{i1} + z_{i1}\hat{\psi}_{i1}\theta_{i1}^{\mathrm{T}}\theta_{i1}\right) - \frac{\dot{K}}{K}z_{i1}, \tag{14}$$

$$\dot{\hat{\psi}}_{i1} = \frac{z_{i1}^2\theta_{i1}^{\mathrm{T}}\theta_{i1}}{2\left(K^2 - z_{i1}^2\right)^2} - \delta_{i1}\hat{\psi}_{i1}, \tag{15}$$

where $c_{i1} > 0$ and $\delta_{i1} > 0$ are designed parameters.

Then, $\dot{V}_{i1}$ can be rewritten as

$$\dot{V}_{i1} \leq -\frac{c_{i1}z_{i1}^2}{K^2 - z_{i1}^2} - \frac{\delta_{i1}\tilde{\psi}_{i1}^2}{2} + \frac{z_{i2}^2}{2} + \frac{\delta_{i1}\psi_{i1}^2}{2} + \frac{1}{2} + \frac{\bar{L}_{i1}^2}{2}.$$

Step 2. The Lyapunov function is established as

$$V_{i2} = V_{i1} + \frac{1}{2}\left(z_{i2}^{(0\sim1)}\right)^{\mathrm{T}} P_i z_{i2}^{(0\sim1)} + \frac{1}{2}\tilde{\psi}_{i2}^2. \tag{16}$$

By differentiating (15), one obtains

$$\dot{V}_{i2} = \dot{V}_{i1} + \left(z_{i2}^{(0\sim1)}\right)^{\mathrm{T}} P_i\tilde{\Phi}_i(0)\, z_{i2}^{(0\sim1)} \tag{17}$$

$$+ \left(z_{i2}^{(0\sim1)}\right)^{\mathrm{T}} P_{iL}\left(G_i u_i + F_{i2}\right) - \tilde{\psi}_{i2}\dot{\hat{\psi}}_{i2}, \tag{18}$$

where $P_{iL} = P_i^{\mathrm{T}} \in \mathbb{R}^{2\times1}$ and $F_{i2} = \bar{f}_i + d_i - \dot{\alpha}_i$. Similarly, the nonlinear function F_{i2} can also be approximated by RBF NNs as $F_{i2}(\cdot) = \omega_{i2}^{\mathrm{T}}\theta_{i2}(\cdot) + L_{i2}(\cdot)$ with $|L_{i2}(\cdot)| < \bar{L}_{i2}$ and $\psi_{i2} = \|\omega_{i2}\|^2$.

By using Young's inequality, the $\dot{V}_{i2}$ can be expressed as

$$\begin{aligned}\dot{V}_{i2} = & -\frac{c_{i1}z_{i1}^2}{K^2 - z_{i1}^2} - \frac{\delta_{i1}\tilde{\psi}_{i1}^2}{2} + \left(z_{i2}^{(0\sim1)}\right)^{\mathrm{T}} P_i\hat{\Phi}_i(0)\, z_{i2}^{(0\sim1)} \\ & + \left(z_{i2}^{(0\sim1)}\right)^{\mathrm{T}} P_{iL}\Bigg(G_i u_i + \frac{\hat{\psi}_{i2}\theta_{i2}\theta_{i2}^{\mathrm{T}}P_{iL}^{\mathrm{T}}z_{i2}^{(0\sim1)}}{2} \\ & + \frac{P_{iL}^{\mathrm{T}}}{2}z_{i2}^{(0\sim1)}\Bigg) + \frac{z_{i2}^2}{2} + \frac{\delta_{i1}\psi_{i1}^2}{2} + 1 + \frac{\bar{L}_{i1}^2}{2} + \frac{\bar{L}_{i2}^2}{2} \\ & + \tilde{\psi}_{i2}\left(\frac{\left(z_{i2}^{(0\sim1)}\right)^{\mathrm{T}} P_{iL}\theta_{i2}\theta_{i2}^{\mathrm{T}}P_{iL}^{\mathrm{T}}z_{i2}^{(0\sim1)}}{2} - \dot{\hat{\psi}}_{i2}\right).\end{aligned} \tag{19}$$

The Nussbaum-type function is selected as $\mathcal{N}(\xi_i) = \xi_i^2\cos\left(\frac{\pi}{2}\xi_i\right)$, then the actual controller u_i and the adaptive law $\dot{\hat{\psi}}_{i2}$ can be constructed as follows:

$$u_i = \mathcal{N}(\xi_i)\bar{u}_i, \tag{20}$$

$$\dot{\hat{\psi}}_{i2} = \frac{\left(z_{i2}^{(0\sim1)}\right)^{\mathrm{T}} P_{iL}\theta_{i2}\theta_{i2}^{\mathrm{T}} P_{iL}^{\mathrm{T}} z_{i2}^{(0\sim1)}}{2} - \delta_{i2}\hat{\psi}_{i2}, \tag{21}$$

where

$$\dot{\xi}_i = -\left(z_{i2}^{(0\sim1)}\right)^{\mathrm{T}} P_{iL}\bar{u}_i,$$

$$\bar{u}_i = A_i^{(0\sim1)} z_{i2}^{(0\sim1)} + \frac{\hat{\psi}_{i2}\theta_{i2}\theta_{i2}^{\mathrm{T}} P_{iL}^{\mathrm{T}} z_{i2}^{(0\sim1)}}{2} + \frac{P_{iL}^{\mathrm{T}} z_{i2}^{(0\sim1)}}{2},$$

and $\delta_{i2} > 0$ is a constant.

On the basis of (20)-(21), the time-derivative of $\dot{V}_{i2}$ can be depicted as

$$\begin{aligned} \dot{V}_{i2} \leq & -\frac{c_{i1} z_{i1}^2}{K^2 - z_{i1}^2} - \mu\left(z_{i2}^{(0\sim1)}\right)^{\mathrm{T}} P_i^{(0\sim1)} z_{i2}^{(0\sim1)} + \frac{z_{i2}^2}{2} \\ & - \frac{\delta_{i1}\tilde{\psi}_{i1}^2}{2} - \frac{\delta_{i2}\tilde{\psi}_{i2}^2}{2} + \left(\bar{G}_i\mathcal{N}(\xi_i) + 1\right)\dot{\xi}_i + \frac{\delta_{i1}\psi_{i1}^2}{2} \\ & + \frac{\delta_{i2}\psi_{i2}^2}{2} + 1 + \frac{\bar{L}_{i1}^2}{2} + \frac{\bar{L}_{i2}^2}{2}. \end{aligned} \tag{22}$$

4 Stability Analysis

Theorem 1. *For the vehicle platoon system (1), the virtual controller is shown in (14), the adaptive laws are designed according to (15) and (21), and the actual controller is developed as in (20), which ensure that all the signals of the system are semi-globally ultimately uniformly bounded.*

Proof. Let $V = V_{i2}$ be defined as the total Lyapunov function. By computing the time derivative of V, we obtain:

$$\begin{aligned} \dot{V} \leq & \sum_{i=1}^{N}\Bigg(-\frac{c_{i1} z_{i1}^2}{K^2 - z_{i1}^2} - \mu\left(z_{i2}^{(0\sim1)}\right)^{\mathrm{T}} P_i z_{i2}^{(0\sim1)} + \frac{z_{i2}^2}{2} \\ & - \sum_{j=1}^{2}\frac{\delta_{ij}}{2}\tilde{\psi}_{ij}^2 + \left(G_i\mathcal{N}\left(\xi_i\right) + 1\right)\dot{\xi}_i\Bigg) + \Delta \\ \leq & -CV + \sum_{i=1}^{N}\left(\bar{G}_i\mathcal{N}(\xi_i) + 1\right)\dot{\xi}_i + \Delta, \end{aligned} \tag{23}$$

where $C = \min\{2c_{i1}, 2\mu - \frac{1}{\lambda_{\max}(P_i)}, \delta_{i1}, \delta_{i2}\}$ and $\Delta = \frac{\delta_{i1}\psi_{i1}^2}{2} + \frac{\delta_{i2}\psi_{i2}^2}{2} + 1 + \frac{\bar{L}_{i1}^2}{2} + \frac{\bar{L}_{i2}^2}{2}$.

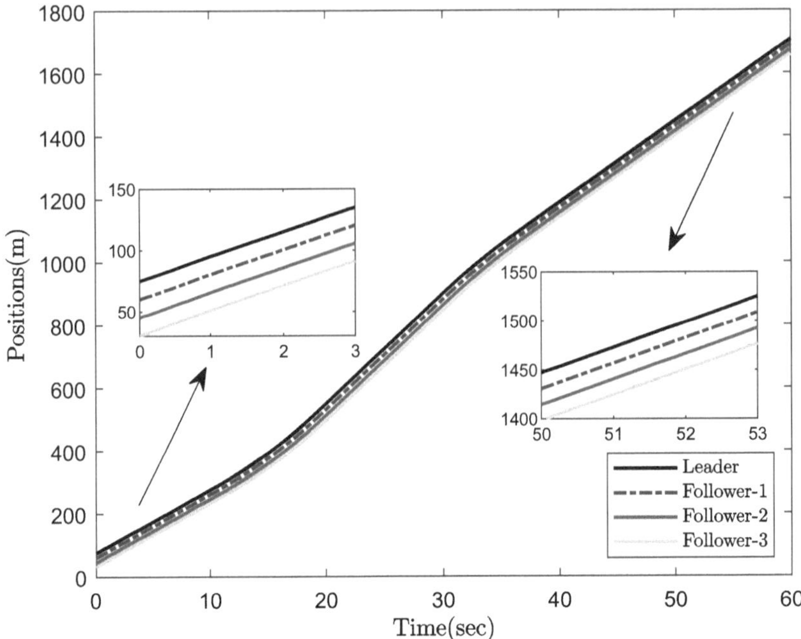

Fig. 1. The position for platoon.

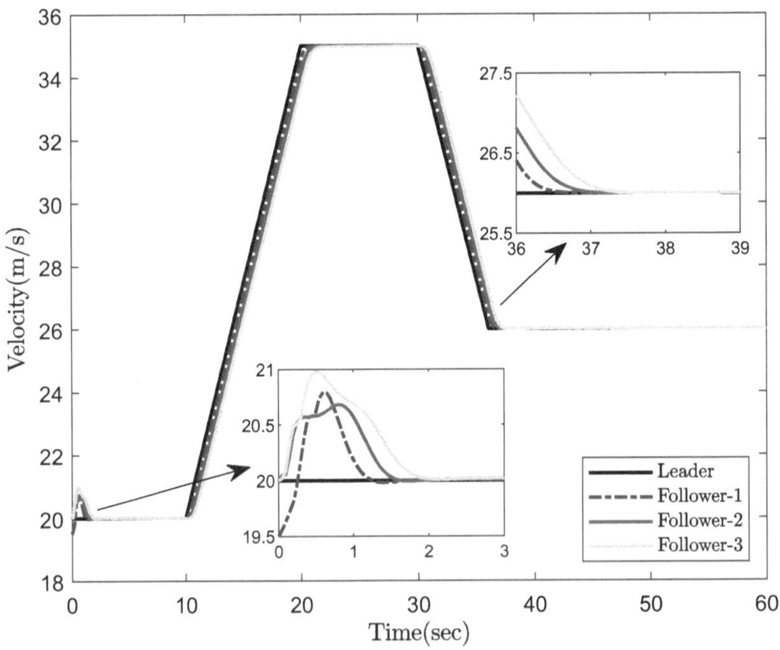

Fig. 2. The velocity for vehicle platoon.

A further conclusion can be drawn that

$$0 \leq V(t) \leq \frac{\Delta}{C} + \left(V(0) - \frac{\Delta}{C}\right) e^{-Ct} + e^{-Ct} \int_0^t \sum_{i=1}^{N} \left(\bar{G}_i \mathcal{N}(\xi_i) + 1\right) \dot{\xi}_i e^{Cv} dv. \tag{24}$$

Therefore, the proof is completed.

5 Simulation Results

For a vehicular platoon consisting of a leader and three followers. The system parameters are selected as $m_i = 1505\,\text{kg}$, $c_i = 0.924$, $\tau_i = 0.51\,\text{s}$, $\mathcal{L}_i = 4.1\,\text{m}$, $\mathcal{D}_i = 6.2\,\text{m}$, $\lambda = 0.1$, $\varsigma = 12.9\,\text{m/s}^2$, $\beta = 5.2$ and $h = 19.9$. The other related parameters are shown as $u_{i,max} = 35$, $u_{i,min} = -30$, $\phi_i = 0.5$, $K(0) = 0.6$, $\kappa = 1.7$, $T_M = 1.5\,\text{s}$, $a = 0.15$, $b_1 = 0.75$, $b_2 = 0.01$, $c_{i1} = 0.1$, $\delta_{i1} = \delta_{i2} = 0.6$, $A_i^{(0\sim1)} = [10, 20]$ and $P_{iL} = [0.0050\ 0.0275]^{\text{T}}$, and the initial values are chosen as $[x_{0p}(0), x_{1p}(0), x_{2p}(0), x_{3p}(0)] = [75, 60, 45, 30]$, $[\dot{x}_{0p}(0), \dot{x}_{1p}(0), \dot{x}_{2p}(0), \dot{x}_{3p}(0)] = [20, 19.5, 20, 20]$ and $[\ddot{x}_{0p}(0), \ddot{x}_{1p}(0), \ddot{x}_{2p}(0), \ddot{x}_{3p}(0)] = [0, 1, 0, 0]$. $\hat{\psi}_{i1} = \hat{\psi}_{i2} = 0.1$ and $\xi_i = 2$, respectively. The external disturbance $d_i = e^{-0.1t}\sin(t)$ and the reference velocity of the leader is selected as

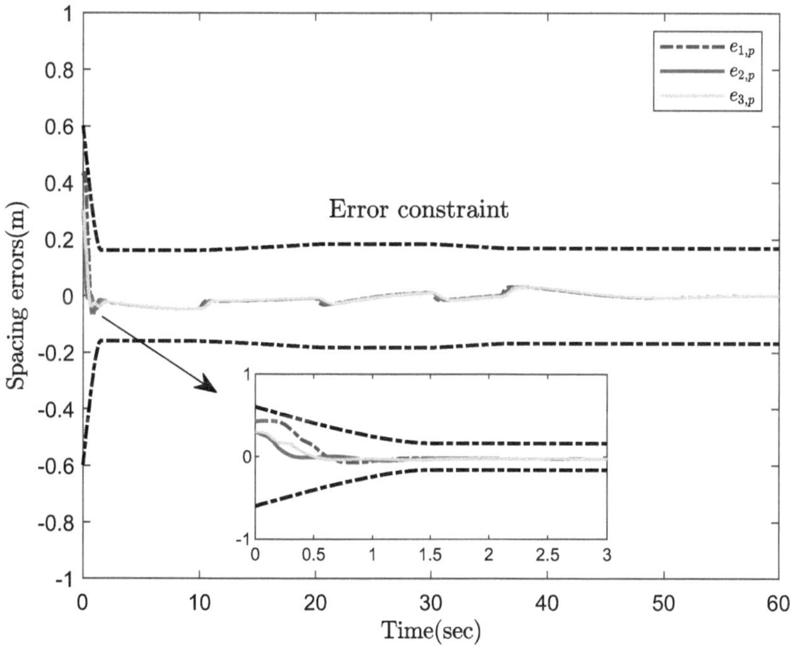

Fig. 3. The spacing error for platoon.

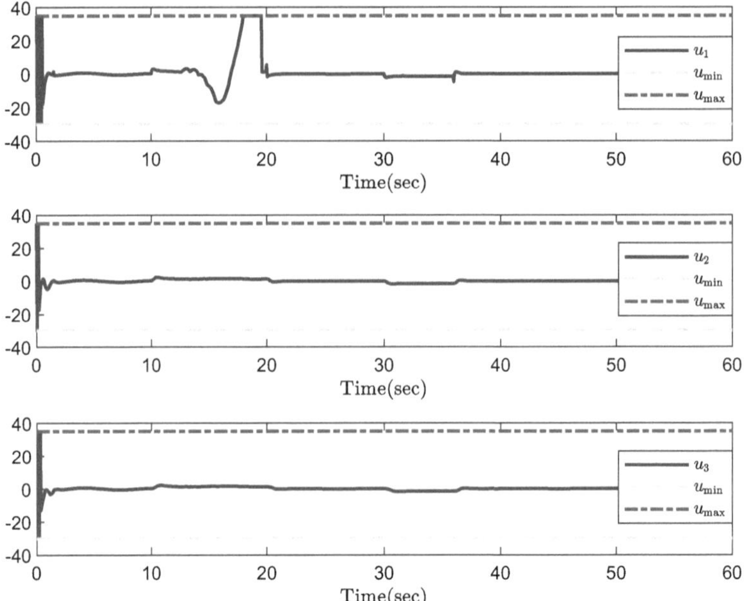

Fig. 4. The control input for platoon.

$$x_{0v} = \begin{cases} 20\,\mathrm{m/s}, & 0\,\mathrm{s} \leq t < 10\,\mathrm{s} \\ (1.5t + 5)\,\mathrm{m/s}, & 10\,\mathrm{s} \leq t < 20\,\mathrm{s} \\ 35\,\mathrm{m/s}, & 20\,\mathrm{s} \leq t < 30\,\mathrm{s} \\ (80 - 1.5t)\,\mathrm{m/s}, & 30\,\mathrm{s} \leq t < 40\,\mathrm{s} \\ 26\,\mathrm{m/s}, & \text{otherwise} \end{cases} \tag{25}$$

Simulation results are shown in Figs. 1, 2, 3 and 4. Figures 1 and 2 demonstrate that the followers maintain well tracking of the leader trajectory and velocity. For the construction of PPC, the spacing errors are limited to a prescribed bounded region within finite time T_M in Fig. 3. Figure 4 shows the trajectory of the control input signal u_i.

6 Conclusion

This paper proposed a novel PPC scheme for vehicle platoon systems subject to input saturation, which was developed using the HOFA system approach. By incorporating an improved performance function, the spacing error converges within a predefined time to an adaptive boundary. RBF NNs were employed to address unknown nonlinear dynamics in the design process. The stability of the resulting closed-loop system was rigorously established via Lyapunov analysis.

Simulation results further demonstrated the effectiveness of the proposed control scheme.

Acknowledgments. This work was partially supported by the Basic Research Project of the Educational Department of Liaoning Province (LJ212410167043, LJ232410167028), the Revitalization of Liaoning Talents Program (XLYC2203201), the National Natural Science Foundation of China (62473060) and the Natural Science Foundation of Liaoning Province (2025-YQ-18, 2025-BS-0825).

Disclosure of Interests. The authors have no competing interests to declare that are relevant to the content of this article.

References

1. Guo, G., Li, D.D.: Adaptive sliding mode control of vehicular platoons with prescribed tracking performance. IEEE Trans. Veh. Technol. **68**(8), 7511–7520 (2019)
2. Qi, H.N., Cao, L., Ren, H.R., Zhao, M.: Fixed-time NN-based adaptive fault-tolerant control for heterogeneous vehicular platoon system with improved exponential spacing policy. Commun. Nonlinear Sci. Numer. Simul. **141**, 108454 (2025)
3. Duan, G.R.: High-order fully actuated system approaches: part I. Models and basic procedure. Int. J. Syst. Sci. **52**(2), 422–435 (2021)
4. Duan, G.R.: High-order system approaches: I. fully-actuated systems and parametric designs. Acta Autom. Sinica **46**(7), 1333–1345 (2020)
5. Liu, Y., Chen, X.B., Mei, Y.F., Wu, Y.L.: Observer-based boundary control for an asymmetric output-constrained flexible robotic manipulator. Sci. China Inf. Sci. **65**(3), 139203 (2022)
6. Gao, Y., Sun, W., Xie, X.P.: Adaptive fuzzy prescribed-time control of high-order nonlinear systems with actuator faults. Inf. Sci. **667**, 120484 (2024)
7. Li, D.D., Guo, G.: Prescribed performance concurrent control of connected vehicles with nonlinear third-order dynamics. IEEE Trans. Veh. Technol. **69**(12), 14793–14802 (2020)
8. Duan, G.R.: High-order fully actuated system approaches: part V. robust adaptive control. Int. J. Syst. Sci. **52**(10), 2129–2143 (2021)

A Novel Memristor-Based Analog Multiplier

Zaiyang Tao[1(✉)], Le Yang[1], Zhanhui Jiang[2], and Xinrui Zhang[1]

[1] School of Electrical and Information Engineering, Wuhan Institute of Technology, Wuhan 430000, China
1691435899@qq.com

[2] Chemical Machinery Equipment Manufacturing and Installation Co., Ltd., Hubei Yihua Group, Yichang 443000, China

Abstract. Most existing analog multipliers use current or voltage as input variables. This paper innovatively proposes a memristor-based analog multiplier, which uses voltage as one input variable and memristance as the other input variable, with a relatively simple circuit structure and a relatively large input-output range. The proposed memristor-based analog multiplier is mainly composed of a memristor control circuit that changes the memristance and a multiplication operation circuit that builds the multiplication relationship between the input voltage and the memristance. In implementing the multiplication relationship, the input voltage is transferred as the partial by the resistor R1 at the input terminal. According to the principle of the Cascode current mirror, the input terminal current is equal to the output terminal current, thereby multiplying the current of output terminal by the memristance can achieve the multiplication of input voltage and memristance. DC analysis and PSPICE simulation indicate that the proposed memristor-based analog multiplier exhibits relatively superior performance.

Keywords: Memristor-based circuit design · Memristor-based analog multiplier · Cascode mirror current source · Analog multiplier

1 Introduction

The memristor possesses advantages such as small size, low energy consumption, easy integration, and non-volatility [9,18]. It can be applied in fields such as storage circuits and neural networks, etc. [10,21]. Furthermore, memristor-based circuits can simulate the biological features [5,7,17]. Additionally, combining memristors with traditional circuits can increase circuit advantages. For example, a continuously adjustable mirrored current source circuit is presented by combining the memristor with the Wilson current mirror circuit [4]. A memristor-based A/D converter with strong anti-interference capability and high resolution is proposed in [23]. In this paper, memristors are combined with analog multipliers to design a novel memristor-based analog multiplier circuit.

C. Li et al. (Eds.): ICNC 2025, CCIS 2946, pp. 270–280, 2026.
https://doi.org/10.1007/978-981-92-1599-7_23

As a fundamental operational unit, the analog multiplier can be applied in the fields of communication systems, signal processing, and neuromorphic circuit design, etc. [13,14]. Early implementations of the analog multiplier faced notable challenges in terms of circuit scale and resource consumption In 1968, B. Gilbert designed a double-signal analog multiplier [6], which requires a large number of logic gates and flip-flops to perform multiplication. Based on the MOS version of Gilbert's six-transistor cell (GSTC), a circuit for analog multiplication capable of handling larger input voltages is designed in [2]. Two types of analog multipliers, one using 12 MOS transistors with 2 resistors and the other using only four MOS transistors are proposed in [16,20]. A unified generation of multiplier architecture which primarily composed of MOS transistors are proposed in [8]. By utilizing the operating characteristics of MOS transistors to establish a multiplication relationship, a circuit structure with a dual translinear-loop configuration is employed simultaneously in [12,15]. An analog multiplier with larger input-output range but relatively complex structure is proposed in [3]. Recently, in the research on current-mode analog multipliers, a new type of current-mode analog multiplier using CMOS technology is presented in [11]. Based on current-mode analog building blocks, an analog multiplier operating at low supply voltage with high linearity is designed in [1]. A multi-input current-mode analog multiplier is proposed in [19], which consists of current splitter and a one-quad multiplier module. However, the range of these circuits' input and output signals is relatively small and the operation process is relatively complex. Moreover, to the best of the author's knowledge, there is currently no circuit that uses memristance as input to an analog multiplier.

Compared to traditional analog multipliers, the novel memristor-based analog multiplier has the following advantages:

(1) The memristance is used as one of the input variable of the analog multiplier, the multiplication relationship between input voltage, input memristance, and output voltage are established.
(2) The complete circuit is mainly composed of 10 MOS transistors, the circuit structure is simple relatively.
(3) The range of input voltage and input memristance is large relatively.

The rest content of this paper is as follow: Sect. 2 introduces the memristor control circuit. Section 3 details the mathematical relationships satisfied by the analog multiplier. Section 4 proposes the complete circuit structure. Section 5 provides an analysis of the DC characteristics of the complete circuit. Section 6 summarizes the work of this paper.

2 Memristor Control Circuit

This paper uses the Ag/AgInSbTe/Ta (AIST) memristor model, which controlled by positive and negative threshold voltages [24]. The memristance variation of the memristor depends on the applied voltage and time.

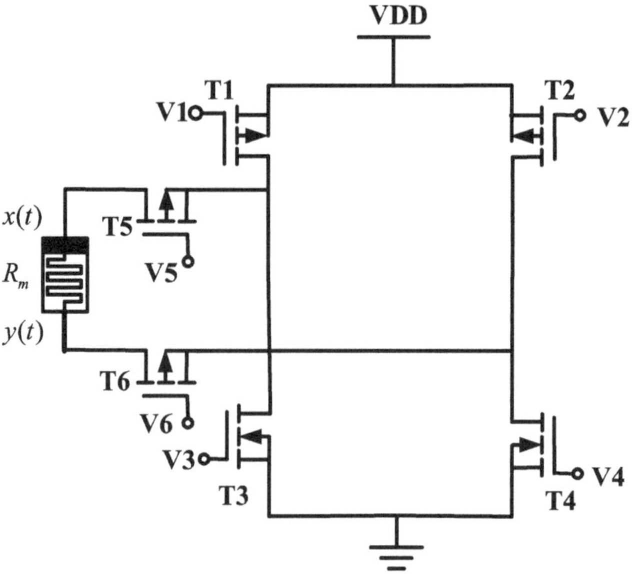

Fig. 1. Memristor control circuit.

Based on the characteristics of the memristor, this paper uses the memristance as a multiplication input variable, the memristor control circuit is employed [22]. The memristor control circuit is shown in Fig. 1, which consists of six MOS transistors and the memristor Rm. T1, T2, T5 and T6 are PMOS transistors, their threshold voltage is -1.5V. T3 and T4 are NMOS transistors, the threshold voltage is 1.5V. VDD=15V, which is the power supply. The MOS transistors here are in the on-off state, with the gate voltage serving as the control voltage. When the NMOS transistor is at on-state, the gate-source voltage satisfies $V_{GS} > V_{TN}$, it can be seen as a conductive wire. When the NMOS transistor work at off-state, the gate voltage satisfies $V_{GS} < V_{TN}$, it can be seen as a tremendous resistor. The control of PMOS transistors is similar to NMOS transistors. By changing the values of the gate voltages V_1, V_2, V_3, V_4, V_5 and V_6, the memristance of the memristor can be controlled to increase, decrease, or remain constant over time.

The control process of the circuit is as follows:

When T1, T4, T5 and T6 work at on-state, T2 and T3 are at off-state. The direction of the current flow from x(t) to y(t). At this time, the voltage of the memristor is greater than the positive threshold voltage $V_{th+} = 12V$, the memristance decreases with time. The mathematical equation that the change in memristance satisfies is

$$v(t)ln(R(t)) - i_0R(t) = (R_{on} - R_{off})ki_{off}t + (v(t)ln(R_s) - i_0R_s) \tag{1}$$

Where $i_0 = 1e{-}3$, $i_{off} = 1e{-}5$ and $R_s = 4k\Omega$. A comparison of results obtained from PSPICE simulations and those calculated using Eq. (3) is presented in Fig. 2(a).

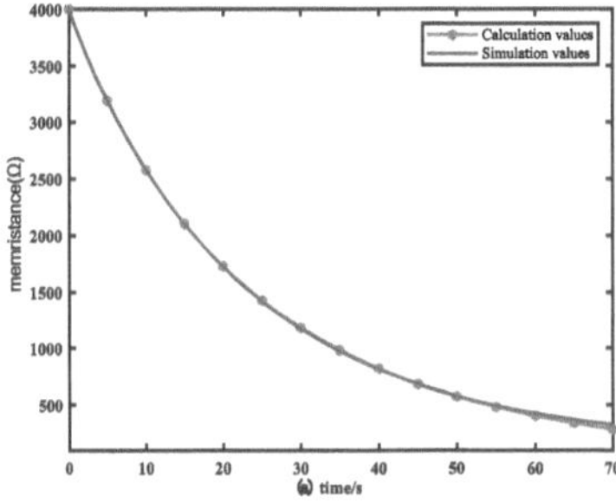

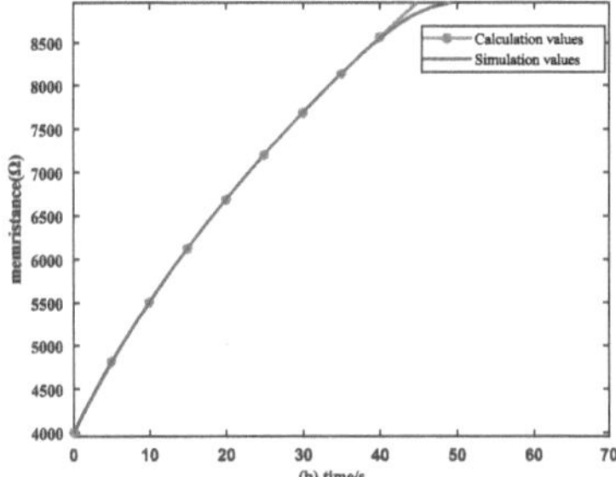

Fig. 2. Simulation results of the memristor control circuit: (a) the memristance decreased; (b) the memristance increased;

Similarly, when T1 and T4 are at off-state, T2, T3, T5 and T6 work at on-state. The current flows from y(t) to x(t). The voltage of the memristor is less than the negative threshold voltage $V_{th-} = -0.5V$, and the memristance increases with time according to the mathematical equation:

$$R(t) = (2(R_{on} - R_{off})k\frac{v(t)}{i_{on}}t + R_s^2)^{1/2}. \tag{2}$$

Figure 2(b) shows a comparison between the PSPICE simulation values and the calculated values from Eq. (4).

When T1–T6 are all off-state. The voltage of the memristor is 0, and no current flows through the memristor, so the memristance remains unchanged.

From the simulation results, the simulation values and the calculation values have a high fit degree. Therefore, the required memristance can be obtained through Eqs. (1) and (2), and the memristor control circuit mentioned above can be used to change the memristance.

3 Multiplication Circuit

3.1 Principle of Multiplication Operation Circuit

This paper utilizes the principle of the Cascode current mirror to establish a multiplication relationship between the voltage at the input terminal and the memristance at the output terminal. The multiplication operation circuit is shown in Fig. 3. Four NMOS transistors, T7–T10, form the Cascode current mirror, which have identical characteristics and operate in the saturation region. $VDD2 = 2V_{GS} = 3V$, R1 = 10KΩ, V_{in} and V_{out} denote the input voltage and output voltage of the multiplication operation circuit, respectively. $VDD1 = 10V$, Rm is the memristor, R2 = R3 = 500Ω.

Due to the characteristic of the current mirror circuit where the current at the input terminal is equal to the current at the output terminal. Thus, the current at the output terminal satisfies the following mathematical equation:

$$I_{Rm} = I_D = \frac{V_{in}}{R1}. \tag{3}$$

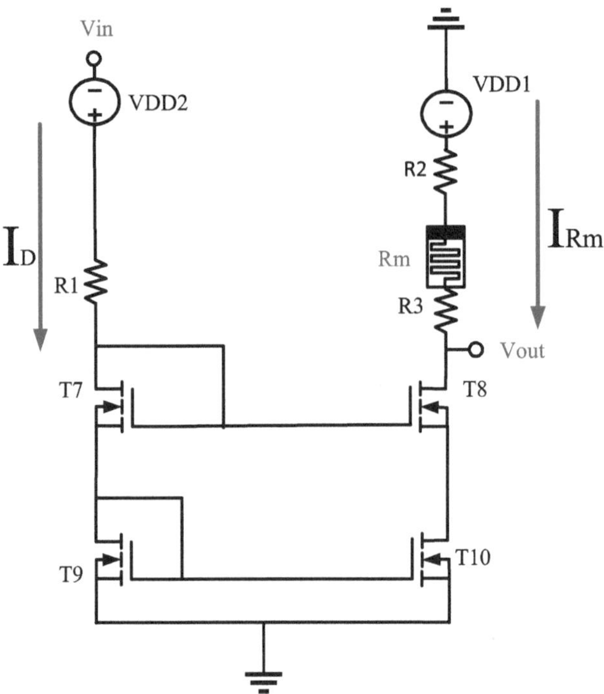

Fig. 3. Multiplication operation circuit: V_{in} is the one input variable, R_m is the other input variable and V_{out} is the output voltage.

In the voltage at the output terminal of the Cascode current mirror, When Rm is fixed, the voltage of the memristor is

$$V_m = R_m.I_{Rm} = \frac{V_{in}}{R1}.R_m. \tag{4}$$

The voltage of fixed resistors R2 and R3 is

$$V_R = (R2 + R3).I_{Rm} = \frac{V_{in}}{R1}.(R2 + R3). \tag{5}$$

So the V_{out} is

$$V_{out} = VDD1 - V_m - V_R = VDD1 - \frac{V_{in}}{R1}.(R_m + R2 + R3), \tag{6}$$

where the voltage V_{in} is one input variable, R_m is the other input variable and output voltage V_{out} is the result of the multiplication operation. They satisfy a linear relationship.

3.2 Simulation of Multiplication Operation Circuit

The PSPICE is used to perform a DC sweep of the input voltage to verify the linear relationship between the input voltage and the output voltage. $R_m = 4k\Omega$, according to the specified data, the Eq. (6) can be simplified as

$$V_{out} = 10 - \frac{V_{in}}{2}. \tag{7}$$

Output voltages are fitted based on PSPICE circuit simulation data and Eq. (7), and the corresponding results are presented in Fig. 4. Here, the red line denotes the simulation results, while the blue line denotes the calculation results from Eq. (7). It can be seen from the figure that the data fitting between the two is relatively high and as the input voltage V_{in} increases from 0 to 10V, the output voltage V_{out} decreases from 10V to 0V. The maximum relative error is 0.248%. Therefore, the mathematical relationship can be used to construct multiplication operation circuits.

4 Complete Circuit

4.1 Principle of Complete Circuit

The novel memristor-based analog multiplier as shown in Fig. 5. The input terminal V_{in} of the cascode current mirror in the multiplication operation circuit is used as one input variable for the analog multiplier, and the memristance R_m at the output terminal is used as the other input variable, whose value is

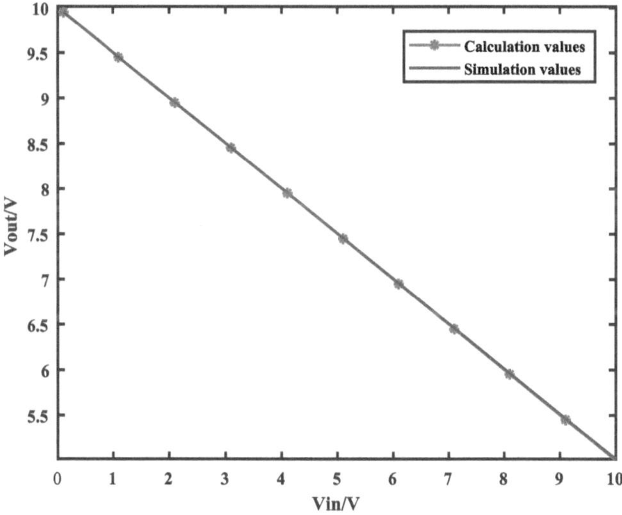

Fig. 4. Relationship between input voltage and output voltage.

adjusted by the memristor control circuit. According to the quantity relationship of Eq. (6) yields the output voltage.

The specific implementation steps of the complete circuit are as follows:

Step 1: First, the memristor control circuit is connected to the multiplication operation circuit in the manner shown in Fig. 5. In order to prevent the adjustment of the memristance from being affected by the output terminal voltage VDD1 of the cascode current mirror, VDD1 is set to 0V. Then, by changing the voltage of MOS transistors V1-V6, the memristance can be controlled to increase, decrease, or remain unchanged.

For example, when the required memristance is greater than the initial memristance, V2, V4, V5 and V6 are set to values less than the threshold voltage. V1 and V3 are set to values greater than the threshold voltage. T2, T3, T5 and T6 work at on-state, T1 and T4 are at off-state, causing the current to flow in reverse through the memristor, and R_m increases with time. Then, according to Eq. (2), the time t at which the memristance needs to change is determined. Changing T1-T6 work at off-state at t-th second to prevent further changes in the memristance. Similarly, when the required memristance is less than the initial value, adjusting the switching state of the T1-T6 to reverse the memristance of R_m.

Step 2: Set VDD1 to 10V. The input V_{in} is multiplied by the adjusted memristance R_m to obtain the output voltage V_{out} according to Eq. (6).

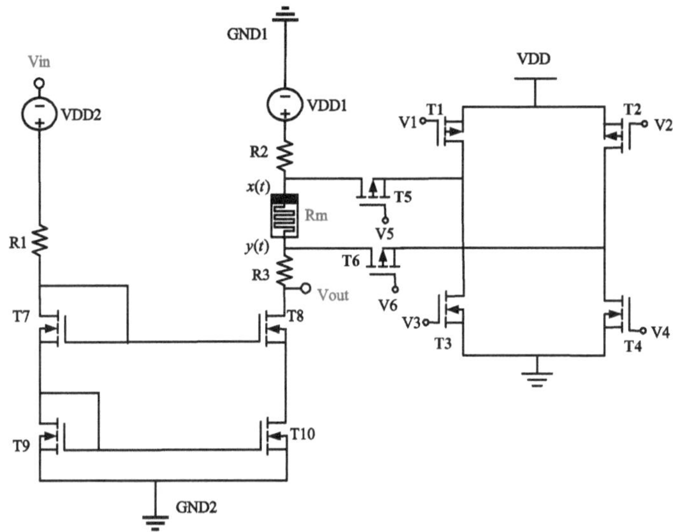

Fig. 5. Overall circuit of analog multiplier.

4.2 Simulation of Complete Circuit

To validate the feasibility of the complete circuit and its capability to multiply input voltages using different memristance, PSPICE is used to simulate the circuit. Two simulation scenarios are set up, with initial values either greater than or less than the required memristance. When $V_{in} = 5V$, R_m is a variable value, substituting into Eq. (6), getting the equation is

$$V_{out} = 10 - \frac{1}{2k}.(Rm + 1000). \tag{8}$$

According to the working principle of the complete circuit, the voltage sources VDD1 and V1–V6 are configured as time-varying voltage sources during simulation. This configuration ensures two key conditions: First, when the memristor control circuit controls the memristor by adjusting its resistance R_m, VDD1 is set to 0. Second, during the operation of the multiplication circuit, all MOS transistors T1–T6 remain in the off-state. In the simulation, the memristor control circuit adjusts the memristance of the controlled memristor every 3 s, while the multiplication circuit operates for 1 s per cycle. The total simulation duration is 81 s, and the corresponding simulation results are presented in Fig. 6.

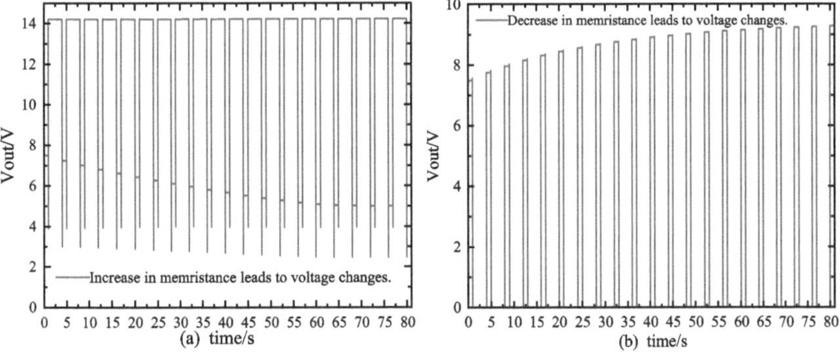

Fig. 6. Output voltage changes caused by different input memristance:(a) Increase in memristance leads to voltage changes; (b) Decrease in memristance leads to voltage changes.

The simulation results in Fig. 6(a) and (b) show that as the memristance of the input memristor increases, the output voltage decreases gradually. Conversely, as the memristance decreases, the output voltage increases gradually. Subsequently, the output voltage values obtained in these two scenarios and their corresponding input memristance is linearly fitted. The fitting results are compared with the calculated results from Eq. (8), as presented in Fig. 7.

It can be observed from Fig. 7 that the simulated values are in high agreement with the calculated values, with a maximum error of less than 0.1%. Therefore, this part of the simulation validates the effectiveness of the memristor-based analog multiplier in working steps and the accuracy of the output voltage.

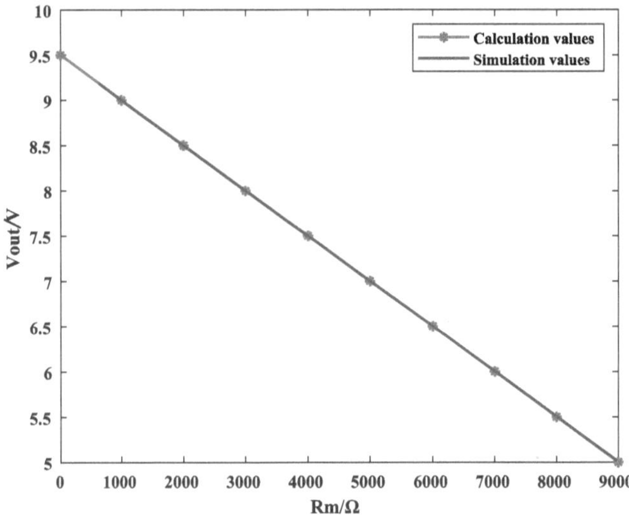

Fig. 7. The simulation results of relationship between input memristance and output voltage.

5 Discussion

5.1 DC Analysis

The section conducts simulation analysis on the DC characteristic of the memristor-based analog multiplier circuit. The simulation results in Fig. 8

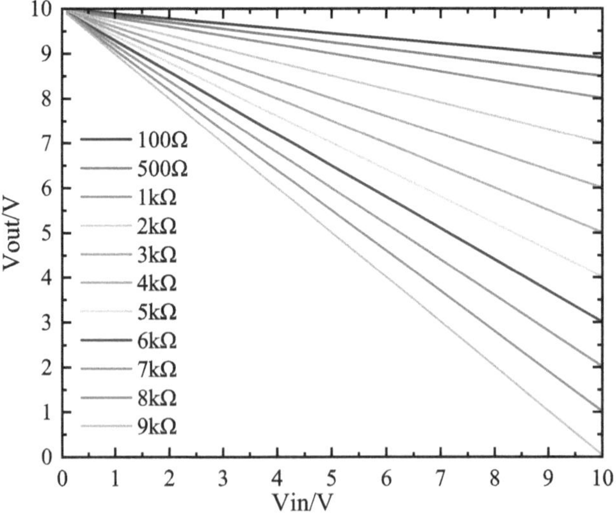

Fig. 8. DC analysis

demonstrate that when both the input voltage and input memristance vary simultaneously, the output still exhibits a linear variation, which verifies the stability and effectiveness of the circuit.

6 Conclusion

This paper presents a novel memristor-based analog multiplier, which has the simple circuit structure, the lower linear error and larger input range relatively. Firstly, the memristor model is introduced, and the control method of the memristor control circuit is verified using PSPICE. Then, the implementation principle of the multiplication operation circuit, the mathematical relationship between the input and output signals, and the verification simulations of PSPICE are given. Additionally, the working principle, steps, and simulation of the complete circuit are presented. Finally, The DC characteristic of the circuit is analyzed. Based on the PSPICE simulations above, it can be concluded that the error of the input voltage does not exceed 0.5% under different input voltages or input memristance, indicating the high accuracy of the output voltage.

Acknowledgments. This work was supported by the National Natural Science Foundation of China under Grants 62106181, Graduate Innovative Fund of Wuhan Institute of Technology, China CX2024571.

Disclosure of Interests. The authors declare no competing interests relevant to the content of this article.

References

1. Agrawal, D., Maheshwari, S.: High performance current-mode four quadrant analog multiplier circuit. In: IMPACT 2022, pp. 1–4. IEEE (2022)
2. Babanezhad, J.N., Temes, G.C.: A 20-v four-quadrant cmos analog multiplier. IEEE J Solid-State Circ. **20**(6), 1158–1168 (1985)
3. Cheng, M., Yang, L., Ding, Z.: Circuit design of a novel analog multiplier. In: ICCSSE 2023, pp. 386–390. IEEE (2023)
4. Cheng, M., Yang, L., Ding, Z.: An improved memristive current mirror circuit for continuous adjustable current output. AEU-Int. J. Electron. Commun. 154765 (2023)
5. Dou, G., Guo, W., Kong, L.: Operant conditioning neuromorphic circuit with addictiveness and time memory for automatic learning. IEEE Trans. Biomed. Circuits Syst (2024)
6. Gilbert, B.: A precise four-quadrant multiplier with subnanosecond response. IEEE J Solid-State Circ. **3**(4), 365–373 (1968)
7. Guo, M., Kong, L., Dou, G.: Neuromorphic circuit of classical and operant conditioning based on tunable neural circuitry motifs. IEEE Trans. Circuits Syst. I: Regul. Pap. (2024)
8. Han, G., Sanchez-Sinencio, E.: Cmos transconductance multipliers: a tutorial. IEEE Trans. Circuits Syst. II **45**(12), 1550–1563 (1998)

9. Joglekar, Y.N., Wolf, S.J.: The elusive memristor: properties of basic electrical circuits. Eur. J. Phys. **30**(4), 661 (2009)
10. Liu, X., Zeng, Z., Donald, C.: Memristor-based lstm network with in situ training and its applications. Neural Netw. **131**, 300–311 (2020)
11. Maryan, M.M., Ghanaatian, A., Azhari, S.J.: Low-power high-speed analog multiplier/divider based on a new current squarer circuit. Arab. J. Sci. Eng. **43**, 2909–2918 (2018)
12. Naderi, A., Mojarrad, H., Ghasemzadeh, H.: Four-quadrant cmos analog multiplier based on new current squarer circuit with high-speed. In: IEEE EUROCON, pp. 282–287. IEEE (2009)
13. Raj, A., Bhaskar, D.R., Kumar, P.: Novel architecture of four quadrant analog multiplier/divider circuit employing single cfoa. Analog Integr. Circ. S. **108**(3), 689–701 (2021)
14. Rajpoot, J., Maheshwari, S.: High performance four-quadrant analog multiplier using dxccii. Circ. Syst. Stgnal. Pr. **39**, 54–64 (2020)
15. Saatlo, A.N., Özoğuz, İS.: Design of a high-linear, high-precision analog multiplier, free from body effect. Turk. J. Electr. Eng. Comput. Sci. **24**(3), 820–832 (2016)
16. Song, H.J., Kim, C.K.: An mos four-quadrant analog multiplier using simple two-input squaring circuits with source followers. IEEE J Solid-State Circ. **25**(3), 841–848 (1990)
17. Sun, J., Yue, Y., Wang, Y.: Memristor-based operant conditioning neural network with blocking and competition effects. IEEE Trans. Industr. Inform (2024)
18. Tetzlaff, R.: Memristors and Memristive Systems, vol. 1. Springer, New York (2014)
19. Unuk, T., Arslanalp, R., Tez, S.: Design of current-mode versatile multi-input analog multiplier topology. AEU-Int. J. Electron. Commun. **160**, 154493 (2023)
20. Wang, Z.: A four-transistor four-quadrant analog multiplier using mos transistors operating in the saturation region. IEEE Trans. Instrum. Meas. **42**(1), 75–77 (1993)
21. Xiao, P., Hong, Q., Du, S.: Analog-in-memory accelerator design based on memristive arrays for opposite directional interference alignment algorithm. IEEE Trans. Industr. Inform (2023)
22. Yang, L., Ding, Z., Zeng, Z.: Memristor crossbar-based pavlov associative memory network for dynamic information correlation. AEU-Int. J. Electron. Commun. **159**, 154472 (2023)
23. Yang, L., Zeng, Z., Ma, Z.: A memristive dual-slope a/d converter. Int. J. Circ. Theor. App. **48**(1), 42–55 (2020)
24. Zhang, Y., Li, Y., Wang, X.: Synaptic characteristics of ag/aginsbte/ta-based memristor for pattern recognition applications. IEEE Trans. Electron Devices **64**(4), 1806–1811 (2017)

Dual Asynchronous Switching Control for Fuzzy Semi-markov Jump Neural Networks with Multi-weighting Event-Triggered Scheme

Yiteng Zhang[1], Qingyi Zhao[2], and Linchuang Zhang[2](✉)

[1] College of Mathematical Sciences, Bohai University, Jinzhou 121013, China
[2] College of Computer Science and Artificial Intelligence, Bohai University, Jinzhou 121013, Liaoning, China
zhanglinchuang123@163.com

Abstract. This paper studies the event-triggered dual asynchronous switching control issue of fuzzy semi-Markov jump neural networks (FSMJNNs). First, a multi-weighting event-triggered scheme is designed to save network resources. Considering the mismatch premise variables and modes, a new membership functions-dependent Lyapunov-Krasovskii functional analysis method is proposed to derive the sufficient conditions for determining the dual asynchronous switching controller gains and guaranteeing the stability of FSMJNNs. Finally, a simulation example is given to illustrate the effectiveness of the proposed method.

Keywords: Fuzzy semi-Markov jump neural networks (FSMJNNs) · Dual asynchronous switching control · Membership functions-dependent Lyapunov-Krasovskii functional analysis method · Multi-weighting event-triggered scheme

1 Introduction

Neural networks (NNs) have been widely studied in the fields such as image encryption [1] and signal processing [2]. Some existing studies focus on the linear systems [3,4]. However, some nonlinear phenomena often occur in the actual environment, such as wind power forecasting in [5–7] and multi-agent systems in [8]. Takagi-Sugeno (T-S) fuzzy model has been used to approximate nonlinear systems and caused a lot of research. For example, a learning-based interval type-2 fuzzy NNs was used to identify nonlinear systems in [9]. In addition, there are often sudden changes in parameters or structures during the system operation. Since the semi-Markov (SM) jump system [10] can be used to describe these abrupt changes, it has become a research hotspot.

To conserve network resources, research has been conducted on the event-triggered scheme (ETS) in [11,12]. Meanwhile, due to the influence of ETS and external disturbances, the mismatch premise variables and modes between the

C. Li et al. (Eds.): ICNC 2025, CCIS 2946, pp. 281–290, 2026.
https://doi.org/10.1007/978-981-92-1599-7_24

system and controller are very common. In addition, most of the Lyapunov functional analysis methods are only related to the system mode, which will cause some conservatism. A membership functions (MFs)-dependent Lyapunov-Krasovskii functional (LKF) was proposed in [13], which reduced conservatism by introducing a switching scheme. However, there are few studies on the multi-weighting event-triggered dual asynchronous switching control of fuzzy semi-Markov jump NNs (FSMJNNs), which promotes the research of this paper.

This paper studies the problem of dual asynchronous switching control for FSMJNNs by using MFs-dependent LKF analysis method. New multi-weighting switching ETS and dual asynchronous switching controller are designed to enhance network resource efficiency while reducing the conservatism. Sufficient conditions are derived to guarantee the system satisfies H_∞ performance.

2 Problem Statement

2.1 FSMJNNs Construction

Take into account the following FSMJNNs:

SYSTEM RULE s **: IF** $\wp_1^{x(t)}$ is $\urcorner_1^s$, $\wp_2^{x(t)}$ is $\urcorner_2^s$, ..., and $\wp_p^{x(t)}$ is $\urcorner_p^s$, **THEN**

$$\begin{cases} \dot{x}(t) = -\mathrm{A}_{\alpha(t),s}x(t) + \mathrm{B}_{\alpha(t),s}g(x(t)) + \mathrm{D}_{\alpha(t),s}u(t) + \mathrm{C}_{\alpha(t),s}\omega(t), \\ z(t) = \mathrm{E}_{\alpha(t),s}x(t), \end{cases} \tag{1}$$

where s is the fuzzy rules number, $\urcorner_a^s$ is the fuzzy set of premise variables $\wp_a^{x(t)}$ $(a = 1, 2, \ldots, p, s = 1, 2, \ldots, n)$, $x(t) \in \mathbb{R}^n$, $u(t) \in \mathbb{R}^{n_u}$ and $z(t) \in \mathbb{R}^{n_z}$ are the state vector, the control input and the output signal, respectively. $\omega(t) \in \mathbb{R}^{n_\omega}$ is the disturbance which belongs to $\mathfrak{L}_2[0, \infty)$. $\alpha(t) \in \mathbb{N} = \{1, 2, \ldots, N\}$ is an SM process. $g(x(t)) \in \mathbb{R}^{n_x}$ denotes the activation function of neurons with $\mathcal{H}_c^- \leq \frac{g_c(m_1)-g_c(m_2)}{m_1-m_2} \leq \mathcal{H}_c^+, \forall m_1, m_2 \in \mathcal{R}, m_1 \neq m_2$, $c = 1, 2, \ldots, n$. The transition probability matrix $\Gamma = [\psi_{\iota\lambda}^h]_{\mathrm{N}\times\mathrm{N}}$ is subject to the probability $\mathrm{P_r}\{\alpha(t+\epsilon) = \lambda \mid \alpha(t) = \iota\} = \psi_{\iota\lambda}^h\epsilon + o(\epsilon)$, when $\iota \neq \lambda$ and $\mathrm{P_r}\{\alpha(t+\epsilon) = \lambda \mid \alpha(t) = \iota\} = 1 + \psi_{\iota\iota}^h\epsilon + o(\epsilon)$, when $\iota = \lambda$. $\lim_{\epsilon\to 0} o(\epsilon)/\epsilon = 0$ and $\psi_{\iota\iota}^h = -\sum_{\lambda=1,\iota\neq\lambda}^{N} \psi_{\iota\lambda}^h$ with $\psi_{\iota\lambda}^h$ being the transition rate from mode ι to mode λ. Then, the FSMJNNs can be expressed as follows:

$$\begin{cases} \dot{x}(t) = \sum_{s=1}^{n} \chi_s(\wp^{x(t)})[-\mathrm{A}_{\alpha(t),s}x(t) + \mathrm{B}_{\alpha(t),s}g(x(t)) \\ \qquad + \mathrm{D}_{\alpha(t),s}u(t) + \mathrm{C}_{\alpha(t),s}\omega(t)], \\ z(t) = \sum_{s=1}^{n} \chi_s(\wp^{x(t)})\mathrm{E}_{\alpha(t),s}x(t), \end{cases} \tag{2}$$

where $\chi_s(\wp^{x(t)}) = \frac{\xi_s(\wp^{x(t)})}{\sum_{s=1}^{n}\xi_s(\wp^{x(t)})}$, $\xi_s(\wp^{x(t)}) = \Pi_a^p \urcorner_a^s(\wp_a^{x(t)})$, $\urcorner_a^s(\wp_a^{x(t)})$ is the membership grade of $\wp_a^{x(t)}$ in $\urcorner_a^s$. $\xi_s(\wp^{x(t)}) \geq 0$, $\chi_s(\wp^{x(t)}) \geq 0$, $\sum_{s=1}^{n}\xi_s(\wp^{x(t)}) > 0$ and $\sum_{s=1}^{n}\chi_s(\wp^{x(t)}) = 1$.

2.2 Event-Triggered Asynchronous Controller Design

t_ib is the latest release instant and the next triggered instant $t_{i+1}b$ can be determined by

$$\begin{aligned} t_{i+1}b = t_ib + \min_{j>0}\{kb|[x(t_ib) - x(t_ib + kb)]^T \Phi_{1\alpha_{(t)}} \\ \times [x(t_ib) - x(t_ib + kb)] > \kappa_{\alpha_{(t)}} x^T(t_ib)\Phi_{2\alpha_{(t)}} x(t_ib)\}, \end{aligned} \tag{3}$$

where $\Phi_{1\alpha_{(t)}} > 0$ and $\Phi_{2\alpha_{(t)}} > 0$ are weighting matrices to be determined. $\kappa_{\alpha(t)} \in [0, 1)$. The holding time is within $[t_ib + \ell_{t_i}, t_{i+1}b + \ell_{t_{i+1}}) = \bigcup_{\hat{i}=0}^{k-1} \Lambda_{\hat{i}}$ for zero-order holder, where $\Lambda_{\hat{i}} = [t_ib + \hat{i}b + \ell_{t_i+\hat{i}},\ t_ib + (\hat{i} + 1)b + \ell_{t_i+(\hat{i}+1)})$ and $\ell_{t_i} \in [0, \bar{\ell}]$. Define $e(t) = x(t_ib) - x(t_ib + kb)$ and $t_{i+1}b = t_ib + kb$. The time delay is defined as $\ell(t) = t - t_ib - \hat{i}b, t \in \Lambda_{\hat{i}}$. Then, one has $0 \le \ell(t) \le b + \bar{\ell} = \ell_M$.

The asynchronous controller is expressed as

CONTROLLER RULE q **: IF** $\wp_1^{x(t)}$ is $\urcorner_1^q$, $\wp_2^{x(t)}$ is $\urcorner_2^q$, ..., and $\wp_p^{x(t)}$ is $\urcorner_p^q$, **THEN**

$$u(t) = \mathrm{K}_{\rho(t),q} x(t_ib), \tag{4}$$

where $\mathrm{K}_{\rho(t)}$ is the controller gain and $(q = 1, 2, \ldots, n)$. $\rho(t)$ is the controller mode taking values in $\mathbb{M} = \{1, 2, \ldots, M\}$ and asynchronous with $\alpha(t)$. The condition probability matrix $\Sigma = [\sigma_{\iota\varrho}]_{\mathbb{N}\times\mathbb{M}}$ is subject to $\Pr\{\rho(t) = \varrho|\, \alpha(t) = \iota\} = \sigma_{\iota\varrho}$, $0 \le \sigma_{\iota\varrho} \le 1$ and $\sum\limits_{\varrho=1}^{M} \sigma_{\iota\varrho} = 1$.

Then, the final asynchronous controller is expressed as:

$$u(t) = \sum_{q=1}^{n} \chi_q(\wp^{x(t_ib)}) \mathrm{K}_{\rho(t),q} x(t_ib). \tag{5}$$

The following conditions are used to deal with the asynchronous MFs:

$$\begin{cases} \chi_q(\wp^{x(t_ib)}) = \theta_q \chi_q(\wp^{x(t)}), \\ |\chi_q(\wp^{x(t)}) - \chi_q(\wp^{x(t_ib)})| \ \le \upsilon_q, \theta_q > 0, \upsilon_q \ge 0, \end{cases} \tag{6}$$

in which

$$\theta_q^{\min} = 1 - \frac{\upsilon_q}{\chi_q(x(t))} \le \theta_q \le 1 + \frac{\upsilon_q}{\chi_q(x(t))} = \theta_q^{\max}. \tag{7}$$

Thus, one has $\frac{\theta_s^{\min}}{\theta_q^{\max}} = \frac{\min\{\theta_s\}}{\max\{\theta_q\}} \le \frac{\theta_s}{\theta_q} \le \frac{\max\{\theta_s\}}{\min\{\theta_q\}} = \frac{\theta_s^{\max}}{\theta_q^{\min}}$. Denoting $\theta_1 = \min \theta_s^{\min}$ and $\theta_2 = \max \theta_s^{\max}$, one has $\varphi_1 = \frac{\theta_1}{\theta_2} \le \frac{\theta_s}{\theta_q} \le \frac{\theta_2}{\theta_1} = \varphi_2$. Then, the closed-loop system is formulated as

$$\begin{cases} \dot{x}(t) = \sum\limits_{s=1}^{n}\sum\limits_{q=1}^{n} \theta_q \chi_s(\wp^{x(t)}) \chi_q(\wp^{x(t)}) [-\mathrm{A}_{\iota s} x(t) + \mathrm{B}_{\iota s} g(x(t)) \\ \qquad + \mathrm{C}_{\iota s}\omega(t) + \mathrm{D}_{\iota s}\mathrm{K}_{\varrho s} x(t - \ell(t)) \\ + \mathrm{D}_{\iota s}\mathrm{K}_{\varrho s} e(t)], z(t) = \sum\limits_{s=1}^{n}\sum\limits_{q=1}^{n} \theta_q \chi_s(\wp^{x(t)}) \chi_q(\wp^{x(t)}) \mathrm{E}_{\iota s} x(t). \end{cases} \tag{8}$$

3 Main Results

Sufficient conditions for guaranteeing the stochastic stability of the system (8) are presented in Theorems 1 and 2.

Theorem 1. *For the known positive constants* h, ℓ_M, κ_ι, φ_1, φ_2, *the system (8) is stochastically stable with* H_∞ *performance, if there exist matrices* $\mathrm{P}_{\iota s} > 0$, $\mathrm{W}_s > 0$, $\mathrm{V}_s > 0$, $\Phi_{1\iota} > 0$, $\Phi_{2\iota} > 0$, $\mathcal{L} > 0$, $\mathrm{K}_{\varrho q}$ *and constant* $\gamma > 0$, *for* $\iota \in \mathbb{N}$, $\varrho \in \mathbb{M}$, $s < q$ *such that*

$$\Theta_{ss\iota} < 0, \tag{9}$$

$$\Theta_{sq\iota} + \varphi_1 \Theta_{qs\iota} < 0, \tag{10}$$

$$\Theta_{sq\iota} + \varphi_2 \Theta_{qs\iota} < 0, \tag{11}$$

where

$$\Theta_{sq\iota} = \begin{bmatrix} \Theta^1_{sq\iota} & \Theta^2_{sq\iota} \\ * & -\mathrm{V}_s^{-1} \end{bmatrix},$$

$$\Theta^1_{sq\iota} = \begin{bmatrix} \Theta^{11}_{sq\iota} & \Theta^{12}_{sq\iota} & 0 & \Theta^{14}_{sq\iota} & \Theta^{15}_{sq\iota} & \Theta^{16}_{sq\iota} \\ * & \Theta^{22}_{sq\iota} & \mathrm{V}_s & \Theta^{24}_{sq\iota} & 0 & 0 \\ * & * & \Theta^{33}_{sq\iota} & 0 & 0 & 0 \\ * & * & * & \Theta^{44}_{sq\iota} & 0 & 0 \\ * & * & * & * & \Theta^{55}_{sq\iota} & 0 \\ * & * & * & * & * & -\gamma^2 I \end{bmatrix},$$

$$\Theta^{11}_{sq\iota} = 2\mathrm{P}_{\iota s}\mathrm{A}_{\iota s} + \sum_{\lambda=1}^{N} \bar{\psi}_{ij}\mathrm{P}_{\iota s} - \mathcal{H}_1\mathcal{L} + \mathrm{E}^T_{\iota s}\mathrm{E}_{\iota s} + \mathrm{W}_s - \mathrm{V}_s,$$

$$\Theta^{12}_{sq\iota} = \sum_{\varrho=1}^{M} \sigma_{\iota\varrho}\mathrm{P}_{\iota s}\mathrm{D}_{\iota s}\mathrm{K}_{\varrho q} + \mathrm{V}_s, \Theta^{14}_{sq\iota} = \sum_{\varrho=1}^{M} \sigma_{\iota\varrho}\mathrm{P}_{\iota s}\mathrm{D}_{\iota s}\mathrm{K}_{\varrho q}, \Theta^{15}_{sq\iota} = \mathrm{P}_{\iota s}\mathrm{B}_{\iota s} + \mathcal{H}_2\mathcal{L},$$

$$\Theta^{16}_{sq\iota} = \mathrm{P}_{\iota s}\mathrm{C}_{\iota s}, \Theta^{22}_{sq\iota} = \kappa_\iota \Phi_{2\iota} - 2\mathrm{V}_s, \Theta^{24}_{sq\iota} = \kappa_\iota\Phi_{2\iota}, \Theta^{33}_{sq\iota} = -\mathrm{W}_s - \mathrm{V}_s,$$

$$\Theta^{44}_{sq\iota} = \kappa_\iota\Phi_{2\iota} - \Phi_{1\iota}, \Theta^{55}_{sq\iota} = -\mathcal{L},$$

$$\Theta^2_{sq\iota} = \ell_M \left[-\mathrm{A}_{\iota s} \ \sum_{\varrho=1}^{M} \sigma_{\iota\varrho}\mathrm{D}_{\iota s}\mathrm{K}_{\varrho q} \ 0 \ \sum_{\varrho=1}^{M} \sigma_{\iota\varrho}\mathrm{D}_{\iota s}\mathrm{K}_{\varrho q} \ \mathrm{B}_{\iota s} \ \mathrm{C}_{\iota s} \right]^T.$$

Proof. Consider the following MFs-dependent LKF: $\mathbb{V}(t) = x^T(t)\mathrm{P}_{i\chi}x(t) + \int_{t-\ell_M}^{t} x^T(v)\mathrm{W}_\chi x(v)\mathrm{d}v + \ell_M \int_{-\ell_M}^{0}\int_{t+v}^{t} \dot{x}^T(u)\mathrm{V}_\chi \dot{x}(u)\mathrm{d}u\mathrm{d}v$.

By applying infinitesimal operator to $\mathbb{V}(t)$, one has

$$\begin{aligned}\boldsymbol{E}\{\mathbb{L}\mathbb{V}(t)\} &= 2x^T(t)\mathrm{P}_{i\chi}\dot{x}(t) + x^T(t)\sum_{\lambda=1}^{N}\bar{\psi}_{\iota\lambda}\mathrm{P}_{i\chi}x(t) \\ &+ x^T(t)\mathrm{W}_{\chi}x(t) - x^T(t-\ell_M)\mathrm{W}_{\chi}x(t-\ell_M) \\ &+ \ell_M^2\dot{x}^T(t)\mathrm{V}_{\chi}\dot{x}(t) - \ell_M\int_{t-\ell_M}^{t}\dot{x}^T(v)\mathrm{V}_{\chi}\dot{x}(v)\mathrm{d}v \\ &+ x^T(t)\dot{\mathrm{P}}_{i\chi}x(t) + \int_{t-\ell_M}^{t}x^T(v)\dot{\mathrm{W}}_{\chi}x(v)\mathrm{d}v \\ &+ \ell_M\int_{-\ell_M}^{0}\int_{t+v}^{t}\dot{x}^T(u)\dot{\mathrm{V}}_{\chi}\dot{x}(u)\mathrm{d}u\mathrm{d}v,\end{aligned}$$

where $\sum\limits_{\lambda=1}^{N}\bar{\psi}_{\iota\lambda} = \boldsymbol{E}\{\psi_{\iota\lambda}^h\} = \int_0^\infty \psi_{\iota\lambda}^h\vartheta_\iota^h\mathrm{d}h$, ϑ_ι^h is the probability density function. For $\chi_s = \chi_s(\wp^{x(t)})$, $\mathrm{P}_{\iota\chi} = \sum_{s=1}^n \chi_s\mathrm{P}_{\iota\chi}$, $\mathrm{W}_\chi = \sum_{s=1}^n \chi_s\mathrm{W}_s$, and $\mathrm{V}_\chi = \sum_{s=1}^n \chi_s\mathrm{V}_s$. Moreover, for $c = 1, 2, \ldots, n$, one has $(g_c(x_c(t)) - \mathcal{H}_c^- x_c(t)) \times (g_c(x_c(t)) - \mathcal{H}_c^+ x_c(t)) \leq 0$, which is equivalent to

$$\begin{bmatrix} x(t) \\ g(x(t)) \end{bmatrix}^T \begin{bmatrix} -\mathcal{H}_1\mathcal{L} & \mathcal{H}_2\mathcal{L} \\ \star & -\mathcal{L} \end{bmatrix} \begin{bmatrix} x(t) \\ g(x(t)) \end{bmatrix} \geq 0,$$

where $\mathcal{H}_1 = \mathrm{diag}\left\{\mathcal{H}_1^-\mathcal{H}_1^+, \ldots, \mathcal{H}_n^-\mathcal{H}_n^+\right\}$, $\mathcal{H}_2 = \mathrm{diag}\left\{\frac{\mathcal{H}_1^- + \mathcal{H}_1^+}{2}, \ldots, \frac{\mathcal{H}_n^- + \mathcal{H}_n^+}{2}\right\}$ and $\mathcal{L} > 0$ is a diagonal matrix. Due to $\sum_{s=1}^n \dot{\chi}_s = 0$, the time derivatives are $\dot{\mathrm{P}}_{\iota\chi} = \sum_{s=1}^n \dot{\chi}_s\mathrm{P}_{\iota s} = \sum_{y=1}^{n-1}\dot{\chi}_y(\mathrm{P}_{\iota y} - \mathrm{P}_{\iota n})$, $\dot{\mathrm{W}}_\chi = \sum_{s=1}^n \dot{\chi}_s\mathrm{W}_s = \sum_{y=1}^{n-1}\dot{\chi}_y(\mathrm{W}_y - \mathrm{W}_n)$, $\dot{\mathrm{V}}_\chi = \sum_{s=1}^n \dot{\chi}_s\mathrm{V}_s = \sum_{y=1}^{n-1}\dot{\chi}_y(\mathrm{V}_y - \mathrm{V}_n)$. The following conditions are given to ensure the negative values of $\dot{\mathrm{P}}_{\iota\chi}$, $\dot{\mathrm{W}}_\chi$ and $\dot{\mathrm{V}}_\chi$. (1) : If $\dot{\chi}_y < 0$, then $\mathrm{P}_{\iota y} - \mathrm{P}_{\iota n} > 0$, $\mathrm{W}_y - \mathrm{W}_n > 0$, $\mathrm{V}_y - \mathrm{V}_n > 0$. (2) : If $\dot{\chi}_y \geq 0$, then $\mathrm{P}_{\iota y} - \mathrm{P}_{\iota n} \leq 0$, $\mathrm{W}_y - \mathrm{W}_n \leq 0$, $\mathrm{V}_y - \mathrm{V}_n \leq 0$. There are 2^{n-1} possible permutations. Then, one has $\boldsymbol{E}\{\mathbb{L}\mathbb{V}(t) + z^T(t)z(t) - \gamma^2\omega^T(t)\omega(t)\} \leq \sum\limits_{s=1}^{n}\sum\limits_{q=1}^{n}\theta_q\chi_s\chi_q\boldsymbol{E}\{\aleph^T(t)\Theta_{sq\iota}\aleph(t)\}$, which is equivalent to $\sum\limits_{s=1}^{n-1}\sum\limits_{s<q}^{n}\theta_q\chi_s\chi_q\aleph^T(t)\left[\Theta_{sq\iota} + \frac{\theta_s}{\theta_q}\Theta_{qs\iota}\right]\aleph(t) + \sum\limits_{s=1}^{n}\theta_s\chi_s^2\aleph^T(t)\left[\Theta_{ss\iota}\right]\aleph(t)$, where $\aleph^T(t) = \left[x^T(t)\ x^T(t-\ell(t))\ x^T(t-\ell_M)\ e^T(t)\ g^T(x(t))\ \omega^T(t)\right]$.

Based on (9)–(11), the following inequality holds: $\boldsymbol{E}\{\mathbb{L}\mathbb{V}(t) + z^T(t)z(t) - \gamma^2\omega^T(t)\omega(t)\} \leq 0$. Integrating this inequality from 0 to t on both sides and it yields that $\boldsymbol{E}\{\int_0^\infty z^T(t)z(t)\mathrm{d}t\} \leq \gamma^2\boldsymbol{E}\{\int_0^\infty \omega^T(t)\omega(t)\mathrm{d}t\}$. Thus, the H_∞ performance is satisfied with the help of [14, Definition 1]. When $\omega(t) = 0$, it can be easily obtained that the system (8) is stochastically stable and is omitted here. The proof is completed.

Remark 1. For $\partial = 1, 2, \ldots, 2^{n-1}$, a switching double asynchronous controller is designed as $u^\partial(t) = \sum\limits_{q=1}^{n}\chi_q(\wp^{x(t_ib)})\mathrm{K}_{\rho(t),q}^\partial x(t_ib)$, and the weighting matrices of ETS $\Phi_{1\alpha(t)}^\partial$ and $\Phi_{2\alpha(t)}^\partial$ are also the switching forms.

Theorem 2. *For the known positive constants h, ℓ_M, κ_ι, φ_1, φ_2, μ, $\ni$, the system (8) is stochastically stable with H_∞ performance, if there exist matrices* $\mathrm{P}_{\iota s} > 0$, $\mathrm{W}_s > 0$, $\mathrm{V}_s > 0$, $\Phi_{1\iota} > 0$, $\Phi_{2\iota} > 0$, $\mathcal{L} > 0$, $\mathrm{X}_{\varrho q}$, $\mathrm{Y}_{\varrho q}$ *and constant* $\gamma > 0$, *for* $\iota \in \mathbb{N}$, $\varrho \in \mathbb{M}$, $s < q$ *such that*

$$\Xi_{ss\iota} < 0, \tag{12}$$

$$\Xi_{sq\iota} + \varphi_1 \Xi_{qs\iota} < 0, \tag{13}$$

$$\Xi_{sq\iota} + \varphi_2 \Xi_{qs\iota} < 0, \tag{14}$$

where

$$\Xi_{sq\iota} = \begin{bmatrix} \mho_{sq\iota} & \Xi^1_{sq\iota} \\ * & \Xi^2_{sq\iota} \end{bmatrix}, \mho_{sq\iota} = \begin{bmatrix} \mho^1_{sq\iota} & \mho^2_{sq\iota} \\ * & \mho^3_{sq\iota} \end{bmatrix},$$

$$\mho^1_{sq\iota} = \begin{bmatrix} \Theta^{11}_{sq\iota} & \mho^{12}_{sq\iota} & 0 & \mho^{14}_{sq\iota} & \Theta^{15}_{sq\iota} & \Theta^{16}_{sq\iota} \\ * & \Theta^{22}_{sq\iota} & \mathrm{V}_s & \Theta^{24}_{sq\iota} & 0 & 0 \\ * & * & \Theta^{33}_{sq\iota} & 0 & 0 & 0 \\ * & * & * & \Theta^{44}_{sq\iota} & 0 & 0 \\ * & * & * & * & \Theta^{55}_{sq\iota} & 0 \\ * & * & * & * & * & -\gamma^2 I \end{bmatrix},$$

$$\mho^{12}_{sq\iota} = \sum_{\varrho=1}^{M} \sigma_{\iota\varrho} \mathrm{D}_{\iota s} \mathrm{Y}_{\varrho q} + \mathrm{V}_s, \ \mho^{14}_{sq\iota} = \sum_{\varrho=1}^{M} \sigma_{\iota\varrho} \mathrm{D}_{\iota s} \mathrm{Y}_{\varrho q},$$

$$\mho^2_{sq\iota} = \ell_M \left[-\mathrm{P}_{\iota s} \mathrm{A}_{\iota s} \ \mho_1 \ 0 \ \mho_2 \right]^T, \mho_1 = \sum_{\varrho=1}^{M} \sigma_{\iota\varrho} \mathrm{D}_{\iota s} \mathrm{Y}_{\varrho q},$$

$$\mho_2 = \left[\mho_1 \ \mathrm{P}_{\iota s} \mathrm{B}_{\iota s} \ \mathrm{P}_{\iota s} \mathrm{C}_{\iota s} \right], \mho^3_{sq\iota} = \ni^2 \mathrm{V}_s - \ni 2 \mathrm{P}_{\iota s},$$

$$\Xi^1_{sq\iota} = \begin{bmatrix} \sigma_{\iota 1}(\Xi^{11}_{sq\iota}) & \ldots & \sigma_{\iota m}(\Xi^{12}_{sq\iota}) \\ \mu(\mathrm{Y}^T_{1q}) & \ldots & \mu(\mathrm{Y}^T_{mq}) \\ 0 & \ldots & 0 \\ \mu(\mathrm{Y}^T_{1q}) & \ldots & \mu(\mathrm{Y}^T_{mq}) \\ 0 & \ldots & 0 \\ 0 & \ldots & 0 \\ \ell_M \sigma_{\iota 1}(\Xi^{11}_{sq\iota}) & \ldots & \ell_M \sigma_{\iota m}(\Xi^{12}_{sq\iota}) \end{bmatrix},$$

$$\Xi^{11}_{sq\iota} = \mathrm{P}_{\iota s} \mathrm{D}_{\iota s} - \mathrm{D}_{\iota s} \mathrm{X}_{1q}, \ \Xi^{12}_{sq\iota} = \mathrm{P}_{\iota s} \mathrm{D}_{\iota s} - \mathrm{D}_{\iota s} \mathrm{X}_{ms},$$

$$\Xi^2_{sq\iota} = \mathrm{diag}\left\{ -\mu 2 \mathrm{X}_{1s}, \ldots, -\mu 2 \mathrm{X}_{ms} \right\},$$

and the controller gains are $\mathrm{K}_{\varrho q} = \mathrm{X}^{-1}_{\varrho q} \mathrm{Y}_{\varrho q}$.

Proof. Firstly, Pre- and post multiplying $\Theta_{sq\iota}$ by $\mathrm{diag}\{\mathrm{I}, \mathrm{I}, \mathrm{I}, \mathrm{I}, \mathrm{I}, \mathrm{I}, \mathrm{P}_{\iota s}\}$. Then, one has

$$\breve{\Theta}_{sq\iota} = \begin{bmatrix} \Theta^1_{sq\iota} & \breve{\Theta}^2_{sq\iota} \\ * & -\mathrm{P}_{\iota s} \mathrm{V}_s^{-1} \mathrm{P}_{\iota s} \end{bmatrix}, \breve{\Theta}^2_{sq\iota} = \Theta^2_{sq\iota} \mathrm{P}_{\iota s}.$$

For a given constant $\ni$, there exists $-\mathrm{P}_{\iota s} \mathrm{V}_s^{-1} \mathrm{P}_{\iota s} \leq \ni^2 \mathrm{V}_s - 2 \ni \mathrm{P}_{\iota s}$. Then, define $\mathrm{Y}_{\varrho q} = \mathrm{X}_{\varrho q} \mathrm{K}_{\varrho q}$. $\breve{\Theta}_{sq\iota} < 0$ is equivalent to $\hat{\Xi}_{sq\iota} < 0$, $\hat{\Xi}_{sq\iota} = \mho_{msi} + 2 \hat{\Xi}^1_{sq\iota} \hat{\Xi}^2_{sq\iota} \hat{\Xi}^3_{sq\iota}$, where

$$\hat{\Xi}_{sq\iota}^{1} = \begin{bmatrix} \sigma_{\iota 1}(\Xi_{sq\iota}^{11}) & \dots & \sigma_{\iota m}(\Xi_{sq\iota}^{12}) \\ 0 & \dots & 0 \\ 0 & \dots & 0 \\ 0 & \dots & 0 \\ 0 & \dots & 0 \\ 0 & \dots & 0 \\ \ell_M \sigma_{\iota 1}(\Xi_{sq\iota}^{11}) & \dots & \ell_M \sigma_{\iota m}(\Xi_{sq\iota}^{12}) \end{bmatrix}, \hat{\Xi}_{sq\iota}^{3} = \begin{bmatrix} 0 \ \mathrm{Y}_{1q} \ 0 \ \mathrm{Y}_{1q} \ 0 \ 0 \ 0 \\ \vdots \ \ \vdots \ \ \vdots \ \ \vdots \ \ \vdots \ \vdots \ \vdots \\ 0 \ \mathrm{Y}_{mq} \ 0 \ \mathrm{Y}_{mq} \ 0 \ 0 \ 0 \end{bmatrix},$$

$$\hat{\Xi}_{sq\iota}^{2} = \mathrm{diag}\left\{\mathrm{X}_{1q}^{-1}, \dots, \mathrm{X}_{mq}^{-1}\right\}.$$

According to [15, Lemma 1], one has (12)–(14). This finishes the proof.

4 Simulation Results

Consider FSMJNNs with the following parameters:

$$\mathrm{A}_{11} = \begin{bmatrix} 0.4 & 0 \\ 0 & 0.1 \end{bmatrix}, \mathrm{A}_{12} = \begin{bmatrix} 0.1 & 0 \\ 0 & 0.2 \end{bmatrix}, \mathrm{A}_{21} = \begin{bmatrix} 0.5 & 0 \\ 0 & 0.6 \end{bmatrix}, \mathrm{A}_{22} = \begin{bmatrix} 0.8 & 0 \\ 0 & 0.7 \end{bmatrix}, \kappa_1 = 0.5,$$

$$\mathrm{B}_{\iota s} = \begin{bmatrix} 3 & 0.2 \\ 0.1 & 0.1 \end{bmatrix}, \mathrm{C}_{\iota s} = \begin{bmatrix} 1.5 \\ -1.5 \end{bmatrix}, \mathrm{D}_{\iota s} = \begin{bmatrix} -1.5 & 0.3 \\ 0.5 & -1 \end{bmatrix}, \mathrm{E}_{\iota s} = \begin{bmatrix} 0.2 & 0 \\ 0 & 0.1 \end{bmatrix}, \kappa_2 = 0.3,$$

$$\mathcal{H}_1 = \begin{bmatrix} 0 & 0 \\ 0 & 0 \end{bmatrix}, \mathcal{H}_2 = \begin{bmatrix} 0.5 & 0 \\ 0 & 0.5 \end{bmatrix}, \ell_M = 0.1, \backepsilon = 0.1, \mu = 0.01, \varphi_1 = 0.6, \varphi_2 = 1.$$

The expectation of the transition probability matrix Γ and the conditional probability matrix Σ are given as

$$\Gamma = \begin{bmatrix} -1.7725 & 1.7725 \\ 2.7082 & -2.7082 \end{bmatrix}, \Sigma = \begin{bmatrix} 0.3 & 0.7 \\ 0.4 & 0.6 \end{bmatrix}.$$

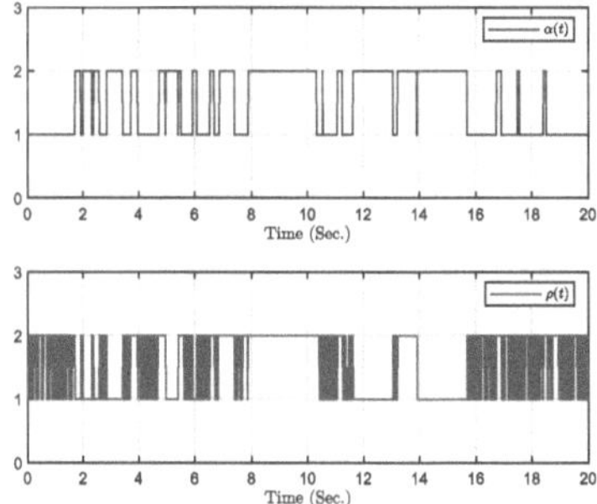

Fig. 1. Mode changes of system and controller.

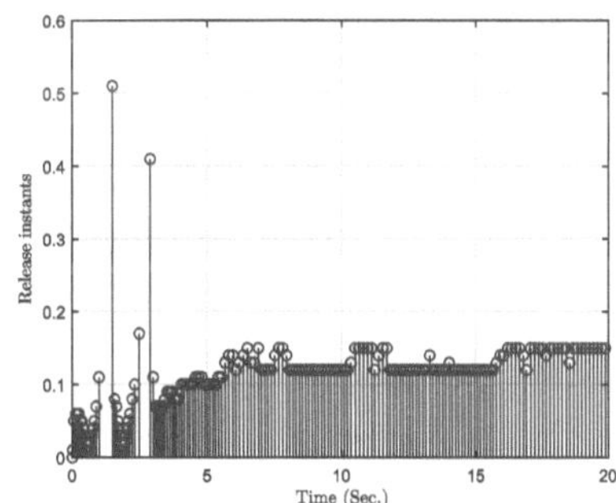

Fig. 2. Triggering release instants and intervals.

Define $g_1(x(t)) = g_2(x(t)) = \tanh(x(t))$. The disturbance is $\omega(t) = \sin(0.02t)$, $0 < t < 3$ and $\omega(t) = 0$ for other times. MFs are given as $\chi_1(x_1(t)) = \frac{1}{1+e^{-2x_1(t)}}$ and $\chi_2(x_1(t)) = 1 - \chi_1(x_1(t))$. The controller gains are determined as follows:

$$\mathrm{K}_{11}^{1} = \begin{bmatrix} 2.5195 & 0.4972 \\ 0.2598 & 1.0579 \end{bmatrix}, \mathrm{K}_{12}^{1} = \begin{bmatrix} 2.9993 & 0.6707 \\ 0.3671 & 1.2927 \end{bmatrix},$$
$$\mathrm{K}_{21}^{1} = \begin{bmatrix} 4.3507 & 0.7216 \\ 1.0755 & 2.1670 \end{bmatrix}, \mathrm{K}_{22}^{1} = \begin{bmatrix} 4.3549 & 0.6478 \\ 1.1255 & 2.2075 \end{bmatrix},$$
$$\mathrm{K}_{11}^{2} = \begin{bmatrix} 1.8601 & -0.2253 \\ 0.4642 & 1.3379 \end{bmatrix}, \mathrm{K}_{12}^{2} = \begin{bmatrix} 2.2595 & 0.2798 \\ 0.3569 & 1.3887 \end{bmatrix},$$
$$\mathrm{K}_{21}^{2} = \begin{bmatrix} 5.1602 & 1.2980 \\ 1.2380 & 2.2445 \end{bmatrix}, \mathrm{K}_{22}^{2} = \begin{bmatrix} 4.7881 & 0.8021 \\ 1.2223 & 2.0534 \end{bmatrix}.$$

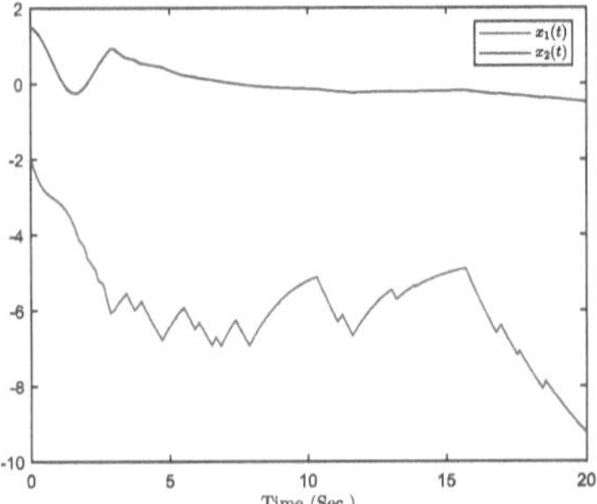

Fig. 3. State responses of open-loop system.

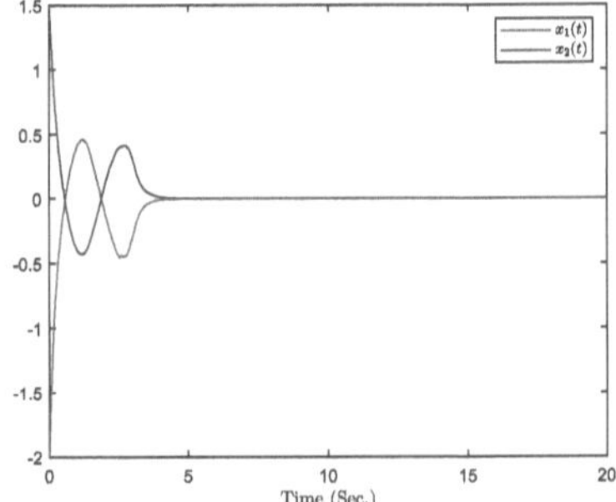

Fig. 4. State responses of closed-loop system.

The initial condition is $x(0) = \begin{bmatrix} -2 & 1.5 \end{bmatrix}^T$. Figure 1 is the modes changes of system and controller. Figure 2 is the release instants and intervals of ETS. We can see that only 196 times are transmitted and the transmission rate is 9.8%, which means the multi-weighting ETS can efficiently save the network resources. Figures 3–4 show the state responses of open-loop system and closed-loop system, respectively. Figure 5 shows that the controller switches dynamically, where $X1-X4$ are the switching points. It is clearly that the proposed dual asynchronous switching controller can ensure the stability of the system.

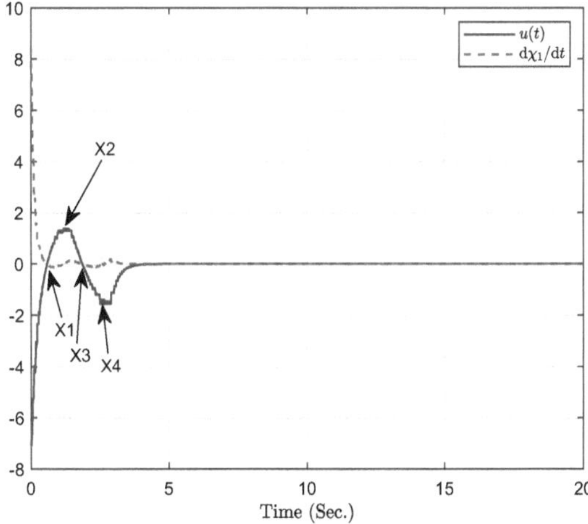

Fig. 5. Trajectories of $d\chi_1/dt$ and $u(t)$.

5 Conclusion

In this paper, the event-triggered dual asynchronous switching control problem has been investigated for FSMJNNs. A multi-weighting ETS has been designed to improve transmission efficiency, and an MFs-dependent LKF analysis method has been used to reduce conservatism. Sufficient conditions have been derived to ensure the stochastic stability of the system. A simulation example has been presented to verify the validity of this analysis method.

Acknowledgments. This work was supported by the National Natural Science Foundation of China (62403071), the Natural Science Foundation of Liaoning Province (2025-BS-0823) and the Innovation Fund Project For Graduate Student of Bohai University (YJC2025-012).

Disclosure of Interests. The authors have no competing interests to declare that are relevant to the content of this article.

References

1. Li, F., Wang, Y.-N., Shen, H.: Dual event-triggered synchronization of two-time-scale jumping neural networks and its application in image encryption and decryption. IEEE Trans. Netw. Sci. Eng. **12**(2), 943–954 (2025)
2. Liu, Z.H., Zhang, J.Y., Zhu, Y.Y., Shi, E.Y., Ai, B.: Robust multidimensional graph neural networks for signal processing in wireless communications with edge-graph information bottleneck. IEEE Trans. Signal Process. (2025). https://doi.org/10.1109/TSP.2025.3574005
3. Lei, Y., Wang, Y.-W., Morărescu, I.-C., Postoyan, R.: Event-triggered fixed-time stabilization of two time scales linear systems. IEEE Trans. Autom. Control **68**(3), 1722–1729 (2023)

4. Lei, Y., Wang, Z., Wang, X., Li, H.Y., Morărescu, I.-C.: Hybrid formation control for heterogeneous uncertain linear two-time-scale systems. Automatica **176**, 112267 (2025)
5. Wang, S., Zhang, W.J., Sun, Y.H., Trivedi, A., Chung, C.Y., Srinivasan, D.: Wind power forecasting in the presence of data scarcity: a very short-term conditional probabilistic modeling framework. Energy **291**, 130305 (2024)
6. Wang, S., Sun, Y.H., Zhang, W.J., Srinivasan, D.: Optimization of deterministic and probabilistic forecasting for wind power based on ensemble learning. Energy **319**, 134884 (2025)
7. Wang, S., Sun, Y.H., Zhang, W.J., Chung, C.Y., Srinivasan, D.: Very short-term wind power forecasting considering static data: an improved transformer model. Energy **312**, 133577 (2024)
8. Fan, S., Meng, M., Fu, Y.K., Deng, C.: Distributed adaptive tracking control for fuzzy nonlinear MASs under Round-Robin protocol. IEEE Trans. Fuzzy Syst. **33**(5), 1488–1498 (2025)
9. Sun, C.X., Wu, X.L., Yang, H.Y., Han, H.G., Zhao, D.Z.: Multi-modal learning-based interval type-2 fuzzy neural network. IEEE Trans. Fuzzy Syst. **32**(11), 6409–6423 (2024)
10. Zhang, L.C., Sun, Y.H., Wu, Z.-G., Shen, M.Q., Pan, Y.N.: Two-layer asynchronous control for a class of nonlinear jump systems: an interval segmentation approach. IEEE Trans. Cybern. **54**(12), 7307–7319 (2024)
11. Kao, Y.G., Feng, X.X., Xie, J., Kao, B.H.: Event-triggered fault-tolerant sliding mode control for discrete Markov jump systems via guaranteed cost control approach. IEEE Trans. Circuits Syst. II Express Briefs **71**(7), 3363–3367 (2024)
12. Zhang, Y.T., Zhang, L.C., Wang, W., Zhu, Z.C.: Co-design of observer-based multiple weighting event-triggered and quadruple asynchronous control for fuzzy semi-Markov jump systems. Inf. Sci. **703**, 121935 (2025)
13. Wang, L.K., Zhao, Y.H., Xie, X.P., Lam, H.-K.: A switching asynchronous control approach for Takagi-Sugeno fuzzy Markov jump systems with time-varying delay. IEEE Trans. Fuzzy Syst. **32**(2), 595–606 (2024)
14. Zhang, L.C., Sun, Y.H., Li, H.Y., Liang, H.J., Wang, J.X.: Event-triggered fault detection for nonlinear semi-Markov jump systems based on double asynchronous filtering approach. Automatica **138**, 110144 (2022)
15. Fang, M., Shi, P., Dong, S.L.: Sliding mode control for Markov jump systems with delays via asynchronous approach. IEEE Trans. Syst. Man Cybern. Syst. **51**(5), 2916–2925 (2021)

MRI-Based Anterior Cruciate Ligament Injury Detection Using Transfer Learning

Junxiu Liu[1,2], Mingxing Li[1,2], Qiang Fu[1,2](✉), Zhaoxin Zhang[1,2], Sheng Qin[1,2], Yuling Luo[1,2], and Xue Ouyang[1,2]

[1] Guangxi Key Lab of Brain-inspired Computing and Intelligent Chips, School of Electronic and Information Engineering, Guangxi Normal University, Guilin, China
qiangfu@gxnu.edu.cn

[2] Key Laboratory of Nonlinear Circuits and Optical Communications (Guangxi Normal University), Education Department of Guangxi Zhuang Autonomous Region, Guilin, China

Abstract. It is critical for the timely discovery of abnormalities or pathologies from radiographic scans, such as magnetic resonance images. Unfortunately, such a task is often rather cumbersome, time-consuming, and error-prone. With the development of machine learning, computer-aided diagnoses systems have been increasingly applied to the field of medical imaging. These auxiliary diagnostic systems are designed to help physicians for better diagnoses. In this work, a neural model for human Anterior Cruciate Ligament (ACL) injury detection is proposed. The network incorporates transfer learning technique with automatic deep layer feature extraction from 3D sagittal knee magnetic resonance images to detect the possibility of mild ACL injury that does not necessitate surgery, and complete ACL rupture that requires surgical intervention. Compared to traditional machine learning methods, the area under the curve has increased by 3%, i.e., the performance of the proposed model has reached a higher classification accuracy rate, which can reduce the misdiagnosis of ACL injury detection. Therefore, the proposed method can be applied for medical diagnosis, such as diabetic retinopathy and pneumonia detection, and can also be used as an early warning system for patients and hospitals to reduce medical costs.

Keywords: knee MRI · transfer learning · neural networks · 3D images

1 Introduction

The Anterior Cruciate Ligament (ACL) is the most commonly injured ligament in the human body, especially among athletes in sports like football or basketball, often requiring surgical intervention [1]. In the United States, around 200,000 ACL tears occur each year, with over 100,000 reconstructions performed. While ACL injuries are frequent in athletes, they can affect anyone. Magnetic Resonance Imaging (MRI) is commonly used to diagnose knee injuries, providing

C. Li et al. (Eds.): ICNC 2025, CCIS 2946, pp. 291–301, 2026.
https://doi.org/10.1007/978-981-92-1599-7_25

detailed images of knee structures and injury severity, which are typically analyzed by radiologists to determine the extent of damage [2].

However, manual diagnosis by radiologists can be time-consuming, labor-intensive, and prone to error. Computer-aided diagnosis (CAD) systems help address these issues by using computer vision, machine learning, and deep learning to analyze medical images [3]. These systems face challenges such as accurately segmenting regions of interest and detecting anomalies within them. While traditional machine learning methods have shown promise in knee injury detection, they often require manual feature extraction and lack full automation [4]. With the rapid development of mobile internet technology, the performance of deep learning in many fields exceeds the traditional machine learning methods [5–7]. In this work, a neural network model based on deep learning methods is proposed for knee injury detection, which is more intelligent and has better performance than traditional machine learning methods.

The main contributions of this work are as follows:

(1) The 3D knee MRI data are pre-processed to make them suitable for deep learning training.
(2) A neural network model is established using transfer learning, and different feature extraction modules are validated, with the AlexNet module demonstrated to be the optimal one.

2 Related Works

Knee MRI serves as an invaluable basis for assessing knee injuries and post-operative care, significantly improving diagnostic accuracy for meniscus and cruciate ligament injuries [8,9]. As a non-invasive method, it causes minimal negative impact on patients [10]. While the images in the MRI examination of the knee can help the physician in making a relevant diagnosis, MRI often has a large number of complicated details. Even if the diagnosis is performed by a professionally certified radiologist, the accurate interpretation of MRI is very time-consuming, and there will inevitably be errors [11]. Therefore, it is necessary to develop an automated computer system to assist professional physicians in interpreting knee MRI. Computer-aided systems have many potential applications in many fields, such as helping radiologists quickly determine the severity of a patient's condition when diagnosing a patient [3]. However, to date, due to the multi-dimensional and multi-planar characteristics of MRI, this has limited the applicability of traditional machine learning image analysis methods to knee MRI [3,12], especially the more complex 3D MRI.

Deep learning methods can automatically learn and extract deep-level features of samples. Therefore, the deep learning method is very suitable for automatically extracting the deep features of medical images and establishing a neural network model with complex interpretations [13,14]. Recently, deep learning methods have surpassed many traditional image processing methods in image analysis tasks. In particular, deep learning methods have made breakthroughs in the field of medical imaging, including the classification of skin cancer medical images [15], detection of diabetic retina medical images [16], and detection of tuberculosis medical images [17].

Because training deep learning models requires huge computing resources, including a large number of datasets. To reduce the computing resources required to train the model and reduce the dependence on large datasets, a transfer learning method based on deep learning was proposed for MRI analysis [18]. In this paper, a fully automatic deep learning model based on transfer learning is proposed to detect knee MRI data. The performance of the model proposed in this paper is compared with that of general radiologists and traditional machine learning methods. Experiments show that the model proposed in this paper has better performance.

3 Methodology

This section introduces the related methods proposed in this work, including transfer learning, feature extraction, and related methods for training neural network models.

3.1 CNN Model

As a multi-layer neural network, CNN is particularly effective when dealing with scenes involving a large number of images. The basic structure of a classical CNN consists of one convolutional layer, one pooling layer, followed by one fully connected layer. In detail, the convolutional layer is designed to extract different features of the image. The pooling layer further abstracts the original features, which greatly reduces the training parameters and eases the over-fitting of the model. In summary, CNN allows a collection of features through the convolution kernel's filtering mechanism, which decreases the number of network parameters through convolutional weight sharing and pooling activity.

3.2 Transfer Learning

For small datasets, training the neural network from the beginning through back-propagation, the final performance of the model will be relatively low. To solve this problem and reduce the computer resources required to train the network, pre-trained networks on other large datasets are used to build models. In transfer learning, the weights of the pre-trained network are pre-set, and only the last several layers belonging to the linear classifier will be retrained. In this work, the feature extraction module of the pre-trained networks is used to extract the deep features of MRI. The feature extraction module used in this paper is introduced in Sect. 3.3.

3.3 Feature Extraction Module

Feature extraction involves deriving high-level features from raw pixels for classifier use, performed here through unsupervised methods without category labels. The two feature extraction modules are described below.

AlexNet, introduced in [19], won the 2012 ILSVRC with a top-5 error of 16.4%, significantly outperforming the runner-up at 26.2%. The network consists of eight layers—five convolutional and three fully-connected—containing approximately 60 million parameters. It employs convolutional kernels of sizes 11×11, 5×5, and 3×3, each followed by ReLU activation and max-pooling layers. This work uses an ImageNet pre-trained AlexNet as one feature extractor.

GoogLeNet [20], winner of ILSVRC 2014, reduced the top-5 error to 6.67%. With 22 layers, it uses only half the parameters of AlexNet. To enable deeper architectures while mitigating overfitting and computational cost, GoogLeNet introduced the Inception module, which utilizes sparse connections. Our model also integrates a pre-trained Inception module for feature extraction.

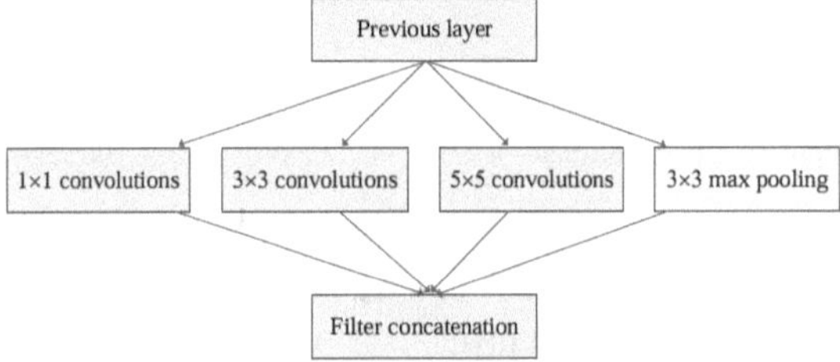

Fig. 1. The Inception module is used to extract the deep features of MRI.

The structure of the Inception module is shown in Fig. 1. It consists of four parts: one 1×1 convolutional layer, one 3×3 convolutional layer, one 5×5 convolutional layer, and one 3×3 max pooling layer. The Inception module combines the operation results of the four components on the channel, with the information of different scales of the image through multiple convolution kernels extracted, and finally performs fusion to obtain a better representation of the image.

3.4 Neural Network Model Based on Transfer Learning

The comprehensive architecture of our proposed transfer learning-based model for 3D knee MRI analysis is detailed in Fig. 2. The processing pipeline begins with a 3D input volume of dimensions $s \times 3 \times 256 \times 256$, where s represents the variable number of sagittal slices in each MRI scan, and the original single-channel grayscale images are replicated across three channels to meet the input requirements of the pre-trained CNN backbones. In the initial feature extraction phase, each 2D slice is processed independently through either an AlexNet or Inception feature extractor, preserving the spatial relationships within each slice while transforming the raw pixel values into high-level semantic features. This process converts the input volume into an intermediate feature representation of size $s \times 256 \times 7 \times 7$, where 256 denotes the number of feature channels extracted by the final convolutional layer, and 7×7 represents the spatially reduced feature map size after successive convolution and pooling operations. Following this, an average pooling layer with a kernel size of 7×7 is applied to each feature map, effectively collapsing the spatial dimensions to 1×1 while maintaining the 256 feature channels, resulting in a tensor of size $s \times 256 \times 1 \times 1$. The model then performs a crucial dimensionality reduction step through max pooling applied along the slice dimension (s), which scans across all s slices and selects the maximum activation value for each of the 256 feature channels, thereby aggregating the most discriminative features from the entire 3D volume into a consolidated 256-dimensional feature vector. This compact representation, which encapsulates the most salient characteristics relevant for ACL injury detection across the complete MRI sequence, is finally propagated through a fully connected layer with sigmoid activation to produce a probabilistic output between 0 and 1, indicating the likelihood of complete ACL rupture versus mild ACL injury. The cross-entropy function is used as the objective function to optimize the model in this paper. The cross-entropy function is calculated by

$$C = -\frac{1}{n} \sum_{i=1}^{n} \left[y_i \ln a_i^L + (1 - y_i) \ln \left(1 - a_i^L\right) \right] \tag{1}$$

where n is the total number of training samples, L denotes the output layer of the network, $y_i \in \{0, 1\}$ is the true label of the i-th sample (1 for complete ACL rupture, 0 for mild ACL injury), and $a_i^L \in (0, 1)$ is the predicted probability output by the sigmoid activation function at layer L for the i-th sample.

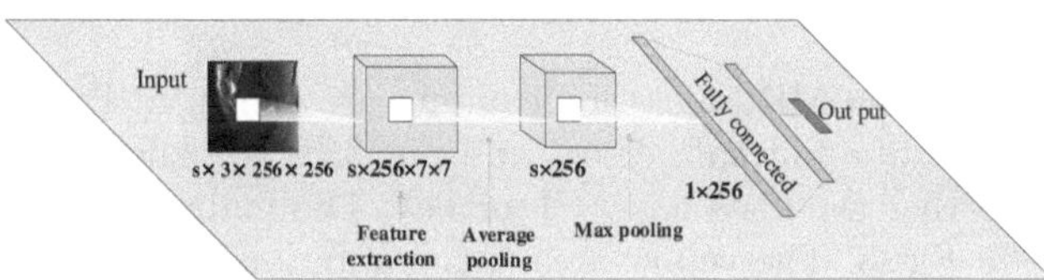

Fig. 2. The neural network model is based on transfer learning for detecting knee ligament injuries.

4 Results

This section first introduces the dataset used in this work and then analyzes the results of different feature extraction modules in transfer learning. A series of comparisons is provided: comparing the results of using different feature extraction modules with other related algorithms. Finally, the contribution of the transfer learning model proposed in this paper in the field of knee MRI is summarized, and the future work prospects are discussed.

4.1 Dataset

The 3D knee MRI dataset used in this study was sourced from the "Rieca" Clinical Hospital archive [21], comprising 969 12-bit grayscale MRI samples. These were acquired between 2007 and 2014 using a Siemens 1.5T scanner with proton density-weighted suppression. Each slice has a thickness of 3–3.6 mm and an inter-slice distance of 0.36 mm. The X-Y plane provides high-resolution information, whereas the Z-axis view is less distinct.

During pre-processing, 25 samples were excluded due to missing DICOM slices or abnormal anatomy, resulting in 944 valid volumes. Their dimensions range from 290 × 300 × 21 to 320 × 320 × 60, with the third value indicating slice count, and each sample is stored as a 3D matrix of different sizes.

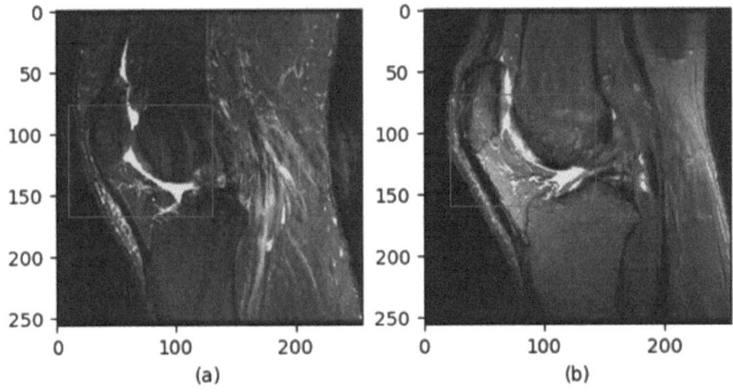

Fig. 3. Processed knee MRI slices for training.

The processed knee MRI slices for training are shown in Fig. 3. Figure 3(a) is a knee MRI slice of a normal person, and Fig. 3(b) is a knee MRI slice with a slight ACL. Whether the knee has ACL or not. The main feature is focused on the red frame in Fig. 3. The yellow area is the anterior cruciate ligament of the knee. However, when the yellow area is partially or completely broken, it means that the knee has an ACL injury.

4.2 Model Performance Evaluation Indicator

In this paper, the binary classification model is used to identify whether the patient's knee injury is a mild ACL injury (without surgical treatment) or a complete ACL rupture (with surgical treatment). The powerful tool for analyzing the binary classification model is the Receiver Operating Characteristic Curve (ROC), and the model with the best performance can be selected by analyzing the ROC.

The binary classification model for assessing knee injury severity can be defined through four diagnostic outcomes: True Positive (TP) occurs when a complete ACL rupture is correctly identified; False Positive (FP) when a rupture is incorrectly diagnosed; True Negative (TN) when the absence of rupture is correctly identified; and False Negative (FN) when an actual rupture is missed by the model.

These four cases can be summarized into a 2 × 2 confusion matrix, as shown in Table 1. Among them, p represents the actual sample is a complete ACL rupture, n represents the actual sample is not a complete ACL rupture, p' represents the diagnosis is a complete ACL rupture, and n' represents the diagnosis is not a complete ACL rupture.

Table 1. Confusion matrix for binary classification

	Actual Positive	Actual Negative
Predicted Positive	True Positive (TP)	False Positive (FP)
Predicted Negative	False Negative (FN)	True Negative (TN)

From the confusion matrix, the False Positive Rate (FPR) and True Positive Rate (TPR) of the binary classification model can be calculated. The calculation formula is shown in (2) and (3):

$$TPR = \frac{TP}{(TP + FN)} \tag{2}$$

$$FPR = \frac{FP}{(FP + TN)} \tag{3}$$

ROC analyzes the performance of the binary classification model through FPR and TPR. The X-axis of the ROC coordinate system is FPR, and the Y-axis is TPR. To calculate the X and Y axes of the ROC coordinate system, the threshold of the binary classification model needs to be set in advance.

The ROC coordinate system divides the coordinate space into two halves by the coordinates (0, 0) and (1, 1). If the coordinates fall in the upper left corner area, it means that the classification effect of the model is better; if the coordinates fall in the lower right corner area, it means that the classification effect of the model is not good. If the FPR and TPR corresponding to all thresholds of the binary classification model are plotted on the ROC coordinate system, a ROC curve can be obtained. The ROC curve is not the final analysis indicator. Each point on the curve can only reflect the classification of each sample, and cannot analyze the overall performance of the model. Only by analyzing all the points on the ROC curve can we judge the quality of a model. At this time, calculating the Area Under Curve (AUC) of the ROC and the coordinate system is the most effective method, and AUC is an index for evaluating the binary classification model. By comparing the AUC of different models, the pros and cons of different model can be analyzed. The larger the AUC value, the better the performance of the model; otherwise, the worse the performance.

4.3 Experimental Results and Analysis

The AlexNet feature extraction module and Inception feature extraction module are used to extract the features of 3D MRI in this paper. The AUC performance of the model is shown in Fig. 4. The models using different feature extraction modules are trained for 25 iterations, respectively. As can be seen from Fig. 4, AlexNet is better than Inception. It uses five times cross-validation when training the model, and also uses GPU acceleration. Its configuration is Nvidia GeForce GTX 1050 Ti with 4GB memory. Due to the computational constraints and the primary focus of this work being on validating the feasibility of the transfer learning approach for 3D MRI analysis, we employed standard and widely adopted training configurations to establish a robust baseline performance, rather than conducting extensive hyperparameter optimization. It can be seen from the experimental results that the model convergence speed and approximation effect of the feature extraction module using AlexNet are better than those using the Inception module as the feature extraction module. Therefore, in the field of knee injury detection, the AlexNet feature extraction module can better extract the deep-level features of 3D MRI than the Inception module.

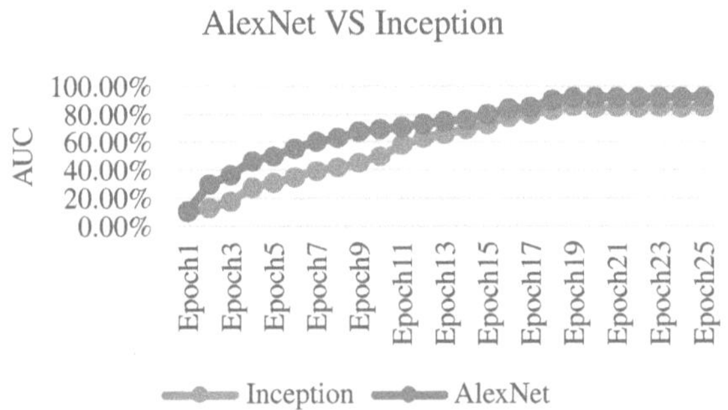

Fig. 4. Performance comparison of feature extraction module using AlexNet and Inception based on transfer learning.

As shown in Table 2, all the methods use the same 3D MRI dataset. The machine learning (ML) model uses two traditional ML algorithms, the support vector machines (SVM) algorithm and the random forest algorithm, which are combined to extract the features from MRI images. The ML model also needs to manually cut the region of interest, and then use the ML model combined with the two ML algorithms for classification training, which makes the identification and detection of knee injuries quite cumbersome. The MRI ligament damage detection model based on transfer learning proposed in this paper overcomes the drawbacks of manual image segmentation, making the detection steps more concise and intelligent. Since the feature extraction module of the pre-trained AlexNet model is used to extract MRI features, the performance of the model is better, with a 3% improvement in AUC.

The 3D MRI data format is a 3D sequence. In this form, most machine learning algorithms cannot directly process them. As a consequence, machine learning methods, such as SVM and RAM, are used in combination with Histogram of oriented gradient (HOG) [22] and Gist [23] when extracting image features. Both SVM and RF are traditional machine learning methods. The process is relatively cumbersome when it comes to detecting knee injuries. The method proposed in this paper can greatly simplify the detection process and is superior to other methods in terms of performance evaluation metrics.

Table 2. Performance comparison with other methods

Methods	AUC
HOG + linSVM [24]	0.894
HOG + RF [24]	0.884
GIST + rbfSVM [24]	0.889
GIST + RF [24]	0.880
This work	**0.924**

5 Conclusion

In this study, a transfer learning-based neural network model was proposed for the automatic detection of Anterior Cruciate Ligament injuries from 3D knee MRI data. The model was designed to leverage pre-trained convolutional neural networks for deep feature extraction, effectively distinguishing between mild ACL injuries and complete ACL ruptures. Comprehensive experiments were conducted on a clinical dataset, demonstrating that the proposed method, particularly with the AlexNet feature extractor, achieved superior performance. Compared to traditional machine learning approaches that rely on manual feature engineering, our model yielded a 3% improvement in the AUC, underscoring its potential for enhancing diagnostic accuracy and reducing clinical misdiagnosis. This work established a more intelligent and automated framework for knee injury detection, which can potentially serve as an early warning system for patients and hospitals to optimize treatment plans and reduce medical costs.

For future work, several directions are recommended to further advance this research. First, the model's performance and generalizability should be validated on larger and more diverse multi-center datasets. Second, other advanced pre-trained models (e.g., ResNet, DenseNet) and feature fusion strategies could be explored to improve the robustness and accuracy of the model. Finally, the applicability of the proposed framework should be investigated for other critical medical imaging tasks, such as the detection of diabetic retinopathy from fundus images and pneumonia from chest X-rays, to assess its broader impact in the field of medical computer-aided diagnosis.

Acknowledgments. This research is supported by the Guangxi Science and Technology Projects under Grant GuiKeAD24010047, the Guangxi Natural Science Foundation under Grants 2022GXNSFFA035028 and 2025GXNSFBA069292, the Guangxi New Engineering and Technical Disciplines Research and Practice Projects under grant XGK2022005, the Basic Ability Enhancement Program for Young and Middle-aged Teachers of Guangxi under Grant 2024KY0074, and a grant (No. BCIC-24-Z7) from Guangxi Key Laboratory of Brain-inspired Computing and Intelligent Chips.

References

1. Lohmander, L.S., Roemer, F.W., Frobell, R.B., Roos, E.M.: Treatment for acute anterior cruciate ligament tear in young active adults. NEJM Evidence **2**(8), EVIDoa2200287 (2023)
2. Chen, C., et al.: High-resolution oblique coronal MRI at optimal flexed-knee angle: a novel imaging method for enhanced anterior cruciate ligament tear diagnosis. J. Orthop. Surg. Res. **19**(1), 456 (2024)
3. Huang, E.P., et al.: Criteria for the translation of radiomics into clinically useful tests. Nat. Rev. Clin. Oncol. **20**(2), 69–82 (2023)
4. Farhadi, F., Barnes, M.R., Sugito, H.R., Sin, J.M., Henderson, E.R., Levy, J.J.: Applications of artificial intelligence in orthopaedic surgery. Front. Med. Technol. **4**, 995526 (2022)
5. Luo, Y., Wan, L., Liu, J., Harkin, J., Cao, Y.: An efficient, low-cost routing architecture for spiking neural network hardware implementations. Neural Process. Lett. **48**(3), 1777–1788 (2018)
6. Liu, J., Huang, Y., Luo, Y., Harkin, J., McDaid, L.: Bio-inspired fault detection circuits based on synapse and spiking neuron models. Neurocomputing **331**, 473–482 (2019)
7. Liu, J., Li, M., Luo, Y., Yang, S., Qiu, S.: Human body posture recognition using wearable devices. In: International Conference on Artificial Neural Networks, pp. 326–337 (2019)
8. Mundal, K., Geeslin, A.G., Solheim, E., Inderhaug, E.: Differences between traumatic and degenerative medial meniscus posterior root tears: a systematic review. Am. J. Sports Med. **53**(1), 228–233 (2025)
9. Arnold, T.C., Freeman, C.W., Litt, B., Stein, J.M.: Low-field MRI: clinical promise and challenges. J. Magn. Reson. Imaging **57**(1), 25–44 (2023)
10. Cance, N., et al.: Delaying anterior cruciate ligament reconstruction increases the rate and severity of medial chondral injuries: a retrospective cohort study. Bone Joint J. **105**(9), 953–960 (2023)
11. Zhang, L., Li, M., Zhou, Y., Lu, G., Zhou, Q.: Deep learning approach for anterior cruciate ligament lesion detection: evaluation of diagnostic performance using arthroscopy as the reference standard. J. Magn. Reson. Imaging **52**(6), 1745–1752 (2020)
12. Oakden-Rayner, L., Carneiro, G., Bessen, T., Nascimento, J.C., Bradley, A.P., Palmer, L.J.: Precision radiology: predicting longevity using feature engineering and deep learning methods in a radiomics framework. Sci. Rep. **7**(1), 1648 (2017)
13. LeCun, Y., Bengio, Y., Hinton, G.: Deep learning. Nature **521**(7553), 436–444 (2015)
14. Ahmed, S.F., et al.: Deep learning modelling techniques: current progress, applications, advantages, and challenges. Artif. Intell. Rev. **56**(11), 13521–13617 (2023)

15. Wei, M.L., Tada, M., So, A., Torres, R.: Artificial intelligence and skin cancer. Front. Med. **11**, 1331895 (2024)
16. Dai, L., et al.: A deep learning system for predicting time to progression of diabetic retinopathy. Nat. Med. **30**(2), 584–594 (2024)
17. Jasmine Pemeena Priyadarsini, M., et al.: Lung diseases detection using various deep learning algorithms. J. Healthc. Eng. **2023**(1), 3563696 (2023)
18. Chen, S., Ma, K., Zheng, Y.: Med3D: Transfer learning for 3D medical image analysis. arXiv preprint arXiv:1904.00625 (2019)
19. Krizhevsky, A., Sutskever, I., Hinton, G.E.: ImageNet classification with deep convolutional neural networks. Adv. Neural Inf. Process. Syst. **25** (2012)
20. Zeng, G., He, Y., Yu, Z., Yang, X., Yang, R., Zhang, L.: Going deeper with convolutions Christian. In: Proceedings of IEEE Conference on Computer Vision and Pattern Recognition (CVPR), pp. 1–9 (2015)
21. Prasoon, A., et al.: Deep feature learning for knee cartilage segmentation using a Triplanar convolutional neural network. In: Medical Image Computing and Computer-Assisted Intervention, pp. 246–253 (2013)
22. Dalal, N., Triggs, B.: Histograms of oriented gradients for human detection. In: 2005 IEEE Computer Society Conference on Computer Vision and Pattern Recognition (CVPR'05). vol. 1, pp. 886–893 (2005)
23. Oliva, A., Torralba, A.: Modeling the shape of the scene: a holistic representation of the spatial envelope. Int. J. Comput. Vision **42**(3), 145–175 (2001)
24. Štajduhar, I., Mamula, M., Miletić, D., Uenal, G.: Semi-automated detection of anterior cruciate ligament injury from MRI. Comput. Methods Programs Biomed. **140**, 151–164 (2017)

Observer-Based Dynamic Event-Triggered Tracking Control for Nonlinear MASs with Unknown Time-Varying Sensor Sensitivity

Hongxuan Song, Guangliang Liu(✉), Yingnan Pan, and Wen Yang

Bohai University, Jinzhou 121013, China
liuguangliang17@163.com

Abstract. This paper addresses the consensus tracking control problem for nonlinear multi-agent systems with unknown and time-varying sensor sensitivity. To reconstruct the real system state from the inaccurate system input signal, a state observer is designed by using the damaged input signal itself. Then, a dynamic event-triggered strategy is proposed, which incorporates the amplitude of the controller and the hyperbolic tangent function into its triggering conditions. Its advantage lies in avoiding unnecessary information transmission when implementing consistent control, thereby reducing the communication burden. Based on Lyapunov theory, it is rigorously proven that all signals in the closed-loop system are semi-globally uniformly ultimately bounded, effectively eliminating Zeno behavior. Finally, the practical value of the method is proved by an example simulation study.

Keywords: Consensus tracking control · state observer · nonlinear multiagent systems · time-varying sensor sensitivity

1 Introduction

With the in-depth application of cooperative control theory in scenarios, the stable operation of multi-agent systems (MASs) increasingly relies on reliable state feedback. However, aging sensors and drifting components cause unknown, time-varying sensitivity, making the true state unobtainable and endangering closed-loop stability. To solve this influence in nonlinear MASs, Du *et al.* [1] completed adaptive compensation with Nussbaum gain technology. Li *et al.* [2] further embedded the Nussbaum function to achieve consensus control in the case of sensitive deviations in output measurements. However, a systematic treatment of sensors with unknown time-varying gains remains largely unexplored.

It is a major challenge to ensure the stability of multi-agent systems while reducing the communication load. Dynamic event-triggered control [3] satisfies it by adjusting the trigger threshold online to reduce bandwidth while maintaining convergence. In [4], the upper bound of sampling error was updated for nonlinear

C. Li et al. (Eds.): ICNC 2025, CCIS 2946, pp. 302–311, 2026.
https://doi.org/10.1007/978-981-92-1599-7_26

MASs in real time by using the intermittent transmission of state variables. In [5], the trigger threshold of MASs was adaptively corrected according to the deviation between the estimated state of the neighbor at the trigger time and its quantized value. However, the above strategies do not include the amplitude of control input signal into the event-triggered threshold function. Inspired by the above research, this paper addresses the consensus control strategy of nonlinear MASs with unknown time-varying sensor sensitivity via a dynamic ETM. The advantage is that the sensitivity of unknown time-varying sensor in the system is eliminated, and the number of event triggering is reduced.

2 Graph Theory

The information interaction topology is described by a directed graph $\mathcal{G} = (\mathcal{V}, \mathcal{E})$, where $\mathcal{V} = (\mathcal{V}_1, \ldots, \mathcal{V}_N)$ is a nonempty set of nodes, and $\mathcal{E} \subseteq \mathcal{V} \times \mathcal{V}$ is the set of edges. The edge from node k to node m is denoted as $(\mathcal{V}_m, \mathcal{V}_k) \in \mathcal{E}$. If the agent m can receive the data of the agent k, then the adjacent weight $a_{m,k} > 0$, otherwise $a_{m,k} = 0$. The in-degree matrix $\mathcal{D} = \text{diag}\{d_1, \ldots, d_N\}$ is defined with $d_m = \sum_{k=1}^{N} a_{m,k}$, and then the Laplacian matrix $\mathcal{L} = \mathcal{D} - \mathcal{A}$ is obtained.

Assumption 1. *[6] If every follower has a directed path to the leader, $\mathcal{G}$ contains a spanning tree rooted at node **0**, with y_0 and $\ddot{y}_0$ bounded.*

Lemma 1. *[6] Define the leader adjacency matrix $\mathcal{B} = \text{diag}\{a_{m,0}\} \in R_+^{N\times N}$, where $a_{m,0} = 0$ indicates that agent m cannot get information from leader, or else, $a_{m,0} = 1$. Then, $\mathcal{L} + B$ is nonsingular.*

3 System Description

3.1 System Model

Define the dynamics of the mth agent $(m = 1, \ldots, N)$

$$\begin{cases} \dot{x}_{m,p} = \varsigma_{m,p}(\bar{x}_{m,p}) + x_{m,p+1} \\ \dot{x}_{m,n} = \varsigma_{m,n}(\bar{x}_{m,n}) + u_m(t) \\ \quad y_m = \varrho_m x_{m,1}, \end{cases} \tag{1}$$

where $u_m(t)$ is the control input signal. $\bar{x}_{m,p} = [x_{m,1}, ..., x_{m,p}]^T$ and $\bar{x}_{m,n} = [x_{m,1}, ..., x_{m,n}]^T$ are state vectors with $1 \leq p \leq n-1$. Unknown nonlinear functions are represented as $\varsigma_{m,p}(\bar{x}_{m,p})$ and $\varsigma_{m,n}(\bar{x}_{m,n})$. y_m denotes an inaccurate system output signal in case of sensor sensitivity decline. ϱ_m denotes the unknown time-varying sensor sensitivity. Finally, we have $x_{m,1} = \gamma_m(t) y_m$ with $\gamma_m(t) = 1/\varrho_m$.

3.2 Error Transformation

The asymptotic error can be defined as

$$\begin{aligned} z_{m,1} &= \sum_{k=1}^{N} a_{mk}(x_{m,1} - x_{k,1}) + a_{m,0}(x_{m,1} - y_0) \\ z_{m,p} &= \hat{x}_{m,p} - o_{m,p},\ \partial_{m,p} = o_{m,p} - \alpha_{m,p-1}, \end{aligned} \tag{2}$$

where $y_0 \in \mathbb{R}$ indicates the output of the leader. $o_{m,p}$ is the first-order filter output. $\alpha_{m,p-1}$ is the virtual design control function. $\partial_{m,p}$ is the filter output error. Then, one has

$$\tau_{m,p}\dot{o}_{m,p} + o_{m,p} = \alpha_{m,p-1}, \quad o_{m,p}(0) = \alpha_{m,p-1}(0) \tag{3}$$

where $\tau_{m,p} > 0$ is the design constant.

Lemma 2. *[7] Since $\varsigma(x)$ is continuous on the compact set Ω, for any $\varepsilon > 0$ there exists an FLS that satisfies*

$$\sup_{x\in\Omega}|\varsigma(x) - \Theta^T\Phi(x)| \leq \varepsilon, \tag{4}$$

where ε is a fuzzy approximated error and κ is the number of fuzzy rules. $\Phi(x)$ and $\Theta^T = [\Theta_1, ..., \Theta_\kappa]$ represent fuzzy basis function and ideal weight vector respectively, where $0 < \Phi(x)^T\Phi(x) \leq 1$ and $\Phi(x) = [\Phi_1(x), ..., \Phi_\kappa(x)]^T / \sum_{k=1}^{N}\Phi_k(x)$.

3.3 Dynamic ETM

The dynamic ETM is designed as follows:

$$u_m(t) = \bar{u}_m(t_k) \quad \forall t \in [t_k, t_{k+1}) \tag{5}$$

$$t_{k+1} = \inf\left\{t > t_k : |e_m(t)| \geq \frac{\Pi_m + \beta_{m,1}\beta_{m,2}}{\beta_{m,1}(\beta_{m,3} + |\alpha_{m,n}|)}\right\} \tag{6}$$

$$\bar{u}_m(t) = -\frac{\Pi_m + \beta_{m,1}\beta_{m,2}}{\beta_{m,1}(\beta_{m,3} + |\alpha_{m,n}|)} tanh[\frac{z_{m,n}(\Pi_m + \beta_{m,1}\beta_{m,2})}{\beta_{m,1}(\beta_{m,3} + |\alpha_{m,n}|)\varkappa}] + \alpha_{m,n}, \tag{7}$$

where $\varkappa$, $\beta_{m,1}$, $\beta_{m,2}$ and $\beta_{m,3}$ are positive constants. $\alpha_{m,n}$ is a virtual control signal. As $|\alpha_{m,n}|$ decreases, the threshold increases, thus reducing the trigger frequency. Define the event-sampling error as $e_m(t) = \bar{u}_m(t) - u_m(t)$. Π_m can be decided as:

$$\dot{\Pi}_m = -\beta_{m,4}\Pi_m + \beta_{m,1}(\beta_{m,2} - (\beta_{m,3} + |\alpha_{m,n}|)|e_m|), \tag{8}$$

where $\beta_{m,4}$ is positive constant. $\Pi_m(0)$ denotes the positive initial value of Π_m.

Lemma 3. *[8] Define $z_1 = (z_{1,1}, z_{2,1}, ..., z_{N,1})^T$, $y = (y_1, y_2, ..., y_N)^T$ and $\bar{y}_r = (y_r, y_r, ..., y_r)^T$, then one gets*

$$||y - \bar{y}|| \leq ||z_1||/\bar{\sigma}(\mathcal{L} + \mathcal{B}), \tag{9}$$

where $\bar{\sigma}(\mathcal{L} + \mathcal{B})$ is the minimum singular value of $\mathcal{L} + \mathcal{B}$.

4 Observer Design and Analysis

Due to the influence of unknown time-varying sensor sensitivity, we design the following state observer

$$\begin{cases} \dot{\hat{x}}_{m,p} = \hat{x}_{m,p+1} + \hat{\Theta}_{m,p}^T \Phi_{m,p}(\hat{x}_{m,p+1}) + h_{m,p}(\hat{\gamma}_m y_m - \hat{x}_{m,1}) \\ \dot{\hat{x}}_{m,n} = u_m(t) + \hat{\Theta}_{m,n}^T \Phi_{m,n}(\hat{x}_{m,n}) + h_{m,n}(\hat{\gamma}_m y_m - \hat{x}_{m,1}), \end{cases} \tag{10}$$

where $\hat{x}_{m,p}$, $\hat{\Theta}_{m,p}$, and $\hat{\gamma}_m$ are the estimations of $x_{m,p}$, $\Theta_{m,p}$, and γ_m, respectively. $h_{m,p}$ $(p = 1, \ldots, n)$ is design parameter.

Define the following error variables as: $\tilde{x}_m = [\tilde{x}_{m,1}, ..., \tilde{x}_{m,n}]^T$, $\tilde{\Theta}_{m,p} = \Theta_{m,p} - \hat{\Theta}_{m,p}$ and $\tilde{\gamma}_m = \gamma_m - \hat{\gamma}_m$ with $\tilde{x}_{m,p} = x_{m,p} - \hat{x}_{m,p}$ $(p = 1, 2, ..., n)$. Then, we get

$$\begin{aligned} \dot{\tilde{x}}_m =& A_m \tilde{x}_m + H_m \tilde{\gamma}_m y_m + \sum_{p=1}^{n-1} I_{m,p} \tilde{\Theta}_{m,p}^T \Phi_{m,p} \left(\hat{\tilde{x}}_{m,p+1}\right) \\ &+ I_{m,n} \tilde{\Theta}_{m,n}^T \Phi_{m,n} \left(\hat{\tilde{x}}_{m,n}\right) + \varepsilon, \end{aligned} \tag{11}$$

where $A_m = \begin{bmatrix} -h_{m,1} & & \\ -h_{m,2} & I_{m,n-1} & \\ \cdots & & \\ -h_{m,n} & 0 \quad \ldots & 0 \end{bmatrix}$, $\Theta_{m,n} = [0, 0, ...1]^T$, $\varepsilon = [\varepsilon_{m,1}, \varepsilon_{m,2}, ..., \varepsilon_{m,n}]^T$

$H_m = [h_{m,1}, ..., h_{m,n}]^T$, $\Theta_{m,n-1} = [\underbrace{0, ...1}_{k}, ..., 0]^T$.

Choose the Lyapunov function as

$$V_0 = \frac{1}{2} \sum_{m=1}^{N} \tilde{x}_m^T R_m \tilde{x}_m. \tag{12}$$

By using Young's inequality, we can derive

$$\begin{aligned} \dot{V}_{m,0} \leq& - \sum_{m=1}^{N} (\lambda_{m,\min}(Q_m) - C_0) \|\tilde{x}_m\|^2 + \sum_{m=1}^{N} \sum_{p=1}^{n} \frac{1}{2} \|R_m\|^2 \|\tilde{\Theta}_{m,p}\|^2 \\ &+ \sum_{m=1}^{N} \frac{1}{2} \|R_m\|^2 \|H_m\|^2 \tilde{\gamma}_m^2 y_m^2 + \frac{1}{2} \bar{\varepsilon}^2, \end{aligned} \tag{13}$$

where $C_0 = 1 + (1/2)\|R_m\|^2$.

5 Controller Design

Step 1: Select the Lyapunov candidate function as:

$$V_{m,1} = \frac{1}{2} z_{m,1}^2 + \frac{1}{2l_{m,1}} \tilde{\Theta}_{m,1}^T \tilde{\Theta}_{m,1} + \frac{1}{3q_{m,1}} |\tilde{\gamma}_m|^3. \tag{14}$$

where $l_{m,1}$ and $q_{m,1}$ stand for positive constants.

Employing Young's inequality and Lemma 2, we design the virtual controller and adaptive law as follows:

$$\alpha_{m,1} = \frac{1}{q_m}\left(a_{m,0}\dot{y}_0 - q_m\hat{\Theta}_{m,1}^T\Phi_{m,1}(\hat{x}_{m,1})\right) - c_{m,1}\hat{z}_{m,1} \tag{15}$$

$$\dot{\hat{\Theta}}_{m,1} = q_m l_{m,1}\hat{z}_{m,1}\Phi_{m,1}(\hat{x}_{m,1}) - \eta_{m,1}\hat{\Theta}_{m,1}, \tag{16}$$

where $c_{m,1}$ and $\eta_{m,1}$ are positive design parameters. $\hat{z}_{m,1} = q_m\hat{\gamma}_m y_m - d_m\hat{\gamma}_k y_k - a_{m,0}y_0$. $q_m = \sum_{k=1}^{N} a_{mk} + a_{m,0}$.

The compensator signal $\hat{\gamma}_m$ is represented as

$$\dot{\hat{\gamma}}_m = \begin{cases} \mathfrak{S}_m, & \mathfrak{S}_m > 0 \\ 0, & \mathfrak{S}_m \leq 0, \end{cases} \tag{17}$$

where $\mathfrak{S}_m = q_{m,1}\Big((1/2)\|R_m\|^2\|H_m\|^2 y_m^2 + (1/2+c_{m,1})(q_m^2 y_m^2 + d_m^2 y_k^2 - a_{m,1}\hat{\bar{\gamma}}_m\Big)$ with $\hat{\bar{\gamma}}_m = \max\{\hat{\gamma}_m, \hat{\gamma}_k\}$. $a_{m,1}$ is a positive design parameter.

Then, we can get

$$\begin{aligned} \dot{V}_{m,1} \leq & -\bar{c}_{m,1}z_{m,1}^2 + \frac{q_m^2}{2}\left(z_{m,2}^2 + \tilde{x}_{m,2}^2 + \partial_{m,2}^2 + 2\|\tilde{\Theta}_{m,1}\|^2\right) + D_{m,1} \\ & + \frac{\mu_{m,1}}{f_{m,1}}\tilde{\Theta}_{m,1}^T\hat{\Theta}_{m,1} - \frac{1}{2}\|R_m\|^2\|H_m\|^2\tilde{\bar{\gamma}}_m^2(x_{m,1})^2 - \bar{a}_{m,1}|\tilde{\bar{\gamma}}_m|^3, \end{aligned} \tag{18}$$

where $D_{m,1} = q_m^2(\varepsilon_{m,1}^*)^2/2 + q_m^2\|\Theta_{m,1}\|^2 + (a_{m,1}/3)\breve{\bar{\gamma}}_m^3 + (1/3)(\breve{\gamma}_m/q_{m,1})^3$ and $\bar{a}_{m,1} = (a_{m,1}-2)/3 > 0$.

Step p (p=2,...,n-1) : Select the Lyapunov candidate function as:

$$V_{m,p} = V_{m,p-1} + \frac{1}{2}z_{m,p}^2 + \frac{\tilde{\Theta}_{m,p}^T\tilde{\Theta}_{m,p}}{2l_{m,p}} + \frac{1}{2}\partial_{m,p}^2, \tag{19}$$

where $l_{m,p}$ is a positive design parameters.

Employing Young's inequality and Lemma 2, we gets

$$\begin{aligned} \alpha_{m,p} = & -c_{m,p}z_{m,p} - h_{m,p}(\hat{\gamma}_m y_m - \hat{x}_{m,1}), \\ & + \dot{o}_{m,p} - \hat{\Theta}_{m,p}^T\Phi_{m,p}(\hat{x}_{m,p}), \end{aligned} \tag{20}$$

$$\dot{\hat{\Theta}}_{m,p} = l_{m,p}z_{m,p}\Phi_{m,p}(\hat{x}_{m,p}) - \mu_{m,p}\hat{\Theta}_{m,p} \tag{21}$$

where $c_{m,p}$ and $\mu_{m,p}$ are positive design parameters.

Then, we can get

$$\begin{aligned} \dot{V}_{m,p} \leq & -\sum_{s=1}^{p}\bar{c}_{m,s}z_{m,s}^2 + \frac{q_m^2}{2}\tilde{x}_{m,2}^2 + \frac{1}{2}\partial_{m,p+1}^2 - \sum_{s=3}^{p}g_{m,s}\partial_{m,s}^2 + D_{m,p} \\ & + \sum_{s=1}^{p}\frac{\eta_{m,s}}{l_{m,s}}\tilde{\Theta}_{m,s}^T\hat{\Theta}_{m,s} - \bar{a}_{m,1}\|\tilde{\bar{\gamma}}_m\|^3 - \frac{1}{2}\|R_m\|^2\|H_m\|^2\tilde{\bar{\gamma}}_m^2 y_m^2 \\ & + \sum_{s=2}^{p}\frac{1}{2}\|\tilde{\Theta}_{m,s}\|^2 + q_m^2\|\tilde{\Theta}_{m,1}\|^2 + \sum_{s=1}^{p}\frac{1}{2}z_{m,s+1}^2. \end{aligned} \tag{22}$$

where $\bar{c}_{m,s} = c_{m,s} - (5/2)$, $g_{m,s} = -1/2 + (1/\tau_{m,s}) - (M_{m,p-1}^2/2\sigma_m)$, $g_{m,2} = -1/2 + (1/\tau_{m,2}) - (M_{m,1}^2/2\sigma_m) - (q_m^2/2)$, and $D_{m,p} = D_{m,p-1} + \|P_{m,p}\|^2 + \sigma_m/2$. $M_{m,p-1}$ and σ_m are positive parameters with $|\dot{\alpha}_{m,p-1}| \leq M_{m,p-1}$.

Step n : Select the Lyapunov candidate function as:

$$V_{m,n} = V_{m,n-1} + \frac{1}{2} z_{m,n}^2 + \frac{1}{2 l_{m,n}} \tilde{\Theta}_{m,n}^T \tilde{\Theta}_{m,n} + \frac{1}{2} \partial_{m,n}^2, \tag{23}$$

where $l_{m,n}$ is a positive design parameter.

Employing Young's inequality and Lemma 2, we get

$$\alpha_{m,n} = -\left(c_{m,n} + 1\right) z_{m,n} - h_{m,n}(\hat{\gamma}_m y_m - \hat{x}_{m,1}), \tag{24}$$

$$\dot{\hat{\Theta}}_{m,n} = -l_{m,n} z_{m,n} \Phi_{m,n}\left(\hat{x}_{m,n}\right) - \eta_{m,n} \hat{\Theta}_{m,n}, \tag{25}$$

where $c_{m,n}$ and $\eta_{m,n}$ are positive design parameters.

Then, one gets

$$\dot{V}_{m,n} \leq \dot{V}_{m,n-1} - c_{m,n} z_{m,n}^2 + \frac{\mu_{m,n}}{f_{m,n}} \tilde{\Theta}_{m,n}^T \hat{\Theta}_{m,n} - g_{m,n} \partial_{m,n}^2 + D_{m,n}, \tag{26}$$

where $D_{m,n} = (\|P_{m,n}\|^2/2) + (\sigma_n/2) + 0.2785\varkappa$. $g_{m,n} = (1/\tau_{m,n}) - (M_{m,n-1}^2/2\sigma_n) - (1/2\tau_{m,n}^2) - 1/2$.

6 Stability Analysis

Theorem 1. *For the nonlinear MASs (1) subject to unknown time-varying sensor sensitivity, and under Assumptions 1-2, the event-triggered controller (24), adaptive update laws (16), (21) and (25) ensure that all signal in the closed-loop system is semi-globally uniformly ultimately bounded (SGUUB) and that Zeno behaviour is excluded. By properly tuning the design parameters, the following inequality can be made to hold for any prescribed tolerance* $\varepsilon^* > 0$*:*

$$\lim_{t \to \infty} \|y - \bar{y}_r\| \leq \varepsilon^*. \tag{27}$$

Proof. Design the entire Lyapunov function as

$$V = \sum_{m=1}^{N} V_{m,n} + V_0 = \sum_{m=1}^{N} V_m. \tag{28}$$

Then, one has

$$\begin{aligned} \dot{V}_m \leq & -\sum_{s=1}^{n} \tilde{c}_{m,s} z_{m,s}^2 - \sum_{s=1}^{n} J_{m,s} \tilde{\Theta}_{m,s}^T \tilde{\Theta}_{m,s} + \bar{D}_{m,n} \\ & - Q^* \|\tilde{x}_m\|^2 - \bar{a}_{m,1} \|\bar{\tilde{\gamma}}_m\|^3 - \sum_{s=2}^{n} g_{m,s} \partial_{m,s}^2, \end{aligned} \tag{29}$$

Choose a positive definite matrix Q_m such that $\lambda_{m,\min}(Q_m) - C_0 - (q_m^2/2) = Q^*$, where Q^* is a positive constant. Define $J_{m,s} = \eta_{m,s}/(2l_{m,s}) - 1/2 - \|R_m\|^2/2 - q_m^2$, $\bar{C}_m = \min\{2Q^*, 2\tilde{c}_{m,s}, 2g_{m,s}, 2J_{m,s}, 2\bar{a}_{m,1}\}$ and $\bar{D}_m = D_{m,p} + D_{m,n} + \sum_{s=1}^{n}(\eta_{m,s}/2l_{m,s})\|P_{m,s}\|^2 + \bar{\varepsilon}^2/2$.

Finally, we get

$$\frac{\bar{D}_m}{\bar{C}_m} \leq \frac{\varepsilon}{2}(\bar{\sigma}(\mathcal{L} + \mathcal{B})^2. \tag{30}$$

According to (30) and Lemma 3, one can get (27).

Now we prove that there exists $t^* > 0$ such that $\forall k \in \mathbf{z}^+$, $\{t_{k+1} - t_k\} \geq t^*$. To this end, by recalling $e_m(t) = \bar{u}_m(t) - u_m(t)$, $\forall t \in [t_k, t_{k+1})$, we have

$$\frac{d}{dt}|e| = \frac{d}{dt}(e * e)^{\frac{1}{2}} = \text{sign}(e)\dot{e} \leq |\dot{\bar{u}}_m(t)|. \tag{31}$$

We obtain that

$$\dot{\bar{u}}_m = \dot{\alpha}_n + \frac{\bar{u}_m}{\mathbf{I}}\dot{\mathbf{I}} + \frac{\bar{u}_m}{\omega_m}\dot{\omega}_m. \tag{32}$$

Because every closed-loop signal that makes up the vector $\bar{u}_m$ is bounded, there is a constant $\kappa > 0$ such that $|\bar{u}_m| \leq \kappa$. Since $e(t_k) = 0$ and the minimum value of $e(t)$ approaches ℓ as $t \to t_{k+1}^-$, the shortest possible inter-execution interval t^* must obey $t^* \geq \ell/\kappa$, which rules out Zeno behaviour.

7 Simulation Result

In order to verify the effectiveness of the proposed control scheme, we carry out the following simulation. Firstly, the communication topology is described in Fig. 1. The dynamics of each agent is modeled as :

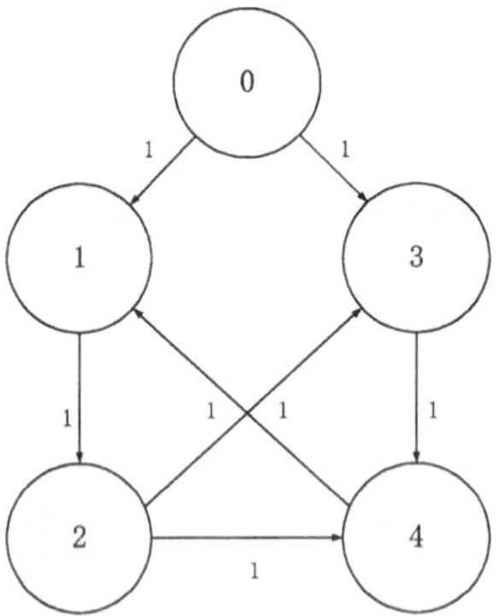

Fig. 1. Topology.

$$\begin{cases} \dot{x}_{m,1} = x_{m,2} + 0.1\cos(0.1x_{m,1}) - 0.1 \\ \dot{x}_{m,2} = u_m + 0.1x_{m,2}\cos(0.2x_{m,1}) \\ y_m = (1 - 0.1|sin(0.1t)|)x_{m,1}. \end{cases} \tag{33}$$

Defined an adjacency matrix $\mathcal{A}$ and a Laplacian matrix $\mathcal{L}$ as follows, respectively.

$$\mathcal{A} = \begin{bmatrix} 0\,0\,0\,1 \\ 1\,0\,0\,0 \\ 0\,1\,0\,0 \\ 0\,1\,1\,0 \end{bmatrix}, \ \mathcal{L} = \begin{bmatrix} 1 & 0 & 0 & -1 \\ -1 & 1 & 0 & 0 \\ 0 & -1 & 1 & 0 \\ 0 & -1 & -1 & 2 \end{bmatrix}$$

Some initial values are selected as: $x_{1,1}(0) = x_{3,1}(0) = 0.5$, $x_{2,1}(0) = x_{4,1}(0) = -0.5$, $x_{m,2}(0) = 0.5$, $\hat{w}_{m,1}(0) = \hat{w}_{m,2}(0) = 0.5$, $\hat{x}_{m,1}(0) = \hat{x}_{m,2}(0) = 0$, $\hat{\gamma}_m(0) = 1.2$ and $\varpi_m(0) = 0$. Some constant parameters are selected as: $c_{m,1} = 8$, , $c_{m,2} = 6.5$, $\sigma_{m,1} = 0.5$, $h_{m,1} = 1.5$, $h_{m,2} = 1$, $l_{m,1} = l_{m,2} = 0.1$, $\eta_{m,1} = \mu_{m,2} = 1$, $\tau_{m,2} = 0.02$, $q_{m,1} = 0.004$, $a_{1,1} = a_{2,1} = a_{3,1} = 8$, $a_{4,1} = 12$, $\beta_{m,1} = \beta_{m,2} = 0.1$, $\beta_{m,3} = \beta_{m,4} = 1$, $\varkappa = 5$, and $m = 1, 2, 3, 4$.

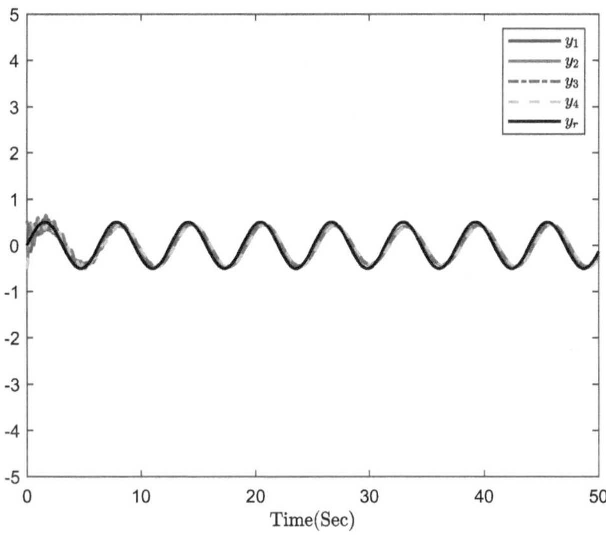

Fig. 2. Trajectories of the system output y_m and reference signal y_r.

Figure 2 illustrates the tracking performance of four followers relative to a leader. As revealed in Fig. 3, the proposed state observer accurately reconstructs the true states $x_{m,1}$ and $x_{m,2}$ despite unknown time-varying sensor sensitivity. The triggering intervals of the dynamic ETM are depicted in Fig. 4. When the system approaches equilibrium, the proposed scheme markedly lowers the trigger frequency.

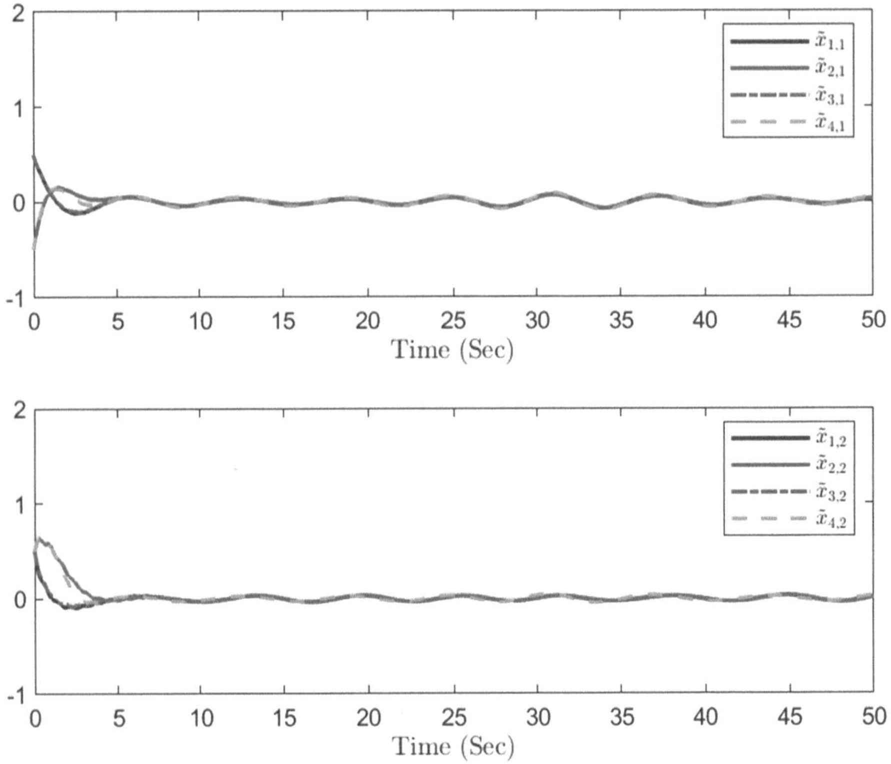

Fig. 3. Observed errors $\tilde{x}_{m,1}$ and $\tilde{x}_{m,2}$.

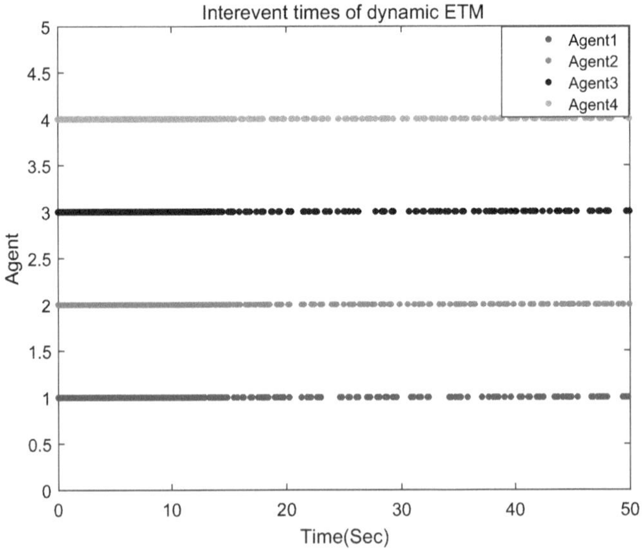

Fig. 4. Interevent times of dynamic ETM.

8 Conclusion

We address the adaptive consensus tracking control problem for nonlinear MASs with unknown time-varying sensor sensitivity. A state observer has been designed from the damaged output signal of the system to reconstruct the unmeasurable state. A dynamic ETM has been designed whose threshold is shaped by both the amplitude of the controller and a hyperbolic-tangent function. Lyapunov analysis indicates that all signals are bounded, and numerical simulations confirm these theoretical findings.

Acknowledgments. This work was supported by the Natural Science Foundation of Liaoning Province under Grant 2025-MS-280, in part by the National Natural Science Foundation of China under Grant 62503062, and also by the Natural Science Foundation of Liaoning Province (2025-YQ-18) and the Basic Research Project of the Educational Department of Liaoning Province (LJ232410167028).

Disclosure of Interests. The authors have no competing interests to declare that are relevant to the content of this article.

References

1. Du, Z., Liang, H., Ahn, C.K.: Adaptive fuzzy control for multi-agent systems with unknown measurement sensitivity via a simplified backstepping approach. IEEE Trans. Circuits Syst. II, Exp. Briefs **69**(6), 2862–2866 (2022)
2. Li, K., Hua, C., You, X., Guan, X.: Distributed output-feedback consensus control of multiagent systems with unknown output measurement sensitivity. IEEE Trans. Autom. Control **66**(7), 3303–3310 (2020)
3. Lv, C., Liu, G., Pan, Y., Hu, Z., Lei, Y.: Event-based distributed cooperative neural learning control for nonlinear multiagent systems with time-varying output constraints. Neural Netw. **187**, 107383 (2025)
4. Xu, B., Li, Y., Hou, Z., Ahn, C.K.: Dynamic event-triggered reinforcement learning-based consensus tracking of nonlinear multi-agent systems. IEEE Trans. Circuits Syst. I, Reg. Papers **70**(5), 2120–2132 (2023)
5. Wang, Y., Zhu, F.: Distributed dynamic event-triggered control for multi-agent systems with quantization communication. IEEE Trans. Circuits Syst. II, Exp. Briefs **71**(4), 2054–2058 (2024)
6. Liang, H., Wang, W., Pan, Y., Lam, H.K., Sun, J.: Synchronous MDADT-based fuzzy adaptive tracking control for switched multiagent systems via modified self-triggered mechanism. IEEE Trans. Fuzzy Syst. **32**(5), 2876–2889 (2024)
7. Liang, H., Li, D., Pan, Y., Li, T.: Adaptive predictor-based event-triggered tracking control for nonlinear multiagent systems with fault detect-switch-compensate mechanism. IEEE Trans. Syst., Man, Cybern., Syst. **54**(2), 1276–1287 (2023)
8. Shang, Y., Chen, B., Lin, C.: Consensus tracking control for distributed nonlinear multiagent systems via adaptive neural backstepping approach. IEEE Trans. Syst., Man, Cybern., Syst. **50**(7), 2436–2444 (2018)

Energy Management for Ship-Integrated Energy System Considering Carbon Emission Intensity Grading Mechanism with EEOI

Qitong Tang[1], Yuxin Zhang[1], Xiaoyang Gao[2(✉)], and Liang'en Yuan[1]

[1] Dalian Maritime University, Navigation College, Dalian 116026, China
[2] Dalian Maritime University, School of Maritime Economics and Management, Dalian 116026, China
xiaoyang_gao@yeah.net

Abstract. In the context of escalating global carbon emissions and rising costs of traditional fossil fuels, the singular reliance on fossil fuel-based equipment for energy supply is no longer sustainable. Consequently, an integrated energy system incorporating both renewable energy sources and conventional fossil fuel equipment, known as the Ship-Integrated Energy System (S-IES), has emerged. As a novel ship energy architecture, the S-IES enables the efficient synergistic utilization of heterogeneous energy sources. To ensure the safe, reliable, and green operation of the S-IES, an energy management model is first established. This model facilitates the synergistic optimization of economic and environmental performance during voyages, namely, the minimization of operational costs and total carbon emissions. Furthermore, to quantify the carbon reduction effectiveness of the constructed energy management model and to align with the carbon emission reduction regulations for the shipping industry drafted by the International Maritime Organization (IMO), this paper establishes a ship carbon emission intensity grading mechanism based on the Energy Efficiency Operational Indicator (EEOI). This mechanism quantifies the carbon emission intensity throughout the entire voyage, contributing to reduced total ship carbon emissions while, to a certain extent, safeguarding the economic benefits of shipowner. Finally, the effectiveness of the proposed S-IES energy management model and the carbon emission intensity grading mechanism is validated using the fundamental data of a bulk carrier as a case study.

Keywords: S-IES · EMP · Energy Management · EEOI · Ship Carbon Emission Grading

1 Introduction

In recent years, the environmental crisis triggered by global climate change has become increasingly severe, and how to reduce carbon emissions has become a core issue of international concern. The shipping industry, a crucial enabler of global trade, is responsible for transporting approximately 80% of international trade by volume [1]. According to

C. Li et al. (Eds.): ICNC 2025, CCIS 2946, pp. 312–325, 2026.
https://doi.org/10.1007/978-981-92-1599-7_27

statistics from the International Maritime Organization (IMO), the annual carbon emissions from shipping currently exceed 1 billion tonnes, accounting for about 2.8% of global carbon emissions [2]. Without effective control, this proportion is projected to rise with the continued development of the shipping industry. To address the conflict between the industry's growth and its escalating greenhouse gas (GHG) emissions, the IMO has introduced regulations and quantitative techniques, including amendments to MARPOL Annex VI [3]and the Carbon Intensity Indicator (CII) [4, 5]. These measures explicitly mandate a 40% reduction in total annual ship CO_2 emissions by 2030 compared to 2008 levels [6], and a 50% reduction in total annual GHG emissions by 2050 [7]. Consequently, the low-carbon transition of ship energy systems has become an inevitable trend for the shipping industry.

As the core carrier of global trade, maritime shipping plays an irreplaceable role in promoting world economic integration. The ship energy system, essential for ensuring safe and reliable navigation, must consistently provide a continuous and high-quality power supply for critical loads such as propulsion. Traditionally, economic has been prioritized in voyage planning and operation, with decision-making processes often centered solely on the core constraint of supply-demand balance, adjusting shipping capacity to balance operational costs and revenue [8, 9]. For instance, one study focuses on fleet adjustment optimization considering supply-demand balance and time window constraints to minimize operational costs and maximize revenue [10]. However, this economic benefit-dominated energy management objective has resulted in significant GHG emissions, exacerbating marine environmental pollution. Exhaust emissions, oil spills, noise pollution, and other impacts from ship operations have caused persistent and far-reaching damage to marine ecosystems and the atmosphere, threatening biodiversity and posing potential risks to the human environment. Against this backdrop, achieving the synergistic optimization of economic and environmental benefits during ship navigation [11] has become a critical challenge in the maritime field. Therefore, this article moves beyond the traditional focus on economic alone by incorporating environmental benefits into the framework, aiming to construct a model that balances both, thereby providing simulation tools and theoretical support for promoting the sustainable development of the shipping industry.

The global shipping industry faces dual pressures to balance environmental sustainability and economic viability, yet strategies that simultaneously enhance ecological and economic performance lack effective implementation pathways. Current mainstream measures-including slow steaming and alternative fuel applications-often pursue environmental goals in isolation, neglecting their systematic trade-offs with operational costs and revenues [12]. proposes a novel biomimetic method utilizing favorable hydrodynamic interactions within vessels; another emphasizes the widespread electrification of maritime transport and the full electrification of ships [13]. However, these approaches fail to optimize evaluation metrics through a carbon rating mechanism to achieve environmental protection targets. To address these issues, this paper designs a CII grading mechanism based on the ship carbon intensity indicator, utilizing the Energy Efficiency Operational Indicator (EEOI) as the foundational data, which describes the carbon emission intensity per unit of transport work (carrying one tonne of cargo over one nautical mile) [14]. This mechanism quantifies a ship's environmental performance while exploring the applicability of energy management schemes. The CII framework classifies ships from A to E based on their carbon emissions per unit of transport work [15], providing a quantitative basis for coordinating environmental compliance and economic incentives [16]. A referenced study clearly and systematically explains the core of CII regulatory requirements, namely, the calculation of carbon intensity based on unit transport work and the resulting A-E grading system. In light of this, the present study breaks from the traditional research paradigm focused solely on economic by integrating environmental considerations, constructing a dual-objective model to support the sustainable development of shipping.

Based on the above discussion, this paper proposes a S-IES energy management model that comprehensively considers economic and environmental benefits and designs CII grading mechanism with EEOI. The energy management model provides quantitative decision, which support for the low-carbon transition of shipping. The main contributions of this article are as follows:

(1) Construction of an Energy Management Model Balancing Operational Performance and Environmental Protection:

A multi-objective energy management model is developed for the Ship-Integrated Energy System (S-IES), aiming to minimize both consumption costs and carbon emissions. This model characterizes the trade-off relationship between economic and environmental benefits.

(2) Establishment of a Ship Carbon Intensity Grading-Driven Environmental Assessment Mechanism: Centered on EEOI, which describes the carbon emission intensity per unit of transport work over a voyage period, a grading system classifying performance into five levels (A-E) is established. The proposed grading system quantifies the coupling relationship between optimized energy management strategies and economic/environmental benefits, enabling dynamic assessment of environmental compliance.

The remainder of this article is structured as follows: Sect. 2 analyzes the energy management problem for S-IES and constructs the energy management model. Section 3 establishes the CII grading system based on EEOI to quantitatively assess the ship's carbon emission rating. Section 4 verifies the effectiveness of the proposed model and method through numerical simulation case studies. Section 5 concludes the article.

2 Energy Management Model for S-IES

S-IES is defined as a next-generation shipboard energy platform centered around an energy optimization scheduling system. It deeply integrates traditional ship energy systems with renewable energy sources, aiming to better align modern ships with the demands of greening and intelligent development. Breaking away from the traditional single-energy supply mode, S-IES achieves unified dispatch of shipboard energy resources, establishing a novel ship energy architecture characterized by multi-energy complementarity and intelligent coordination. It facilitates the deep integration and optimization of thermal and electrical loads. The fundamental model framework of S-IES is depicted in Fig. 1.

As illustrated in Fig. 1, under the green shipping objectives advocated by the IMO, the S-IES ensures high-quality energy supply by establishing a "supply-consumption-load" collaboration mechanism. The S-IES can be categorized into the energy supply side and the load side based on energy provision and consumption. The supply side enables the collaboration operation of renewable energy devices and conventional fossil-fuel-based equipment, which can reduce the demand for fossil fuels and enhance the utilization of renewable energy. The load side employs distributed scheduling methods to coordinately control various electrical/thermal load equipment aboard ships, such as heating systems, communication devices, and domestic power appliances. Currently, when executing navigation missions, ships must simultaneously consider the operational costs of energy equipment and the total greenhouse gas emissions from fossil fuels, thereby holistically addressing both economic and environmental benefits. Consequently, inspired by [17], this paper constructs the operational functions for the S-IES energy equipment using the operational cost functions and carbon dioxide emission calculation formulas presented therein.

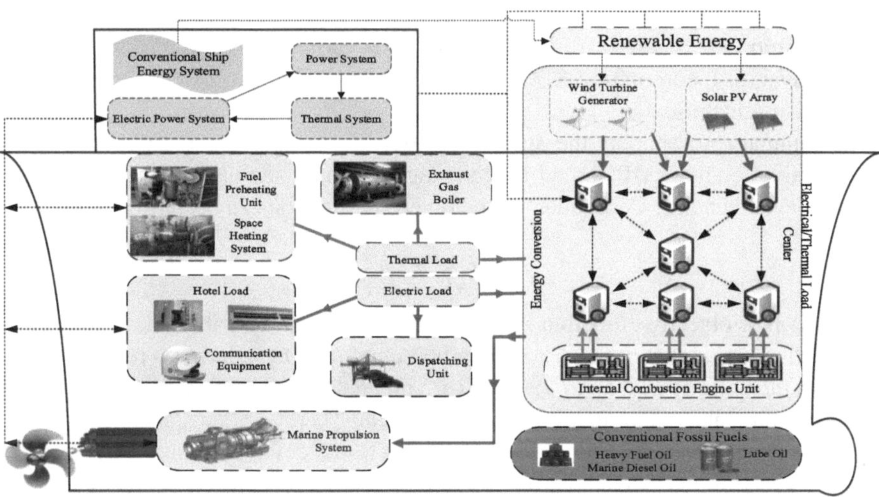

Fig. 1. Ship-integrated Energy System

Specifically, the operational cost functions for fossil-fuel-based equipment (FE) and renewable energy equipment (RE) within the S-IES can be defined as:

$$C_i^{\mathrm{RE}} = a_i(\phi_i)^2 + c_i, i \in \Omega_{\mathrm{RE}} \tag{1}$$

$$C_i^{\mathrm{FE}} = a_i(\phi_i)^2 + b_i\phi_i + c_i, i \in \Omega_{\mathrm{FE}} \tag{2}$$

where $\phi_i \in \{p_i, h_i\}$, denoting the power output and heating output of the relevant energy equipment. Ω_{RE} , Ω_{FE} representing set of the conventional FE and RE. i is the ordinal index of each device within the set; C_i^{RE} representing the operational costs of re aboard ships; C_i^{FE} represent the operational costs of FE.

Therefore, the total operational cost of S-IES can be defined as:

$$F_{EC} = \sum_{i \in \Omega_{\mathrm{FE}}} C_i^{\mathrm{FE}} + \sum_{i \in \Omega_{\mathrm{RE}}} C_i^{\mathrm{RE}} \tag{3}$$

where, F_{EC} denotes the economic benefit in the energy management problem (EMP) of S-IES, which corresponds to the total operational cost of energy consumption. Conversely, EEOI is introduced as a critical measure of the ship's carbon emission intensity for the optimization objectives. Its specific definition is as follows:

$$F_{\mathrm{EO}} = \left(I^{\mathrm{FE}}\right)/(\mathrm{Cap} \cdot \mathrm{Dist}) \tag{4}$$

$$I^{\mathrm{FE}} = \sum_{i \in \Omega_{\mathrm{FE}}} \left(\alpha_i(\phi_i)^2 + \beta\phi_i + \gamma_i\right) \tag{5}$$

where, F_{EO} represents EEOI. Dist represents the voyage distance per time segment; Cap represent the passenger capacity determined by vessels; I^{FE} represents the total carbon emissions from shipboard energy equipment.

IMO proposes the use of EEOI to quantify the carbon emission intensity of ships during voyages. Incorporating EEOI into EMP of S-IES enhances accountability for maritime sustainability.

In summary, to support the sustainable development of the marine ecosystem, the objective for the EMP should be constructed by considering both economic and environmental benefits. Consequently, the objective function can be defined as:

$$\min\{F_{EC}, F_{EO}\} \tag{6}$$

where (6) is a objective function of the S-IES's EMP. It describes the compromise between the operational costs F_{EC} and carbon emissions F_{EO} during a round-voyage. However, the units of F_{EC} and F_{EO} in (6) are inherently incompatible, making it difficult to compute and determine an optimal schedule. Therefore, normalization of these objectives is necessary. The purpose of (6) is not to select the minimum value of each objective function independently, but rather to construct a compromise solution based on both functions. It reformulates the multi-objective EMP as a single-objective one. This method aims to find an equilibrium solution that balances the performance across different objectives. Consequently, (6) can be reformulated using a normalized form, and defined as:

$$\min\left\{\underbrace{\frac{F_{EC} - F_{\mathrm{EC}}^{\mathrm{m}}}{F_{\mathrm{EC}}^{\mathrm{M}} - F_{\mathrm{EC}}^{\mathrm{m}}}}_{\mathcal{G}_1}, \underbrace{\frac{F_{\mathrm{EO}} - F_{\mathrm{EO}}^{\mathrm{m}}}{F_{\mathrm{EO}}^{\mathrm{M}} - F_{\mathrm{EO}}^{\mathrm{m}}}}_{\mathcal{G}_2}\right\} \tag{7}$$

The operational integrity and safety of the vessel are ensured by adhering to the following constraints, which complete the energy management model of S-IES:

$$\sum_{i\in\Omega_E} p_{i,k}^{\mathrm{FE}} + p_{i,k}^{\mathrm{RE}} - p_{i,k}^{\mathrm{LE}} = 0 \tag{8}$$

$$\sum_{i\in\Omega_E} h_{i,k}^{\mathrm{FE}} + h_{i,k}^{\mathrm{RE}} - h_{i,k}^{\mathrm{LE}} = 0 \tag{9}$$

$$\phi_i^m \le \phi_i \le \phi_i^M, \quad i \in \Omega_{\mathrm{FE}} \cup \Omega_{\mathrm{RE}} \cup \Omega_{\mathrm{LE}}, \phi_i \in \{p_i, h_i\} \tag{10}$$

$$\sum_{i\in\Omega_{\mathrm{RE}}} \phi_i{}^{\mathrm{RE}} + \sum_{i\in\Omega_{\mathrm{FE}}} \phi_i{}^{\mathrm{FE}} - \max\left\{\phi_i{}^{\mathrm{RE}}, \phi_i{}^{\mathrm{FE}}\right\} > \sum_{i\in\Omega_{\mathrm{LE}}} \phi_i{}^{\mathrm{LE}} \tag{11}$$

here, Ω_E denotes the node set of energy equipment, and LE represents the power and heating load equipment, (8) and (9) represent the power supply-demand balance and thermal supply-demand balance constraints, respectively. (10) imposes limits on the energy output of the equipment. Furthermore, to ensure the normal operation of various equipment (the electrical and thermal networks) during ship navigation, (11) defines the ramping constraints for reliable and safe operation.

Based on the analysis, the EMP for the S-IES can be formulated as the optimization of the objective function in (6), subject to the constraints defined in (8) to (11).

$$\begin{aligned} &\min\ \{\kappa_1\mathcal{G}_1 + \kappa_2\mathcal{G}_2\} = \sum_s f_s(\phi_s), \forall s \in \Omega_E \\ &\mathrm{s.t.}(8)\text{-}(11) \end{aligned} \tag{12}$$

where $\phi_i \in \{p_i, h_i\}$, $\kappa_1 + \kappa_2 = 1$, $\Omega_E = \Omega_{\mathrm{FE}} \cup \Omega_{\mathrm{RE}}$. The weighting coefficients must satisfy $\kappa_1 + \kappa_2 = 1$. f_s is a composite function defined to holistically consider the ship's economic and environmental benefits.

To achieve collaboration of the S-IES,the inherent trade-off between F_{EO} and F_{EC} must be addressed, as simultaneously achieving the optimum for both economic and environmental objectives is infeasible. Consequently, the functions representing ship operational cost and voyage emission intensity are scalarized into a single-objective optimization problem using dynamic weighting coefficients.

The selection of the weighting coefficients κ_1 and κ_2 is critical as they reflect the decision-maker's preference regarding the trade-off between cost and emissions. In [18], a sensitivity analysis was conducted to investigate the impact of different weight combinations on the optimization results. A range of values for κ_1 (and correspondingly $\kappa_2 = 1 - \kappa_1$) were tested. The specific weights used in the simulation (e.g., $\kappa_1 = 0.5$, $\kappa_2 = 0.5$ for a balanced scenario, or other values that lead to the presented results) were chosen to achieve a pre-defined target. Such as a specific EEOI grade improvement while accepting a reasonable increase in operational cost. This approach provides a pragmatic methodology for weight selection, moving beyond purely empirical choices and linking them directly to the regulatory and economic goals defined in this paper.

As (12), this problem is solved subject to the operational constraints imposed during the ship's voyage.

3 Ship Carbon Emission Grading Mechanism

The establishment of the EMP's model for S-IES serves as a technical approach; however, a method is required to verify its effectiveness in reducing CO_2 emission levels. Moreover, an carbon emission grading mechanism is constructed with the utilization of EEOI. It quantifies the performance of the carbon reduction during the ship's sailing. This mechanism quantifies ship carbon emission intensity based on EEOI. In accordance with current IMO regulations, ship carbon emission intensity is divided into five grades, ranging from A to E. To assign a carbon intensity grade to a ship, the reduction factor and grade boundaries for that specific ship type must first be calculated. This section considers bulk carriers and tanker vessels as the research subjects. this section details the methodology for calculating these reduction factors and grade boundaries.

$$\mathrm{EEOI}_{\mathrm{ref}} = a \cdot \mathrm{Deadweight}^{-c} \tag{13}$$

where, $\mathrm{EEOI}_{\mathrm{ref}}$ represents the reference baseline value for carbon emissions in 2019. Deadweight denotes the cargo carrying capacity of the respective ship. Parameters a and c are estimated via median regression fitting; specifically, for bulk carriers, a is 4745 and c is 0.622, while for tanker vessels, a is 5247 and c is 0.610. Furthermore, the carbon intensity assessment standards evolve annually, with the reduction factors for the years 2023, 2024, 2025, and 2026 set at 5%, 7%, 9%, and 11%, respectively. The carbon intensity grade can be determined by comparing its value against the benchmark known as the ship carbon intensity baseline, which is specifically defined based on data from 2019.

$$\text{Required } \mathrm{EEOI}_k = (1 - T_k) \cdot \mathrm{EEOI}_{\mathrm{ref}} \tag{14}$$

where Required $\mathrm{EEOI}_{\mathrm{k}}$ denotes the carbon emission standard value for the k year, and T_k represents the reduction factor for k year.

In accordance with IMO regulations, four boundaries are calculated using (13), and the ship carbon emission intensity is categorized into five distinct grades: the Excellent boundary (B_1), the Good boundary (B_2), the Moderate boundary (B_3), and the Poor boundary (B_4), as illustrated in Fig. 2.

Specifically, the carbon intensity boundary for the k year is defined as follows:

$$B_1 = \exp(L_1)\text{Required } \mathrm{EEOI}_k \tag{15}$$

$$B_2 = \exp(L_2)\text{Required } \mathrm{EEOI}_k \tag{16}$$

$$B_3 = \exp(L_3)\text{Required } \mathrm{EEOI}_k \tag{17}$$

$$B_4 = \exp(L_4)\text{Required } \mathrm{EEOI}_k \tag{18}$$

In (15–18), parameters L_1, L_2, L_3, and L_4 represent the rating boundary parameters. These parameters are estimated using the quantile regression method, with carbon intensity data from different ship types in 2019 serving as the sample data set, as detailed in the table (Table 1).

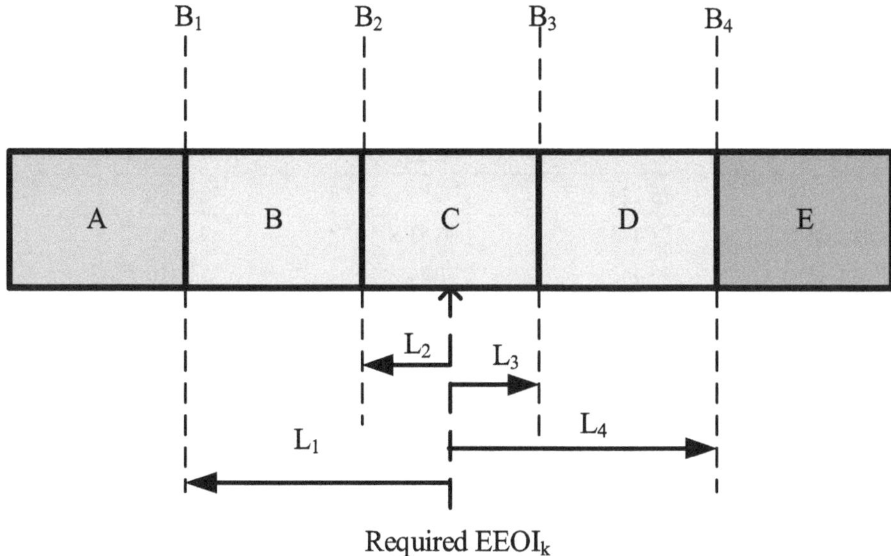

Fig. 2. CII Rating Boundaries Based on the EEOI Calculation Methodology.

Table 1. Boundary parameters in the CII rating for bulk carriers and tanker vessels

Type	$\exp(L_1)$	$\exp(L_2)$	$\exp(L_3)$	$\exp(L_4)$
Bulk Carrier	0.86	0.94	1.06	1.18
Tank Vessel	0.82	0.93	1.08	1.28

A total of five bulk carriers and four tanker vessels are selected as the research subjects. Taking the year 2023 as an example, the EEOI for each subject vessel can be derived based on (4). Subsequently, the Required $EEOI_k$ for the year 2025 is calculated using (13) and (14), as summarized in Table 2.

Based on the results presented in Table 2 and utilizing (15) to (18), the corresponding carbon emission rating boundaries for each vessel can be calculated. The results are summarized in Table 3.

Using the data from Tables 2 and 3, the actual operational EEOI of each vessel is compared against the calculated boundaries to determine its carbon emission rating. Furthermore, accounting for the distinct annual reduction factors from 2023 to 2026, the ratings for both bulk carriers and tanker vessels are assessed on a yearly basis. The specific rating results are illustrated in the Figs. 3 and 4.

As observed from Figs. 3 and 4, when applying EEOI for rating bulk carriers and tank vessels, the annual boundaries for both vessel types exhibit a progressively decreasing trend from 2023 to 2026. For bulk carriers assessed using EEOI, approximately 27% fall within Grades A, C, D and about 10% within Grades B, E, indicating a relatively poor overall compliance rate. In contrast, tank vessels demonstrate significantly improved

Table 2. Carbon Emission Intensity of Vessels in 2025

Type	Deadweight(t)	Carbon Emission Intensity Indicator/ [t $CO_2 \cdot (t \cdot n\ mile)^{-1}$]	
		EEOI	Required $EEOI_k$
Bulk Carrier 1	76 124	4.56×10^{-6}	3.78×10^{-6}
Bulk Carrier 2	74 443	6.53×10^{-6}	3.82×10^{-6}
Bulk Carrier 3	46 462	1.34×10^{-5}	5.35×10^{-6}
Bulk Carrier 4	38 338	7.56×10^{-5}	5.78×10^{-6}
Bulk Carrier 5	21 651	8.80×10^{-6}	8.24×10^{-6}
Tank Vessel 1	7 060	1.62×10^{-5}	2.04×10^{-5}
Tank Vessel 2	4 607	3.38×10^{-5}	2.64×10^{-5}
Tank Vessel 3	3 676	3.46×10^{-5}	3.03×10^{-5}
Tank Vessel 4	3 268	9.64×10^{-5}	3.26×10^{-5}

Table 3. CII rating thresholds for distinct vessel types in 2025 /$g \cdot CO_2 \cdot (t \cdot n\ mile)^{-1}$

Type	B_1	B_2	B_3	B_4
Bulk Carrier 1	3.25	3.55	4.00	4.46
Bulk Carrier 2	3.29	3.59	4.05	4.51
Bulk Carrier 3	4.60	5.03	5.67	6.31
Bulk Carrier 4	4.97	5.43	6.13	6.82
Bulk Carrier 5	7.09	7.75	8.73	9.72
Tank Vessel 1	16.70	18.94	22.00	26.07
Tank Vessel 2	21.66	24.57	28.53	33.82
Tank Vessel 3	24.86	28.20	32.75	38.81
Tank Vessel 4	26.71	30.30	35.19	41.70

compliance, with each grade containing roughly 20% of the ships.. Furthermore, analysis of Figs. 3 and 4 reveals a pronounced downward trend in the carbon intensity ratings for bulk carriers over time, while the trend for tank vessels more gradually. This pattern reflects the increasingly stringent global regulatory landscape for ship carbon emissions and the evolving rigor of the ship energy efficiency rating system. Specifically, a vessel currently rated as grade A can be reasonably projected to deteriorate to grades D or E in subsequent years if no corrective measures are implemented. Therefore, the analysis of EMP for S-IES is critically important.

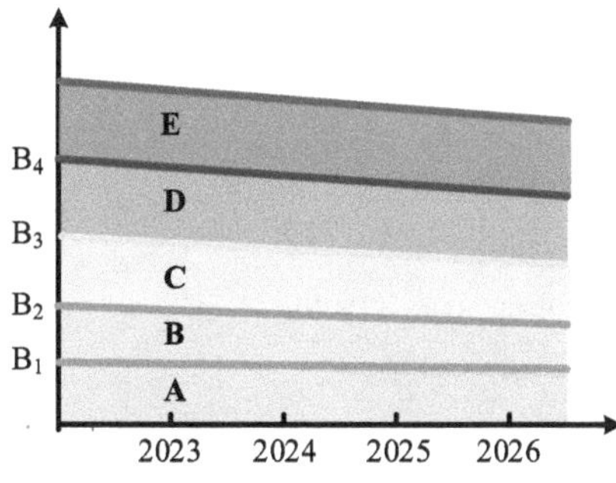

Fig. 3. Trends of Carbon Intensity Boundaries for Bulk Carriers (2023–2026).

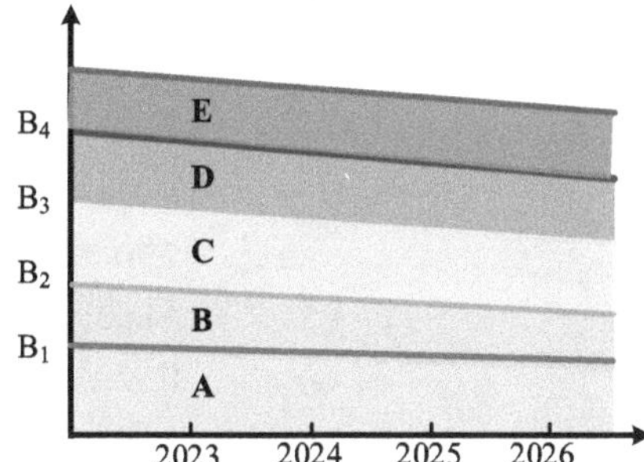

Fig. 4. Trends of Carbon Intensity Boundaries for Tank Vessels (2023–2026).

4 Simulation Analysis

This section validates the effectiveness of the proposed ship carbon intensity grading mechanism and the energy management model of S-IES through numerical simulations. Bulk Carrier 1 from Table 2 is selected as the case study vessel, with a deadweight tonnage of 76,124 t. The voyage from Ningbo Port, China, to the Port of Oakland, USA, spanning approximately 5,900 nautical miles, is used as the test scenario. The entire navigation mission is divided into 38 time intervals, with the load demand profiles for each time slot illustrated in Fig. 5. The vessel is assumed to be equipped with seven power generators and five heating generators, whose operational parameters are provided in Tables 4 and 5.

Table 4. Operational cost and carbon emission parameters for seven power output equipments

	P-1	P-2	P-3	P-4	P-5	P-6	P-7
$a_i/\ \$ \cdot p^{-2}$	0.56	0.61	0.65	0.71	0.76	0.75	0.72
$b_i/\ \$ \cdot p^{-1}$	27.00	13.00	20.00	-	-	-	-
$c_i/\ \$$	110.00	74.00	112.00	132.00	122.00	113.00	125.00
$\alpha_i/\ kg \cdot p^{-2}$	0.32	0.34	0.28	-	-	-	-
$\beta_i/\ kg \cdot p^{-1}$	30.00	25.00	36.00	-	-	-	-
$\alpha_i/\ kg$	47.00	50.00	37.00	-	-	-	-

To validate the effectiveness of the proposed ship energy management model, two simulation scenarios are established: Scenario 1: Considers economic benefit only, Scenario 2: Simultaneously considers economic and environmental benefits.

In Scenario 1, the power outputs of the seven power supply devices and five heating supply devices across the 38 time intervals are shown in Figs. 6 and 7, respectively.

In Scenario 2, the power outputs of the seven power supply devices and five heating supply devices across the 38 time intervals are depicted in Figs. 8 and 9, respectively.

Table 5. Operational cost and carbon emission parameters for five heating output equipments

	P-1	P-2	P-3	Pt-4	P-5
a_{il} $\$ \cdot h^{-2}$	0.054	0.061	0.072	0.074	0.071
b_{il} $\$ \cdot h^{-1}$	2.70	1.30	-	-	-
c_{il} $\$$	14.00	7.40	11.20	13.20	12.20
α_{il} $kg \cdot h^{-2}$	0.32	0.34	-	-	-
β_{il} $kg \cdot h^{-1}$	30.00	25.00	-	-	-
α_{il} kg	47.00	50.00	-	-	-

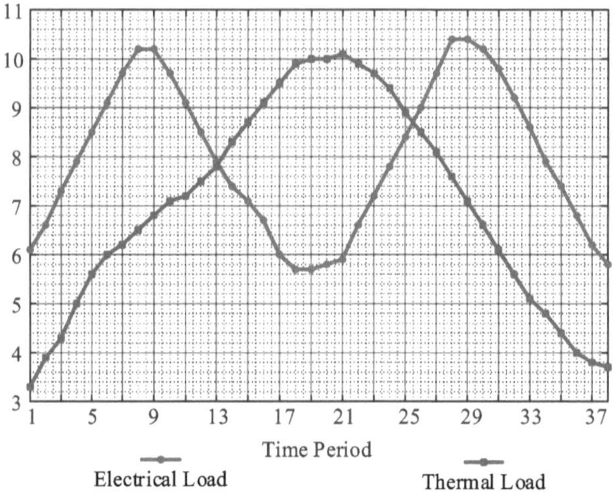

Fig. 5. Equipment Load Profiles Across 38 Time Periods

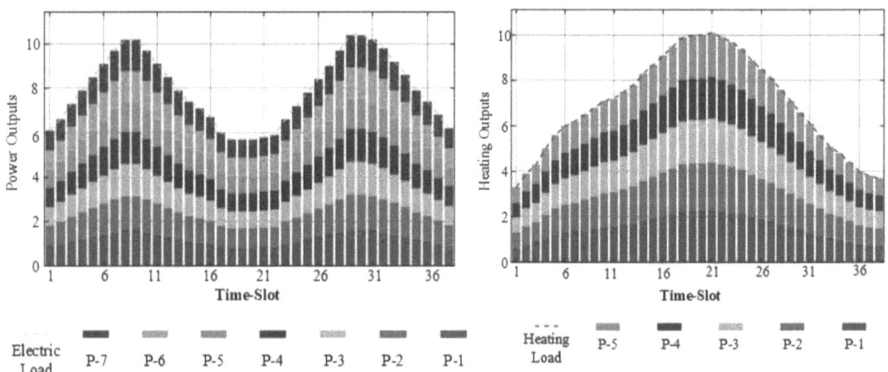

Fig. 6. Outputs of power generators under scenario-1

Fig. 7. Outputs of heating generators under scenario-1

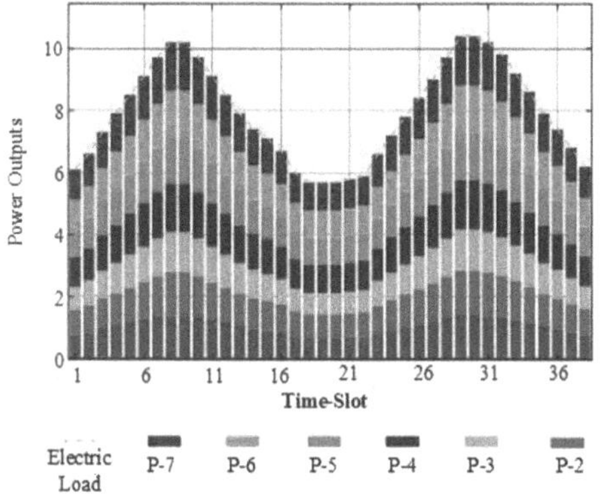

Fig. 8. Outputs of power generators under scenario-2

Fig. 9. Outputs of heating generators under scenario-2

Figures 6, 8, 7 and 9 depict the power and heating outputs under the two scenarios during the voyage, respectively. The results demonstrate distinct power output profiles for both the ship's electrical and thermal networks across the 38 time intervals under the different operational scenarios. Furthermore, the corresponding profiles for total ship carbon emissions and equipment operational costs are presented in Figs. 10 and 11.

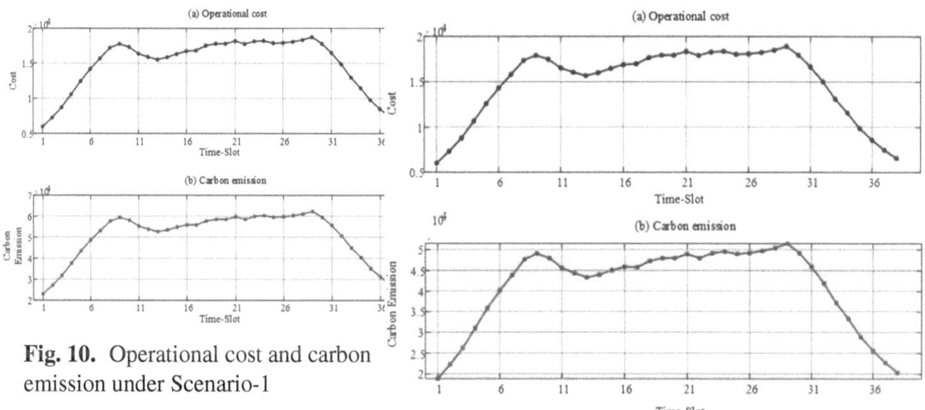

Fig. 10. Operational cost and carbon emission under Scenario-1

Fig. 11. Operational cost and carbon emission under Scenario-2

As observed from Figs. 9 and 10 under Scenario 2, the operational cost of the ship energy system exhibits a minor increase, remaining comparable to that of Scenario 1. However, the total carbon emissions in Scenario 2 demonstrate a significant reduction compared to Scenario 1.

Furthermore, based on the energy management results from Scenarios 1 and 2, the carbon reduction effectiveness is evaluated using the ship carbon intensity grading mechanism established in Sect. 3. Specifically, B_1, B_2, B_3, B_4 for Bulk Carrier 1 are [3.25, 3.55,

4.00, 4.46] ($\times 10^{-3}$kg $CO_2 \cdot (t \cdot n\,mile)^{-1}$). The EEOI values for Scenarios 1 and 2 are calculated as 4.24×10^{-3} (kg $CO_2 \cdot (t \cdot n\,mile)^{-1}$) and 3.50×10^{-3} (kg $CO_2 \cdot (t \cdot n\,mile)^{-1}$), respectively, based on the tested vessel's fundamental information. Therein, the EEOI in Scenario 2 is reduced by 17.45% compared to Scenario 1. Consequently, the carbon intensity grades for Scenarios 1 and 2 fall within Grade D and Grade B, respectively.

In summary, although the proposed energy management model leads to a moderate increase in operational cost, it substantially reduces the ship's carbon emissions.

5 Conclusions

This paper addresses the EMP of the S-IES to facilitate green and secure shipping. To mitigate the environmental impact associated with aggregate carbon emissions, an energy management model for S-IES is established, enabling the synergistic optimization of economic and environmental benefits throughout the entire voyage. Furthermore, a ship carbon emission intensity grading mechanism is constructed based on the EEOI. The effectiveness of the proposed energy management model and the grading mechanism is validated through numerical simulations using a bulk carrier operating on the route between Ningbo Port, China, and the Port of Oakland, USA. The results demonstrate that the proposed approach achieves a significant reduction of 338.4 t in total carbon emissions for the complete voyage, while the operational cost increases by merely 0.96%. Consequently, the ship's carbon intensity rating improves from an initial Grade D to Grade B. Future work will focus on analyzing the impacts of renewable energy uncertainty and volatility on the operational reliability of the S-IES, aiming to further enhance navigation safety performance.

Acknowledgments. This work was supported in part by the National Natural Science Foundation of China under Grants 52301418, 52471376; Sichuan Science and Technology Program under Grants 2024NSFSC0878; the Fundamental Research Funds for the Central Universities under Grant 3132025295;

Disclosure of Interests. The authors have no competing interests to declare that are relevant to the content of this article.

References

1. International Maritime Organization: Low Carbon GIA Explores Book And Claim Concept. https://greenvoyage2050.imo.org/1916-2/. Accessed 15 Oct 2025
2. International Maritime Organization: IMO Strategy on Reduction of GHG Emissions from Ships (2023). https://www.imo.org/en/ourwork/environment/pages/2023-imo-strategy-on-reduction-of-ghg-emissions-from-ships.aspx. Accessed 15 Oct 2025
3. International Maritime Organization: MARPOL Annex VI Amendments (Including CII). https://www.imo.org/zh/mediacentre/pressbriefings/pages/cii-and-eexi-entry-into-force.aspx. Accessed 15 Oct 2025
4. International Maritime Organization: Adoption of the Carbon Intensity Indicator (CII) for Ships. https://wwwcdn.imo.org/localresources/en/KnowledgeCentre/IndexofIMOResolutions/MEPCDocuments/MEPC.346(78).pdf. Accessed 15 Oct 2025

5. International Maritime Organization: Guidelines for the Carbon Intensity Indicator (CII) for Ships (2022). https://wwwcdn.imo.org/localresources/en/KnowledgeCentre/IndexofIMOResolutions/MEPCDocuments/MEPC.346(78).pdf. Accessed 15 Oct 2025
6. International Maritime Organization: IMO Strategy on Reduction of GHG Emissions from Ships (2023). https://wwwcdn.imo.org/localresources/en/KnowledgeCentre/IndexofIMOResolutions/MEPCDocuments/MEPC.377(80).pdf, 5. Accessed 15 Oct 2025
7. International Maritime Organization: Signed an Agreement to Halve Carbon Emissions of Shipping Industry in (2050). https://www.imo.org/en/MediaCentre/PressBriefings/Pages/06GHGinitialstrategy.aspx,6. Accessed 15 Oct 2025
8. Sree, K.D., Gunti, G., Neelima, N., Nagaraja, K.V.: Optimizing cargo ship operational costs: a comprehensive algorithmic approach. In: 2024 4th International Conference on Intelligent Technologies, CONIT, pp. 1–6, Bangalore, India (2024)
9. Haliem, M., Aggarwal, V., Bhargava, B.: AdaPool: a diurnal-adaptive Fleet management framework using model-free deep reinforcement learning and change point detection. IEEE Trans. Intell. Transp. Syst. **23**(3), 2471–2481 (2022)
10. Varelas, T., Plitsos, S.: Real-time ship management through the lens of big data. In: 2020 IEEE Sixth International Conference on Big Data Computing Service and Applications, BigDataService, pp. 142–147, Oxford, UK (2020)
11. Zhang, Y., Xiao, Y., Teng, F., Li, T.: Distributed energy Management for Ship-Integrated Energy System: secure and green sailing. IEEE Trans. Transp. Electrification. **11**(3), 8189–8201 (2025)
12. He, Y., et al.: Ship emission reduction via energy-saving formation. IEEE Trans. Intell. Transp. Syst. **25**(3), 2599–2614 (2024)
13. Fang, S., Wang, Y., Gou, B., Xu, Y.: Toward future green maritime transportation: an overview of seaport microgrids and all-electric ships. IEEE Trans. Veh. Technol. **69**(1), 207–219 (2020)
14. Chen, X., Lv, S., Wu, B., Li, C., Xian, J., Wu, H.: Ship energy efficiency operation index variation analysis via empirical voyage data. In: 2023 7th International Conference on Transportation Information and Safety, ICTIS, pp. 1957–1961, China (2023)
15. Zhang, Y., Xiao, Y., Teng, F., Li, T.: Distributed optimal energy management strategy for S-IES with IMO's regulation on CII. Ocean Eng. **340**(3), 122397 (2025)
16. D'Agostino, F., Gallo, M., Saviozzi, M., Silvestro, F.: A model predictive control-based energy management strategy for secure operations in shipboard power systems. IEEE Trans. Transp. Electrification. **11**(1), 4818–4829 (2025)
17. Zhang, Y., Xiao, Y., Teng, F., Li, T.: Distributed Energy Management Method With EEOI Limitation for the Ship-Integrated Energy System. IEEE Syst. J. **18**(2), 1332–1343 (2024)
18. Hottenroth, H., et al.: Viere: beyond climate change. Multi-attribute decision making for a sustainability assessment of energy system transformation pathways. Renew. Sust. Energ. Rev. **156**, 1364–0321 (2022)

Enhancing Few-Shot Class Incremental Learning with Local Expert-Guided Class Separation

Qining Ren and Zhigang Zeng(✉)

School of Artificial Intelligence and Automation, Huazhong University of Science and Technology, Wuhan, China
{qiningren,zgzeng}@hust.edu.cn

Abstract. Continuous learning is a hallmark of biological intelligence, enabling lifelong adaptation. Inspired by this capability, few-shot class incremental learning (FSCIL) seeks to endow models with a similar facility: the ability to continually learn new classes from limited data while retaining previously acquired knowledge. However, existing methods face significant challenges, such as catastrophic forgetting and overfitting to new, limited-data tasks. In this work, we adopt the common paradigm of finetuning the model during the base session and employing a prototypical classifier. Our approach focuses on enhancing the inter-class separation in the feature space. Strong separation is critical as it creates a more distinct and spacious representation for accommodating new classes in subsequent incremental sessions, thereby mitigating catastrophic forgetting. We propose a two-stage, local expert-guided learning framework to progressively enhance inter-class separation. In the first stage, we apply prefix-tuning to help the model acquire domain-specific knowledge and achieve an initial class separation. In the second stage, we introduce local experts, each focusing on a specific region in the feature space. To obtain discriminative expert features that complement the original ones, we insert additional prompts and propose two prompt fusion modules based on the intermediate patch-level features from the backbone. Furthermore, to effectively train the experts, we propose the Selective Anchor-based Supervised Contrastive loss, Prior-Guided Class Separation loss, and Matching Score Reverse Prediction loss. Experimental results on three datasets demonstrate that our method outperforms existing state-of-the-art approaches.

Keywords: Class incremental learning · Catastropic forgetting · Prompt tuning

1 Introduction

Emulating the continuous learning capabilities of biological intelligence represents a major challenge for deep neural networks. While powerful on static

C. Li et al. (Eds.): ICNC 2025, CCIS 2946, pp. 326–337, 2026.
https://doi.org/10.1007/978-981-92-1599-7_28

datasets, these models struggle when knowledge must be integrated sequentially from non-stationary data streams, as is common in real-world applications. This has spurred research in class incremental learning (CIL) [5,15,18,19,22,27,42], and more recently, few-shot class incremental learning (FSCIL) [34], which directly addresses the biological reality of learning new concepts from sparse data.

The primary challenges in FSCIL are catastrophic forgetting [9] and overfitting. Catastrophic forgetting causes models to lose previously learned knowledge, while overfitting–aggravated by limited data–hinders the direct application of CIL methods. To address these issues, a common strategy involves training the model on sufficient base session data and freezing the backbone during incremental sessions. Prototype-based classification is often used to incorporate new classes with minimal retraining, prioritizing stability and mitigating overfitting.

Recent methods [1,31,49] further improve adaptability by enhancing feature space structure during base training. Increasing inter-class separation and intra-class compactness facilitates the integration of new classes and reduces feature overlap. For example, supervised contrastive loss [13] is widely adopted to strengthen class discrimination, while methods like FACT [49] and SVAC [31] introduce pseudo-classes to reserve feature space for future classes. While effective, these approaches primarily compress the original feature space and are often based on shallow CNNs.

In this work, we also focus on feature space enhancement but build upon a pretrained Vision Transformer (ViT) [6] and introduce additional features to boost class separation. Improved separation simplifies the incorporation of new classes, thus alleviating forgetting and overfitting. Since the original feature space is often uneven, with crowded regions hindering discrimination, we propose local experts, each specializing in a local region. Expert features emphasize separating nearby classes, complementing the original features and facilitating incremental learning.

Our framework consists of two stages. First, we apply prefix-tuning to shallow ViT layers, training the model with cross-entropy and supervised contrastive losses to establish a well-separated base feature distribution. We then perform clustering to identify centroids in dense regions as expert centers. In the second stage, we insert learnable prompts in deeper transformer layers to extract expert-specific features without interfering with the original forward pass. Two fusion modules–Expert Prompt Fusion Module (EPFM) and Cross-Layer Prompt Fusion Module (CPFM)–integrate information across experts and layers. We also propose three loss functions–Selective Anchor-based Supervised Contrastive (SASC) loss, Matching Score Reverse Prediction (MSRP) loss, and Prior-Guided Class Separation (PGCS) loss–to train the experts for better local class separation.

In summary, our main contributions are:

1) A two-stage FSCIL framework that progressively enhances inter-class separation in the feature space.

2) Local experts constructed via prompt-based fusion modules, trained with three novel losses to improve discrimination among local classes.
3) Extensive experiments on three benchmarks showing that our method outperforms state-of-the-art FSCIL approaches.

2 Related Works

2.1 Few-Shot Learning

Few-shot Learning (FSL) enables models to adapt to new classes with limited samples, a crucial capability for real-world applications. Existing approaches primarily follow two paradigms: optimization-based methods [3,4,8,24] that enhance generalization through meta-learning (e.g., MAML [8] and Reptile [24]), and metric-based methods [20,30,32,37,44,47] that learn feature spaces for similarity-based classification of novel samples.

2.2 Class Incremental Learning

Class Incremental Learning (CIL) addresses the challenge of learning new classes while mitigating catastrophic forgetting. Existing methods can be categorized into four main approaches. Replay-based methods [5,27,28,36] preserve previous knowledge by storing or generating samples from past tasks. Regularization-based approaches [7,12,15,19,46] constrain model updates through techniques like distillation or parameter importance preservation. Architecture-based methods [18,21,42,45] dynamically expand network structures or allocate subnetworks for new tasks. More recently, prompt-based techniques [11,29,33,40,41] have adapted pretrained Vision Transformers for CIL, including methods that maintain prompt pools (e.g., L2P [41], DualPrompt [40]) and instance-level prompt generation approaches (e.g., APG [33], DAP [11]).

2.3 Few-Shot Class Incremental Learning

Few-Shot Class Incremental Learning (FSCIL) extends CIL with the added challenge of learning new classes from limited data per task. Dynamic structure-based methods adapt the network architecture incrementally [34,35,48], such as neural gas networks in TOPIC [35]. Feature/space-based methods like FACT [49], WaRP [14], and SAVC [31] aim to structure or reserve regions in the feature space for future classes. Prototype-based approaches [10,23,43] rely on class prototypes for classification, e.g., using quasi-orthogonal prototypes as in C-FSCIL [10]. While direct application of prompt-based CIL methods to FSCIL often performs poorly, recent adaptations like PriViLege [25] incorporate language guidance and distillation to achieve strong results.

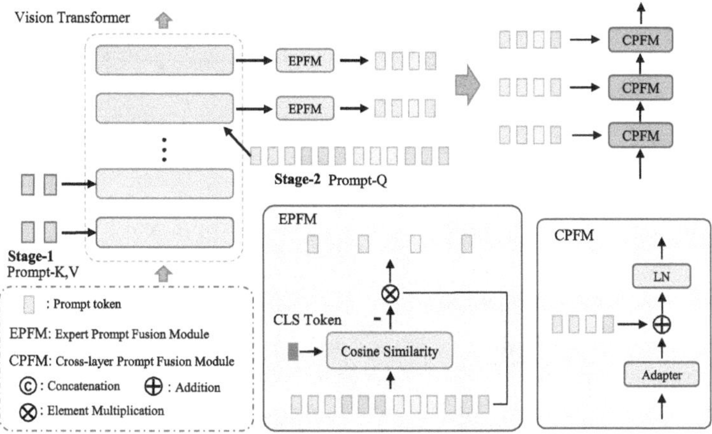

Fig. 1. In Stage 1, prompts are inserted into shallow layers via prefix-tuning to capture domain knowledge and establish initial class separation. Stage 2 constructs local experts by incorporating additional prompts to query intermediate patch features. Expert features are then extracted through Expert Prompt Fusion (EPFM) and Cross-Layer Fusion (CPFM) modules. Colors indicate prompt tokens from the same expert.

3 Methodology

3.1 Preliminary

Problem Formulation and Prototypical Network. In FSCIL, models learn sequentially from tasks $\{D_0, ..., D_T\}$, where only the base session D_0 has abundant labeled samples. Subsequent sessions follow an N-way K-shot setup with disjoint class sets $\{C_t\}$. Performance is evaluated on all seen classes up to session t.

We adopt prototypical networks [30] where a backbone f extracts features and computes class prototypes $p_c = \frac{1}{N_c} \sum f(x_i)$. These prototypes serve as classifier weights, enabling new class incorporation without retraining. The classifier output uses cosine similarity: $g(f(x_i)) = \frac{f(x_i)}{\|f(x_i)\|} \frac{P^T}{\|P\|}$.

Prompt Tuning with Vision Transformers. We use prefix-tuning [17] on pretrained ViTs, keeping original parameters ϕ frozen while learning prompts p. For an MSA layer with input X, prefix-tuning modifies the forward pass to: $f_{Pre-T}(X; p) = \text{MSA}(X_Q, [p_K; X_K], [p_V; X_V])$.

3.2 Overall Framework

Our two-stage framework (Fig. 1) enhances inter-class separation progressively. Stage 1 inserts prompts into shallow ViT layers, optimized with cross-entropy and supervised contrastive loss [13]:

$$\mathcal{L}_{SupCon}^{i} = -\frac{1}{|P_i|}\sum_{z_j \in P_i} \log \frac{\exp(z_i \cdot z_j/\tau)}{\sum_{z_k \in N_i} \exp(z_i \cdot z_k/\tau)} \tag{1}$$

In Stage 2, we cluster features to identify dense regions and assign local experts to each cluster. Experts leverage patch-level features from deeper layers via query-only prompts: $f(X;p) = \text{MSA}([p;X_Q], X_K, X_V)$. This preserves original features while extracting complementary expert features.

3.3 Building Local Experts

The [CLS] token in ViT's final layer captures global information, while our local experts extract complementary details from patch-level intermediate features. We insert learnable prompts into selected layers, but unlike prefix-tuning which modifies keys and values, we only concatenate prompts to queries:

$$f(X;p) = \text{MSA}([p;X_Q], X_K, X_V) \tag{2}$$

Each expert receives n_p prompt tokens per layer, with $p_l^e \in \mathbb{R}^{n_p \times d}$ denoting the output of expert e at layer l.

Expert Prompt Fusion Module (EPFM). We guide prompt fusion using the current layer's [CLS] feature g_l. Prompts with higher similarity to g_l receive lower weights to encourage diversity:

$$c_l^e = \text{softmax}(-\frac{g_l}{||g_l||} \cdot \frac{p_l^{eT}}{||p_l^e||}) \in \mathbb{R}^{n_p} \tag{3}$$

$$q_l^e = \sum_{i=1}^{n_p} c_{l,i}^e \cdot p_{l,i}^e \tag{4}$$

where c_l^e are prompt weights and q_l^e is expert e's output prompt.

Cross-Layer Prompt Fusion Module (CPFM). We fuse prompts across layers using adapters and layer normalization:

$$\boldsymbol{h}_i = \text{LN}(\boldsymbol{q}_i + \mathcal{A}_i(\boldsymbol{h}_{i-1})) \tag{5}$$

with adapter $\mathcal{A}(\boldsymbol{x}) = \alpha W_u \text{GeLU}(W_d \boldsymbol{x})$. Final expert features combine with global features:

$$h_{final}^e = \text{Linear}(\text{concat}(g \times h^e, h^e)) \tag{6}$$

Our approach offers three advantages: (1) comprehensive cross-layer information integration, (2) efficient parameter usage without additional attention modules, and (3) single-pass inference for all experts. We use top-p expert selection based on cosine similarity scores $s_{i,e}$, with final output:

$$\boldsymbol{O}_i = \sum_{e=1}^{N_e} s_{i,e} \cdot O_i^e \tag{7}$$

3.4 Training Loss

Selective Anchor-Based Supervised Contrastive Loss. Only local-class samples serve as anchors, with non-local samples as negatives:

$$\mathcal{L}_{SASC} = \sum_{i=1}^{N} \sum_{e=1}^{N_e} s_{i,e} \mathcal{L}_{SASC}^{i,e} \tag{8}$$

$$\mathcal{L}_{SASC}^{i,e} = -\frac{1}{|P_i|} \sum_{z_j^e \in P_i} \log \frac{\exp(z_i^e \cdot z_j^e / \tau)}{\sum_{z_k^e \in N_i} \exp(z_i^e \cdot z_k^e / \tau)} \tag{9}$$

Matching Score Reverse Prediction Loss. Experts predict sample matching using binary cross-entropy:

$$\mathcal{L}_{MSRP} = \frac{1}{N_e N} \sum_{i=1}^{N} \sum_{e=1}^{N_e} \mathcal{L}_{MSRP}^{i,e} \tag{10}$$

$$\mathcal{L}_{MSRP}^{i,e} = -[y_i^e \log(s_{i,e}) + (1 - y_i^e) \log(1 - s_{i,e})] \tag{11}$$

Prior-Guided Class Separation Loss. We enhance separation using semantic priors from Stage 1 prototypes:

$$\mathcal{L}_{PGCS}^{e} = \frac{1}{|C_e|} \sum_{i \in C_e} \log \sum_{j \in C_e} e^{\lambda(i,j) \cdot \omega_i \cdot \omega_j / \tau} \tag{12}$$

$$\lambda(i,j) = k \cdot \frac{p_i^{ori}}{||p_i^{ori}||} \cdot \frac{p_j^{ori\,T}}{||p_j^{ori}||} \tag{13}$$

Total Loss. Base session:

$$\mathcal{L}_{Base} = \mathcal{L}_{CE} + \mathcal{L}_{SASC} + \lambda_1 \mathcal{L}_{MSRP} + \lambda_2 \mathcal{L}_{PGCS} \tag{14}$$

Incremental sessions (excluding PGCS):

$$\mathcal{L}_{Inc} = \mathcal{L}_{CE} + \mathcal{L}_{SASC} + \lambda_1 \mathcal{L}_{MSRP} \tag{15}$$

4 Experiments

4.1 Experimental Setup

We evaluate our method on CIFAR-100 [2], CUB200 [38], and miniImageNet [26] following standard FSCIL protocols. CIFAR-100 and miniImageNet use a 60-class base session with seven 5-way 10-shot incremental sessions; CUB200 uses a 100-class base session with 10-way 5-shot sessions. Performance is measured by base session accuracy (A_{Base}), last session accuracy (A_{Last}), and average accuracy (A_{Avg}).

We use ViT-B/16 pretrained on ImageNet21K. Images are resized to 224×224. Training uses 5 epochs for Stage 1, 5 for Stage 2, and 3 for incremental sessions, with 8 experts, batch size 128, and Adam optimizer. Prefix-tuning (length=5) is applied to the first 6 layers in Stage 1; expert prompts (length=10) are inserted in the last 3 layers in Stage 2.

4.2 Comparison with State-of-the-Art

As shown in Table 1, our method achieves the best performance across all datasets and metrics. On CIFAR-100, we improve A_{Last} by 1.96% and A_{Avg} by 1.55% over the prior best method. Figure 2 shows consistent performance gains across sessions.

Table 1. Comparison with state-of-the-art methods.

Dataset	CUB200			CIFAR-100			miniImageNet		
Models	A_{Base}	A_{Last}	A_{Avg}	A_{Base}	A_{Last}	A_{Avg}	A_{Base}	A_{Last}	A_{Avg}
L2P [41]	77.60	38.26	54.10	90.23	51.60	68.76	93.34	64.20	77.58
DualPrompt [40]	77.97	42.37	58.43	90.87	52.71	69.14	92.33	66.56	78.30
CODA-Prompt [29]	78.80	49.30	61.12	91.20	54.93	72.01	92.45	69.46	80.06
CEC [48]	77.43	63.11	68.27	90.34	63.60	76.40	92.71	82.12	86.57
TEEN [39]	78.28	69.11	72.13	91.02	78.23	84.20	93.13	88.08	91.42
PriViLege [25]	82.20	75.08	77.50	90.88	86.06	88.08	96.68	94.10	95.27
ours	**85.67**	**79.54**	**80.97**	**92.25**	**88.02**	**89.63**	**96.92**	**94.43**	**95.57**

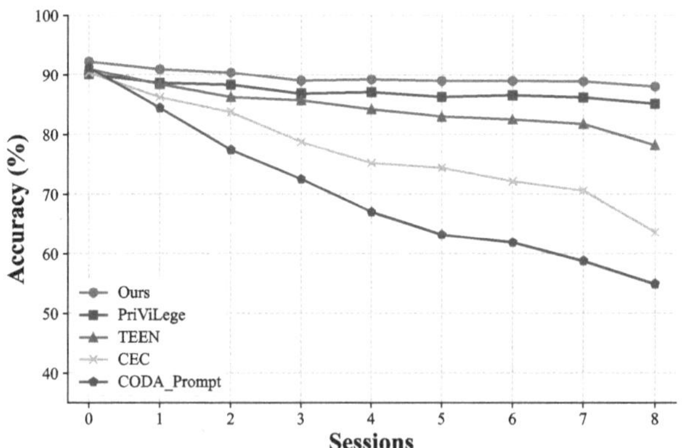

Fig. 2. Accuracy per session on CIFAR-100.

4.3 Ablation Studies

Table 2 shows the contribution of each loss component. The full model achieves 88.02% A_{Last} on CIFAR-100. Removing $\mathcal{L}_{MSRP}$ causes the largest performance drop (1.98%), highlighting its importance for expert specialization.

Table 3 validates our fusion modules. Combining EPFM and CPFM improves A_{Avg} by 3.13% over the baseline, demonstrating the effectiveness of cross-layer feature integration.

Table 2. Ablation study of loss functions on CIFAR-100.

$\mathcal{L}_{CE}$	$\mathcal{L}_{SASC}$	$\mathcal{L}_{PGCS}$	$\mathcal{L}_{MSRP}$	A_{Last}
✓	✓	✓		86.37
✓	✓	✓	✓	**88.02**
✓	✓		✓	87.54

Table 3. Ablation study of prompt fusion modules.

Methods	A_{Base}	A_{Last}	A_{Avg}
Baseline	88.82	84.20	86.50
+ EPFM	89.38	85.57	87.28
+ EPFM + CPFM	**92.25**	**88.02**	**89.63**

Table 4. Expert number ablation on CIFAR-100.

Experts	A_{Base}	A_{Last}	A_{Avg}
0	90.76	85.17	87.61
2	91.41	86.75	88.74
4	91.45	87.23	88.95
6	92.17	87.98	89.54
8	92.25	**88.02**	**89.63**
10	**92.57**	87.79	89.57

Table 5. Prefix-tuning configurations.

Layers, Length	A_{Last}	PD
0, –	82.34	6.30
2, 5	87.12	4.56
4, 5	87.75	4.43
6, 5	**88.02**	**4.23**
6, 10	87.93	4.57
6, 15	87.01	5.89

4.4 Further Analysis

Number of Local Experts. We evaluate expert count impact on CIFAR-100 (Table 4). The baseline (0 experts) uses Stage 1 features directly. Increasing experts improves base accuracy, but beyond 8 experts, last-session accuracy drops due to overfitting from reduced local class allocation.

Prefix-Tuning in First Stage. We analyze prefix-tuning intensity (Table 5). Skipping Stage 1 reduces last-session accuracy by 5.68%, highlighting its importance. Optimal results (6 layers, prompt length 5) achieve 88.02% A_{Last} with minimal performance drop (4.23%). Longer prompts increase overfitting risk.

5 Conclusions

In this paper, we propose a novel two-stage learning framework for the few-shot class incremental learning (FSCIL) scenario using pretrained ViTs. To adapt the model to incremental classes with limited samples, we progressively enhance class separation with local experts. In the first stage, the model is tuned on all base classes to realize an initial class separation. We then construct local experts, each focusing only on a specific local region in the feature space. By integrating discriminative information within local classes, these experts expand the original features to further improve the inter-class separation. Our method achieves superior results across all benchmarks compared to previous approaches, demonstrating the effectiveness of our method.

Acknowledgments. The work was supported by the National Key R&D Program of China under Grant 2021ZD0201300, the National Natural Science Foundation of China under Grant 61936004, the Innovation Group Project of the National Natural Science Foundation of China under Grant 61821003, and the 111 Project on Computational Intelligence and Intelligent Control under Grant B18024.

References

1. Ahmed, N., Kukleva, A., Schiele, B.: OrCo: towards better generalization via orthogonality and contrast for few-shot class-incremental learning. In: Proceedings of the IEEE/CVF Conference on Computer Vision and Pattern Recognition, pp. 28762–28771 (2024)
2. Alex, K.: Learning multiple layers of features from tiny images. https://www.cs.toronto.edu/~kriz/learning-features-2009-TR.pdf (2009)
3. Antoniou, A., Edwards, H., Storkey, A.: How to train your MAML. In: International Conference on Learning Representations (2018)
4. Arnold, S., Iqbal, S., Sha, F.: When MAML can adapt fast and how to assist when it cannot. In: International Conference on Artificial Intelligence and Statistics, pp. 244–252. PMLR (2021)
5. Belouadah, E., Popescu, A.: IL2M: class incremental learning with dual memory. In: Proceedings of the IEEE/CVF International Conference on Computer Vision, pp. 583–592 (2019)
6. Dosovitskiy, A.: An image is worth 16×16 words: Transformers for image recognition at scale. arXiv preprint arXiv:2010.11929 (2020)
7. Douillard, A., Cord, M., Ollion, C., Robert, T., Valle, E.: PODNet: pooled outputs distillation for small-tasks incremental learning. In: Computer vision–ECCV 2020: 16th European conference, Glasgow, UK, August 23–28, 2020, proceedings, part XX 16, pp. 86–102. Springer (2020)
8. Finn, C., Abbeel, P., Levine, S.: Model-agnostic meta-learning for fast adaptation of deep networks. In: International Conference on Machine Learning, pp. 1126–1135. PMLR (2017)
9. French, R.M.: Catastrophic forgetting in connectionist networks. Trends Cogn. Sci. **3**(4), 128–135 (1999)
10. Hersche, M., Karunaratne, G., Cherubini, G., Benini, L., Sebastian, A., Rahimi, A.: Constrained few-shot class-incremental learning. In: Proceedings of the IEEE/CVF Conference on Computer Vision and Pattern Recognition, pp. 9057–9067 (2022)
11. Jung, D., Han, D., Bang, J., Song, H.: Generating instance-level prompts for rehearsal-free continual learning. In: Proceedings of the IEEE/CVF International Conference on Computer Vision, pp. 11847–11857 (2023)
12. Kang, M., Park, J., Han, B.: Class-incremental learning by knowledge distillation with adaptive feature consolidation. In: Proceedings of the IEEE/CVF Conference on Computer Vision and Pattern Recognition, pp. 16071–16080 (2022)
13. Khosla, P., et al.: Supervised contrastive learning. Adv. Neural. Inf. Process. Syst. **33**, 18661–18673 (2020)
14. Kim, D.Y., Han, D.J., Seo, J., Moon, J.: Warping the space: weight space rotation for class-incremental few-shot learning. In: The Eleventh International Conference on Learning Representations (2023)
15. Kirkpatrick, J., et al.: Overcoming catastrophic forgetting in neural networks. Proc. Natl. Acad. Sci. **114**(13), 3521–3526 (2017)
16. Lester, B., Al-Rfou, R., Constant, N.: The power of scale for parameter-efficient prompt tuning. arXiv preprint arXiv:2104.08691 (2021)
17. Li, X.L., Liang, P.: Prefix-tuning: Optimizing continuous prompts for generation. arXiv preprint arXiv:2101.00190 (2021)
18. Li, X., Zhou, Y., Wu, T., Socher, R., Xiong, C.: Learn to grow: a continual structure learning framework for overcoming catastrophic forgetting. In: International Conference on Machine Learning, pp. 3925–3934. PMLR (2019)

19. Li, Z., Hoiem, D.: Learning without forgetting. IEEE Trans. Pattern Anal. Mach. Intell. **40**(12), 2935–2947 (2017)
20. Liu, B.: Negative margin matters: understanding margin in few-shot classification. In: Vedaldi, A., Bischof, H., Brox, T., Frahm, J.-M. (eds.) ECCV 2020. LNCS, vol. 12349, pp. 438–455. Springer, Cham (2020). https://doi.org/10.1007/978-3-030-58548-8_26
21. Mallya, A., Davis, D., Lazebnik, S.: Piggyback: adapting a single network to multiple tasks by learning to mask weights. In: Proceedings of the European Conference on Computer Vision (ECCV), pp. 67–82 (2018)
22. Masana, M., Liu, X., Twardowski, B., Menta, M., Bagdanov, A.D., Van De Weijer, J.: Class-incremental learning: survey and performance evaluation on image classification. IEEE Trans. Pattern Anal. Mach. Intell. **45**(5), 5513–5533 (2022)
23. Mazumder, P., Singh, P., Rai, P.: Few-shot lifelong learning. In: Proceedings of the AAAI Conference on Artificial Intelligence. vol. 35, pp. 2337–2345 (2021)
24. Nichol, A.: On first-order meta-learning algorithms. arXiv preprint arXiv:1803.02999 (2018)
25. Park, K.H., Song, K., Park, G.M.: Pre-trained vision and language transformers are few-shot incremental learners. In: Proceedings of the IEEE/CVF Conference on Computer Vision and Pattern Recognition, pp. 23881–23890 (2024)
26. Ravi, S., Larochelle, H.: Optimization as a model for few-shot learning. In: International Conference on Learning Representations (2017)
27. Rebuffi, S.A., Kolesnikov, A., Sperl, G., Lampert, C.H.: iCaRL: incremental classifier and representation learning. In: Proceedings of the IEEE Conference on Computer Vision and Pattern Recognition, pp. 2001–2010 (2017)
28. Rolnick, D., Ahuja, A., Schwarz, J., Lillicrap, T., Wayne, G.: Experience replay for continual learning. Adv. Neural Inf. Process. Syst. **32** (2019)
29. Smith, J.S., et al.: CODA-Prompt: continual decomposed attention-based prompting for rehearsal-free continual learning. In: Proceedings of the IEEE/CVF Conference on Computer Vision and Pattern Recognition, pp. 11909–11919 (2023)
30. Snell, J., Swersky, K., Zemel, R.: Prototypical networks for few-shot learning. Adv. Neural Inf. Process. Syst. **30** (2017)
31. Song, Z., Zhao, Y., Shi, Y., Peng, P., Yuan, L., Tian, Y.: Learning with fantasy: semantic-aware virtual contrastive constraint for few-shot class-incremental learning. In: Proceedings of the IEEE/CVF Conference on Computer Vision and Pattern Recognition, pp. 24183–24192 (2023)
32. Sung, F., Yang, Y., Zhang, L., Xiang, T., Torr, P.H., Hospedales, T.M.: Learning to compare: relation network for few-shot learning. In: Proceedings of the IEEE Conference on Computer Vision and Pattern Recognition, pp. 1199–1208 (2018)
33. Tang, Y.M., Peng, Y.X., Zheng, W.S.: When prompt-based incremental learning does not meet strong pretraining. In: Proceedings of the IEEE/CVF International Conference on Computer Vision, pp. 1706–1716 (2023)
34. Tao, X., Hong, X., Chang, X., Dong, S., Wei, X., Gong, Y.: Few-shot class-incremental learning. In: Proceedings of the IEEE/CVF Conference on Computer Vision and Pattern Recognition, pp. 12183–12192 (2020)
35. Tao, X., Hong, X., Chang, X., Dong, S., Wei, X., Gong, Y.: Few-shot class-incremental learning. In: Proceedings of the IEEE/CVF Conference on Computer Vision and Pattern Recognition, pp. 12183–12192 (2020)
36. Tiwari, R., Killamsetty, K., Iyer, R., Shenoy, P.: GCR: gradient coreset based replay buffer selection for continual learning. In: Proceedings of the IEEE/CVF Conference on Computer Vision and Pattern Recognition, pp. 99–108 (2022)

37. Vinyals, O., Blundell, C., Lillicrap, T., Wierstra, D., et al.: Matching networks for one shot learning. Adv. Neural Inf. Process. Syst. **29** (2016)
38. Wah, C., Branson, S., Welinder, P., Perona, P., Belongie, S.: The Caltech-UCSD birds-200-2011 dataset (2011)
39. Wang, Q.W., Zhou, D.W., Zhang, Y.K., Zhan, D.C., Ye, H.J.: Few-shot class-incremental learning via training-free prototype calibration. Adv. Neural Inf. Process. Syst. **36** (2024)
40. Wang, Z., et al.: DualPrompt: complementary prompting for rehearsal-free continual learning. In: European Conference on Computer Vision, pp. 631–648. Springer (2022)
41. Wang, Z., et al.: Learning to prompt for continual learning. In: Proceedings of the IEEE/CVF Conference on Computer Vision and Pattern Recognition, pp. 139–149 (2022)
42. Yan, S., Xie, J., He, X.: DER: dynamically expandable representation for class incremental learning. In: Proceedings of the IEEE/CVF Conference on Computer Vision and Pattern Recognition, pp. 3014–3023 (2021)
43. Yang, Y., Yuan, H., Li, X., Lin, Z., Torr, P., Tao, D.: Neural collapse inspired feature-classifier alignment for few-shot class incremental learning. arXiv preprint arXiv:2302.03004 (2023)
44. Ye, H.J., Zhan, D.C., Jiang, Y., Zhou, Z.H.: Heterogeneous few-shot model rectification with semantic mapping. IEEE Trans. Pattern Anal. Mach. Intell. **43**(11), 3878–3891 (2020)
45. Yoon, J., Yang, E., Lee, J., Hwang, S.J.: Lifelong learning with dynamically expandable networks. arXiv preprint arXiv:1708.01547 (2017)
46. Zenke, F., Poole, B., Ganguli, S.: Continual learning through synaptic intelligence. In: International Conference on Machine Learning, pp. 3987–3995. PMLR (2017)
47. Zhang, C., Cai, Y., Lin, G., Shen, C.: DeepEMD: few-shot image classification with differentiable earth mover's distance and structured classifiers. In: Proceedings of the IEEE/CVF Conference on Computer Vision and Pattern Recognition, pp. 12203–12213 (2020)
48. Zhang, C., Song, N., Lin, G., Zheng, Y., Pan, P., Xu, Y.: Few-shot incremental learning with continually evolved classifiers. In: Proceedings of the IEEE/CVF Conference on Computer Vision and Pattern Recognition, pp. 12455–12464 (2021)
49. Zhou, D.W., Wang, F.Y., Ye, H.J., Ma, L., Pu, S., Zhan, D.C.: Forward compatible few-shot class-incremental learning. In: Proceedings of the IEEE/CVF Conference on Computer Vision and Pattern Recognition, pp. 9046–9056 (2022)

Prompt-Based Hypernetwork in Rehearsal-Free Class Incremental Learning

Yang Liu and Zhigang Zeng(✉)

Huazhong University of Science and Technology, Wuhan 430074, China
{m202373667,zgzeng}@hust.edu.cn

Abstract. Among the three scenarios of continual learning, task incremental learning is considered an ideal scenario under certain assumptions, where the model is assumed to have access to task identifiers during the inference phase. In this scenario, numerous advanced methods have emerged, among which hypernetwork-based approaches stand out. These methods use an auxiliary network to generate weights or masks for the main network, serving to protect the weights and achieving favorable results. However, a limitation of this approach is that, due to the task-specific nature of the hypernetwork, task identifiers are essential. To address this limitation, we introduce prompt matching mechanism into the hypernetwork, enabling this approach to be applied in class-incremental learning without recourse to a task identifier. Furthermore, we combine the prompt-based hypernetwork with prompt-tuning methods to explore additional functionalities of prompts. Experimental results show that our prompt-based hypernetwork experiences almost no performance degradation in the class incremental learning scenario. Additionally, when our hypernetwork is used as a plug-in for prompt-tuning methods in class-incremental learning, it improves accuracy by 1.0–1.5% and achieves state-of-the-art (SOTA) results on several continual learning benchmarks.

Keywords: Continual learning · Hypernetwork · Prompt tuning

1 Introduction

Methods for learning on static data distributions are well established and perform well. However, in real-world scenario, data distributions are often dynamic and require models to adapt continually. Therefore, this requires us to continuously update the model to accommodate the integration of knowledge from old and new tasks in order to avoid degradation of model performance, which is referred to as continual learning or incremental learning or lifelong learning [1–6].

The primary objective of continual learning is to enable a single model to learn a series of tasks without experiencing catastrophic forgetting. A straightforward approach to achieving this is to store some data from previous tasks and

C. Li et al. (Eds.): ICNC 2025, CCIS 2946, pp. 338–352, 2026.
https://doi.org/10.1007/978-981-92-1599-7_29

combine it with new data into a comprehensive dataset for model updates. This strategy is known as rehearsal [7–14]. However, as data accumulates, training becomes increasingly expensive. And in some cases privacy concerns with the data prevent it from being stored for long periods of time, further limiting the feasibility of rehearsal. Generative rehearsal [15–17] is an alternative to it, using generative models to get old data for new tasks, but this approach is highly dependent on the effectiveness of the generative models. As a result, rehearsal-free continual learning methods have attracted more attention. The dynamic architecture-based approach [18–22] is a well-established strategy in rehearsal-free continual learning methods. Its core idea is to assign a separate architecture for each task. This approach is typically explored from two perspectives: one is to expand the network by adding new components for each task, and the other is to compress the network by pruning the model after learning each task to get a more streamlined but similarly performing subnetwork, which is based on the lottery hypothesis, so that the remaining parameters can be used to learn a new task.

Hypernetwork-based [23–31] method can be viewed as a variant of dynamic architecture-based approaches. Unlike traditional scaling methods, hypernetworks mitigate storage requirements associated with network expansion and alleviate limitations on the number of tasks that can be learned in compressed networks compared with pruning method. Specifically, although hypernetwork also need to add another additional architecture [25], this type of model requires adding only once in the training phase, and the architecture being added is called a hypernetwork. Rather than directly performing tasks, the hypernetwork generates weights for a main task-performing network. It leverages the task identification information for weight generation in forward propagation and is optimized with the main network from end-to-end back propagation optimization. By generating weights and protect them in hypernetwork, this kind of model helps mitigate catastrophic forgetting in the main network.

Despites its strength, hypernetworks are generally restricted to task incremental learning due to their reliance on task identifiers. We propose incorporating a matching mechanism into the hypernetwork, thereby extending its applicability to class-incremental learning scenarios.

Furthermore, since prompt-tuning has been used in continual learning with exceptional performance [32–36], this motivated us to dig into whether the prompt-match mechanism can be introduced to hypernetwork, and whether one prompt can be utilized for not only fine-tuning but also other tasks such as weight generating. Our study shows that the inclusion of a prompt matching mechanism overcomes the need for additional task information of the hypernetwork and experience little performance degradation. And without adding additional prompt parameters, the prompt components employed in the prompt-tuning approach can be used to guide the hypernetwork for weight generation at the same time and can improve the performance of the model by 1.0–1.5%.

To summarize, our work makes the following contributions:

- We introduce the prompt matching mechanism into the hypernetwork, which makes the hypernetwork-based method applicable to the field of class incremental learning.
- Besides, we propose a simple but effective orthogonal update method for hypernetworks, which is not only applicable to the method in this paper but also to hypernetwork-based methods in general.
- We introduce the prompt-based hypernetwork method as a plug-in to the prompt-tuning-based continual learning method, and propose P-Hnet architecture, which is a generalized method that can help existing prompt-tuning methods to improve the performance from the perspective of expanding the functionality of the prompts.
- We conducted experiments on Split-CIFAR-100, Split-ImageNet-R, and Split-DomainNet benchmarks in a no-rehearsal continual learning setup, which improved the performance of the basic prompt-tuning method we used in P-Hnet by about 1.0–1.5% and achieved SOTA on some benchmarks.

2 Related Work

Continual learning aims to overcome the catastrophic forgetting problem that arises when learning from dynamical data distributions. Depending on the problem setting and application scenario, methods in continual learning can be categorized into several pipelines. Among them, regularization-based methods [37–40] constructing a regular loss penalty that constrains the model itself or the model prediction results by limiting the drastic transformation of previously learned knowledge. Rehearsal-based methods are suitable for scenarios when old task data can be stored [14,41,42] or generated [15–17,25]. These methods are effective in mitigating the forgetting of prior knowledge. Additionally, dynamic architecture-based approaches, which have been previously discussed, are another common approach in continual learning and will not be revisited here.

2.1 Hypernetwork-Based Method.

Hypernetwork was proposed by Ha et al. [43] and introduced to continual learning by Von et al. [25]. In continual learning, it belongs to the dynamic architecture-based method, but unlike general methods that require dynamically expanding or compressing the architecture, hypernetwork methods generate weights for the main network, i.e., the network used for classification or other tasks by the way called mapping, which serves to protect the weights of that part. Previous hypernetwork-based methods for continual learning typically require additional task-specific information, such as task identifiers or task identity embeddings. As a result, these methods are mainly applicable in task-incremental learning scenarios. Since the parameters of the main network are generally large, directly generating all the weights will result in the parameters of the hypernetwork being much larger than the main network. There has been some work done to address this issue. Von et al. [25] proposes to generate the weights in chunks, which reduces the number of parameters of the hypernetwork by introducing additional layer embeddings information. In terms of the type of hypernetwork, Chandra et al. [29] and Jin et al. [28] use LSTM and attention block instead of MLP which is used in the original method to generate weights respectively. Kamil et al. [31] do not use the hypernetwork to generate the weights directly, but instead use them to generate soft masks for a pretrained network, which balances the importance of old and new knowledge when learning a new task. As fine-tuning has been widely used and excelled in various fields, some scholars have also combined hypernetworks with fine-tuning models. Ding et al. [27] utilizes the hypernetwork to generate fine-tuning parameters such as prompts, adapter parameters, and lora matrices for the pre-trained main network. While this approach integrates the hypernetwork with prompts, it is fundamentally differs from our method. The prompt in this method is a product of the hypernetwork and is used to fine-tune the main network, which essentially still utilizes the task identification information as the source of information for the hypernetwork and generates part of the weights for the main network. Whereas in our method, the prompt serves as an information source for the hypernetwork and does not require additional task information.

2.2 Prompt-Based Method

Continual learning methods based on prompt-tuning are one of the most effective classes of methods in recent years. Such methods are often based on a pre-trained ViT model and introducing prompt-tuning techniques [49,50] from natural language processing into continual learning. L2P [32] proposes to learn a pool of prompts and uses a query-key matching mechanism to select a prompt, and most of the existing work is based on a modification of the prompting mechanism or the training mechanism. Dualprompt [33] proposes e-prompt, which represents task-specific knowledge, and g-prompt, which represents task-shared knowledge. Codaprompt [34] modifies the prompts into prompt components and introduces a method for selecting prompt selection based on the attention mechanism. Cprompt [35] investigates the consistency between prompts and classifiers, and OS-prompt [36] turns the training process from two phases into a one phase.

3 Preliminaries

3.1 Continual Learning with Hypernetwork

In the continual learning setting, the model learns T tasks sequentially with dataset $\mathcal{D} = \{\mathcal{D}_1, \ldots, \mathcal{D}_T\}$. $\mathcal{D}_t = \{(x_i, y_i)\}_t$ is provided to the model at the

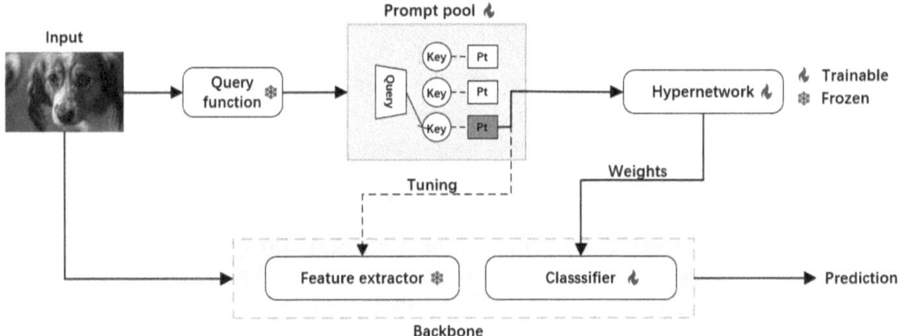

Fig. 1. The illustration of the proposed method P-Hnet. It consists of two main modules, a prompt-based hypernetwork model, and a prompt-tuning model. In this architecture, the hypernetwork can be used as a plug-in for the prompt-tuning model to co-utilise the prompt information.

t-th stage. Our problem is based on an image classification task so x_i denotes an image and y_i is the image label. We focus on class incremental learning, a scenario where task identification information is unknown at inference, which is more challenging compared to domain incremental learning and task incremental learning.

Our purpose is to learn a meta network named hypernetwork and optimize the following loss function

$$\psi^* = min \frac{1}{n_t} \sum_{\psi} \mathcal{L}(f(h(\psi), x_{i,t})) \tag{1}$$

where $\mathcal{L}(\cdot)$ is a classification objective loss such as the cross-entropy loss. $h(\cdot)$ denotes the hypernetwork model, ψ is the weights of the hypernetwork. $f(\cdot)$ denotes the main network and $h(\psi)$ is the weights of the main network, which is generated by the hypernetwork. n_t is the number of samples for task t.

3.2 Prompt-Based Method

Prompt-tuning methods typically employ a pre-trained ViT model as the backbone, including a feature extractor $f_e(\cdot)$ and a classifier $f_c(\cdot)$. Besides, the model has a set of additional parameters called the prompt pool (P, K). The prompts denoted as $P \in R^{L_P \times D}$, are a set of trainable vectors or matrices , where D denotes the embedding dimension and L_P indicates the length of the prompt. The Keys are a small set of vectors with the same dimensions as the input feature vectors, which are used to match the similarity with the input and get the corresponding prompt components, denoted as $K \in R^D$. During trainig, the feature extractor is frozen, the classifier, the keys and prompts are trainable.

4 Method

4.1 Prompt-Based Hypernetwork

Existing hypernetwork-based methods use the task identification embedding $e = \{e_1, \ldots, e_T\}$ as input information, and at the time of learning the t-th task, e_t is used as the model input to learn the weights of hypernetwork and the corresponding task embedding through a two-stage approach [25]. During inference, both the image and the image's corresponding task embedding are needed. Specifically, as shown in the following equations, the task embedding is used to generate a task-specific weight for the test image, which is then used to perform the task of classification.

$$y_i = f(\theta, x_i) \tag{2}$$

$$\theta = h(\psi, e_t) \tag{3}$$

We introduce the prompt pool for replacing the task identification information embedding. As shown in Fig. 1, the prompt pool consists of a set of keys and corresponding prompts. When the inputs get feature representations through the pre-trained encoder, the representations are sent to the prompt pool to do similarity matching with each key in the set of keys, and the prompt corresponding to the key with the highest similarity is taken as the hypernetwork input. Then hypernetwork generates weights for the main network that performs the target task. Since the pre-trained ViT is used as a feature extractor, we only generate weights for the classifier. Detailed process is shown in the following equation.

$$\theta = h(\psi, p_t) \tag{4}$$

$$p_t = topk(q, k) \tag{5}$$

$$q = f_e(x) \tag{6}$$

where q denotes the query obtained through pre-trained query function. Similar to earlier setting, we use the frozen feature extractor of the backbone as the query function. Topk returns the prompt components corresponding to the top-k keys that are most similar to query.

4.2 Orthogonal Penalty Loss

Using a hypernetwork to generate weights for the main network does not directly overcome forgetting. Forgetting still occurs in the hypernetwork itself. Most of the previous continual learning methods based on hypernetworks use a two-stage training method with regularization loss to ensure the effectiveness of the hypernetwork, which is a compromise. Instead, We use an orthogonal gradient updating strategy to replace it.

In continual learning, the orthogonal gradient updating based approach [44–48] allows the knowledge of the new task to be updated in the orthogonal direction of the knowledge of the old task, thus ideally not having an impact on the old

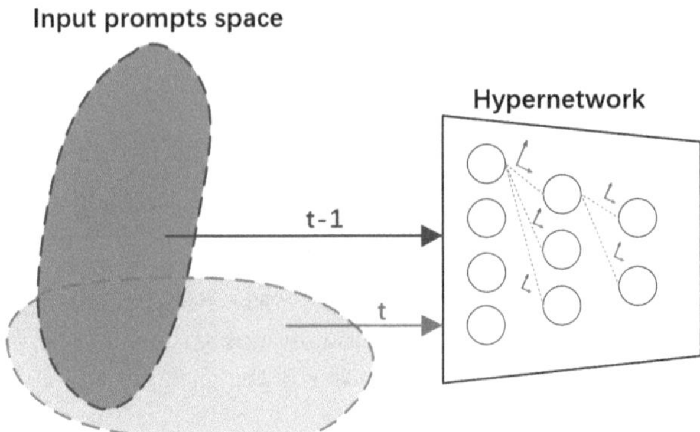

Fig. 2. When the network is an MLP or a convolutional neural network, the direction of update of the weights lies in the space tensed by the inputs. We discard the cumbersome multi-step orthogonal gradient update optimisation process and directly impose an orthogonal penalty loss term on the input prompt space.

task. In the existing representative orthogonal projection methods, OWM [47] makes the update direction of the weights of each layer orthogonal to the inputs of that layer, and OGD [48] makes it orthogonal to the gradient of the weights of the old task. However, in reality, the capacity of the orthogonal space, and the preservation of the gradient of the knowledge of the old tasks are inevitable factors that limit the performance of this kind of methods. Methods such as GPM [45] and DualGPM [46] can preserve the gradient information of the old tasks through learning a matrix. Although GPM and DualGPM have achieved great performance, there is still a need to maintain an additional matrix of gradient information. Since the hypernetwork we use is a simple MLP model, according to existing studies, the updated gradient directions of the linear and convolutional layers are located in the space tensored by the input vectors. Since the inputs to the hypernetwork consist of trainable prompts or task embeddings, it motivates us to explore orthogonal gradient updates within the input space of the hypernetwork, specifically in the prompt space, as shown in Fig. 2. We propose an orthogonal penalty loss item for the input prompt space:

$$\mathcal{L}_{ortho} = \sum_{i \neq j} ||p_i^T p_j||_2 + \sum_i ||1 - p_i^T p_i||_2 \tag{7}$$

where p represents the normalized prompt vector for each task. $||\cdot||_2$ is the Frobenius norm of the vector or the matrix. The loss ensures that the input prompts corresponding to each task is orthogonal to the prompts of the other tasks, so that the direction of the hypernetwork gradient update of the corresponding task is also orthogonal to the direction of the hypernetwork gradient of the other tasks. Moreover, this loss function can be applied not only in prompt-based hypernetworks, but also in general MLP or convolutional hypernetwork-based methods, with the only modification of replacing the prompt p with task-specific information embedding e in the loss term.

$$\mathcal{L}_{ortho} = \sum_{i \neq j} ||e_i^T e_j||_2 + \sum_{i} ||1 - e_i^T e_i||_2 \tag{8}$$

4.3 A Plug-in to Prompt-Tuning Method

We integrate the prompt-based hypernetwork proposed above as a plug-in to an existing prompt-tuning method. Since the role of the hypernetwork is to generate weights for the target main task-network, in this case we chose to generate weights for the classification header. The choice of the prompt-tuning method is not our focus: this is a generic plug-in for most prompt-tuning methods. The usage of our plug-in is shown in Fig. 1. In the prompt-tuning method, the query generated by the pre-trained query function will not only fine-tune the pre-trained main network, but it will also serve as the input to the hypernetwork plug-in. The weights of the main network classifier are no longer used, instead, the classifier weights generated by the hypernetwork are employed to classify the images. We refer to this framework that combines the plugin and prompt-tuning methods as P-hnet. In this framework, there are various choices of hypernetwork types, including the current mainstream architectures. However, when orthogonal penalty loss terms are used, convolutional or MLP hypernetworks are a better choice. In addition, the type of prompt sent to hypernetwork vary as well, depending on the chosen basic prompt-tuning method. It is worth noting that the query function, the feature extractor in the main network is frozen in our framework, as same as in the prompt-tuning based approach. The hypernetwork, the prompt components and the keys, are trainable.

4.4 Full Loss Function

Our loss function consists of three parts, cross-entropy loss $\mathcal{L}_{en}$, orthogonal penalty loss $\mathcal{L}_{orth}$ and prompt matching loss $\mathcal{L}_{match}$, and the final loss is as follows:

$$\mathcal{L}_{total} = \mathcal{L}_{en} + \lambda_1 \mathcal{L}_{orth} + \lambda_2 \mathcal{L}_{match} \tag{9}$$

where λ_1, λ_2 represent the trade-off coefficients. $\mathcal{L}_{match}$ is a loss function that improves matching accuracy by penalising the similarity between keys and query within the same task [33].

5 Experiments

5.1 Experiments Details

We utilize the Split CIFAR-100, Split ImageNet-R, and Split-DomainNet benchmarks for class-incremental continual learning.

- **Split-CIFAR100** [51] consists of 60,000 32×32 images with 100 classes. This dataset is a widely used learning benchmark in continual learning and is divided into 10 tasks, each containing 10 classes.

- **Split-ImageNet-R** [52] has 30,000 color images with 200 classes. This dataset is considered to be a very challenging benchmark in continual learning. Each task contains 40 classes when divided into 5 tasks and 20 classes each when divided into 10.
- **Split-DomainNet** [53] is a large-scale dataset containing about 600,000 images in 345 categories. We use images from its five different domains as continual learning tasks, each consisting of 69 classes.

Implementation Details. Similar to earlier PCL methods, we use VIT-B/16 pre-trained on ImageNet-1k as the backbone for our experiments. And we chose Dualprompt as the base prompt-tuning method when used with a hypernetwork plug-in. For the type of hypernetwork, a simple two-layer MLP is chosen as our model. In addition, we used Adam as our optimiser and all experiments were run on the A800.

Evaluation Metrics. We evaluate the performance using two evaluation metrics [6] which are widely used in continual learning: the average accuracy (denoted as Acc) and the average forgetting (denoted as FF).

Competitors. We compare our approach with the current mainstream prompt-based methods such as L2P [32], dualprompt [33], CodaPrompt [34], ESN [54], and state-of-the-art methods such as OSprompt [36]. For a fair comparison, all models were reproduced in the same experimental setting using the same pre-trained model with the best experimental parameters set in the original paper. The upper-bound (UB) denotes the upper bound of the experiment, obtained by learning with all data collectively.

Table 1. Result (%) on ImageNet-R under 10-tasks (20 classes per task) and 20-tasks (10 classes per task) setting. Best results are marked in bold.

	10 Tasks		20 Tasks	
Method	Acc(↑)	FF(↓)	Acc(↑)	FF(↓)
UB	80.27	-	80.27	-
L2P	74.60 ± 0.90	$\mathbf{4.52 \pm 0.50}$	72.09 ± 1.12	$\mathbf{4.86 \pm 1.37}$
ESN	75.11 ± 0.36	5.68 ± 0.77	70.57 ± 0.62	6.84 ± 0.36
CodaPrompt	75.51 ± 0.81	5.91 ± 1.36	72.25 ± 0.78	6.65 ± 0.31
OS-Prompt	75.36 ± 0.67	4.95 ± 0.60	73.18 ± 0.49	4.89 ± 0.59
DualPrompt	74.87 ± 0.85	7.18 ± 0.15	71.69 ± 1.06	7.68 ± 0.96
Ours	$\mathbf{75.81 \pm 0.72 (+1.0)}$	6.03 ± 0.26	$\mathbf{73.2 \pm 0.84 (+1.5)}$	6.79 ± 0.59

Table 2. Result (%) on CIFAR-100 under 10-tasks (10 classes per task) setting. Best results are marked in bold.

Method	Acc(↑)	FF(↓)
UB	91.99	-
L2P	86.38 ± 0.31	5.88 ± 0.76
ESN	86.61 ± 0.22	6.08 ± 0.48
CodaPrompt	85.73 ± 0.14	7.13 ± 0.44
OS-Prompt	86.32 ± 0.53	5.01 ± 0.47
DualPrompt	86.01 ± 0.22	5.86 ± 0.62
Ours	**87.37+0.22(+1.3)**	**5.01 ± 0.60**

Table 3. Result (%) on DomainNet under 5-tasks (69 classes per task) setting. Best results are marked in bold.

Method	Acc(↑)	FF(↓)
UB	89.15	-
L2P	81.17 ± 0.83	8.98 ± 1.25
ESN	79.22 ± 2.04	10.62 ± 2.12
CodaPrompt	80.04 ± 0.79	10.16 ± 0.35
OS-Prompt	82.71 ± 0.38	8.95 ± 1.01
DualPrompt	81.70 ± 0.78	8.04 ± 0.31
Ours	**83.15 ± 0.61(+1.4)**	**7.12 ± 0.47**

5.2 Main Results and Discussion

Overall Performance Evaluation. Table 1 shows the comparison of our method (P-Hnet) with the SOTA competitors on the Split-ImageNet benchmark. Our method does not use the SOTA prompt-tuning method in our framework, what is more noteworthy is the improvement of our method over the Dual-prompt method(denoted by the figures in parentheses). As mentioned earlier, the prompt-tuning model chosen for our method is Dualprompt, which is a relatively basic model compared to other more advanced methods. We chose it because improving on a simpler model better demonstrates the effectiveness of our method and reduces the boosting effect from coupling with other improvements. We improve the performance over Dualprompt by 1.0% on 10-tasks, and 1.5% on 20-task settings.

Table 2 shows the comparative results of our approach on CIFAR100. Due to the excellent foundation of Dualprompt and our great improvement, we achieved SOTA results on this dataset.

Table 3 shows the results on Domainnet, which is a very challenging dataset due to its large number. Nonetheless, our method still improves 1.4% on Dual-prompt and outperforms other methods.

Discussion. We note that several existing studies share similarities with our study, and we compare and discuss these studies below.

Ding et al.citeding2024bridging and He et al. [55] leverage the hypernetwork to generate prompts or other fine-tuning components, which are subsequently used to fine-tune the pre-trained model.

As mentioned earlier, these studies still rely on task-specific information. And in terms of the role of the hypernetwork, which serves as a protector of the weights, these works shifted the protection from the backbone weights to the fine-tuning parameters or weights of fine-tuning components. Essentially, they remain within the traditional task-specific hypernetwork framework. In contrast, our method does not require any task-specific information. Instead, we use prompts to eliminate the need for explicit task-specific data, which moves beyond the traditional framework's focus on protection.

Additionally, Ni et al. [56] also employs non-task-specific information for weight generation guidance, but a language model is utilized in their framework. By embedding category names into prompt templates and inputting them into a large language model (LLM), the model utilizes the LLM's robust semantic processing capabilities to guide weight generation for an image classification networks. This study emphasizes the impact of linguistic information on visual processing, requiring additional pre-trained language models as support. In contrast to that study, our work focuses on exploring additional information from prompts, using the existing prompt components within fine-tuning models as inputs to the hypernetwork, without the need for additional known information.

Ablation Study. We further analyze our approach through ablation experiments. First we compared the two orthogonal approaches(DualGPM and ours) and compared the experimental results without applying orthogonal gradient updating (i.e. using the two-stage method). We perform experiments on the Split-CIFAR100 dataset. As shown in Table 4, the two orthogonal methods are almost equally effective, with our method being slightly lower by a little bit, but given its simplicity, this reduction is perfectly acceptable. Both orthogonal methods are much higher than the one using the two-stage method.

Table 4. Analysis of orthogonal gradient updating in hypernetwork.

Method	Acc(↑)	FF(↓)
Two-stage	80.15 ± 0.97	10.88 ± 0.91
DualGPM	87.41 ± 0.25	5.23 ± 0.58
Ours	87.37 ± 0.22	5.01 ± 0.60

In addition, we explored the ways in which prompts act on both hypernetworks and prompt-tuning. We explored three implementations: taking a portion of the promtps used in prompt-tuning for weight generation; taking all the prompts weighted and summed and taking all the prompts concatenated. Results are in Table 5. As shown in it, the results of the three approaches do not differ much, and we analyse that this is because the amount of input information required to guide the hypernetwork is small. In general hypernetwork-based approaches, the dimensions of the input vectors are usually 8 to 32, while the dimension of prompts is 768, which is far more than enough to satisfy the needs of the hypernetwork. Therefore all the three approaches are able to satisfy the needs.

Table 5. Analysis of how prompts are used in both models.

Method	Acc(↑)	FF(↓)
Partial	87.37 ± 0.22	5.01 ± 0.60
Sum	87.37 ± 0.22	5.01 ± 0.60
Cat	87.55 ± 0.20	4.97 ± 0.59

We also conducted experiments on the impact of the prompt matching mechanism on the CIFAR dataset. As shown in Table 6. Where prompt-matching denotes the use of the prompt pool we introduced above, and task-id denotes that the task identification information is known in the inference phase. The results show that the P-hnet architecture with the prompt matching mechanism does not show much degradation after the loss of task identification information.

Table 6. Analysis of the impact of prompt-matching mechanism compared with task-id.

Method	Acc(↑)	FF(↓)
Prompt-matching	87.37 ± 0.22	5.01 ± 0.60
Task-ID	87.93 ± 0.12	4.79 ± 0.44

6 Conclusion

In this paper, additional applications of prompts in continual learning were revealed and discussed in depth for the first time. We introduced prompts into existing work on hypernetwork, enabling their use for parameter protection in the scenario of class incremental learning. Moreover, we proposed a hypernetwork-specific orthogonal gradient updating method, which was simple to implement but effective. In addition, we dug deeper into the role of prompts and proposed P-Hnet, a framework in which prompts were used for both fine-tuning and weight generation. We validated its effectiveness and generality on mainstream benchmarks and conducted an analysis.

Despite its demonstrated effectiveness, our approach has several limitations. The incorporation of the hypernetwork introduces additional parameters, leading to higher computational costs. Although the efficacy of combining the hypernetwork with prompts has been validated empirically, it remains to be tested with a wider range of advanced techniques. Future work will focus on optimizing the hypernetwork architecture and integrating more sophisticated methods into the framework to enhance its performance and broaden its applicability.

Acknowledgement. The work was supported by the National Key R&D Program of China under Grant 2021ZD0201300, the Foundation for Outstanding Research Groups of Hubei Province of China under Grant 2025AFA012, the 111 Project on Computational Intelligence and Intelligent Control under Grant B18024, and the Postdoctoral Fellowship Program of CPSF under Grant Numbers GZB20250428 and 2025M771702.

References

1. Belouadah, E., Popescu, A., Kanellos, I.: A comprehensive study of class incremental learning algorithms for visual tasks. Neural Netw. **135**, 38–54 (2021)
2. De Lange, M., et al.: A continual learning survey: defying forgetting in classification tasks. IEEE TPAMI **44**(7), 3366–3385 (2021)
3. Masana, M., Liu, X., Twardowski, B., Menta, M., Bagdanov, A.D., Van De Weijer, J.: Class-incremental learning: survey and performance evaluation on image classification. IEEE TPAMI **45**(5), 5513–5533 (2022)
4. Sodhani, S., et al.: An introduction to lifelong supervised learning. arXiv preprint arXiv:2207.04354 (2022)
5. Van de Ven, G.M., Tolias, A.S.: Three scenarios for continual learning. arXiv preprint arXiv:1904.07734 (2019)
6. Wang, L., Zhang, X., Su, H., Zhu, J.: A comprehensive survey of continual learning: theory, method and application. IEEE TPAMI (2024)
7. Aljundi, R., et al.: Online continual learning with maximal interfered retrieval. In: NeurIPS, vol. 2 (2019)
8. Aljundi, R., Lin, M., Goujaud, B., Bengio, Y.: Gradient based sample selection for online continual learning. In: NeurIPS, vol. 32 (2019)
9. Bang, J., Kim, H., Yoo, Y., Ha, J., Choi, J.: Rainbow memory: continual learning with a memory of diverse samples. In: CVPR, pp. 8218–8227 (2021)
10. Buzzega, P., Boschini, M., Porrello, A., Abati, D., Calderara, S.: Dark experience for general continual learning: a strong, simple baseline. NeurIPS **33**, 15920–15930 (2020)
11. Chaudhry, A., Ranzato, M., Rohrbach, M., Elhoseiny, M.: Efficient lifelong learning with a-gem. arXiv preprint arXiv:1812.00420 (2018)
12. Chaudhry, A., et al.: Continual learning with tiny episodic memories. In: Workshop on Multi-Task and Lifelong Reinforcement Learning (2019)
13. Rebuffi, S., Kolesnikov, A., Sperl, G., Lampert, C.H., et al.: Incremental classifier and representation learning. In: CVPR, pp. 5533–5542 (2001)
14. Lopez-Paz, D., Ranzato, M.: Gradient episodic memory for continual learning. In: NeurIPS, vol. 30 (2017)
15. Rolnick, D., Ahuja, A., Schwarz, J., Lillicrap, T., Wayne, G.: Experience replay for continual learning. In: NeurIPS, vol. 32 (2019)

16. Robins, A.: Catastrophic forgetting, rehearsal and pseudorehearsal. Connect. Sci. **7**(2), 123–146 (1995)
17. Gepperth, A., Karaoguz, C.: Incremental learning with self-organizing maps. In: WSOM, pp. 1–8 (2017)
18. Ebrahimi, S., Meier, F., Calandra, R., Darrell, T., Rohrbach, M.: Adversarial continual learning. In: ECCV, pp. 386–402 (2020)
19. Lee, S., Ha, J., Zhang, D., Kim, G.: A neural Dirichlet process mixture model for task-free continual learning. arXiv preprint arXiv:2001.00689 (2020)
20. Lomonaco, V., Maltoni, D.: Core50: a new dataset and benchmark for continuous object recognition. In: CoRL, pp. 17–26 (2017)
21. Maltoni, D., Lomonaco, V.: Continuous learning in single-incremental-task scenarios. Neural Netw. **116**, 56–73 (2019)
22. Rusu, A.A., et al.: Progressive neural networks. arXiv preprint arXiv:1606.04671 (2016)
23. Hemati, H., Lomonaco, V., Bacciu, D., Borth, D.: Partial hypernetworks for continual learning. In: Conference on Lifelong Learning Agents, pp. 318–336 (2023)
24. Chauhan, V.K., Zhou, J., Lu, P., Molaei, S., Clifton, D.A.: A brief review of hypernetworks in deep learning. Artif. Intell. Rev. **57**(9), 250 (2024)
25. Von Oswald, J., Henning, C., Grewe, B.F., Sacramento, J.: Continual learning with hypernetworks. arXiv preprint arXiv:1906.00695 (2019)
26. Ehret, B., Henning, C., Cervera, M.R., Meulemans, A., Von Oswald, J., Grewe, B.F.: Continual learning in recurrent neural networks. arXiv preprint arXiv:2006.12109 (2020)
27. Ding, F., Xu, C., Liu, H., Zhou, B., Zhou, H.: Bridging pre-trained models to continual learning: a hypernetwork based framework with parameter-efficient fine-tuning techniques. Inf. Sci. **674**, 120710 (2024)
28. Jin, H., Kim, G., Ahn, C., Kim, E.: Growing a brain with sparsity-inducing generation for continual learning. In: ICCV, pp. 18961–18970 (2023)
29. Chandra, D.S., Varshney, S., Srijith, P.K., Gupta, S.: Continual learning with dependency preserving hypernetworks. In: WACV, pp. 2339–2348 (2023)
30. Krukowski, P., Bielawska, A., Książek, K., Wawrzyński, P., Batorski, P., Spurek, P.: HyperInterval: hypernetwork approach to training weight interval regions in continual learning. arXiv preprint arXiv:2405.15444 (2024)
31. Książek, K., Spurek, P.: HyperMask: adaptive hypernetwork-based masks for continual learning. arXiv preprint arXiv:2310.00113 (2023)
32. Wang, Z., et al.: Learning to prompt for continual learning. In: CVPR, pp. 139–149 (2022)
33. Wang, Z., et al.: DualPrompt: complementary prompting for rehearsal-free continual learning. In: ECCV, pp. 631–648 (2022)
34. Smith, J.S., et al.: Coda-Prompt: continual decomposed attention-based prompting for rehearsal-free continual learning. In: CVPR, pp. 11909–11919 (2023)
35. Gao, Z., Cen, J., Chang, X.: Consistent prompting for rehearsal-free continual learning. In: CVPR, pp. 28463–28473 (2024)
36. Kim, Y., Li, Y., Panda, P.: One-stage prompt-based continual learning. In: ECCV, pp. 163–179 (2025)
37. Aljundi, R., Babiloni, F., Elhoseiny, M., Rohrbach, M., Tuytelaars, T.: Memory aware synapses: Learning what (not) to forget. In: ECCV, pp. 139–154 (2018)
38. Kirkpatrick, J., et al.: Overcoming catastrophic forgetting in neural networks. PNAS **114**(13), 3521–3526 (2017)

39. Titsias, M.K., Schwarz, J., Matthews, A.G., Pascanu, R., Teh, Y.W.: Functional regularisation for continual learning with gaussian processes. arXiv preprint arXiv:1901.11356 (2019)
40. Zenke, F., Poole, B., Ganguli, S.: Continual learning through synaptic intelligence. In: ICML, pp. 3987–3995 (2017)
41. Hou, S., Pan, X., Loy, C.C., Wang, Z., Lin, D.: Learning a unified classifier incrementally via rebalancing. In: CVPR, pp. 831–839 (2019)
42. Kemker, R., McClure, M., Abitino, A., Hayes, T., Kanan, C.: Measuring catastrophic forgetting in neural networks. In: AAAI, vol. 32 (2018)
43. Ha, D., Dai, A., Le, Q.V.: Hypernetworks. arXiv preprint arXiv:1609.09106 (2016)
44. Guo, Y., Hu, W., Zhao, D., Liu, B.: Adaptive orthogonal projection for batch and online continual learning. In: AAAI, vol. 36, pp. 6783–6791 (2022)
45. Saha, G., Garg, I., Roy, K.: Gradient projection memory for continual learning. arXiv preprint arXiv:2103.09762 (2021)
46. Liang, Y., Li, W.: Adaptive plasticity improvement for continual learning. In: CVPR, pp. 7816–7825 (2023)
47. Zeng, G., Chen, Y., Cui, B., Yu, S.: Continual learning of context-dependent processing in neural networks. Nat. Mach. Intell. **1**(8), 364–372 (2019)
48. Farajtabar, M., Azizan, N., Mott, A., Li, A.: Orthogonal gradient descent for continual learning. In: AISTATS, pp. 3762–3773 (2020)
49. Liu, P., Yuan, W., Fu, J., Jiang, Z., Hayashi, H., Neubig, G.: Pre-train, prompt, and predict: a systematic survey of prompting methods in natural language processing. ACM Comput. Surv. **55**(9), 1–35 (2023)
50. Lester, B., Al-Rfou, R., Constant, N.: The power of scale for parameter-efficient prompt tuning. arXiv preprint arXiv:2104.08691 (2021)
51. Krizhevsky, A., Hinton, G. et al.: Learning multiple layers of features from tiny images (2009)
52. Hendrycks, D., et al.: The many faces of robustness: a critical analysis of out-of-distribution generalization. In: ICCV, pp. 8340–8349 (2021)
53. Peng, X., Bai, Q., Xia, X., Huang, Z., Saenko, K., Wang, B.: Moment matching for multi-source domain adaptation. In: ICCV, pp. 1406–1415 (2019)
54. Wang, Y., Ma, Z., Huang, Z., Wang, Y., Su, Z., Hong, X.: Isolation and impartial aggregation: a paradigm of incremental learning without interference. In: AAAI, vol. 37, pp. 10209–10217 (2023)
55. He, Y., et al.: HyperPrompt: prompt-based task-conditioning of transformers. In: ICML, pp. 8678–8690 (2022)
56. Ni, B., et al.: Enhancing visual continual learning with language-guided supervision. In: CVPR, pp. 24068–24077 (2024)
57. Hayes, T.L., Cahill, N.D., Kanan, C.: Memory efficient experience replay for streaming learning. In: ICRA, pp. 9769–9776 (2019)

Active Disturbance Rejection Method for Positioning and Payload Swing Suppression of Tower Cranes

Siyan Song[1], Jing Zhao[2], Naiqi Wu[3], Menghua Zhang[4], Ming Li[5], and Wei Peng[1(✉)]

[1] The School of Information and Electrical Engineering, Shandong Jianzhu University, Jinan 250101, China
Pengwei19@sdjzu.edu.cn

[2] School of Mechanical Engineering and Automation, Northeastern University, Shenyang, China
zhaoj@mail.neu.edu.cn

[3] Macau Institute of Systems Engineering, Macau University of Science and Technology, Macau, China
nqwu@muct.edu.mo

[4] School of Electrical Engineering, University of Jinan, Jinan, China
zhangmenghua@mail.sdu.edu.cn

[5] Shandong Inspur Database Technology Co., Ltd, Jinan, China
Liming2017@inspur.com

Abstract. Tower cranes, as key lifting equipment in construction and ports, face challenges in precise positioning and payload anti-sway control due to their under-actuation, nonlinearity and strong coupling characteristics. To address those challenges, this paper proposes a controller method combining active disturbance rejection control and differential flatness theory. Firstly, the differential flatness of the 4°-of-freedom(4-DOF) tower crane system was proved. The system state was expressed as the algebraic combination of the flat output and its finite-order derivatives, thereby transforming the control problem into a tracking problem of the flat output.Secondly, a tracking differential is designed to arrange the transition process to suppress overshoot, and an extended state observer is utilized to estimate the total disturbance of the system and higher-order state quantities, achieving feedforward compensation and state reconstruction. Finally, the effectiveness of the proposed method is verified through simulation. The results show that this method can effectively suppress the load swing while achieving precise positioning of the jib and the trolley.

Keywords: Tower Crane · Differential Flatness · Active Disturbance Rejection Control

1 Introduction

Tower cranes are indispensable for material handling in construction and real estate due to their high efficiency, extensive reach, and strong load capacity [1, 2]. However, as underactuated systems [3], they face inherent control challenges: their limited control

C. Li et al. (Eds.): ICNC 2025, CCIS 2946, pp. 353–365, 2026.
https://doi.org/10.1007/978-981-92-1599-7_30

inputs must simultaneously drive jib and trolley positioning while suppressing payload swing. These difficulties are compounded by nonlinear dynamics, strong coupling, and external disturbances, making controller design and stability analysis highly complex. Currently, the operation of tower cranes still mainly relies on manual control, demanding skilled operators while introducing safety risks from fatigue-induced errors [4]. Consequently, the development of automatic crane controllers presents considerable theoretical importance and practical value.

The control of crane systems has an extensive history of research. In [5], an enhanced energy-coupled nonlinear anti-swing control strategy was introduced, addressing the key issues of low positioning accuracy and large payload swing in underactuated tower cranes, with demonstrated strong robustness. In [6], an adaptive robust sliding mode control approach was developed, integrating nonlinear dynamic modeling with adaptive parameter estimation to effectively counteract unknown friction and wind disturbances, enabling precise positioning and swing suppression. For 4-DOF tower cranes facing underactuation, external disturbances, and parameter uncertainties, an adaptive tracking control scheme was proposed in [4], achieving accurate payload tracking while effectively suppressing swing. In [5], an adaptive integral sliding mode control technique was adopted, which eliminates chattering issues common in traditional sliding mode control while balancing positioning accuracy, swing suppression, and system robustness. Based on the adaptive backstepping principle, a fault-tolerant controller was designed in [7]. By refining the traditional backstepping structure, the system maintains precise payload positioning and swing suppression even under actuator faults. Targeting 3D offshore boom cranes, a novel model predictive control method accounting for input and output constraints was proposed in [8]. Using a discretized system model with only unactuated states as system variables to reduce computational load, the method achieves precise payload positioning and effective anti-swing under ship rolling and heaving disturbances. In [9], a neural network-based adaptive anti-swing control strategy was presented, compensating for ship rolling disturbances and input dead zones through neural networks to realize finite-time positioning of the jib and rope along with payload swing suppression for underactuated offshore cranes. A unified online adaptive near-optimal control framework was constructed in [10], where an auxiliary system approximates and compensates for parametric uncertainties, simplifying the dynamic programming process and enabling stable control of uncertain systems. This framework has been validated on underactuated surface vessels, offering a valuable reference for control design of tower cranes and similar underactuated systems.

Despite ongoing research, tower cranes control is still constrained by its inherent underactuation, which critically affects positioning accuracy and anti-sway performance. Furthermore, strong coupling among system states increases modeling and control difficulties, leading existing methods to compromise between structural complexity and performance. Based on this, in order to balance control performance and engineering practicality, we attempt to develop a control method with a relatively simple structure, fewer control parameters that need to be adjusted, and the ability to efficiently suppress load swing, in order to specifically solve the above technical problems and provide a better solution for the automatic control of tower cranes.

In order to deal with the uncertainties and external disturbances in tower crane systems, this paper adopts the idea of the active disturbance rejection control(ADRC). The ADRC is mainly composed of TD and ESO [11, 12], TD is designed to arrange the transition process [13, 14], extract the input signal with random noise and its differential signal, and resolve the contradiction between overshoot and rapidity, ESO is used to observe the high-order terms of the system and the total error, and compensate for the total error [15, 16]. In this article, the differential flatness of the tower crane is proved to solve the underactuation problem of the system. In general, the main contributions of this paper are summarized as follows:

(1) The differential flatness of the 4-DOF tower crane is proved. Then, the control problem of the system is transformed into a control problem of flat output and the problem of state coupling is solved.
(2) By using an extended state observer, higher-order time derivatives with flat outputs are not required, which is also difficult to measure in practice.

2 Problem Statement

2.1 Tower Crane Dynamics

In this section, the control problem of accurate positioning for trolley and jib, the swing elimination of payload is considered for 4-DOF (see Fig. 1).

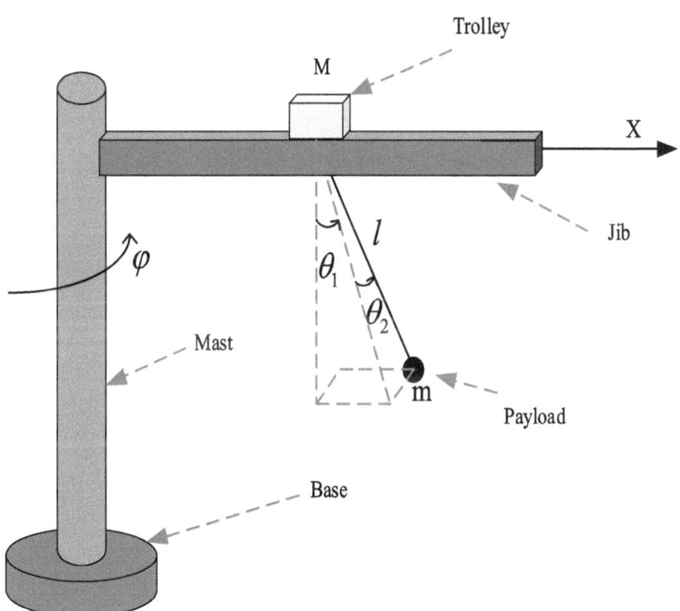

Fig. 1. An underactuated tower crane

The dynamics can be obtained by Lagrange method as follows [17–19], the state variables and parameters are defined in Table 1.

$$\begin{aligned}&\left[m\left(S_1^2C_2^2+S_2^2\right)l^2+2mxlC_2S_1+J+(M+m)x^2\right]\ddot{\varphi}-mlS_2\ddot{x}-ml^2C_1C_2S_2\ddot{\theta}_1\\&+ml(C_2x+lS_1)\ddot{\theta}_2+2(M+m)x\dot{x}\dot{\varphi}+2mlC_1C_2x\dot{\varphi}\dot{\theta}_1-mlS_2\left(2\dot{\varphi}S_1+\dot{\theta}_2\right)x\dot{\theta}_2\\&+2mlS_1C_2\dot{x}\dot{\varphi}+ml^2S_{21}C_2^2\dot{\varphi}\dot{\theta}_1+ml^2S_1S_2C_2\dot{\theta}_1^2\\&+ml^2C_1^2S_{22}\dot{\varphi}\dot{\theta}_2+2ml^2C_1S_2^2\dot{\theta}_1\dot{\theta}_2=F_\varphi\end{aligned}\tag{1}$$

$$\begin{aligned}&-mlS_2\ddot{\varphi}+(M+m)\ddot{x}+mlC_1C_2\ddot{\theta}_1-mlS_1S_2\ddot{\theta}_2-(M+m)x\dot{\varphi}^2-2mlC_1S_2\dot{\theta}_1\dot{\theta}_2\\&-mlC_2\left[S_1\left(\dot{\varphi}^2+\dot{\theta}_1^2+\dot{\theta}_2^2\right)+2\dot{\varphi}\dot{\theta}_2\right]=F_x\end{aligned}\tag{2}$$

$$\begin{aligned}&-ml^2C_1C_2S_2\ddot{\varphi}+mlC_1C_2\ddot{x}+ml^2C_2^2\ddot{\theta}_1-mlC_1C_2(x+lS_1C_2)\dot{\varphi}^2+mglS_1C_2\\&-2ml^2C_2\left(\dot{\varphi}C_1C_2+\dot{\theta}_1S_2\right)\dot{\theta}_2=0\end{aligned}\tag{3}$$

$$\begin{aligned}&ml(C_2x+lS_1)\ddot{\varphi}-mlS_1S_2\ddot{x}+ml^2\ddot{\theta}_2+2mlC_2\dot{x}\dot{\varphi}+ml\left(xS_1S_2-lC_1^2S_2C_2\right)\dot{\varphi}^2\\&+2ml^2C_1C_2^2\dot{\varphi}\dot{\theta}_1+ml^2\dot{\theta}_1^2S_2C_2+mglC_1S_2=0\end{aligned}\tag{4}$$

2.2 Control Objectives

The control goal is to deliver the payload to the desired position while completely eliminating the swing of the payload, that is

$$\lim_{t\to\infty}\varphi=\varphi_d,\ \lim_{t\to\infty}x=x_d\tag{5}$$

$$\lim_{t\to\infty}\theta_1=\lim_{t\to\infty}\theta_2=0\tag{6}$$

Table 1. Parameters and variables.

Symbols	Meaning
φ	Jib slow angle
x	Trolley displacement
θ_1, θ_2	Payload swing angles
l	Cable length
M	Trolley mass
m	Payload mass
J	Jib ineritial moment
g	Gravitational constane
S_1, S_2, C_1, C_2	Abbreviations of $\sin\theta_1,\ \sin\theta_2,\ \cos\theta_1,\ \cos\theta_2$
S_{21}, S_{22}	Abbreviations of $\sin 2\theta_1,\ \sin 2\theta_2$
F_φ	Slew control torque
F_x	Translation control force

3 Controller Design

3.1 Flat Output Structure

During the operation of the tower crane, the position of the payload is [20]:

$$x_p = x\cos\varphi + l\theta_1\cos\varphi - l\theta_2\sin\varphi \tag{7}$$

$$y_p = x\sin\varphi + l\theta_1\sin\varphi + l\theta_2\cos\varphi \tag{8}$$

Convert (7) and (8) into matrix form

$$\begin{pmatrix} x_p \\ y_p \end{pmatrix} = \begin{pmatrix} \cos\varphi & -\sin\varphi \\ \sin\varphi & \cos\varphi \end{pmatrix} \begin{pmatrix} x + l\theta_1 \\ l\theta_2 \end{pmatrix} \tag{9}$$

Using the inverse property of the rotation matrix, it can be obtained that:

$$\begin{pmatrix} x + l\theta_1 \\ l\theta_2 \end{pmatrix} = \begin{pmatrix} \cos\varphi & \sin\varphi \\ -\sin\varphi & \cos\varphi \end{pmatrix} \begin{pmatrix} x_p \\ y_p \end{pmatrix} \tag{10}$$

Then (10) can be expressed as follows:

$$x + l\theta_1 = x_p\cos\varphi + y_p\sin\varphi \tag{11}$$

$$l\theta_2 = -x_p\sin\varphi + y_p\cos\varphi \tag{12}$$

After some simplifications, another form of (11) and (12) can be obtained

$$\theta_1 = \frac{1}{l}\left(x_p\cos\varphi + y_p\sin\varphi - x\right) \tag{13}$$

$$\theta_2 = \frac{1}{l}\left(y_p\cos\varphi - x_p\sin\varphi\right) \tag{14}$$

Taking the second derivative of (7) and (8):

$$\begin{aligned} \ddot{x}_p = {} & \ddot{x}\cos\varphi - 2\dot{x}\dot{\varphi}\sin\varphi - x\dot{\varphi}^2\cos\varphi - x\ddot{\varphi}\sin\varphi + l\ddot{\theta}_1\cos\varphi \\ & - l\ddot{\theta}_2\sin\varphi - 2l\dot{\theta}_1\dot{\varphi}\sin\varphi - 2l\dot{\theta}_2\sin\varphi - 2l\dot{\theta}_2\dot{\varphi}\cos\varphi - l\theta_1\ddot{\varphi}\sin\varphi \\ & - l\theta_1\dot{\varphi}^2\cos\varphi - l\theta_2\ddot{\varphi}\cos\varphi + l\theta_2\dot{\varphi}^2\sin\varphi \end{aligned} \tag{15}$$

$$\begin{aligned} \ddot{y}_p = {} & \ddot{x}\sin\varphi + 2\dot{x}\dot{\varphi}\cos\varphi - x\dot{\varphi}^2\sin\varphi + x\ddot{\varphi}\cos\varphi + l\ddot{\theta}_1\sin\varphi \\ & + l\ddot{\theta}_2\cos\varphi + 2l\dot{\theta}_1\dot{\varphi}\cos\varphi - 2l\dot{\theta}_2\dot{\varphi}\sin\varphi + l\theta_1\ddot{\varphi}\cos\varphi \\ & - l\theta_1\dot{\varphi}^2\sin\varphi - l\theta_2\ddot{\varphi}\sin\varphi - l\theta_2\dot{\varphi}^2\cos\varphi \end{aligned} \tag{16}$$

Integrating the results of multiplying (15) and (16) by cosφ, sin φ respectively, one obtains

$$\ddot{x}_p\cos\varphi + \ddot{y}_p\sin\varphi = \ddot{x} - x\dot{\varphi}^2 + l\ddot{\theta}_1 - 2l\dot{\theta}_2\dot{\varphi} - l\theta_1\dot{\varphi}^2 - l\theta_2\ddot{\varphi} \tag{17}$$

$$\ddot{x}_p \sin\varphi - \ddot{y}_p \cos\varphi = -2\dot{x}\dot{\varphi} - x\ddot{\varphi} - l\ddot{\theta}_2 - 2l\dot{\theta}_1\dot{\varphi} - l\theta_1\ddot{\varphi} + l\theta_2\dot{\varphi}^2 \tag{18}$$

Substituting (3) and (4) into (17) and (18), it can be obtained that

$$\ddot{x}_p \cos\varphi + \ddot{y}_p \sin\varphi = -g\theta_1 \tag{19}$$

$$\ddot{x}_p \sin\varphi - \ddot{y}_p \cos\varphi = g\theta_2 \tag{20}$$

Based on (19) and (20), another correlation can be deduced as

$$\theta_1 = -\frac{\ddot{x}_p \cos\varphi + \ddot{y}_p \sin\varphi}{g} \tag{21}$$

$$\theta_2 = \frac{\ddot{x}_p \sin\varphi - \ddot{y}_p \cos\varphi}{g} \tag{22}$$

Combining (13) and (21), it can be obtained that

$$\ddot{x}_p \cos\varphi + \ddot{y}_p \sin\varphi = -\frac{g}{l}\left(x_p \cos\varphi + y_p \sin\varphi - x\right) \tag{23}$$

Combining (14) and (22), it can be obtained that

$$\ddot{x}_p \sin\varphi - \ddot{y}_p \cos\varphi = \frac{g}{l}\left(y_p \cos\varphi - x_p \sin\varphi\right) \tag{24}$$

Integrating the results of multiplying (23) and (24) by cosφ, sin φ respectively:

$$\ddot{x}_p = -\frac{g}{l}x_p + \frac{g}{l}x\cos\varphi \tag{25}$$

$$\ddot{y}_p = -\frac{g}{l}y_p + \frac{g}{l}x\sin\varphi \tag{26}$$

According to (25) and (26), another relationship can be inferred as

$$x\cos\varphi = x_p + \frac{l}{g}\ddot{x}_p \tag{27}$$

$$x\sin\varphi = y_p + \frac{l}{g}\ddot{y}_p \tag{28}$$

Squaring and adding (27) and (28) yields:

$$x^2\left(\cos^2\varphi + \sin^2\varphi\right) = \left(x_p + \frac{l}{g}\ddot{x}_p\right)^2 + \left(y_p + \frac{l}{g}\ddot{y}_p\right)^2 \tag{29}$$

Therefore

$$x = \sqrt{\left(x_p + \frac{l}{g}\ddot{x}_p\right)^2 + \left(y_p + \frac{l}{g}\ddot{y}_p\right)^2} \tag{30}$$

From (27) and (28), it can be obtainted that

$$\tan\varphi = \frac{x\sin\varphi}{x\cos\varphi} = \frac{y_p + \frac{l}{g}\ddot{y}_p}{x_p + \frac{l}{g}\ddot{x}_p} \tag{31}$$

Furthermore, the following equationes can be obtained

$$\varphi = \arctan\left(\frac{y_p + \frac{l}{g}\ddot{y}_p}{x_p + \frac{l}{g}\ddot{x}_p}\right) \tag{32}$$

$$\cos\varphi = \frac{x_p + \frac{l}{g}\ddot{x}_p}{\sqrt{\left(x_p + \frac{l}{g}\ddot{x}_p\right)^2 + \left(y_p + \frac{l}{g}\ddot{y}_p\right)^2}} \tag{33}$$

$$\sin\varphi = \frac{y_p + \frac{l}{g}\ddot{y}_p}{\sqrt{\left(x_p + \frac{l}{g}\ddot{x}_p\right)^2 + \left(y_p + \frac{l}{g}\ddot{y}_p\right)^2}} \tag{34}$$

Combining (21) and (22), it can be obtainted that

$$\theta_1 = -\frac{1}{g}\left(\ddot{x}_p\frac{x_p + \frac{l}{g}\ddot{x}_p}{\sqrt{\left(x_p + \frac{l}{g}\ddot{x}_p\right)^2 + \left(y_p + \frac{l}{g}\ddot{y}_p\right)^2}} + \ddot{y}_p\frac{y_p + \frac{l}{g}\ddot{y}_p}{\sqrt{\left(x_p + \frac{l}{g}\ddot{x}_p\right)^2 + \left(y_p + \frac{l}{g}\ddot{y}_p\right)^2}}\right) \tag{35}$$

$$\theta_2 = \frac{1}{g}\left(\ddot{x}_p\frac{y_p + \frac{l}{g}\ddot{y}_p}{\sqrt{\left(x_p + \frac{l}{g}\ddot{x}_p\right)^2 + \left(y_p + \frac{l}{g}\ddot{y}_p\right)^2}} - \ddot{y}_p\frac{x_p + \frac{l}{g}\ddot{x}_p}{\sqrt{\left(x_p + \frac{l}{g}\ddot{x}_p\right)^2 + \left(y_p + \frac{l}{g}\ddot{y}_p\right)^2}}\right) \tag{36}$$

Accordingly, the tower crane system exhibits the property of differential flatness, with the payload coordinates x_p and y_p serving as the flat outputs. As a result, all state variables can be expressed in terms of the payload position and its finite-order derivatives. The relationship between the state variables and the flat outputs, established by Eqs. (30), (32),(35, 36), allows the control problem to be reformulated as the control of the flat outputs. Thus, the control objective reduces to ensuring that the flat outputs asymptotically converge to their desired target positions.Through (5) and (6), the target values of the flat outputs can be calculated as follows

$$x_{pd} = x_d\cos\varphi_d \tag{37}$$

$$y_{p_d} = x_d\sin\varphi_d \tag{38}$$

3.2 Controller Design

At first, in order to arrange the transition process, a tracking differentiator(TD) is designed as follows:

$$\begin{cases} \dot{v}_1 = v_2, \\ \dot{v}_2 = v_3, \\ \dot{v}_3 = v_4, \\ \dot{v}_4 = -r(r(r(r(v_1 - v_0) + 4v_2) + 6v_3) + 4v_4) \end{cases} \tag{39}$$

where r is the control parameter for tracking differentiator performance according to system requirements, $v_0 = [x_{pd}\ y_{pd}]^T$ is the vector composed of flat output target values, and $v_1 \in \mathbb{R}^2$ is the tracking signal of v_0, $v_2, v_3, v_4 \in \mathbb{R}^2$ are the first to third order time derivatives of v_1 respectively.

Secondly, an extended state observer is designed as

$$\begin{cases} e = z_1 - f \\ \dot{z}_1 = z_2 - \beta_1 e \\ \dot{z}_2 = z_3 - \beta_2 e \\ \dot{z}_3 = z_4 - \beta_3 e \\ \dot{z}_4 = -\beta_4 e + u \end{cases} \tag{40}$$

where $f = [x_p\ y_p]^T \in \mathbb{R}^2$, $z_1 \in \mathbb{R}^2$ is the estimation vector of f and z_2, z_3, z_4 are the estimation vectors of the first to third order time derivatives of f respectively. $\beta_1, \beta_2, \beta_3$ and β_4 are positive definite control parameter matrics as

$$\beta_j = \mathrm{diag}\{\beta_{jx}, \beta_{jy}\}, j = 1, \cdots, 4. \tag{41}$$

Thirdly, the following feedback control law is proposed:

$$\begin{cases} e_1 = v_1 - z_1 \\ e_2 = v_2 - z_2 \\ e_3 = v_3 - z_3 \\ e_4 = v_4 - z_4 \\ u = \sigma_1 e_1 + \sigma_2 e_2 + \sigma_3 e_3 + \sigma_4 e_4 \end{cases} \tag{42}$$

where $\sigma_1, \sigma_2, \sigma_3$ and σ_4 are positive definite control parameter matrices as

$$\sigma_j = \mathrm{diag}\{\sigma_{jx}, \sigma_{jy}\}, j = 1, \cdots, 4. \tag{43}$$

Finally, by integrating the extended state observer with Eqs. (30), (32), (35), and (36), the estimated results of the state variables and their high-order derivatives can be derived through calculation. Using the estimated values of the state variables and their derivatives obtained above, the final controller can be further computed as follows:

$$F_\varphi = \left[2m\hat{x}l\hat{\theta}_1 + J + (M+m)\hat{x}^2\right]\ddot{\hat{\varphi}} - ml\hat{\theta}_2\ddot{\hat{x}} + m\hat{x}l\ddot{\hat{\theta}}_2 + 2(M+m)\hat{x}\dot{\hat{x}}\dot{\hat{\varphi}} + 2m\hat{x}l\dot{\hat{\varphi}}\dot{\hat{\theta}}_1 + 2ml\hat{\theta}_1\dot{\hat{x}}\dot{\hat{\varphi}} \tag{44}$$

$$F_x = -ml\hat{\theta}_2\ddot{\hat{\varphi}} + (M+m)\ddot{\hat{x}} + ml\ddot{\hat{\theta}}_1 - (M+m)\hat{x}\dot{\hat{\varphi}}^2 - 2ml\dot{\hat{\varphi}}\dot{\hat{\theta}}_2 \tag{45}$$

where $\hat{x}$, $\hat{\varphi}$, $\hat{\theta}_1$ and $\hat{\theta}_2$ are the estimated value of x, φ, θ_1 and θ_2.

4 Simulation Results and Analysis

In this section, to verify the performance of the proposed controller, a dynamic model is first established in MATLAB and Simulink. Then, the proposed controller is compared with the PD controller and the EE controller, the control performance of the controllers is evaluated from three aspects: positioning error, maximum swing angle and residual swing angle.

The system parameters of the simulation experiment platform are chosen as $M = 3.5\text{kg}$, $\text{m} = 1\text{kg}$

$$l = 0.6\text{m}, \text{g} = 9.8\ \text{m/s}^2, J = 6.8\text{kg} \cdot \text{m}^2.$$

The initial slew angle of the jib, initial trolley displacement, and initial payload swing angles are all set to zero, which means that

$$\varphi(0) = 0\,\text{deg}, x(0) = 0\text{m}, \theta_1(0) = 0\,\text{deg}, \theta_2(0) = 0\,\text{deg}$$

Additionally, the desired jib slew angle and desired trolley position are configured as

$$\varphi_d = 45\,\text{deg}, x_d = 1\text{m}$$

The formulaic representation of the comparative method PD is as follows:

$$F_x = -k_{p1}e_x - k_{d1}\dot{x} \tag{46}$$

$$F_\varphi = -k_{p2}e_\varphi - k_{d2}\dot{\varphi} \tag{47}$$

The formulaic representation of the comparative method EE is as follows:

$$F_x = -p_{11}(e_x - k_a \sin\theta_1) - p_{12}(\dot{e}_x - k_a\dot{\varphi}\cos\theta_1) \tag{48}$$

$$F_\varphi = -p_{21}\left(e_\varphi - k_b \sin\theta_2\right) - p_{22}\left(\dot{e}_\varphi - k_b\dot{x}\cos\theta_1\right) \tag{49}$$

where $e_x = x - x_d$, $e_\varphi = \varphi - \varphi_d$ and $k_{p1}, k_{p2}, k_{d1}, k_{d2}, p_{11}, p_{12}, p_{21}, p_{22}, k_a, k_b$ all stand for positive constant.

The control gains for the PD and EE methods are determined using the trial-and-error approach:

$$\begin{aligned} &k_{p1} = 16, k_{d1} = 23, k_{p2} = 8, k_{d2} = 12, p_{11} = 6, \\ &p_{12} = 13, \text{p}_{21} = 5, \text{p}_{22} = 8, k_a = 2.3, k_b = 1.2 \end{aligned}$$

The control gains for proposed control method are chosen as:

$$\begin{aligned} &r = 1.2, \omega_0 = 6.2, \overline{\omega}_0 = 5.4, \omega_c = 2.3, \overline{\omega}_c = 2 \\ &\beta_{1x} = 4\omega_0, \beta_{2x} = 6\omega_0^2, \beta_{3x} = 4\omega_0^3, \beta_{4x} = \omega_0^4 \\ &\beta_{1y} = 4\overline{\omega}_0, \beta_{2y} = 6\overline{\omega}_0^2, \beta_{3y} = 4\overline{\omega}_0^3, \beta_{4y} = \overline{\omega}_0^4 \\ &\sigma_{1x} = 4\omega_c, \sigma_{2x} = 6\omega_c^2, \sigma_{3x} = 4\omega_c^3, \sigma_{4x} = \omega_c^4 \\ &\sigma_{1y} = 4\overline{\omega}_c, \sigma_{2y} = 6\overline{\omega}_c^2, \sigma_{3y} = 4\overline{\omega}_c^3, \sigma_{4y} = \overline{\omega}_c^4 \end{aligned}$$

The simulation results of the PD control method, EE control method the proposed control method are plotted in Fig. 2, and the corresponding quantitative results are provided in Table 2.

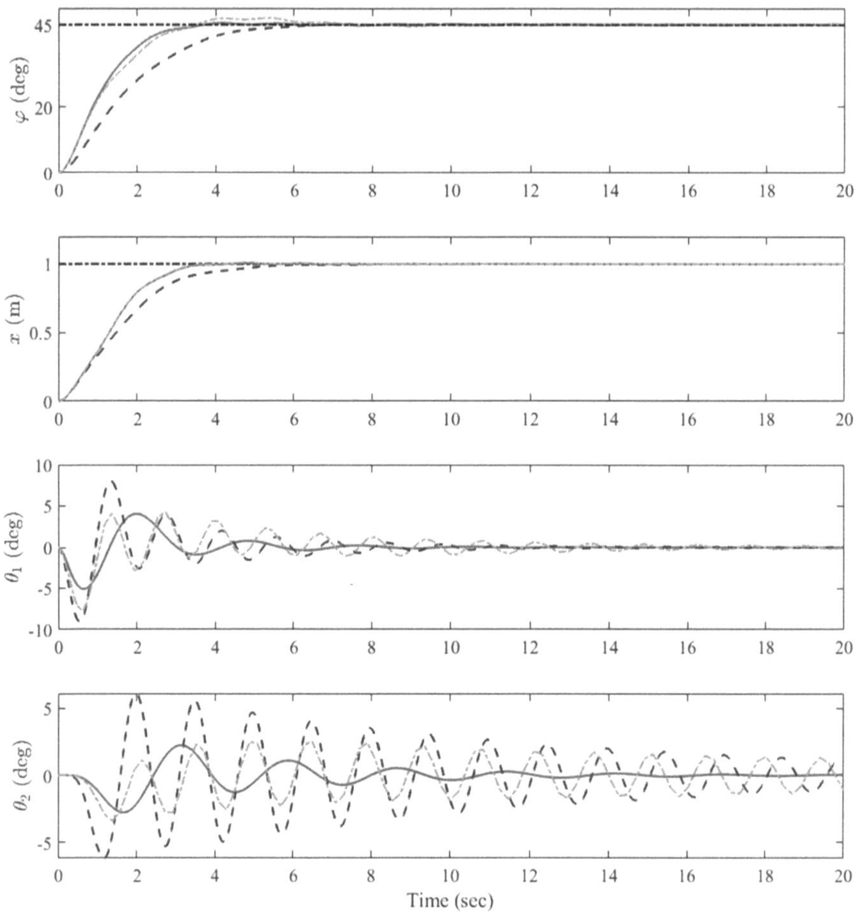

Fig. 2. The control performance of the methods

As illustrated in Fig. 3, the control forces/torques of all methods eventually converge to zero. The proposed approach converges fast and smoothly.

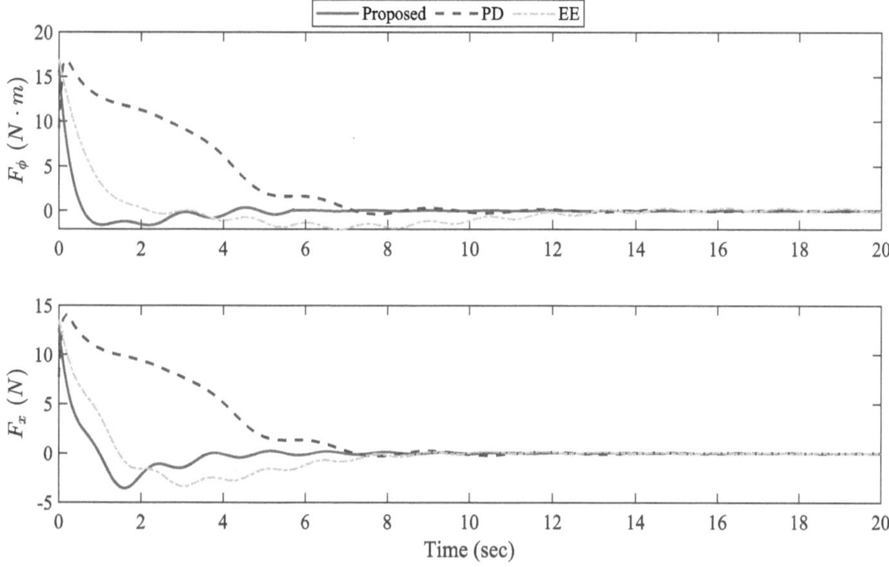

Fig. 3. The control force of the methods

Table 2. Quantative results of simulation.

controllers	$\Delta\varphi$(deg)	Δx(m)	θ_1(max)deg	θ_2(max)deg	$\theta_1(res)$deg	$\theta_2(res)$deg
PD controller	0.0379	0.0046	7.9995	6.1199	0.0298	1.1241
EE controller	0.0543	0.0035	4.1136	2.3249	0.0324	1.0714
Proposed controller	0.0055	0.0012	4.0651	2.2295	0.0007	0.0345

From Fig. 2 and Table 2, there is obviously under the same transportation time, the designed control method (4.25 s, 4.42 s) reaches the target position faster than the PD control method (7.1 s, 6.1 s), and the EE control method (5.9 s,4.67 s) the payload swing angle of the proposed control method (θ_1(max) : 4.0651 deg ; θ_2(max) : 2.2295 deg ; $\theta_1(res)$: 0.0007 deg ; $\theta_2(res)$: 0.0345 deg) is smaller than that of the PD control method (θ_1(max) : 7.9995 deg ; θ_2(max) : 6.1199 deg ; $\theta_1(res)$: 0.0298 deg ; $\theta_2(res)$: 1.1241 deg) and the EE control method(θ_1(max) : 4.1136 deg ; θ_2(max) : 2.3249 deg ; $\theta_1(res)$:

0.0324 deg ; $\theta_2(res)$: 1.0714 deg)These simulation results directly demonstrate that the proposed method can achieve superior control performance.

5 Conlusion

To effectively control underactuated tower cranes, this paper proposed a method based on active disturbance rejection control on the basis of proving the differential flatness of the system. The problem of system state coupling was successfully solved by introducing flat output. Subsequently, a tracking differentiator was designed based on the flat output, effectively arranging the transition process from the initial value to the target value. Moreover, an extended state observer was utilized to avoid the need for higher-order derivatives that are difficult to measure in practice. The simulation results show that the designed controller can achieve precise positioning of the trolley and the jib, effectively suppress the payload swing at the same time, and its performance was superior to the PD controller.

In the future, non-ideal factors such as input dead zones and friction forces will be taken into account, and more universal control methods will be sought to reduce reliance on system models. At the same time, more precise and stable state observation schemes will be studied.

Acknowledgments. This study is partly supported by the National Natural Science Foundation of China (No. 62573272, 62173216, 62076150, 61903226, 62273163), the Taishan Scholar Foundation of Shandong Province under Grant (No. tsqn202312212), the Outstanding Youth Foundation of Shandong Province under Grant (No.ZR2023YQ056), the Key Research and Development Program of Shandong Province (No. 2021CXGC011205), the Natural Science Foundation of Shandong Province (ZR2023QF020).

References

1. Zhang, M., Jiang, X., Zhou, Z., et al.: Rapid and restricted swing control via adaptive output feedback for 5-DOF tower crane systems. Mech. Syst. Signal Process. **212**, 111283 (2024)
2. Ur Rehman, S.M.F., Mohamed, Z., Husain, A.R., et al.: Adaptive input shaper for payload swing control of a 5-DOF tower crane with parameter uncertainties and obstacle avoidance. Autom. Constr. **154**, 104963 (2023)
3. Liu, Z., Ma, X.: Piecewise time polynomials-based control methods for obstacle avoidance and precision positioning of tower crane systems with varying cable lengths. Machines. **12**(11), 775 (2024)
4. Chen, H., Fang, Y., Sun, N.: An adaptive tracking control method with swing suppression for 4-DOF tower crane systems. Mech. Syst. Signal Process. **123**, 426–442 (2019)
5. Shi, H.T., Huang, J.Q., Bai, X., et al.: Nonlinear anti-swing control of underactuated tower crane based on improved energy function. Int. J. Control. Autom. Syst. **19**(12), 3967–3982 (2021)
6. Hou, C., Liu, C., Li, Z., et al.: Tower crane systems modeling and adaptive robust sliding mode control design under unknown frictions and wind disturbances. Trans. Inst. Meas. Control. **47**(4), 795–809 (2025)
7. Xia, J.Y., Ouyang, H.M., Zhang, M.H.: Fault-tolerant controller design based on adaptive backstepping for tower cranes with actuator faults. ISA Trans. **146**, 463–471 (2024)
8. Lin, J., Fang, Y., Lu, B., et al.: Constrained model predictive control for 3-D offshore boom cranes. Control. Eng. Pract. **142**, 105741 (2024)
9. Yang, T., Sun, N., Chen, H., et al.: Neural network-based adaptive antiswing control of an underactuated ship-mounted crane with roll motions and input dead zones. IEEE Trans. Neural Netw. Learn. Syst. **31**(3), 901–914 (2019)
10. Zhang, Y., Li, S., Liu, X.: Adaptive near-optimal control of uncertain systems with application to underactuated surface vessels. IEEE Trans. Control Syst. Technol. **26**(4), 1204–1218 (2017)

11. Kang, X., Chai, L., Liu, H.: Anti-swing and positioning for double-pendulum tower cranes using improved active disturbance rejection controller. Int. J. Control. Autom. Syst. **21**(4), 1210–1221 (2023)
12. Li, Z., Chen, H., Che, L.: Antiswing control of offshore cranes under ship rolling disturbances: an active disturbance rejection control based approach. Nonlinear Dyn. **112**(23), 21097–21116 (2024)
13. Guo, Q., Chai, L., Liu, H.: Anti-swing sliding mode control of three-dimensional double pendulum overhead cranes based on extended state observer. Nonlinear Dyn. **111**(1), 391–410 (2023)
14. Tian, Y., Ma, H., Ma, L., et al.: Path tracking control of commercial vehicle emergency obstacle avoidance based on MPC and active disturbance rejection control. IEEE Trans. Transp. Electrification. (2024)
15. Du, Y., Cao, W., She, J.: Analysis and design of active disturbance rejection control with an improved extended state observer for systems with measurement noise. IEEE Trans. Ind. Electron. **70**(1), 855–865 (2022)
16. Tan, P., Liu, J., Sun, M., et al.: Disturbance compensation-based deep reinforcement learning control strategy for underactuated overhead crane systems: design and experiments. IEEE Trans. Intell. Transp. Syst. (2025)
17. Yang, T., Sun, N., Chen, H., et al.: Observer-based nonlinear control for tower cranes suffering from uncertain friction and actuator constraints with experimental verification. IEEE Trans. Ind. Electron. **68**(7), 6192–6204 (2020)
18. Zhang, M., Zhang, Y., Ouyang, H., et al.: Adaptive integral sliding mode control with payload sway reduction for 4-DOF tower crane systems. Nonlinear Dyn. **99**(4), 2727–2741 (2020)
19. Sun, N., Fang, Y., Chen, H., et al.: Slew/translation positioning and swing suppression for 4-DOF tower cranes with parametric uncertainties: design and hardware experimentation. IEEE Trans. Ind. Electron. **63**(10), 6407–6418 (2016)
20. Liu, Z., Yang, T., Sun, N., et al.: An antiswing trajectory planning method with state constraints for 4-DOF tower cranes: design and experiments. IEEE Access. **7**, 62142–62151 (2019)

Stability of Nonlinear Impulsive Systems Under Hybrid Event-Triggered Control

Weisen Hu[1], Cheng Hu[1,2(✉)], and Juan Yu[1,2]

[1] College of Mathematics and System Sciences, Xinjiang University, Urumqi 830017, China
hucheng@xju.edu.cn

[2] Xinjiang Key Laboratory of Applied Mathematics (XJDX1401), Urumqi 830017, China

Abstract. This paper investigates the stability of nonlinear delayed impulsive systems by hybrid event-triggered control. Firstly, by integrating event-triggered control with impulsive control, a novel hybrid event-triggered mechanism is designed, and the Zeno behavior is excluded by means of the method of classification discussion. Secondly, based on the proposed hybrid event-triggered control scheme, quantitative relationships among impulsive gains, event-triggering parameters, and exponential convergence rates are established, which rigorously guarantee system stability. Finally, the effectiveness of the theoretical findings is verified through a numerical example.

Keywords: Stability · nonlinear impulsive systems · event-triggered control

1 Introduction

Impulsive systems, a class of hybrid dynamic models characterized by continuous dynamics and discrete state jumps, are widely employed in various fields such as neural networks and smart grids [1,2]. Typical application examples of impulsive systems include neuronal spiking behaviors and transient failures in power grids. Given the inherent difficulties in implementing full-time continuous control for such systems, impulsive control has gained significant academic interest as an efficient alternative that relies solely on instantaneous state jumps at discrete instants. Nowadays, impulsive control strategy has been successfully incorporated into various advanced control frameworks, including intermittent control [3,4], and event-triggered control [5–7].

Event-triggered (ET) control is a resource-efficient approach that employs discontinuous, on-demand actions. By replacing the fixed periodic sampling of time-triggered control with an event-driven mechanism, it activates control updates only when specific event conditions are met, thereby fundamentally avoiding unnecessary resource consumption.

C. Li et al. (Eds.): ICNC 2025, CCIS 2946, pp. 366–376, 2026.
https://doi.org/10.1007/978-981-92-1599-7_31

ET impulsive control merges the advantages of both event-triggered and impulsive control. In this framework, the impulsive controller is activated exclusively by the event-triggering function, with no control transmissions occurring between triggering instants. By leveraging the properties of impulsive system theory and Lyapunov methods, the ET impulsive control strategy [8,9] has been applied to various nonlinear systems involving impulses, and the issue of global exponential stability was extensively investigated.

Although ET impulsive control can significantly reduce communication and computational costs, it also carries inherent risks. A primary concern is Zeno behavior, where an infinite number of events occur in a finite time if the triggering function is ill-designed. In particular, in [10], Borgers and Heemels demonstrated that the Zeno behavior may occur in the presence of external disturbances or noise. In addition, ET impulsive control may also exhibit excessively long inter-impulse intervals, i.e., no triggering occurs for a long time. To mitigate its impact on the system stability, a minimal inter-event interval [11] was introduced, which ensures the minimum control frequency and further guarantees the stability of the addressed system.

Based on the above analysis, this paper will develop an event-triggered control strategy to explore the stability problem of nonlinear impulsive systems. The main innovations of this paper are mainly in the following two aspects:

(1) Unlike existing ET control designs [12,13], a kind of hybrid event-triggered (HET) impulsive control scheme is developed by introducing composite triggering conditions that integrate both event-based and time-dependent triggers. This design not only avoids resource waste by preventing excessively long intervals but also intrinsically guarantees a minimum control frequency, enhancing overall system reliability.
(2) Based on the proposed HET impulsive control strategy, the method of classification discussion is used to eliminate the Zeno behavior, and a quantitative relationship between the ET parameters and the exponential convergence rate is established, which ensures the stability of the addressed impulsive system.

Table 1. Notations

Symbol	Explanation
$\mathbb{Z}^+$, $\mathbb{R}$, $\mathbb{R}^+$	$\mathbb{Z}^+ = \{1, 2, 3, \ldots\}$, $\mathbb{R} = (-\infty, +\infty)$, $\mathbb{R}^+ = [0, +\infty)$.
$\mathbb{R}^n$	The n-dimensional Euclidean space.
$\mathbb{R}^{n\times n}$	The set of square matrices of order n.
$C(s_1, s_2)$	$C(s_1, s_2) = \{\varphi : s_1 \to s_2 \text{ is continuous}\}$, $s_1 \subseteq \mathbb{R}$, $s_2 \subseteq \mathbb{R}^n$.

2 Preliminaries

Consider the following impulsive delayed system:

$$\begin{cases} \dot{x}(t) = \Upsilon(t, x(t), x(t-\tau(t))), & t \neq t_\theta,\ t \geq t_0, \\ x(t) = F(x(t^-)), & t = t_\theta,\ \theta \in \mathbb{Z}^+, \\ x_{t_0} = \phi \in C([-\tau, 0], \mathbb{R}^n), \end{cases} \tag{1}$$

where $x(t) \in \mathbb{R}^n$ is the system state, continuous function $\Upsilon : \mathbb{R}^+ \times \mathbb{R}^n \times \mathbb{R}^n \to \mathbb{R}^n$ and $F : \mathbb{R}^n \to \mathbb{R}^n$ satisfy $\Upsilon(t, \mathbf{0}, \mathbf{0}) = \mathbf{0}$ with $t \in \mathbb{R}^+$ and $F(\mathbf{0}) = \mathbf{0}$, $\tau(t)$ is the time-varying delay satisfying $\tau(t) \in [0, \tau]$ with $\tau > 0$, $x_{t_0} = x(t_0 + s)$, $s \in [-\tau, 0]$. The time sequence $\{t_\theta, \theta \in \mathbb{Z}^+\}$ denotes the impulse instants determined by an ET mechanism to be designed later. Let $\lim_{s\to t^+} x(s) = x(t^+)$ and $\lim_{s\to t^-} x(s) = x(t^-)$. Let $x(t_\theta^+) = x(t_\theta)$ at each impulse time t_θ, ensuring the right-continuity of the solutions to system (1), here $\mathbb{R}, \mathbb{R}^+, \mathbb{Z}^+, \mathbb{R}^n$ are defined in Table 1.

Definition 1. *For an impulse sequence $\{t_\theta, \theta \in \mathbb{Z}^+\}$, let $x(t) = x(t, t_0, \phi) \in \mathbb{R}^n$ represent the trajectory of system (1) initiated at (t_0, ϕ). The origin of system (1) is asymptotically stable, defined as being both stable and globally attractive, that is, $\|x(t)\| \to 0$ as $t \to +\infty$ for any initial state $\phi \in C([-\tau, 0], \mathbb{R}^n)$ and $t_0 \in \mathbb{R}^+$.*

Definition 2. *[14] The function $V : [t_0 - \tau, +\infty) \times \mathbb{R}^n \to \mathbb{R}^+$ belongs to class υ_0 if*

(i) V is continuous on each interval $[t_{\theta-1}, t_\theta)$, $\lim\limits_{(t,v)\to(t_\theta^-,x)} V(t, v) = V(t_\theta^-, x)$ exists;

(ii) $V(t, x)$ is locally Lipschitzian and $V(t, \mathbf{0}) \equiv 0$.

Definition 3. *[5] Let $V \in \upsilon_0$, then the upper right-hand Dini derivative of V with respect to system (1) for $\psi \in C([-\tau, 0], \mathbb{R}^n)$ is defined as*

$$D^+V(t, \psi(0)) = \limsup_{h\to 0^+} \frac{1}{h} \{V(t + h, \psi(0) + h\Upsilon) - V(t, \psi(0))\}.$$

Definition 4. *[15] A continuous function $\alpha : [0, a) \to \mathbb{R}^+$ is said to be of class $\mathbf{K}$ if α is strictly increasing and satisfies $\alpha(0) = 0$. If $\alpha \in \mathbf{K}$, $a = +\infty$ and α is radially unbounded, i.e., $\lim\limits_{r\to+\infty} \alpha(r) = +\infty$, then α is said to be of class $\mathbf{K}_\infty$.*

Lemma 1. *[11] Let $\upsilon_1(t)$ and $\upsilon_2(t) \in C([t_\theta, t_{\theta+1}), \mathbb{R}^+)\ (\theta \in \mathbb{Z}^+)$. Suppose that for constants ϑ_1, $\vartheta_2 > 0$, $0 < \bar{\mu} < 1$, it has that*

$$\begin{cases} D^+\upsilon_1(t) \leq \vartheta_1\upsilon_1(t) + \vartheta_2\upsilon_1(t - \tau(t)),\ t \in [t_\theta, t_{\theta+1}), \\ \upsilon_1(t_\theta) \leq \bar{\mu}\upsilon_1(t_\theta^-), \end{cases}$$

$$\begin{cases} D^+\upsilon_2(t) > \vartheta_1\upsilon_2(t) + \vartheta_2\upsilon_2(t - \tau(t)),\ t \in [t_\theta, t_{\theta+1}), \\ \upsilon_2(t_\theta) = \bar{\mu}\upsilon_2(t_\theta^-), \end{cases}$$

where $0 \leq \tau(t) \leq \tau$, then, if $\upsilon_1(t) \leq \upsilon_2(t)$, $t \in [t_0 - \tau, t_0]$, it is derived that $\upsilon_1(t) \leq \upsilon_2(t)$ for all $t \geq t_0$.

3 Main Results

In this paper, based on HET control, some sufficient conditions are established for ensuring the stability of the addressed impulsive system. The HET mechanism will be considered in the following form

$$\begin{aligned}&(2a)\ t_\theta = \min\{t^*_\theta, t_{\theta-1}+\hat{\tau}\},\ \theta \in \mathbb{Z}^+,\\ &(2b)\ t^*_1 = \inf\{t \geq t_0 | g(t) \geq e^\sigma g_{t_0}\},\\ &(2c)\ t^*_\theta = \inf\{t \geq t_{\theta-1} | g(t) \geq e^\sigma g(t_{\theta-1})\},\ \theta \geq 2,\end{aligned} \tag{2}$$

where $\sigma > 0$, $\hat{\tau} > 0$, $g(t) := g(t, x(t))$ serves as the Lyapunov function associated with the system state in system (1), $g_{t_0} = \sup_{s\in[-\tau,0]} g(s + t_0)$.

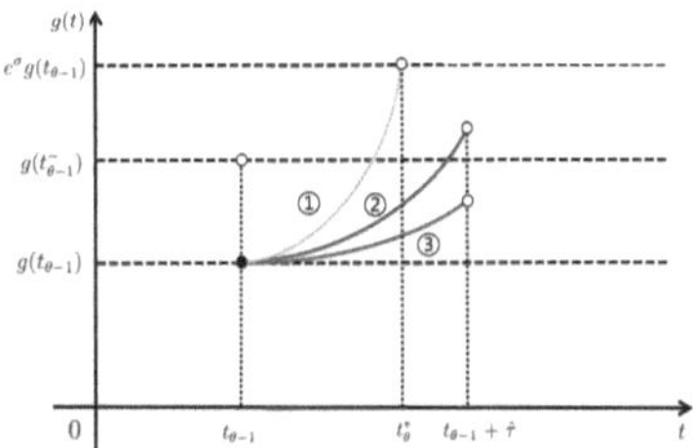

Fig. 1. Three triggering situations.

Based on the definition of t_θ in (2a), there are certain moments in sequence $\{t_\theta,\ \theta \in \mathbb{Z}^+\}$ that satisfy the condition $t_\theta = t_{\theta-1} + \hat{\tau}$. Then, these moments are defined as τ_θ, i.e., $\tau_\theta = t_\theta = t_{\theta-1} + \hat{\tau}$. Therefore, we have $g(\tau^-_\theta) < e^\sigma g(t_{\theta-1})$, which generates two cases (see Figure. 1): (I) $g(t^-_{\theta-1}) < g(\tau^-_\theta) < e^\sigma g(t_{\theta-1})$, (II) $g(\tau^-_\theta) \leq g(t^-_{\theta-1})$. On this basis, τ_θ satisfying case (I) is denoted by $\tilde{\tau}_\theta$, and τ_θ satisfying case (II) is represented by $\bar{\tau}_\theta$. To express it more clearly, two types of parameters $\tilde{\lambda}_\theta > 0$ and $\bar{\lambda}_\theta \geq 0$ are introduced to denote $g(\tilde{\tau}^-_\theta) = e^{\tilde{\lambda}_\theta} g(t^-_{\theta-1}) < e^\sigma g(t_{\theta-1})$ and $g(\bar{\tau}^-_\theta) = e^{-\bar{\lambda}_\theta} g(t^-_{\theta-1})$, that is, $\tilde{\lambda}_\theta$ and $\bar{\lambda}_\theta$ are determined by the forced triggering moment in HET mechanism (2).

Theorem 1. *Assume that there exist a function* $g \in v_0$ *and constants* $\alpha > 0$, $\beta > 0$, $\sigma > 0$, $0 < \mu < 1$, $\delta \geq 1$ *satisfying* $\ln \delta < \ln \mu + \sigma$ *such that*
(1) $D^+g(t, x(t)) \leq \alpha g(t, x(t)) + \beta g(t+s, x(t+s))$, $t \neq t_\theta$, $t \geq t_0$, $s \in [-\tau, 0]$,
(2) $g(t, g(x(t))) = \mu g(t^-, x(t^-)), t = t_\theta,\ \theta \in \mathbb{Z}^+$.
Therefore, Zeno behavior is eliminated in system (1) under HET mechanism (2).

Proof. Let $x(t) = x(t, t_0, \phi)$ denote the solution of system (1) under the initial condition (t_0, ϕ). Based on the properties of the HET mechanism (2), three scenarios should be discussed.

Case 1: The triggered time sequence t_θ, $\theta \in \mathbb{Z}^+$ is composed exclusively of forced impulse instants, that is $t_\theta = t_{\theta-1} + \hat{\tau}$ for all $\theta \in \mathbb{Z}^+$. In this case, it is easy to see that the Zeno behavior is excluded.

Case 2: The triggered impulse sequence t_θ, $\theta \in \mathbb{Z}^+$ is composed exclusively of event-triggered instants, that is, $t_\theta = t^*_\theta$, for all $\theta \in \mathbb{Z}^+$. By the definition of the initial value, one can obtain

$$g(t) \leq g_{t_0} \leq \delta g_{t_0} < e^\sigma g_{t_0}, t \in [t_0 - \tau, t_0]. \tag{3}$$

According to HET mechanism (2),

$$g(t) < e^\sigma g_{t_0},\ t \in [t_0, t_1),\ g(t_1^-) = e^\sigma g_{t_0}. \tag{4}$$

Combining inequalities (3) and (4), one has

$$g(t) < e^\sigma g_{t_0}, t \in [t_0 - \tau, t_1). \tag{5}$$

By virtue of the formula (4), there exists $t_* := \sup\{t \in [t_0, t_1) | g(t) \leq g_{t_0}\}$ satisfying $g(t_*) = g_{t_0}$, $g(t) > g_{t_0}$, $t \in (t_*, t_1)$, which combines with inequality (5), one has

$$e^\sigma g(t) > e^\sigma g_{t_0} \geq g(s),\ t \in [t_*, t_1),\ s \in [t_0 - \tau, t_1).$$

So $D^+ g(t) < (\alpha + \beta e^\sigma) g(t)$, $t \in [t_*, t_1)$. Then integrating the above inequality on the interval $[t_*, t_1)$, one has $t_1 - t_0 \geq t_1 - t_* > \frac{\ln \delta}{\alpha + \beta e^\sigma}$. Repeating such procedure, we finally get $t_\theta - t_{\theta-1} > \frac{\ln \delta}{\alpha + \mu\beta e^\sigma}$, $\theta \in \mathbb{Z}^+$. It is easy to see that the Zeno beavior is excluded.

Case 3: The triggered impulse sequence $t_\theta, \theta \in \mathbb{Z}^+$ comprises forced impulse instants as well as ET instants. Reduction to absurdity will be employed to demonstrate that this scenario precludes Zeno behavior. In this case, the occurrence of Zeno behavior in system (1) would imply the existence of infinitely many impulses within a finite time. Let $[t_0, T]$ be the finite interval exhibiting Zeno behavior, and T be the Zeno time. Therefore, the interval $[T - \frac{\hat{\tau}}{2}, T]$ contains infinitely many impulse instants. If there exists a forced triggering time in the interval $[T - \frac{\hat{\tau}}{2}, T)$, then according to the definition of $\hat{\tau}$, there must be one and only one forced triggering time in the interval $[T - \frac{\hat{\tau}}{2}, T)$, and then the forced trigger moments in $[t_0, T]$ are finite in number, which denoted as $\{\tau_{q_i},\ i = 1, 2, \ldots, r\}$, $q_i \in \{j^i + 1, j^i + 2, \ldots, j^i + n_1^i, j_1^i + 1, \ldots, j_1^i + n_2^i, \ldots, j_{m_i-1}^i + 1, \ldots, j_{m_i}^i\}$, where $j_k^i = j^i + n_1^i + n_2^i + \ldots + n_k^i$, $j_{m_i}^i < j^{i+1} + 1$, n_k^i represents the number of distinct forced-triggering types, $\tau_{j_{m_r}^r} \in [T - \frac{\hat{\tau}}{2}, T)$ is the last forced triggering instant, with all subsequent triggers being event-triggered. For convenience, we denote $\tilde{\Delta}_{j_k^i} = \tilde{\lambda}_{j_{k-1}^i+1} + \tilde{\lambda}_{j_{k-1}^i+2} + \ldots + \tilde{\lambda}_{j_k^i}$ and $\bar{\Delta}_{j_k^i} = \bar{\lambda}_{j_{k-1}^i+1} + \bar{\lambda}_{j_{k-1}^i+2} + \ldots + \bar{\lambda}_{j_k^i}$, $k \in \{1, 2, \ldots, m_i\}$, $i \in \{1, 2, \ldots, r\}$. Then four scenarios should be discussed.

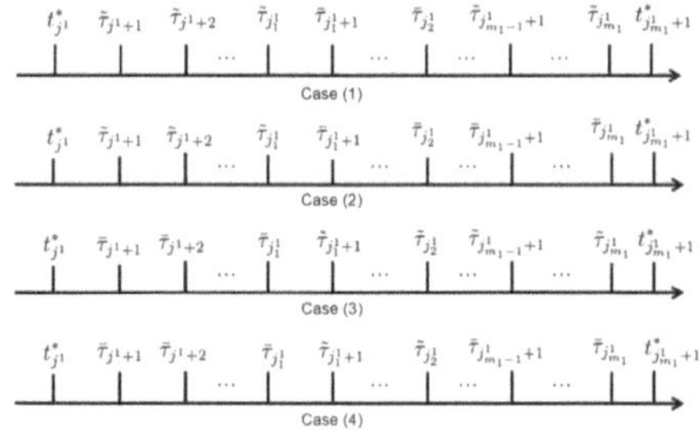

Fig. 2. Four cases of forced-triggering classification.

Subcase 1: If $r = 1$ and $j^1 \neq 0$, for $\theta = 1, 2, \ldots, j^1$, $t_\theta = t^*_\theta$, that is, $\{t_1, \ldots, t_{j^1}\}$ are the set of event-triggered instants and for $\theta = j^1+1, j^1+2, \ldots, j^1_{m_1}$, $t_\theta = \tau_\theta$, that is $\{t_{j^1+1}, \ldots, t_{j^1_{m_1}}\}$ are the set of forced-triggered instants, it should be noted that there are four possible cases to be considered in Fig. 2.

Case (1): $t_{j^1+1} = \tilde{\tau}_{j^1+1}$ and $t_{1+j^1_{m_1-1}} = \tilde{\tau}_{j^1_{m_1-1}+1}$. Because $t = t^*_{j^1_{m_1}+1}$, $t^*_{j^1}$ are ET instants, one can obtain

$$g(t) < g(t^-_{j^1}) < \delta g(t^-_{j^1}) < \mu e^\sigma g(t^-_{j^1}),\ t \in [t_0 - \tau, t_{j^1}). \tag{6}$$

In light of HET mechanism (2),

$$g(t) \leq e^\sigma g(t_{j^1}) = \mu e^\sigma g(t^-_{j^1}), \quad t \in [t_{j^1}, t_{j^1+1}), g(t^-_{j^1+1}) = e^{\tilde{\lambda}_{j^1+1}} g(t^-_{j^1}). \tag{7}$$

By using of inequalities (6) and (7), we have

$$g(t) \leq \mu e^\sigma g(t^-_{j^1}) < \mu e^{\sigma+\tilde{\lambda}_{j^1+1}} g(t^-_{j^1}),\ t \in [t_0 - \tau, t_{j^1+1}).$$

Repeating such procedure, one has

$$g(t) \leq \mu e^{\sigma+\tilde{\Delta}_{j^1_1}} g(t^-_{j^1}), \quad t \in [t_0 - \tau, t_{j^1_1+1}). \tag{8}$$

Leveraging HET mechanism (2),

$$g(t) \leq e^\sigma g(t_{j^1_1+1}) = \mu e^{\sigma+\tilde{\Delta}_{j^1_1}-\bar{\lambda}_{j^1_1+1}} g(t^-_{j^1}),\ t \in [t_{j^1_1+1}, t_{j^1_1+2}), \tag{9}$$

Repeat inequalities (8)–(9), we have

$$g(t) < \mu e^{\sigma+\tilde{\Delta}_{j^1_1}+\tilde{\Delta}_{j^1_3}+\ldots+\tilde{\Delta}_{j^1_{m_1}}} g(t^-_{j^1}), \quad t \in [-\tau + t_0, t_{j^1_{m_1}+1}). \tag{10}$$

By parity of reasoning with Case 2, so one has

$$t_{j^1_{m_1}+1} - t_{j^1_{m_1}} \geq t_{j^1_{m_1}+1} - t_* > \frac{\ln \delta}{\alpha + \mu\beta e^{\sigma+\bar{\Delta}_{j^1_2}+\bar{\Delta}_{j^1_4}+\ldots+\bar{\Delta}_{j^1_{m_1-1}}}}.$$

By reproducing the analytical steps as in Case (1), it can be readily observed that Cases (2)–(4) yield analogous conclusions. Hence, it can be concluded that there exists a constant $c_1 > 0$, such that $t_{j^1_{m_1}+1} - t_{j^1_{m_1}} > c_1$ always holds.

Reproducing the analytical steps from Case 2, it is easy to see $t_\theta - t_{\theta-1} \geq c_1$, when $\theta \in \{j^1_{m_1}+1,\ j^1_{m_1}+2, \ldots\}$. It then can be deduced that $t_{j^1_{m_1}+s} > sc_1 + t_{j^1_{m_1}}$, one obtains that $t_{j^1_{m_1}+s} \to +\infty$ as $s \to +\infty$, which gives rise to a contradiction.

Subcase 2: If $r = 1$ and $j^1 = 0$, then on the interval $[t_0, \tau_{j^1_{m_1}}]$, no event-triggered sequence exists, and the forced-triggered instants are given by $\{\tau_\theta, \theta = 1, 2, \ldots, j^1_{m_1}\}$. By applying similar arguments as in Subcase 1, one can deduce that $t_{j^1_{m_1}+s} > sc_1 + t_{j^1_{m_1}}$. It then follows that $t_{j^1_{m_1}+s} \to +\infty$ as $s \to +\infty$, leading to a contradiction.

Subcase 3: If $r > 1$ and $j^1 \neq 0$, it follows that on the interval $[t_0, \tau_{j^r_{m_r}}]$, for $\theta \in \{1, 2, \ldots, j^1\} \cup (\bigcup_{i=1}^{r-1}\{j^i_{m_i}+1, j^i_{m_i}+2, \ldots, j^{i+1}\})$, $t_\theta = t^*_\theta$. Repeating the steps as in Subcase 1, we can obtain $t_{j^i_{m_i}+1} - t_{j^i_{m_i}} > c_i$, $i \in \{1, 2, \ldots, r\}$. Using arguments similar to those in Case 2, one can obtain $t_{j^r_{m_r}+s+1} - t_{j^r_{m_r}+s} > c_r$, $s \geq 1$. It then can be deduced that $t_{j^r_{m_r}+s} > sc_r + t_{j^r_{m_r}}$, and $t_{j^r_{m_r}+s} \to +\infty$ as $s \to +\infty$, which leads to a contradiction.

Subcase 4: If $r > 1$ and $j^1 = 0$, for $\theta \in \bigcup_{i=1}^{r-1}\{j^i_{m_i}+1, j^i_{m_i}+2, \ldots, j^{i+1}\}$, $t_\theta = t^*_\theta$. Reproducing the analytical steps from Subcase 3, one can deduce that $t_{j^r_{m_r}+s} > sc_1 + t_{j^r_{m_r}}$. It then follows that $t_{j^r_{m_r}+s} \to +\infty$ as $s \to +\infty$, leading to a contradiction.

In the absence of forced triggering times in $[T - \frac{\hat{\tau}}{2}, T]$, namely $\tau_{j^r_{m_r}} \notin [T - \frac{\hat{\tau}}{2}, T]$. Then, mirroring the analysis in Subcase 1–4, we conclude that $t_{j^r_{m_r}+s} \to +\infty$ as $s \to +\infty$, which gives rise to a contradiction. It follows that Zeno behavior does not occur in Case 3.

The proof is completed.

Theorem 2. *Under conditions in Theorem 1, the origin of system (1) is asymptotically stable if there exist functions* $\varkappa \in \mathbf{K}_\infty$ *such that*

(1) $\varkappa(|x|) \leq g(t, x(t))$, *for all* $x \in \mathbb{R}^n$, $t \geq t_0 - \tau$,

(2) $\dfrac{\ln \mu}{\hat{\tau}} + \alpha \leq -\gamma$,

(3) $-\mu\gamma + \beta < 0$.

Proof. For any $\varepsilon > 0$, define $v(t)$ as an unique solution of the system

$$\begin{cases} D^+v(t) = \alpha v(t) + \beta v(t - \tau(t)) + \varepsilon, & t \in [t_\theta, t_{\theta+1}), \\ v(t_\theta) = \mu v(t_\theta^-), & \theta \in \mathbb{Z}^+, \\ v(t) = g_{t_0}, & t \in [t_0 - \tau, t_0]. \end{cases} \tag{11}$$

From Lemma 1 and $g(t) \leq v(t)$, $t \in [-\tau + t_0, t_0]$, We conclude that $g(t) \leq v(t)$ for $t \geq t_0$. Applying the variation of parameters formula, $v(t)$ can be written as

$$v(t) = v(t_0)\bar{v}(t, t_0) + \int_{t_0}^{t} \varepsilon + \bar{v}(t, s)\left(\beta v(s - \tau(s))\right) \mathrm{d}s, \tag{12}$$

where $\bar{v}(t, s),\ 0 \leq s \leq t$, satisfying

$$\bar{v}(t, s) = e^{\alpha(t-s)} \prod_{s \leq t_\theta < t} \mu \leq e^{(-r-\frac{\ln \mu}{\tilde{\tau}})(t-s)} \mu^{(\frac{t-s}{\tilde{\tau}}-1)} \leq \frac{1}{\mu} e^{-r(t-s)}. \tag{13}$$

Next, we will prove

$$v(t) \leq \Pi e^{-\eta(t-t_0)} + \Sigma,\ t \geq t_0, \tag{14}$$

in which $\eta > 0$ is a unique solution satisfying $\mu(\eta - \gamma) + \beta e^{\eta\tau} = 0$, $\Sigma = \frac{\varepsilon}{\mu\gamma - \beta}$, and $\Pi = \frac{1}{\mu} g_{t_0}$. Let $\tilde{t} = \inf\{t > t_0 :\ v(t) \geq \Pi e^{-\eta(t-t_0)} + \Sigma\}$, so one can obtain

$$v(t) < \Pi e^{-\eta(-t_0+t)} + \Sigma,\ \text{for } t \in (t_0, \tilde{t}) \text{ and } v(\tilde{t}) = \Pi e^{-\eta(\tilde{t}-t_0)} + \Sigma. \tag{15}$$

Based on inequalities (13) and (14), the following estimation could be conducted

$$\begin{aligned} v(\tilde{t}) <& \exp\{-\gamma(-t_0 + \tilde{t})\}\Pi + \frac{\beta \Pi e^{\eta\tau}}{(\gamma - \eta)\mu}\left[-e^{-\gamma(-t_0+\tilde{t})} + e^{-\eta(-t_0+\tilde{t})}\right] \\ &+ \frac{\beta\Sigma + \varepsilon}{\mu\gamma}\left[1 - e^{-\gamma(-t_0+\tilde{t})}\right] \\ \leq& \Pi e^{-\eta(-t_0+\tilde{t})} + \Sigma, \end{aligned}$$

which leads to contradiction with the formula (15), Hence, it can be concluded that inequality (14) is true. Then we can obtain that $g(t) \leq \Pi e^{-\eta(t-t_0)} + \Sigma$, $t \in [t_0, +\infty)$, and $\Sigma \to 0$ as $\varepsilon \to 0$, so

$$|x(t)| \leq \varkappa^{-1}(\Pi e^{-\eta(t-t_0)}),\ \text{fort} \in [t_0, +\infty), \tag{16}$$

which indicates that system (1) is asymptotically stable.

4 Numerical Simulations

The following model is considered

$$\begin{cases} \dot{x}(t) = Ax(t) + Bf(x(t-\tau)),\ t \neq t_\theta, \\ x(t_\theta) = \sqrt{\mu} x(t_\theta^-),\ \theta \in \mathbb{Z}^+, \end{cases} \tag{17}$$

where $x(t) = [x_1(t), x_2(t)]^\top \in \mathbb{R}^2$, $\tau = 1$, $\mu = 0.69$, $A = \begin{pmatrix} -1.5 & -0.6 \\ -0.8 & -1.5 \end{pmatrix}$, $B = \begin{pmatrix} -0.6 & -0.1 \\ -0.2 & -0.8 \end{pmatrix}$, $f(x(t)) = \begin{pmatrix} \tanh(x_1(t)) \\ \tanh(x_2(t)) \end{pmatrix}$.

Choose the Lyapunov function $V(t) = (x(t))^\top x(t)$, then

$$\begin{cases} D^+V(t) \le aV(t) + bV(t-\tau), & t \ne t_\theta, \\ V(t_\theta) = \sqrt{\mu} V(t_\theta^-), & \theta \in \mathbb{Z}^+, \end{cases} \tag{18}$$

where $a = 0.5$, $b = 0.2$. Then conditions (2), (3) in Theorem 2 are satisfied with $\hat{\tau} < 0.4698$ and condition $\ln \delta < \ln \mu + \sigma$ in Theorem 1 is satisfied with $\sigma > 0.3711$. Choose $\hat{\tau} = 0.43$, $\sigma = 0.38$, for system (17), the following HET mechanism (19) is given by

$$\begin{aligned} t_\theta &= \min\{t_\theta^*, t_{\theta-1} + 0.43\},\ \theta \in \mathbb{Z}^+, \\ t_1^* &= \inf\{t \ge V(t) \ge e^{0.38} \sup_{s\in[-1,0]} \|x(s)\|^2\}, \\ t_\theta^* &= \inf\{t \ge t_{\theta-1} \big| V(t) \ge e^{0.38} V(t_{\theta-1})\},\ \theta \ge 2. \end{aligned} \tag{19}$$

According to Theorem 1, the stability of system (17) under the given parameters and controller is guaranteed, as shown in Fig. 3 for different initial values. Besides, the trigger types under controller (19) is illustrated in Figure.4.

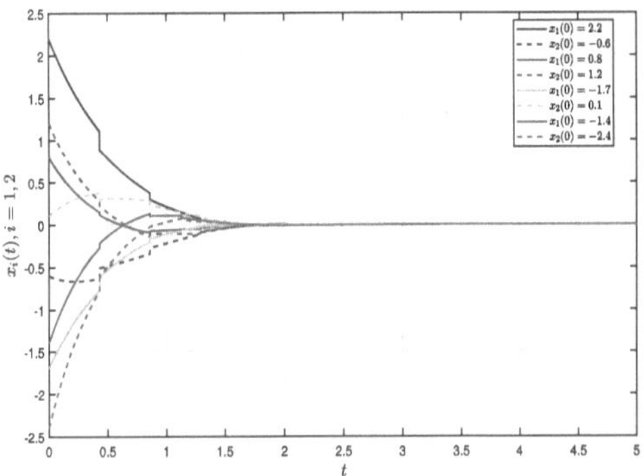

Fig. 3. Stability of system (17) under the HET impulsive control with different initial values.

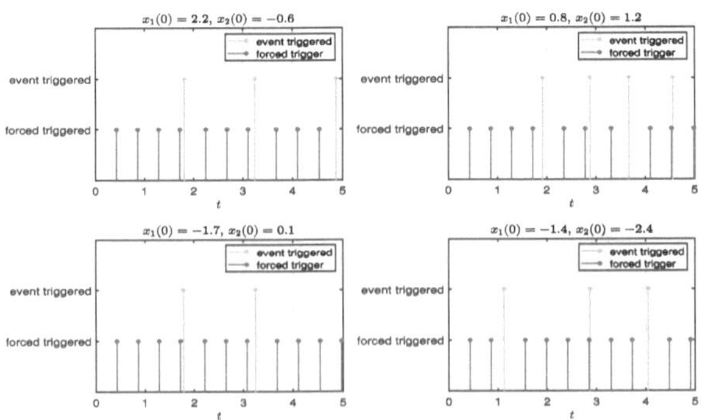

Fig. 4. Trigger types.

5 Conclusion

This paper develops a HET impulsive control scheme to tackle the stability problem of nonlinear impulsive systems. Firstly, an HET impulsive control strategy incorporating dual triggering conditions was designed, with rigorous theoretical analysis demonstrating the exclusion of Zeno behavior. Subsequently, several effective criteria ensuring system stability were derived based on the Lyapunov method. Finally, the effectiveness of the proposed control approach was verified through numerical simulations.

Acknowledgments. This work was supported by Tianshan Talent Training Program (2022TSYCCX0013), by National Natural Science Foundation of China (Grant Nos. 62373317, 62263029), by the Key Project of Natural Science Foundation of Xinjiang (2021D01D10), and by Intelligent Control and Optimization Research Platform at Xinjiang University.

References

1. Varshney, L.R., Chen, B.L., Paniagua, E., Hall, D.H., Chklovskii, D.B.: Structural properties of the caenorhabditis elegans neuronal network. PLoS Comput. Biol. **7**(2), e1001066 (2011)
2. Adhikari, B.M., Prasad, A., Dhamala, M.: Time-delay-induced phase-transition to synchrony in coupled bursting neurons. Chaos Interdisc. J. Nonlinear Sci. **21**(2), 023116 (2011)
3. Hu, H.X., Xu, L.G.: Finite-time and fixed-time stability for nonlinear systems via aperiodically intermittent control. Nonlinear Dyn. **113**(8), 8475–8490 (2025)
4. Zhang, J.R., Wang, J.A., Zhang, J., Li, M.J., Zhao, Z.C., Wen, X.Y.: Fixed-time event-triggered pinning synchronization of complex network via aperiodically intermittent control. Neurocomputing **614**, 128818 (2025)
5. Li, X.D., Liu, W.L., Gorbachev, S., Cao, J.D.: Event-triggered impulsive control for input-to-state stabilization of nonlinear time-delay systems. IEEE Trans. Cybern. **54**(4), 2536–2544 (2023)
6. Yang, N., Gu, X.L., Su, H.: Event-triggered delayed impulsive control for functional differential systems on networks. Commun. Nonlinear Sci. Numer. Simul. **131**, 107850 (2024)
7. Liu, X., Chen, L.L., Zhao, Y.F., Li, H.L.: Event-triggered hybrid impulsive control for synchronization of fractional-order multilayer signed networks under cyber attacks. Neural Netw. **172**, 106124 (2024)
8. Sun, H.B., Kong, X.L., Yang, J., Hou, L.L., Yang, D.: Self-adjustable and flexible performance-based event-triggered asymptotic tracking control of nonlinear systems with unknown control directions. IEEE Trans. Cybern. **55**(8), 3960–3973 (2025)
9. Zhang, T.P., Feng, C.J., Xia, X.N.: Adaptive event-triggered control of switched nonlinear systems with unknown input dead zones and unmodeled dynamics. J. Franklin Inst. **362**(2), 107453 (2025)
10. Li, H., Hua, C.C., Li, K.: Global dynamic double side event-triggered adaptive control for interconnected nonlinear systems via intermittent output feedback. IEEE Trans. Cybern. **55**(3), 1083–1092 (2025)

11. Xie, X., Wei, T.D., Li, X.D.: Hybrid event-triggered approach for quasi-consensus of uncertain multi-agent systems with impulsive protocols. IEEE Trans. Circuits Syst. I Regul. Pap. **69**(2), 872–883 (2022)
12. Lin, S.R., Liu, X.W., Huang, Y.L.: Event-triggered pinning control for passivity and synchronization of directly coupled delayed reaction-diffusion neural networks. Int. J. Adapt. Control Signal Process. **38**(2), 558–579 (2024)
13. Wang, S.T., Shi, K.B., Cao, J.D., Wen, S.P.: Fuzzy adaptive event-triggered synchronization control mechanism for T-S fuzzy RDNNs under deception attacks. Commun. Nonlinear Sci. Numer. Simul. **134**, 107985 (2024)
14. Li, X.D., Wu, J.H.: Stability of nonlinear differential systems with state-dependent delayed impulses. Automatica **64**, 63–69 (2016)
15. Khalil, H.K., Grizzle, J.W.: Nonlinear Systems. Prentice Hall Upper Saddle River, New Jersey, USA (2002)

MPC-Based Resilient Cooperative Target Tracking of Autonomous Surface Vehicles Under False Data Injection Attacks

Nan Gu[1,2](✉), Yiwen Li[1,2], Ronghui Li[1,2], Aijun Li[1,2], and Zhouhua Peng[1,2]

[1] School of Marine Electrical Engineering, Dalian Maritime University, Dalian 116026, China
{ngu,yiwenli,ronghuili,aijunli,zhpeng}@dlmu.edu.cn

[2] Dalian Key Laboratory of Swarm Control and Electrical Technology for Intelligent Ships, Dalian 116026, China

Abstract. This paper addresses the cooperative target tracking control problem for multiple tracking autonomous surface vehicles (ASVs) under false data injection (FDI) attacks from the target ASV. A model predictive control (MPC) - based resilient cooperative target tracking control scheme is proposed for mitigating the effect of the attacks and restoring the tracking performance. Specifically, a nominal cooperative target tracking control law is designed to achieve cooperative target tracking without FDI attacks where a finite-time extended state observer (ESO) is proposed to estimate the model uncertainties. Next, the FDI attacks from the target ASV are modeled by using a MPC method to hinder the cooperative tracking of multiple ASVs. Then, optimal resilient signals are developed via the MPC approach to counteract the attack effects and thereby recover cooperative tracking performance. The input-to-state stability (ISS) of the proposed finite-time ESO error dynamics is analyzed by employing a homogeneous Lyapunov function, besides in the entire closed-loop control system, all tracking error signals are uniformly ultimately bounded. The efficacy of the proposed resilient control method is confirmed through simulation results.

Keywords: Cooperative Target tracking · Autonomous Surface Vehicle · False Data Injection attacks · Resilient control

1 Introdution

In recent years, autonomous surface vehicles (ASVs) have emerged as a key component of intelligent ocean systems, with applications spanning oceanographic research, environmental monitoring, search and rescue missions, and military surveillance [4,14]. Cooperative target tracking [10,15] is one of the most fundamental and essential cooperative control tasks for ASVs, which allows ASVs to autonomously track designated maritime targets such as vessels, buoys, or moving objects of interest. Effective cooperative target tracking is essential for

C. Li et al. (Eds.): ICNC 2025, CCIS 2946, pp. 377–387, 2026.
https://doi.org/10.1007/978-981-92-1599-7_32

maintaining situational awareness, ensuring mission success, and enabling collaborative operations in complex maritime environments.

Current research in cooperative target tracking control of ASVs has achieved promising results. In [3], a hierarchical reinforcement learning control is designed for ASVs to achieve cooperative target tracking, where the follower ASVs can receive the message from the neighbor ASV via cloud-supported communication. In [6], a decentralized leader-follower control is designed, enabling USVs to achieve target tracking and collision avoidance, where follower USV know the message of its leader USV through network. However, to achieve the cooperative target tracking task, modern ASVs increasingly rely on networked communication systems for sensor data transmission, control command delivery. This cyber-physical system renders ASVs easier to be cyber-attacked, which leads to ASVs deviating from the desired trajectory [9,12]. The FDI attacks are aiming to modify the data of sensor measurements or control signals during the data transmission [11,16]. Thus, it is significant to investigate resilient control strategies for cyber-physical systems under FDI attacks.

Several promising results are put forward to address the attacks [1,2,8,13]. In [1], a nonlinear observer is proposed to estimate the FDI attacks in real-time, and a secure nonlinear controller is designed such that FDI attacks can be mitigated. In [2], an attack detector is equipped and a deep reinforcement learning (DRL) solution is proposed such that the secure control problem of connected and automated vehicles under FDI attacks is effectively solved. In [8], a data transmission mechanism and an ESO is developed such that the FDI attacks can be identified and compensated, a secure control scheme is proposed such that the multi-ASV system to accurately track the desired reference trajectory. However, the mentioned methods rely on accurate attack estimation and precise compensation.

Based on the above observations, this paper investigates the multi-ASV system target tracking problem under false data injection attacks, and proposes a resilient control law under optimal attacks which is generated by MPC optimization problem. At first, nominal cooperative target tracking control law is proposed based on a finite-time ESO with relative line-of-sight distance. Then, the attacks and the optimal compensations is designed based on MPC optimization approach. The proposed control method adopts the perspective of defenders. It utilizes the current and nominal states to guide the follower ASVs back to their nominal states in order to mitigate the effects of the attacks. Simulation results show the effectiveness of the resilient control law.

2 Preliminaries and Problem Formation

The motion of ASV moving in a horizontal plane is described in earth-fixed reference frame $X_E - X_E$ and a body-fixed reference frame $X_B - X_B$. The kinematic of the target ASV is constructed as

$$\begin{cases} x_0(k+1) = x_0(k) + T_s[u_0(k)\cos\psi_0(k) - v_0(k)\sin\psi_0(k)] \\ y_0(k+1) = y_0(k) + T_s[u_0(k)\sin\psi_0(k) + v_0(k)\cos\psi_0(k)] \\ \psi_0(k+1) = \psi_0(k) + T_s r_0(k) \end{cases} \tag{1}$$

where $x_0(k), y_0(k), \psi_0(k)$ are the position and orientation of the target ASV in earth-fixed reference frame; $u_0(k), v_0(k), r_0(k)$ represent surge velocity, sway velocity and yaw rate of target ASV in body-fixed reference frame, T_s is the discretization period.

The kinematic model of ith tracking ASVs is constructed as

$$\begin{cases} x_i(k+1) = x_i(k) + T_s[u_i(k)\cos\psi_i(k) - v_i(k)\sin\psi_i(k)] \\ y_i(k+1) = y_i(k) + T_s[u_i(k)\sin\psi_i(k) + v_i(k)\cos\psi_i(k)] \\ \psi_i(k+1) = \psi_i(k) + T_s r_i(k) \end{cases} \tag{2}$$

where $x(k), y(k), \psi(k)$ are the position and orientation of the ith follower ASV in earth-fixed reference frame; $u(k), v(k), r(k)$ represent surge velocity, sway velocity and yaw rate of the i-th tracking ASV in body-fixed reference frame.

The objective of this paper is to develop resilient cooperative target tracking control law under FDI attacks, enabling the underactuated tracking ASVs to track the target ASV with a desired surrounding formation.

3 Controller Design

3.1 Nominal Cooperative Target Tracking Control Law

At first, the cooperative target tracking errors are defined as

$$\begin{cases} e_{\rho i}(k) = \rho_i(k) - l_i \\ e_{\beta i}(k) = \beta_i(k) - \psi_i(k) - \beta_{si}(k) \end{cases} \tag{3}$$

where $\rho_i(k) = \sqrt{(y_0(k) - y_i(k) + \Delta_{yi})^2 + (x_0(k) - x_i(k) + \Delta_{xi})^2}$ is the relative range between the ith follower ASV and ith virtual point; $\beta_i(k) = \text{atan2}(y_0(k) - y_i(k) + \Delta_{yi}, x_0(k) - x_i(k) + \Delta_{xi})$ is the angle between the ith follower ASV and ith virtual point; Δ_{xi} and Δ_{yi} denote the expected position error between the ith virtual point and the ith tracking ASV; l_i denotes the desired distance; $\beta_{si}(k) = \text{atan2}(vi(k), ui(k))$ is the sideslip angle. Considering the discrete-time dynamics of $e_{\rho i}(k)$ and $e_{\beta i}(k)$ and using (1) and (2), we have

$$\begin{cases} e_{\rho i}(k+1) = e_{\rho i}(k) + T_s[u_0(k)\cos(\beta_i(k) - \psi_0(k)) + v_0(k)\sin(\beta_i(k) - \psi_0(k)) \\ \qquad - u_i(k) - v_i(k)\sin(\beta_i(k) - \psi_i(k)) + 2u_i(k)\sin^2\frac{\beta_i(k) - \psi_i(k)}{2}] \\ e_{\beta_i}(k+1) = e_{\rho i}(k) + \frac{T_s}{\rho_i}[u_0(k)\sin(\psi_0(k) - \beta_i(k)) + v_0(k)\cos(\psi_0(k) - \beta_i(k)) \\ \qquad - v_i(k)\cos(\beta_i(k) - \psi_i(k)) + u_i(k)\sin(\beta_i(k) - \psi_i(k))] - T_s r_i(k) \end{cases} \tag{4}$$

which can be simplified as

$$\begin{cases} e_{\rho i}(k+1) - e_{\rho i}(k) = T_s\zeta_{ui}(k) - T_s u_i(k) \\ e_{\beta i}(k+1) - e_{\beta i}(k) = T_s\zeta_{ri}(k) - T_s r_i(k) \end{cases} \tag{5}$$

where u_i, r_i are the cooperative target tracking resilient control law which will be designed later. The unknown functions $\zeta_{ui}(k)$ and $\zeta_{ri}(k)$ are obtained as $\zeta_{ui}(k) = u_0(k)\cos(\beta_i(k) - \psi_0(k)) + v_0(k)\sin(\beta_i(k) - \psi_0(k)) - v_i(k)\sin(\beta_i(k) - \psi_i(k)) + 2u_i(k)\sin^2\frac{\beta_i(k)-\psi_i(k)}{2}$, $\zeta_{ri}(k) = \frac{1}{\rho_i}(u_0(k)\sin(\psi_0(k) - \beta_i(k)) + v_0(k)\cos(\psi_0(k) - \beta_i(k)) - v_i(k)\cos(\beta_i(k) - \psi_i(k)) + u_i(k)\sin(\beta_i(k) - \psi_i(k)))$

To estimate $\zeta_{ui}(k)$ and $\zeta_{ri}(k)$, a finite-time ESO is developed as follows

$$\begin{cases} \hat{e}_{\rho i}(k+1) = \hat{e}_{\rho i}(k) + T_s[\hat{\zeta}_{ui}(k) - u_i(k) - k_{1\rho} sig^{\alpha_1}(\hat{e}_{\rho i}(k) - e_{\rho i}(k))] \\ \hat{\zeta}_{ui}(k+1) = \hat{\zeta}_{ui}(k) + T_s[-k_{2\rho} sig^{2\alpha_1 - 1}(\hat{e}_{\rho i}(k) - e_{\rho i}(k))] \\ \hat{e}_{\beta i}(k+1) = \hat{e}_{\beta i}(k) + T_s[\hat{\zeta}_{ri}(k) - r_i(k) - k_{1\beta} sig^{\alpha_2}(\hat{e}_{\beta i}(k) - e_{\beta i}(k))] \\ \hat{\zeta}_{ri}(k+1) = \hat{\zeta}_{ri}(k) + T_s[-k_{2\beta} sig^{2\alpha_2 - 1}(\hat{e}_{\beta i}(k) - e_{\beta i}(k))] \end{cases} \tag{6}$$

where the symbol $\mathrm{sig}^{\alpha}(\cdot)$ is defined as $\mathrm{sig}^{\alpha}(\cdot) = \mathrm{sign}(\cdot)|\cdot|^{\alpha}$, $\alpha > 0$; $\hat{e}_{\rho i}(k)$, $\hat{e}_{\beta i}(k)$ are the estimated values of $e_{\rho i}(k)$, $e_{\beta i}(k)$, $\hat{\zeta}_{ui}(k)$, $\hat{\zeta}_{ri}(k)$ are the estimated values of $\zeta_{ui}(k)$, $\zeta_{ri}(k)$ respectively; the positive gains of the finite-time ESO are represented as $k_{1\rho}$, $k_{2\rho}$, $k_{1\beta}$ and $k_{2\beta}$; α_1, $\alpha_2 \in (\frac{1}{2}, 1)$ are design parameters.

Assumption 1. Defining the difference of the $\zeta_{ui}(k)$ and $\zeta_{ri}(k)$ are $\Delta_{\zeta ui}$, $\Delta_{\zeta ri}(k)$. $\Delta_{\zeta ui}(k)$, $\Delta_{\zeta ri}(k)$ are bounded, satisfying $\|\Delta_{\zeta ui}\| \le \zeta_{ui}^*$, $\|\Delta_{\zeta ri}\| \le \zeta_{ri}^*$, with ζ_{ui}^*, ζ_{ri}^* are positive constants.

Defining the estimation errors $\tilde{e}_{\rho i}(k) = \hat{e}_{\rho i}(k) - e_{\rho i}(k)$, $\tilde{\zeta}_{ui}(k) = \hat{\zeta}_{ui}(k) - \zeta_{ui}(k)$, $\tilde{e}_{\beta i}(k) = \hat{e}_{\beta i}(k) - e_{\beta i}(k)$, $\tilde{\zeta}_{ri}(k) = \hat{\zeta}_{ri}(k) - \zeta_{ri}(k)$, it follows that

$$\begin{cases} \tilde{e}_{\rho i}(k+1) = \tilde{e}_{\rho i}(k) + T_s[\tilde{\zeta}_{ui}(k) - k_{1\rho} sig^{\alpha_1}(\tilde{e}_{\rho i}(k))] \\ \tilde{\zeta}_{ui}(k+1) = \tilde{\zeta}_{ui}(k) + T_s[-k_{2\rho} sig^{2\alpha_1 - 1}(\tilde{e}_{\rho i}(k)) - \Delta_{\zeta ui}(k)] \\ \tilde{e}_{\beta i}(k+1) = \tilde{e}_{\beta i}(k) + T_s[\tilde{\zeta}_{ri}(k) - k_{2\rho} sig^{\alpha_2}(\tilde{e}_{\beta i}(k))] \\ \tilde{\zeta}_{ri}(k+1) = \tilde{\zeta}_{ri}(k) + T_s[-k_{2\rho} sig^{2\alpha_2 - 1}(\tilde{e}_{\beta i}(k)) - \Delta_{\zeta ri}(k)] \end{cases} \tag{7}$$

The following presents the nominal control law for cooperative target tracking

$$\begin{cases} u_{ni}(k) = c_{\rho i} e_{\rho i}(k)/\Pi_{\rho i} + \hat{\zeta}_{ui}(k) \\ r_{ni}(k) = c_{\beta i} e_{\beta i}(k)/\Pi_{\beta i} + \hat{\zeta}_{ri}(k) \end{cases} \tag{8}$$

where $c_{\rho i}, c_{\beta i}$ are positive constants, $\Pi_{\rho i} = \sqrt{e_{\rho i}(k)^2 + \Delta_{\rho i}^2}$, $\Pi_{\beta i} = \sqrt{e_{\beta i}(k)^2 + \Delta_{\beta i}^2}$, and $\Delta_{\rho i}, \Delta_{\beta i}$ are positive constants.

3.2 MPC-Based Cooperative Target Tracking Resilient Control Law

Before the FDI attacks injecting, the control law of tracking ASVs is (8). The target ASV executes FDI attacks to prevent being caught up when the nominal cooperative target tracking control signals are transmitted to the tracking ASVs. To address FDI attacks, a resilient cooperative target tracking control framework for ASVs is proposed, which is given by

$$\begin{cases} u_{di}(k) = u_{ai}(k) + u_{ni}(k) \\ r_{di}(k) = r_{ai}(k) + r_{ni}(k) \end{cases} \tag{9}$$

where $u_{ai}(k)$ and $r_{ai}(k)$ are the false data injection attacks of the ith tracking ASV, which will be developed later.

From the perspective of the target ASV, there are two primary objectives, avoiding being caught up by the tracking ASVs, and minimizing the cost of attack execution. Then, the attacks are generated by solving the MPC optimization problem

$$\begin{aligned} \boldsymbol{U}_a(k) &= \text{argmax} J_{adv} \\ &= \text{argmax} \sum_{k=1}^{K} [\boldsymbol{\eta}_d(k) - \boldsymbol{\eta}_n(k)]^{\mathrm{T}} \mathbf{Q}_a [\boldsymbol{\eta}_d(k) - \boldsymbol{\eta}_n(k)] - \boldsymbol{U}_a(k)^{\mathrm{T}} \mathbf{R}_a \boldsymbol{U}_a(k) \end{aligned} \tag{10}$$

subject to

$$\begin{cases} \boldsymbol{\eta}_d(k+1) = f_p(\boldsymbol{\eta}_d(k), u_{di}(k), r_{di}(k), T_s) \\ \underline{\mathbf{B}}_a \le \boldsymbol{U}_a(k) \le \overline{\mathbf{B}}_a \end{cases} \tag{11}$$

where $\boldsymbol{\eta}_d(k) = [x_d(k), y_d(k), \psi_d(k)]^{\mathrm{T}}$ is the state of a tracking ASV under attacks; $\boldsymbol{\eta}_n(k) = [x_n(k), y_n(k), \psi_n(k)]^{\mathrm{T}}$ is the nominal state of a tracking ASV; $\overline{\mathbf{B}}_a = [\bar{u}_a, \bar{r}_a]^{\mathrm{T}}$ and $\underline{\mathbf{B}}_a = [-\bar{u}_a, -\bar{r}_a]^{\mathrm{T}}$ are constraints of the attacks; $\boldsymbol{U}_a(k) = [u_{ai}(k), r_{ai}(k)]^{\mathrm{T}}$ is the attacks; $\mathbf{Q}_a$, $\mathbf{R}_a$ are positive definite matrices.

The tracking performance of tracking ASVs is undesirable with the control law (9). As tracking ASVs, it is necessary to design an optimal compensation signal to mitigate the false data injection attack, and recover the tracking performance, then the resilient control law is designed as follows

$$\begin{cases} u_i(k) = u_{ci}(k) + u_{ai}(k) + u_{ni}(k) \\ r_i(k) = r_{ci}(k) + r_{ai}(k) + r_{ni}(k) \end{cases} \tag{12}$$

where $u_{ci}(k)$ and $r_{ci}(k)$ are the compensations, which will be designed as follows.

Similar to the target ASV, the tracking ASVs need to achieve two goals, effectively mitigating the impact of the attack and minimizing the corresponding compensation cost. Based on the objective, the optimal compensation is obtained through the solution of the following MPC optimization problem

$$\begin{aligned} \boldsymbol{U}_c(k) &= \text{argmin} J_{sec} \\ &= \text{argmin} \sum_{k=1}^{K} [\boldsymbol{\eta}_s(k) - \boldsymbol{\eta}_n(k)]^{\mathrm{T}} \mathbf{Q}_c [\boldsymbol{\eta}_s(k) - \boldsymbol{\eta}_n(k)] + \boldsymbol{U}_c(k)^{\mathrm{T}} \mathbf{R}_c \boldsymbol{U}_c(k) \end{aligned} \tag{13}$$

subject to

$$\begin{cases} \boldsymbol{\eta}_s(k+1) = f_p(\boldsymbol{\eta}_s(k), u_i(k), r_i(k), T_s) \\ \qquad \underline{\mathbf{B}}_c \leq \boldsymbol{U}_c(k) \leq \overline{\mathbf{B}}_\mathbf{c} \end{cases} \tag{14}$$

where $\boldsymbol{\eta}_s(k) = [x_s(k), y_s(k), \psi_s(k)]^{\mathrm{T}}$ are the state of a tracking ASV under secure control; $\overline{\mathbf{B}}_c = [\bar{u}_c, \bar{r}_c]^{\mathrm{T}}$ and $\underline{\mathbf{B}}_c = [-\bar{u}_c, -\bar{r}_c]^{\mathrm{T}}$ are constraints of the compensations; $\boldsymbol{U}_c(k) = [u_c(k), r_c(k)]^{\mathrm{T}}$ is the compensations; $\mathbf{Q}_c$, $\mathbf{R}_c$ are positive definite matrices; $\boldsymbol{\eta}_n(k) = [x_n(k), y_n(k), \psi_n(k)]^{\mathrm{T}}$ is the nominal state of a tracking ASV.

By using (5) and (12), the error dynamics can be express as

$$\begin{cases} e_{\rho i}(k+1) - e_{\rho i}(k) = T_s[-c_{\rho i} e_{\rho i}(k)/\Pi_{\rho i} - \tilde{\zeta}_{ui}(k) - u_{ai}(k) - u_{ci}(k)] \\ e_{\beta i}(k+1) - e_{\beta i}(k) = T_s[-c_{\beta i} e_{\beta i}(k)/\Pi_{\beta i} - \tilde{\zeta}_{ri}(k) - r_{ai}(k) - r_{ci}(k)] \end{cases} \tag{15}$$

4 Stability Analysis

Lemma 1. *[5] Under Assumption 1, the finite-time ESO error dynamic in (7): $[\Delta_{\zeta ui}, \Delta_{\zeta ri}] \to [\tilde{\zeta}_{ui}(k), \tilde{\zeta}_{ri}(k)]$ is input-to-state stable (ISS) in a finite time.*

Lemma 2. *The tracking errors system (15): $[\tilde{\zeta}_{ui}(k), \tilde{\zeta}_{ri}(k), u_{ai}(k), r_{ai}(k), u_{ci}(k), r_{ci}(k)] \to [e_{\rho i}(k), e_{\beta i}(k)]$ is ISS.*

Proof. Construct a Lyapunov function as follows

$$V_{ei}(k) = \frac{1}{2} \sum_{i=1}^{N} (e_{\rho i}(k)^2 + e_{\beta i}(k)^2). \tag{16}$$

The time derivative of $V_{ei}(k)$ is given by

$$\begin{aligned} \Delta_{V_e} &= \frac{V_{ei}(k+1) - V_{ei}(k)}{T_s} \\ &= -\sum_{i=1}^{N} \Big(\frac{c_{\rho i} e_{\rho i}(k)^2}{\Pi_\rho} + \frac{c_{\beta i} e_{\beta i}(k)^2}{\Pi_\beta} - e_{\rho i}(k)(\tilde{\zeta}_{ui}(k) + u_{ai}(k) + u_{ci}(k)) \\ &\quad - e_{\beta i}(k)(\tilde{\zeta}_{ri}(k) + r_{ai}(k) + r_{ci}(k)) \\ &\leq -\frac{c_{\min} \|\boldsymbol{E}_1\|^2}{\sqrt{\|\boldsymbol{E}_1\|^2 + \Delta_{\max}^2}} + \|\boldsymbol{\varpi}\| \|\boldsymbol{E}_1\| + \|\boldsymbol{a}\| \|\boldsymbol{E}_1\| + \|\mathbf{c}\| \|\boldsymbol{E}_1\| \end{aligned} \tag{17}$$

where $c_{\min} = \min\{\|\boldsymbol{c}_\rho\|, \|\boldsymbol{c}_\beta\|\}$, $\boldsymbol{c}_\rho = [c_{\rho 1}, \cdots, c_{\rho N}]$, $\boldsymbol{c}_\beta = [c_{\beta 1}, \cdots, c_{\beta N}]$; $\boldsymbol{E}_1 = [e_\rho, e_\beta]^{\mathrm{T}}$, $\boldsymbol{e}_\rho = [e_{\rho 1}, \cdots, e_{\rho N}]$, $\boldsymbol{e}_\beta = [e_{\beta 1}, \cdots, e_{\beta N}]$; $\Delta_{\max} = \max\{\|\boldsymbol{\Delta}_\rho\|, \|\boldsymbol{\Delta}_\beta\|\}$,

$\boldsymbol{\Delta}_{\rho} = [\Delta_{\rho 1}, \cdots, \Delta_{\rho N}]$, $\boldsymbol{\Delta}_{\beta} = [\Delta_{\beta 1}, \cdots, \Delta_{\beta N}]$; $\boldsymbol{\varpi} = [\|\tilde{\boldsymbol{\zeta}}_u\|, \|\tilde{\boldsymbol{\zeta}}_r\|]^{\mathrm{T}}$, $\tilde{\boldsymbol{\zeta}}_u = [\tilde{\zeta}_{u1}, \cdots, \tilde{\zeta}_{uN}]^{\mathrm{T}}$, $\tilde{\boldsymbol{\zeta}}_r = [\tilde{\zeta}_{r1}, \cdots, \tilde{\zeta}_{rN}]^{\mathrm{T}}$, $\boldsymbol{a} = [\|\boldsymbol{u}_a\|, \|\boldsymbol{r}_a\|]^{\mathrm{T}}$, $\boldsymbol{c} = [\|\boldsymbol{u}_c\|, \|\boldsymbol{r}_c\|]^{\mathrm{T}}$, $\boldsymbol{u}_a = [u_{a1}, \cdots, u_{aN}]^{\mathrm{T}}$, $\boldsymbol{r}_a = [r_{a1}, \cdots, r_{aN}]^{\mathrm{T}}$, $\boldsymbol{c}_a = [u_{c1}, \cdots, u_{cN}]^{\mathrm{T}}$, $\boldsymbol{r}_c = [r_{c1}, \cdots, r_{cN}]^{\mathrm{T}}$.

Noting that as $\|\boldsymbol{E}_1\| \geq \frac{\|\tilde{\boldsymbol{\zeta}}_u\|}{\xi c_{\min}} + \frac{\|\tilde{\boldsymbol{\zeta}}_r\|}{\xi c_{\min}} + \frac{\|\boldsymbol{u}_a\|}{\xi c_{\min}} + \frac{\|\boldsymbol{r}_a\|}{\xi c_{\min}} + \frac{\|\boldsymbol{u}_c\|}{\xi c_{\min}} + \frac{\|\boldsymbol{r}_c\|}{\xi c_{\min}} \geq \frac{\|\boldsymbol{\varpi}\|}{\xi c_{\min}} + \frac{\|\boldsymbol{a}\|}{\xi c_{\min}} + \frac{\|\boldsymbol{c}\|}{\xi c_{\min}}$, $\Delta_{V_e} \leq -c_{\min}(1-\xi)\|\boldsymbol{E}_1\|^2$, where $0 < \xi < 1$. Then it can be concluded that the system (15) is ISS under *Assumption 1*, and $\|\boldsymbol{E}_1\| \leq \max\{\gamma^{\boldsymbol{E}_1}(\|\boldsymbol{E}_1(0)\|, t)\}, \sigma^{\boldsymbol{\varpi}_1}(\|\boldsymbol{\varpi}_1\|)$, where $\gamma^{\boldsymbol{E}_1}$ is $\mathcal{KL}$ function and $\sigma^{\boldsymbol{\varpi}_1}(s) = \frac{\|\boldsymbol{\varpi}_1\|}{\xi c_{\min}} s$ is $\mathcal{K}$ function.

Theorem 1. *Consider the multi-ASVs system consists of the target ASV* (1) *and the tracking ASV* (2)*, combining with the finite-time ESO* (6)*, the nominal cooperative target tracking control law* (8)*. Under the Assumption 1, the overall closed-loop cooperative target tracking system described by* (7) *and* (15) *is ISS, and all error signals of the mentioned system are uniformly ultimately bounded.*

Proof. Lemma 1 show that the subsystem (7) with states $\tilde{\boldsymbol{\zeta}}_u$, $\tilde{\boldsymbol{\zeta}}_r$ and inputs $\Delta_{\zeta ui}, \Delta_{\zeta ri}$ is ISS within a finite time. Lemma 2 shows that the subsystem (15) with states $\boldsymbol{e}_\rho$, $\boldsymbol{e}_\beta$ and inputs $\tilde{\boldsymbol{\zeta}}_u$, $\tilde{\boldsymbol{\zeta}}_r$, $\boldsymbol{u}_a$, $\boldsymbol{r}_a$, $\boldsymbol{u}_c$, $\boldsymbol{r}_c$ is ISS. By [7], it proves that the entire closed-loop cooperative control system consisted by subsystem (7) and (15) with states $\tilde{\boldsymbol{\zeta}}_u, \tilde{\boldsymbol{\zeta}}_r, \boldsymbol{e}_\rho, \boldsymbol{e}_\beta$ and inputs $\Delta_{\zeta ui}, \Delta_{\zeta ri}$ is ISS, i.e., there exists class $\mathcal{KL}$ function γ, and $\mathcal{K}$ function σ^{ϕ}, such that $\boldsymbol{E}(t) \geq \max\{\gamma(\|\boldsymbol{E}(0)\|\, t), \sigma^{\phi}(\|\boldsymbol{\phi}\|)\}$, where $\boldsymbol{E} = [\tilde{\boldsymbol{\zeta}}_u, \tilde{\boldsymbol{\zeta}}_r, \boldsymbol{e}_\rho, \boldsymbol{e}_\beta]^{\mathrm{T}}$ and $\boldsymbol{\phi} = [\Delta_{\zeta ui}, \Delta_{\zeta ri}, \tilde{\boldsymbol{\zeta}}_u, \tilde{\boldsymbol{\zeta}}_r, \boldsymbol{a}, \boldsymbol{c}]^{\mathrm{T}}$. Note that $\Delta_{\zeta ui}$, $\Delta_{\zeta ri}$ are bounded by ζ_{ui}^*, ζ_{ri}^*; $\boldsymbol{a}$ and $\boldsymbol{c}$ are constrained by $\underline{\mathbf{B}}_a, \overline{\mathbf{B}}_a, \underline{\mathbf{B}}_c, \overline{\mathbf{B}}_c$. Then, the entire closed-loop cooperative target tracking system error signals $\boldsymbol{e}_\rho$, $\boldsymbol{e}_\beta$ are all bounded.

5 Results of Simulation

This section presents a simulation example to verify the effectiveness of the proposed resilient cooperative target tracking control law against the FDI attacks. Consider a cooperative multi-ASVs system consisting of a target ASV and six tracking ASVs. The initial state of the target ASV is set to $(x_0, y_0, \psi_0) = (10, 8, \pi/4), (u_0, v_0, r_0) = (1.0, 0.0, 0.0)$. The initial speed $(u_i(0), v_i(0), r_i(0))$ of tracking ASVs are zero, and the initial states of tracking ASVs are initialized as.

$$\begin{cases} (x_1(0), y_1(0), \psi_1(0)) = (18, 8, \pi/4), & (x_2(0), y_2(0), \psi_2(0)) = (12, 16, \pi/6) \\ (x_3(0), y_3(0), \psi_3(0)) = (12, 0, 0), & (x_4(0), y_4(0), \psi_4(0)) = (6, 14, \pi/2) \\ (x_5(0), y_5(0), \psi_5(0)) = (6, -2, \pi/2), & (x_6(0), y_6(0), \psi_6(0)) = (-2, 8, 0) \end{cases} \tag{18}$$

The desired tracking distance for the ASV to follow the virtually transformed reference point is set as $l_1 = l_2 = l_3 = l_4 = l_5 = l_6 = 1m$. The distance tracking deviation from the virtual point to the follower ASV is set as:

$$\begin{cases} (\Delta_{x1}, \Delta_{y1}) = (10, 0), & (\Delta_{x2}, \Delta_{y2}) = (4, 8), & (\Delta_{x3}, \Delta_{y3}) = (4, -8) \\ (\Delta_{x4}, \Delta_{y4}) = (-4, 8), & (\Delta_{x5}, \Delta_{y5}) = (-4, -8), & (\Delta_{x6}, \Delta_{y6}) = (-10, 0) \end{cases} \tag{19}$$

During the simulation, the finite-time ESO gains are chosen as follows: $k_{1\rho} = k_{2\rho} = 3.0, k_{1\beta} = k_{2_\beta} = 5.0, \alpha_1 = \alpha_2 = 0.8$; the nominal control law gains are chosen as: $c_{\rho i} = c_{\beta i} = 0.6$; the parameters are set as $K = 25$, $T_s = 0.1$; the attacks and compensations constrains are chosen as: $[\bar{u}_a, \bar{r}_a] = [\bar{u}_c, \bar{r}_c] = [0.3, 0.1]$; the following weighting matrices are adopted: $\mathbf{Q}_a = \mathbf{Q}_c = \text{diag}\{1.0, 1.0, 0.1\}$, $\mathbf{Q}_a = \mathbf{Q}_c = \text{diag}\{0.5, 0.05\}$.

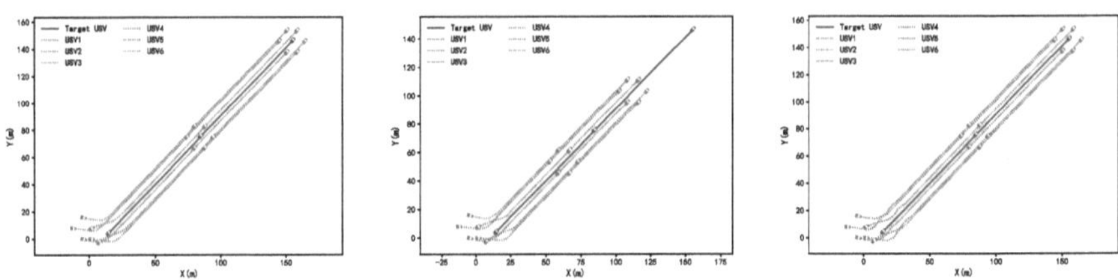

(a) Tracking Performance without FDI Attacks. (b) Tracking Performance under FDI Attacks. (c) Tracking Performance with Compensations.

Fig. 1. Tracking Performances by using different control law.

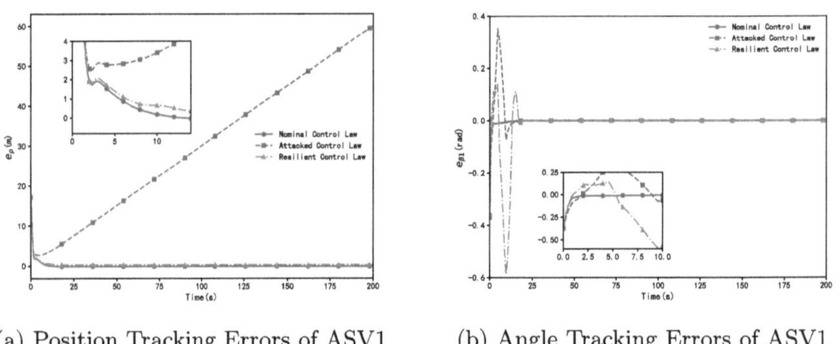

(a) Position Tracking Errors of ASV1 (b) Angle Tracking Errors of ASV1

Fig. 2. Tracking errors of ASV1 by using different control laws.

Figures 1, 2,3 and 4 present the simulation results. Figure 1a gives the tracking trajectory with nominal cooperative target tracking control law. Figure 1b shows the tracking performance under the FDI attack. Figure 1c shows the tracking performance with the FDI attack can be recovered by using the resilient control law, six tracking ASVs catch up with the target ASV, and surround it, which demonstrates the effectiveness of the resilient compensation control law. Figure 2 shows the position and angle tracking errors of ASV1. Figure 3 reveals the nominal control inputs u_{n1}, r_{n1}, the control inputs under the attacks u_{d1}, r_{d1}, and the control inputs under the compensations u_{s1}, r_{s1} of ASV1. Figure 4a shows the

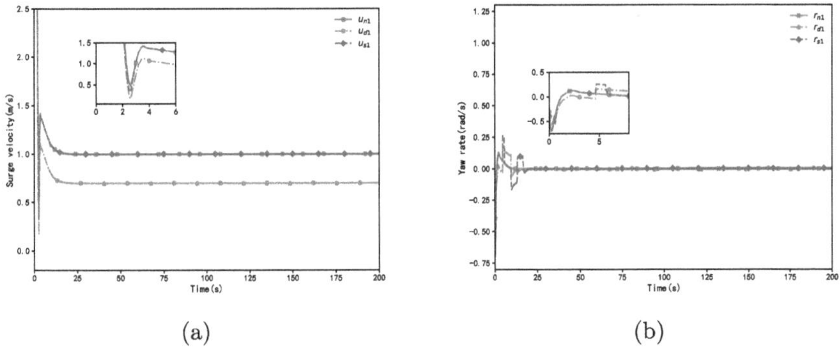

(a) (b)

Fig. 3. Control Inputs of ASV1.

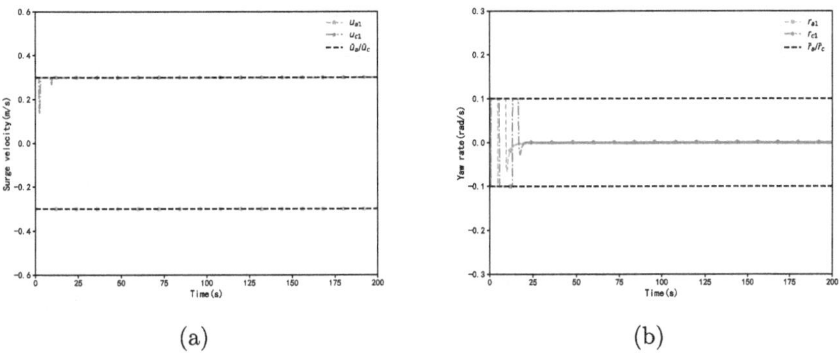

(a) (b)

Fig. 4. Attacks and Optimal Compensations of ASV1.

optimal false data injection attacks u_{a1}, r_{a1} of ASV1 and Fig. 4b shows the optimal compensations u_{c1}, r_{c1} of ASV1, it is obvious that the MPC-based resilient control generates an optimal compensations that counters the attacks.

6 Conclusion

This paper focuses on the cooperative target tracking of multiple ASVs under FDI attacks. The MPC-based resilient target tracking control law is proposed for recovering the tracking performance of tracking ASVs under the FDI attacks. The proposed method enables the multi-ASVs to cooperatively track the target and preserve the expected formation under FDI attacks. Finally, the cascade stability analysis demonstrates that the closed-loop cooperative tracking system is ISS. The simulation results validate the effectiveness of the proposed approach in mitigating FDI attacks and restoring the tracking performance of follower ASVs.

Acknowledgments. This work was supported in part by the National Science and Technology Major Project under Grant 2022ZD0119902, in part by the National Natural Science Foundation of China under Grants 52501374 and 52471372, in part by the Doctoral Scientific Research Foundation of Liaoning Province under Grant 2024-BS-012, in part by the Liaoning Revitalization Leading Talents Program under Grant XLYC2402054, in part by the Key Basic Research of Dalian under Grant 2023JJ11CG008, and in part by the Open Project of State Key Laboratory of Maritime Technology and Safety under Grant SKLMTS-DMU-2024-05.

References

1. Ansari-Bonab, P., Holland, J.C., Cunningham-Rush, J., Noei, S., Sargolzaei, A.: Secure control design for cooperative adaptive cruise control under false data injection attack. IEEE Trans. Intell. Transp. Syst. **25**(8), 9723–9732 (2024)
2. Chen, G., Wu, T., Li, X., Zhang, Y.: Secure and safe control of connected and automated vehicles against false data injection attacks. IEEE Trans. Intell. Transp. Syst. **25**(9), 12347–12360 (2024)
3. Ding, T.F., Ge, M.F., Liu, Z.W., Wang, L., Liu, J.: Reinforcement learning formation tracking of networked autonomous surface vehicles with bounded inputs via cloud-supported communication. IEEE Trans. Intell. Veh. **9**, 469–480 (2024)
4. Fossen, T.I.: Handbook of Marine Craft Hydrodynamics and Motion Control. John (2011)
5. Gao, S., Peng, Z., Wang, H., Liu, L., Wang, D.: Long duration coverage control of multiple robotic surface vehicles under battery energy constraints. IEEE/CAA J. Automatica Sin. **11**, 1695–1698 (2024)
6. He, S., Wang, M., Dai, S.L., Luo, F.: Leader-follower formation control of USVs with prescribed performance and collision avoidance. IEEE Trans. Industr. Inf. **15**(1), 572–581 (2018)
7. Krstic, M., Kokotovic, P.V., Kanellakopoulos, I.: Nonlinear and Adaptive Control Design. John, Inc (1995)
8. Liu, Z.Q., Ge, X., Han, Q.L., Wang, Y.L., Zhang, X.M.: Secure cooperative path following of autonomous surface vehicles under cyber and physical attacks. IEEE Trans. Intell. Veh. **8**(6), 3680–3691 (2023)
9. Lu, A.Y., Yang, G.H.: False data injection attacks against state estimation without knowledge of estimators. IEEE Trans. Autom. Control **67**(9), 4529–4540 (2022)
10. Ma, L., Wang, Y.L., Han, Q.L.: Cooperative target tracking of multiple autonomous surface vehicles under switching interaction topologies. IEEE/CAA J. Automatica Sin. **10**(3), 673–684 (2022)
11. Pang, Z.H., Fan, L.Z., Dong, Z., Han, Q.L., Liu, G.P.: False data injection attacks against partial sensor measurements of networked control systems. IEEE Trans. Circ. Syst. II Express Briefs **69**(1), 149–153 (2021)
12. Pang, Z.H., Liu, G.P., Zhou, D., Hou, F., Sun, D.: Two-channel false data injection attacks against output tracking control of networked systems. IEEE Trans. Industr. Electron. **63**(5), 3242–3251 (2016)
13. Pasqualetti, F., Dörfler, F., Bullo, F.: Attack detection and identification in cyber-physical systems. IEEE Trans. Autom. Control **58**(11), 2715–2729 (2013)
14. Peng, Z., Wang, J., Wang, D., Han, Q.L.: An overview of recent advances in coordinated control of multiple autonomous surface vehicles. IEEE Trans. Industr. Inf. **17**(2), 732–745 (2020)

15. Yu, Y., Peng, S., Dong, X., Li, Q., Ren, Z.: UIF-based cooperative tracking method for multi-agent systems with sensor faults. Sci. China Inf. Sci. **62**(1), 10202 (2019)
16. Zhao, R., Zuo, Z., Shi, Y., Wang, Y., Zhang, W.: Dos and stealthy deception attacks for switched systems: a cooperative approach. IEEE Trans. Autom. Control **69**(7), 4396–4410 (2023)

Synchronization Control of Time-Delay Inertial Memristive Neural Networks with Unknown Parameters

Ning Wu and Jiemei Zhao(✉)

Wuhan Polytechnic University, Wuhan, China
jiemeizhao@163.com

Abstract. This study investigates the synchronization control problem of drive-response systems in inertial memristive neural networks with unknown parameters and time delays. Initially, nonsmooth analysis theory is employed to analyze and process both the drive and response systems, and a drive-response error system is established using a non-order-reduction approach. Subsequently, a state feedback controller and corresponding weight update rules for the unknown parameters are designed. By applying Lyapunov stability theory, the error system is analyzed, and synchronization control criteria for inertial memristive neural networks with unknown parameters and time delays are derived.

Keywords: Memristive neural networks · Unknown parameters · Time-delays · Synchronization

1 Introduction

The memristor, first proposed by Professor Chua in 1971 [1], is recognized as the fourth basic circuit element alongside the resistor, capacitor, and inductor. Due to its capability to emulate brain synapses, research on memristive neural networks (MNNs) continues to be a prominent area of study [2,3]. MNNs have been developed and successfully applied across various fields, including signal processing, image processing, associative memory, pattern recognition, robotics, control, and other engineering and scientific fields.

In recent years, MNNs have attracted considerable attention from researchers across various disciplines. Currently, MNNs are classified into two categories based on their differential order: first-order MNNs and second-order MNNs that include inertial terms [4]. Compared to first-order MNNs, those with inertial terms offer significant advantages, such as increased storage capacity and improved fault tolerance. As a result, the dynamic characteristics of IMNNs have become a prominent focus of scholarly research [5,6]. There are two common approaches to studying IMNNs [7]. The first involves reducing the second-order MNNs system to a first-order system for analysis [2,8]. However, this method increases the system's dimensionality, leading to higher computational complexity. The second approach uses non-order-reduction method, which retains the

C. Li et al. (Eds.): ICNC 2025, CCIS 2946, pp. 388–398, 2026.
https://doi.org/10.1007/978-981-92-1599-7_33

system's higher-order dynamic information and avoid the computational burden caused by dimensional expansion. Nonetheless, these methods pose greater research challenges.

Currently, numerous researchers have independently initiated investigations into MNNs. This paper investigates the fixed-time synchronization problem of time-delay MNNs with mixed pulse effects, as discussed in [9–11]. The synchronization problem of multiple reaction-diffusion MNNs with known/unknown parameters and switching topologies using a distributed control approach are discussed in [12,13]. The network model incorporates both unbounded discrete time-varying delays and bounded distributed time-varying delays, as described in [14]. The exponential synchronization of IMNNs was addressed in [15]. However, most existing research on synchronization control of MNNs are assumes that the parameters are known and fixed. In practical applications, obtaining precise model parameters is often challenging. Therefore, this study focuses on the synchronization control problem of time-delayed IMNNs under conditions of parameter uncertainty.

This article proposed the problem of synchronization control of time-delay IMNNs with unknown parameters. This paper is summarized in three key aspects

1) The synchronization for IMNNs with time delays and parameter uncertainties is investigated.
2) A novel weight update rule and synchronization controller are developed to ensure that the drive-response system achieves synchronization despite parameter uncertainties.
3) A non-order-reduction method is used to analyze IMNNs, thereby avoiding the increase in model dimensionality that typically accompanies order-reduction techniques.

2 Problem Formulation and Preliminaries

Consider time-delay IMNNs with unknown parameters

$$\begin{aligned}\ddot{s}_i(t) = &- c_i s_i(t) - d_i \dot{s}_i(t) + \sum_{j=1}^{n} a_{ij}(s_i(t)) f_j(s_j(t)) \\ &+ \sum_{j=1}^{n} b_{ij}(s_i(t)) g_j(s_j(t - h_j(t))) + I_i, \end{aligned} \tag{1}$$

where $s_i(t)$ represents the state of the i neuron in the network at time t. c_i is the capacitance-related coefficient for the i neuron, d_i is the damping coefficient of the i neuron. $a_{ij}(s_i(t))$ is the connection weight from the i neuron to the j neuron, $b_{ij}(s_i(t))$ is the dynamic connection weight representing the delay from the i neuron to the j neuron and $a_{ij}(s_i(t)), b_{ij}(s_i(t))$ is unknown. $f_j(\cdot)$ and $g_j(\cdot)$ denotes activation functions for neurons with and without time delay. $h_j(t)$ denotes time-varying time delay, in which $0 \leq h_j(t) \leq h$ and $\dot{h}_j(t) \leq \theta \leq 1$, ($h, \theta$ is a constant), I_i is an external disturbance.

Based on a simplified model of the memristor, the weight of the memristor is considered as follows:

$$a_{ij}(s_i(t)) = \begin{cases} a_{ij}^*, & |s_j(t)| < \xi_i, \\ a_{ij}^* + \Delta a_{ij}, & |s_j(t)| \geq \xi_i, \end{cases}$$

$$b_{ij}(s_i(t)) = \begin{cases} b_{ij}^*, & |s_j(t)| < \xi_i, \\ b_{ij}^* + \Delta b_{ij}, & |s_j(t)| \geq \xi_i. \end{cases}$$

Let $A_{ij} =\mid a_{ij}^* \mid + \mid \Delta a_{ij} \mid$, $B_{ij} =\mid b_{ij}^* \mid + \mid \Delta b_{ij} \mid$, where $a_{ij}^*, b_{ij}^*, \Delta a_{ij}, \Delta b_{ij}$ all are constants, switch jump ξ_i is positive number.

Assumption 1. The neural activation functions $f_j(\cdot)$ and $g_j(\cdot)$ are globally Lipschitz continuous, this implies that the following condition holds for any s and v such that

$$|f_j(s) - f_j(v)| \leq F_j|s - v|,$$
$$|g_j(s) - g_j(v)| \leq G_j|s - v|,$$

where F_j and G_j is a positive number.

Assumption 2. The neural activation functions $f_j(\cdot)$ and $g_j(\cdot)$ are bounded and satisfy the following conditions:

$$f_j(\cdot) \leq M_1, g_j(\cdot) \leq M_2,$$

where constants $M_1, M_2 > 0, j = 1, 2, \ldots, n$.

Assumption 3. The neuron activation function satisfies $f_j(\pm\xi_j) = 0$ and $g_j(\pm\xi_j) = 0$.

Since IMNNs are right-discontinuous differential equations, solutions are considered in the Filippov sense. Using differential inclusion theory and set-valued mapping theory, model (1) can be represented as follows

$$\begin{aligned} \ddot{s}_i(t) \in & - c_i s_i(t) - d_i \dot{s}_i(t) + \sum_{j=1}^{n} co[a_{ij}(s_i(t))] f_j(s_j(t)) \\ & + \sum_{j=1}^{n} co[b_{ij}(s_i(t))] g_j(s_j(t - h_j(t))) + I_i, \end{aligned} \tag{2}$$

where

$$\mathrm{co}\big[a_{ij}(s_i(t))\big] = \begin{cases} a_{ij}^*, & |s_j(t)| < \xi_j, \\ a_{ij}^* + \theta_{ij}\Delta a_{ij}, & |s_j(t)| = \xi_j, \\ a_{ij}^* + \Delta a_{ij}, & |s_j(t)| > \xi_j. \end{cases}$$

$$\mathrm{co}\big[b_{ij}(s_i(t))\big] = \begin{cases} b_{ij}^*, & |s_j(t)| < \xi_j, \\ b_{ij}^* + \bar{\theta}_{ij}\Delta b_{ij}, & |s_j(t)| = \xi_j, \\ b_{ij}^* + \Delta b_{ij}, & |s_j(t)| > \xi_j, \end{cases}$$

with $\theta_{ij}, \bar{\theta}_{ij} \in [0, 1]$.

Remark 1. In fact, when the activation function satisfies conditions $f_j(\pm\xi_j) = 0$ and $g_j(\pm\xi_j) = 0$, solutions in the Filippov sense are unnecessary. Suppose that for each $j = 1, 2, \ldots, n$, state variable $s_j(t)$ becomes 0 after a certain point in time and remains 0. It readily follows that the differential equation (1) is continuous at any given moment. Therefore, the conventional methods for solving differential equations can be applied.

Considering drive-response synchronization, with model (1) as the drive system, the response system is described as follows:

$$\begin{aligned} \ddot{w}_i(t) = &- \hat{c}_i(t)w_i(t) - \hat{d}_i(t)\dot{w}_i(t) + \sum_{j=1}^{n} \hat{a}_{ij}(w_i(t))f_j(w_j(t)) \\ &+ \sum_{j=1}^{n} \hat{b}_{ij}(w_i(t))g_j(w_j(t-h_j(t))) + I_i + u_i(t), \end{aligned} \tag{3}$$

where $w_i(t)$ represents the state of the i neuron in the network at time t. $f_j(\cdot)$ and $g_j(\cdot)$ denote the activation functions for neurons without and with delay. Here $\hat{c}_i(t)$, $\hat{d}_i(t)$, $\hat{a}_{ij}(w_i(t))$, $\hat{b}_{ij}(w_i(t))$ are unknown parameters, and $u_i(t)$ is the control input used for drive-response synchronization.

Define $e_i(t) = w_i(t) - s_i(t)$ as the error of the drive-response system. If Assumption 3 holds, combining the drive system (1) with the response system (3) yields the synchronization error system

$$\begin{aligned} \ddot{e}_i(t) = &\sum_{i=1}^{n}\big(-c_i(t)e_i(t) - (\hat{c}_i(t) - c_i)w_i(t) - (\hat{d}_i(t) - d_i)\dot{w}_i(t) \\ &- d_i\dot{e}_i(t)\big) + \sum_{j=1}^{n}\big(\hat{a}_{ij}(w_i(t)) - a_{ij}(s_i(t))\big)f_j(w_j(t)) \\ &+ \sum_{j=1}^{n} a_{ij}(s_i(t))\bar{f}_j(e_j(t)) \\ &+ \sum_{j=1}^{n}\big(\hat{b}_{ij}(w_i(t)) - b_{ij}(s_i(t))\big)g_j(w_j(t-h_j(t))) \\ &+ \sum_{j=1}^{n} b_{ij}(s_i(t))\bar{g}_j(e_j(t-h_j(t))) + u_i(t), \end{aligned} \tag{4}$$

where

$$\begin{aligned} \bar{f}_j(e_j(t)) &= f_j(w_i(t)) - f_j(s_i(t)), \\ \bar{g}_j(e_j(t-h(t))) &= g_j(w_j(t-h_j(t))) - g_j(s_j(t-h_j(t))). \end{aligned}$$

Next, a controller is designed to synchronize time-delay IMNNs with unknown parameters. The controller is defined as follows

$$u_i(t) = -\alpha_i e_i(t) - \beta_i \dot{e}_i(t) - \delta_i sign(\dot{e}_i(t)), \tag{5}$$

where $\alpha_i > 0, \beta_i > 0, \delta_i > 0$ are the gain of the controller.

Theorem 1. *If Assumptions 1–2 hold, there exist constants $\alpha_i, \beta_i, \gamma_i, F_i, G_i, M_1$ and M_2 such that the model parameter update rule satisfies*

$$\begin{aligned}\dot{\hat{c}}_i(t) &= w_i(t)\dot{e}_i(t),\\ \dot{\hat{a}}_{ij}(t) &= -f_j(w_j(t))\dot{e}_i(t),\\ \dot{\hat{b}}_{ij}(t) &= -g_j(w_j(t-h_j(t)))\dot{e}_i(t),\\ \dot{\hat{d}}_i(t) &= \dot{w}_i(t)\dot{e}_i(t),\end{aligned}$$

then the drive system (1) *and the response system* (3), *both containing uncertain parameters, can achieve synchronization under the controller* (6).

Proof. Constructing the Lyapunov functional

$$v(t) = v_1(t) + v_2(t) + v_3(t), \tag{6}$$

where

$$\begin{aligned}v_1(t) =& \frac{1}{2}\sum_{i=1}^{n} e_i^2(t) + \frac{1}{2}\sum_{i=1}^{n} \dot{e}_i^2(t),\\ v_2(t) =& \frac{1}{2(1-\theta)}\sum_{i=1}^{n}\sum_{j=1}^{n}\int_{t-h_j(t)}^{t} B_{ij}G_j e_j^2(s)\,ds,\\ v_3(t) =& \frac{1}{2}\sum_{i=1}^{n}(\hat{c}_i(t)-c_i)^2 + \frac{1}{2}\sum_{i=1}^{n}(\hat{d}_i(t)-d_i)^2\\ &+ \frac{1}{2}\sum_{i=1}^{n}\left(\hat{a}_{ij}^2(t) + \hat{b}_{ij}^2(t)\right).\end{aligned}$$

Taking the derivative (8) along the trajectory of the error system (4) yields

$$\begin{aligned}\dot{v}_1(t) =& \sum_{i=1}^{n} e_i(t)\dot{e}_i(t) + \sum_{i=1}^{n}\big(-c_i e_i(t)\dot{e}_i(t) - (\hat{c}_i(t)-c_i)\big)w_i(t)\dot{e}_i(t)\\ &- \big(\hat{d}_i(t)-d_i\big)\dot{w}_i(t)\dot{e}_i(t) - d_i(t)\dot{e}_i^2(t)\\ &+ \sum_{i=1}^{n}\sum_{j=1}^{n}\big[\hat{a}_{ij}(w_i(t)) - a_{ij}(s_i(t))\big]f_j(w_j(t))\dot{e}_i(t)\\ &+ \sum_{i=1}^{n}\sum_{j=1}^{n} a_{ij}(s_i(t))\dot{e}_i(t)\bar{f}_j(e_j(t))\\ &+ \sum_{i=1}^{n}\sum_{j=1}^{n}\big[\hat{b}_{ij}(w_i(t)) - b_{ij}(s_i(t))\big]g_j(w_j(t-h_j(t)))\dot{e}_i(t)\\ &+ \sum_{i=1}^{n}\sum_{j=1}^{n} b_{ij}(s_i(t))\dot{e}_i(t)\bar{g}_j(e_j(t-h_j(t))) + \sum_{i=1}^{n} u_i(t)\dot{e}_i(t).\end{aligned} \tag{7}$$

$$\dot{v}_2(t) = \sum_{i=1}^{n}\sum_{j=1}^{n}\frac{B_{ij}G_j}{2(1-\theta)}\left[e_j^2(t) - e_j^2(t-h_j(t))(1-\dot{h}_j(t))\right]$$
$$\leq \sum_{i=1}^{n}\sum_{j=1}^{n}\frac{B_{ij}G_j}{2(1-\theta)}e_j^2(t) - \frac{B_{ij}G_j}{2}e_j^2(t-h_j(t)). \quad (8)$$

$$\dot{v}_3(t) = \sum_{i=1}^{n}(\hat{c}_i(t) - c_i)w_i(t)\dot{e}_i(t) + \sum_{i=1}^{n}(\hat{d}_i(t) - d_i)\dot{w}_i(t)\dot{e}_i(t)$$
$$- \sum_{i=1}^{n}\hat{a}_{ij}(t)f_j(w_j(t))\dot{e}_i(t)$$
$$- \sum_{i=1}^{n}\hat{b}_{ij}(t)g_j(w_j(t-h_j(t)))\dot{e}_i(t). \quad (9)$$

According to (9)–(11), we obtain

$$\dot{v}(t) \leq \sum_{i=1}^{n}(e_i(t)\dot{e}_i(t) - c_ie_i(t)\dot{e}_i(t) - d_i\dot{e}_i^2(t))$$
$$+ \sum_{i=1}^{n}\sum_{j=1}^{n}A_{ij}f_j(w_j(t))\dot{e}_i(t) + \frac{1}{2}\sum_{i=1}^{n}\sum_{j=1}^{n}F_jA_{ij}(s_i(t))\dot{e}_i^2(t)$$
$$+ \frac{1}{2}\sum_{i=1}^{n}\sum_{j=1}^{n}F_jA_{ij}(s_i(t))e_j^2(t) + \sum_{i=1}^{n}\sum_{j=1}^{n}\frac{G_jB_{ij}}{2}(s_i(t))e_j^2(t-h_j(t))$$
$$+ \sum_{i=1}^{n}\sum_{j=1}^{n}\frac{G_jB_{ij}}{2}(s_i(t))\dot{e}_i^2(t) - \delta_ie_i(t)\big) + \sum_{i=1}^{n}\sum_{j=1}^{n}B_{ij}g_j(w_j(t-h_j(t)))\dot{e}_i(t)$$
$$+ \sum_{i=1}^{n}\sum_{j=1}^{n}\frac{B_{ij}G_j}{2(1-\theta)}e_j^2(t) - \sum_{i=1}^{n}\sum_{j=1}^{n}\frac{B_{ij}G_j}{2}e_j^2(t-h_j(t))$$
$$- \sum_{i=1}^{n}\big(\alpha_ie_i(t)\dot{e}_i(t) - \beta_i\dot{e}_i^2(t)$$
$$\leq \sum_{i=1}^{n}(1-c_i-\alpha_i)e_i(t)\dot{e}_i(t) + \sum_{i=1}^{n}\sum_{j=1}^{n}A_{ij} \mid f_j(w_j(t)) \mid\mid \dot{e}_i(t) \mid$$
$$\leq \sum_{i=1}^{n}(1-c_i-\alpha_i)e_i(t)\dot{e}_i(t) + \sum_{i=1}^{n}\sum_{j=1}^{n}A_{ij} \mid f_j(w_j(t)) \mid\mid \dot{e}_i(t) \mid$$
$$- \sum_{i=1}^{n}d_i(t)\dot{e}_i^2(t) + \frac{1}{2}\sum_{i=1}^{n}\sum_{j=1}^{n}F_j \mid A_{ij}(s_i(t)) \mid \dot{e}_i^2(t)$$
$$+ \frac{1}{2}\sum_{i=1}^{n}\sum_{j=1}^{n}F_j \mid A_{ij}(s_i(t)) \mid e_j^2(t) \quad (10)$$

$$
\begin{aligned}
&+\sum_{i=1}^{n}\sum_{j=1}^{n}B_{ij}\mid g_j(w_j(t-h_j(t)))\mid\mid \dot{e}_i(t)\mid \\
&+\frac{1}{2}\sum_{i=1}^{n}\sum_{j=1}^{n}G_j\mid B_{ij}(s_i(t))\mid e_i^2(t-h_j(t)) \\
&+\frac{1}{2}\sum_{i=1}^{n}\sum_{j=1}^{n}G_j\mid B_{ij}(s_i(t))\mid \dot{e}_i^2(t)+\sum_{i=1}^{n}\sum_{j=1}^{n}\frac{B_{ij}G_j}{2(1-\theta)}e_j^2(t) \\
&-\sum_{i=1}^{n}\sum_{j=1}^{n}\frac{B_{ij}G_j}{2}e_j^2(t-h_j(t))-\beta_i\dot{e}_i^2(t)-\delta_i\mid \dot{e}_i(t)\mid .
\end{aligned}
$$

According to the mean-value inequality theory, we obtain

$$
\begin{aligned}
\mid \dot{e}_i(t)e_j(t)\mid &\leq \frac{1}{2}(\dot{e}_i^2(t)+e_j^2(t)), \\
\mid \dot{e}_i(t)e_j(t-h_j(t))\mid &\leq \frac{1}{2}(\dot{e}_i^2(t)+e_j^2(t-h_j(t))).
\end{aligned}
$$

Based on Assumption 1 and 2, (12) can be written as

$$
\begin{aligned}
\dot{v}(t)\leq &\sum_{i=1}^{n}(1-c_i-\alpha_i)e_i^2(t)+\frac{1}{2}\sum_{i=1}^{n}(1-c_i-\alpha_i)\dot{e}_i^2(t) \\
&-\sum_{i=1}^{n}d_i\dot{e}_i^2(t)+\sum_{i=1}^{n}\sum_{j=1}^{n}M_1A_{ij}|\dot{e}_i(t)|-\sum_{i=1}^{n}\delta_i|\dot{e}_i(t)| \\
&+\frac{1}{2}\sum_{i=1}^{n}\sum_{j=1}^{n}F_jA_{ij}\dot{e}_i^2(t)+\frac{1}{2}\sum_{i=1}^{n}\sum_{j=1}^{n}F_jA_{ij}e_j^2(t) \\
&+\sum_{i=1}^{n}\sum_{j=1}^{n}M_2B_{ij}|\dot{e}_i(t)|+\frac{1}{2}\sum_{i=1}^{n}\sum_{j=1}^{n}G_jB_{ij}e_i^2(t-h_j(t)) \\
&+\frac{1}{2}\sum_{i=1}^{n}\sum_{j=1}^{n}G_jB_{ij}\dot{e}_i^2(t)+\sum_{i=1}^{n}\sum_{j=1}^{n}\frac{B_{ij}G_j}{2(1-\theta)}e_j^2(t) \\
&-\sum_{i=1}^{n}\sum_{j=1}^{n}\frac{B_{ij}G_j}{2}e_j^2(t-h_j(t))-\sum_{i=1}^{n}\beta_i\dot{e}_i^2(t) \\
\leq &\sum_{i=1}^{n}\sum_{j=1}^{n}(\frac{1}{2}A_{ij}F_j+\frac{1}{2}B_{ij}G_j-d_i-\beta_i+\frac{1}{2}(1-c_i-\alpha_i))\dot{e}_i^2(t) \\
&+\sum_{i=1}^{n}\sum_{j=1}^{n}(\frac{B_{ij}G_j}{2(1-\theta)}+\frac{1}{2}A_{ij}F_j+\frac{1}{2}(1-c_i-\alpha_i)))e_i^2(t) \\
&+\sum_{i=1}^{n}\sum_{j=1}^{n}(M_1A_{ij}+M_2B_{ij}-\delta_i)\mid \dot{e}_i(t)\mid .
\end{aligned}
$$

Let

$$\alpha_i = 1 - c_i + A_{ij}F_j + \frac{B_{ji}G_i}{(1-\theta)},$$
$$\beta_i = 1 - d_i - \frac{B_{ji}G_i}{2(1-\theta)} + \frac{1}{2}B_{ij}G_j,$$
$$\delta_i = M_1 A_{ij} + M_2 B_{ij}.$$

Then

$$\dot{v}(t) \leq \sum_{i=1}^{n} -\dot{e}_i^2(t) \leq 0.$$

Therefore, the drive-response systems (1) and (3) can achieve synchronization under with unknown parameters.

3 Numerical Example

Now, we provide a numerical example to demonstrate the effectiveness of the proposed criteria.

Example 1. consider the IMNNs with unknown parameters

$$\begin{aligned}\ddot{s}_i(t) =& - c_i s_i(t) - d_i \dot{s}_i(t) + \sum_{j=1}^{n} a_{ij}(s_i(t)) f_j(s_j(t)) \\ &+ \sum_{j=1}^{n} b_{ij}(s_i(t)) g_j(s_j(t - h_j(t))) + I_i, i = 1, 2 \end{aligned} \tag{11}$$

where the time-varying delay $h_j(t) = \frac{e^t}{1+e^t}$ and satisfies $0 < h_j(t) < \frac{1}{2}$, $\dot{h}_j(t) \leq \frac{1}{4}$, and the initial condition of (11) is $s_1(\kappa) = 1.3, s_2(\kappa) = -1.0, \kappa \in [-0.5, 0)$, and $c_1 = 1.2, c_2 = 1.6$. The other parameters are give follows:

$$a_{11}(s_1(t)) = \begin{cases} -2.2, & |s_1(t)| < 0.8, \\ -1.8, & |s_1(t)| \geq 0.8, \end{cases}$$

$$a_{12}(s_2(t)) = \begin{cases} 1.4, & |s_2(t)| < 0.8, \\ 2.0, & |s_2(t)| \geq 0.8, \end{cases}$$

$$a_{21}(s_1(t)) = \begin{cases} 2.1, & |s_1(t)| < 0.8, \\ 2.4, & |s_1(t)| \geq 0.8, \end{cases}$$

$$a_{22}(s_2(t)) = \begin{cases} -1.2, & |s_2(t)| < 0.8, \\ -2.2, & |s_2(t)| \geq 0.8, \end{cases}$$

$$b_{11}(s_1(t-\tau(t))) = \begin{cases} -2.8, & |s_1(t-h_j(t))| < 0.8, \\ -3.2, & |s_1(t-\tau(t))| \geq 0.8, \end{cases}$$

$$b_{12}(s_2(t-\tau(t))) = \begin{cases} 1.6, & |s_2(t-h_j(t))| < 0.8, \\ 2.0, & |s_2(t-\tau(t))| \geq 0.8, \end{cases}$$

$$b_{21}(s_1(t-\tau(t))) = \begin{cases} 1.0, & |s_1(t-h_j(t))| < 0.8, \\ 1.6, & |s_1(t-\tau(t))| \geq 0.8, \end{cases}$$

$$b_{22}(s_2(t-\tau(t))) = \begin{cases} -0.2, & |s_2(t-h_j(t))| < 0.8, \\ -0.5, & |s_2(t-\tau(t))| \geq 0.8. \end{cases}$$

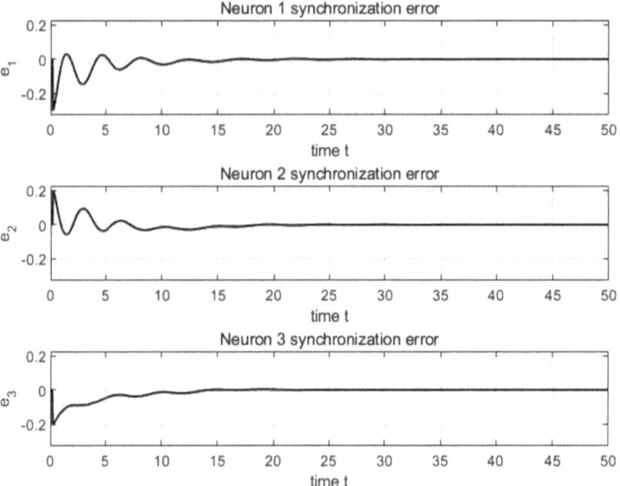

Fig. 1. Error trajectories of e_1, e_2, and e_3.

The corresponding response system

$$\ddot{w}_i(t) = -\hat{c}_i(t)w_i(t) - \hat{d}_i(t)\dot{w}_i(t) + \sum_{j=1}^{n} \hat{a}_{ij}(w_i(t))f_j(w_j(t)) + \sum_{j=1}^{n} \hat{b}_{ij}(w_i(t))g_j(w_j(t-h_j(t))) + I_i + u_i(t), i = 1, 2.$$

The initial condition of response system is given by $w_1(\kappa) = -2.2, w_1(\kappa) = 0.5, \kappa \in [-0.5, 0)$, and the initial values of weights are $a_{ij} = b_{ij} = -1, i, j = 1, 2$. $M_1 = M_2 = 1$. In this numerical example, activation functions are chosen as

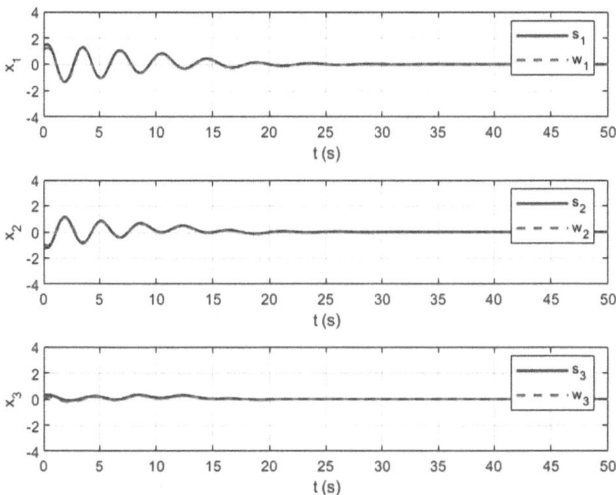

Fig. 2. Trajectories of (s_1, w_1), (s_2, w_2), and (s_3, w_3).

$f_j = tan(\pi s)$ and $g_j = tan(2\pi s)$, satisfying Theorem 1. Figure 1 illustrates different states of error e_1, e_2, e_3 respectively. Figure 2 displays the trajectories of nodes s_1, w_1, s_2, w_2, s_3, w_3.

4 Conclusion

This study investigates the synchronization control problem of drive-response system in IMNNs with unknown parameters and time delays. To tackle these uncertainties, a novel weight update rule and synchronization controller are developed, facilitating synchronization between the drive and response systems despite the presence of uncertain parameters. Furthermore, a non-order-reduction approach is utilized to deal with the IMNNs, effectively circumventing the dimensionality explosion problem commonly associated with decaying methods.

Acknowledgement. This work was supported by the Natural Science Foundation of Wuhan (Chenguang Project) under Grant 2024040801020332.

References

1. Chua, L.: Memristor-the missing circuit element. IEEE Trans. Circ. Theory **18**(5), 507–519 (1971)
2. Shen, Y., Zhao, J., Yu, L.: Reachable set estimation of delayed second-order memristive neural networks. Appl. Math. Comput. **484**, 128994 (2025)

3. Zhao, J., Shen, Y., Wang, L., Yu, L.: Reachable set estimation of inertial complex-valued memristive neural networks. Express Briefs. IEEE Trans. Circ. Syst. II (2024)
4. Ji, X., Dong, Z., Zhou, G., Lai, C.S., Yan, Y., Qi, D.: Memristive system based image processing technology: a review and perspective. Electronics **10**(24), 3176 (2021)
5. Luo, D., Wang, C., Deng, Q., Sun, Y.: Dynamics in a memristive neural network with three discrete heterogeneous neurons and its application. Nonlinear Dyn. **113**(6), 5811–5824 (2025)
6. Zhao, J., Wang, F.: Synchronization of reaction-diffusion delayed inertial memristive neural networks via adaptive pinning control. Neural Netw., 108183 (2025)
7. Sheng, Y., Huang, T., Zeng, Z., Li, P.: Exponential stabilization of inertial memristive neural networks with multiple time delays. IEEE Trans. Cybern. **51**(2), 579–588 (2019)
8. Wang, X., Jian, J.: Multi-type synchronization for second-order memristive neural networks with mixed time-varying delays. Neural Process. Lett. **55**(2), 1759–1781 (2023)
9. Wang, D., Li, L.: Fixed-time synchronization of delayed memristive neural networks with impulsive effects via novel fixed-time stability theorem. Neural Netw. **163**, 75–85 (2023)
10. Peng, L., Li, X., Bi, D., Xie, X., Xie, Y.: Multiple μ-stable synchronization control for coupled memristive neural networks with unbounded time delays. IEEE Trans. Syst. Man Cybern. Syst. **52**(2), 990–1002 (2020)
11. Li, R., Cao, J., Li, N.: Stabilization of reaction-diffusion fractional-order memristive neural networks. Neural Netw. **165**, 290–297 (2023)
12. Liu, N., Cheng, J., Chen, Y., Yan, H., Zhang, D., Qi, W.: Adaptive protocol-based control for reaction-diffusion memristive neural networks with semi-Markov switching parameters. Inf. Sci. **677**, 120947 (2024)
13. Cao, Y., Liu, N., Zhang, C., Zhang, T., Luo, Z.: Synchronization of multiple reaction-diffusion memristive neural networks with known or unknown parameters and switching topologies. Knowl.-Based Syst. **254**, 109595 (2022)
14. Cui, Y., Cheng, P.: Exponential synchronization of stochastic time-delayed memristor-based neural networks via pinning impulsive control. Int. J. Control Autom. Syst. **22**(7), 2283–2292 (2024)
15. Cao, Y., Cao, Y., Guo, Z., Huang, T., Wen, S.: Global exponential synchronization of delayed memristive neural networks with reaction-diffusion terms. Neural Netw. **123**, 70–81 (2020)

SNN-Based Lightweight Denoising Method for Event Cameras

Hongzhi Huang[1,2,3], Xiaoping Wang[1,2,3](✉), Weibin Feng[1,2,3], Xin Zhan[1,2,3], and Jiang Li[1,2,3]

[1] School of Artificial Intelligence and Automation, Huazhong University of Science and Technology, Wuhan 430074, China
wangxiaoping@hust.edu.cn

[2] Key Laboratory of Image Processing and Intelligent Control of Education Ministry of China, Huazhong University of Science and Technology, Wuhan 430074, China

[3] Hubei Key Laboratory of Brain-Inspired Intelligent Systems, Huazhong University of Science and Technology, Wuhan 430074, China

Abstract. Compared to image frames, the event stream captured by event cameras possesses the advantages of being high temporal resolution, sparsity and asynchronism. However, these characteristics also make it highly susceptible to noise which negatively affects the performance of downstream tasks. Existing event denoising methods are either too straightforward for varying noise-ratio scenes, or too complex, resulting in low efficiency and difficult to use in practice.

In search of a denoising method that combines both performance and efficiency, we propose a lightweight and real-time event denoising algorithm based on the Spiking Neural Networks (SNN). Specifically, we introduce the Threshold-Limited PLIF neuron model, which leverages membrane potentials to capture the spatio-temporal correlations essential for effective denoising. With this neuron as the fundamental component, we design a frame-by-frame denoising network, called DeSNN. The proposed architecture utilizes the SNN architecture to integrate the spatio-temporal information from input event frames, and subsequently generates denoised outputs. Our method allows the entire processing pipeline to be maintained in the spiking data format, thereby fully exploiting the strengths of both dynamic vision sensors (DVS) and SNNs. Extensive experimental results demonstrated that our method achieves both high performance and high efficiency, which is beneficial for real-time downstream tasks. Furthermore, we have implemented our method in several simple real-world scenarios, illustrating its practical applicability and potential for deployment.

Keywords: Event Cameras · Event Denoising · Spiking Neural Networks

1 Introduction

Event camera, also known as the Dynamic Vision Sensor (DVS), are an asynchronous sensing device inspired by the mechanism of the human retina [1].

C. Li et al. (Eds.): ICNC 2025, CCIS 2946, pp. 399–414, 2026.
https://doi.org/10.1007/978-981-92-1599-7_34

In contrast to RGB camera which captures image from entire scene in a synchronous frame-based manner, event camera generates data independently and asynchronously from each pixel. This distinctive sensing paradigm offers several advantages, including high resolution, high dynamic range and low power consumption [2,3]. Consequently, event camera is currently employed in a number of fields, such as object and gesture recognition [4–7], optical flow and depth estimation [8,9], image reconstruction and motion deblurring [10,11] and simultaneous localization and mapping (SLAM) [12,13].

Despite significant advancements in the field of event camera, its integration into traditional image field remains limited. One major reason is the substantial noise present in the event stream. The high sensitivity of event camera enables it to capture fast or subtle motion, but also makes it prone to noise. Background disturbances or stroboscopic lighting will generate noise in the event camera, which are more noticeable under low-light conditions. The large amount of noise not only increases energy consumption but also drowns out useful events, which seriously affects downstream tasks [14–16]. Therefore, the event stream pre-denoising is a crucial step to improve the performance.

Existing event denoising methods can be broadly classified into two categories: filter-based methods and deep learning-based methods. Filter-based methods often perform poorly in varying noise-ratio scenarios due to their relatively simple structures. Deep learning-based methods are difficult to deploy in real-world applications because of their complex network structures, though achieving high accuracy [3]. Moreover, most denoising methods are limited to event-by-event assessment, necessitating the encoding of both the event itself and the surrounding information. This results in an non-negligible increase in computation cost while simultaneously hindering parallelism. To be truly beneficial for downstream tasks, it is essential to develop a denoising method with both high performance and efficiency.

In recent years, a novel artificial neural networks—Spiking Neural Networks (SNNs)—has attracted the attention of researchers. Different from traditional neural networks which transmit continuous-valued signals, SNN communicates via discrete spikes between neurons, where each spike signifies the generation of an event. The fundamental principles and characteristics of SNN are closely aligned with those of event camera—the event stream generated from event camera can be interpreted as a spike stream, which is the native data format processed throughout SNN. In terms of performance, the spatio-temporal correlation analysis required for event denoising constitutes a core strength of SNN. In terms of efficiency, SNN also share key advantages with event camera, including low power consumption and sparsity. Therefore, it is a natural and logical approach to utilize SNN for event denoising [17].

In this work, we propose a simple yet effective and real-time event denoising algorithm. We design Threshold-Limited PLIF (ThPLIF) to translates the basis of denoising—the spatio-temporal correlation of events—into membrane potentials. Leveraging ThPLIF as the fundamental component, we construct a real-time event denoising network called Denoising SNN (DeSNN). Event stream is input in a frame-by-frame manner while maintaining spike data format throughout the entire processing, thus combining the advantages of both fields. Extensive evaluations on denoising datasets and real-world assessments has demonstrated that our solution concurrently achieves high accuracy, minimal power consumption and low latency, thereby making it a promising solution for real-world applications. The main contributions of this work are given as follows.

- We provide the theoretical rationale for event denoising using frame-based methods and SNN model.
- We design Threshold-limited PLIF (ThPLIF) neuron model for characterizing the spatio-temporal correlation of events.
- Utilizing ThPLIF as the fundamental component, we propose Denoising SNN (DeSNN), a real-time and lightweight event denoising network.

This article is structured as follows. In Sect. 2, we explore the methods and rationale for frame-by-frame event denoising, and then develop neurons and real-time lightweight denoising algorithm based on SNN that are well-suited to event denoising tasks. Section 3 assesses the denoising algorithm from multiple perspectives, including its performance and efficiency. Finally, the conclusions are drawn in Sect. 4.

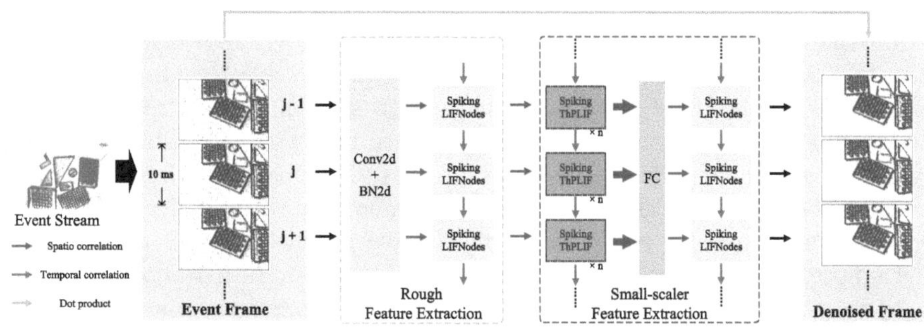

Fig. 1. Framework of our DeSNN. Event stream is encoded into event frames at fixed time intervals (10 ms in this study) and fed into the network, and the denoised event frames are outputted synchronously. Spatial features are extracted by convolution, while temporal features are conveyed by ThPLIF in DeSNN. The denoised frames can be utilized either directly as a downstream task or in a manner analogous to a funnel to filter the event stream. n in small-scaler feature extraction is set to 4.

2 Methodology

2.1 Theoretical Basis of Event Data

In contrast to RGB camera, event camera does not capture the absolute value of light intensity, but rather the changes that occur in it. An event stream containing n events can be represented by a $n \times 4$ matrix e:

$$e[k] = (t_k, p_k, x_k, y_k). \tag{1}$$

Each pixel in the event camera detects the changes independently. At the pixel location (x_k, y_k), when the change of light intensity $I(t_k)$ at time t_k reaches a threshold $\Delta\theta$ compared to the light intensity $I(t_{pre})$ at the time of the last previously generated event, an event is generated and can be encoded as a 1×4 vector, as shown in the following equation:

$$|\log I(t_k) - \log I(\max(t_{pre}))| > \Delta\theta. \tag{2}$$

p_k in Eq. 1 indicates the polarity of the event, which is positive in the case of light intensity increases (i.e. $\Delta I(t_k) > 0$), denoted by $+1$, and negative in the case of the opposite, denoted by -1 [18].

2.2 Frame-Based Denoising

The raw event stream can not convey image information directly because of two problems. First, its microsecond-level temporal resolution results in an extremely large number of time steps when treated as a time series. Second, the asynchronous and sparse nature of the event stream results in a significant lack of event points in most regions. To address these issues, various event representations have been proposed such as time surface [19], voxel grid [20], etc. The most widely adopted is the event frame representation [6,21,22], which integrates events into frames based on a predefined temporal resolution or frame number.

Building upon this strategy, we reformulate the event denoising task from an event-by-event to a frame-by-frame process. Our goal is to design a frame-based denoising model that takes event frame as input and generates denoised event frame as output, effectively remove noise in this frame. Specifically, we first ignore the polarity and discuss only whether the event exists or not within each frame. The possible performance gains from considering polarity is not worth the almost doubled computational cost, while the dataset we used does not contain polarity information. Subsequently, the event stream is compressed into event frames at a fixed time interval of 10 milliseconds (the selection of time interval is detailed in Sect. 3.6). The data with a duration of L ms is then divided into $\lceil L/10 \rceil$ frames. This process can be equated to a reduction in the temporal resolution of the event camera to 10 ms, or sampling per 10 ms. The entire process can be represented as the following equation:

$$E[j, x, y] = \begin{cases} 1, & \text{if} \sum\limits_{t_0 + j\Delta T \le t_i \le t_0 + (j+1)\Delta T} e(p_i, x_i, y_i) \ge 1 \\ 0, & \text{otherwise,} \end{cases} \tag{3}$$

where ΔT equals to 10 ms. The visualization of the process is shown in Fig. 2.

We argue that frame-by frame-denoising strategy is one of a key factor contributing to the success of our work. It not only allow us to leverage successful experiences from the image domain, but also enables the integration of event

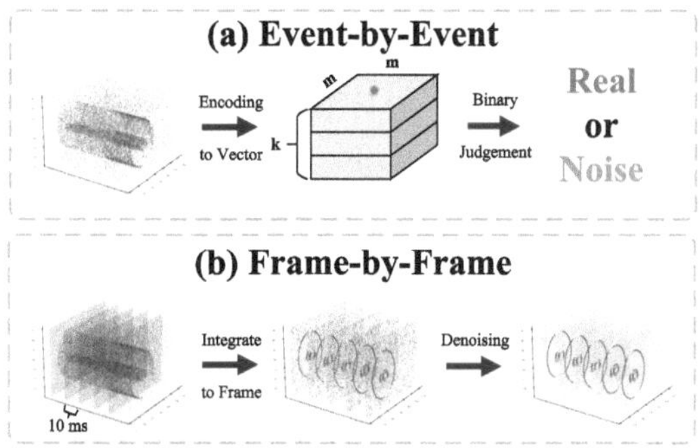

Fig. 2. The two different denoising sequences using deep learning-based method. (a) Each event is encoded as a vector with the k nearest events in $m \times m$ neighborhood, which requires additional computation and storage space. (b) Our denoising method bases on the event frame, which can get the labels simultaneously.

stream with neural networks while preserving the spiking and temporal properties. Some methods [3,23,24] divide the event stream into multiple windows or stacks, only allowing parallel judgments within each window. However, these approaches are still event-by-event denoising and incapable of exchanging information across different windows. In contrast, we propose the first frame-based denoising framework where the information can be conveyed between consecutive frames through neuron and SNN architecture—just like video tasks.

2.3 Threshold-Limited PLIF

For an event, without considering other auxiliary information (e.g., RGB frames and IMU information), the key to judge whether it is noise is the spatio-temporal correlation [25]. Events that appear in isolation and sparsely distribute in space or time are more likely to be background activity or thermal noise. Conversely, a cluster of temporally continuous events in a region are more likely to represent real motion. Similar properties exist in neuroscience. When a neuron is subjected to sustained stimulation, it becomes increasingly sensitive to subsequent stimuli and less difficult to activate. By associating the biological mechanism of "stimulus, activation, action potential" with events, we propose to use spiking neurons to the noise judgment.

We propose a Threshold-Limited Parametric Leaky-Integrate and Fire spiking neuron model (ThPLIF), designed for deep learning to illustrate the trigger of an event in terms of neuronal activation. ThPLIF builds upon the PLIF model [6], which is a version of standard Leaky Integrate-and-Fire (LIF) neuron model with learnable membrane time constant. ThPLIF's iterative version of neuronal dynamics can be shown in the following equation:

$$\begin{cases} H[t] & = V[t-1] - \frac{1}{\tau}(V[t-1] - V_{reset}) + X[t], \\ S[t] & = \Theta(H[t] - V_{th}), \\ V[t] & = H[t](1 - S[t]) + V_{th}S[t]. \end{cases} \tag{4}$$

$H[t]$ represents charging process of the neuron, which incorporates the current input $X[t]$ and the residual potential of previous time step $V[t-1]$ controlled by the membrane time constant τ. $S[t]$ represents the firing process, where $\Theta(x)$ denotes the Heaviside step function. If the membrane potential reaches the threshold V_{th}, the neuron will fire and generate a spike. $V[t]$ represents the resetting process, which is modified for event denoising purposes. In the original PLIF model, the membrane potential will reset to V_{reset} (typically 0) whenever a spike is emitted. In contrast, our ThPLIF will reset to V_{th}, ensuring that the potential carried over to the next time step remains at or below the threshold.

We aim to preserve spatio-temporal information in neurons across time steps, but the resetting step in traditional models would set the membrane potential to zero, leads to the loss of embedded information. To address this issue, we replace it with a threshold-limited resetting strategy. The structure of ThPLIF with a schematic representation of membrane potential changes is shown in Fig. 3. By negating the reset property (and the accompanying refractory period), ThPLIF neuron becomes more sensitivity to temporally correlated events, thereby successfully integrates spatio-temporal correlation.

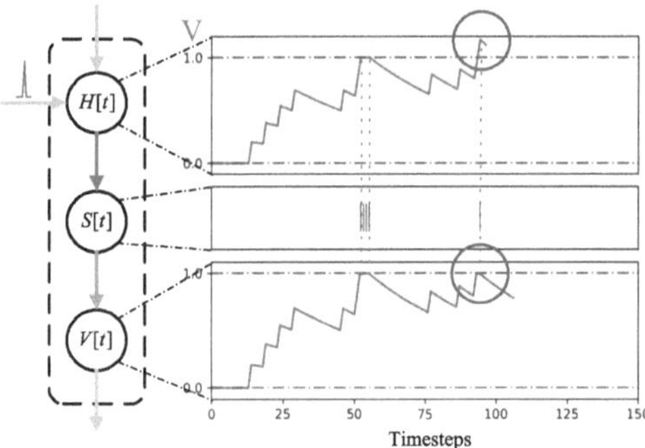

Fig. 3. The structure and schematic diagram of membrane potential changes of ThPLIF. The optimization of the resetting step ensures that the neuron is able to respond positively to dense spikes (during the period of timesteps 50–75).

2.4 DenoisingSNN

Event camera employs a dedicated circuit at each pixel to detect the light intensity changes, while our denoising network adopts a similar strategy by equipping each pixel location with a ThPLIF neuron. The neuron operates in a binary mode—either activated or deactivated—to determine whether the events at that pixel location within a frame should be passed. Neuronal activation is governed by the membrane potential, which evolves based on spatio-temporal correlations. When a significant number of events occur in a spatio-temporal region, the corresponding neurons maintain high membrane potential, thereby enabling them to activate and passing all events. The modification of the resetting step ensured that the membrane potentials correspond to the spatio-temporal correlations without any false breaks. With ThPLIF neurons, we combine the event denoising task as well as biological mechanisms, making our approach intuitive and interpretable.

The complete denoising model is shown in Fig. 1. It mainly consists of two convolutional modules for feature extraction at different scales, along with ThPLIF for representing spatio-temporal correlation in place of activation function. Event frames are sequentially inputted into the network frame-by-frame in time steps. The first component is a large-scale rough feature extraction. This process captures coarse spatial features, which are subsequently combined with small-scale neighborhood features. The integration of these features results in the spatial excitation of the pixel location at this time step, which is subsequently fed into the corresponding ThPLIF neuron. The ThPLIF neuron integrates the excitation with its own membrane potential and activates if the threshold is reached. The output frame will be combined with the original input frame via dot product, preventing false spikes at event-free locations due to surrounding excitations. The output spike can represent a part of output event frame, or indicate that all event at that location are passed as real events within that frame-like a funnel. Spiking data format is maintained throughout the process, keeping temporal precision and computational efficiency. Notably that the denoising process does not involve any down-sampling step, ensures that each pixel corresponds to an independent ThPLIF neuron, preserves the same frame sizes between input and output.

Our method seems relatively straightforward, comprising a limited number of modules and artificially defined features from deep learning perspective. However, this simplicity is not a limitation—it is a deliberate design choice, which can maintain the combined advantages of low power consumption and sparsity of SNN and event camera. The convolution-neuron structure effectively extracts spatio-temporal correlations while preserving the independence and asynchrony of each pixel.

3 Results

3.1 Experiments Setup

Dataset Pre-processing. The neuromorphic datasets processing method in SpikingJelly is utilized, with the events stream being integrated at fixed 10 ms. A frame matrix and a label matrix are obtained after integration, both of which have the shape $[L/10, H, W]$. Frame matrix indicates the location of the events within each frame, while label matrix indicates the location of real events.

Implementation Details. Our model is implemented with the open-source framework SpikingJelly [26]. The kernel size are set to 13 for rough feature extraction and 3 for small-scale feature extraction. The threshold V_{th} and the rest value V_{rest} of all neurons are set to 1 and 0. V_{reset} is equal to V_{rest} for LIF while V_{th} for ThPLIF. The membrane time constant τ of LIF is set to 2, which is learnable in ThPLIF. The network employs sigmoid function with $\alpha = 4.0$ as surrogate of gradient function. Given the binary nature of spiking neuron outputs (0 or 1), we adopt BCEWithLogitsLoss as the training objective.

Evaluation Methods. Most previous studies use the Signal-to-Noise Ratio (SNR) only to evaluate denoising performance, as shown below:

$$\mathrm{SNR} = 20 \times \log_{10} \frac{M}{N}. \tag{5}$$

It indicates the ratio of real events M to noise events N after denoising. This measure is not comprehensive: an overly aggressive filter may also yield a high SNR simply by removing most events, including many real ones [1]. Therefore, we selected precision, recall, and F1-score metrics in deep learning field to better assess both noise suppression and preservation of true events.

3.2 Evaluation in DVSCLEAN

The density of event stream can reach millions of events per second, making the manual labeling impractical and inefficient. Alternative methods must be employed, and DVSCLEAN [2] with a simulated dataset and a real dataset proposed a solution. The simulated part is generated by the ESIM algorithm [27] with artificial noise injected, so that we can get a dataset with reliable labels. The simulated dataset comprises two noise ratios—50% and 100%—with each ratio containing 49 scenes. We divide the specified 39 scenes as the training set, and the remaining 10 scenes are test set [2].

To evaluate our method's performance, we select 4 SOTA event denoising methods. Two filter-based methods, Background Activity Filter (BAF [25]) and Probabilistic Undirected Graph Model (PUGM [23]), and two deep learning-based methods: Event Denoising Convolutional Neural Network (EDnCNN [14]) and Asynchronous Event Denoising Network (AEDNet [2]). Experimental results are presented in Table 1, shows that our method achieved SOTA in both 50% and 100% noise ratio scenes. Noted that our method attains the highest SNR, indicating that filtered events stream contains almost only real events within the blank area and near real events, as visualized in Fig. 4.

In addition to the original dataset, we also tested all methods on dataset with reduced resolution to 10 ms, which our model was trained and validated. We performed the reduced resolution test using DVSCLEAN with 100% noise ratio. The experimental results presented in Table 3 shows that the performance of denoising methods does not vary much when using reduced resolution datasets,

Table 1. Performance on DVSCLEAN.

Noise Ratio	Method	Metrics				
		Acc	Pre	Rec	F1	SNR
50%	BAF	92.7%	97.5%	91.5%	94.2%	32.82
	PUGM	87.1%	96.1%	83.8%	89.2%	28.30
	EDnCNN	98.1%	98.8%	97.1%	97.9%	38.58
	AEDNet	97.2%	95.3%	**98.9%**	97.1%	26.13
	Ours	**98.2%**	**99.0%**	98.3%	**98.6%**	**43.02**
100%	BAF	91.3%	91.4%	91.5%	91.2%	21.26
	PUGM	85.8%	93.9%	76.0%	83.3%	24.16
	EDnCNN	97.2%	97.4%	97.0%	97.2%	31.78
	AEDNet	96.6%	95.0%	**98.3%**	96.6%	25.56
	Ours	**97.7%**	**98.5%**	97.0%	**97.7%**	**41.72**

Table 2. Performance on ED-KoGTL.

Light Condition	Method	Metrics				
		Acc	Pre	Rec	F1	SNR
750 lux	BAF	97.3%	99.0%	98.2%	98.6%	40.20
	PUGM	96.2%	**99.4%**	96.7%	98.0%	**44.83**
	EDnCNN	91.7%	92.4%	98.7%	95.4%	21.66
	AEDNet	96.7%	98.6%	98.0%	98.3%	37.27
	Ours	**98.5%**	98.6%	**99.8%**	**99.2%**	37.24
	-in DC	98.3%	98.7%	99.7%	99.2%	37.47
5 lux	BAF	92.8%	97.8%	94.3%	96.0%	32.88
	PUGM	92.1%	98.0%	93.3%	95.6%	33.88
	EDnCNN	90.2%	94.8%	93.6%	94.2%	25.29
	AEDNet	95.3%	96.8%	98.2%	97.5%	29.49
	Ours	**97.1%**	**98.2%**	**98.6%**	**98.4%**	**34.95**
	-in DC	97.3%	97.9%	99.2%	98.5%	33.38

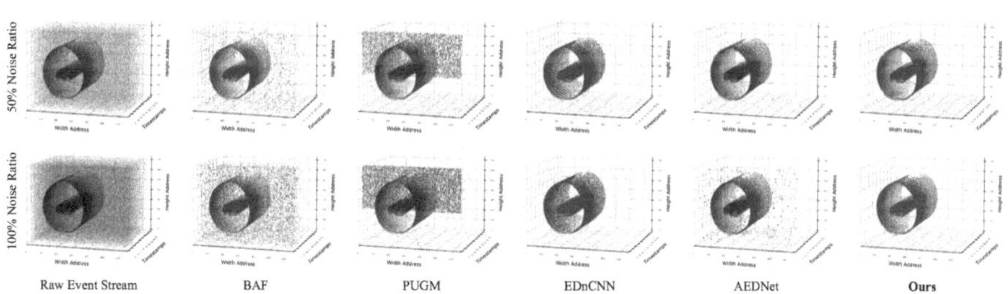

Fig. 4. Visualization of denoising result of DVSCLEAN. Blue points are real events while the green are noise events. (Color figure online)

especially ours. The result can be explained by the sparsity of the event stream: the sparsity ensures that the event stream does not carry excessive information within 10 ms, and the reduction in resolution does not lead to significant loss of relevant events or information. The average loss of events in frame compression is approximately 1.4% in our experience, which is acceptable for practical applications. These findings support one of our points: maintaining the original high resolution and integrity of the event stream is not essential in event tasks, e.g., EDnCNN and AEDNet. Irrespective of their performance, they will not offer significant advantages in terms of efficiency and actually diminish the low-power advantage of event camera.

Table 3. Comparison of performance with reduced resolution dataset in DVSCLEAN with 100% ratio.

Dataset	Original	Reduce Resolution to 10 ms
BAF	91.29%	91.49%
PUGM	85.79%	85.10%
EDnCNN	97.23%	97.66%
AEDNet	96.57%	95.70%
Ours	**97.72%**	**97.71%**

3.3 Evaluation in ED-KoGTL

ED-KoGTL [1] is a dataset comprising identical motion trajectories of objects under different lighting conditions. The dataset is labeled by the Known-object Ground-Truth Labeling (KoGTL) method, using synchronized APS frames to extract object edges so that events can be judged by their distance from the edges. KoGTL is grounded in logical principles without elaborate mathematical derivations, resulting in a real-world dataset with reliable ground-truth labels.

We tested our method on two publicly available event streams in ED-KoGTL: very good light condition (750 lux) and low light condition (5 lux). The total duration time of dataset is inadequate for our frame-based method, so we employed the data augmentation [28], which is only effective for the frame-based method. The experimental results are presented in Table 2, demonstrates that our method achieved SOTA in both 750 lux and 5 lux light conditions. The visualization of performance in Fig. 5 shows that our method successfully removes isolated noise while preserving the original event edges.

In addition to train our model directly on ED-KoGTL, we also validate its performance with the network pre-trained on DVSCLEAN with 100% noise ratio. The results demonstrate a minor discrepancy in performance and a slight improvement in outcomes compared to direct training in 5 lux light condition scenes. This can prove the generalization of our method and has the potential to be applied in real-world scenarios.

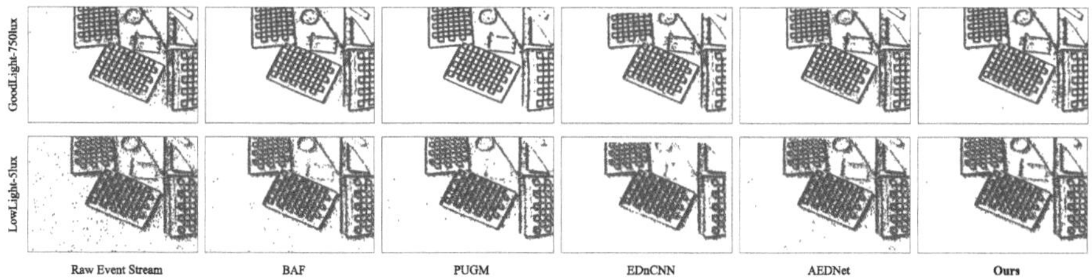

Fig. 5. Visualization of denoising result of ED-KoGTL.

3.4 Ablation Study

Effect of Each Components. We first evaluate the contribution of each component of our model, including the proposed ThPLIF neuron and the rough feature extraction module. We performed several ablation experiments on DVS-CLEAN with 100% noise ratio, as shown in Table 4. The result indicates that both ThPLIF neuron as well as the rough feature extraction contribute positively to the denoising performance. As ThPLIF neurons are removed, a more significant drop in Recall is observed compared to Precision, reflecting the role of ThPLIF in preventing consecutive events from being erroneously filtered due to membrane potential resets. It is noteworthy that the simple network without the rough feature extraction module can be interpreted as a simplified version of BAF-the easiest denoising method [25]. Our ablation analysis reflects an evolutionary enhancement of BAF, incorporating learnable components while preserving its efficiency-driven design principles.

Table 4. Ablation study of components in DVSCLEAN with 100% ratio.

Rough Feature Ex.	Neuron	Metrics			
		Acc	Pre	Rec	F1
✗	PLIF	93.7%	96.1%	91.2%	93.6%
✗	**ThPLIF**	94.2%	95.3%	93.3%	94.3%
✓	$\Theta(x)$	94.4%	97.8%	91.0%	94.3%
✓	LIF	97.4%	98.5%	96.5%	97.4%
✓	PLIF	97.6%	98.7%	96.5%	97.6%
✓	**ThPLIF**	**97.7%**	98.5%	**97.0%**	**97.7%**

Effect on Downstream Tasks. To validate the practical benefit of our method, we evaluate the denoising impact on DVS128 Gesture [29], a neuromorphic dataset taken with dynamic vision sensor and widely used for gesture recognition tasks. We use two baseline classification methods for evaluation: HATS [19] and PLIF [6]. The performance of HATS and PLIF on DVS128 Gesture before denoising are 83.3% and 88.2%, while on the denoised dataset are 85.2% and 91.3%, respectively. The visualization before and after applying our denoising method is shown in Fig. 6. Since our DeSNN doesn't training from a denoising dataset with polarity information, we also removed the polarity in DVS128 Gesture, resulted in slight drop in accuracy for PLIF. We hypothesize that the dataset with polarity information would allow our method to learn and perform better at denoising.

Fig. 6. Visualization of comparison in DVS128 Gesture whether denoised or not.

3.5 Practical Application

Efficiency Analysis. To evaluate the practicality of our method, we conduct an efficiency analysis on DVSCLEAN with 100% noise ratio, comparing the number of parameters, inference runtime and energy consumption of each method. Following Kim et al. [30], the energy consumption of deep learning-based methods can be quantified by MAC operations and the number of neuron activations, while the activation number of ANNs is equal to the number of neurons. For EDnCNN and AEDNet, the energy consumption is calculated event-by-event while our method operates in a frame-by-frame manner, so we calculate the average energy consumption per frame, then multiplied it by the number of frames and normalize it by the total number of events. All experiments were conducted on a PC with NVIDIA GEFORCE RTX 3070 GPU, using the event stream MAH00490_100 from DVSCLEAN. As shown in Table 5, our method exhibits the lowest energy consumption, the fewest number of parameters and the shortest running time, indicating the potential for practical applications. It is noteworthy that SNN can achieve higher efficiency on the neuromorphic hardware platform than on GPUs.

Practical Application. Our denoising algorithm has been successfully deployed in a real-time system integrated with event camera. We use DAVIS346 Color to capture event streams in real time and transfer them to GPU for processing. The data is read and synchronize denoised combining the dv-processing library in Python. The time stamp, raw event frames and corresponding denoised

Table 5. Efficiency comparison in DVSCLEAN with 100% ratio.

Type	Method	Params (k)	Inference Time (s)	Preprocess Time (s)	Energy (mJ)
Filter-based	BAF	–	70.2	–	–
	PUGM	–	117.8	–	–
Learning-based	EDnCNN	614.5	6.7	176.3	8.44e−2
	AEDNet	52663.2	692.4	1044.0	3.62
	Ours	**1.4**	**5.5**	**12.2**	**6.99e−5**

outputs are displayed on the screen, as shown in Fig. 7. Through a series of trials, the average time consumption of the denoising network is approximately 0.5 ms per frame. While additional overhead arises from data communication and preprocessing, the total processing time remains around 5 ms on average, well within the 10 ms temporal resolution we selected. We expect that the communication and processing latency could be significantly reduced with further optimization, such as hardware acceleration or low-level code tuning.

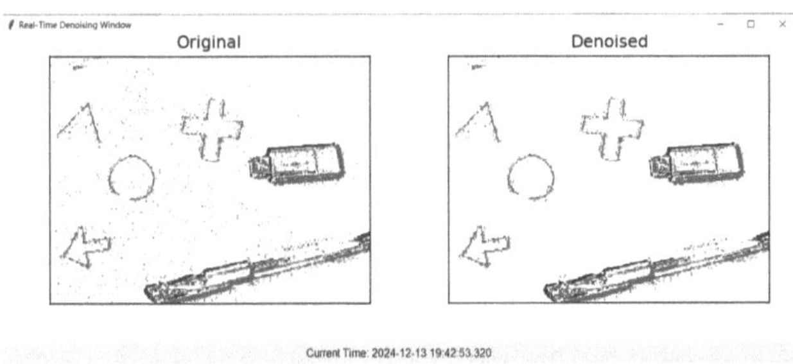

Fig. 7. The real-time synchronize denoising.

3.6 Discussion

Selection of Temporal Resolution. The decision of 10 ms resolution is not only a result of task requirements and experience in the image domain, but also a combination of network training and practical application. On one hand, there currently lacks an efficient parallel algorithm that supports event-wise forward inference or back propagation. Enforcing such an event-driven approach will result in increased memory accesses and computational complexity, which ultimately undermining system efficiency [31]. On the other hand, downstream tasks such as gesture detection and object tracking-which are applied in real world scenarios-are typically executed at around 30 FPS (i.e., 33 ms per frame), which is significantly coarser than our 10 ms framing interval. This ensures that our preprocessing pipeline remains backward compatible with existing applications and systems.

Unlike previous denoising methods that attempt to evaluate each event independently, we argue that such fine-grained processing may be unnecessary. As discussed in Sect. 2.2, meaningful patterns emerge only when events are aggregated into structured frames. Furthermore, as most downstream pipelines already convert event stream into frames for further processing, it becomes more effective to perform denoising at the frame level rather than on a per-event basis.

4 Conclusion

In this work, we propose a real-time lightweight denoising method for event cameras. Our approach is based on event frame representation, which minimizes information loss and eliminates the need for additional coding. The spatio-temporal correlation of events is explicitly modeled through the threshold-limited spiking neuron ThPLIF, which is used as the core component for the proposed DeSNN. DeSNN extracts the spatial information through feature extraction at two scales and transmits the temporal information through ThPLIF, achieving effective denoising result with a minimal number of modules—thus maximizing efficiency. Extensive evaluations of both performance and efficiency show that our denoising method achieves high accuracy and low consumption simultaneously comparing with other SOTA algorithms. Examples of application in real-world scenarios demonstrate the practical potential of our method. In conclusion, our work provides an engineering solution for real-time denoising of event cameras.

Acknowledgement. This work was supported in part by the National Natural Science Foundation of China under Grant 62236005, the Foundation for Outstanding Research Groups of Hubei Province of China under Grant 2025AFA012, and the 111 Project on Computational Intelligence and Intelligent Control under Grant B18024.

Disclosure of Interests. The authors have no competing interests to declare that are relevant to the content of this article.

References

1. Alkendi, Y., Azzam, R., Ayyad, A., Javed, S., Seneviratne, L., Zweiri, Y.: Neuromorphic camera denoising using graph neural network-driven transformers. IEEE Trans. Neural Netw. Learn. Syst. **35**(3), 4110–4124 (2022)
2. Fang, H., Wu, J., Li, L., Hou, J., Dong, W., Shi, G.: AEDNet: asynchronous event denoising with spatial-temporal correlation among irregular data. In: Proceedings of the 30th ACM International Conference on Multimedia, pp. 1427–1435 (2022)
3. Fang, H., Wu, J., Hou, Q., Dong, W., Shi, G.: Fast window-based event denoising with spatiotemporal correlation enhancement, arXiv preprint arXiv:2402.09270 (2024)
4. Calabrese, E., et al.: DHP19: dynamic vision sensor 3D human pose dataset. In: Proceedings of the IEEE/CVF Conference on Computer Vision and Pattern Recognition Workshops, pp. 0–0 (2019)

5. Jin, X., Zhang, M., Yan, R., Pan, G., Ma, D.: R-SNN: region-based spiking neural network for object detection. IEEE Trans. Cogn. Dev. Syst. (2023)
6. Fang, W., Yu, Z., Chen, Y., Masquelier, T., Huang, T., Tian, Y.: Incorporating learnable membrane time constant to enhance learning of spiking neural networks. In: Proceedings of the IEEE/CVF International Conference on Computer Vision, pp. 2661–2671 (2021)
7. Ren, H., et al.: Rethinking efficient and effective point-based networks for event camera classification and regression: Eventmamba, arXiv preprint arXiv:2405.06116 (2024)
8. Liu, M., Delbruck, T.: Adaptive time-slice block-matching optical flow algorithm for dynamic vision sensors. BMVC (2018)
9. Muglikar, M., Bauersfeld, L., Moeys, D.P., Scaramuzza, D.: Event-based shape from polarization. In: Proceedings of the IEEE/CVF Conference on Computer Vision and Pattern Recognition, pp. 1547–1556 (2023)
10. Zhang, P., Liu, H., Ge, Z., Wang, C., Lam, E.Y.: Neuromorphic imaging with joint image deblurring and event denoising. IEEE Trans. Image Process. (2024)
11. Duan, P., Wang, Z.W., Zhou, X., Ma, Y., Shi, B.: EventZoom: learning to denoise and super resolve neuromorphic events. In: Proceedings of the IEEE/CVF Conference on Computer Vision and Pattern Recognition, pp. 12824–12833 (2021)
12. Peng, X., Wang, Y., Gao, L., Kneip, L.: Globally-optimal event camera motion estimation. In: Vedaldi, A., Bischof, H., Brox, T., Frahm, J.-M. (eds.) ECCV 2020. LNCS, vol. 12371, pp. 51–67. Springer, Cham (2020). https://doi.org/10.1007/978-3-030-58574-7_4
13. Chamorro, W., Sola, J., Andrade-Cetto, J.: Event-based line slam in real-time. IEEE Robot. Autom. Lett. **7**(3), 8146–8153 (2022)
14. Baldwin, R., Almatrafi, M., Asari, V., Hirakawa, K.: Event probability mask (EPM) and event denoising convolutional neural network (EDNCNN) for neuromorphic cameras. In: Proceedings of the IEEE/CVF Conference on Computer Vision and Pattern Recognition, pp. 1701–1710 (2020)
15. Wang, Z., Yuan, D., Ng, Y., Mahony, R.: A linear comb filter for event flicker removal. In: 2022 International Conference on Robotics and Automation (ICRA), pp. 398–404. IEEE (2022)
16. Czech, D., Orchard, G.: Evaluating noise filtering for event-based asynchronous change detection image sensors. In: 6th IEEE International Conference on Biomedical Robotics and Biomechatronics (BioRob), pp. 19–24. IEEE (2016)
17. Roy, K., Jaiswal, A., Panda, P.: Towards spike-based machine intelligence with neuromorphic computing. Nature **575**(7784), 607–617 (2019)
18. Zhang, P., Ge, Z., Song, L., Lam, E.Y.: Neuromorphic imaging with density-based spatiotemporal denoising. IEEE Trans. Comput. Imaging **9**, 530–541 (2023)
19. Sironi, A., Brambilla, M., Bourdis, N., Lagorce, X., Benosman, R.: HATS: histograms of averaged time surfaces for robust event-based object classification. In: Proceedings of the IEEE Conference on Computer Vision and Pattern Recognition, pp. 1731–1740 (2018)
20. Deng, Y., Chen, H., Chen, H., Li, Y.: EVVGCNN: a voxel graph CNN for event-based object classification. arXiv preprint arXiv:2106.00216 (2021). **1**(2), 6
21. Cordone, L., Miramond, B., Ferrante, S.: Learning from event cameras with sparse spiking convolutional neural networks. In: 2021 International Joint Conference on Neural Networks (IJCNN), pp. 1–8. IEEE (2021)
22. Samadzadeh, A., Far, F.S.T., Javadi, A., Nickabadi, A., Chehreghani, M.H.: Convolutional spiking neural networks for spatio-temporal feature extraction. Neural Process. Lett. **55**(6), 6979–6995 (2023)

23. Wu, J., Ma, C., Li, L., Dong, W., Shi, G.: Probabilistic undirected graph based denoising method for dynamic vision sensor. IEEE Trans. Multimedia **23**, 1148–1159 (2020)
24. Feng, W., Wang, X., Zhan, X., Huang, H.Z.: Event denoising for dynamic vision sensor using residual graph neural network with density-based spatial clustering. Neurocomputing **636**, 130026 (2025)
25. Delbruck, T., et al.: Frame-free dynamic digital vision. In: Proceedings of Intl. Symp. on Secure-Life Electronics, Advanced Electronics for Quality Life and Society, vol. 1, pp. 21–26. Citeseer (2008)
26. Fang, W., et al.: Spikingjelly: an open-source machine learning infrastructure platform for spike-based intelligence. Sci. Adv. **9**(40), eadi1480 (2023)
27. Gehrig, D., Gehrig, M., Hidalgo-Carrió, J., Scaramuzza, D.: Video to events: recycling video datasets for event cameras. In: Proceedings of the IEEE/CVF Conference on Computer Vision and Pattern Recognition, pp. 3586–3595 (2020)
28. Li, Y., Kim, Y., Park, H., Geller, T., Panda, P.: Neuromorphic data augmentation for training spiking neural networks. In: European Conference on Computer Vision, pp. 631–649. Springer (2022)
29. Amir, A., et al.: A low power, fully event-based gesture recognition system. In: Proceedings of the IEEE Conference on Computer Vision and Pattern Recognition, pp. 7243–7252 (2017)
30. Kim, Y., Chough, J., Panda, P.: Beyond classification: directly training spiking neural networks for semantic segmentation. Neuromorphic Comput. Eng. **2**(4), 044015 (2022)
31. Dalgaty, T., et al.: The CNN vs. SNN event-camera dichotomy and perspectives for event-graph neural networks. In: 2023 Design, Automation & Test in Europe Conference & Exhibition (DATE), pp. 1–6. IEEE (2023)

Safety-Critical Distributed Containment Maneuvering of Underactuated Autonomous Surface Vehicles via Local LiDAR Perception

Nan Gu[1,2(✉)], Aijun Li[1,2], Yiwen Li[1,2], Lu Liu[1,2], and Zhouhua Peng[1,2]

[1] School of Marine Electrical Engineering, Dalian Maritime University, Dalian 116026, China
{ngu,aijunli,yiwenli,luliu,zhpeng}@dlmu.edu.cn

[2] Dalian Key Laboratory of Swarm Control and Electrical Technology for Intelligent Ships, Dalian 116026, China

Abstract. Ensuring collision-free navigation is essential for the safe operation of multiple autonomous surface vehicles. This paper addresses the problem of safe containment maneuvering of multiple underactuated ASVs in unknown environments. A distributed containment maneuvering controller based on local LiDAR perception is proposed, which allows the ASVs to move into the convex hull formed by multiple virtual leaders following parameterized paths, achieving coordinated formations while maintaining collision-free behavior. First, the containment maneuvering control law is developed based on the information of the neighbor information. Then, by introducing path parameter deviations among multiple virtual leaders, a path parameter update law is designed. Finally, obstacle information is obtained through LiDAR perception, and a Gaussian process is employed to construct a control barrier function-based safety constraint. The designed containment maneuvering control law is optimized via quadratic programming to ensure real-time safety. Lyapunov stability analysis demonstrates that the overall control system is asymptotically stable, ensuring the safety of the system. The simulation results confirm the effectiveness of the proposed approach.

Keywords: Collision Avoidance · Distributed Containment Maneuvering · Control Barrier Function · Autonomous Surface Vehicles

1 Introduction

In recent years, autonomous surface vehicle (ASV) technologies have seen significant development potential [2,3,9,13], and containment maneuvering has received widespread attention. In [5], a graph-theory-based distributed mechanism is proposed, through which formation containment maneuvering is achieved via parameterized paths. In [11], driven barrier Lyapunov functions-based LOS

C. Li et al. (Eds.): ICNC 2025, CCIS 2946, pp. 415–425, 2026.
https://doi.org/10.1007/978-981-92-1599-7_35

guidance algorithm is proposed to enhance the convergence speed of containment formation errors and path tracking. In [12], a potential field-based coordinated maneuvering guidance is proposed to generate safe trajectories for each ASV, enabling effective convergence to the convex hull defined by the virtual leaders. In [15] a noncooperative-game-based fault-tolerant fuzzy containment control method is proposed, enabling a balance between individual and group objectives while ensuring that multiple vessels converge to the convex hull and respect individual preferences and constraints. However, the aforementioned studies [5,11,12,15] do not tackle the problem of obstacle avoidance.

To address the obstacle avoidance problem in ASV formations, several methods have been proposed that integrate formation control with collision avoidance strategies. In [17], a variable-size artificial potential field strategy is proposed, using a virtual structure for formation organization and a dynamic window method for obstacle avoidance. In [4], a hierarchical control framework integrating formation control and obstacle avoidance is proposed based on an adaptive neural potential field method. In [16], a heading-constrained control barrier function (CBF) is proposed to design a yaw-rate control law for obstacle avoidance. In [7], a barrier-certified distributed model predictive control method is proposed to enable safe obstacle avoidance for a formation of multiple ASVs. In [14], a high-order control barrier function-based method is proposed to achieve collision avoidance. In [1], a backstepping-based control law is designed using a relaxed control barrier function and an analytical convex optimization method, enabling faster obstacle avoidance for the ASVs. However, obstacle avoidance in the aforementioned studies typically relies on prior knowledge of the obstacles.

Based on the above discussion, existing studies have rarely addressed the simultaneous realization of local-perception-based obstacle avoidance and containment maneuvering control for ASV fleets operating in unknown environments. To address this gap, this paper proposes a LiDAR-based containment maneuvering control approach that enables the fleet of ASVs to avoid both dynamic and static obstacles without requiring prior environmental knowledge.

2 Preliminaries

2.1 Gaussian Process

A Gaussian Process (GP) is a Bayesian regression method defined by a mean function and a covariance (kernel) function, eliminating the need for explicit basis functions. Any finite subset of a GP follows a joint Gaussian distribution. A commonly used kernel is the squared exponential (SE) function $k(c_i, c_j) = \sigma_{SE}^2 e^{-\frac{1}{2}(c_i-c_j)^T(\boldsymbol{L}^T\boldsymbol{L})^{-1}(c_i-c_j)}$ where $\boldsymbol{L}$ and σ_{SE} denoting the length scale matrix and signal variance, respectively.

Given a set of N_c data points with inputs $\{c_i\}_{i=1}^{N_c}$ and corresponding outputs $\boldsymbol{Y} = \{y_i\}_{i=1}^{N_c}$, the Gaussian Process can be used to obtain a prediction at a query point x^q. At a given query point x^q, the predicted output y_* follows a normal distribution, and is given by $y_* \mid x^q \sim \mathcal{N}\left(\mu(x^q), \sigma^2(x^q)\right)$, where $\mu(x^q) =$

$\boldsymbol{k}_*^T \boldsymbol{K}_{SEN}^{-1} \boldsymbol{Y}$, $\sigma^2(x^q) = k(x^q, x^q) - \boldsymbol{k}_*^T \boldsymbol{K}_{SEN}^{-1} \boldsymbol{k}_*$. $\boldsymbol{k}_*$ links the query point x^q to the training set $\{c_i\}_{i=1}^{N_c}$, and $\mathcal{N}$ representing the normal distribution.

2.2 Problem Formulation

Consider a networked ASVs system including N_f ASVs and $N_l - N_f$, $(N_l > N_f)$ virtual leaders. The kinematics model of ASVs can be expressed as follows

$$\begin{cases} \dot{x}_i = U_i \cos \psi_i \\ \dot{y}_i = U_i \sin \psi_i \\ \dot{\psi}_i = r_i + \beta_{id} \end{cases} \tag{1}$$

where $U_i = \sqrt{u_i^2 + v_i^2}$ represents the resultant speed; $\boldsymbol{p}_i = [x_i, y_i]^{\mathrm{T}}$ denotes the position; $\psi_i = \phi_i + \beta_i$ denotes the course angle with $\beta_i = \arctan(v_i/u_i)$, and $\beta_{id} = \dot{\beta}_i$; ϕ_i represents the heading angle; β_i denotes the sideslip angle; r_i denotes the yaw rate; u_i, v_i are the surge and sway velocities, respectively.

Define $\boldsymbol{p}_i^p = [x_i^p, y_i^p]^{\mathrm{T}}$ as the point obtained by moving a distance $l_i \in \mathbb{R}_+$ from $\boldsymbol{p}_i$ along the direction of the resultant velocity. It can be expressed as $\boldsymbol{p}_i^p = \boldsymbol{p}_i + l_i \boldsymbol{\eta}_i$, with $\boldsymbol{\eta}_i = [\cos \psi_i, \sin \psi_i]^T$, the time derivative of $\boldsymbol{p}_i^p$ is given as follows

$$\begin{bmatrix} \dot{x}_i^p \\ \dot{y}_i^p \end{bmatrix} = \begin{bmatrix} U_i \cos \psi_i \\ U_i \sin \psi_i \end{bmatrix} + l_i \dot{\psi}_i \begin{bmatrix} -\sin \psi_i \\ \cos \psi_i \end{bmatrix} = \boldsymbol{R}_l(\psi_i) \begin{bmatrix} U_i \\ r_i + \beta_{id} \end{bmatrix} \tag{2}$$

where rotation matrix $\boldsymbol{R}_l(\psi_i)$ is given by $\boldsymbol{R}_l(\psi_i) = \begin{bmatrix} \cos \psi_i & -l_i \sin \psi_i \\ \sin \psi_i & l_i \cos \psi_i \end{bmatrix}$.

The ASVs, indexed from 1 to N_f, are required to track the convex hull formed by multiple virtual leaders (numbered from $N_f + 1$ to N_l), with only a subset of ASVs directly communicating with the leaders.

The virtual leaders are assumed to move along a parameterized path, with the path information specified as $\boldsymbol{X}_{kr}(\theta_k^l) = [x_{kr}(\theta_k^l), y_{kr}(\theta_k^l)]^{\mathrm{T}} \in \mathbb{R}^2, k = N_f + 1, \ldots, N_l$, where θ_k^l is a path variable. The partial derivative of $\boldsymbol{X}_{kr}(\theta_k^l)$ with respect to the path parameter is defined as $\boldsymbol{X}_{kr}^{\theta_k^l}(\theta_k^l) = \partial \boldsymbol{X}_{kr}(\theta_k^l)/\partial \theta_k^l$. It is assumed that $\boldsymbol{X}_{kr}(\theta_k^l)$ and $\boldsymbol{X}_{kr}^{\theta_k^l}(\theta_k^l)$ are bounded.

A graph $\mathcal{G} = \{\mathcal{V}, \mathcal{E}\}$ is used to describe the communication topology between the ASVs and the virtual leaders, where $\mathcal{V} = \{1, 2, \ldots, N_l\}$ denotes the set of ASVs and $\mathcal{E} \subseteq \mathcal{V} \times \mathcal{V}$ represents the information flow. Its associated Laplacian matrix is defined as $\boldsymbol{L}^s = \boldsymbol{D}^s - \boldsymbol{A}^s \in \mathbb{R}^{N_l \times N_l}$. It can be expressed as follows $\boldsymbol{\mathcal{L}}^s = \begin{bmatrix} \boldsymbol{L}_1^s & \boldsymbol{L}_2^s \\ \boldsymbol{L}_3^s & \boldsymbol{L}_0^s \end{bmatrix}$, where $\boldsymbol{L}_1^s \in \mathbb{R}^{N_f \times N_f}$, $\boldsymbol{L}_2^s \in \mathbb{R}^{N_f \times (N_l - N_f)}$, $\boldsymbol{L}_3^s \in \mathbb{R}^{(N_l - N_f) \times N_f}$ and $\boldsymbol{L}_0^s \in \mathbb{R}^{(N_l - N_f) \times (N_l - N_f)}$. The adjacency matrix $\boldsymbol{A}^s = [a_{ij}] \in \mathbb{R}^{N_l \times N_l}$ is defined as $a_{ij} = 1$ if $(i, j) \in \mathcal{E}$ (i.e., ASV i receives information from ASV j) and $a_{ij} = 0$ otherwise. The in-degree matrix is defined as $\boldsymbol{D}^s = \mathrm{diag}\{d_1, d_2, \ldots, d_{N_l}\}$.

Assumption 1. At least one virtual leader can reach every follower through a directed path.

Assumption 2. The virtual leaders form a connected, undirected communication graph, with at least one leader receiving information from the super leader. Followers communicate over a undirected topology.

To obtain the obstacle positions, the unsafe positions $\boldsymbol{B}_{iq}$ detected by the qth laser beam are expressed as follows $\boldsymbol{B}_{iq} = \begin{bmatrix} x_i + D_{iq}\cos(\theta_i + q \cdot \theta_{\text{res}}) \\ y_i + D_{iq}\sin(\theta_i + q \cdot \theta_{\text{res}}) \end{bmatrix}$, where $\boldsymbol{B}_{iq}$ denotes the obstacle detected by the qth beam of the ith ASV. D_{iq} denotes the distance from the qth LiDAR beam to the obstacle and θ_{res} denotes the angular resolution.

At time t, once the edge of an obstacle is detected by the qth laser beam of the LiDAR, the associated unsafe position (i.e., the location of the point on edge of the obstacle) $\boldsymbol{B}_{iq}$ will be used in the Gaussian process model when the following condition is satisfied, $\min_{j=1,\ldots,N_{oe}} \|\boldsymbol{B}_{iq} - \boldsymbol{B}_j\| > d_{\text{sample}}$, where d_{sample} represents the minimum sampling distance and N_{oe} denotes the number of points sampled by the lidar. Upon satisfying this condition, the corresponding output value is set to -1, which prevents overly dense sampling points.

Assumption 3. [6] At the lidar sampling instant t_{s+1}, the states are restricted to the intersection of the zero super-level sets corresponding to t_s and t_{s+1}.

This papers aims to design a control method enables a fleet of under-actuated ASVs to perform containment maneuvering with local LiDAR perception such that the fleet of ASVs is safe in unknown environments.

3 Design and Analysis

3.1 Controller Design

The containment maneuvering errors are defined as

$$\boldsymbol{Z}_{i1} = \boldsymbol{R}_i^{\mathrm{T}}(\psi_i)\left\{\sum_{j=1}^{N_f} a_{ij}(\boldsymbol{p}_i^p - \boldsymbol{p}_j^p) + \sum_{k=N_f+1}^{N_l} a_{ik}(\boldsymbol{p}_i^p - \boldsymbol{X}_{kr}(\theta_k^l))\right\} \tag{3}$$

where $\boldsymbol{Z}_{i1}$ is the containment maneuvering error in the body-fixed frame, with two components. The first component $\sum_{j=1}^{N_f} a_{ij}(\boldsymbol{p}_i^p - \boldsymbol{p}_j^p)$ represents the error of the ith ASV relative to its neighbors. The second component $\sum_{k=N_f+1}^{N_l} a_{ik}(\boldsymbol{p}_i^p - \boldsymbol{X}_{kr}(\theta_k^l))$ describes the path maneuvering error of those unmanned surface vehicles that obtain path information, i.e., $a_{ik} = 1$, for all i, k,the system will adopt a centralized architecture, which demands substantial communication resources to acquire the path information.

Defining $\dot{\theta}_k^l = v_s - \omega_k$. According to (2), differentiating (3) with respect to time yields

$$\begin{aligned}\dot{\boldsymbol{Z}}_{i1} = -r_i \boldsymbol{S}\boldsymbol{Z}_{i1} + \boldsymbol{G}_i \boldsymbol{V}_{ic} - \sum_{j=1}^{N_f} a_{ij} \boldsymbol{R}_i^{\mathrm{T}}(\psi_i)\boldsymbol{R}_l(\psi_j) \begin{bmatrix} U_j \\ r_j + \beta_{jd} \end{bmatrix} \\ - \sum_{k=N_f+1}^{N_l} a_{ik} \boldsymbol{R}_i^{\mathrm{T}}(\psi_i) \boldsymbol{X}_{kr}^{\theta_k^l}(\theta_k^l)(v_s - \omega_k)\end{aligned} \tag{4}$$

where ω_k is a coordination variable and $\boldsymbol{G}_i$ is given by $\boldsymbol{G}_i = \begin{bmatrix} d_i & 0 \\ 0 & d_i/l_i \end{bmatrix}$ where $d_i = \sum_{j=1}^{N_l} a_{ij}$.

To stabilize (4), the following distributed containment maneuvering guidance law is introduced.

$$\boldsymbol{V}_{ic} = \boldsymbol{G}_i^{-1} \left\{ -\boldsymbol{K}_{ic}\boldsymbol{Z}_{i1} + \boldsymbol{R}_i^{\mathrm{T}}(\sum_{j=1}^{N_f} a_{ij}(\psi_i)\boldsymbol{R}_l(\psi_j) \begin{bmatrix} U_j \\ r_j + \beta_{jd} \end{bmatrix} + \sum_{k=N_f+1}^{N_l} a_{ik} \boldsymbol{X}_{kr}^{\theta_k^l}(\theta_k^l) v_s) \right\} \tag{5}$$

where $\boldsymbol{K}_{ic} \in \mathbb{R}^{2\times 2}$ is a diagonal gain matrix.

Substituting (5) and (4), the error dynamics are given by

$$\dot{\boldsymbol{Z}}_{i1} = -r_i \boldsymbol{S}\boldsymbol{Z}_{i1} - \boldsymbol{K}_{ic}\boldsymbol{Z}_{i1} - \sum_{k=N_f+1}^{N_l} a_{ik} \boldsymbol{R}_i^{T}(\psi_i) \boldsymbol{X}_{kr}^{\theta_k^l}(\theta_k^l)\omega_k \tag{6}$$

To ensure coordination among virtual leaders and enable feedback from followers, ω_k is defined as follows:

$$\begin{cases} \omega_k = l_{k1}\vartheta_k + \varpi_k \\ \dot{\varpi}_k = -l_{k2}(\varpi_k - \mu_k \vartheta_k) \end{cases} \tag{7}$$

where $l_{k1}, l_{k2}, \mu_k \in \mathbb{R}_+$ are positive parameter; $\vartheta_k = -\sum_{i=1}^{N_f} a_{ki} (\boldsymbol{X}_{kr}^{\theta_k^l}(\theta_k^l))^T \boldsymbol{R}_i(\psi_i)\boldsymbol{Z}_{i1} + (\theta_k^l - \theta_l - \mathcal{P}_{kl})$, $l \in \{N_f+1, ..., N_l\}$, in which $\theta_k^l - \theta_l^l - \mathcal{P}_{kl}$ represents the path coordination error between the kth virtual leader and its neighboring virtual leaders, $\mathcal{P}_{kl} = \mathcal{P}_{k0} - \mathcal{P}_{l0}$ represents the path coordination error associated with the k-th virtual leader and the super leader. The boundedness of the closed-loop signals is ensured by the inclusion of l_{k1} and ϖ_k.

The global path coordination error is defined as $\theta_{k\epsilon}^l = \theta_k^l - \theta_0^l - \mathcal{P}_{k0}$ and it follows that

$$\dot{\boldsymbol{\theta}}_e^l = -\boldsymbol{\omega} = -\boldsymbol{\mu}\boldsymbol{\mathcal{H}}\boldsymbol{\theta}_e^l \tag{8}$$

where $\boldsymbol{\omega} = [\omega_{N_f+1}, ..., \omega_{N_l}]^T$, $\boldsymbol{\mu} = \mathrm{diag}\{\mu_{N_f+1}, ..., \mu_{N_l}\}$, $\boldsymbol{\theta}_e^l = [\theta_{(N_f+1)e}^l, ..., \theta_{Ne}^l]^T$; and $\boldsymbol{\mathcal{H}} = \boldsymbol{L}_0 + \mathrm{diag}\{b_1, ..., b_{(N_l-N_f)}\}$, where $b_i = 1$ if the ith virtual leader has access to the super leader, $b_i = 1$; otherwise $b_i = 0$. All the eigenvalues of $\boldsymbol{\mathcal{H}}$ are positive [8], which guarantees the global asymptotic stability of the leader system.

The dynamics of the errors θ_{ke}^{l} and ϖ_k are given by

$$\begin{cases} \dot{\theta}_{ke}^{l} = -l_{k1}\vartheta_k - \varpi_k \\ \dot{\varpi}_k = -l_{k2}\left(\varpi_k - \mu_k\vartheta_k\right) \end{cases} \tag{9}$$

Based on the GP mean prediction, the safety function is expressed as

$$h_i(\pi(\boldsymbol{X}_i^q)) = m_i(\pi(\boldsymbol{X}_i^q)) + \boldsymbol{k}_{i*}^{\mathrm{T}}\boldsymbol{K}_{i,SEN}^{-1}(\boldsymbol{Y}_i - m_i(\pi(\boldsymbol{X}_i^q))) \tag{10}$$

$$\boldsymbol{K}_{i,SEN}(\pi(\boldsymbol{B}_n), \pi(\boldsymbol{B}_m)) = \exp\left(-\frac{\|\boldsymbol{B}_n^b - \boldsymbol{B}_m^b\|^2}{2\ell^2}\right) + \boldsymbol{\sigma}_N^2\boldsymbol{I} \tag{11}$$

where $\boldsymbol{X}_i^q$ is defined as the query point, and $m_i(\pi(\boldsymbol{X}_i^q)) = 1$. To ensure efficient use of data points, the GP assumes the environment to be safe by default if no obstacle points are detected; $\boldsymbol{\sigma}_N^2$ represents the noise variance. ℓ denotes a scalar length and $\boldsymbol{I}$ is the identity matrix. The function $\pi(\cdot)$ is used to transform the position coordinates into the body-fixed coordinate frame; $(\cdot)^b$ represents the position expressed in the body frame.

The data set must include at least one sample, and the nominal control input is used when no obstacles are detected. Then, a safety controller is designed based on the CBF-constrained control space, with the optimal input $\boldsymbol{V}_i^{opt}$ computed from the following quadratic program

$$\boldsymbol{V}_i^{opt}(t) = \arg\min_{\boldsymbol{V}_i^{opt}\in\mathbb{R}^2} \|\boldsymbol{V}_i^{opt} - \boldsymbol{V}_{ic}(t)\|^2 \tag{12a}$$

$$L_f h_i(\pi(\boldsymbol{X}_i^q)) + L_g h_i(\pi(\boldsymbol{X}_i^q))\boldsymbol{V}_i^{opt} + \gamma(h_i(\pi(\boldsymbol{X}_i^q))) \geq 0 \tag{12b}$$

where $L_f h_i(\pi(\boldsymbol{X}_i^q)) = \frac{\partial h_i(\pi(\boldsymbol{X}_i^q))}{\partial \pi(\boldsymbol{X}_i^q)}\boldsymbol{f}$ and $L_g h_i(\pi(\boldsymbol{X}_i^q)) = \frac{\partial h_i(\pi(\boldsymbol{X}_i^q))}{\partial \pi(\boldsymbol{X}_i^q)}\boldsymbol{g}$ are denoted the Lie derivatives of $h_i(\pi(\boldsymbol{X}_i^q))$ with respect to the vector fields $\boldsymbol{f} = 0$ and $\boldsymbol{g} = \mathrm{diag}\{1, l_i\}$, with $\frac{\partial h_i(\pi(\boldsymbol{X}_i^q))}{\partial \pi(\boldsymbol{X}_i^q)} = [\boldsymbol{Y}_i^{\mathrm{T}}\boldsymbol{K}_{i,SEN}^{-1}\frac{\partial k_{i,n*}}{\partial \boldsymbol{X}_i^q}]^{\mathrm{T}}$ and $\frac{dk_{i,n*}}{\partial \boldsymbol{X}_i^q} = -\frac{2K_{i,SE}(\boldsymbol{X}_i^q, B_n)(\boldsymbol{X}_i^q - \boldsymbol{B}_n)^{\mathrm{T}}}{\ell}$; $\boldsymbol{f}$ denotes the intrinsic dynamics of the system and $\boldsymbol{g}$ denotes the input matrix. $\boldsymbol{B}_n, \boldsymbol{B}_m \in \mathbb{O}$, where $\mathbb{O}$ represents the boundary point dataset obtained by lidar sampling.

3.2 Stability Analysis

Lemma 1. *Under Assumptions 1 and 2, the systems described by* (6) *and* (9) *are asymptotically stable.*

Proof. The Lyapunov function is defined as

$$V_2 = \frac{1}{2}\sum_{i=1}^{N_f} \boldsymbol{Z}_{i1}^{\mathrm{T}}\boldsymbol{Z}_{i1} + \frac{1}{2}{\boldsymbol{\theta}_e^l}^{\mathrm{T}}\mathcal{H}\boldsymbol{\theta}_e^l + \sum_{K=N_f+1}^{N_l} \frac{\varpi_k^2}{2l_{k2}\mu_k} \tag{13}$$

by differentiating V_c along (6), (7) and (8), it follows that

$$\dot{V}_2 \leq \sum_{i=1}^{N_f} -\lambda_{\min}(\boldsymbol{K}_{ic})\|\boldsymbol{Z}_{i1}\|^2 - \sum_{k=N_f+1}^{N_l} (l_{k1}\vartheta_k^2 + \frac{\varpi_k^2}{\mu_k}) \tag{14}$$

Defining $\boldsymbol{E}_2^{'} = [\boldsymbol{Z}_1^{\mathrm{T}}, \boldsymbol{\vartheta}^{\mathrm{T}}, \boldsymbol{\varpi}^{\mathrm{T}}]^{\mathrm{T}}$, where $\boldsymbol{Z}_1 = [\boldsymbol{Z}_{11}, \ldots, \boldsymbol{Z}_{N_f 1}]^{\mathrm{T}}, \boldsymbol{\vartheta} = [\vartheta_{N_f+1}, \ldots, \vartheta_{N_l}]^{\mathrm{T}}$, and $\boldsymbol{\varpi} = [\varpi_{N_f+1}, \ldots, \varpi_{N_l}]^{\mathrm{T}}, \boldsymbol{E}_2 = [\boldsymbol{Z}_1^{\mathrm{T}}, \boldsymbol{\varpi}^{\mathrm{T}}, \boldsymbol{\theta}_e^{l\,\mathrm{T}}]^{\mathrm{T}}$, it follows that $\boldsymbol{E}_2^{'} = \boldsymbol{\Phi}\boldsymbol{E}_2$, where

$$\boldsymbol{\Phi} = \begin{bmatrix} \boldsymbol{I}_{N_f} \otimes \boldsymbol{I}_2 & 0 & 0 \\ 0 & \boldsymbol{I}_{N_l - N_f} \otimes \boldsymbol{I}_2 & 0 \\ -\boldsymbol{\Gamma} \otimes \boldsymbol{I}_2 & 0 & \boldsymbol{\mathcal{H}} \otimes \boldsymbol{I}_2 \end{bmatrix}$$

and

$$\boldsymbol{\Gamma} = \begin{bmatrix} a_{(N_f+1)1}(\boldsymbol{X}_{(N_f+1)r}^{\theta^l_{N_f+1}})^{\mathrm{T}}\boldsymbol{R}_1(\psi_1) & \cdots & a_{(N_f+1)N_f}(\boldsymbol{X}_{(N_f+1)r}^{\theta^l_{N_f+1}})^{\mathrm{T}}\boldsymbol{R}_{N_f}(\psi_{N_f}) \\ \vdots & \ddots & \vdots \\ a_{N_l 1}(\boldsymbol{X}_{N_l r}^{\theta^l_{N_l}})^{\mathrm{T}}\boldsymbol{R}_1(\psi_1) & \cdots & a_{N_l N_f}(\boldsymbol{X}_{N_l r}^{\theta^l_{N_l}})^{\mathrm{T}}\boldsymbol{R}_{N_f}(\psi_{N_f}) \end{bmatrix}.$$

Then $\dot{V}_2 \leq -c_2\lambda^2(\boldsymbol{\Phi})\|\boldsymbol{E}_2\|^2$, where $c_2 = \min_{i=1,\ldots,N_f,\, k=N_f+1,\ldots,N_l} \{\lambda_{\min}(\boldsymbol{K}_{ic}), l_{k1}, 1/\mu_k\}$. Therefore, global asymptotic stability of the closed-loop system is guaranteed.

Lemma 2. *[6] A GP-based safety function is constructed as in* (10)*, which employs a smooth kernel inversely proportional to the Euclidean distance and marking unsafe positions as* -1*. The resulting quadratic program is feasible with a nontrivial inequality constraint.*

Theorem 1. *Under Assumptions 1–3 and Lemmas 1–2, the closed-loop system is asymptotically stable. Moreover, any locally Lipschitz continuous controller satisfying* (12) *guarantees that the switching zero-super-level set remains forward invariant within the safe set.*

Proof. According to Lemma 1, the closed-loop system is asymptotically stable. At each sampling instant t_s ensures the safety function is smooth and locally Lipschitz, guaranteeing forward invariance within each interval (t_s, t_{s+1}). Assumption 3 ensures the system is thus maintained within the safe set after switching, and the switching zero-super-level set is consequently forward invariant.

It implies that the closed-loop system is asymptotically stable and its states always remain within the safe set.

4 Simulation

This section presents simulation results to demonstrate the effectiveness of the proposed distributed containment and obstacle avoidance controller for multiple underactuated ASVs with local LiDAR perception. A multi-ASV network system composed of five underactuated ASVs, three virtual leaders, and one super leader is considered. The communication topology, shown in Fig. 1, satisfies Assumptions 1 and 2. The LiDAR scanning schematic is shown in Fig. 3.

The dynamic model of each ASV adopts the Cybership II parameters from [10]. The trajectory of virtual leader 6 is defined as $\boldsymbol{p}_{6r}(\theta_6^l) = [0.2\theta_6^l; 0.2\theta_6^l]$ and it receives information from the super leader.

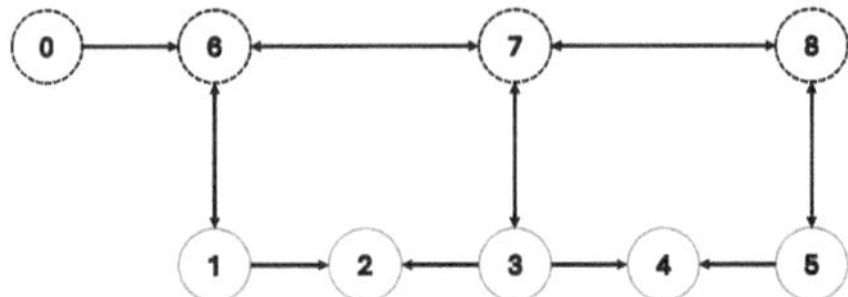

Fig. 1. The communication topology employed in the simulation.

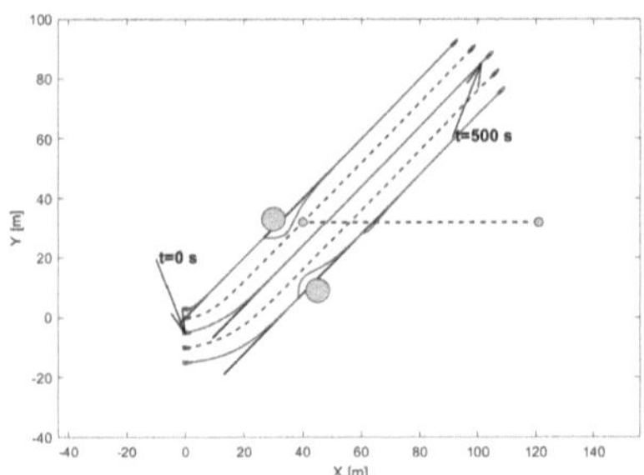

Fig. 2. Paths of the five follower ASVs.

The parameters are selected as $\mathcal{P}_{60} = -20$, $\mathcal{P}_{70} = 0$, $\mathcal{P}_{80} = -20$, $\boldsymbol{K}_{ic} = \text{diag}\{0.1, 0.1\}$, $l_{k1} = 0.01$, $v_s = 1$, $l_{k2} = 10$, $\mu_k = 10$, $\ell = 2$, $\sigma_N^2 = 0.01$, $\sigma_{SE}^2 = 1$, $d_{\text{sample}} = 0.05$, $\gamma(h_i(\pi(\boldsymbol{X}_i^q))) = 5 \times h_i(\pi(\boldsymbol{X}_i^q))^3$. The LiDAR measurements are performed at 5 Hz, featuring an angular resolution of $\theta_{res} = 1°$, and a detection range extending to $d_{\max} = 6$. The initial positions of the five vessels are $\boldsymbol{p}_1(0) = [0; 3]$, $\boldsymbol{p}_2(0) = [0; 0]$, $\boldsymbol{p}_3(0) = [0; -5]$, $\boldsymbol{p}_4(0) = [0; -10]$, $\boldsymbol{p}_5(0) = [0; -15]$. The initial heading angels are all 0. The static obstacles are centered at [30; 33] and [45; 9] with a radius of 4, while the dynamic obstacle moves from [120; 32] to [40; 32] with a radius of 2.

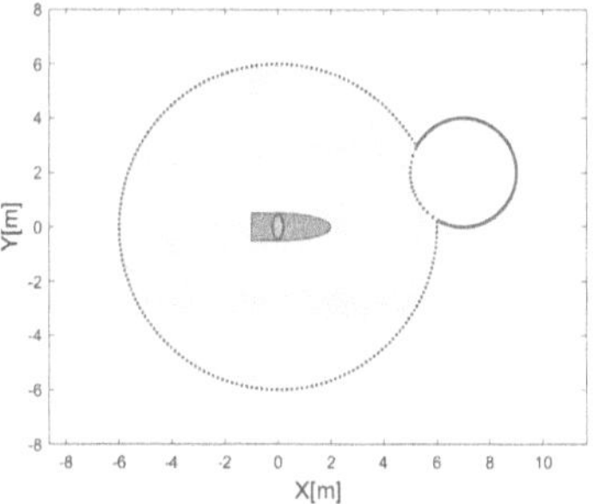

Fig. 3. Schematic of LiDAR scanning of obstacles.

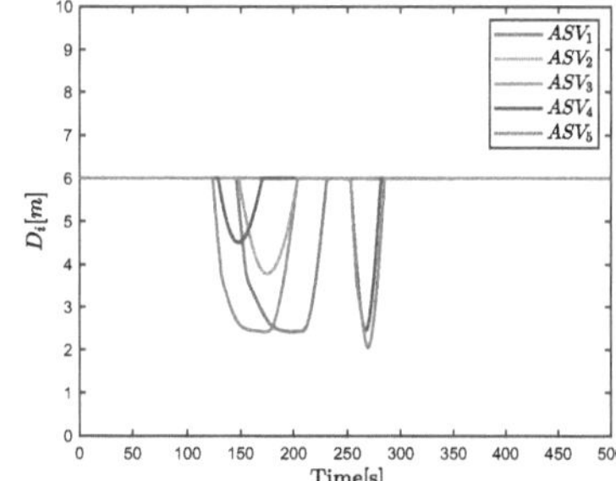

Fig. 4. The distance between the ASVs and the obstacle.

The simulation results are shown in Fig. 2 and 5. As illustrated in Fig. 2 the proposed control approach enables five ASVs to converge to the convex hull of three virtual leaders and form a triangular formation with collision-free navigation around both static and dynamic obstacles. Figure 4 shows that all ASVs maintain safe distances from obstacles. Figure 5 shows that the along-track and cross-track errors converge to zero.

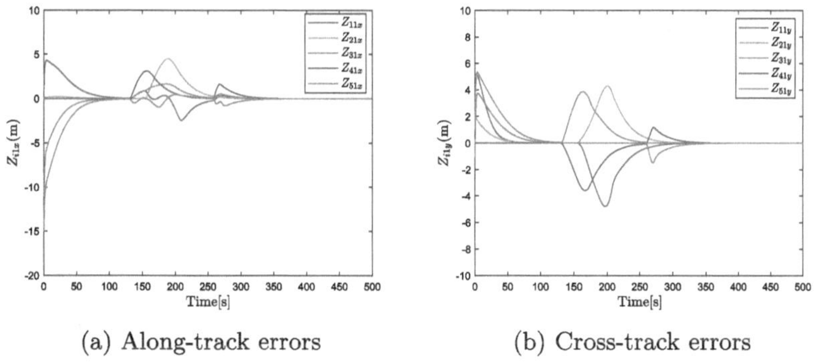

(a) Along-track errors

(b) Cross-track errors

Fig. 5. Simulation results of along-track and cross-track errors.

5 Conclusions

This paper addresses the containment maneuvering and obstacle avoidance control problem for formations of underactuated ASVs. A LiDAR-based local perception obstacle avoidance containment maneuvering controller is proposed. Using the proposed control approach, underactuated ASVs converge to the convex hull of the virtual leaders and form the desired formation while avoiding collisions with static and dynamic obstacles. The stability analysis confirms that error signals remain bounded near zero, with the LiDAR-based safety function maintaining the system within the safe set. The simulation results further validate the effectiveness of the proposed control approach.

Acknowledgments. This work was supported in part by the National Science and Technology Major Project under Grant 2022ZD0119902, in part by the National Natural Science Foundation of China under Grants 52501374 and 52471372, in part by the Doctoral Scientific Research Foundation of Liaoning Province under Grant 2024-BS-012, in part by the Liaoning Revitalization Leading Talents Program under Grant XLYC2402054, in part by the Key Basic Research of Dalian under Grant 2023JJ11CG008, and in part by the Open Project of State Key Laboratory of Maritime Technology and Safety under Grant SKLMTS-DMU-2024-05.

References

1. von Ellenrieder, K.D., Camurri, M.: Relaxed control barrier function based control for closest approach by underactuated USVs. IEEE J. Oceanic Eng. (2024)
2. Er, M.J., Gong, H., Liu, Y., Liu, T.: Intelligent trajectory tracking and formation control of underactuated autonomous underwater vehicles: a critical review. IEEE Trans. Syst. Man Cybern. Syst. **54**(1), 543–555 (2023)
3. Fossen, T.I.: Handbook of Marine Craft Hydrodynamics and Motion Control. Wiley (2011)
4. Ghommam, J., Saad, M., Mnif, F., Zhu, Q.M.: Guaranteed performance design for formation tracking and collision avoidance of multiple USVs with disturbances and unmodeled dynamics. IEEE Syst. J. **15**(3), 4346–4357 (2020)
5. Gu, N., Wang, D., Peng, Z., Liu, L.: Distributed containment maneuvering of uncertain under-actuated unmanned surface vehicles guided by multiple virtual leaders with a formation. Ocean Eng. **187**, 105996 (2019)
6. Keyumarsi, S., Atman, M.W.S., Gusrialdi, A.: LiDAR-based online control barrier function synthesis for safe navigation in unknown environments. IEEE Robot. Autom. Lett. **9**(2), 1043–1050 (2023)
7. Lv, G., Peng, Z., Liu, L., Wang, J.: Barrier-certified distributed model predictive control of under-actuated autonomous surface vehicles via neurodynamic optimization. IEEE Trans. Syst. Man Cybern. Syst. **53**(1), 563–575 (2022)
8. Mei, J., Ren, W., Ma, G.: Distributed containment control for Lagrangian networks with parametric uncertainties under a directed graph. Automatica **48**(4), 653–659 (2012)
9. Peng, Z., Wang, J., Wang, D., Han, Q.L.: An overview of recent advances in coordinated control of multiple autonomous surface vehicles. IEEE Trans. Industr. Inf. **17**(2), 732–745 (2020)
10. Skjetne, R., Fossen, T.I., Kokotović, P.V.: Adaptive maneuvering, with experiments, for a model ship in a marine control laboratory. Automatica **41**(2), 289–298 (2005)
11. Wang, H., Zheng, J., Fu, J., Wang, Y.: Bounded containment maneuvering protocols for marine surface vehicles with quantized communications and tracking errors constrained guidance: theory and experiment. IEEE Trans. Cybern. (2024)
12. Wang, Y., Qu, Y., Zhao, S., Fu, H.: Adaptive neural containment maneuvering of underactuated surface vehicles with prescribed performance and collision avoidance. Ocean Eng. **297**, 116779 (2024)

13. Wei, H., Sun, Q., Chen, J., Shi, Y.: Robust distributed model predictive platooning control for heterogeneous autonomous surface vehicles. Control. Eng. Pract. **107**, 104655 (2021)
14. Wen, G., Fu, J., Lu, H., Sun, J., Shen, H.: Robust collision avoidance and path-following of USVs with reduced conservativeness: a control barrier function-based approach. J. Field Robot. **42**(4), 1388–1400 (2025)
15. Wu, W., Zhang, Y., Jia, Z., Lu, J.G., Zhang, W.: Adaptive fault-tolerant fuzzy containment control for networked autonomous surface vehicles: a noncooperative game approach. IEEE Trans. Fuzzy Syst. **32**(7), 4192–4204 (2024)
16. Xu, Y., Liu, L., Gu, N., Wang, D., Peng, Z.: Multi-ASV collision avoidance for point-to-point transitions based on heading-constrained control barrier functions with experiment. IEEE/CAA J. Automatica Sinica **10**(6), 1494–1497 (2022)
17. Yan, X., Jiang, D., Miao, R., Li, Y.: Formation control and obstacle avoidance algorithm of a multi-USV system based on virtual structure and artificial potential field. J. Mar. Sci. Eng. **9**(2), 161 (2021)

A Multi-graph Fusion Attention Network for Traffic Flow Forecasting

Wenlin He[1], Dawen Xia[1,2](✉), Jian Zhang[1], Han Zhang[1], Yang Hu[3], and Huaqing Li[4]

[1] College of Data Science and Information Engineering, Guizhou Minzu University, Guiyang 550025, China
{wenlinhe,jianzhang,hanzhang}@stu.gzmu.edu.cn

[2] Engineering Research Center of Micro-nano and Intelligent Manufacturing of Ministry of Education, Kaili University, Kaili 556011, China
dwxia@gzmu.edu.cn

[3] State Key Laboratory of Public Big Data, Guizhou University, Guiyang 550025, China
gs.yhu23@gzu.edu.cn

[4] College of Electronic and Information Engineering, Southwest University, Chongqing 400715, China
huaqingli@swu.edu.cn

Abstract. Traffic flow forecasting constitutes a fundamental task in Intelligent Transportation Systems (ITSs). Although existing methods capture spatiotemporal dependencies from traffic data, most rely solely on a single graph structure, making it difficult to jointly model long-term topological priors and short-term dynamics. To this end, this paper proposes a Multi-Graph Fusion Attention Network (MGFAN) for traffic flow forecasting, including a Periodic Attention Feature Selection Module (PAFSM) and a Multi-Graph Fusion Attention Gated Recurrent Unit (MGFA-GRU). Specifically, PAFSM adopts attention mechanisms to extract periodic patterns across different temporal scales. Moreover, MGFA-GRU integrates a multi-graph fusion module that dynamically constructs adjacency matrices from periodic features and historical data, where adaptive fusion weights are utilized to jointly model static and dynamic graphs to preserve stable topological priors while exploring dynamic variations. Additionally, a dual attention mechanism is combined with the graph convolution network to model long-term dependencies and multi-scale spatial representations. Extensive experiments across four real-world datasets demonstrate that MGFAN outperforms comparable baselines.

Keywords: Traffic flow forecasting · Multi-graph fusion attention · Spatiotemporal dependencies · Multi-scale periodic features

1 Introduction

In the process of urban intelligentization, Intelligent Transportation Systems (ITSs) have achieved unprecedented development with the advancement of tech-

C. Li et al. (Eds.): ICNC 2025, CCIS 2946, pp. 426–440, 2026.
https://doi.org/10.1007/978-981-92-1599-7_36

nologies such as big data, cloud computing, and artificial intelligence [3]. As a core component of ITSs, accurate traffic flow forecasting relies on spatiotemporal data from the Transportation Internet of Things to alleviate traffic congestion, optimize scheduling, and enhance the operational efficiency and sustainability of transportation systems.

Early studies mainly relied on traditional statistical and machine learning methods, such as HA [14] and VAR [20]. While these approaches captured linear features from traffic data, they depended on linearity and stationarity assumptions, making it difficult to extract the dynamic and nonlinear characteristics of traffic conditions. With the advancement of deep learning, RNN [21], LSTM [17], and GRU [7] were widely utilized for capturing temporal dependencies. Afterwards, researchers combined Graph Convolutional Networks (GCNs) with temporal forecasting models to jointly model spatiotemporal features, including T-GCN [23], DCRNN [10], ASTGCN [8], and STSGCN [15]. Furthermore, MST-ATG [9] employed a hybrid framework based on the Transformer, which effectively captured multi-scale temporal dependencies and spatial heterogeneities through an adaptive attention mechanism. However, most of approaches still relied on static graph structures, limiting their ability to adapt to the dynamic changes in traffic networks. Therefore, AGCRN [2], DGCRN [12], and DDGCRN [19] introduced dynamic graph generation mechanisms based on hypergraphs. Recently, FDGT [1] further advanced this paradigm by integrating differential graph learning with the attention mechanism, enabling dynamically modeling future-oriented spatiotemporal dependencies. IEDSFAN [22] uncovered complex and dynamic hidden relationships within road networks by integrating static connectivity graphs with dynamic adaptive graphs. Nevertheless, most approaches overlooked the inherent spatial connectivity graph, failing to fully capture the authentic spatial dependencies. In particular, most methods lack explicit modeling of multi-period temporal dependencies and cross-node attention interactions, and rely solely on a single graph structure, making them difficult to jointly model long-term topological priors and short-term dynamics.

To address these challenges, this paper proposes a Multi-Graph Fusion Attention Network (MGFAN) for traffic flow forecasting. By incorporating a Periodic Attention Feature Selection Module (PAFSM), the MGFAN adaptively models multi-scale temporal dependencies. It generates dynamic graphs based on periodic features and traffic signals. Moreover, a fusion weight matrix is designed to dynamically weight static and dynamic graphs, collaboratively modeling long-term topological priors and short-term dynamic variations. This approach overcomes the limitations of traditional methods that rely on a single graph. In addition, to enhance modeling capabilities of spatial dependencies, a dual attention mechanism is introduced into the GCN.

The main contributions of this paper are summarized as follows:

- We construct a Multi-Graph Fusion Attention Network (MGFAN) to overcome the limitation of existing methods constrained to a single-graph representation. MGFAN collaboratively integrates static topological graphs with dynamic graphs to jointly model long-term topological priors and short-term dynamic variations. Meanwhile, an attention mechanism is incorporated to enhance information exchange among nodes, thereby enabling more comprehensive modeling of complex spatiotemporal dependencies.

- We design a Multi-Graph Fusion Module (MGFM) that generates dynamic graphs based on multi-scale periodic features and real-time traffic flow to capture dynamic inter-node relationships. Unlike existing approaches that construct dynamic graphs solely from historical traffic flow, MGFM explicitly incorporates daily and weekly periodic features as prior knowledge. Furthermore, an adaptive fusion weighting mechanism is introduced further to dynamically balance static and dynamic graphs, achieving effective integration of complementary spatial information.
- We propose a GCN with a dual attention mechanism to enhance the modeling of complex spatial dependencies, where multi-head attention is applied both before and after graph convolution. This design enriches node representations through contextual information and adaptively fuses outputs from multiple diffusion orders, enhancing the capability to capture non-local spatial dependencies and multi-scale structural patterns.

The remainder of this paper is organized as follows: Sect. 2 introduces preliminaries, Sect. 3 details the MGFAN model, Sect. 4 presents experimental results and analyses, and Sect. 5 concludes the paper.

2 Preliminaries

This section first defines relevant concepts on the traffic network and traffic flow, and then formally describes the traffic flow forecasting task.

Definition 1 (Traffic Network): A traffic network can be represented as $G = (V, E, A)$, where V is the set of road nodes and the number of nodes is N, and E is the edges between nodes. We employ two types of adjacency matrices: A_t^{in} and A_t^{out}. A_t^{in} denotes the traffic flow inputted into this node from other nodes at time step t, while A_t^{out} represents the outflow characteristics from this node to other nodes at time step t.

Definition 2 (Traffic Flow): Traffic flow can be defined as $X \in \mathbb{R}^{N \times L \times F}$, where L represents the total length of the input sequence, and F is the number of traffic signals. Furthermore, the input traffic signal $X^{(t)} \in \mathbb{R}^{D \times N}$ represents data collected at time step t from all observations in the traffic network G, where D represents the initial dimension of the feature channel.

Formalization: The goal of traffic flow forecasting is to predict future traffic flow $Y = \left[Y^{(t+1)}, Y^{(t+2)}, \cdots, Y^{(t+Q)}\right]$ based on historical observations $X = \left[X^{(t-q+1)}, X^{(t-q+2)}, \cdots, X^{(t)}\right]$. Formally, the task is expressed as $Y = f(X; G)$, where q and Q denote the lengths of the historical and future horizons, respectively, and $f(\cdot)$ represents the mapping function that captures the relationship between past and future traffic states.

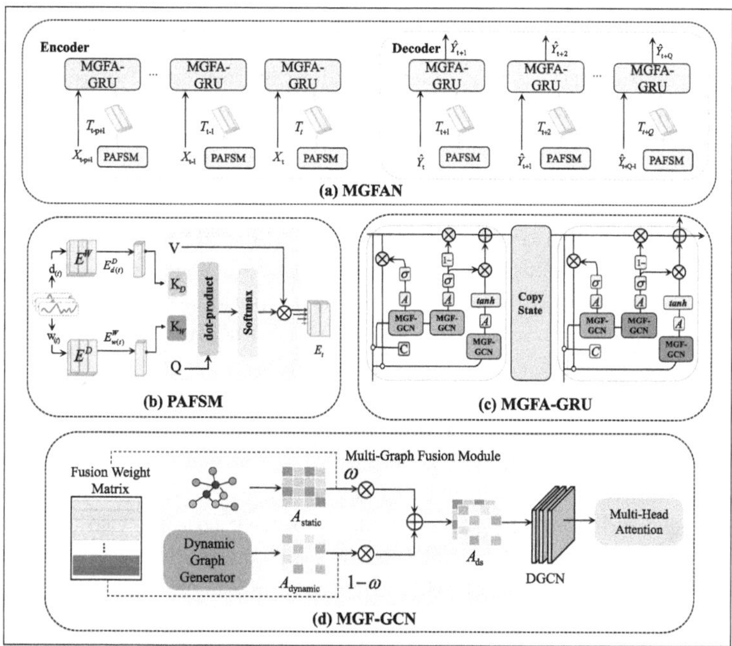

Fig. 1. Overview of MGFAN. (a) MGFAN adopts an encoderdecoder architecture with a Multi-Graph Fusion Attention Gated Recurrent Unit (MGFA-GRU) and a Periodic Attention Feature Selection Module (PAFSM), (b) PAFSM extracts multi-scale periodic features, (c) MGFA-GRU captures spatiotemporal dependencies via a stacked Multi-Graph Fusion Graph Convolutional Network (MGF-GCN), and (d) MGF-GCN fuses static and dynamic graphs and enhances node features with multi-head attention mechanisms.

3 MGFAN Model

This section describes the proposed MGFAN model in details.

3.1 Model Overview

The overall framework of MGFAN is illustrated in Fig. 1. This model adopts an encoder-decoder architecture designed to comprehensively model the complex spatiotemporal dependencies and periodic characteristics inherent within traffic flow. In the encoder, the model incorporates a Periodic Attention Feature Selection Module (PAFSM) and a Multi-Graph Fusion Attention Gated Recurrent Unit (MGFA-GRU). The former adaptively extracts temporal features such as daily and weekly periods, while the latter captures spatial dependencies across dynamic and static graphs. In the decoder, the model progressively generates multi-step forecasting based on the previous forecasting results and periodic features, thereby facilitating regression-based traffic flow forecasting.

3.2 Periodic Attention Feature Selection Module

The periodic characteristics of urban traffic flow are influenced by residents' daily travel behaviors and lifestyles, including morning or evening peaks and weekday or weekend variations. The periodic characteristics exhibit significant variations across different periods per day, which is important for forecasting traffic flow. To enhance the model's perception on these features, we introduce the Periodic Attention Feature Selection Module (PAFSM). This module adaptively models the dependencies between the current time step and periodic features, dynamically adjusting periodic embeddings. This allows periods with greater predictive influence, such as peak hours, to play a more significant role in the forecasting process.

Specifically, as illustrated in Fig. 1(b), we construct the daily feature pool, $E^D \in \mathbb{R}^{N_d \times p}$, and the weekly feature pool, $E^W \in \mathbb{R}^{N_w \times p}$, to learn corresponding multi-scale features, where N_d denotes the number of time slots in a day, and $N_w = 7$ represents the number of days in a week. To obtain the periodic features corresponding to the current time step, we introduce two periodic index functions, $d(t)$ and $w(t)$, to extract the periodic embeddings, $E^D_{d(t)}$ and $E^W_{w(t)}$, from the feature pools. These are then used to model the periodic features at the current time step. Next, a multi-head attention mechanism calculates the correlations between the current traffic state input X_t and the two periodic embeddings, thereby adoptively capturing the influence of different periodic patterns on the current time step. Finally, outputs of the attention mechanism from the daily and weekly features are fused to obtain the enhanced representations.

$$Z_D = \text{Softmax}(\frac{Q(X_t) \cdot K(E^D_{d(t)})^T}{\sqrt{d}}) \cdot V(E^D_{d(t)}), \tag{1}$$

$$Z_W = \text{Softmax}(\frac{Q(X_t) \cdot K(E^W_{w(t)})^T}{\sqrt{d}}) \cdot V(E^W_{w(t)}), \tag{2}$$

$$Z = Z_D \odot Z_W, \tag{3}$$

where d denotes the feature dimension. After performing dot products on $Q(X_t)$ and $K(\cdot)$ values and applying scaling, the attention weights are obtained through normalization with the Softmax function, which is defined as $\text{Softmax}(\mathbf{z})_i = \frac{e^{z_i}}{\sum_j e^{z_j}}$. Subsequently, multiplying these weights by $V(\cdot)$ yields the corresponding attention representations. The symbol $\odot$ represents the element-wise multiplication or broadcast dot product. Finally, the fused periodic features $Z \in \mathbb{R}^{N \times F}$ are concatenated with the original traffic flow X_t to obtain the enhanced features for input.

$$\tilde{X}_t = \left[X_t \parallel Z\right]. \tag{4}$$

3.3 Dynamic Graph Generator

The underlying spatial dependencies among traffic flow on road segments vary dynamically over time. However, topology- and distance-based graphs remain static, making them incapable of capturing temporal variations in spatial relationships. To address this, we design a dynamic graph generation module (see Fig. 2), to adaptively capture the spatiotemporal dependencies between traffic nodes. Specifically, we extract data-driven dynamic flow from traffic data to characterize the latent dynamic relationships, given as follows:

$$F_t = \mathrm{MLP}(\tilde{X}_t), \tag{5}$$

where $F_t \in \mathbb{R}^{N \times P}$ denotes the dynamic features extracted from $\tilde{X}_t$. The Multi-Layer Perceptron (MLP) is implemented as a two-layer fully connected network with a ReLU activation function.

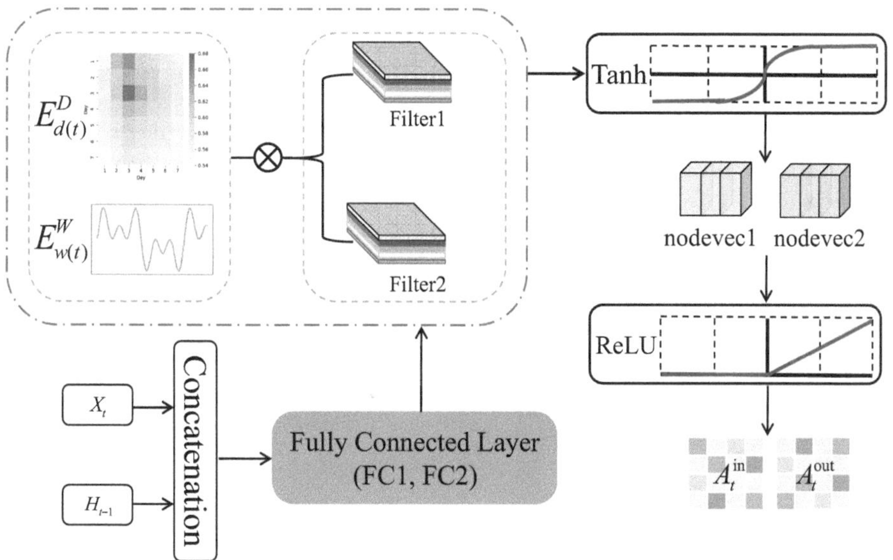

Fig. 2. Dynamic graph generator.

In fact, traffic flow is usually bidirectional, exhibiting contrasting inflow and outflow characteristics. Therefore, we denote the dynamic signals of inflow and outflow as F_t^{in} and F_t^{out}, respectively. To generate accurate forecasting, we consider the dynamic patterns of traffic flow across different roads, as well as the temporal evolution of similar traffic flow patterns at different time steps. Combining the extracted dynamic signals with multi-period embedding yields a dynamic periodic graph embedding, expressed as follows:

$$E_t^{in} = \tanh(F_t^{in} \odot E_{d(t)}^D \odot E_{w(t)}^W), \tag{6}$$

$$E_t^{out} = \tanh(F_t^{out} \odot E_{d(t)}^D \odot E_{w(t)}^W), \tag{7}$$

where $\odot$ denotes the element-wise Hadamard product. The hyperbolic tangent function $\tanh(\cdot)$ is defined as $\tanh(x) = \frac{e^x - e^{-x}}{e^x + e^{-x}}$, which maps input values to the range $(-1, 1)$. Subsequently, a dynamic periodic graph is generated according to the similarity among nodes, formulated as follows:

$$A_t^{in} = \text{ReLU}(E_t^{in} E_t^{out^T} - E_t^{out} E_t^{in^T}), \tag{8}$$

$$A_t^{out} = \text{ReLU}(E_t^{out} E_t^{in^T} - E_t^{in} E_t^{out^T}), \tag{9}$$

where ReLU() denotes the activation function with $\text{ReLU}(x) = \max(0, x)$, and A_t^{in} and A_t^{out} represent the dynamic graph adjacency matrices for the inflow and outflow directions, respectively, at the current time step t. The generated periodic dynamic graph is asymmetric, better capturing the directional dependencies of traffic flow.

3.4 Multi-graph Fusion Module

In traffic flow modeling, a single graph structure struggles to simultaneously accommodate long-term stable topologies and short-term dynamic characteristics. Static graphs (e.g., road topology maps) provide stable structural priors, while dynamic graphs reflect real-time traffic variation patterns. To effectively integrate the strengths of static and dynamic graphs, we propose a multi-graph fusion mechanism that adaptively combines static adjacency matrices with data-driven dynamic graphs via an attention mechanism. As illustrated in Fig. 1(d), to effectively fuse information from static and dynamic graphs, we design a multi-graph fusion mechanism, defined as follows:

$$\omega_t = \sigma([X_t \parallel H_{t-1}]), \tag{10}$$

$$A_{\text{ds}} = \omega_t \cdot A_{\text{static}} + (1 - \omega_t) \cdot A_{\text{dynamic}}, \tag{11}$$

where σ represents the activation function, Sigmoid defined as $\sigma(x) = \frac{1}{1+e^{-x}}$, which calculates the fusion weight $\omega_t \in \mathbb{R}^{B \times N \times 1}$ for the current input state. This enables adaptive modeling of the fusion ratio for each node. The final output of the multi-graph fusion module is denoted by A_{ds}. Obviously, through this fusion mechanism, the model can adaptively select static or dynamic graph information at different time steps, enabling the collaborative modeling of long-term topological priors and short-term dynamics.

3.5 Multi-graph Fusion Graph Convolutional Network

A main challenge in traffic flow forecasting lies in the complex and dynamically changing spatial dependencies among traffic nodes. To enhance the global perception capability, a multi-head attention mechanism is introduced to deal with node features $H \in \mathbb{R}^{B \times N \times D}$. This contextually models the original node features by treating each node as a *Query* and comparing it with the *Key* and *Value* of all other nodes to obtain an attention matrix. The resulting feature representation is globally contextually enhanced as follows:

$$H' = MHA(H). \tag{12}$$

When constructing the adjacency structure for the graph convolution, a static adjacency graph originated from the road topology is fused with a dynamic periodic graph based on historical periodic traffic flow relationships to generate the fusion graph A^f_{ds}. By separately constructing the inflow graph A^f_{in} and outflow graph A^f_{out}, and then normalizing them, the diffusion probability matrix is obtained as follows:

$$P^{(k)}_{in} = (D^f_{in})^{-1}(A^{in}_{ds} + I)^k, \tag{13}$$

$$P^{(k)}_{out} = (D^f_{out})^{-1}(A^{out}_{ds} + I)^k, \tag{14}$$

where I denotes the identity matrix, $k \in [0, K]$ signifies the diffusion order, and D^f_{in} and D^f_{out} represent the degree matrix. The diffusion graph convolution operation performed on the fusion graph is expressed as follows:

$$Z = \sum_{k=0}^{K}(P^{(k)}_{in}H' \odot W^{(k)}_{in} + P^{(k)}_{out}H' \odot W^{(k)}_{out}), \tag{15}$$

$$Z' = MHA([Z^{(0)}; Z^{(1)}; \cdots ; Z^{(K)}]), \tag{16}$$

where $\odot$ denotes the element-wise multiplication, and $W^{(k)}_{in}$ and $W^{(k)}_{out}$ represent learnable parameters. All weights are shared with the node embedding matrix E to enhance parameter efficiency. To improve the sensitivity of different diffusion orders, an attention fusion module is incorporated after graph convolution for modeling the interrelationships among multi-order information.

3.6 Multi-graph Fusion Attention Gated Recurrent Unit

To model temporal features from traffic flow, we employ the GRU [5] to capture temporal characteristics. Specifically, a multi-graph fusion graph convolutional layer replaces the MLP layer in the GRU. Subsequently, as shown in Fig. 1(c), an attention mechanism is introduced to further enhance the ability to select key moments, resulting in a novel architecture termed the Multi-Graph Fusion Attention Gated Recurrent Unit (MGFA-GRU). The MGFA-GRU is defined as follows:

$$r_t = \sigma(W_r * G(x_t \parallel H_{t-1})), \tag{17}$$

$$u_t = \sigma(W_u * G(x_t \parallel H_{t-1})), \tag{18}$$

$$\tilde{H}_t = \tanh(W_r * G(x_t \parallel (u_t \odot H_{t-1}))), \tag{19}$$

$$H_t = r_t \odot H_{t-1} + (1 - r_t) \odot \tilde{H}_t, \tag{20}$$

$$\alpha_t = \mathrm{Softmax}\left(q \cdot \tanh\left(W_\alpha H_t + b_\alpha\right)\right), \tag{21}$$

$$H_{\alpha tt} = \sum_{t=1}^{T} \alpha_t \cdot H_t, \tag{22}$$

where X_t represents the current input, H_{t-1} denotes the hidden state from the previous time step, $*G$ is the graph convolution operation, W_r, W_u, and W_h are the learnable parameters, $\odot$ denotes the Hadamard product, and $\|$ represents the concatenation operation. The graph convolution component integrates static adjacency graphs with dynamic periodic graphs, while sharing node embeddings to enhance the sensitivity to spatial structures in time-dependent modeling. q, W_α, and b_α denote learnable parameters. The mechanism adaptively aggregates historical temporal information by weighted aggregation, focusing on critical temporal segments to effectively mitigate information decay in long sequences.

4 Experiments

This section experimentally validates the proposed MGFAN model on four real-world datasets.

4.1 Datasets

We evaluate the performance of MGFAN on four real-world datasets, including PEMS03, PEMS04, PEMS07, and PEMS08. These datasets are collected by the California Performance Measurement System (PEMS) [15] with five-minute intervals. To reduce the training complexity and enhance the computational efficiency, all raw data are normalized by the Z-scores.

4.2 Baselines

To validate the performance of MGFAN, we compare it with nine baselines:

(1) VAR [20]: A vector autoregressive model based on statistical properties of time series.
(2) FC-LSTM [16]: A fully connected LSTM-based recurrent network.
(3) GRU-ED [16]: A GRU encoder-decoder network with dual RNNs.
(4) DCRNN [10]: A diffusion graph convolutional recurrent network.
(5) AGCRN [2]: An adaptive graph convolutional recurrent network with node-specific parameters and data-driven graph construction.
(6) DSTAGNN [11]: A dynamic spatiotemporal perception graph neural network with spatiotemporal attention and convolution for capturing dynamic dependencies.
(7) STG-NCDE [4]: A spatiotemporal graph neural network with neural control differential equations.
(8) STIDGCN [13]: A spatiotemporal dynamic graph convolutional network with interactive learning and downsampling mechanisms.
(9) PDG2Seq [6]: A neural network based on periodic dynamic graphs for better modeling spatial dependencies.

4.3 Parameter Settings and Evaluation Metrics

The objective of hyperparameter selection is to identify the dimension p of periodic embeddings from the set of $\{4, 5, \ldots, 20\}$, the dimension d of node embedding matrices within the set of $\{1, 4, 8, 12\}$, the depth K of graph convolutions from the set of $\{1, 2, 3\}$, and the number of attention heads $h \in \{1, 2, 4, 8, 16\}$. Optimal parameter values are determined through experimental validations. During the training phase, the model employs the MultiStepLR scheduler to adjust the learning rate. Table 1 presents the specific parameter settings used for each dataset.

Furthermore, the experimental datasets are divided into three subsets: training (60%), validation (20%), and test (20%). All experiments are conducted on a server platform equipped with an NVIDIA A40 GPU (48 GB VRAM) with Python 3.8, PyTorch 1.13, and CUDA 11.8. The proposed MGFAN model uses the Adam optimizer with Mean Absolute Error (MAE) as the loss function.

In addition, to evaluate the forecasting performance, three metrics are employed as follows [18]: mean absolute error (MAE), root mean square error (RMSE), and mean absolute percentage error (MAPE).

Table 1. Parameter settings.

Datasets	p	h	d	K	Batch Size	Learning Rate
PEMS03	8	8	4	1	64	0.003
PEMS04	16	8	8	1	64	0.003
PEMS07	20	4	10	1	16	0.00075
PEMS08	16	8	8	2	64	0.003

4.4 Experimental Results

As illustrated in Table 2, the performance of MGFAN is compared with baselines across four PEMS datasets, where values in bold denote the best performance and underlined values indicate the second-best performance. From Table 2 with the average values, the experimental results demonstrate that MGFAN attains the best performance on the majority of metrics on all datasets. Specifically, on the PEMS07 and PEMS08 datasets, MGFAN achieves a decrease of 2.3%–62.5% in MAE, 3.1%–75.7% in MAPE, and 0.3%–57.2% in RMSE on PEMS07; a decrease of 1.3%–42.5% in MAE, 0.5%–41.8% in MAPE, and 1.1%–36.4% in RMSE on PEMS08. Notably, the RMSE values on PEMS03 and PEMS04 are slightly higher than those of STIDGCN because the downsampling operation in STIDGCN smooths short-term fluctuations, whereas MGFAN preserves more temporal details, slightly increasing RMSE despite better overall accuracy. Among graph-based models, data-driven dynamic graph approaches, AGCRN and DSTAGNN, outperform static graph methods, including DCRNN and STG-NCDE. The experimental results indicate that dynamic modeling can facilitate identifying latent spatial dependencies. However, these models typically overlook the pronounced periodicity of traffic flow and fail to fully exploit temporal contextual information when extracting long-term dependencies. In contrast, MGFAN reduces MAE, MAPE, and RMSE by 5.8%–25.4%, 0.3%–33.8%, and 2.4%–16.1%, respectively, across the four datasets. This is because MGFAN integrates periodic features with dynamic signals to generate a dynamic graph and achieves effective fusion of static and periodic dynamic graphs via a multi-graph fusion module, thereby making fuller use of spatial and temporal information.

Table 2. Performance comparisons between MGFAN and baselines on four PEMS datasets.

Models	PEMS03			PEMS04			PEMS07			PEMS08		
	MAE	MAPE (%)	RMSE	MAE	MAPE (%)	RMSE	MAE	MAPE (%)	RMSE	MAE	MAPE (%)	RMSE
VAR	23.65	24.50	38.26	24.54	17.24	38.61	50.22	32.22	75.63	19.19	13.10	29.81
FC-LSTM	21.33	23.33	35.11	26.77	18.23	40.65	29.98	13.20	45.94	23.09	14.99	35.17
GRU-ED	19.12	19.31	32.85	23.68	16.44	39.27	27.66	12.20	43.49	22.00	13.33	36.22
DCRNN	17.99	18.34	30.31	21.22	14.17	33.44	25.22	11.82	38.61	16.82	10.92	26.36
AGCRN	15.98	15.23	28.25	19.83	12.97	32.26	22.37	9.12	36.55	15.95	10.09	25.20
DSTAGNN	15.57	14.68	27.21	19.30	12.70	31.46	21.42	9.01	34.51	15.67	9.94	24.77
STG-NCDE	15.50	14.90	27.06	19.13	12.68	30.94	20.45	8.65	33.73	15.32	8.90	24.72
STIDGCN	14.76	15.28	**24.59**	18.16	12.24	**29.77**	19.26	8.11	32.51	13.45	8.77	23.28
PDG2Seq	14.62	14.88	25.47	18.24	12.09	30.08	19.28	8.07	33.04	13.60	8.99	23.37
MGFAN	**14.39**	**14.64**	25.43	**18.02**	**11.97**	30.19	**18.81**	**7.82**	**32.41**	**13.27**	**8.73**	**23.02**

To demonstrate the predictive performance of all models across different horizons, from Fig. 3, the experimental results show that PDG2Seq effectively models spatiotemporal dependencies by incorporating periodic features into dynamic signals and outperforms STG-NCDE, DSTGANN, and AGCRN. MGFAN integrates further periodic features with spatial information in dynamic graphs. Therefore, it demonstrates the superiority in both short- and long-term forecasting.

4.5 Ablation Experiments

To verify the effectiveness of the components of MGFAN, we design four MGFAN variants as followings.

(1) w/o MPA: This variant removes the periodic attention feature selection module.
(2) w/o MGF: This variant removes the multi-graph fusion module and uses only dynamic graph.
(3) w/o GCN: This variant replaces the GCN modules with MLP layers.
(4) w/o DAM: This variant removes the attention mechanism both before and after the GCN.

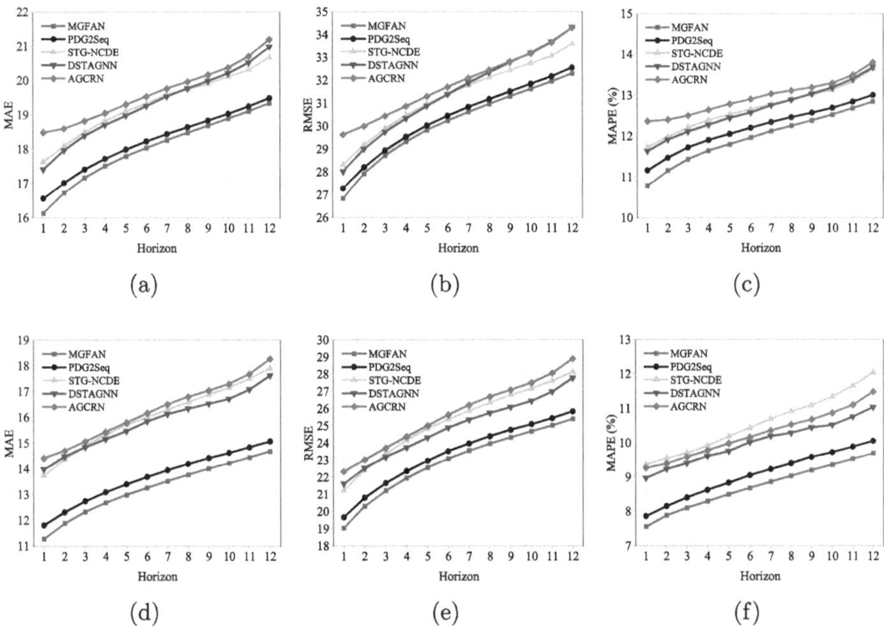

Fig. 3. Evaluation metrics of all models on PEMS04 and PEMS08. (a) MAE on PEMS04, (b) RMSE on PEMS04, (c) MAPE on PEMS04, (d) MAE on PEMS08, (e) RMSE on PEMS08, and (f) MAPE on PEMS08.

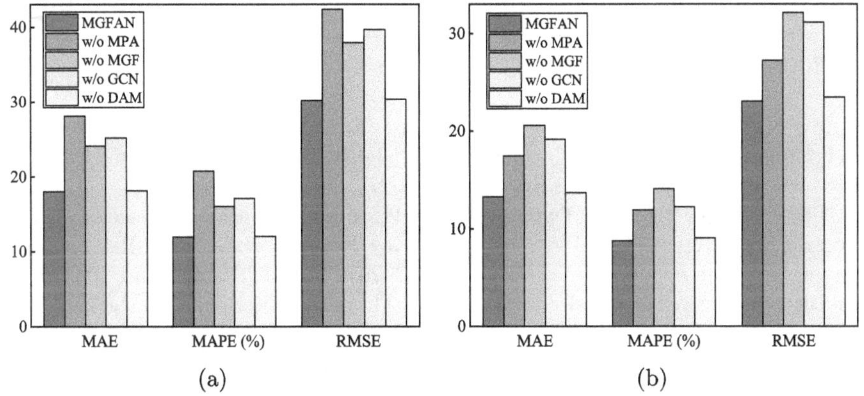

Fig. 4. Ablation experiments of MGFAN on PEMS04 and PEMS08. (a) PEMS04 and (b) PEMS08.

As shown in Fig. 4, the four variant models exhibit different performance across various metrics on the PEMS04 and PEMS08 datasets. However, MGFAN demonstrates significantly superior performance compared to all variants, indicating that each module within MGFAN contributes positively. Notably, the performance of the w/o MPA variant declines markedly, particularly on the PEMS04 dataset, highlighting the critical role of multi-period dependency modeling in traffic flow forecasting. Compared to MGFAN, the w/o MGF variant performs worse, demonstrating that the stable topological prior provided by static graphs significantly enhances the short-term dependencies modeled by dynamic graphs. In the w/o GCN variant, replacing graph convolutions with an MLP markedly reduces the predictive performance, further demonstrating the irreplaceable role of graph convolutions in extracting spatial features. Although w/o DAM shows smaller declines across metrics, its MAPE degradation is particularly pronounced, indicating that attention mechanisms before and after graph convolutions enhance spatial contextual expression between nodes, enhancing the capability to prioritize critical dependencies.

5 Conclusion

To overcome the limitations of most existing methods that rely solely on a single graph structure, while collaboratively modeling long-term topological priors and short-term dynamic variations, we proposed a Multi-Graph Fusion Attention Network (MGFAN) for traffic flow forecasting. MGFAN incorporated a multi-graph fusion module that comprehensively considered the periodicity of traffic data, the hidden dynamic signals, and static road network graphs, achieving dynamic fusion to fully extract spatiotemporal features. Furthermore, to capture latent dependencies between road nodes, we introduced attention networks before and after graph convolutions. Extensive experiments on four real-world traffic datasets demonstrated that MGFAN outperforms the comparable baselines in predictive performance.

Acknowledgments. This work was supported in part by the National Natural Science Foundation of China (No. 62462013), the High-Level Innovative Talent Project of Guizhou Province (No. QKHPTRC-GCC2023027), the Science and Technology Innovation Talent Team Project of Data Science and Computational Intelligence of Guizhou Province (No. QKHRC-CXTD2025038), the Key Project of the Basic Research Program of Guizhou Province (No. QKHJCZD2026144), the Science and Technology Breakthrough Project of Hundred Schools and Thousand Enterprises of the Education Department of Guizhou Province (No. QJJ2025011), the Key Project of Engineering Research Center of Micro-nano and Intelligent Manufacturing of Ministry of Education (No. WZG202501), and the Scientific Research Platform (Key Laboratory and Engineering Research Center) Project of Kaili University (Nos. YTH-PT202501 and YTH-PT202602).

Disclosure of Interests. The authors declare that they have no known competing financial interests or personal relationships that could have appeared to influence the work reported in this paper.

References

1. Bai, D., et al.: Future-heuristic differential graph transformer for traffic flow forecasting. Inf. Sci. **701**, 121852 (2025)
2. Bai, L., Yao, L., Li, C., Wang, X., Wang, C.: Adaptive graph convolutional recurrent network for traffic forecasting. In: Proceedings of the Advances in Neural Information Processing Systems, vol. 33, pp. 17804–17815 (2020)
3. Chen, B., Cheng, H.H.: A review of the applications of agent technology in traffic and transportation systems. IEEE Trans. Intell. Transp. Syst. **11**(2), 485–497 (2010)
4. Choi, J., Choi, H., Hwang, J., Park, N.: Graph neural controlled differential equations for traffic forecasting. In: Proceedings of the AAAI Conference on Artificial Intelligence, vol. 36, pp. 6367–6374 (2022)
5. Dey, R., Salem, F.M.: Gate-variants of gated recurrent unit (GRU) neural networks. In: Proceedings of the IEEE 60th International Midwest Symposium on Circuits and Systems (MWSCAS), pp. 1597–1600. IEEE (2017)
6. Fan, J., Weng, W., Chen, Q., Wu, H., Wu, J.: PDG2Seq: periodic dynamic graph to sequence model for traffic flow prediction. Neural Netw. **183**, 106941 (2025)
7. Fu, R., Zhang, Z., Li, L.: Using LSTM and GRU neural network methods for traffic flow prediction. In: Proceedings of the 31st Youth Academic Annual Conference of Chinese Association of Automation (YAC), pp. 324–328. IEEE (2016)
8. Guo, S., Lin, Y., Feng, N., Song, C., Wan, H.: Attention based spatial-temporal graph convolutional networks for traffic flow forecasting. In: Proceedings of the AAAI Conference on Artificial Intelligence, vol. 33, pp. 922–929 (2019)
9. Hu, Y., Li, S., Xia, D., Zhang, W., Yuan, P., Wu, F., Li, H.: A multi-view spatial-temporal adaptive Transformer-GRU framework for traffic flow prediction. IEEE Internet Things J. **12**(6), 7114–7132 (2025)
10. Huang, Y., Weng, Y., Yu, S., Chen, X.: Diffusion convolutional recurrent neural network with rank influence learning for traffic forecasting. In: Proceedings of the 18th IEEE International Conference on Trust, Security and Privacy in Computing and Communications/13th IEEE International Conference on Big Data Science and Engineering (TrustCom/BigDataSE), pp. 678–685. IEEE (2019)
11. Lan, S., Ma, Y., Huang, W., Wang, W., Yang, H., Li, P.: DSTAGNN: dynamic spatial-temporal aware graph neural network for traffic flow forecasting. In: Proceedings of the International Conference on Machine Learning, pp. 11906–11917. PMLR (2022)
12. Li, F., et al.: Dynamic graph convolutional recurrent network for traffic prediction: benchmark and solution. ACM Trans. Knowl. Discov. Data **17**(1), 1–21 (2023)
13. Liu, A., Zhang, Y.: Spatial-temporal dynamic graph convolutional network with interactive learning for traffic forecasting. IEEE Trans. Intell. Transp. Syst. **25**(7), 7645–7660 (2024)
14. Liu, J., Guan, W.: A summary of traffic flow forecasting methods. J. Highw. Transp. Res. Develop. **21**(3), 82–85 (2004)
15. Song, C., Lin, Y., Guo, S., Wan, H.: Spatial-temporal synchronous graph convolutional networks: a new framework for spatial-temporal network data forecasting. In: Proceedings of the AAAI Conference on Artificial Intelligence, vol. 34, pp. 914–921. (2020)
16. Sutskever, I., Vinyals, O., Le, Q.V.: Sequence to sequence learning with neural networks. Adv. Neural. Inf. Process. Syst. **27** (2014)

17. Van Houdt, G., Mosquera, C., Nápoles, G.: A review on the long short-term memory model. Artif. Intell. Rev. **53**(8), 5929–5955 (2020). https://doi.org/10.1007/s10462-020-09838-1
18. Wang, X., et al.: Traffic flow prediction via spatial temporal graph neural network. In: Proceedings of the Web Conference 2020, pp. 1082–1092 (2020)
19. Weng, W., et al.: A decomposition dynamic graph convolutional recurrent network for traffic forecasting. Pattern Recogn. **142**, 109670 (2023)
20. Williams, B.M., Hoel, L.A.: Modeling and forecasting vehicular traffic flow as a seasonal Arima process: theoretical basis and empirical results. J. Transp. Eng. **129**(6), 664–672 (2003)
21. Wu, Y., Tan, H., Qin, L., Ran, B., Jiang, Z.: A hybrid deep learning based traffic flow prediction method and its understanding. Transp. Res. Part C Emerg. Technol. **90**, 166–180 (2018)
22. Yu, L., Wang, Z., Yang, W., Qu, Z., Ren, C.: IEDSFAN: information enhancement and dynamic-static fusion attention network for traffic flow forecasting. Complex Intell. Syst. **11**(1), 28 (2025)
23. Zhao, L., Song, Y., Zhang, C., Liu, Y., Wang, P., Lin, T., Deng, M., Li, H.: T-GCN: a temporal graph convolutional network for traffic prediction. IEEE Trans. Intell. Transp. Syst. **21**(9), 3848–3858 (2019)

Adaptive Platoon Control of Heterogeneous Vehicles With Extended State Observer and Modified Constant Time Headway Policy

Shanshan Tian, Liang Cao(✉), and Meng Zhao

College of Mathematical Sciences, Bohai University, Jinzhou 121013, China
caoliang0928@163.com

Abstract. This paper focuses on the problem of adaptive platoon tracking control in the presence of uncertainties and external disturbances. A distributed platoon control strategy with active disturbance rejection is developed, utilizing an extended state observer (ESO) along with a modified constant time headway policy (CTHP). Firstly, an ESO is constructed to estimate external disturbances. Subsequently, a modified CTHP is employed to mitigate excessive transient response induced by non-zero initial spacing. This approach thereby eliminates the need for the zero-velocity assumption and resolves the issue of initial spacing errors. Finally, a series of simulation experiments are conducted to evaluate the efficacy of the proposed control scheme.

Keywords: Extended state observer · Spacing policy · Vehicle platoon

1 Introduction

In recent years, the accelerating pace of global urbanization has led to increasingly prevalent traffic congestion, drawing widespread attention to Intelligent Transportation Systems (ITS). However, the growing complexity of the traffic environment demands greater control capabilities. As a key cooperative control technology within ITS, vehicle platoon control coordinates speed and spacing among multiple vehicles [1]. It holds significant potential to improve road efficiency, enhance traffic flow stability and increase driving safety.

As road traffic demand continues to grow, driving safety has become a critical issue. Notably, in practical vehicle platoon, state constraints are critical to ensuring both platoon stability and driving safety, for example limitations on inter-vehicle spacing [2]. The most widely adopted policies include the constant spacing policy (CSP) [3] and the constant time headway policy (CTHP) [4]. The CSP has been applied in practice due to its ability to maintain fixed inter-vehicle spacing and enhance road capacity. However, under high-speed driving conditions, vehicles require longer safety distances, which the fixed spacing in

C. Li et al. (Eds.): ICNC 2025, CCIS 2946, pp. 441–450, 2026.
https://doi.org/10.1007/978-981-92-1599-7_37

CSP fails to accommodate dynamically. To address this limitation, the CTHP was introduced, which linearly adjusts the desired spacing according to vehicle speed. Nevertheless, the CTHP leads to significantly larger inter-vehicle spacing at high velocities and typically suffers from issues of non-zero initial spacing. To overcome this drawback, a modified CTHP was developed that ensures zero initial spacing, effectively mitigating excessive transient responses and initial deviation problems in [5]. Although this method performs well in handling transient response and initial state issues, its control performance in the presence of external disturbances requires further validation. Therefore, integrating the modified CTHP with disturbance rejection techniques presents a research direction worthy of exploration.

Rejecting external disturbances is crucial for ensuring stable and cooperative operation in vehicle platoon. Without effective suppression, these uncertainties can directly disrupt the overall stability of vehicle platoon. The extended state observer (ESO) offers an effective solution by estimating and compensating for external disturbances, making it a highly favored method for engineering control applications. Deng *et al.* [6] effectively reduced the noise sensitivity of the observer through DC-ESO. Zhang *et al.* [7] employed an NESO to observe and compensate for the composite disturbances in the current prediction equation.

Building upon the above discussions, this paper addresses the cooperative control challenges of heterogeneous vehicle platoon under dynamic disturbances and modified CTHP. The proposed approach employs the modified CTHP to address excessive transient response while maintaining zero initial spacing. Then, an ESO is utilized to estimate and compensate for external disturbances. In contrast to existing studies, this paper integrates the ESO with the modified CTHP to develop robustness for vehicle platoon. The proposed scheme demonstrates stable performance, guarantees minimal spacing errors and incorporates acceleration considerations.

2 Problem Description

2.1 Model Design

We consider a vehicle platoon comprising one leader and N followers. For the i-th vehicle, its states are defined with respect to position, velocity and acceleration, denoted by $x_{i,p}$, $x_{i,v}$ and $x_{i,a}$, respectively.

$$\begin{aligned}
\dot{x}_{i,p} &= x_{i,v} \\
\dot{x}_{i,v} &= x_{i,a} \\
\dot{x}_{i,a} &= -\frac{x_{i,a}}{\tau_i} - \frac{\bar{c}_i(x_{i,v}^2 + 2\tau_i x_{i,v} x_{i,a}) + \lambda_i}{m_i \tau_i} + \frac{u_i}{m_i \tau_i} + W_i \\
y_i &= x_{i,p}
\end{aligned} \tag{1}$$

where $\tau_i > 0$ represents the engine time constant of the i-th follower. u_i and y_i denote the control input and output, respectively. The term $\bar{c}_i$ is given by

$\frac{1}{2}\rho_{ai}\bar{A}_i C_{di}$, with the constituting parameters namely the air density ρ_{ai}, the vehicle cross-sectional area $\bar{A}_i$ and the air drag coefficient C_{di}. m_i is the i-th vehicle mass and g is the gravity acceleration. λ_i is given by $\lambda_i = m_i g \sin(\delta_i) + r_i m_i \cos(\delta_i)$, where δ_i represents the road slope.

2.2 Modified Constant Time Headway Policy (CTHP)

A modified CTHP policy is proposed to address these challenges, particularly the excessive transients and non-zero initial spacing errors. Its formulation is presented below:

$$e_i(t) = \tilde{e}_i(t) - \Gamma_i(t)$$

where

$$\tilde{e}_i(t) = d_i(t) - \Delta_{i-1,i} - h(x_{i,v}(t) - x_{0,v}(t))$$
$$\Gamma_i(t) = \{\tilde{e}_i(0) + [\pi_i \tilde{e}_i(0) + \dot{\tilde{e}}_i(0)]t + \frac{1}{2}[\pi_i^2 \tilde{e}_i(0) + 2\pi_i \dot{\tilde{e}}_i(0) + \ddot{\tilde{e}}_i(0)]t^2\}e^{-\pi_i t}$$

The inter-vehicle spacing for consecutive vehicles is denoted by $d_i(t) = x_{i-1,p}(t) - x_{i,p}(t) - H_i$, where H_i, $\Delta_{i-1,i}$ and h represent the physical length of vehicle, the desired inter-vehicle distance and CTHP, respectively. $\pi_i > 0$ and $\Gamma_i(t)$ is an exponential term introduced to satisfy the condition.

$$e_i(0) = 0, \dot{e}_i(0) = 0, \ddot{e}_i(0) = 0$$

As evidenced by the results, the proposed modified policy is effective in driving the initial state deviations to zero.

2.3 Neural Networks (NNs)

NNs are presented to estimate a continuous unknown function $F_i(x)$ as follows:

$$F_i(x) = \varpi_i^T \varphi_i(x) + \varepsilon_i(x), \forall x \in \Xi$$

where ϖ_i is the ideal weight vector, $\Psi_{i,1} = \|\varphi_{i,1}\|^2$, $\varphi_i(x) = [\varphi_1(x), \varphi_2(x), ..., \varphi_p(x)]^T$ means the basis function vector with $p > 1$ being the number of NNs nodes, φ_i represents the Gaussian function, $\varepsilon_{i,1}(x)$ is denoted as the approximation error satisfying $|\varepsilon_{i,1}| \le \bar{\varepsilon}_{i,1}$, where $\bar{\varepsilon}_{i,1}(x) > 0$ is an unknown constant, and Ξ is a compact set.

2.4 Extended State Observer (ESO)

To estimate the total disturbance in the systems (1), an ESO is designed.

$$\begin{cases} \tilde{x}_{i,a} = x_{i,a} - \hat{x}_{i,a} \\ \dot{\hat{x}}_{i,a} = \dfrac{u_i}{m_i \tau_i} + \hat{G}_i + \rho_{i,p}\mu_i \tilde{x}_{i,a} \\ \dot{\hat{G}}_i = \rho_{i,v}\mu_i^2 \tilde{x}_{i,a} - \rho_{i,a}\hat{G}_i \end{cases} \tag{2}$$

where the estimation values of $x_{i,a}$ and G_i are denoted as $\hat{x}_{i,a}$ and $\hat{G}_i$, respectively. $G_i = -\frac{x_{i,a}}{\tau_i} - \frac{\bar{c}_i(x_{i,v}^2+2\tau_i x_{i,v}x_{i,a})+\lambda_i}{m_i\tau_i} + W_i$. The estimation errors are defined as $\tilde{x}_{i,a} = x_{i,a} - \hat{x}_{i,a}$ and $\tilde{G}_i = G_i - \hat{G}_i$. The design parameters μ_i, $\rho_{i,p}$, $\rho_{i,v}$ and $\rho_{i,a}$ are selected such that the characteristic multinomial $s^2 + (\rho_{i,p}\mu_i - \rho_{i,a})s + (\rho_{i,v}\mu_i^2 - \rho_{i,p}\mu_i\rho_{i,a})$ is the Hurwitz polynomial.

To facilitate the subsequent analysis, we first state a key assumption and essential lemmas.

Assumption 1. *[8] The external disturbance is considered to be bounded, satisfying $|W(t)| \leq W^*$.*

Lemma 1. *[9] With appropriate selection of $\rho_{i,p}$, $\rho_{i,v}$, $\rho_{i,a}$, and μ_i, the ESO (2) ensures that $\hat{x}_{i,a}$ and $\hat{G}_i$ converge to bounded compact sets of $x_{i,a}$ and G_i, respectively, for system (1).*

Lemma 2. *[10] We construct the tracking differentiator as*

$$\begin{cases} \dot{\phi}_1 = \phi_2 \\ \dot{\phi}_2 = -\sigma^2\Big[\kappa_1 \mathrm{sgn}(\phi_1 - \alpha_d)|\phi_1 - \alpha_d|^{\frac{1}{2}} \\ \qquad + \kappa_2 \mathrm{sgn}\left(\dfrac{\phi_2}{\sigma}\right)\left|\dfrac{\phi_2}{\sigma}\right|^{\frac{2}{3}}\Big] \end{cases} \tag{3}$$

where ϕ_1 and ϕ_2 represent the state variables, α_d denotes input signal, and $\sigma > 0$. If the signal α_d satisfies $\sup_{t\in[0,\infty)}\left|\alpha_d^{(i)}\right| < \infty$ for $i = 0, 1, 2$, and $\kappa_1 > 0$, $\kappa_2 > 0$. Then, there exist $\xi_d^\star > 0$ and $\xi_d^* > 0$ such that $|\phi_1 - \alpha_d| \leq \xi_d^\star$, $|\phi_2 - \dot{\alpha}_d| \leq \xi_d^*$.

3 Main Results

Consider the following change of coordinates:

$$\begin{aligned} \zeta_{i,p} &= e_{i,p} \\ \zeta_{i,v} &= x_{i,v} - \alpha_{i,p} \\ \zeta_{i,a} &= x_{i,a} - \alpha_{i,v} \end{aligned} \tag{4}$$

where $\alpha_{i,p}$ and $\alpha_{i,v}$ are the virtual controllers.

Step 1: Differentiating $\zeta_{i,p}$ with respect to time using (4) yields

$$\dot{\zeta}_{i,p} = x_{i-1,v} - \zeta_{i,v} - \alpha_{i,p} - h(x_{i,a} - x_{0,a}) - \dot{\Gamma}_i \tag{5}$$

Accordingly, a Lyapunov function is constructed as follows:

$$V_{i,p} = \frac{1}{2}\zeta_{i,p}^2 + \frac{1}{2d_{i,p}}\tilde{\Psi}_{i,p}^2 \tag{6}$$

where $d_{i,p} > 0$. $\tilde{\Psi}_{i,p} = \Psi_{i,p} - \hat{\Psi}_{i,p}$, $\hat{\Psi}_{i,p}$ is the estimation value of $\Psi_{i,p}$.

Using Young's inequality, we derive the virtual control and adaptive law as

$$\alpha_{i,p} = c_{i,p}\zeta_{i,p} + x_{i-1,v} + \frac{\zeta_{i,p}\Psi_{i,p}\varpi_{i,p}^T\varpi_{i,p}}{2b_i^2} + \frac{\zeta_{i,p}}{2} - \dot{\Gamma}_i \tag{7}$$

$$\dot{\hat{\Psi}}_{i,p} = \frac{r_{i,p}\zeta_{i,p}^2\varpi_{i,p}^T\varpi_{i,p}}{2b_i^2} - \bar{d}_{i,p}\hat{\Psi}_{i,p} \tag{8}$$

where $c_{i,p} > 0$, $b_i > 0$, and $\bar{d}_{i,p} > 0$.

The substitution of (7) and (8) into (6) gives

$$\dot{V}_{i,p} \leq -c_{i,p}\zeta_{i,p}^2 - \zeta_{i,p}\zeta_{i,v} + \frac{\bar{d}_{i,p}}{d_{i,p}}\tilde{\Psi}_{i,p}\hat{\Psi}_{i,p} + \Xi_{i,1} \tag{9}$$

where $\Xi_{i,1} = \frac{b_i^2}{2} + \frac{\bar{\varepsilon}_{i,p}^2}{2}$.

Step 2: Differentiating $\zeta_{i,v}$ yields

$$\dot{\zeta}_{i,v} = \zeta_{i,a} + \alpha_{i,v} - \dot{\alpha}_{i,p} \tag{10}$$

Accordingly, we employ a Lyapunov function constructed as follows:

$$V_{i,v} = V_{i,p} + \frac{1}{2}\zeta_{i,v}^2 \tag{11}$$

To avoid the computational explosion, we get

$$\begin{cases} \dot{\alpha}_{i,p} = \dot{\phi}_{p,i,1} + q_{i,p}, \ |q_{i,p}| \leq \bar{q}_{i,p} \\ -\zeta_{i,v}q_{i,p} \leq \frac{p_{i,v}^2}{2}\zeta_{i,v}^2 + \frac{\bar{q}_{i,p}^2}{2p_{i,v}^2} \end{cases} \tag{12}$$

where $p_{i,v} > 0$ and $\bar{q}_{i,p} > 0$.

Based on the preceding derivation, the virtual controller is constructed as

$$\alpha_{i,v} = -c_{i,v}\zeta_{i,v} + \zeta_{i,p} + \dot{\phi}_{p,i,1} - \frac{p_{i,v}^2}{2}\zeta_{i,v} \tag{13}$$

where $c_{i,v} > 0$.

Then, we get

$$\dot{V}_{i,v} \leq -\sum_{j=1,2} c_{i,j}\zeta_{i,j}^2 + \zeta_{i,v}\zeta_{i,a} + \frac{\bar{d}_{i,p}}{d_{i,p}}\tilde{\Psi}_{i,p}\hat{\Psi}_{i,p} + \Xi_{i,2} \tag{14}$$

where $\Xi_{i,2} = \frac{\bar{q}_{i,p}^2}{2p_{i,v}^2} + \Xi_{i,1}$.

Step 3: The derivative of $\zeta_{i,a}$ is given by

$$\dot{\zeta}_{i,a} = \frac{u_i}{m_i\tau_i} + G_i - \dot{\alpha}_{i,v} \tag{15}$$

where $G_i = -\frac{x_{i,a}}{\tau_i} - \frac{\bar{c}_i(x_{i,v}^2+2\tau_i x_{i,v}x_{i,a})+\lambda_i}{m_i\tau_i} + W_i$.

Based on the established stability conditions, the following Lyapunov function is constructed:

$$V_{i,a} = V_{i,v} + \frac{1}{2}\zeta_{i,a}^2 + \frac{1}{2}\tilde{x}_{i,a}^2 + \frac{1}{2}\tilde{G}_i^{\ 2} \tag{16}$$

Successive applications of Lemma 2 and Young+'s inequality yield

$$\begin{cases} \dot{\alpha}_{i,v} = \dot{\phi}_{v,i,1} + q_{i,v},\ |q_{i,v}| \leq \bar{q}_{i,v} \\ -\zeta_{i,a}q_{i,v} \leq \frac{p_{i,a}^2}{2}\zeta_{i,a}^2 + \frac{\bar{q}_{i,v}^2}{2p_{i,a}^2} \end{cases} \tag{17}$$

where $p_{i,a} > 0$ and $\bar{q}_{i,v} > 0$.

Therefore, the control law is

$$u_i = -m_i\tau_i(c_{ia}\zeta_{i,a} + \zeta_{i,v} + \hat{G}_i - \dot{\phi}_{v,i,1} + \frac{p_{i,a}^2}{2}\zeta_{i,a}) \tag{18}$$

where $c_{i,a} > 0$.

The following relationship is established as

$$\begin{aligned} \dot{V}_{i,a} \leq & -\sum_{j=p,v} c_{i,j}\zeta_{i,j}^2 - (c_{i,a} - l_{i,5}^2)\zeta_{i,a}^2 - (\rho_{i,p}\mu_i - \frac{1}{4l_{i,1}^2} - l_{i,3}^2\rho_{i,2}\mu_i^4)\tilde{x}_{i,a}^2 - (\rho_{i,a} \\ & - \frac{\rho_{i,a}}{4l_{i,4}^2} - \frac{\rho_{i,v}\mu_i^4}{4l_{i,3}^2} - \frac{1}{4l_{i,2}^2} - l_{i,1}^2 - \frac{1}{4l_{i,5}^2})\tilde{G}_i^{\ 2} + \frac{\bar{d}_{i,p}}{d_{i,p}}\tilde{\Psi}_{i,p}\hat{\Psi}_{i,p} + \Xi_{i,3} \end{aligned} \tag{19}$$

where $\Xi_{i,3} = \frac{\bar{q}_{i,v}^2}{2p_{i,a}^2} + l_{i,4}^2\rho_{i,a}G_i^2 + \bar{G}_i + \Xi_{i,2}$. $l_{i,1}$, $l_{i,2}$, $l_{i,3}$, and $l_{i,4}$ are positive constants.

Theorem 1. *Given that Assumption 1 is satisfied, the closed-loop system (1) under virtual controllers (7, 13) and tracking differentiators (12, 17) achieves the control objectives with all signals remaining ultimately bounded.*

Proof. The Lyapunov function for the vehicle platoon is formulated as follows:

$$V_i = \sum_{i=1}^{N} V_{i,a} \tag{20}$$

Based on (9), (14) and (19), $\dot{V}_i$ is expressed as

$$\begin{aligned} \dot{V}_i \leq & \sum_{i=1}^{N}[-c_{i,p}\zeta_{i,p}^2 - c_{i,v}\zeta_{i,v}^2 - (c_{i,a} - l_{i,5}^2)\zeta_{i,a}^2 - (\rho_{i,p}\mu_i - \frac{1}{4l_{i,1}^2} - l_{i,3}^2\rho_{i,v}\mu_i^4)\tilde{x}_{i,a}^2 \\ & - (\rho_{i,a} - \frac{\rho_{i,a}}{4l_{i,4}^2} - \frac{\rho_{i,v}\mu_i^4}{4l_{i,3}^2} - \frac{1}{4l_{i,2}^2} - l_{i,1}^2 - \frac{1}{4l_{i,5}^2})\tilde{G}_i^{\ 2} + l_{i,4}^2\rho_{i,a}G_i^2 + \bar{G}_i + \frac{\bar{q}_{i,p}^2}{2p_{i,v}^2} \\ & + \frac{\bar{q}_{i,v}^2}{2p_{i,a}^2} + \frac{b_i^2}{2} + \frac{\bar{\varepsilon}_{i,1}^2}{2} + \frac{\bar{d}_{i,p}}{2d_{i,p}}\Phi_{i,p}^2 - \frac{\bar{d}_{i,p}}{2d_{i,p}}\tilde{\Phi}_{i,p}^2] \end{aligned} \tag{21}$$

From the above derivation, it follows that

$$\dot{V}_i \leq -C_i V_i + \Xi_i \tag{22}$$

where $C = \min\{2c_{i,p}, 2c_{i,v}, 2(c_{i,a} - l_{i,5}^2), 2(\rho_{i,p}\gamma_i - l_{i,3}^2\rho_{i,v}\mu_i^4 - \frac{1}{4l_{i,1}^2}), 2(\rho_{i,a} - \frac{\rho_{i,3}}{4l_{i,4}^2} - \frac{\rho_{i,2}\gamma_i^4}{4l_{i,3}^2} - \frac{1}{4l_{i,2}^2} - l_{i,1}^2 - \frac{1}{4l_{i,5}^2}), \bar{d}_{i,p}\}$ and $\Xi = \sum_{i=1}^{N}[l_{i,4}^2\rho_{i,a}G_i^2 + \bar{G}_i + \frac{1}{2}b_i^2 + \frac{\bar{q}_{i,p}^2}{2p_{i,v}^2} + \frac{\bar{q}_{i,v}^2}{2p_{i,a}^2} + \frac{b_i^2}{2} + \frac{\bar{\varepsilon}_{i,1}^2}{2} + \frac{\bar{d}_{i,p}}{2d_{i,p}}\Psi_{i,p}^2 - \frac{\bar{d}_{i,p}}{2d_{i,p}}\tilde{\Psi}_{i,p}^2]$. The proof is completed.

4 Simulation Results

The simulation involves the vehicle platoon comprising a leader and three followers ($N = 3$), with a series of parameters specified as follows.

Table 1. The Parameters of Heterogeneous Vehicles

Parameters ($i = p, v, a$)
$m_0 = 1000$, $m_1 = 1200$, $m_2 = 1400$, $m_3 = 1500$, $H_0 = 4$, $H_1 = 3$, $H_2 = 5$, $H_3 = 5$
$\bar{c}_i = 0.5$, $\bar{c}_i = 0.26$, $\bar{c}_i = 0.3$, $\bar{c}_i = 0.774$, $\tau_0 = 0.5$, $\tau_1 = 0.45$, $\tau_2 = 0.55$, $\tau_3 = 0.6$
$b_i = 1$, $\rho_{i,1} = 4$, $\rho_{i,2} = 850$, $\rho_{i,3} = 8$, $\mu_i = 0.3$, $c_{i,1} = c_{i,2} = 30$, $c_{i,3} = 150$
$l_{i,1} = 1$, $l_{i,2} = 10$, $l_{i,3} = 0.25$, $l_{i,4} = 10$, $l_{i,5} = 10$, $\rho_{ai} = 1.29$, $\bar{A}_i = 2$, $C_{di} = 0.3$
$\sigma_i = 1$, $k_{11} = k_{12} = 0.1$, $k_{21} = k_{22} = 0.1$, $g = 9.8$, δ_i, $h = 0.2$, $\Delta_{i-1,i} = 6$

Table 2. Initial States and Initial Values of Vehicle platoon

Initial States	$i = 0,\ p,\ v,\ a$
Position	$x_{i,p}(0) = [60,\ 40,\ 20,\ 0]\,\mathrm{m}$
Velocity	$x_{i,v}(0) = [15,\ 15,\ 15,\ 15]\,\mathrm{m/s}$
Acceleration	$x_{i,a}(0) = [0,\ 0,\ 0.5,\ 1]\,\mathrm{m/s^2}$
Initial Values	$i = 0,\ p,\ v,\ a$
Tracking differentiators	$\phi_{i,1,1}(0) = \phi_{i,1,2}(0) = \phi_{i,2,1}(0) = \phi_{i,2,2}(0) = 0$
Adaptive values	$\dot{\hat{\Psi}}_1(0) = \dot{\hat{\Psi}}_2(0) = \dot{\hat{\Psi}}_3(0) = 0.1$

The parameters for heterogeneous vehicles are listed in Tables 1 and 2. All parameters and initial state settings are consistent across comparative studies. Moreover, the disturbance terms are $W_1(t) = 10\cos(t)$, $W_2(t) = 20\sin(t)$, and

$W_3(t) = 5\sin(2t)$, respectively. Then, $\zeta_{0,v}(t)$ is designed as follows:

$$\zeta_{0,v}(t) = \begin{cases} 15\,\text{m/s}, & t < 5\,\text{s} \\ 2t + 5\,\text{m/s}, & 5\,\text{s} \leq t < 10\,\text{s} \\ 25\,\text{m/s}, & 10\,\text{s} \leq t < 20\,\text{s} \\ -0.5t + 35\,\text{m/s}, & 20\,\text{s} \leq t < 30\,\text{s} \\ 20\,\text{m/s}, & 30\,\text{s} \leq t < 60\,\text{s}. \end{cases}$$

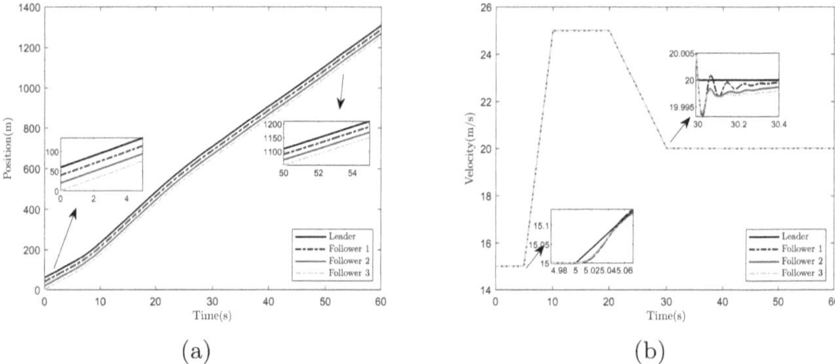

Fig. 1. (a) The vehicle positions with modified CTHP. (b) The vehicle velocities of with modified CTHP.

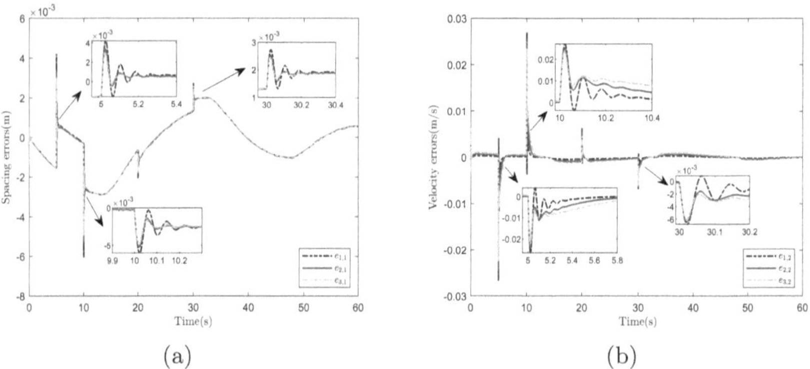

Fig. 2. (a) The spacing errors of vehicle platoon with modified CTHP. (b) The velocity errors of vehicle platoon with modified CTHP.

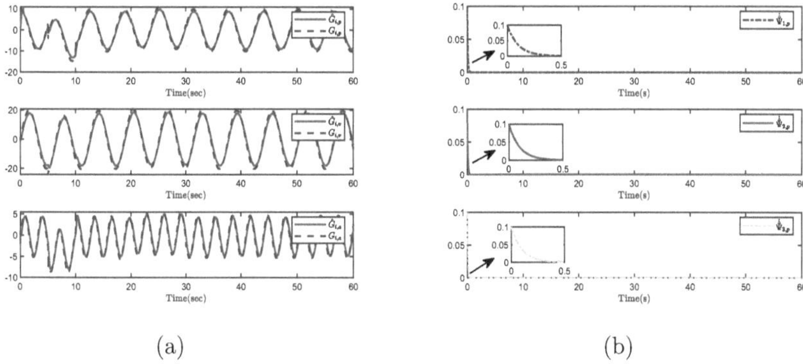

(a) (b)

Fig. 3. (a) The ESO signals of vehicle platoon. (b) The trajectories of adaptive laws $\dot{\hat{\Psi}}_{1,p}$, $\dot{\hat{\Psi}}_{2,p}$ and $\dot{\hat{\Psi}}_{3,p}$.

Simulation results for vehicle platoon with the modified CTHP are shown in Figs. 1a–2b. Figure 1a shows the position trajectories of the leader and followers, confirming that the followers accurately track the trajectory of the leader after a brief transient. Figure 1b illustrates the velocity profiles, indicating that all vehicles attain velocity consensus. The spacing and relative velocity errors, shown in Figs. 2a and 2b respectively, converge to a small neighborhood of zero. Furthermore, Fig. 3a demonstrates that the ESO effectively estimates uncertain dynamics with ultimately bounded errors, while Fig. 3b presents the convergent adaptation of control parameters $\dot{\hat{\Psi}}_{1,p}$, $\dot{\hat{\Psi}}_{2,p}$ and $\dot{\hat{\Psi}}_{3,p}$. These results validate the effectiveness of the proposed control scheme.

Acknowledgments. This work was partially supported by the National Natural Science Foundation of China (62473059), the Revitalization of Liaoning Talents Program (XLYC2403188), the Jinzhou Talents Program (JXYC240102) and the Natural Science Foundation of Liaoning Province (2024MS183, 2024BS235).

Disclosure of Interests.. The authors have no competing interests to declare that are relevant to the content of this article.

References

1. Zhang, M., Wang, C., Zhao, W., Liu, J., Zhang, Z.: A multi-vehicle self-organized cooperative control strategy for platoon formation in connected environment. IEEE Trans. Intell. Transp. Syst. **26**(3), 4002–4018 (2025)
2. Yang, B., Yan, S., Wang, Z., Nakano, K.: Prediction based trajectory planning for safe interactions between autonomous vehicles and moving pedestrians in shared spaces. IEEE Trans. Intell. Transp. Syst. **24**(10), 10513–10524 (2023)
3. Li, P., Lam, J., Lu, R.: Robust switched velocity-dependent path-following control for autonomous ground vehicles. IEEE Trans. Intell. Transp. Syst. **24**(5), 4815–4826 (2023)

4. Rezaee, H., Parisini, T., Polycarpou, M.M.: Leaderless cooperative adaptive cruise control based on the constant time-gap spacing policy. IEEE Trans. Autom. Control **69**(1), 659–666 (2024)
5. Boo, J., Chwa, D.: Integral sliding mode control-based robust bidirectional platoon control of vehicles with the unknown acceleration and mismatched disturbance. IEEE Trans. Intell. Transp. Syst. **24**(10), 10881–10894 (2023)
6. Deng, J., Xue, W., Zhang, L., Bao, Q., Mao, Y.: Disturbance-compression extended state observer with noise insensitivity: application to electro-optical tracking system. IEEE Trans. Autom. Sci. Eng. **22**, 17761–17777 (2025)
7. Zhang, Z., Wang, X., Xu, J.: Robust amplitude control set model predictive control with low-cost error for SPMSM based on nonlinear extended state observer. IEEE Trans. Power Electron. **39**(6), 7016–7028 (2024)
8. Qi, H., Cao, L., Ren, H., Zhao, M.: Fixed-time NN-based adaptive fault-tolerant control for heterogeneous vehicular platoon system with improved exponential spacing policy. Commun. Nonlinear Sci. Numer. Simul. **141**, 108454 (2025)
9. Chen, Q., Zhou, Y., Ahn, S., Xia, J., Li, S., Li, S.: Robustly string stable longitudinal control for vehicle platoons under communication failures: a generalized extended state observer-based control approach. IEEE Trans. Intell. Veh. **8**(1), 159–171 (2023)
10. Zhang, Y., Wang, D., Peng, Z.: Consensus maneuvering for a class of nonlinear multivehicle systems in strict-feedback form. IEEE Trans. Cybern. **49**(5), 1759–1767 (2019)

Neural-Network-Based Self-triggered Observed Platoon Control for Autonomous Vehicles

Zihan Li[1], Ziming Wang[2,3], Chenning Liu[1], and Xin Wang[1(✉)]

[1] Southwest University, Chongqing 400715, China
xinwangswu@163.com
[2] Swinburne University of Technology, Melbourne, VIC 3122, Australia
[3] Tsinghua University, Beijing 100084, China

Abstract. This paper investigates autonomous vehicle (AV) platoon control under uncertain dynamics and intermittent communication, which remains a critical challenge in intelligent transportation systems. To address these issues, this paper proposes an adaptive consensus tracking control framework for nonlinear multi-agent systems (MASs). The proposed approach integrates backstepping design, a nonlinear sampled-data observer, radial basis function neural networks, and a self-triggered communication mechanism. The radial basis function neural networks approximate unknown nonlinearities and time-varying disturbances, thereby enhancing system robustness. A distributed observer estimates neighboring states based on limited and intermittent measurements, thereby reducing dependence on continuous communication. Moreover, self-triggered mechanism is developed to determine triggering instants, guaranteeing a strictly positive minimum inter-event time and preventing Zeno behavior. The theoretical analysis proves that all closed-loop signals are uniformly ultimately bounded (UUB), and tracking errors converge to a compact set. Simulation results demonstrate that the proposed approach achieves high robustness, adaptability, and communication efficiency, making it suitable for real-world networked vehicle systems.

Keywords: platoon control · self-triggered control · radial basis function neural networks · autonomous vehicles

1 Introduction

Autonomous vehicle platoon control has become a key research direction in intelligent transportation systems, aiming to improve traffic efficiency, reduce fuel consumption, and enhance road safety through coordinated vehicle motion [1]. Vehicle-to-vehicle (V2V) and vehicle-to-infrastructure (V2I) communications enable real-time information exchange, thereby facilitating the formation and maintenance of stable platoons under predefined communication topologies.

C. Li et al. (Eds.): ICNC 2025, CCIS 2946, pp. 451–463, 2026.
https://doi.org/10.1007/978-981-92-1599-7_38

However, traditional control strategies often rely on idealized assumptions about vehicle dynamics and communication reliability, overlooking practical issues such as modeling uncertainties [2,3], unknown disturbances [4], and intermittent communication failures [5]. These challenges motivate the development of robust and adaptive frameworks capable of maintaining performance under dynamic conditions.

In practical platoons, packet loss, time-varying delays, and bandwidth constraints are inevitable. To cope with these problems, observer-based designs have been introduced to estimate the unmeasured or unreliable states of neighboring vehicles using limited sensor data. The observer-based self-triggered adaptive control framework reduces dependence on continuous communication, improving robustness and practicality under real-world constraints [6–9]. Moreover, integrating adaptive techniques with observer design enables compensation for modeling errors and external disturbances, ensuring accurate estimation and stable cooperative motion.

Traditional time-triggered control schemes execute control tasks periodically at fixed intervals [10–12], leading to unnecessarily high-frequency updates and resource consumption. Event-triggered control [13–15] has been proposed to address this by updating control signals only when specific conditions are violated, significantly reducing communication load while maintaining closed-loop stability. However, ETC typically requires continuous monitoring of the triggering condition, which, despite reducing communication, still imposes a substantial burden on the agent's sensing and computational resources. To overcome this limitation, self-triggered control (STC) has emerged as an advanced strategy. STC eliminates continuous monitoring by calculating the precise next triggering instant t_{k+1} at the current event time t_k. This allows the control input to be held constant across the interval $t \in [t_k, t_{k+1})$. Existing research on STC has demonstrated its ability to maintain stability and performance in various multi-agent systems, particularly in tackling complex issues like distributed tracking control and ensuring resource efficiency under strict network constraints [6,9,12,17]. This prediction-based update mechanism drastically enhances the practicality of implementing cooperative control in resource-constrained environments.

Motivated by these considerations, it is with this goal that the work, specifically this paper, proposes an adaptive consensus tracking framework for nonlinear high-order multi-agent systems based on backstepping, filtering techniques, radial basis function neural networks, and distributed observers. The main contributions are summarized as follows (Fig. 1):

- Adaptive neural networks compensation: neural networks are employed to approximate and compensate for unknown nonlinearities and disturbances, enhancing robustness and adaptability without requiring precise model knowledge.
- A distributed observer estimates neighboring agents' full states using limited and intermittent output data, reducing reliance on continuous communication and perfect measurements [6,16].

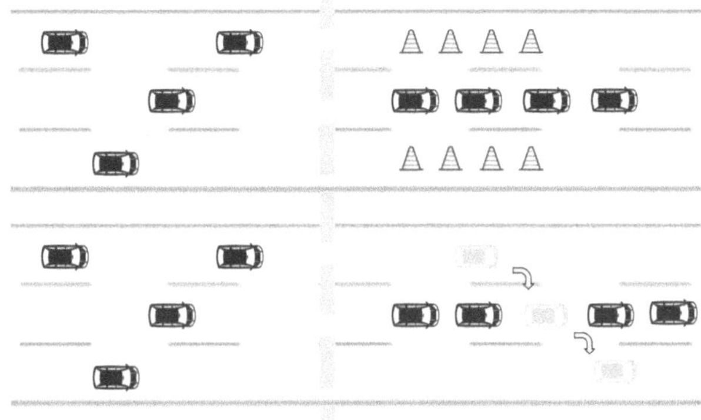

Fig. 1. AVs formation control in linear formation and linear-queue formation.

- A self-triggered control mechanism determines the next triggering instant based solely on the current control signal and its rate of change, avoiding continuous state monitoring [12,17]. It guaranties a strictly positive minimum inter-event time and allows adjustable trade-offs between control accuracy and communication burden, enhancing practicality for networked systems.

The remainder of this paper is organized as follows. Section 2 reviews preliminaries on MASs, observer design, and graph theory, and formulates the problem. Section 3 presents the self-triggered control mechanism. Section 4 analyzes stability. Section 5 provides simulation results. Finally, Sect. 6 concludes the paper.

2 Preliminaries and Problem Formulation

2.1 Graph Theory

In the AV platoon, communication interactions are modeled by a directed graph $\mathcal{G} = (\mathcal{V}, \mathcal{E})$, where $\mathcal{V}$ is the set of follower vehicles and $\mathcal{E}$ represents the directed communication links. The communication topology is described by the adjacency matrix $\mathcal{A} = [a_{ij}] \in \mathbb{R}^{N \times N}$, where $a_{ij} = 1$ if follower i receives information from vehicle j, and $a_{ij} = 0$ otherwise. The in-degree matrix $\mathcal{D} = \text{diag}\{d_i\}$ satisfies $d_i = \sum_{j=1}^{N} a_{ij}$, and the Laplacian matrix is $\mathcal{L} = \mathcal{D} - \mathcal{A}$. The leader-follower coupling is defined by the pinning gain matrix $\mathcal{B} = \text{diag}\{b_i\}$, where $b_i = 1$ if the follower i can access the leader's state and $b_i = 0$ otherwise. The matrix $\mathcal{H} = \mathcal{L} + \mathcal{B}$ plays a key role in stability analysis; if the communication graph contains a directed spanning tree rooted at the leader, all eigenvalues of $\mathcal{H}$ possess positive real parts. Notably, the proposed control scheme does not require explicit knowledge of $\mathcal{H}$, instead, it relies only on locally sampled information, thereby enhancing robustness against communication uncertainties.

2.2 Systems Model

The longitudinal dynamics of each autonomous vehicle (AV) in the platoon are considered. The tenth vehicle is described by a second-order nonlinear system:

$$\dot{p}_i = v_i \tag{1}$$

$$\dot{v}_i = \frac{1}{m_i}(F_i - f_i(v_i)) + d_i \tag{2}$$

where $i \in \mathcal{N} = 1, 2, \ldots, N, p_i \in \mathbb{R}$ and $v_i \in \mathbb{R}$ represent the position and velocity of the vehicle, respectively $m_i > 0$ denotes vehicle mass, and $F_i \in \mathbb{R}$ is the control input force. The term $f_i(v_i)$ encapsulates unknown non-linear resistance forces, including aerodynamic drag and rolling resistance. The term $d_i \in \mathbb{R}$ represents unknown external disturbances.

2.3 Radial Basis Function Neural Networks

According to research [18,19], radial basis function neural networks are employed to approximate the unknown nonlinearities $\Delta_i(v_i, d_i)$ inherent in vehicle dynamics, leveraging their universal approximation capability to model any continuous function $F(Z) = W^{*T}\Psi(Z) + \epsilon(Z)$ over a compact set. Here, Z denotes the input vector, W^* represents the ideal weight vector, $\Psi(Z)$ consists of a set of Gaussian basis functions $\psi_k(Z) = \exp\left(-\|Z - \mu_k\|^2/\eta_k^2\right)$, and $\epsilon(Z)$ signifies the bounded approximation error. The online estimate of the uncertainty is constructed as $\hat{\Delta}_i(Z_i) = \hat{W}_i^T\Psi_i(Z_i)$, where $\hat{W}_i$ is the adaptive estimate of the ideal weights, and $\tilde{W}_i = W_i^* - \hat{W}_i$ denotes the weight estimation error. This approach facilitates the development of robust and self-learning control capabilities without reliance on precise prior models.

2.4 Self-triggered Mechanism Design

In self-triggered control, the control input is held constant between consecutive triggering instants and is updated at the current trigger time t_k. The control hold law is given by:

$$u_i(t) = u_i(t_k), \quad \forall t \in [t_k, t_{k+1}) \tag{3}$$

The core of this mechanism involves computing the next triggering instant t_{k+1} at the current time t_k. The proposed triggering control mechanism is designed as follows:

$$t_{k+1} = t_k + \min\left\{\frac{s_\sigma|u_i(t_k)| + s_D}{\max(|\dot{\hat{u}}_i(t_k)|, s_\Lambda)}, \ T_{\max}\right\} \tag{4}$$

where $T_{\max}$ specifies the maximum inter-event time to ensure the minimum update frequency. Constants include: $s_\sigma \in (0, 1)$ scaling the threshold with control magnitude; $s_D > 0$ providing a fixed threshold to prevent unbounded intervals; $s_\Lambda > 0$ maintaining a minimum update rate when $\dot{\hat{u}}_i(t_k)$ is small. In contrast to event-triggered control, this self-triggered control mechanism (4) eliminates the need for continuous state measurement. The computation is executed only at each triggering instant, thereby reducing the agent's sensing workload.

3 Main Result

To facilitate the controller design, the model is rewritten by defining the control input as $u_i = F_i/m_i$. The uncertainty, which includes internal resistance and external disturbances, is denoted as $\Delta_i = -f_i(v_i)/m_i + d_i$. With these new definitions, the vehicle dynamic model can therefore be transformed into the following presented compact form:

$$\begin{aligned} \dot{p}_i &= v_i \\ \dot{v}_i &= u_i + \Delta_i\left(v_i, d_i\right) \end{aligned} \tag{5}$$

In this paper, it is the uncertainty $\Delta_i(\cdot)$ that is an unknown non-linear function. To handle this, neural network is employed to approximate it over a compact set Ω with arbitrary accuracy, leading to the following expression:

$$\Delta_i = W_i^{*T}\Psi_i\left(Z_i\right) + \epsilon_i\left(Z_i\right) \tag{6}$$

where $Z_i = [v_i, d_i]^T \in \Omega \subset \mathbb{R}^2$ is the input vector to the neural networks, $W_i^* \in \mathbb{R}^\ell$ is the ideal optimal weight vector, and $\Psi_i(Z_i) = [\psi_{i1}(Z_i), \psi_{i2}(Z_i), \ldots, \psi_{i\ell}(Z_i)]^T \in \mathbb{R}^\ell$ is the Gaussian basis function vector. The term $\epsilon_i(Z_i)$ denotes the bounded approximation error satisfying $|\epsilon_i(Z_i)| \leq \bar{\epsilon}_i$, where $\bar{\epsilon}_i > 0$ is a designed constant. Substituting (6) into (5) yields the final model used for controller synthesis:

$$\begin{aligned} \dot{p}_i &= v_i \\ \dot{v}_i &= u_i + W_i^{*T}\Psi_i\left(Z_i\right) + \epsilon_i\left(Z_i\right) \end{aligned} \tag{7}$$

It is crucial to note that the control algorithm developed in this paper does not require precise knowledge of the vehicle mass m_i, the nonlinear function $f_i(v_i)$, or the disturbance d_i. Instead, the adaptive laws and the neural network estimator will identify and compensate for the uncertainty Δ_i, enhancing the system's robustness and adaptability.

Given the limitations and the need to reduce communication load, a sampling-based state observer is proposed. It estimates the full state of both the preceding and ego vehicles using sampled position data. Let t_k denote the k-th sampling instant with fixed interval $T = t_{k+1} - t_k$. The measured position of the preceding vehicle $(i-1)$, affected by sensor noise and perception errors, is given as:

$$\tilde{p}_{i-1}(t_k) = p_{i-1}(t_k) + \zeta_{i-1}(t_k) \tag{8}$$

where $\zeta_{i-1}(t_k)$ is the bounded measurement noise that satisfies $|\zeta_{i-1}(t_k)| \leq \bar{\zeta}_{i-1}$.A Luenberger-type nonlinear sampled-data observer is proposed for the ego vehicle i to estimate the unmeasurable states:

$$\dot{\hat{p}}_i = \hat{v}_i + L_{i1}(\tilde{p}_{i-1}(t_k) - \hat{p}_i) \tag{9}$$

$$\dot{\hat{v}}_i = u_i + \hat{\Delta}_i(\hat{v}_i) + L_{i2}(\tilde{p}_{i-1}(t_k) - \hat{p}_i) \tag{10}$$

Here, $\hat{p}_i$ and $\hat{v}_i$ are the estimated positions and velocities of the ego vehicle, respectively. $L_{i1} > 0$ and $L_{i2} > 0$ are the observer gain constants to be designed. $\hat{\Delta}_i(\hat{v}_i) = \hat{W}_i^T \Psi_i(\hat{v}_i)$ is the online estimate of the lumped uncertainty of the neural network.The dynamics of the observer error are defined as $e_{p_i} = p_i - \hat{p}_i$ and $e_{v_i} = v_i - \hat{v}_i$.

Assumption 1. Based on research [20,21], the vehicle velocity dynamics are Lipschitz continuous, meaning there exists a designed constant $\gamma_i > 0$ such that:

$$|\dot{v}_i(t_1) - \dot{v}_i(t_2)| \leq \gamma_i |t_1 - t_2|, \quad \forall t_1, t_2 \in [t_k, t_{k+1}) \tag{11}$$

Assumption 2. Based on research [20,21], the rate of change of vehicle position is bounded; that is, the velocity satisfies $|v_i(t)| \leq v_i^{\max}$ for all $t \geq 0$, where $v_i^{\max} > 0$ is an unknown constant.

Under these assumptions and with appropriately chosen gains L_{i1}, L_{i2}, it is guaranteed that the dynamics of the observation error are ultimately uniformly bounded. The key innovation lies in the structure of the observer, which uses the sampled position of the preceding vehicle $\tilde{p}_{i-1}(t_k)$as a corrective input rather than requiring continuous state information from other vehicles or infrastructure.

In control design, the formulation of the tracking error is central to achieving control objectives. For bipartite tracking control with competitive-cooperative relationships, the definition of the tracking error differs from that of traditional consensus. The bipartite tracking error e_{i1} for agent i is typically defined as:

$$e_{i1} = \sum_{j=1}^{N} |a_{ij}|(y_i - \text{sgn}(a_{ij})y_j) + |b_i|(y_i - \text{sgn}(b_i)y_r) \tag{12}$$

where a_{ij} is the element of the adjacency matrix representing the cooperative-competitive relationship between agents, b_i represents the connection to the leader, y_i and y_j are the outputs of the agents, y_r is the output of the leader, and $\text{sgn}(\cdot)$ is the signum function. This formulation ensures that, under competitive interactions, the agents asymptotically converge to values with identical magnitudes and opposite signs.

In vehicle formation control, the tracking error is more directly related to the system states and the desired trajectory. For agent i, its position tracking error $z_{i,1}$ and velocity tracking error $z_{i,2}$ are typically defined as:

$$z_{i,1} = \hat{x}_i - q_i \tag{13}$$

$$z_{i,2} = \hat{v}_i - \dot{q}_i - \alpha_i \tag{14}$$

Here, $\hat{x}_i$ and $\hat{v}_i$ represent the position and velocity of agent i estimated by the observer, q_i and $\dot{q}_i$ denote its desired position and velocity trajectories, and α_i is the virtual control law designed in the backstepping procedure.

4 Stability Analysis

Theorem 1: Consider the MASs (5) with uncertainties. With the self-triggered mechanism defined by (3) and (4), the following properties hold:

1. All signals in the closed-loop system, including the tracking error e, the neural network weight estimates $\hat{W}$, and the disturbance constant estimates $\hat{\Theta}$, are uniformly ultimately bounded (UUB).
2. The tracking error e converges to a compact set around the origin, the size of which can be adjusted by the design constants s_σ, s_D, s_Λ, and the adaptive gains.
3. The inter-event times are strictly positive, i.e., there exists a constant $\tau > 0$ such that $\inf_k(t_{k+1} - t_k) \geq \tau$; thus, Zeno behavior is excluded.

Proof: Consider the Lyapunov function candidate:

$$V = \frac{1}{2}e^T Pe + \frac{1}{2}\left(\tilde{W}^T \Gamma^{-1}\tilde{W}\right) + \frac{1}{2}\tilde{\Theta}^T \Xi^{-1}\tilde{\Theta} \tag{15}$$

where e is the tracking error vector, $\tilde{W}$ and $\tilde{\Theta}$ are the weight estimation error matrices for the neural network and the disturbance constants, respectively, and P, Γ, Ξ are positive definite matrices. The time derivative of V along the trajectories of the system is given by:

$$\dot{V} = e^T P\dot{e} + \left(\tilde{W}^T \Gamma^{-1}\dot{\tilde{W}}\right) + \tilde{\Theta}^T \Xi^{-1}\dot{\tilde{\Theta}} \tag{16}$$

Substituting the system dynamics, the control law $u_i(t) = u_i(t_k)$, and the adaptive update laws, we obtain an expression that contains a negative definite term $-e^T Qe$ and cross terms involving the control input error $e_u(t) = u_i(t_k) - \bar{u}_i(t)$, where $\bar{u}_i(t)$ denotes the ideal continuous control signal. The design of the self-triggered mechanism ensures that, for $t \in [t_k, t_{k+1})$, the control input error $e_u(t)$ satisfies

$$|e_u(t)| \leq s_\sigma |u_i(t_k)| + s_D \tag{17}$$

where s_σ and s_D are positive design constants. Using Young's inequality and standard norm inequalities, the derivative $\dot{V}$ can then be upper bounded as:

$$\begin{aligned} \dot{V} &\leq -\lambda_{\min}(Q)\|e\|^2 + c_1\|e\| \cdot |e_u(t)| + c_2 \\ &\leq -\alpha V + \beta \end{aligned} \tag{18}$$

A key step in the analysis is to bound the control input error. The self-triggered mechanism ensures that, for $t \in [t_k, t_{k+1})$, the control input error $e_u(t)$ satisfies

$$|e_u(t)| \leq s_\sigma |u_i(t_k)| + s_D \tag{19}$$

where s_σ and s_D are positive design constants. Using Young's standard norm properties, the derivative of the Lyapunov function V can be upper bounded as

$$\begin{aligned} \dot{V} &\leq -\lambda_{\min}(Q)\|e\|^2 + c_1\|e\| \cdot |e_u(t)| + c_2 \\ &\leq -\alpha V + \beta \end{aligned} \tag{20}$$

for some positive constants $c_1, c_2, \alpha, \beta > 0$. α characterizes the exponential convergence rate of the closed-loop system, indicating how rapidly the tracking and observer errors decay in the absence of triggering-induced accumulation. In contrast, β quantifies the residual steady-state bound resulting from sampling effects, estimation inaccuracies, and triggering-induced input perturbations. Then integrating both sides gives

$$0 \leq V(t) \leq \frac{\beta}{\alpha} + \left(V(0) - \frac{\beta}{\alpha}\right) e^{-\alpha t}, \tag{21}$$

which shows that $V(t)$ exponentially converges to a bounded residual set of radius.

Therefore, according to the properties of positive definite matrices, it is corroborated that the observer error e, tracking errors $z_{i,1}$ and $z_{i,2}$, and constant estimation errors $\tilde{W}_{i,j}$ and $\tilde{\sigma}_i$ are all bounded, i.e.,

$$\|e\|^2 \leq \frac{2V}{\phi_{\min}(P)}, \quad \|z_{i,1}\|^2 \leq 2V, \quad \|z_{i,2}\|^2 \leq 2V$$

$$\|\tilde{W}_{i,j}\|^2 \leq \frac{2V}{\phi_{\min}(\mathcal{O}_{i,j})}, \quad \|\tilde{\sigma}_i\|^2 \leq \frac{2V}{\phi_{\min}(\Delta_{i,j})}$$

Now we show that there exists a positive constant $t^* > 0$ such that for all $k \in \mathbb{Z}^+$, the triggering intervals satisfy $\{t_{k+1} - t_k\} \geq t^*$. To this end, by recalling $e_i(t) = w_i(t) - u_i(t)$, during the period $t_i^k \leq t_i < t_i^{k+1}$, we obtain

$$\frac{d}{dt}\|e_i\| = \frac{d}{dt}(e_i^T e_i)^{\frac{1}{2}} = [\mathrm{sgn}(e_{i,1}), \mathrm{sgn}(e_{i,2})]^T e_i = \|w_i\| \tag{22}$$

where $e_{i,j}$ is the j-th component of the vector e_i, $j = 1, 2$.

This result indicates that the growth rate of the control error is bounded by the control magnitude, thereby ensuring a strictly positive minimum inter-event interval t^* and excluding Zeno behavior. Hence, all closed-loop signals remain bounded, and overall system stability is guaranteed. Finally, the inequality $\dot{V} \leq -\alpha V + \beta$ proves that all signals in the closed-loop system, including the tracking error e and the estimation errors $\tilde{W}$, $\tilde{\Theta}$, are UUB.

To exclude Zeno behavior, we prove the existence of a positive lower bound for the inter-event times. The time derivative of the control signal $\bar{u}_i(t)$ can be shown to be bounded. Suppose there exists a constant $\Upsilon > 0$ such that:

$$\left|\frac{d}{dt}(\bar{u}_i(t))\right| \leq \Upsilon \tag{23}$$

The error $e_u(t)$ evolves from 0 at t_k to the threshold $s_\sigma|u_i(t_k)| + s_D$ at t_{k+1}. The time required for the process of this growth is at least:

$$\tau = \frac{s_\sigma|u_i(t_k)| + s_D}{\Upsilon} > 0 \tag{24}$$

Therefore, the inter-event time is bounded below by a positive constant, $t_{k+1} - t_k \geq \tau > 0$. This result explicitly excludes Zeno behavior,as the bounded rate of change of the ideal continuous control law $\bar{u}_i(t)$ ensures that the time required for the control error to grow from zero to the triggering threshold possesses a positive lower bound τ, thereby guaranteeing a strictly positive interval between consecutive triggering instants.

5 Illustrative Example

5.1 Simulation Setup and Scenario Design

To validate the proposed observer-based self-triggered adaptive control mechanism, numerical simulations are conducted in MATLAB for a heterogeneous vehicle platoon comprising one leader and three followers with distinct masses $m_i = [1800, 1900, 1850, 1760]$ kg. The simulation uses a sampling interval of $T = 0.001$ s over a 50 s duration.

Each vehicle $i \in \{1, 2, 3, 4\}$ (vehicle 1 as leader) has a longitudinal-lateral position $p_i = [p_{xi}, p_{yi}]^\top$ and velocity $v_i = [v_{xi}, v_{yi}]^\top$, with control inputs u_{xi}, u_{yi} resisting unknown nonlinear forces f_{xi}, f_{yi} and small disturbances d_{xi}, d_{yi}. The dynamics follow:

$$\dot{v}_{xi} = (u_{xi} + f_{xi} + d_{xi})/m_i, \quad \dot{v}_{yi} = (u_{yi} + f_{yi} + d_{yi})/m_i.$$

Defining $x_i = [p_{xi}, p_{yi}]^\top$ and $\dot{x}_i = v_i$, the observer estimates $\hat{x}_i = [\hat{p}_{xi}, \hat{p}_{yi}]^\top$ and $\hat{v}_i = [\hat{v}_{xi}, \hat{v}_{yi}]^\top$.

To introduce nonzero initial estimation errors, heterogeneous initial states are assigned. For illustration, the leader is initialized at $x_1(0) = [50,\ 6.0]^\top$m with $v_1(0) = [15,\ 0]^\top$m/s, while its observer estimates start from $\hat{x}_1(0) = [48,\ 5.5]^\top$m and $\hat{v}_1(0) = [14,\ 0.2]^\top$m/s. The remaining followers are positioned sequentially behind the leader, with comparable longitudinal velocities and similarly perturbed observer initializations. This ensures a non-negligible initial mismatch between the true and estimated states for evaluating observer convergence and self-triggered control performance.

The leader's longitudinal velocity $v_{x1}(t)$ follows: 12 m/s for $0 \leq t < 25$; a smooth S-curve deceleration to 6 m/s during $25 \leq t < 31$; and 6 m/s for $t \geq 31$. The lateral position remains constant at $p_{y1}(t) = 5.4$ m. Controller constants are chosen as $K_1 = \text{diag}(1.5, 1.0)$, $K_2 = \text{diag}(35, 5)$, $C_1 = C_2 = 8$. The neural network employs $l = 5$ basis functions with centers in $[-12, 12]$ and width $\sigma = 2.5$. The self-triggered mechanism involves several constants $(s_\sigma, s_D, s_\Lambda, T_{\max})$, which balance control accuracy and communication efficiency. Specifically, s_σ adjusts the triggering sensitivity relative to the control magnitude (smaller values yield higher accuracy but require more updates), while s_D prevents excessive triggering when the control signal is small, typically set as $s_D \approx 0.01u_{\max}$. The term s_Λ ensures numerical stability when the control rate is nearly constant and can be chosen based on noise levels. The upper bound $T_{\max}$ avoids overly long inter-event times and is selected according to system dynamics, satisfying

$T_{\max} < 1/\gamma_i$. These constants are obtained through analytical constraints and limited simulation tuning, ensuring both practical implementability and theoretical consistency.

Two simulation scenarios are designed to evaluate the performance of the proposed self-triggered control algorithm.

In linear formation scenario, vehicles maintain the same lane with a desired inter-vehicle spacing of 10 m behind their immediate predecessor. This setup represents a standard highway cruising condition with longitudinal coordination.

In linear-queue scenario, a gap twice the standard inter-vehicle distance is reserved between the second and third vehicles to allow surrounding traffic to change lanes safely. This prevents unnecessary road occupation and reduces interference with vehicles merging from either side, while testing the controller's ability to maintain formation under such a specific spacing constraints. Communication efficiency is quantified using the triggering ratio:

$$\text{Triggering Ratio} = \frac{N_{\text{trigger}}}{N_{\text{total}}} \times 100\%$$

where N_{trigger} is the number of control updates for each AV, and N_{total} is the total simulation step count.

Table 1. Self-triggered control statistics

Vehicle	Self-Triggered	Continuous	Ratio (%)
AV1	18185	50000	63.63
AV2	44405	50000	11.19
AV3	44439	50000	11.12
AV4	44702	50000	10.60

The communication update counts for each vehicle under both the self-triggered and continuous control modes, along with the calculated triggering ratios, are summarized in Table 1. To improve visibility, only a subset of the triggering instants is displayed in Figure 3; the statistical results in Table 1 are based on the full triggering data. The results demonstrate that the self-triggered mechanism significantly reduces the number of required communication updates compared to continuous control.

5.2 Simulation Results and Analysis

From the longitudinal position plots in Fig. 2 (a), the followers respond with minimal delay, preserving the target 10 m spacing with minor fluctuations. The smooth transitions and bounded spacing errors confirm that the PD control with adaptive neural network compensation suppresses overshoot and enhances disturbance rejection. The lateral trajectories in Fig. 2 (b), the linear formation

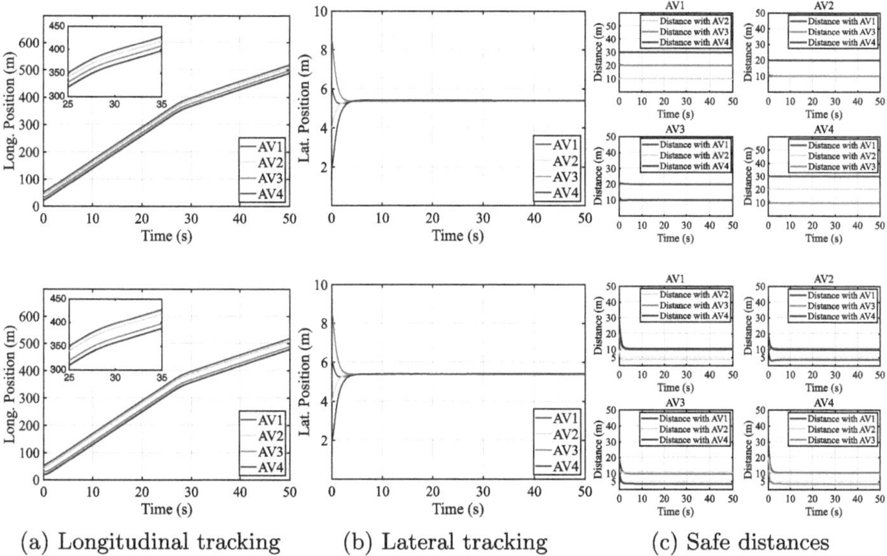

(a) Longitudinal tracking (b) Lateral tracking (c) Safe distances

Fig. 2. Longitudinal and lateral tracking performance of the vehicle formation control and safe distances in control process.

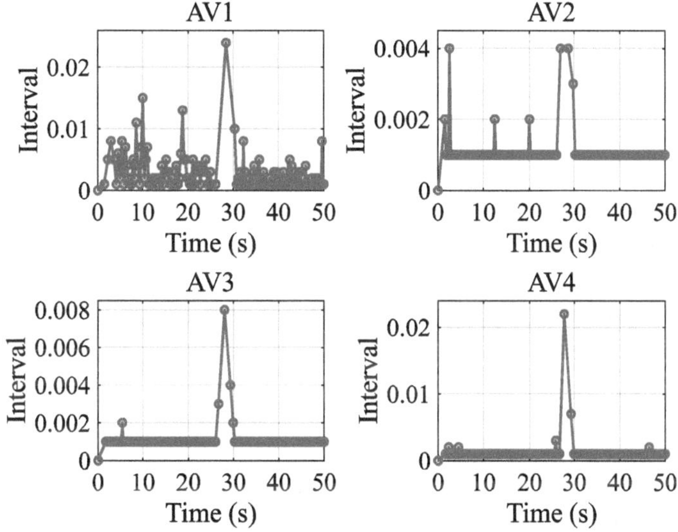

Fig. 3. The Longitudinal update time interval of self-triggered control law for the AVs in linear formation.

maintains nearly constant lateral offsets, while in the linear-queue formation, followers converge smoothly to the leader's path within 15 s without overshoot, indicating well-damped and stable lateral dynamics.

The safe-distance results in Fig. 2 (c) further confirm collision-free formation maintenance. Pairwise distances remain above safety thresholds with smooth temporal profiles, demonstrating robust spacing regulation. The evolution of self-triggered intervals in Fig. 3 shows that the control-update frequency adapts to system dynamics: intervals shorten during transients and lengthen in steady states. Simulation statistics confirm that the proposed scheme lowers update events compared to continuous control while maintaining high tracking accuracy.

Overall, the results demonstrate that the proposed adaptive neural network controller with self-triggered updates achieves accurate trajectory tracking, stable formation maintenance, and reduced communication demands. The system exhibits fast transient response, negligible steady-state error, and reliable safety margins, validating its potential for AV platooning and cooperative formation control.

6 Conclusion

This paper introduces an observer-based self-triggered adaptive control framework to tackle the consensus tracking problem of nonlinear MASs under communication constraints and uncertain dynamics. By integrating neural networks, a sampled-data observer, and a self-triggered control mechanism, the proposed approach alleviates the reliance on continuous communication and precise model constants. The theoretical analysis guaranties the uniform ultimate boundedness of all system signals while ensuring the asymptotic convergence of tracking errors to a neighborhood around the origin. The self-triggered mechanism maintains a strictly positive minimum inter-event time, preventing Zeno behavior and optimizing communication efficiency. Simulation results confirm the robustness and practicality of the approach under real-world conditions.

Acknowledgments. This study did not receive funding.

Disclosure of Interests. The authors have no competing interests to declare that are relevant to the content of this article.

References

1. Chen, C.L.P., Wen, G.X., Liu, Y.J., Liu, Z.: Observer-based adaptive backstepping consensus tracking control for high-order nonlinear semi-strict-feedback multiagent systems. IEEE Trans. Cybern. **46**(7), 1591–1601 (2016)
2. Wang, Z., Gao, Y., Rikos, A.I., Pang, N., Ji, Y.: Fixed-relative-switched threshold strategies for consensus tracking control of nonlinear multiagent systems. In: Proceedings of the IEEE International Conference Control Automation (ICCA), pp. 899–905 (2025)
3. Wang, X., Yin, Z., Lei, Y., Huang, T., Kurths, J.: Secure consensus for switched multiagent systems under DoS attacks: hybrid event-triggered and impulsive control approach. IEEE Trans. Cybern. **55**(5), 2400–2410 (2025)

4. Wang, Z., Piao, S., Ji, Y., Wang, X., Tsung, F.: An efficient dual-observer method for leader-following consensus control of multiagent systems. In: 2025 IEEE 21st International Conference Automation Science Engineering (CASE), pp. 3468–3473 (2025)
5. Wang, L., Wang, X., Wang, Z.: Event-triggered optimal tracking control for strict-feedback nonlinear systems with non-affine nonlinear faults. Nonlinear Dyn. **112**(17), 15413–15426 (2024)
6. Pang, N., Wang, X., Wang, Z.: Observer-based event-triggered adaptive control for nonlinear multiagent systems with unknown states and disturbances. IEEE Trans. Neural Netw. Learn. Syst. **34**(9), 6663–6669 (2023)
7. Wang, Z.M.: Hybrid event-triggered control of nonlinear system with full state constraints and disturbance. In: 36th Chinese Control Decision Conference, pp. 2122–2127 (2024)
8. Ni, Z., Li, F., Wang, Y., Shen, H.: Sliding-mode control for 2-D hidden Markov jump Roesser systems with partial information and its application in metal rolling process. IEEE Trans. Autom. Sci. Eng. **22**, 6851–6859 (2024)
9. Li, F., Ni, Z., Su, L., Xia, J., Shen, H.: Passivity-based finite-region control of 2-D hidden Markov jump Roesser systems with partial statistical information. Nonlinear Anal. Hybrid Syst **51**, 101433 (2024)
10. Åström, K.J., Bernhardsson, B.: Comparison of periodic and event based sampling for first-order stochastic systems. IFAC Proc. Volumes **32**(2), 5006–5011 (1999)
11. Xing, L.T., Wen, C.Y., Liu, Z.T., Su, H.Y., Cai, J.P.: Event-triggered adaptive control for a class of uncertain nonlinear systems. IEEE Trans. Auto. Cont. **62**(4), 2071–2076 (2017)
12. Wang, Z., Wang, X., Pang, N.: Adaptive fixed-time control for full state-constrained nonlinear systems: switched-self-triggered case. IEEE Trans. Circ. Syst. II Express Briefs **71**(2), 752–756 (2024)
13. Tabuada, P.: Event-triggered real-time scheduling of stabilizing control tasks. IEEE Trans. Autom. Control **52**(9), 1680–1685 (2007)
14. Theodosis, D., Dimarogonas, D.V.: Event-triggered control of nonlinear systems with updating threshold. IEEE Control Syst. Lett. **3**(3), 655–660 (2019)
15. Meng, L., Wang, X., Wang, Z.: Event-triggered optimized control for nonlinear multiagent systems via reinforcement learning strategy. Cogn. Comput. **17**(5), 1–9 (2025)
16. Elena, P., Romain, P., Daniele, A., Dragan, N., Maurice, H.W.: Decentralized event-triggered estimation of nonlinear systems. Automatica **160**, 111414 (2024)
17. Anta, A., Tabuada, P.: To sample or not to sample: self-triggered control for nonlinear systems. IEEE Trans. Autom. Control **55**(9), 2030–2042 (2010)
18. Wang, Z.M., Wang, H., Wang, X., Pang, N., Shi, Q.: Event-triggered adaptive neural control for full state-constrained nonlinear systems with unknown disturbances. Cogn. Comput. **16**(2), 717–726 (2023)
19. Wang, X., Zhang, S., Li, H., Zhang, W., Li, H., Huang, T.: Neuroadaptive containment control for nonlinear multiagent systems with input saturation: an event-triggered communication approach. IEEE Trans. Syst. Man Cybern. Syst. **55**(5), 3163–3173 (2025)
20. Wang, Z., Zhang, Y., Zhao, C., Yu, H.: Adaptive event-triggered formation control of autonomous vehicles. arXiv preprint arXiv:2506.06746 (2025)
21. Xue, Y., Wang, C., Ding, C., Yu, B., Cui, S.: Observer-based event-triggered adaptive platooning control for autonomous vehicles with motion uncertainties. Transp. Res. Part C Emerg. Technol. **159**, 104462 (2024)

Reinforcement Learning-Based Prescribed Performance Formation Control for Unmanned Surface Vehicles

Gengqi Li, Liang Cao, and Meng Zhao(✉)

College of Mathematical Sciences, Bohai University, Jinzhou 121013, China
zhaomengzm157@163.com

Abstract. This paper proposes the reinforcement learning-based prescribed performance optimal formation control approach for a fleet of unmanned surface vehicles (USVs). Firstly, to improve the prescribed performance of formation errors while ensuring collision avoidance and preserving connectivity between two successive USVs, the monotone tube boundary functions are designed. Second, to obtain realistic optimal solutions, a reinforcement learning algorithm based on the actor-critic framework is employed, with the unknown parameters approximated using a identifier neural networks. Lastly, the simulation results clearly demonstrate the viability and efficacy of the suggested control strategy inside the backstepping method.

Keywords: Reinforcement learning · unmanned surface vehicles · monotone tube boundary function

1 Introduction

Unmanned surface vehicles (USVs), as an important branch in the field of unmanned systems, have naturally become a popular focus of research in the field [1]. In spite of the great progress made in the formation control of USVs, there are still some unavoidable problems, such as collisions due to practical environmental factors or exceeding the connection range between two consecutive vehicles.

It is worth noting that in existing research, scholars have conducted preliminary explorations into the state constraint problem in USV formation systems through methods such as designing time-varying Lyapunov functions for obstacles. However, there remains room for further refinement in developing a universal framework that simultaneously addresses transient performance and constraint satisfaction. The monotonic tube boundary function was proposed in [2], which quantitatively predetermines the formation error with defined metrics, including convergence time and overshoot. Nowadays, the improvement of transient and steady state performance by incorporating prescribed performance control (PPC) into USV formation control has received great attention [3,4].

C. Li et al. (Eds.): ICNC 2025, CCIS 2946, pp. 464–473, 2026.
https://doi.org/10.1007/978-981-92-1599-7_39

However, because of the limited energy of USVs, it is a difficult challenge to permit the employment of cheap cost to accomplish the necessary trajectory during platoon formation.

The primary role of optimal control is to achieve minimum performance indices, reach optimal performance and reduce energy consumption. For USVs, the optimization ability to balance performance and control cost is very important. In general, we rely on solving the Hamilton-Jacobi-Bellman (HJB) equation as a means of deriving the optimal solution. Since the HJB equation of the USV system presents its inherent nonlinear dynamics, which greatly increases the difficulty of solving it. Reinforcement learning (RL) has proven to be a powerful tool to effectively address this challenge [5,6]. A simplified optimal backstepping control framework was devised for nonlinear systems with unknown dynamics in [7], greatly reducing the computational complexity. During the learning process, it is crucial to ensure that the formation error has a specific transient performance during the learning process because the convergence speed of the formation error influences the learning speed of the NN weights.

Based on the above discussion, this paper proposes a RL-based prescribed performance optimal formation control strategy. A monotonic tube boundary function is designed in the framework of backstepping optimality to construct a novel formation control strategy. The monotonic tube boundary function not only well constrains the overshooting of the formation error, reduces chattering and allows the formation error to converge to a neighborhood of the origin at predetermined time, but also maintains the monotonicity of the function in the presence of parameter variations.

2 Problem Description

2.1 Model Design

Consider a string of N fully actuated USVs with $\aleph = 1, ..., N$. The kinematics of the ith USV for $i \in \aleph$ is given by

$$\begin{aligned} \dot{x}_i &= u_i \cos\psi_i - v_i \sin\psi_i \\ \dot{y}_i &= u_i \sin\psi_i + v_i \cos\psi_i \\ \dot{\psi}_i &= r_i \end{aligned} \tag{1}$$

Let $\boldsymbol{\eta}_i = [x_i, y_i, \psi_i]^T$ denote position and heading angle of the ith vehicle and $\boldsymbol{\nu}_i = [u_i, v_i, r_i]^T$ express the velocity vector. For $i \in \aleph$, the kinetics of the ith vehicle dynamic can be expressed as

$$\mathbf{M}_i\dot{\boldsymbol{\nu}}_i = -\mathbf{C}(\boldsymbol{\nu}_i)\boldsymbol{\nu}_i - \mathbf{D}(\boldsymbol{\nu}_i)\boldsymbol{\nu}_i + \boldsymbol{\tau}_i + \boldsymbol{\tau}_{wi}(t) \tag{2}$$

where $\mathbf{M}_i = \mathbf{M}_i^T > 0$ stands for the vessel constant inertia matrix, $\mathbf{D}(\boldsymbol{\nu}_i)$ is the damping matrix, $\boldsymbol{\tau}_i = [\tau_{ui}, \tau_{vi}, \tau_{ri}]^T$ is the control input, $\boldsymbol{\tau}_{wi} = [\tau_{wui}, \tau_{wvi}, \tau_{wri}]^T$ is the external disturbance, and $\mathbf{C}(\boldsymbol{\nu}_i)$ is the total Coriolis and centripetal acceleration matrix.

2.2 LOS-Based Optimal Formation Control

Make $\boldsymbol{\eta}_0 = [x_0, y_0, \psi_0]^T$ for the desired reference trajectory. The angle and line-of-sight (LOS) range between two nearby vehicles are defined as follows:

$$\varphi_i(t) = \text{atan2}(\breve{y}_i(t), \breve{x}_i(t)) \tag{3}$$

$$d_i(t) = \sqrt{(\breve{x}_i(t))^2 + (\breve{y}_i(t))^2} \tag{4}$$

and then, one yields

$$\begin{aligned} \breve{y}_i &= y_{i-1} - y_i = d_i \sin\varphi_i \\ \breve{x}_i &= x_{i-1} - x_i = d_i \cos\varphi_i \end{aligned} \tag{5}$$

where atan2 $(\cdot)$ is a smooth function.

2.3 Collision Avoidance and Connectivity Maintenance

To avoid collisions between the follower and its leader, the LOS needs to satisfy the following condition:

$$d_{i,\min} < d_i(t) < d_{i,\max} \tag{6}$$

where $d_{i,\min} > 0$ is a safe distance, and $d_{i,\max}$ ($d_{i,\max} > d_{i,\min} > 0$) indicates the connection range.

The following is the definition of formation error, which consists of angle error and LOS error:

$$e_{\psi i} = \psi_{i-1} - \psi_i \tag{7}$$

$$e_{di} = d_i - d_{i,\text{des}} \tag{8}$$

where $d_{i,\text{des}}$ indicates the desired distance with $0 < d_{i,\min} < d_{i,\text{des}} < d_{i,\max}$. LOS error e_{di} satisfies

$$d_{i,\min} - d_{i,\text{des}} < e_{di} < d_{i,\max} - d_{i,\text{des}} \tag{9}$$

where $d_{i,\min} - d_{i,\text{des}} < 0$ and $d_{i,\max} - d_{i,\text{des}} > 0$.

2.4 Prescribed Performance for Formation Errors

Employing the inherent characteristics of the hyperbolic cosecant function csch$(\cdot)$, at its initial point, we devise a set of viable monotone tube boundary functions as follows:

$$\text{col}(\bar{b}_{ji}, \underline{b}_{ji}) = \text{sign}\big(e_{ji}(0)\big)(\mathfrak{q}_{ji} - \mathfrak{q}_{ji,\infty})\mathbf{1}_2 + \Pi_{ji}\mathfrak{q}_{ji} \tag{10}$$

with $\mathbf{1}_2 = \text{col}(1,1)$, $\Pi_{ji} = \text{diag}(\bar{\lambda}_{ji}, -\underline{\lambda}_{ji})$, $i \in \aleph$, $j = d, \psi$ and

$$\mathfrak{q}_{ji} = \begin{cases} \text{csch}(\mathfrak{q}_{ji,0} + \dfrac{\alpha t}{T_c - t}) + \mathfrak{q}_{ji,\infty}, & 0 \le t < T_c \\ \mathfrak{q}_{ji,\infty}, & t \ge T_c \end{cases} \tag{11}$$

with $T_c > 0$ denotes the time constant, $\alpha \in \mathbb{R}_{>0}$ denotes the rate of convergence of the formation error e_{ji}, $\mathfrak{q}_{ji,0}$ represents the initial value of preselected bounds, and $\bar{\lambda}_{ji}$, $\underline{\lambda}_{ji}$, $\mathfrak{q}_{ji,\infty} \in \mathbb{R}_{>0}$.

2.5 Dynamic Surface Control Technique

Define the error coordinate transformation and the boundary layer error:

$$\mathbf{z}_{2i} = \boldsymbol{\nu}_i - \boldsymbol{\alpha}_{fi},\ \mathbf{e}_{\hat{\alpha}_i^*} = \boldsymbol{\alpha}_{fi} - \hat{\boldsymbol{\alpha}}_i^* \tag{12}$$

where $\mathbf{z}_{2i}$ and $\boldsymbol{\alpha}_{fi}$ denote the filtered virtual control input vector and the optimal virtual control input vector is $\hat{\boldsymbol{\alpha}}_i^* = [\hat{\alpha}_{1i}^*, \hat{\alpha}_{2i}^*, \hat{\alpha}_{3i}^*]^T$. Let $\hat{\boldsymbol{\alpha}}_i^*$ pass through a first-order filter to have $\boldsymbol{\alpha}_{fi}$, that is $\boldsymbol{\mu}_i\dot{\boldsymbol{\alpha}}_{fi} + \boldsymbol{\alpha}_{fi} = \hat{\boldsymbol{\alpha}}_i^*, \boldsymbol{\alpha}_{fi}(0) = \hat{\boldsymbol{\alpha}}_i^*(0)$. Thus, $\dot{\mathbf{e}}_{\hat{\alpha}_i^*}$ with $\mathbf{e}_{\hat{\alpha}_i^*} = [e_{\hat{\alpha}_{1i}^*}, e_{\hat{\alpha}_{2i}^*}, e_{\hat{\alpha}_{3i}^*}]^T$ is explained as $\dot{\mathbf{e}}_{\hat{\alpha}_i^*} = \boldsymbol{\mu}_i^{-1}(\hat{\boldsymbol{\alpha}}_i^* - \boldsymbol{\alpha}_{fi}) - \dot{\hat{\boldsymbol{\alpha}}}_i^* = \boldsymbol{\mu}_i^{-1}\mathbf{e}_{\hat{\alpha}_i^*} - \mathbf{B}_i(\cdot)$ where $\dot{\hat{\boldsymbol{\alpha}}}_i^* \triangleq \mathbf{B}_i(\cdot)$ is bounded, and satisfies $|B_{li}| \leq \bar{b}_i$.

2.6 Constrained Error Transformation

We introduce the error transformation function $T(z_{ji})$ as follows:

$$T(z_{ji}) = \frac{\exp(z_{ji})\bar{b}_{ji} + \underline{b}_{ji}}{1 + \exp(z_{ji})} \tag{13}$$

where z_{ji} is the transformed error. Let transformation error $e_{ji} = T(z_{ji})$, we have

$$\dot{z}_{ji} = p_{ji}\dot{e}_{ji} - q_{ji} \tag{14}$$

It is clear that p_{ji} is bounded, and exists constant $\bar{p}_{ji} > 0$, such that $|p_{ji}| < \bar{p}_{ji}$.

Differentiating Eqs. (7) and (8) by considering Eqs. (4) and (5) and along kinematics model (1) yields

$$\dot{e}_{di} = -u_i \cos\theta_i + v_i \sin\theta_i + \Xi_i \tag{15}$$

$$\dot{e}_{\psi i} = \dot{\psi}_{i-1} - r_i \tag{16}$$

where $\theta_i = \psi_i - \varphi_i$ and $\Xi_i = \dot{x}_{i-1}\cos\varphi_i + \dot{y}_{i-1}\sin\varphi_i$. Substituting Eqs. (15) and (16) into Eq. (14), we have

$$\dot{z}_{di} = p_{di}(-u_i\cos\theta_i + v_i\sin\theta_i + \Xi_i) - q_{di} \tag{17}$$

$$\dot{z}_{\psi i} = p_{\psi i}(\dot{\psi}_{i-1} - r_i) - q_{\psi i} \tag{18}$$

3 Main Results

Step 1: The optimal performance index are represented by the following function:

$$J_{di}^*(z_{di}) = \int_t^\infty h_{di}\big(z_{di}(s), \alpha_{1i}^*(z_{di}), \alpha_{2i}^*(z_{di})\big)ds \tag{19}$$

$$J_{\psi i}^*(z_{\psi i}) = \int_t^\infty h_{\psi i}\big(z_{\psi i}(s), \alpha_{3i}^*(z_{\psi i})\big)ds \tag{20}$$

where $h_{di}(z_{di}, \alpha_{1i}, \alpha_{2i}) = z_{di}^2(t) + \alpha_{1i}^2(z_{di}) + \alpha_{2i}^2(z_{di})$ and $h_{\psi i}(z_{\psi i}, \alpha_{3i}) = z_{\psi i}^2(t) + \alpha_{3i}^2(z_{\psi i})$ are the cost functions.

The HJB equations associated with Eqs. (17) and (18) are obtained as follows:

$$H_{di}\Big(z_{di}, \alpha_{1i}^*, \alpha_{2i}^*, \frac{dJ_{di}^*}{dz_{di}}\Big) = 0,\ H_{\psi i}\Big(z_{\psi i}, \alpha_{3i}^*, \frac{dJ_{\psi i}^*}{dz_{\psi i}}\Big) = 0 \tag{21}$$

By solving $\partial H_{di}/\partial \alpha_{1i}^* = 0$, $\partial H_{di}/\partial \alpha_{2i}^* = 0$ and $\partial H_{\psi i}/\partial \alpha_{3i}^* = 0$, we obtain

$$\alpha_{1i}^* = \frac{p_{di}\cos\theta_i}{2}\frac{dJ_{di}^*(z_{di})}{dz_{di}},\ \alpha_{2i}^* = -\frac{p_{di}\sin\theta_i}{2}\frac{dJ_{di}^*(z_{di})}{dz_{di}},\ \alpha_{3i}^* = \frac{p_{\psi i}}{2}\frac{dJ_{\psi i}^*(z_{\psi i})}{dz_{\psi i}}$$

The term $dJ_{ji}^*(z_{ji})/dz_{ji}$, $j = d, \psi$ is decomposed as

$$\frac{dJ_{di}^*(z_{di})}{dz_{di}} = \frac{2}{p_{di}^2}(c_{z_{di}}z_{di} - q_{di}) + \frac{2}{p_{di}}\Xi_i + V_{NNdi} \tag{22}$$

$$\frac{dJ_{\psi i}^*(z_{\psi i})}{dz_{\psi i}} = \frac{2}{p_{\psi i}^2}(c_{z_{\psi i}}z_{\psi i} - q_{\psi i}) + \frac{2}{p_{\psi i}}\dot{\psi}_{i-1} + V_{NN_{\psi i}} \tag{23}$$

where $c_{z_{di}}$ and $c_{z_{\psi i}}$ are positive design constants and V_{NNdi} and $V_{NN_{\psi i}}$ are continuous functions.

Then, yields

$$\alpha_{1i}^* = \cos\theta_i\Big(\frac{1}{p_{di}}(c_{z_{di}}z_{di} - q_{di}) + \Xi_i\Big) + \frac{p_{di}\cos\theta_i}{2}V_{NNdi}$$

$$\alpha_{2i}^* = -\sin\theta_i\Big(\frac{1}{p_{di}}(c_{z_{di}}z_{di} - q_{di}) + \Xi_i\Big) - \frac{p_{di}\sin\theta_i}{2}V_{NNdi}$$

$$\alpha_{3i}^* = \frac{1}{p_{\psi i}}(c_{z_{\psi i}}z_{\psi i} - q_{\psi i}) + \dot{\psi}_{i-1} + \frac{p_{\psi i}}{2}V_{NN_{\psi i}}$$

The NNs are used to estimate the unknown and continuous function V_{NNji} within the compact set Ω. Then, V_{NNji} is represented as $V_{NNji} = W_{ji}^{*T}S_{J_{ji}}(z_{ji}) + \varepsilon_{ji}(z_{ji})$.

The optimal controllers are unavailable because W_{ji}^* is unknown. To obtain the available controllers, the following critic and actor NNs implement RL. Next, the actor and critic NNs are created as

$$\frac{d\hat{J}_{di}^*(z_{di})}{dz_{di}} = \frac{2}{p_{di}^2}(c_{z_{di}}z_{di} - q_{di}) + \frac{2}{p_{di}}\Xi_i + \hat{W}_{c_{di}}^T S_{J_{di}}$$

$$\frac{d\hat{J}_{\psi i}^*(z_{\psi i})}{dz_{\psi i}} = \frac{2}{p_{\psi i}^2}(c_{z_{\psi i}}z_{\psi i} - q_{\psi i}) + \frac{2}{p_{\psi i}}\dot{\psi}_{i-1} + \hat{W}_{c_{\psi i}}^T S_{J_{\psi i}}$$

$$\hat{\alpha}_{1i}^* = \cos\theta_i\Big(\frac{1}{p_{di}}(c_{z_{di}}z_{di} - q_{di}) + \Xi_i\Big) + \frac{p_{di}\cos\theta_i}{2}\hat{W}_{a_{di}}^T S_{J_{di}}$$

$$\hat{\alpha}_{2i}^* = -\sin\theta_i\Big(\frac{1}{p_{di}}(c_{z_{di}}z_{di} - q_{di}) + \Xi_i\Big) - \frac{p_{di}\sin\theta_i}{2}\hat{W}_{a_{di}}^T S_{J_{di}}$$

$$\hat{\alpha}_{3i}^* = \frac{1}{p_{\psi i}}(c_{z_{\psi i}}z_{\psi i} - q_{\psi i}) + \dot{\psi}_{i-1} + \frac{p_{\psi i}}{2}\hat{W}_{a_{\psi i}}^T S_{J_{\psi i}}$$

where $\hat{W}_{c_{ji}}$ and $\hat{W}_{a_{ji}}$ are the critic and actor NN weight and $d\hat{J}^*_{ji}(z_{ji})/dz_{ji}$ is the estimation of $dJ^*_{ji}(z_{ji})/dz_{ji}$.

The updating law for the critic and actor NN weight are defined as follows:

$$\dot{\hat{W}}_{c_{ji}} = -\gamma_{c_{ji}}(S_{J_{ji}}S^T_{J_{ji}} + \sigma_i I_{p_i})\hat{W}_{c_{ji}} \tag{24}$$

$$\dot{\hat{W}}_{a_{ji}} = -(S_{J_{ji}}S^T_{J_{ji}} + \sigma_i I_{p_i})\big(\gamma_{a_{ji}}(\hat{W}_{a_{ji}} - \hat{W}_{c_{ji}}) + \gamma_{c_{ji}}\hat{W}_{c_{ji}}\big) \tag{25}$$

where $\gamma_{c_{ji}} > 0$, $\sigma_i > 0$, and $I_{p_i} \in \mathbb{R}^{pi}$ is an identity matrix.

These designed parameters $c_{z_{ji}}$, $\gamma_{c_{ji}}$ and $\gamma_{a_{ji}}$ are chosen to satisfy

$$c_{z_{ji}} > \frac{p^2_{ji}}{4} + \bar{p}^2_{ji}, \gamma_{a_{ji}} > \frac{p^2_{ji}}{2}, \gamma_{a_{ji}} > \gamma_{c_{ji}} > \frac{\gamma_{a_{ji}}}{2} \tag{26}$$

Consider the following candidate for the Lyapunov function V_{1i}:

$$V_{1i} = \frac{1}{2}z^2_{di} + \frac{1}{2}z^2_{\psi i} + \frac{1}{2}\tilde{W}^2_{a_{di}} + \frac{1}{2}\tilde{W}^2_{c_{di}} + \frac{1}{2}\tilde{W}^2_{a_{\psi i}} + \frac{1}{2}\tilde{W}^2_{c_{\psi i}}$$

where $\tilde{W}_{c_{ji}} = \hat{W}_{c_{ji}} - W^*_{J_{ji}}$ and $\tilde{W}_{a_{ji}} = \hat{W}_{a_{ji}} - W^*_{J_{ji}}$ are NN weight errors.

Based on the condition (26) and utilize Young's inequality, one gets

$$\begin{aligned}\dot{V}_{1i} \leq &- (c_{z_{di}} - \frac{p^2_{di}}{4})z^2_{di} - (c_{z_{\psi i}} - \frac{p^2_{\psi i}}{4})z^2_{\psi i} - \mathbf{z}^T_{2i}\boldsymbol{\Psi}_i - \mathbf{e}^T_{\hat{\alpha}^*_i}\boldsymbol{\Psi}_i - \frac{\gamma_{c_{di}}}{2}\tilde{W}^T_{c_{di}}(S_{J_{di}}S^T_{J_{di}} \\ &+ \sigma_i I_{p_i})\tilde{W}_{c_{di}} - \frac{\gamma_{c_{di}}}{2}\tilde{W}^T_{a_{di}}(S_{J_{di}}S^T_{J_{di}} + \sigma_i I_{p_i})\tilde{W}_{a_{di}} - \frac{\gamma_{c_{\psi i}}}{2}\tilde{W}^T_{c_{\psi i}}(S_{J_{\psi i}}S^T_{J_{\psi i}} \\ &+ \sigma_i I_{p_i})\tilde{W}_{c_{\psi i}} - \frac{\gamma_{c_{\psi i}}}{2}\tilde{W}^T_{a_{\psi i}}(S_{J_{\psi i}}S^T_{J_{\psi i}} + \sigma_i I_{p_i})\tilde{W}_{a_{\psi i}} + C_{1i}(t)\end{aligned}$$

where $C_{1i}(t) = (\frac{\gamma_{a_{ji}}}{2} + \frac{\gamma_{c_{ji}}}{2})W^{*T}_{J_{ji}}(S_{J_{ji}}S^T_{J_{ji}} + \sigma_i I_{p_i})W^*_{J_{ji}}$.

Step 2: Inspired by the disturbance observer in [8], the observation error is defined as

$$\dot{\tilde{\boldsymbol{\tau}}}_{di} = \mathbf{z}_{2i} - \mathbf{C}_{di}\mathbf{M}^{-1}_i\big(\tilde{\boldsymbol{\tau}}_{di} - \tilde{\mathbf{W}}^T_i\mathbf{S}_i(Z_i)\big) - \dot{\boldsymbol{\tau}}_{di} \tag{27}$$

The NN weight $\hat{W}_{li}$ is updated by

$$\dot{\hat{W}}_{li} = -\boldsymbol{\Gamma}_{li}\big(S_{li}(Z_i)z_{2li} + \zeta_{li}\hat{W}_{li}\big) \tag{28}$$

where $\zeta_{li} > 0$ and $\boldsymbol{\Gamma}_{li} = \boldsymbol{\Gamma}^T_{li} > 0$ is the gain matrix.

Similar to the method employed in the step 1, the optimal controller is derived

$$\begin{aligned}\hat{\boldsymbol{\tau}}^*_i = &- \mathbf{C}_{2i}\mathbf{z}_{2i} + \mathbf{C}(\boldsymbol{\nu}_i)\boldsymbol{\nu}_i + \hat{\mathbf{W}}^T_i\mathbf{S}_i(Z_i) - \hat{\boldsymbol{\tau}}_{di} - \mathbf{M}_i\boldsymbol{\mu}^{-1}_i\mathbf{e}_{\hat{\alpha}^*_i} + \boldsymbol{\Psi}_i \\ &- \frac{\mathbf{M}^{-1}_i}{2}\hat{\mathbf{W}}^T_{a_{2i}}\mathbf{S}_{J_{2i}}\end{aligned} \tag{29}$$

The appropriate Lyapunov function is selected for the purpose of

$$\begin{aligned}V_{2i} = &V_{1i} + \frac{1}{2}\sum_{l=1}^{3}e^2_{\hat{\alpha}^*_{li}} + \frac{1}{2}\sum_{l=1}^{3}\tilde{W}^T_{a_{2li}}\tilde{W}_{a_{2li}} + \frac{1}{2}\sum_{l=1}^{3}\tilde{W}^T_{c_{2li}}\tilde{W}_{c_{2li}} + \frac{1}{2}\sum_{l=1}^{3}\tilde{W}^T_{li}\Gamma^{-1}_{li}\tilde{W}_{li} \\ &+ \frac{1}{2}\mathbf{z}^T_{2i}\mathbf{M}_i\mathbf{z}_{2i} + \frac{1}{2}\tilde{\boldsymbol{\tau}}^T_{di}\tilde{\boldsymbol{\tau}}_{di}\end{aligned}$$

Deriving this equation and according to the Young's inequality, we have

$$\begin{aligned}\dot{V}_{2i} \leq &-(\mathrm{c}_{z_{di}}-\frac{p_{di}^2}{4}-\bar{p}_{di}^2)z_{di}^2-(\mathrm{c}_{z_{\psi i}}-\frac{p_{\psi i}^2}{4}-\frac{\bar{p}_{\psi i}}{2})z_{\psi i}^2-(\frac{1}{\mu_{li}}-1)e_{\hat{\alpha}_{li}^*}^2-(\mathbf{C}_{di}\mathbf{M}_i^{-1}\\ &-I_3)\tilde{\boldsymbol{\tau}}_{di}^T\tilde{\boldsymbol{\tau}}_{di}-\left(\mathbf{C}_{2i}-\frac{\left\|\mathbf{M}_i^{-1}\right\|^2}{4}I_3\right)\mathbf{z}_{2i}^T\mathbf{z}_{2i}-\sum_{l=1}^{3}(\frac{\zeta_{li}}{2}-\frac{\lambda_{W_{li}}^{\max}}{2}\left\|\mathbf{C}_{di}\mathbf{M}_i^{-1}\right\|^2)\\ &\times\tilde{W}_{li}^T\tilde{W}_{li}-\frac{\gamma_{c_{di}}}{2}\sigma_i\tilde{W}_{a_{di}}^T\tilde{W}_{a_{di}}-\frac{\gamma_{c_{di}}}{2}\sigma_i\tilde{W}_{c_{di}}^T\tilde{W}_{c_{di}}-\frac{\gamma_{c_{\psi i}}}{2}\sigma_i\tilde{W}_{a_{\psi i}}^T\tilde{W}_{a_{\psi i}}\\ &-\frac{\gamma_{c_{\psi i}}}{2}\sigma_i\tilde{W}_{c_{\psi i}}^T\tilde{W}_{c_{\psi i}}-\frac{\gamma_{c_{2li}}}{2}\sigma_i\tilde{W}_{a_{2li}}^T\tilde{W}_{a_{2li}}-\frac{\gamma_{c_{2li}}}{2}\sigma_i\tilde{W}_{c_{2li}}^T\tilde{W}_{c_{2li}}+C_{2i}(t)\end{aligned}$$

where $C_{2i}(t)=(\frac{\gamma_{a_{2li}}}{2}+\frac{\gamma_{c_{2li}}}{2})W_{J_{2li}}^{*T}(S_{J_{2li}}S_{J_{2li}}^T+\sigma_i I_{p_i})W_{J_{2li}}^*+\frac{3\bar{b}_i^2}{2}+\frac{\bar{\tau}_{di}^2}{2}+C_{1i}(t)$, and I_3 is the identity matrix.

Theorem 1. *Consider a set of USVs. If the design parameters* $\mathrm{c}_{z_{di}}$, $\gamma_{c_{di}}$, $\gamma_{a_{di}}$, $\mathrm{c}_{z_{\psi i}}$, $\gamma_{a_{\psi i}}$, $\gamma_{c_{\psi i}}$, C_{zli}, $\gamma_{c_{2li}}$, $\gamma_{a_{2li}}$ *are chosen appropriately to satisfy the conditions, then the optimal virtual controllers* $\hat{\alpha}_{1i}^*$, $\hat{\alpha}_{2i}^*$, $\hat{\alpha}_{3i}^*$ *and the optimal controller* $\hat{\boldsymbol{\tau}}_i^*$ *with the NNs updating laws can achieve the following properties:*
1) All signals of closed-loop systems are SGUUB.
2) The transient and steady-state performances specified by (10) are guaranteed.
3) Each vehicle is tracked with the ideal trajectory.

Proof. There exists a constant $\beth_i > 0$ that satisfies $C_{2i}(t) < \beth_i$. Let $\eth_i = \min\Big\{4c_{z_{di}}-p_{di}^2-4\bar{p}_{di}^2, 4c_{z_{\psi i}}-p_{\psi i}^2-2\bar{p}_{\psi i}^2, 4(\frac{1}{\mu_{li}}-1), 4\lambda_{\min}(\mathbf{C}_{di}\mathbf{M}_i^{-1}-I_3), 2\gamma_{c_{di}}\sigma_i, 2\gamma_{c_{\psi i}}\sigma_i, 2\gamma_{c_{2li}}\sigma_i, \frac{\lambda_{\min}\left(4\mathbf{C}_{2i}-\left\|\mathbf{M}_i^{-1}\right\|^2 I_3\right)}{\lambda_{\max}(\mathbf{M}_i)}, \frac{2(\zeta_{li}-\lambda_{W_{li}}^{\max}\left\|\mathbf{C}_{di}\mathbf{M}_i^{-1}\right\|^2)}{\lambda_{\max}(\Gamma_{li}^{-1})}\Big\}$. Then, $\dot{V}_{2i}$ is given by

$$\dot{V}_{2i} \leq -\eth_i V_{2i} + \beth_i$$

Based on Lemma 1 in [9], we obtain

$$V_{2i} \leq e^{-\eth_i t}V_{2i}(0) + \frac{\beth_i}{\eth_i}(1-e^{-\eth_i t})$$

When t tends to infinity, we get

$$\begin{aligned}&|z_{ji}| \leq \sqrt{2\hbar_i},\quad |\tilde{W}_{a_{ji}}| \leq \sqrt{2\hbar_i},\quad |\tilde{W}_{c_{ji}}| \leq \sqrt{2\hbar_i},\quad |\tilde{W}_{a_{2li}}| \leq \sqrt{2\hbar_i},\quad |\tilde{W}_{c_{2li}}| \leq \sqrt{2\hbar_i},\\ &|\tilde{W}_{li}| \leq \sqrt{2\hbar_i},\quad |e_{\hat{\alpha}_{li}^*}| \leq \sqrt{2\hbar_i},\quad ||\mathbf{z}_{2i}|| \leq \sqrt{2\hbar_i/\lambda_{\min}(\mathbf{M}_i)},\quad ||\boldsymbol{\tau}_{di}|| \leq \sqrt{2\hbar_i}\end{aligned}$$

where $\hbar_i = \beth_i/\eth_i$, $j = d, \psi$.

4 Simulation Results

A simulation analysis for five identical USVs, i.e., $N = 5$, is presented in this section. The vehicle weighs 23.8 kg and has a length of 1.255 m. The initial values of the weights are $\hat{W}_{li} = 0$, $\hat{W}_{c_{di}} = \hat{W}_{a_{di}} = \hat{W}_{c_{\psi i}} = \hat{W}_{a_{\psi i}} = \hat{W}_{c_{2li}} = \hat{W}_{a_{2li}} = [0.6, ..., 0.6]^T \in \mathbb{R}^{27\times 1}$. The initial conditions are $\boldsymbol{\delta}_{21}(0) = \boldsymbol{\delta}_{22}(0) = [0.1, 0.1, 0.1]^T$ and $\boldsymbol{\delta}_{2\iota}(0) = [0.5, 0.5, 0.5]^T$, $\iota = 3, ..., 5$. The design parameters of the optimal controller $\hat{\boldsymbol{\tau}}_i^*$, optimal virtual controllers $\hat{\alpha}_{1i}^*$, $\hat{\alpha}_{2i}^*$, $\hat{\alpha}_{3i}^*$, the disturbance observer $\boldsymbol{\delta}_i$ and weight adaptive parameters $\hat{W}_{li}$, $\hat{W}_{c_{di}}$, $\hat{W}_{a_{di}}$, $\hat{W}_{c_{\psi i}}$, $\hat{W}_{a_{\psi i}}$, $\hat{W}_{c_{2li}}$ and $\hat{W}_{a_{2li}}$ are taken as $\gamma_{c_{di}} = \gamma_{c_{\psi i}} = \gamma_{c_{2li}} = 2$, $\gamma_{a_{di}} = \gamma_{a_{\psi i}} = \gamma_{a_{2li}} = 3$, $\Gamma_{li} = 1$, $\zeta_{1i} = 1$, $\zeta_{2i} = 0.5$, $\zeta_{3i} = 0.15$, $c_{z_{di}} = 32$, $c_{z_{\psi i}} = 33$, $\mathbf{C}_{2i} = \text{diag}[33, 33, 30]$, $\mathbf{C}_{di} = \text{diag}[33, 33, 29]$ and $\mu_{li} = 0.01$.

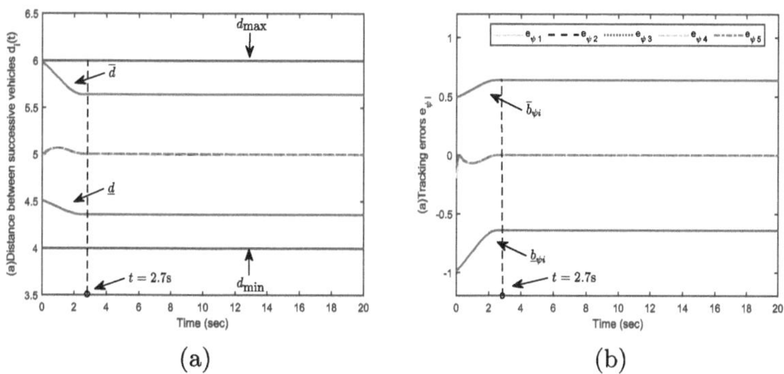

(a) (b)

Fig. 1. (a) The LOS range $d_i(t)$, and the monotonic tube boundary functions. (b) The yaw angle error $e_{\psi i}$, and performance functions.

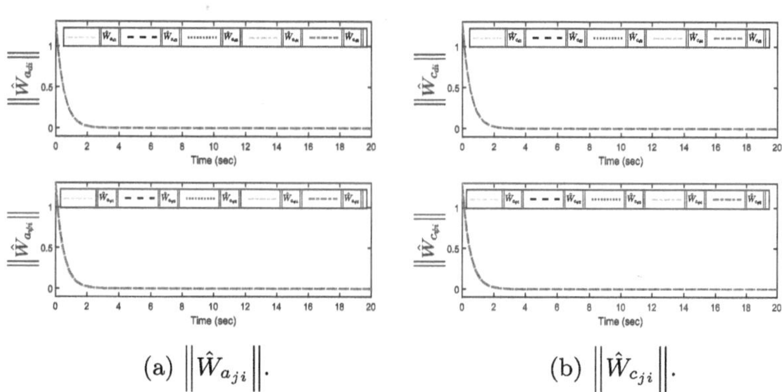

(a) $\left\|\hat{W}_{a_{ji}}\right\|$. (b) $\left\|\hat{W}_{c_{ji}}\right\|$.

Fig. 2. The norms of weight estimates $\left\|\hat{W}_{a_{ji}}\right\|$ and $\left\|\hat{W}_{c_{ji}}\right\|$, $j = d, \psi$, $i = 1, \ldots, 5$.

Simulation results are shown in Figs. 1-4. The evolution of the LOS range $d_i(t)$, the monotonic tube boundary functions $\bar{d} = \bar{b}_{di} + d_{\text{des}}$ and $\underline{d} = \underline{b}_{di} + d_{\text{des}}$, the safety distance for surface navigation $d_{\min}$, and the range of connections between nearby vehicles $d_{\max}$ are all shown in Fig. 1, where $\bar{b}_{di}$ and $\underline{b}_{di}$ are the performance functions. Figure 2 shows that the angle error $e_{\psi i}$ does not extend beyond the region bounded by the monotonic tube boundary. The norm of the NN weight estimates are shown in Figs. 2 and 3. The optimal control input $\hat{\boldsymbol{\tau}}_i^* = [\hat{\tau}_{ui}^*, \hat{\tau}_{vi}^*, \hat{\tau}_{ri}^*]^T$ is presented in Fig. 4.

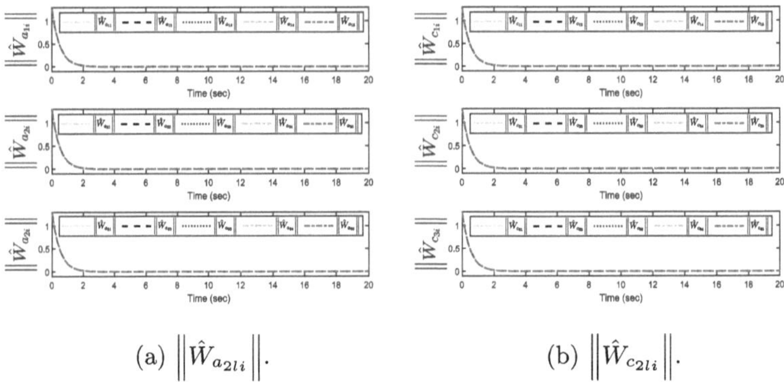

(a) $\left\|\hat{W}_{a_{2li}}\right\|$. (b) $\left\|\hat{W}_{c_{2li}}\right\|$.

Fig. 3. The norms of weight estimates $\left\|\hat{W}_{a_{2li}}\right\|$ and $\left\|\hat{W}_{c_{2li}}\right\|$, $l = 1, 2, 3$.

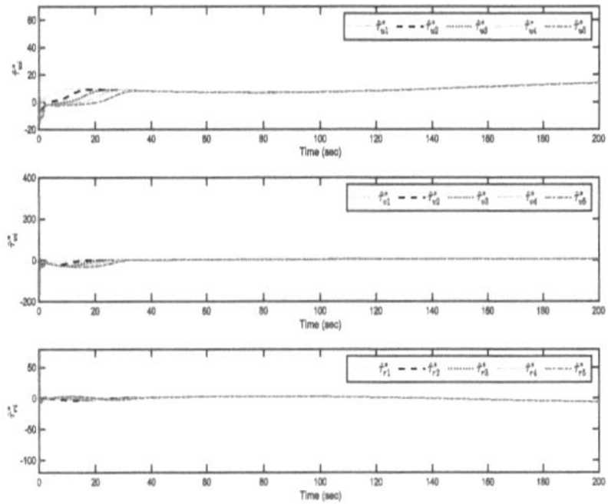

Fig. 4. Control inputs.

5 Conclusions

This research has proposed a prescribed performance optimal backstepping technique based on RL to achieve the preserved connection and collision avoidance of USVs formation. To guarantee that the formation error can precisely and steadily converge to a near-zero state within a predetermined time range, a class of predefined performance boundary functions has been used, greatly increasing the formation accuracy. The control strategy has been built using an actor-critic structure RL method. The adaptive optimization algorithm introduced has ensured optimal performance by guiding each vehicle to seamlessly follow the leader while avoiding collisions and preserving a predefined formation. Simulation results have validated the effectiveness of the proposed approach.

Acknowledgments. This work was partially supported by the National Natural Science Foundation of China (62473059), the Revitalization of Liaoning Talents Program (XLYC2403188), the Jinzhou Talents Program (JXYC240102), and the Natural Science Foundation of Liaoning Province (2024MS183, 2024BS235).

Disclosure of Interests. The authors have no competing interests to declare that are relevant to the content of this article.

References

1. Wang, N., Ahn, C.: Coordinated trajectory-tracking control of a marine aerial-surface heterogeneous system. IEEE/ASME Trans. Mechatron. **26**(6), 3198–3210 (2021)
2. Shi, Y., Yi, B., Xie, W., Zhang, W.: Enhancing prescribed performance of tracking control using monotone tube boundaries. Automatica **159**, 111304 (2024)
3. Li, J., Zhang, G., Cabecinhas, D., Pascoal, A., Zhang, W.: Prescribed performance path following control of USVs via an output-based threshold rule. IEEE Trans. Veh. Technol. **73**(5), 6171–6182 (2024)
4. Li, J., Xiang, X., Zhang, Q., Yang, S.: Robust practical prescribed time trajectory tracking of USV with guaranteed performance. Ocean Eng. **302**, 117622 (2024)
5. Lian, B., Xue, W., Lewis, F., Chai, T.: Inverse reinforcement learning for multi-player noncooperative apprentice games. Automatica **145**, 110524 (2022)
6. Wang, X., Guang, W., Huang, T., Kurths, J.: Optimized adaptive finite-time consensus control for stochastic nonlinear multiagent systems with non-affine nonlinear faults. IEEE Trans. Autom. Sci. Eng. **21**(4), 5012–5023 (2024)
7. Wen, G., Chen, C., Ge, S.: Simplified optimized backstepping control for a class of nonlinear strict-feedback systems with unknown dynamic functions. IEEE Trans. Cybern. **51**(9), 4567–4580 (2020)
8. Cao, L., Pan, Y., Liang, H., Ahn, C.: Event-based adaptive neural network control for large-scale systems with nonconstant control gains and unknown measurement sensitivity. IEEE Trans. Syst. Man Cybern. Syst. **54**(11), 7027–7038 (2024)
9. Guo, X., Wang, C., Liu, L.: Adaptive fault-tolerant control for a class of nonlinear multi-agent systems with multiple unknown time-varying control directions. Automatica **167**, 111802 (2024)

ABODE-VAE: An Autonomous Feature Importance Learning Framework for Age-Stratified Mortality Prediction in Sepsis Patients

Erzhuo Duan[1], You Zhao[1], Shiying Sun[2], Hongjun Yang[2], and Xing He[1(✉)]

[1] Chongqing Key Laboratory of Nonlinear Circuits and Intelligent Information Processing, College of Electronic and Information Engineering, Southwest University, Chongqing 400715, China
hexingdoc@swu.edu.cn

[2] State Key Laboratory of Multimodal Artificial Intelligence Systems, Institute of Automation, Chinese Academy of Sciences, Beijing 100190, China

Abstract. Sepsis is a serious disease that leads to life-threatening organ dysfunction and is associated with a high risk of mortality. Effective early in-hospital mortality prediction is essential for improving patient prognosis. However, most current predictive models suffer from two major limitations: class imbalance in clinical datasets and the insufficient interpretability of the models needed to support clinical decisions. Therefore, we propose a deep learning model named the attention-based ordinary differential equation variational autoencoder (ABODE-VAE), which autonomously assigns importance scores to features, thereby enhancing model interpretability. In addition, the model integrates the weighted representations with minority-class latent features learned by the variational autoencoder, incorporating reconstruction-error signals to enhance minority representations and thus mitigate class imbalance. This study analysed intensive care unit (ICU) patients diagnosed with sepsis from the Medical Information Mart for Intensive Care IV (MIMIC-IV), focusing on two distinct age cohorts: younger adults (18–65 years) and older adults ($\geq$65 years). The experimental results demonstrate that the proposed model achieves outstanding performance, with area under the receiver operating characteristic curves (AUROCs) of 0.8808 and 0.8643, respectively. Moreover, several key mortality predictors consistently identified across both age cohorts include the Glasgow Coma Scale score, oxygen saturation, mechanical ventilation, and body temperature.

Keywords: Deep learning · Hospital mortality prediction · Intensive care unit · Sepsis

1 Introduction

Sepsis is a life-threatening disease, defined as organ dysfunction caused by a dysregulated host response to infection [1]. In intensive care units (ICUs), sepsis

C. Li et al. (Eds.): ICNC 2025, CCIS 2946, pp. 474–483, 2026.
https://doi.org/10.1007/978-981-92-1599-7_40

is one of the most common complications. A study conducted in ICUs across 44 hospitals in China reported a sepsis incidence rate of 20.6%, with 90-day mortality reaching 35.5% [2]. Owing to its high incidence and mortality, sepsis poses a significant healthcare challenge and places a heavy burden on medical resources and the economy. In 2017, sepsis was the most expensive condition to treat in U.S. hospitals, far surpassing the costs of the second-most costly disease. Patients who are diagnosed with sepsis after hospital admission, as well as those whose condition deteriorates, often experience higher mortality and increased economic burden. Similar to acute myocardial infarction, early recognition and appropriate treatment within the first few hours of sepsis onset can improve patient outcomes [3].

Early identification of mortality risk in sepsis patients is crucial for timely intervention and improved clinical outcomes. In clinical practice, various scoring systems are routinely employed to evaluate the severity of sepsis. These scoring systems include the Sequential Organ Failure Assessment (SOFA), the Acute Physiology and Chronic Health Evaluation III (APACHE III), and the Simplified Acute Physiology Score II (SAPS II) [4]. Despite their clinical utility, these scoring systems depend on manually selected features and expert-defined weights, which may fail to capture complex organ interactions or generalize across diverse patient populations. Artificial intelligence has entered a new phase, with machine learning, particularly deep learning, revealing immense potential and value in healthcare applications. Machine learning methods such as XGBoost, random forest (RF), logistic regression (LR), naive Bayes (NB), and neural networks, are capable of analysing vast amounts of data from electronic health records in various formats to predict patient prognosis [5]. Hou et al. [6] applied XGBoost to predict 30-day mortality in sepsis patients. They reported that XGBoost achieved an area under the receiver operating characteristic curve (AUROC) of 0.857, significantly surpassing the SAPS II scores, demonstrating its potential for prognostic support. Tang et al. [7] proposed a deep learning model to improve 30-day all-cause hospital readmission prediction. By integrating imaging and health record data, the model achieved superior performance compared with existing models. Pandey et al. [8] utilized neural ordinary differential equations (neural ODEs) with additional regularization to improve few-shot organ segmentation performance.

In medical applications, a major challenge lies in handling class imbalance. Specifically, there are substantially fewer clinical deaths than survival cases, causing models to bias toward the majority class. To address this issue, data augmentation strategies employ generative models such as variational autoencoders (VAEs), which learn the latent data distribution to generate new synthetic samples [9]. At the same time, understanding model decisions is equally important. Predicting patients' risk of mortality necessitates a focus on interpreting variables, as understanding the contributing factors is critical for clinical decision-making. To improve model interpretability, attention mechanisms have been widely adopted, enabling a more precise identification of the most influential features driving predictive outcomes.

To mitigate data imbalance and improve model interpretability, we propose a deep learning framework named the attention-based ordinary differential equation variational autoencoder (ABODE-VAE). Our main contributions are summarized as follows:

1. We propose the ABODE-VAE framework, which synergistically couples neural ODEs, VAEs, and attention mechanisms. The model first applies a keyless attention mechanism to weight feature importance for improved interpretability, and fuses the resulting weighted representations with VAE-based latent and reconstruction features to enhance minority-class feature representation, thereby alleviating class imbalance.
2. This study conducts stratified analysis across different age cohorts to identify both unique and shared risk factors, thereby constructing personalized and effective predictive models.

2 Methods

2.1 Dataset

We utilized a retrospective cohort design based on the publicly available Medical Information Mart for Intensive Care IV (MIMIC-IV) database. Patients diagnosed with sepsis according to the Third International Consensus Definitions for Sepsis and Septic Shock (Sepsis-3) were included. Patients aged 18 years and older were considered. To account for multiple admissions, we extracted data from each individual' s first hospital admission and first ICU stay. Only patients with an ICU stay longer than 24 h were included, and those with unknown outcomes were excluded. Sepsis can affect anyone, but some populations are at higher risk. An age of 65 years has been identified as the threshold for worse outcomes [10]. The study population was categorized into two cohorts: younger adults (18âĂŞ65 years) and older adults ($\geq$65 years). The inclusion criteria for the study subjects are detailed in Fig. 1. We ultimately obtained 66 features, with 1 feature used to label whether sepsis patients survived during hospitalization and the remaining 65 features serving as predictor variables. These predictor variables span various categories, including demographic data, vital signs, level of consciousness, laboratory test results, and treatment measures.

2.2 Network Architecture

Our objective is to develop a classification model for imbalanced medical datasets and identify key features for predicting in-hospital mortality. We propose a method named ABODE-VAE. It consists of three modules: the keyless attention mechanism module, the VAE module, and the ODE-Net classification module. Figure 2 illustrates the overall framework.

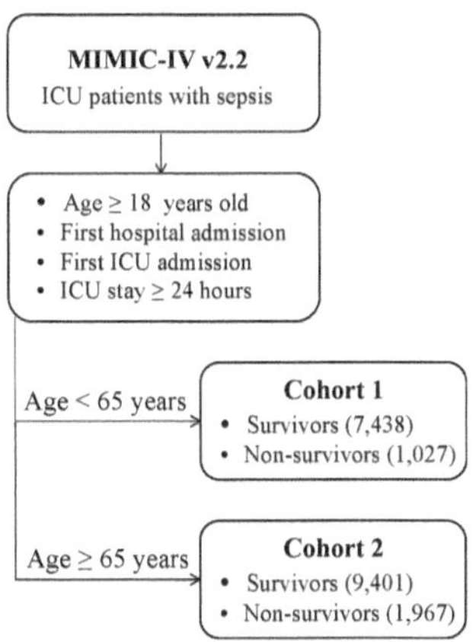

Fig. 1. Flow diagram of patient selection.

Keyless Attention Mechanism Module. At the outset, we introduce an effective approach that leverages the output produced by ODE-Net, which aligns with the input dimensions, and normalizes it via the Softmax function to convert it into a probability distribution for computing the importance weights of each feature among patient samples. Given binary classifications, $\{(x_i, y_i)\}_{i=1}^{N} \subset \mathbb{R}^d \times \{0, 1\}$ represents the medical dataset containing N patient samples. Each sample x_i is a d-dimensional vector, where the j-th feature of the sample x_i is denoted by x_{ij}. The label y_i of each sample belongs to one of the two classes. The network input $\mathbf{X} = [x_1, x_2, \ldots, x_N]^T$ serves as the initial condition at time t_0 for the ODE-Net. The Neural-ODE block, parameterized by ϑ, models the temporal evolution of hidden states g_ϑ, producing the terminal state at time t_1 as output:

$$\mathbf{X}(t_0) = [x_1, x_2, \ldots, x_N]^T \tag{1}$$

$$g_\vartheta(\mathbf{X}(t), t) = \frac{d\mathbf{X}(t)}{dt} \tag{2}$$

$$\mathbf{X}(t_1) = \mathbf{X}(t_0) + \int_{t_0}^{t_1} g_\vartheta(\mathbf{X}(t), t)\, dt \tag{3}$$

The output $\mathbf{X}(t_1) \in \mathbb{R}^{N \times d}$ is further normalized to calculate the attention weights:

$$\mathbf{W} = [\alpha_{ij}] = \frac{\exp(e_{ij})}{\sum_{k=1}^{d} \exp(e_{ik})} \tag{4}$$

By assigning different weights to the input, the model can focus on the important features more effectively. The original input is then multiplied by the attention weights:

$$\mathbf{O} = [o_{ij}] = \alpha_{ij} \odot x_{ij} \tag{5}$$

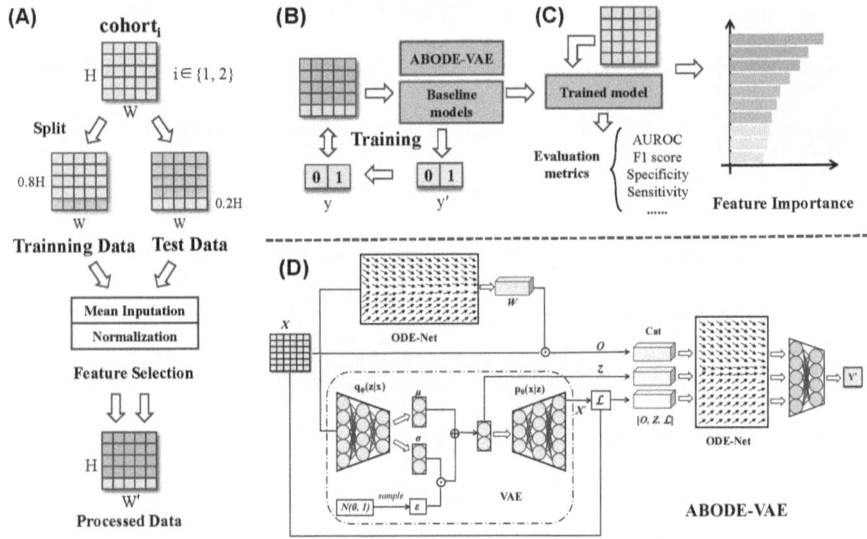

Fig. 2. Architecture and workflow of the ABODE-VAE model for in-hospital mortality prediction. **(A)** Data preprocessing. **(B)** Model training. **(C)** Model evaluation and interpretation. **(D)** Architecture of the proposed ABODE-VAE model.

VAE Module. In the VAE module, latent variables and reconstruction errors are incorporated to mitigate class imbalance and enhance feature discrimination. VAEs learn the data distribution and generate new samples through random sampling in the latent space, which helps capture underrepresented features. During training, the encoder estimates the posterior distribution of the latent variable z_i and samples from it as:

$$z \sim q_\phi(z|x) = \mathcal{N}\left(\mu(x), diag\left(\sigma(x)^2\right)\right) \tag{6}$$

where $\mu(x) \in \mathbb{R}^d$ and $\sigma(x) \in \mathbb{R}^d$ are the mean and standard deviation of the latent distribution, respectively. The decoder then maps z_i back to the data space, producing the reconstructed input x_i'.

The reconstruction error can then be calculated as follows:

$$\mathcal{L} = \frac{1}{N}\sum_{i=1}^{N} \mathcal{L}(x_i, x_i') \tag{7}$$

where $\mathcal{L} \in \mathbb{R}^{N\times c}$ can be calculated via various distance metrics. When training solely on normal data, encoders usually fail to reconstruct anomalous data, resulting in high reconstruction errors, which are considered outliers. In imbalanced classification, reconstruction error serves as an auxiliary signal, encouraging the model to focus on hard-to-reconstruct samples often belonging to the minority class.

ODE-Net Classification Module. Ultimately, we introduce the ODE-Net classification network. As ODE-Net preserves the input dimensionality, a fully connected layer is added to transform the high-dimensional input into a one-dimensional output. Let $\mathbf{O} \in \mathbb{R}^{N\times d}$, $\mathbf{Z} \in \mathbb{R}^{N\times k}$, and $\mathcal{L} \in \mathbb{R}^{N\times c}$ denote the attention-weighted output, latent representation, and reconstruction error, respectively. These matrices are concatenated along the feature dimension to form the final input:

$$\mathbf{X}_f = [\mathbf{O}, \mathbf{Z}, \mathcal{L}] = [f_1, f_2, \ldots, f_N]^T \tag{8}$$

The input of the network is $\mathbf{X}_f(t_0)$, which is used as the initial condition. The Neural-ODE block, parameterized by ψ, models the temporal evolution of hidden states r_ψ, producing the terminal state at time t_1 as output:

$$\mathbf{X}_f(t_0) = [f_1, f_2, \ldots, f_N]^T \tag{9}$$

$$r_\psi(\mathbf{X}_f(t), t) = \frac{d\mathbf{X}_f(t)}{dt} \tag{10}$$

$$\mathbf{X}_f(t_1) = \mathbf{X}_f(t_0) + \int_{t_0}^{t_1} r_\psi(\mathbf{X}_f(t), t)\, dt \tag{11}$$

The resulting $\mathbf{X}_f(t_1)$ is subsequently passed through a fully connected network, culminating in the prediction $\mathbf{Y}'$.

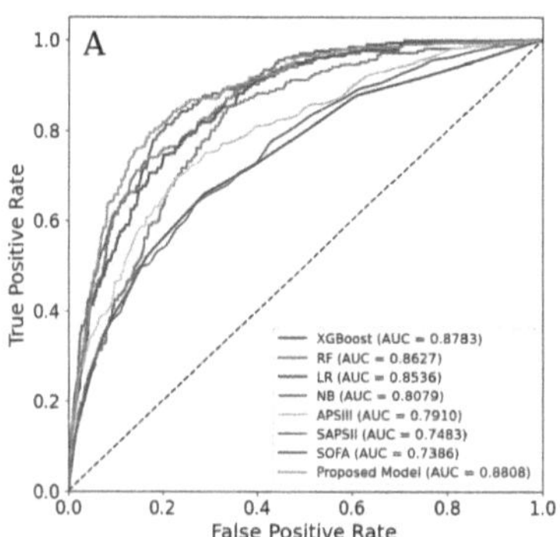

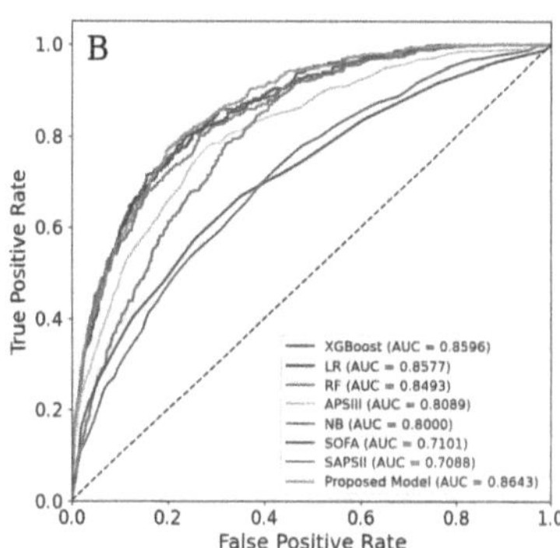

Fig. 3. Comparison of the proposed method with other machine learning methods and scoring systems in predicting in-hospital mortality. (**A**) Cohort 1. (**B**) Cohort 2.

3 Experiments

3.1 Overall Performance Evaluation

In this study, we independently analyzed two cohorts of sepsis patients, including 8,465 younger and 11,368 older individuals. The datasets were partitioned into training and test sets, with 80% of the data allocated for training purposes and the remaining 20% reserved for testing. Notably, we ensured that the proportions of surviving and dead samples remained consistent within the training and test sets for each cohort. To demonstrate the effectiveness of the proposed approach, we compared ABODE-VAE with seven representative and widely used models, including XGBoost, RF, LR, NB, APS III, SAPS II, and SOFA.

In Cohort 1, ABODE-VAE performed well, achieving an AUROC of 0.8808 (95% CI: 0.8569âĂŞ0.9036). XGBoost closely followed, with an AUROC of 0.8783 (95% CI: 0.8558âĂŞ0.9000). RF and LR had AUROC values of 0.8627 (95% CI: 0.8376–0.8871) and 0.8536 (95% CI: 0.8276–0.8780), respectively. Other models, such as NB, APS III, SAPS II, and SOFA, presented AUROC values ranging from 0.7386 to 0.8079, indicating relatively weak performance. The corresponding ROC curves are illustrated in Fig. 3. However, the AUROC may not be sensitive due to the imbalance in the distribution of the data. Therefore, we analysed the sensitivity, specificity, area under the precisionâĂŞrecall curve (AUPRC), and so on. As shown in Table 1, ABODE-VAE also achieved excellent results across other metrics.

Table 1. Performance metrics of different models in Cohort 1

Method	AUROC	Sensitivity	Specificity	Accuracy	F1 score	AUPRC	Brier score
XGBoost	0.8783	0.8146	0.7944	0.7968	0.4926	**0.5304**	0.0829
RF	0.8627	0.7415	**0.8306**	0.8198	0.4992	0.5005	0.0826
LR	0.8536	0.7463	0.7991	0.7927	0.4658	0.4709	0.0833
NB	0.8079	**0.8537**	0.6626	0.6858	0.3968	0.3920	0.1774
APS III	0.7910	0.6927	0.7728	0.7631	0.4146	0.4175	0.0978
SAPS II	0.7483	0.6390	0.7285	0.7177	0.3541	0.3270	0.1094
SOFA	0.7386	0.6585	0.7103	0.7041	0.3502	\	\
ABODE-VAE	**0.8808**	0.7951	0.8293	**0.8252**	**0.5241**	0.5276	**0.0771**

In Cohort 2, the AUROC of ABODE-VAE was 0.8643 (95% CI: 0.8455–0.8825), indicating good predictive ability. The sensitivity of ABODE-VAE in this group was 0.7735, and the specificity was 0.8017, suggesting that the model performs adequately in identifying true high-risk patients and accurately excluding low-risk patients. These results are presented in Fig. 3 and Table 2. Notably, the brier scores for the two cohorts are 0.0771 and 0.1003, which are metrics used to assess the accuracy of probability predictions, with lower values indicating more accurate predictions. The calibration of ABODE-VAE was visually evaluated via a calibration plot, as shown in Fig. 4, and the results indicate satisfactory calibration.

Table 2. Performance metrics of different models in Cohort 2

Method	AUROC	Sensitivity	Specificity	Accuracy	F1 score	AUPRC	Brier score
XGBoost	0.8596	0.8015	0.7634	0.7700	0.5464	**0.6082**	0.1031
RF	0.8493	0.8015	0.7347	0.7463	0.5220	0.5870	0.1082
LR	0.8577	**0.8066**	0.7507	0.7603	0.5377	0.5946	0.1011
NB	0.8000	0.7837	0.6816	0.6992	0.4738	0.4396	0.1935
APS III	0.8089	0.7735	0.7214	0.7304	0.4980	0.5140	0.1126
SAPS II	0.7088	0.7328	0.5726	0.6003	0.3879	0.3470	0.1604
SOFA	0.7101	0.5725	0.7453	0.7155	0.4102	\	\
ABODE-VAE	**0.8643**	0.7735	**0.8017**	**0.7968**	**0.5682**	0.6058	**0.1003**

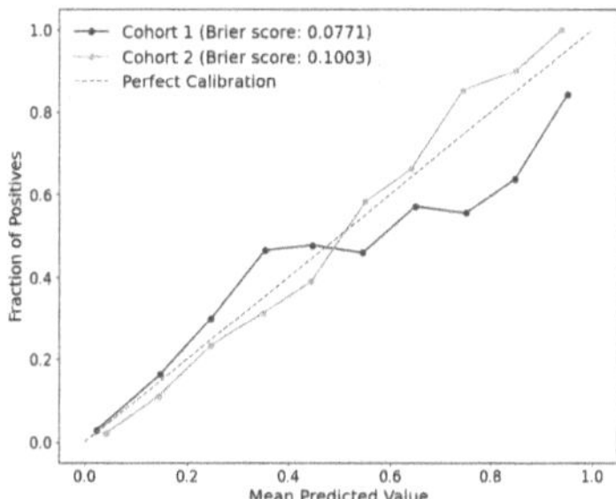

Fig. 4. Calibration plot across two cohorts.

3.2 Model Interpretability

To understand the model's decision-making mechanism and identify the factors that significantly impact mortality prediction, we extracted the top 20 important features determined by ABODE-VAE through the keyless attention mechanism, as shown in Fig. 5. The top 10 important features in both age groups show a high degree of similarity, including Glasgow Coma Scale, pulse oxygen saturation, mechanical ventilation, temperature, pH, and sodium. Moreover, we identified differences between the two cohorts. For younger patients, admission age showed greater relevance in mortality prediction, whereas in older patients, metabolic-related features such as base excess and calcium levels were more significant predictors.

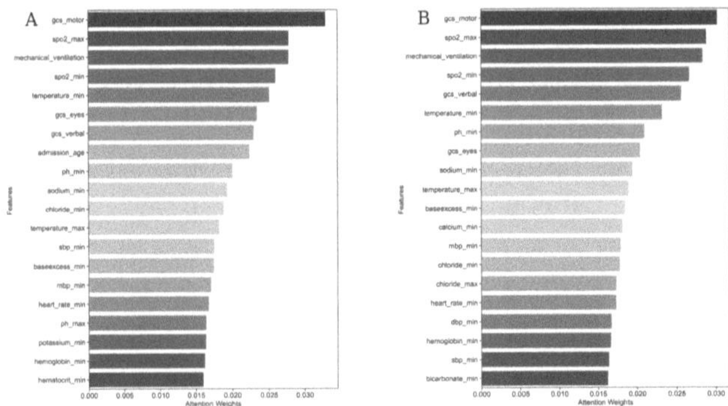

Fig. 5. Top 20 variables contributing to in-hospital mortality prediction using ABODE-VAE. (**A**) Attention weights in Cohort 1. (**B**) Attention weights in Cohort 2.

4 Conclusion

In conclusion, ABODE-VAE offers a reliable predictive solution that combines accuracy and interpretability and aims to meet the demand for efficient risk assessment in critical care environments, thereby maximizing the survival rate of hospitalized sepsis patients. In future work, we plan to extend this framework to multi-center datasets and explore its applicability to other critical conditions.

Acknowledgments. hospitalized sepsis patients

References

1. Singer, M., et al.: The third international consensus definitions for sepsis and septic shock (sepsis-3). JAMA **315**(8), 801–810 (2016)
2. Weng, L., et al.: National incidence and mortality of hospitalized sepsis in China. Critical Care **27**(1), 84 (2023)
3. Rawat, N., et al.: Prospective, multi-site study of patient outcomes after implementation of the trews machine learning-based early warning system for sepsis. Nat. Med. **28**(7), 1455–1460 (2022)
4. Hwang, S.Y., et al.: Prognostic performance of sequential organ failure assessment, acute physiology and chronic health evaluation III, and simplified acute physiology score ii scores in patients with suspected infection according to intensive care unit type. J. Clin. Med. **12**(19), 6402 (2023)
5. Hu, Y., Jin, Y., Jiang, Z., Zheng, Q.: PVT2DNet: polyp segmentation with vision transformer and dual decoder refinement strategy. J. Vis. Commun. Image Represent. **104**, 104304 (2024)
6. Hou, N., et al.: Predicting 30-days mortality for mimic-iii patients with sepsis-3: a machine learning approach using XGBoost. J. Transl. Med. **18**, 1–14 (2020)
7. Tang, S., Tariq, A., Dunnmon, J.A., Sharma, U., Elugunti, P., Rubin, D.L., Patel, B.N., Banerjee, I.: Predicting 30-day all-cause hospital readmission using multimodal spatiotemporal graph neural networks. IEEE J. Biomed. Health Inform. **27**(4), 2071–2082 (2023)

8. Pandey, P., Chasmai, M., Sur, T., Lall, B.: Robust prototypical few-shot organ segmentation with regularized neural-odes. IEEE Trans. Med. Imaging **42**(9), 2490–2501 (2023)
9. Li, Z., et al.: WF-VAE: enhancing video VAE by wavelet-driven energy flow for latent video diffusion model. In: Proceedings of the Computer Vision and Pattern Recognition Conference, pp. 17778–17788 (2025)
10. Marini, C.P., Lieb, D.A.: Sepsis, Septic Shock, and Its Treatment in Geriatric Patients. Springer, Cham, Switzerland (2023). https://doi.org/10.1007/978-3-031-30651-8_53

Acceleration Dual Q-Learning Control for Nonlinear Discrete-Time Systems via Swarm Intelligence

Zeyu Zhou, Peihao Du(✉), Guangdeng Chen, and Qi Zhou

College of Electronic and Information Engineering, Southwest University, No. 2, Tiansheng Road, Beibei District, Chongqing 400715, China
{zhouzeyu,peihaodu}@swu.edu.cn

Abstract. This paper proposes an accelerated dual Q-learning control framework for nonlinear discrete-time systems subject to external disturbances. An acceleration factor $\alpha > 1$ is introduced in the Q-function update law, which substantially improves convergence speed over traditional Q-learning. The method solves a minimax optimal control problem formulated as a discounted performance index with $L2$-gain robustness, iteratively approximating the Hamilton-Jacobi-Bellman equation. We rigorously prove that the Q-value sequence is bounded, non-decreasing, and converges to the optimal solution when the discount factor γ satisfies $\gamma \leq \frac{1}{\alpha}$. The implementation utilizes an actor-critic neural network structure: a critic network approximates the Q-function while action networks handle the optimal control and worst-case disturbance policies. Particle swarm optimization with adaptively scheduled parameters efficiently searches the compressed policy space. Simulation results on a helicopter attitude control model show markedly faster convergence and enhanced disturbance rejection capability compared to standard value iteration methods, confirming the theoretical predictions and practical utility of the proposed approach.

Keywords: Q-learning · Acceleration learning · Particle swarm optimization · Adaptive dynamic programming · Convergence

1 Introduction

Optimal control of nonlinear discrete-time systems subject to external disturbances remains a fundamental challenge in modern control theory and engineering applications [1,2]. Traditional methods, such as dynamic programming [3,4], suffer from the "curse of dimensionality," rendering them computationally intractable for high-dimensional systems. While reinforcement learning (RL), particularly Q-learning, has emerged as a powerful data-driven alternative by approximating solutions to the Hamilton-Jacobi-Bellman (HJB) equation without requiring complete system models [5,6], its practical deployment is hampered by notoriously slow convergence rates and limited robustness guarantees. This

C. Li et al. (Eds.): ICNC 2025, CCIS 2946, pp. 484–498, 2026.
https://doi.org/10.1007/978-981-92-1599-7_41

computational inefficiency becomes particularly critical in safety-sensitive applications like aerospace systems [7,8], where rapid adaptation to disturbances is essential.

Recent advances in adaptive dynamic programming (ADP) have integrated Q-learning with function approximation techniques to mitigate dimensional limitations [9,10]. However, standard Q-learning implementations typically restrict the learning rate to the conservative range of (0, 1), prioritizing stability at the expense of convergence speed [11]. Moreover, existing approaches [12,13] often conflate the value function $\mathcal{V}(x_k)$ with the Q-function $Q(x,u)$, overlooking the core principle of Q-learning: optimizing action-selection for future states rather than merely evaluating current-state values. This conceptual simplification, while computationally convenient, sacrifices the algorithm's innate capacity for anticipatory policy improvement. Additionally, most frameworks [14] lack rigorous theoretical guarantees for convergence under acceleration, particularly in the presence of adversarial disturbances.

To address these limitations, this paper proposes a novel accelerated dual Q-learning control framework that fundamentally restructures the update law through an acceleration factor $\alpha > 1$, embedded within a minimax optimal control formulation. The key innovation lies in decoupling the Q-function update from the conservative stability constraints, enabling aggressive yet theoretically grounded acceleration. We rigorously prove that the resulting Q-value sequence is bounded, monotonically non-decreasing, and converges to the optimal solution when the discount factor γ satisfies $\gamma \leq \frac{1}{\alpha}$: A condition that establishes an explicit trade-off between acceleration and convergence guarantees. This theoretical contribution distinguishes our work from heuristic acceleration attempts by providing necessary and sufficient conditions for stability.

The implementation leverages an actor-critic neural network architecture, where a critic network approximates the Q-function while separate action networks compute optimal control and worst-case disturbance policies. To overcome local optima in high-dimensional policy spaces, we employ particle swarm optimization (PSO) [15,16] with adaptively scheduled parameters that dynamically balance exploration and exploitation. By encoding the entire policy as a single parameter matrix and searching a compressed latent representation, the method achieves order-of-magnitude reductions in iteration counts compared to standard value iteration.

The efficacy of the proposed approach is validated through extensive simulations on a helicopter attitude control model under $L2$-gain bounded disturbances [17,18]. Results demonstrate markedly faster convergence and superior disturbance rejection capability relative to conventional Q-learning, confirming both the theoretical predictions and practical utility for robust aerospace applications.

The main contributions of this paper are highlighted as follows.

(I) Different from most current works simply equate the value function $\mathcal{V}(x_k)$ directly with the Q value $Q_j(x_k, \pi_j(x_k), \omega_k^j)$, the proposed Q-learning focuses on the essence: optimizing the next-state action.

(II) An acceleration factor $\alpha > 1$ is proposed, transcending the conventional $0 < \alpha < 1$ limitation. We rigorously establish a necessary and sufficient convergence condition and prove that the resulting Q-value sequence is bounded, monotonically non-decreasing, and convergent to the optimal solution.

(III) An adaptive particle swarm optimization (PSO) algorithm integrated with adaptive dynamic programming (ADP) is proposed. The algorithm adaptively balances global versus local search based on real-time Q-value fitness feedback, achieving order-of-magnitude iteration reductions.

The remainder of this paper is organized as follows. Section 2 formulates the minimax optimal control problem and derives the accelerated Q-learning update law. Section 3 presents the iterative algorithm and establishes its monotonicity, boundedness, and convergence properties. Section 4 details the actor-critic implementation and PSO-based policy improvement mechanism. Section 5 provides simulation results and comparative analysis. Section 6 concludes the paper and outlines future research directions.

2 Problem Statement and Preliminaries

Consider a class of discrete-time nonlinear systems

$$x_{k+1} = f(x_k) + g(x_k)u_k + \omega_k \tag{1}$$

where $x_k \in \Omega \subset \mathbb{R}^n$ is the state vector, Ω is the compact ensuring that system (1) is controllable and stabilizable within Ω, $u_k \in \mathbb{R}^m$ is the control input vector, $f \in \mathbb{R}^n$ and $g \in \mathbb{R}^{n\times m}$ are the state dynamic vector and input dynamic matrix, respectively, $\omega_k \in \mathbb{R}^q$ is the time-varying disturbance , T is the time interval, and $k \in \mathbb{N}$ is the discrete-time step.

The control purpose is to design a state feedback control pair $\{u_k, \omega_k\}$ ensuring the closed-loop system (1) is asymptotically stable and minimize the performance index

$$\begin{aligned} \mathcal{V}(x_k) &= \frac{1}{2}\sum_{l=k}^{\infty}\gamma^{l-k}\mathcal{U}(x_k, u_k, \omega_k) \\ &= \frac{1}{2}\sum_{l=k}^{\infty}\gamma^{l-k}\left[x_l^T P x_l + u_l^T R u_l - \zeta^2 \omega_l^T \omega_l\right] \end{aligned} \tag{2}$$

where $\gamma \in (0,1)$ is the discount factor, $P \in \mathbb{R}^{n\times n}$ and $R \in \mathbb{R}^{m\times m}$ are positive definite weighting matrices with appropriate dimensions, and ζ is the L_2 gain for (1).

Remark 1. In (2), the discount factor γ guarantees that the current reward of value function is more valuable than future reward. Generally, the performance index is constructed by the quadratic terms of the inputs and states, while the L_2 gain is introduced to enhance the robustness. It is noted that the admissible condition for (1) should be satisfied.

To obtain the optimal control problem, define the Hamilton-Jacobi-Bellman equation as

$$\mathcal{V}^*(x_k) = \min_{u_k} \max_{\omega_k} \left\{ \frac{1}{2}\mathcal{U}(x_k, u_k, \omega_k) + \gamma \mathcal{V}^*(x_{k+1}) \right\} \tag{3}$$

The Q-function is introduced for the state-action reward, which evaluates the expected cost function based on current state and action as

$$Q_j(x_k, \pi_j(x_k), \omega_k^j) = \frac{1}{2}\mathcal{U}(x_k, u_k, \omega_k) + \gamma Q_j(x_{k+1}, \pi_j(x_{k+1}), \omega_{k+1}^j) \tag{4}$$

with the optimal form as

$$Q^*(x_k, u_k^*, \omega_k^*) = \frac{1}{2}\mathcal{U}(x_k, u_k, \omega_k) + \gamma Q^*(x_{k+1}, u_{k+1}^*, \omega_{k+1}^*) \tag{5}$$

where $\pi_j(x_k) = u_k^j$ denotes the control policy for j-th iteration at state x_k.

Based on the Q-learning definition, the learning purpose is to approximate $\mathcal{V}^*(x_k)$ by $Q^*(x_k, u_k^*, \omega_k^*)$, that is

$$\mathcal{V}^*(x_k) = \min_{u_k} Q^*(x_k, u_k, \omega_k) = Q^*(x_k, u_k^*, \omega_k^*) \tag{6}$$

The crucial update formula for Q-learning is defined as

$$\begin{aligned} Q_{j+1}(x_k, \pi_{j+1}(x_k), \omega_k^{j+1}) = Q_j(x_k, \pi_j(x_k), \omega_k^j) + \alpha \Big[\tfrac{1}{2}\mathcal{U}(x_k, u_k, \omega_k) \\ + \gamma \min_{u_k} \max_{\omega_k} Q_j(x_{k+1}, \pi(x_{k+1}), \omega_{k+1}) - Q_j(x_k, \pi_j(x_k), \omega_k^j) \Big] \end{aligned} \tag{7}$$

where α is the learning rate. Generally, α is chosen as a constant between 0 and 1; however, such a strategy tends to slow down convergence. Therefore, this paper sets $\alpha > 1$ as a new acceleration factor and rigorously proves its convergence properties.

It is noted that the temporal difference (TD) is the fundamental for reinforcement learning as

$$\underbrace{\frac{1}{2}\mathcal{U}(x_k, u_k, \omega_k) + \gamma \underbrace{\min_{u_k} \max_{\omega_k} Q_j(x_{k+1}, \pi(x_{k+1}), \omega_{k+1})}_{Optiumal\,Q\,for\,next\,state}}_{TD\,target} - \underbrace{Q_j(x_k, \pi_j(x_k), \omega_k^j)}_{Current\,Q} \tag{8}$$

In (8), the TD error is constructed by TD target and current Q value, and the core of (8) is to obtain the corresponding control pair including the optimal control and the worst-case disturbances based on the optimal Q value for next state and gradient method as

$$\begin{cases} \pi_{j+1}(x_{k+1}) = u_k^{j+1} = u_k^j - \beta \dfrac{\partial Q_j(x_{k+1}, u_{k+1}, \omega_{k+1})}{\partial x_{k+1}} \dfrac{\partial x_{k+1}}{\partial u_k} \\ \omega_k^{j+1} = \omega_k^j + \beta \dfrac{\partial Q_j(x_{k+1}, u_{k+1}, \omega_{k+1})}{\partial x_{k+1}} \dfrac{\partial x_{k+1}}{\partial \omega_k} \end{cases} \tag{9}$$

where β is the learning rate.

Furthermore, the Q value can be updated as

$$\begin{aligned} &Q_{j+1}(x_k,\pi_{j+1}(x_k),\omega_k^{j+1}) = Q_j(x_k,\pi_j(x_k),\omega_k^j) \\ &+\alpha\left[\frac{1}{2}\mathcal{U}(x_k,u_k,\omega_k)+\gamma Q_j(x_{k+1},\pi_{j+1}(x_{k+1}),\omega_{k+1}^{j+1}) - Q_j(x_k,\pi_j(x_k),\omega_k^j)\right] \end{aligned} \tag{10}$$

Compared to traditional update equation of value function, (10) has two major contributions: Most current works simply equate the value function $\mathcal{V}(x_k)$ directly with the Q value $Q_j(x_k,\pi_j(x_k),\omega_k^j)$, which ignores essence of Q-learning: optimizing the next-state action; The learning rate α can accelerate or adjust the learning process to achieve more satisfactory performance.

3 Iterative Q-Learning Algorithm

An iterative Q-learning algorithm framework is proposed in this section, and the corresponding properties are derived.

3.1 Derivation of the Iterative Algorithm

An iterative Q-learning algorithm framework is established to obtain the optimal control policy and the Q value function with a initial semipositive function $Q_0(x_k,\pi_0(x_k),\omega_k^0) = \Upsilon_0(x_k)$. The detailed iterative algorithm processes are illustrated in Algorithm 1.

3.2 Monotonicity and Convergence of the Iterative Algorithm

Theorem 1. *Considering system (1) with performance index (2), let the proposed Algorithm 1 update the control policy, the worst disturbance, and the Q value function.*

Then, following conclusions can be obtained.

(a) ***(Boundedness)*** *There is an upper bound Z such that the inequalities $0 \le Q_j(x_k,\pi_j(x_k),\omega_k^j) \le Z, j = 1,2,\ldots$.*

(b) (Monotonicity) The Q- value sequence $\left\{Q_j(x_k,\pi_j(x_k),\omega_k^j), j=1,2,\ldots\right\}$ is nondecreasing as $Q_j(x_k,\pi_j(x_k),\omega_k^j) \le Q_{j+1}(x_k,\pi_{j+1}(x_k),\omega_k^{j+1})$.

(c) ***(optimality)*** *$\lim_{j\to\infty} Q_j(x_k,\pi_j(x_k),\omega_k^j) = Q^*(x_k,u_k^*,\omega_k^*)$, $\lim_{j\to\infty}\pi_j(x_k) = \pi^*(x_k)$, and $\lim_{j\to\infty}\omega_k^j = \omega_k^*$.*

Proof. **Proof of (a):** Let (υ_k,τ_k) be an admissible control pair, and define an auxiliary Q value function Y updated by

$$\begin{aligned} &Y_{j+1}(x_k,\upsilon_k,\tau_k) = Y_j(x_k,\upsilon_k,\tau_k) \\ &+\alpha\left[\frac{1}{2}\mathcal{U}(x_k,\upsilon_k,\tau_k)+\gamma Y_j(x_{k+1},\upsilon_{k+1},\tau_{k+1}) - Y_j(x_k,\upsilon_k,\tau_k)\right] \end{aligned} \tag{11}$$

Algorithm 1. Iterative Q-learning Algorithm

Input: The initial policies u_k^0 and ω_k^0, the initial Q function $Q_0(x_k, \pi_0(x_k), \omega_k^0) = \Upsilon_0(x_k)$ and the computation threshold $\varepsilon_f > 0$.

Output: u_k^*, ω_k^*, and $Q^*(x_k, u_k^*, \omega_k^*)$.

1: **repeat**

2: **Policy Improvement:**

3: Update the control policy (The optimal control and worst disturbance policies) using the gradient method:

$$\begin{cases} \pi_{j+1}(x_{k+1}) = \arg\min_{u_k} \max Q_j(x_{k+1}, \pi(x_{k+1}), \omega_{k+1}) \\ \omega_k^{j+1} = \arg\min_{u_k} \max Q_j(x_{k+1}, \pi(x_{k+1}), \omega_{k+1}) \end{cases}$$

4: **Value Function Update:**

5: Update the Q-function as

$$\begin{aligned} &Q_{j+1}(x_k, \pi_{j+1}(x_k), \omega_k^{j+1}) = Q_j(x_k, \pi_j(x_k), \omega_k^j) \\ &+\alpha\left[\tfrac{1}{2}\mathcal{U}(x_k, u_k, \omega_k) + \gamma Q_j(x_{k+1}, \pi_{j+1}(x_{k+1}), \omega_{k+1}^{j+1}) - Q_j(x_k, \pi_j(x_k), \omega_k^j)\right] \end{aligned}$$

6: Increment the iteration index $j = j + 1$;

7: **until** $|Q_{j+1}(x_k, \pi_{j+1}(x_k), \omega_k^{j+1}) - Q_j(x_k, \pi_j(x_k), \omega_k^j)| < \varepsilon_f$

where $\mathcal{U}(x_k, \upsilon_k, \tau_k) = x_k^T Q x_k + \upsilon_k^T R \upsilon_k - \zeta^2 \tau_k^T \tau_k$. Therefore, defining $Y_{j+1}^k = Y_{j+1}(x_k, \upsilon_k, \tau_k)$, it can be concluded with the initial condition $Q_0(x_k, u_k, \omega_k) = Y_0(x_k, \upsilon_k, \tau_k) = 0$ that

$$Y_{j+1}^k - (1-\alpha)\, Y_j^k = \frac{\alpha}{2}\mathcal{U}(x_k, \upsilon_k, \tau_k) + \alpha\gamma Y_j^{k+1} \tag{12}$$

Then, it can be obtained that

$$Y_j^k - (1-\alpha)\, Y_{j-1}^k = \frac{\alpha}{2}\mathcal{U}(x_k, \upsilon_k, \tau_k) + \alpha\gamma Y_{j-1}^{k+1} \tag{13}$$

Based on (12) and (13), we have

$$Y_{j+1}^k - (2-\alpha)\, Y_j^k + (1-\alpha)\, Y_{j-1}^k = \alpha\gamma\left(Y_j^{k+1} - Y_{j-1}^{k+1}\right) \tag{14}$$

Similarly, it can be inferred that

$$\begin{cases} Y_j^{k+1} - (1-\alpha)\, Y_{j-1}^{k+1} = \frac{\alpha}{2}\mathcal{U}(x_{k+1}, \upsilon_{k+1}, \tau_{k+1}) + \alpha\gamma Y_{j-1}^{k+2} \\ Y_{j-1}^{k+1} - (1-\alpha)\, Y_{j-2}^{k+1} = \frac{\alpha}{2}\mathcal{U}(x_{k+1}, \upsilon_{k+1}, \tau_{k+1}) + \alpha\gamma Y_{j-2}^{k+2} \end{cases} \tag{15}$$

Then, it can be obtained that

$$Y_j^{k+1} - (2-\alpha)\, Y_{j-1}^{k+1} + (1-\alpha)\, Y_{j-2}^{k+1} = \alpha\gamma\left(Y_{j-1}^{k+2} - Y_{j-2}^{k+2}\right) \tag{16}$$

According to (14) and (16), define a sequence $\{\Theta_j, j = 1, 2, \ldots\}$ as

$$\Theta_1 = Y_{j+1}^k - (2-\alpha) Y_j^k + (1-\alpha) Y_{j-1}^k = \alpha\gamma \left(Y_j^{k+1} - Y_{j-1}^{k+1}\right) \tag{17a}$$

$$\Theta_2 = Y_j^{k+1} - (2-\alpha) Y_{j-1}^{k+1} + (1-\alpha) Y_{j-2}^{k+1} = \alpha\gamma \left(Y_{j-1}^{k+2} - Y_{j-2}^{k+2}\right) \tag{17b}$$

$$\vdots \tag{17c}$$

$$\Theta_j = Y_2^{k+j-1} - (2-\alpha) Y_1^{k+j-1} + (1-\alpha) Y_0^{k+j-1} = \alpha\gamma \left(Y_1^{k+j} - Y_0^{k+j}\right) \tag{17d}$$

To present the relationship of the sequence $\{\Theta_j, j = 1, 2, \ldots\}$, (17b) can be constructed as

$$\begin{aligned}\Theta_2 &= Y_j^{k+1} - (2-\alpha) Y_{j-1}^{k+1} + (1-\alpha) Y_{j-2}^{k+1} \\ &= \left(Y_j^{k+1} - Y_{j-1}^{k+1}\right) - (1-\alpha)\left(Y_{j-1}^{k+1} - Y_{j-2}^{k+1}\right) \\ &= \frac{1}{\alpha\gamma}\Theta_1 - (1-\alpha)\left(Y_{j-1}^{k+1} - Y_{j-2}^{k+1}\right)\end{aligned} \tag{18}$$

Due to (υ_k, τ_k) is an admissible control pair and the condition $Q_0(x_k, u_k, \omega_k) = Y_0(x_k, \upsilon_k, \tau_k) = 0$, $Y_{j-1}^{k+1} \geq Y_{j-2}^{k+1}$ holds, that is $\Theta_2 \geq \frac{1}{\alpha\gamma}\Theta_1$. Similarly, by repeating the above process, there has

$$\Theta_j \geq \left(\frac{1}{\alpha\gamma}\right)^{j-1} \Theta_1 \tag{19}$$

It is noted that $\alpha > 1$ to accelerate the convergence rate. Invoking (17a) into (19) yields

$$\begin{aligned}(\alpha\gamma)^{j-1}\Theta_j &= (\alpha\gamma)^j Y_1^{k+j} \\ &\geq \Theta_1 = Y_{j+1}^k - (2-\alpha) Y_j^k + (1-\alpha) Y_{j-1}^k \\ &= \left(Y_{j+1}^k - Y_j^k\right) - (1-\alpha)\left(Y_j^k - Y_{j-1}^k\right) \\ &\geq Y_{j+1}^k - Y_j^k\end{aligned} \tag{20}$$

Based on (20), the conclusion can be recursively derived

$$(\alpha\gamma)^j Y_1^{k+j} \geq Y_{j+1}^k - Y_j^k \tag{21a}$$

$$(\alpha\gamma)^{j-1} Y_1^{k+j-1} \geq Y_j^k - Y_{j-1}^k \tag{21b}$$

$$\vdots \tag{21c}$$

$$(\alpha\gamma)^1 Y_1^{k+1} \geq Y_2^k - Y_1^k \tag{21d}$$

Add all formulas in (21) as

$$Y_{j+1}^k - Y_1^k \leq (\alpha\gamma)^j Y_1^{k+j} + (\alpha\gamma)^{j-1} Y_1^{k+j-1} + \ldots + (\alpha\gamma) Y_1^{k+1} \tag{22}$$

Selecting the discount rate $\gamma \leq \frac{1}{\alpha}$, we have

$$\begin{aligned}Y_{j+1}^k &\leq (\alpha\gamma)^j Y_1^{k+j} + (\alpha\gamma)^{j-1} Y_1^{k+j-1} + \ldots + (\alpha\gamma) Y_1^{k+1} + Y_1^k \\ &\leq Y_1^{k+j} + Y_1^{k+j-1} + \ldots Y_1^k\end{aligned} \tag{23}$$

Since (υ_k, τ_k) is a stabilizing and admissible control pair ($x_k \rightarrow 0$ when $k \rightarrow \infty$), it can be concluded that

$$\forall j : Y_{j+1}^{k} \leq \sum_{i=0}^{j} Y_1^{k+i} \leq Z \tag{24}$$

Additionally, the designed control pair $\left(\pi_j(x_k), \omega_k^j\right)$ is to minimize the Q-value in Algorithm 1. Therefore, we have

$$\forall j : Q_{j+1}(x_k, \pi_{j+1}(x_k), \omega_k^{j+1}) \leq Y_{j+1}^{k} \leq Z \tag{25}$$

Proof of (b): Define an auxiliary Q function for $j \neq 0$ with initial condition $\Upsilon_0(x_k, \pi_0(x_k), \omega_k^0) = Q_0(x_k, \pi_0(x_k), \omega_k^0)$ as

$$\begin{aligned} &\Upsilon_{j+1}(x_k, \pi_{j+1}(x_k), \omega_k^{j+1}) = \Upsilon_j(x_k, \pi_j(x_k), \omega_k^j) \\ &+ \alpha\left[\frac{1}{2}\mathcal{U}(x_k, u_k, \omega_k) + \gamma\Upsilon_j(x_{k+1}, \pi_j(x_{k+1}), \omega_{k+1}^j) - \Upsilon_j(x_k, \pi_j(x_k), \omega_k^j)\right] \end{aligned} \tag{26}$$

Based on (26), $\Upsilon_1(x_k, \pi_1(x_k), \omega_k^1)$ can be described as

$$\begin{aligned} &\Upsilon_1(x_k, \pi_1(x_k), \omega_k^1) = \Upsilon_0(x_k, \pi_0(x_k), \omega_k^0) \\ &+ \alpha\left[\frac{1}{2}\mathcal{U}(x_k, u_k, \omega_k) + \gamma\Upsilon_0(x_{k+1}, \pi_0(x_{k+1}), \omega_{k+1}^0) - \Upsilon_0(x_k, \pi_0(x_k), \omega_k^0)\right] \\ &\geq Q_0(x_k, \pi_0(x_k), \omega_k^0) \\ &+ \alpha\left[\frac{1}{2}\mathcal{U}(x_k, u_k, \omega_k) + \gamma Q_0(x_{k+1}, \pi_1(x_{k+1}), \omega_{k+1}^1) - Q_0(x_k, \pi_0(x_k), \omega_k^0)\right] \\ &= Q_1(x_k, \pi_1(x_k), \omega_k^1) \end{aligned} \tag{27}$$

Without loss of generality, it can be inferred that $\Upsilon_{j-1}(x_k, \pi_{j-1}(x_k), \omega_k^{j-1}) \geq Q_{j-1}(x_k, \pi_{j-1}(x_k), \omega_k^{j-1})$ holds. Then, we have

$$\begin{aligned} &\Upsilon_j(x_k, \pi_j(x_k), \omega_k^j) = \Upsilon_{j-1}(x_k, \pi_{j-1}(x_k), \omega_k^{j-1}) + \alpha\left[\frac{1}{2}\mathcal{U}(x_k, u_k, \omega_k)\right. \\ &\left.+\gamma\Upsilon_{j-1}(x_{k+1}, \pi_{j-1}(x_{k+1}), \omega_{k+1}^{j-1}) - \Upsilon_{j-1}(x_k, \pi_{j-1}(x_k), \omega_k^{j-1})\right] \\ &\geq Q_{j-1}(x_k, \pi_{j-1}(x_k), \omega_k^{j-1}) + \alpha\left[\frac{1}{2}\mathcal{U}(x_k, u_k, \omega_k)\right. \\ &\left.+\gamma Q_{j-1}(x_{k+1}, \pi_j(x_{k+1}), \omega_{k+1}^j) - Q_{j-1}(x_k, \pi_{j-1}(x_k), \omega_k^{j-1})\right] \\ &= Q_j(x_k, \pi_j(x_k), \omega_k^j) \end{aligned} \tag{28}$$

In (28), the condition $\Upsilon_{j-1}(x_k, \pi_{j-1}(x_k), \omega_k^{j-1}) \geq Q_{j-1}(x_k, \pi_{j-1}(x_k), \omega_k^{j-1})$ can ensure $\Upsilon_j(x_k, \pi_j(x_k), \omega_k^j) \geq Q_j(x_k, \pi_j(x_k), \omega_k^j)$.

Meanwhile, $Q_1(x_k, \pi_1(x_k), \omega_k^1) \geq \Upsilon_0(x_k, \pi_0(x_k), \omega_k^0) = 0$ holds. Then, it can be inferred that

$$\begin{aligned} Q_{j+1}(x_k, \pi_{j+1}(x_k), \omega_k^{j+1}) &\geq Q_j(x_k, \pi_j(x_k), \omega_k^j) + \alpha \left[\frac{1}{2}\mathcal{U}(x_k, u_k, \omega_k) \right. \\ &\left. + \gamma Q_j(x_{k+1}, \pi_j(x_{k+1}), \omega_{k+1}^j) - Q_j(x_k, \pi_j(x_k), \omega_k^j) \right] \\ &\geq \Upsilon_{j-1}(x_k, \pi_{j-1}(x_k), \omega_k^{j-1}) + \alpha \left[\frac{1}{2}\mathcal{U}(x_k, u_k, \omega_k) \right. \\ &\left. + \gamma \Upsilon_{j-1}(x_k, \pi_{j-1}(x_k), \omega_k^{j-1}) - \Upsilon_{j-1}(x_k, \pi_{j-1}(x_k), \omega_k^{j-1}) \right] \\ &= \Upsilon_j(x_k, \pi_j(x_k), \omega_k^j) \end{aligned} \tag{29}$$

Based on (27) and (29), we have

$$Q_j(x_k, \pi_j(x_k), \omega_k^j) \leq \Upsilon_j(x_k, \pi_j(x_k), \omega_k^j) \leq Q_{j+1}(x_k, \pi_{j+1}(x_k), \omega_k^{j+1}) \tag{30}$$

Proof of (c): Defining the$j$-th admissible control pair $\left(\upsilon_k^j, \tau_k^j\right)$, invoking (23) yields

$$Y_{j+1}^k \leq \sum_{i=0}^{j} Y_1^{k+i} \leq \sum_{i=0}^{\infty} Y_1^{k+i} \leq \Xi \tag{31}$$

where Ξ is a sufficiently large number. Based on (24), it can be further obtained that

$$\lim_{j\to\infty} Q_j(x_k, \pi_j(x_k), \omega_k^j) \leq \lim_{j\to\infty} Y_j^k \leq \Xi \tag{32}$$

(32) indicates the existence of the limit as $\lim\limits_{j\to\infty} Q_j(x_k, \pi_j(x_k), \omega_k^j)$. Additionally, in **Proof of (b)**, $Q_j(x_k, \pi_j(x_k), \omega_k^j)$ is nondecreasing. Since the sequence is monotonically increasing and bounded above, the limit $\lim\limits_{j\to\infty} Q_j(x_k, \pi_j(x_k), \omega_k^j) = Q^*(x_k, u_k^*, \omega_k^*)$ exists. Furthermore, we can also obtain that $\frac{\partial Q_j(x_{k+1}, u_{k+1}, \omega_{k+1})}{\partial x_{k+1}}$ converges to optimal when $j \to \infty$, that is, the corresponding control sequence pair $\left(\pi_j(x_k), \omega_k^j\right)$ converges to $(\pi^*(x_k), \omega_k^*)$.

The whole proof is complete. □

4 The Implementation of the Algorithm

In this section, the critic network for Q function and the action networks for control policy and worst disturbance are presented, and the swarm intelligence algorithm to optimize the matrix is introduced.

4.1 The Action-Critic Structure

The critic is to approximate the Q value function as

$$\hat{Q}_j(x_k, u_k) = W_{j,c2}^T \sigma_c \left(W_{j,c1}^T \Theta_k \right) \tag{33}$$

where $\Theta_k = [x_k; u_k] \in \mathbb{R}^{n+m}$, $W_{j,c1} \in \mathbb{R}^{(n+m)\times h_c}$, $W_{j,c2} \in \mathbb{R}^{h_c \times (n+m)}$ are the hidden and the output weight matrices, $\sigma_c \in \mathbb{R}^{h_c}$ is the activation function vector, h_c is the number of neurons. The loss function is defined as

$$E_j^c = \frac{\left\| \hat{Q}_j(x_k, u_k) - Q_j(x_k, \pi_j(x_k), W_k^j) \right\|^2}{2} \tag{34}$$

The gradient descent method to solve the weight matrices can be defined as

$$\begin{aligned} W_{j+1,c1} &= W_{j,c1} - \alpha_{c1} \frac{\partial E_j^c}{\partial W_{j,c1}} \\ W_{j+1,c2} &= W_{j,c2} - \alpha_{c2} \frac{\partial E_j^c}{\partial W_{j,c2}} \end{aligned} \tag{35}$$

where α_{c1} and α_{c2} are the learning rates.

The action is to approximate the worst disturbance as

$$\hat{\omega}_j(x_k) = W_{j,d2}^T \sigma_d \left(W_{j,d1}^T x_k \right) \tag{36}$$

where $W_{j,d1} \in \mathbb{R}^{(q)\times h_d}$, $W_{j,d2} \in \mathbb{R}^{h_d \times (q)}$ are the hidden and the output weight matrices, $\sigma_d \in \mathbb{R}^{h_q}$ is the activation function vector, h_q is the number of neurons. The loss function is defined as

$$E_j^d = \frac{\left\| \hat{\omega}_j(x_k) - \omega_j(x_k) \right\|^2}{2} \tag{37}$$

The gradient descent method to solve the weight matrices can be defined as

$$\begin{aligned} W_{j+1,d1} &= W_{j,d1} - \alpha_{d1} \frac{\partial E_j^d}{\partial W_{j,d1}} \\ W_{j+1,d2} &= W_{j,d2} - \alpha_{d2} \frac{\partial E_j^d}{\partial W_{j,d2}} \end{aligned} \tag{38}$$

where α_{c1} and α_{c2} are the learning rates.

The action to approximate the optimal control policy as

$$\hat{u}_j(x_k) = \mathcal{A}^T \sigma_a (x_k) \tag{39}$$

where $\sigma_a \in \mathbb{R}^{h_a}$ is the activation function vector, h_a is the numbers of neuron, $\mathcal{A} \in \mathbb{R}^{h_a \times m}$ is the policy parameter matrix.

To accelerate convergence, we encode the entire policy as a single parameter matrices $\mathcal{A}$ and search over particles rather than over raw policy vectors u, which shrinks the search space to a single latent representation that spans all data, yielding order-of-magnitude iteration reductions.

4.2 Evolution Policy Improvement

The policy population Ψ_j^u at j-th iteration of PSO with N particles are defined as

$$\Psi_i^u = \{\mathcal{A}_1^i, \mathcal{A}_2^i, \dots \mathcal{A}_N^i\} \tag{40}$$

where $i = 1, 2, \dots I$ is the index of the iteration in PSO

The fitness functions are important for PSO to guide the optimization direction, which can be defined as the Q function $\hat{Q}_j(x_k, u_k)$. The smallest fitness value in the population can be defined as $\mathcal{A}_*^i$. At each iteration, the particles are updated by

$$\mathcal{A}_p^{i+1} = \mathcal{A}_p^i + \vartheta_p^i \tag{41}$$

where $p = 1, 2, \dots N$ denotes the particles, and ϑ_p^i is the velocity as

$$\vartheta_p^i = \nu\vartheta_p^{i-1} + c_1 r_1 \left(\mathcal{B}_p - \vartheta_p^{i-1}\right) + c_2 r_2 \left(\mathcal{B}_g - \vartheta_p^{i-1}\right) \tag{42}$$

where r_1 and r_2 are random variables uniformly distributed in $[0, 1]$, $\mathcal{B}_p$ is the local optimal solution for j-th particle, $\mathcal{B}_g$ is the global optimal solution for whole population, and ν, c_1, and c_2 are positive parameters as

$$c_1 = c_1^{\max} - \left(c_1^{\max} - c_1^{\min}\right)\left(\frac{i}{I}\right)^2 \tag{43a}$$

$$c_2 = c_2^{\min} + \left(c_2^{\max} - c_2^{\min}\right)\left(\frac{i}{I}\right)^2 \tag{43b}$$

$$\nu = \nu_{\min} + \frac{i}{I}\frac{\hat{Q}_j\left(x_k, \mathcal{A}_p^i \sigma_a(x_k)\right) - \hat{Q}_j^{\min}}{\hat{Q}_j^{\max} - \hat{Q}_j^{\min}}\left(\nu_{\max} - \nu_{\min}\right) \tag{43c}$$

where $c_1^{\max}$, $c_1^{\min}$, $c_2^{\max}$, $c_2^{\min}$, $\nu_{\max}$ and $\nu_{\min}$ are the preset constants weights,and $\hat{Q}_j^{\min}$ and $\hat{Q}_j^{\max}$ the minimum and maximum fitness values.

Remark 2. In (42), the parameters c_1 and c_2 are scheduled to favor exploration in the early stage and exploitation in the late stage, while ν adaptively balances global exploration against local refinement: larger ν widens the search, smaller ν accelerates local convergence.

Then, based on the PSO, the control policy is updated by the Q-function as

$$\hat{u}_j(x_k, \mathcal{A}) = \arg\min_{\mathcal{A}} \left\|\hat{Q}_j(x_k, u_k)\right\| \tag{44}$$

5 Simulation Results and Analysis

In this section , the helicopter model is introduced to verify the advantage of the proposed algorithm. According to the flight dynamics and aerodynamics, the discrete-time attitude model of a helicopter is expressed as

$$\begin{cases} \Omega_{k+1} = \Omega_k + T_0 H_k \xi_k \\ \xi_{k+1} = \xi_k - T_0 J^{-1}\xi_k \times J\xi_k + T_0 J^{-1}\Sigma_k + \omega_k \end{cases} \tag{45}$$

where $\Omega_k = [\phi_k, \theta_k, \psi_k]^T$ and $\xi_k = [p_k, q_k, r_k]^T$ are the attitude angle and angular rate vectors in $\Re_b$, respectively, $T_0 = 0.01$ is the sampling period, $\Sigma_k = [\Sigma_{xk}, \Sigma_{yk}, \Sigma_{zk}]^T$ is the torque vector in three axes of $\Re_b$, $J = diag\{J_{xx}, J_{yy}, J_{zz}\} = diag\{358, 778, 601\}$ is the inertia matrix, and H_k is given by

$$H_k = \begin{bmatrix} 1 & \sin\phi_k \tan\theta_k & \cos\phi_k \tan\theta_k \\ 0 & \cos\phi_k & -\sin\phi_k \\ 0 & \sin\phi_k / \cos\theta_k & \cos\phi_k / \cos\theta_k \end{bmatrix} \tag{46}$$

To describe the attitude system briefly, the helicopter model (45) can be rewritten as

$$X_{k+1} = F(X_k) + Bu_k + [0; \omega_k] \tag{47}$$

where $F(X_k) = [\Omega_k + T_0 H_k \xi_k, \xi_k - T_0 J^{-1} \xi_k \times J\xi_k]^T$, $X_k = [\Omega_k, \xi_k]^T$, $B = [0, T_0 J^{-1}]$, and $u_k = \Sigma_k$.

Compared to traditional V value function, the convergence process of Q value is investigated in Fig. 1.

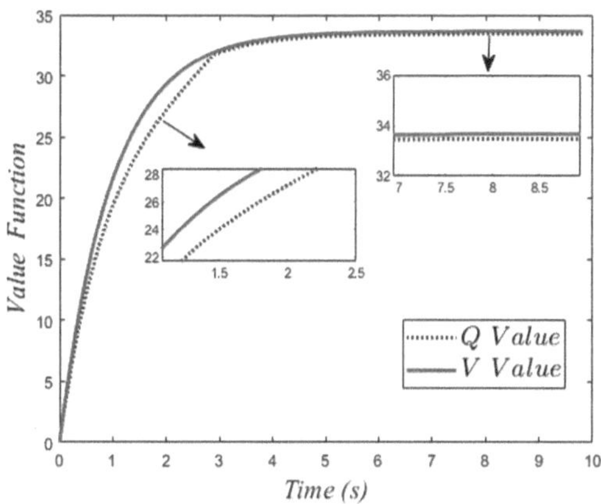

Fig. 1. The convergence process of Q function compared to V value.

It can be seen from Fig. 1 that the Q value is smaller than the V value in the middle of the iteration. This is because the core of Q-learning is to learn the optimal policy for the next state rather than the current state. However, in the later stage of learning, the convergence of the overall algorithm can be guaranteed.

The anti-disturbance capability of Q value is investigated in Fig. 2.

In Fig. 2, the proposed Q-learning algorithm effectively suppresses the impact of disturbances with $L2$-gain on the system.

The different acceleration factor of Q value is presented in Fig. 3 ($\alpha = 1.5$) and Fig. 4 ($\alpha = 0.5$).

It can be seen from Fig. 3 that when $\alpha = 1.5$, the designed optimal control strategy stabilizes the system very well. However, in Fig. 4, when $\alpha = 0.5$, the

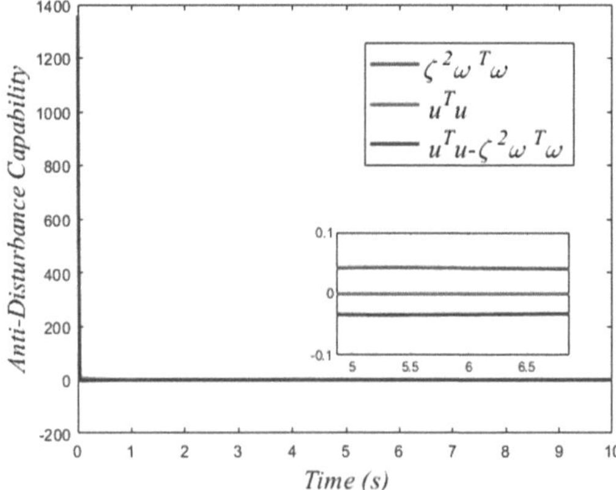

Fig. 2. The anti-disturbance capability of Q function.

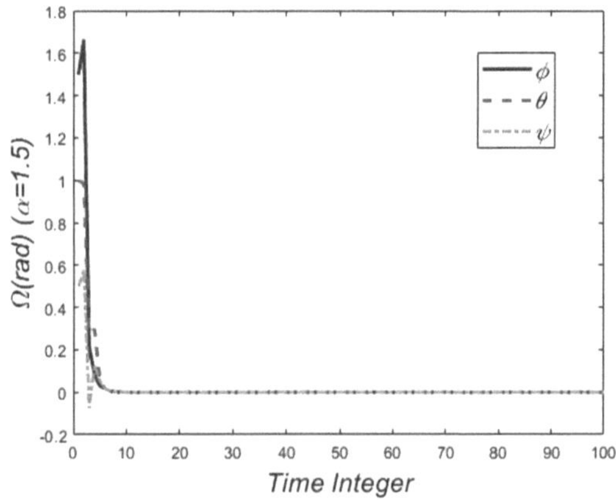

Fig. 3. The state response helicopter. ($\alpha = 1.5$.

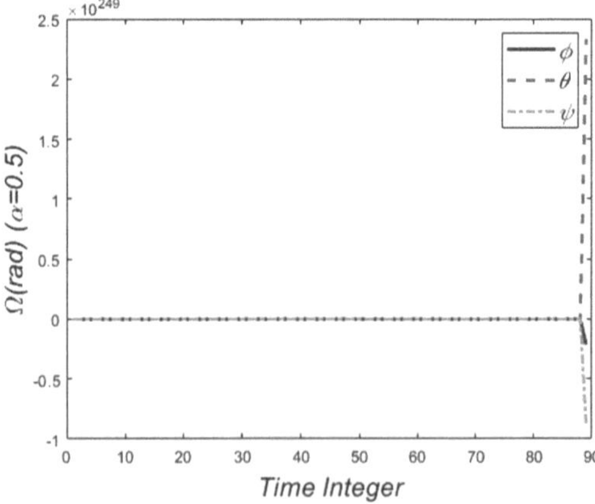

Fig. 4. The state response helicopter. ($\alpha = 0.5$.

system is directly unstable and does not converge. Therefore, for the learning rate in the Q-learning algorithm, it must be set as $\gamma \leq \frac{1}{\alpha}$ to ensure global convergence.

6 Conclusion

This paper has presented an accelerated dual Q-learning framework for nonlinear discrete-time systems subject to $L2$-bounded disturbances. By introducing an acceleration factor$\alpha > 1$ into the Q-function update law, we achieve orders-of-magnitude faster convergence while retaining rigorous stability guarantees. Theoretically, we proved that the generated Q-value sequence is bounded, monotonically non-decreasing, and converges to the optimal solution provided the discount factor satisfies $\gamma \leq \frac{1}{\alpha}$, thereby establishing for the first time a necessary and sufficient condition for aggressive Q-learning acceleration. Practically, an actor-critic neural implementation with separate networks for the critic, optimal control, and worst-case disturbance policies is developed, and a particle-swarm optimizer with adaptively scheduled parameters is employed to escape local optima in the compressed policy space. Comparative simulations on a high-fidelity helicopter attitude model demonstrate that the proposed algorithm converges faster and exhibits superior disturbance rejection than conventional value iteration, confirming the theoretical predictions and its viability for safety-critical aerospace applications. Future work will extend the framework to multi-agent and event-triggered settings, and will explore hardware deployment on embedded flight-control platforms.

Acknowledgments. This study was funded by X (grant number Y).

Disclosure of Interests. The authors have no competing interests to declare that are relevant to the content of this article.

References

1. Jiang, X., Wang, Y., Zhao, D., Shi, L.: Online pareto optimal control of mean-field stochastic multi-player systems using policy iteration. SCIENCE CHINA Inf. Sci. **67**(4), 140202 (2024)
2. Cao, L., Qin, Y., Pan, Y., Liang, H.: Prescribed performance-based optimal formation control for USVs with position constraints and yaw angle time-varying partial constraints. IEEE Trans. Intell. Transp. Syst. **26**(3), 4109–4121 (2025)
3. Guo, S., Pan, Y., Li, H., Cao, L.: Dynamic event-driven adp for n-player nonzero-sum games of constrained nonlinear systems. IEEE Trans. Autom. Sci. Eng. **22**(1), 7657–7669 (2025)
4. Xiong, H., Chen, G., Ren, H., Li, H.: Broad-learning-system-based model-free adaptive predictive control for nonlinear mass under dos attacks. IEEE/CAA J. Autom. Sinica **12**(2), 381–393 (2025)
5. He, Q., Zhang, L., Fang, H., Wang, X., Ma, L., Yu, K., Zhang, J.: Multistage competitive opinion maximization with q-learning-based method in social networks. IEEE Trans. Neural Netw. Learn. Syst. **36**(4), 7158–7168 (2025)

6. Lyu, J., Ma, X., Li, X., Lu, Z.: Mildly conservative q-learning for offline reinforcement learning. Adv. Neural. Inf. Process. Syst. **35**, 1711–1724 (2022)
7. Zhou, Z., Wang, Y., Wu, Q.: Resilient H_∞ control for nonlinear systems with uncertainties and disturbances based on equivalence robust passivity. IEEE Trans. Aerosp. Electron. Syst. **60**(3), 3598–3610 (2024)
8. Lu, C., Meng, D., Li, H.: Data-induced learning control for unknown nonlinear systems with full-tracking performances. SCIENCE CHINA Inf. Sci. **68**(8), 1–14 (2025)
9. Al-Dabooni, S., Wunsch, D.C.: Online model-free n-step HDP with stability analysis. IEEE Trans. Neural Netw. Learn. Syst. **31**(4), 1255–1269 (2019)
10. Zhou, Z., Wang, Y., Wu, Q.: An advanced optimal tracking control for nonlinear discrete-time systems based on (n + 1)-step gradient learning. IEEE Trans. Neural Netw. Learn. Syst. **36**(10), 18696–18710 (2025)
11. Wang, D., Yuan, Z., Liu, A., Lin, Q., Qiao, J.: Model-free neuro-fuzzy q-learning control with swarm intelligence. IEEE Trans. Fuzzy Syst. **33**(9), 3035–3046 (2025)
12. Yang, Y., Kiumarsi, B., Modares, H., Xu, C.: Model-free λ-policy iteration for discrete-time linear quadratic regulation. IEEE Trans. Neural Netw. Learn. Syst. **34**(2), 635–649 (2021)
13. Bian, T., Jiang, Z.: Reinforcement learning and adaptive optimal control for continuous-time nonlinear systems: a value iteration approach. IEEE Trans. Neural Netw. Learn. Syst. **33**(7), 2781–2790 (2021)
14. Hou, J., Wang, D., Liu, D., Zhang, Y.: Model-free H_∞ optimal tracking control of constrained nonlinear systems via an iterative adaptive learning algorithm. IEEE Trans. Syst. Man Cybern. Syst. **50**(11), 4097–4108 (2018)
15. De Oca, M.A.M., Stutzle, T., Birattari, M., Dorigo, M.: Frankenstein's pso: a composite particle swarm optimization algorithm. IEEE Trans. Evol. Comput. **13**(5), 1120–1132 (2009)
16. Regaya, C.B., Hamdi, H., Farhani, F., Marai, A., Zaafouri, A., Chaari, A.: Real-time implementation of a novel MPPT control based on the improved PSO algorithm using an adaptive factor selection strategy for photovoltaic systems. ISA Trans. **146**, 496–510 (2024)
17. Wang, M., Peng, X.J., Li, H., He, Y.: Distributed cooperative learning control for multiagent systems with data protection and disturbance observation. SCIENCE CHINA Inf. Sci. **69**(1), 1–14 (2026)
18. Peng, X.J., He, Y., Liu, Z., You, L., Li, H.: Time-varying formation H_∞ tracking control and optimization for delayed multi-agent systems with exogenous disturbances. IEEE Trans. Autom. Sci. Eng. **22**, 5637–5647 (2025)

Stochastic Disturbance Rejection Tracking for Surface Ships with Thruster Saturation

Yuxin Liu[1], Yifeng Zhang[1(✉)], Kaichen Zhong[1], Hongxuan Wang[2], Xiangjia Meng[3], and Xin Hu[1]

[1] School of Mathematics and Statistics Science, Ludong University, Yantai, China
wfzhangyifeng@126.com
[2] School of Hydraulic and Civil Engineering, Ludong University, Yantai, China
[3] Ulsan Ship and Ocean College, Ludong University, Yantai, China

Abstract. This article describes a composite disturbance rejection tracking control for ships with stochastic disturbances under actuator saturation. The stochastic disturbances are described by the first-order Markov process. The composite control scheme is built by incorporating the stochastic disturbance observer and the auxiliary filter with the vectorial backstepping technique. First, a stochastic disturbance observer is designed to provide stochastic disturbance on-line estimations. Second, an auxiliary filter is employed to generate filtered versions of non-achievable portions of the control derivation between commanded control signals and actual control signals, such that the stochastic anti-disturbance control performance can be preserved under actuator saturation effects. The composite stochastic disturbance rejection control achieves the ship position and heading tracking with ensuring the closed-loop stability. Simultaneous achievement of stochastic disturbance suppression and mitigation of actuator saturation is accomplished. Simulations conducted on a 1:70 scale ship model illustrate the validity of the proposed control strategy.

Keywords: Surface ship · disturbance observer · disturbance rejection · trajectory tracking · actuator saturation

1 Introduction

The ships operating in the varying ocean environment inevitably suffer from stochastic disturbances induced by ocean waves, wind and currents. The stochastic disturbance rejection tracking control for ships is a challenging problem [1,2]. Due to propulsion limits, ship control signals face saturation, reducing control performance. [3]. Improving adaptability and robustness of trajectory tracking under actuator saturation is important.

In 1960s, the early ship motion control system relied on linear proportional-integral-derivative controllers with difficult parameter adjustment procedure [4]. Subsequently, the multivariate optimal control and Kalman filter theory were

C. Li et al. (Eds.): ICNC 2025, CCIS 2946, pp. 499–508, 2026.
https://doi.org/10.1007/978-981-92-1599-7_42

introduced into the wave filtering and state estimation for ship motion control, yet the ship nonlinear kinematics and kinetics are required to be linearized in different operating points [5,6]. In 1990s, the nonlinear vectorial backstepping control was applied to ship motion control systems, which avoided the linearization of ship nonlinear kinematics and kinetics [7]. The above results in [4–7] neglect stochastic characteristics in ocean environmental disturbances. The presence of stochastic disturbances complicates the control design for ship trajectory tracking. The disturbance observer-based control was introduced for the first time to microprocessor-controlled DC motors subject to disturbances in the frequency-domain [8]. For the time-domain nonlinear systems, a nonlinear disturbance observer is designated based on the mathematical model and partially known disturbance knowledge [9]. The disturbance observer-based control lies in that a disturbance observer is constructed to provide the on-line estimations of disturbances such that the disturbance estimations are employed as disturbance compensation terms in the control design, yet it cannot be applicable to stochastic systems [10–14]. Further, considering the stochastic systems with multiple disturbances including partially known disturbances as well as white noises, the disturbance observer-based control is extended to stochastic systems for the first time [15]. This stochastic disturbance observer is promising to ship trajectory tracking control owing to the estimation and rejection capability for stochastic disturbances. Due to the physical limitations of the ship's propulsion system, the actuator saturation is the inherent adverse effects and would decrease the disturbance rejection performance. For ship dynamic positioning under unknown constant disturbances and actuator saturation, proportional-integral control with anti-windup and model predictive control with optimal strategy have been proposed. [16,17] Disturbance observer-based dynamic surface control was developed for time-varying disturbances and actuator saturation [18], but it is ineffective for ship tracking under stochastic disturbances.

In the simultaneous presence of stochastic disturbances and actuator saturation, in this work, a composite stochastic disturbance rejection tracking control is designed for ships based on the stochastic disturbance observer as well as the auxiliary filter. The composite control structure consists of two layers including the inner layer as well as the outer layer. In the outer layer, a disturbance observer is employed to estimate and reject stochastic disturbances. The inner layer includes the auxiliary filer to attenuate actuator saturation. The stochastic disturbance rejection and the actuator saturation attenuation are realized simultaneously.

2 Surface Ship Motion Modeling

As illustrated in Fig. 1, the north-east are given to describe the ship movements. The nonlinear mathematical model of ships in tracking mode can be described as:

$$\dot{\eta} = J(\psi)\nu \tag{1}$$

$$M\dot{\nu} + C(\nu)\nu + D(\nu)\nu = \tau + \tau_d(t) \tag{2}$$

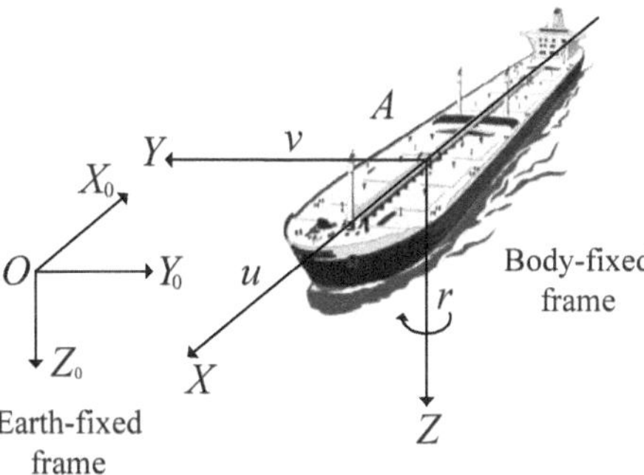

Fig. 1. Coordinate systems for ship movement model [1].

where $\eta = [x, y, \psi]^{\mathrm{T}}$ is the position and heading vector in the north-east coordinate system, which consists of the actual position of ships (x, y) and heading angel ψ. $\nu = [\mu, v, r]^{\mathrm{T}}$ is the velocity vector in the body-fixed coordinate system, which consists of the ship forward speed μ, the lateral speed ν and the yaw rate r. Define the rotation matrix as follows:

$$J(\psi) = \begin{bmatrix} \cos(\psi) & -\sin(\psi) & 0 \\ \sin(\psi) & \cos(\psi) & 0 \\ 0 & 0 & 1 \end{bmatrix} \tag{3}$$

It satisfies $J^{-1}(\psi) = J^{\mathrm{T}}(\psi)$ and $\|J(\psi)\| = 1$ with $\|\cdot\|$ as the 2-norm. M denotes the inertia matrix with added mass, characterized as symmetric and positive-definite, $C(\nu)$ is the Coriolis centripetal force matrix, $D(\nu)$ is the damping matrix. $\tau = [\tau_1, \tau_2, \tau_3]^{\mathrm{T}}$ is the control vector provided by propeller, which consists of the surge control force τ_1, the sway control force τ_2 and the yaw control torque τ_3. Due to the physical limits of ship propellers, the amplitudes of control force and torque are subject to saturation as [18]:

$$\tau_i = \begin{cases} \tau_{i\max}, & \tau_{ci} > \tau_{i\max} \\ \tau_{ci}, & \tau_{i\min} \le \tau_{ci} \le \tau_{i\max} \\ \tau_{i\min}, & \tau_{ci} < \tau_{i\min} \end{cases} \tag{4}$$

where $\tau_{i\max} > 0$ and $\tau_{i\min} > 0$ indicate the maximum and minimum control limits imposed by the propulsion system. $\tau_c = [\tau_{c_1}, \tau_{c_2}, \tau_{c_3}]^{\mathrm{T}}$ is the commanded control vector calculated by the designed control law, which consists of the forward command control force τ_{c_1}, the lateral drift command control force τ_{c_2} and the yaw command control torque τ_{c_3}. $\tau_d(t) = [\tau_{d_1}(t), \tau_{d_2}(t), \tau_{d_3}(t)]^{\mathrm{T}}$ represents the disturbance force and torque vector induced by environment disturbance such as waves, winds and currents, which are expressed by the first-order Markov process:

$$\dot{\tau}_d(t) = -T^{-1}\tau_d(t) + \Psi\zeta(t) \tag{5}$$

where $T \in \mathbb{R}^{3\times 3}$ is the time constant positive matrix, $\Psi \in \mathbb{R}^{3\times 3}$ is the amplitude matrix and $\zeta(t) \in \mathbb{R}^3$ is the Gaussian white noise vector.

To facilitate the stochastic control design later, according to stochastic theory, the ship movement model (1)–(2) can be expressed as follows:

$$d\eta = J(\psi)\nu dt \tag{6}$$
$$M d\nu + C(\nu)\nu dt + D(\nu)\nu dt = \tau dt + \tau_d(t)dt \tag{7}$$

By replacing $\xi(t)$ with $\frac{dW(t)}{dt}$, the disturbance can be rewritten as

$$d\tau_d(t) = -T^{-1}\tau_d dt + \Psi dW. \tag{8}$$

3 Control Objective

A composite disturbance rejection backstepping control is targeted for the trajectory tracking of ships with stochastic disturbances under actuator saturation, such that the ship position and yaw angle track the desired trajectory vector $\eta_d = [x_d, y_d, \psi_d]^{\mathrm{T}}$ and all signals of the ship tracking control system are globally ultimately bounded.

4 Stochastic Disturbance Observer

The stochastic disturbance observer is constructed as:

$$\hat{\tau}_d(t) = q(t) + K_0\,\nu, \tag{9}$$
$$\begin{aligned} \mathrm{d}q(t) = {} & \left[-T^{-1} + K\right] q(t)\,\mathrm{d}t + K_0\,\nu\,\mathrm{d}t \\ & - K\,\tau\,\mathrm{d}t + K\big(C(\nu)\,\nu + D(\nu)\,\nu\big)\,\mathrm{d}t. \end{aligned} \tag{10}$$

where $\hat{d}(t)$ is the disturbance estimation value,$q(t) \in \mathbb{R}^3$ is the auxiliary vector of the observer, $K_0 \in \mathbb{R}^{3\times3}$ and $K \in \mathbb{R}^{3\times3}$ are design matrix satisfying $K = K_0 M^{-1}$.

Defining observer estimation error vector as:

$$\tilde{\tau}_d = \tau_d - \hat{\tau}_d \tag{11}$$

On the basis of (5) and (9)–(10), the dynamics of disturbance estimation error can be obtained as

$$\begin{aligned} d\tilde{\tau}_d &= -T^{-1}\tau_d dt + \Psi dW + (T^{-1} + K)\hat{\tau}_d dt - K\tau_d dt \\ &= -(T^{-1} + K)\tilde{\tau}_d dt + \Psi dW \end{aligned} \tag{12}$$

Choosing the Lyapunov function:

$$V_e = \tilde{\tau}_d^{\mathrm{T}} P \tilde{\tau}_d \tag{13}$$

with the matrix $P = P^{\mathrm{T}}$ satisfying

$$P(T^{-1} + K) + (T^{-1} + K)^{\mathrm{T}} P = -Q \tag{14}$$

where the matrix Q is a positive-definite matrix.

In the light of (12)–(14), it is obtained that

$$\begin{aligned}\mathcal{L}V_e =& \frac{\partial V_e}{\partial \tilde{\tau}_d}[-(T^{-1}+K)\tilde{\tau}_d] + Tr\{\Psi^{\mathrm{T}}P\Psi\} \\ \leq& -\tilde{\tau}_d^{\mathrm{T}}[P(T^{-1}+K)+(T^{-1}+K)^{\mathrm{T}}P]\tilde{\tau}_d + Tr\{\Psi^{\mathrm{T}}P\Psi\} \\ \leq& -\varrho_1 V_e + \beta_1 \end{aligned} \tag{15}$$

Herein, $\varrho_1 = \lambda_{\min}(Q)$ and $\beta_1 = Tr\{\Psi^{\mathrm{T}}P\Psi\}$. Select the function $\kappa = \lambda_{\min}(P)||\tilde{\tau}_d||^p$ and the positive number $p = 2$, such that

$$\kappa(||\tilde{\tau}_d||^p) = \lambda_{\min}(P)||\tilde{\tau}_d||^2 \leq \tilde{\tau}_d^{\mathrm{T}}P\tilde{\tau}_d = V_e \tag{16}$$

$$\lim_{t\to\infty}\sup\mathbb{E}||\tilde{\tau}_d(t;t_0,\tilde{\tau}_d(0))||^2 \leq \kappa^{-1}\Big(\frac{\varrho_1}{\beta_1}\Big) \tag{17}$$

Hence, the disturbance error system is asymptotically stable in mean square. It follows from the expressions of ϱ_1 and β_1 that the disturbance estimation errors can be made small by adjusting the design matrix K_0 together with K satisfying $K = K_0 M^{-1}$.

5 Composite Disturbance Rejection Backstepping Design

To address the actuator saturation effects, the following auxiliary filer is constructed by [22]

$$d\xi_1 = -K_1\xi_1 dt + J(\psi)\xi_2 dt \tag{18}$$

$$d\xi_2 = -M^{-1}K_2\xi_2 dt + M^{-1}\Delta\tau dt \tag{19}$$

where ξ_1 and ξ_2 denotes the state vector produced by the auxiliary filter. $K_1 = K_1^{\mathrm{T}}$ and $K_2 = K_2^{\mathrm{T}}$ represent the 3×3 positive-definite design matrices and $\Delta\tau = \tau - \tau_c$ represents the control derivation induced by actuator saturation (4).

Based on the vectorial backstepping design [21, 24], the composite disturbance rejection control is derived in following two steps.

Step 1. Definite position error vector as:

$$z_1 = \eta - \eta_d - \xi_1 \tag{20}$$

Its time derivatives is

$$dz_1 = J(\psi)\nu dt - d\eta_d + K_1\xi_1 dt - J(\psi)\xi_2 dt \tag{21}$$

Here, let ν be the virtual control for (21) and select the stabilizing function vector for ν as

$$\alpha_1 = -J^{-1}(\psi)(K_1 z_1 + K_1\xi_1 - \dot{\eta}_d). \tag{22}$$

Step 2. Definite velocity error vector as:

$$z_2 = \nu - \alpha_1 - \xi_2 \tag{23}$$

It follows from (21)–(23) that

$$\begin{aligned} dz_1 =& J(\psi)(z_2 + \alpha_1 + \xi_2)dt - d\eta_d + K_1\xi_1 dt - J(\psi)\xi_2 dt \\ =& J(\psi)z_2 dt - K_1 z_1 dt \end{aligned} \tag{24}$$

Choose the Lyapunov function candidate

$$V_1 = z_1^{\mathrm{T}} z_1 \tag{25}$$

In virtue of (24)–(25), it is obtained that

$$\begin{aligned} \mathcal{L}V_1 =& \frac{\partial V_1}{\partial z_1}[J(\psi)z_2 - K_1 z_1] \\ =& -z_1^{\mathrm{T}} K_1 z_1 + z_1^{\mathrm{T}} J(\psi) z_2 \end{aligned} \tag{26}$$

According to (23), the derivation of z_2 is

$$dz_2 = dv - d\alpha_1 - d\xi_2 \tag{27}$$

It follows from (7) that

$$\begin{aligned} dz_2 =& M^{-1}[\tau + \tau_d - C(\nu)\nu - D\nu]dt - d\alpha_1 - d\xi_2 \\ =& M^{-1}[\tau + \tau_d - C(\nu)\nu - D\nu + K_2\xi_2 - \Delta\tau - M\dot{\alpha}_1]dt \\ =& M^{-1}[\tau_c + \tau_d - C(\nu)\nu - D\nu + K_2\xi_2 - M\dot{\alpha}_1]dt \end{aligned} \tag{28}$$

Choose the Lyapunov function candidate

$$V_2 = V_1 + z_2^{\mathrm{T}} M z_2 \tag{29}$$

According to (26) and (28), the derivation of V_2 is

$$\begin{aligned} \mathcal{L}V_2 =& \mathcal{L}V_1 + z_2^{\mathrm{T}}\big(\tau_c + \tau_d - C(\nu)\nu - D\nu + K_2\xi_2 - M\dot{\alpha}_1\big) \\ =& -z_1^{\mathrm{T}} K_1 z_1 + z_1^{\mathrm{T}} J(\psi) z_2 \\ &+ z_2^{\mathrm{T}}\big(\tau_c + \tau_d - C(\nu)\nu - D\nu + K_2\xi_2 - M\dot{\alpha}_1\big). \end{aligned} \tag{30}$$

Design the commanded anti-disturbance tracking control law:

$$\tau_c = -J^{\mathrm{T}}(\psi)z_1 - K_2 z_2 + M\dot{\alpha}_1 + C(\nu)\nu + D\nu - K_2\xi_2 - \hat{\tau}_d \tag{31}$$

where $K_2 = K_2^{\mathrm{T}} \in \mathbb{R}^{3\times 3}$ is a positive-definite design matrix. According to (30) and (31), it follows that

$$\begin{aligned} \mathcal{L}V_2 \le& -z_1^{\mathrm{T}} K_1 z_1 - z_2^{\mathrm{T}} K_2 z_2 + z_2^{\mathrm{T}}(\tau_d - \hat{\tau}_d) \\ \le& -z_1^{\mathrm{T}} K_1 z_1 - z_2^{\mathrm{T}} K_2 z_2 + \frac{1}{2} z_2^{\mathrm{T}} z_2 + \frac{1}{2}||\tilde{\tau}_d||^2 \\ =& -z_1^{\mathrm{T}} K_1 z_1 - z_2^{\mathrm{T}}\left(K_2 - \frac{1}{2} I_{3\times 3}\right) z_2 + \frac{1}{2}||\tilde{\tau}_d||^2 \end{aligned} \tag{32}$$

Theorem 1. *Consider the ship movement model* (1)–(2) *with stochastic disturbances described by first-order Markov process* (8) *under actuator saturation. By designing the commanded anti-disturbance tracking control law based on the disturbance observer* (9)–(10) *and auxiliary filter* (18)–(19)*, the ship's position* (x, y) *and the yaw angle* ψ *can track the desired trajectory* $\eta_d = [x_d, y_d, \psi_d]^{\mathrm{T}}$ *and all signals of the ship closed-loop systems are guaranteed to be uniformly ultimately bounded through properly selecting design parameter matrices* K_0, K, K_1 *and* K_2 *which satisfy* $K = K_0 M^{-1}$ *and* $\lambda_{\min}(K_2) > \frac{1}{2}$.

Proof: Structure the following Lyapunov function:

$$V = z_1^{\mathrm{T}} z_1 + z_2^{\mathrm{T}} M z_2 + \tilde{\tau}_d^{\mathrm{T}} P \tilde{\tau}_d \tag{33}$$

In the light of (15) and (32), the derivative of (33) is:

$$\begin{aligned} \mathcal{L}V \leq & -z_1^{\mathrm{T}} K_1 z_1 - z_2^{\mathrm{T}}\left(K_2 - \tfrac{1}{2} I_{3\times 3}\right) z_2 \\ & - \tilde{\tau}_d^{\mathrm{T}} Q \tilde{\tau}_d + \tfrac{1}{2}\|\tilde{\tau}_d\|^2 + \mathrm{Tr}\{\Psi^{\mathrm{T}} P \Psi\} \\ \leq & -\varrho_2 V + \beta_2. \end{aligned} \tag{34}$$

It can obtain the following result:

$$\mathcal{L}V \leq -\varrho_2 V + \beta_2 \tag{35}$$

Herein, $\varrho_2 = \lambda_{\min}\left\{K_1, (K_2 - \frac{1}{2} I_{3\times 3})\right\}$ and $C = \frac{1}{2}\|\tilde{\tau}_d\|^2 + Tr\{\varPsi^{\mathrm{T}} P \varPsi\}$ are the positive constants with the condition $\lambda_{\min}(K_2) > \frac{1}{2}$. According to Lemma 1, choose the function $\kappa = \lambda_{\min}(P)\|\tilde{\tau}_d\|^p$ and the positive number $p = 2$, such that

$$\begin{aligned} \kappa(\|z_1\|^p, \|z_2\|^p, \|\tilde{\tau}_d\|^p) = & \|z_1\|^2 + \lambda_{\min}(M)\|z_2\|^2 \\ & + \lambda_{\min}(P)\|\tilde{\tau}_d\|^2 \leq V \end{aligned} \tag{36}$$

$$\lim_{t\to\infty} \sup \mathbb{E}\|z_1(t; t_0, z_1(0))\|^2 \leq \kappa^{-1}\left(\tfrac{\varrho_2}{\beta_2}\right). \tag{37}$$

Hence, the tracking error dynamics of the ship are asymptotically stable in the mean-square sense. From the formulations of ϱ_2 and β_2, it can be inferred that the tracking errors can be reduced by tuning the design matrices K_0, K, K_1, and K_2, subject to the conditions $K = K_0 M^{-1}$ and $\lambda_{\min}(K_2) > \frac{1}{2}$.

6 Simulation Study

To validate the composite disturbance rejection tracking control, a 1:70 scaled model ship CyberShip II is taken as the example for simulations, the ship model dynamic parameters are:

$$M = \begin{bmatrix} 25.8 & 0 & 0 \\ 0 & 33.8 & 1.0115 \\ 0 & 1.0115 & 2.76 \end{bmatrix} \tag{38}$$

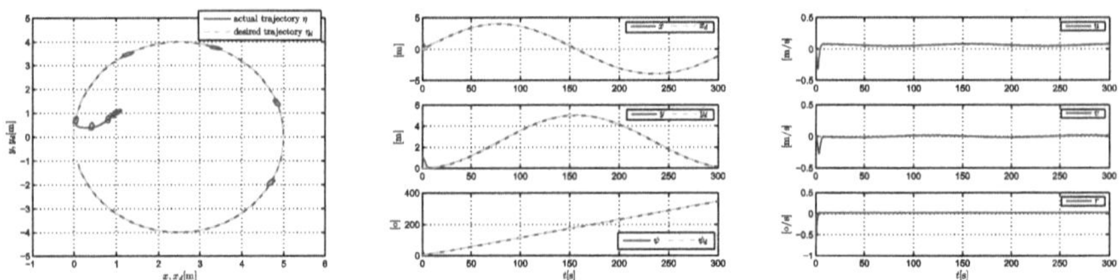

(a) Ship horizontal tracking. (b) Ship position x, y and heading ψ. (c) Ship velocities μ, v, r.

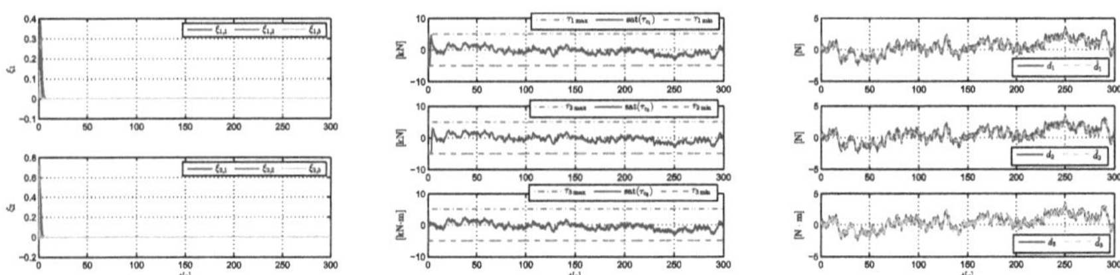

(d) States vectors ξ_1 and ξ_2. (e) Control forces τ_1, τ_2 and yaw torque τ_3. (f) Stochastic disturbances d_1, d_2, d_3 and estimations $\hat{d}_1$, $\hat{d}_2$, $\hat{d}_3$.

Fig. 2. Illustrative results in disturbance.

$$C(\nu) = \begin{bmatrix} 0 & 0 & -(33.8\,v + 1.0115\,r) \\ 0 & 0 & 25.8\,u \\ 33.8\,v + 1.0115\,r & -25.8\,u & 0 \end{bmatrix} \tag{39}$$

$$D = \begin{bmatrix} 0.72 & 0 & 0 \\ 0 & 0.86 & -0.11 \\ 0 & -0.11 & -0.5 \end{bmatrix} \tag{40}$$

The desired trajectory $\eta_d = [x_d,\ y_d,\ \psi_d]^{\mathrm{T}}$ of ship is

$$x_d = 4\sin(0.02\mathrm{t}) \tag{41}$$

$$y_d = 2.5(1 - \cos(0.02\mathrm{t})) \tag{42}$$

$$\psi_d = 0.02\mathrm{t} \tag{43}$$

The initial conditions of the ship are chosen as

$$\eta(0) = [1\,\mathrm{m},\ \ 1\,\mathrm{m},\ \ 45^\circ]^{\mathrm{T}} \tag{44}$$

$$\nu(0) = [0\,\mathrm{m/s},\ \ 0\,\mathrm{m/s},\ \ 0^\circ/s]^{\mathrm{T}} \tag{45}$$

To demonstrate the adaptability and robustness of the composite disturbance rejection control scheme, a disturbance case is taken in simulations.

The disturbance parameters are set as

$$T = diag(10, 10, 10) \tag{46}$$

$$\Psi = \mathrm{diag}(0.05, 0.05, 0.05) \tag{47}$$

The design parameter matrices are $K_1 = diag(1.5, 1.5, 1.5)$, $K_2 = diag(50, 50, 50)$, $K_0 = diag(60, 60, 40)$ and $P_0 = diag(4, 4, 4)$.

The simulation results, plotted in Fig. 2, illustrate that the proposed composite anti-disturbance controller achieves satisfactory tracking performance. The ship successfully navigates the desired trajectory (Fig. 2a, 2b) and respects saturation constraints (Fig. 2e), while the stochastic disturbances are effectively estimated (Fig. 2f). It is verified from these results that all signals, including ship velocities (Fig. 2c) and auxiliary states (Fig. 2d), are bounded, which validate the theoretical results successfully.

7 Conclusion

A composite stochastic disturbance rejection tracking control was presented for ships facing both stochastic disturbances and actuator saturation. The scheme fuses a stochastic disturbance observer with an auxiliary filter via backstepping, creating a hierarchical and implementable control structure. Based on Lyapunov stability, the designed controller makes the ship track the desired trajectory with small errors, and all signals in the closed-loop system are globally uniformly ultimately bounded. Simulations on the 1:70 scale model ship verified the availability of the proposed scheme.

Acknowledgments. This work was supported in part by the National Natural Science Foundation of China under Grant 62273172, Natural Science Foundation of Shandong Province under Grant ZR2024MF055, Yantai Science and Technology Innovation Development Plan Research Project under Grant 2024JCYJ092.

References

1. Fossen, T.I.: Handbook of Marine Craft Hydrodynamics and Motion Control. John Wiley & Sons Ltd, Chichester (2011)
2. Sørensen, A.J.: A survey of dynamic positioning control systems. Annu. Rev. Control. **35**, 123–136 (2002)
3. Zhu, G.B., Ma, Y., Li, Z.X., Malekian, R., Sotelo, M.: Event-triggered adaptive neural fault-tolerant control of underactuated MSVs with input saturation. IEEE Trans. Intell. Transp. Syst. (2021). https://doi.org/10.1109/TITS.2021.3066461
4. Fossen, T.I.: Guidance and Control of Ocean Vehicles. John Wiley & Sons Ltd, Chichester (1994)
5. Balchen, J.G., Jenssen, N.A., Salid, S.: Dynamic positioning using Kalman filtering and optimal control theory. In: Proceedings of IFAC/IFIP Symposium on Automation in Offshore oil Field Operation, pp. 183–186. Bergen, Norway (1976)

6. Saeid, S., Jenssen, N.A., Balchen, J.G.: Design and analysis of a dynamic positioning system based on Kalman filtering and optimal control. IEEE Trans. Autom. Control **28**(3), 331–339 (1983)
7. Fossen, T.I., Grøvlen, A.: Nonlinear output feedback control of dynamically positioned ships using vectorial observer backstepping. IEEE Trans. Control Syst. Technol. **6**(1), 121–128 (1998)
8. Ohishi, K., Nakao, M., Ohnishi, K., Miyachi, K.: Microprocessor-controlled DC motor for load-insensitive position servo system. IEEE Trans. Ind. Electron. **34**(1), 44–49 (1987)
9. Chen, W.H.: Disturbance observer based control for nonlinear systems. IEEE/ASME Trans. Mechatron. **9**(4), 706–710 (2004)
10. Guo, L., Cao, S.Y.: Anti-disturbance Control for Systems with Multiple Disturbances. CRC Press, Boca Raton, FL (2014)
11. Li, S.H., Yang, J., Chen, W.H., Chen, X.S.: Disturbance Observer-Based Control: Methods and Applications. CRC Press, Boca Raton, FL (2014)
12. Zhang, H.F., Wei, X.J., Zhang, L.Y.: Disturbance observer-based control for a class of strict-feedback nonlinear systems with derivative-bounded disturbances. Trans. Inst. Meas. Control. **42**(14), 2601–2610 (2020)
13. Hu, X., Wei, X.J., Kao, Y.G., Han, J.: Robust synchronization for under-actuated vessels based on disturbance observer. IEEE Trans. Intell. Transp. Syst. (2021). https://doi.org/10.1109/TITS.2021.3054177
14. Wei, X.J., Zhang, H.F., Guo, L.: Composite disturbance-observer-based control and variable structure control for non-linear systems with disturbances. Trans. Inst. Meas. Control. **31**(5), 401–423 (2009)
15. Wei, X.J., Wu, Z.J., Karimi, H.R.: Disturbance observer-based disturbance attenuation control for a class of stochastic systems. Automatica **63**, 21–25 (2016)
16. Donaire, A., Perez, T.: Dynamic positioning of marine craft using a port-Hamiltonian framework. Automatica **48**(5), 851–856 (2012)
17. Veksler, A., Johansen, T.A., Borrelli, F., Realfsen, B.: Dynamic positioning with model predictive control. IEEE Trans. Control Syst. Technol. **24**(4), 1340–1353 (2016)
18. Du, J.L., Hu, X., Krstić, M., Sun, Y.Q.: Robust dynamic positioning of ships with disturbances under input saturation. Automatica **73**, 207–214 (2016)
19. Dong, L.W., Wei, X.J., Zhang, H.F.: Anti-disturbance control based on nonlinear disturbance observer for a class of stochastic systems. Trans. Inst. Meas. Control. **41**(6), 1665–1675 (2019)
20. Mao, X.R., Yuan, C.G.: Stochastic Differential Equations with Markovian Switching. Imperial College Press, London (2006)
21. Deng, H., Krstic, M.: Stabilization of stochastic nonlinear systems driven by noise of unknown covariance. IEEE Trans. Autom. Control **46**(8), 1237–1253 (2001)
22. Hu, X., Wei, X.J., Zhang, H.F., Han, J.: Adaptive saturation compensation for strict-feedback systems with unknown control coefficient and input saturation. Int. J. Adapt. Control Signal Process. **35**(6), 1083–1098 (2021)
23. Krstić, M., Kanellakopoulos, I., Kokotović, P.: Nonlinear and Adaptive Control Design. John Wiley & Sons, New York (1995)
24. Li, Y.M., Min, X., Tong, S.C.: Observer-based fuzzy adaptive inverse optimal output feedback control for uncertain nonlinear systems. IEEE Trans. Fuzzy Syst. **29**(6), 1484–1495 (2021)

Neural Network-Based Adaptive Impulsive Control Strategy for Leader-Following Consensus in Uncertain Multiagent Systems

Chengyao Yan, Yiyan Han, and Xin Wang(✉)

College of Electronic and Information Engineering, Southwest University, Chongqing, China
yyhansera@swu.edu.com, xinwangswu@163.com

Abstract. This paper investigates a Neural Network-based (NN-based) adaptive impulsive control approach to address the leader-following consensus problem in uncertain multiagent systems under deception attacks. A novel adaptive impulsive control strategy is developed to mitigate the impact of uncertain nonlinearities and deception attacks. This strategy not only ensures faster convergence but also reduces the consumption of communication resources. Simulation examples are provided to verify the effectiveness of the proposed adaptive impulsive control method.

Keywords: Multiagent Systems · Adaptive impulsive control · Leader-following consensus

1 Introduction

With the rapid advancement of artificial intelligence, data-intensive communication challenges have become increasingly prominent. Multiagent systems (MASs), which excel in distributed information processing, have garnered significant attention and are widely applied in various fields, including aircraft control, UAVs, and power systems [1–3]. Coordination control, which relies on information sharing among neighboring agents, enables cooperative behaviors such as consensus, formation, and containment, with leader-following consensus control being the most fundamental [4,5].

Traditional consensus control methods mainly rely on continuous control schemes, which, despite their effectiveness, can lead to substantial communication burdens. Hybrid impulsive control strategies have been developed to counteract network attacks and external disturbances, but they often focus on isolated attack modes [6,7]. In reality, attacks can simultaneously impact both continuous and impulsive control channels, making the study of consensus under such conditions particularly significant. Adaptive control methods, particularly those incorporating neural networks, have demonstrated success in addressing system

C. Li et al. (Eds.): ICNC 2025, CCIS 2946, pp. 509–520, 2026.
https://doi.org/10.1007/978-981-92-1599-7_43

uncertainties [8–10]. However, the stability challenges posed by deception attacks in hybrid control paradigms have not been extensively explored [11–13].

Deception attacks, especially false data injection (FDI) attacks, have emerged as a significant threat to the stability and security of MASs, given the increasing dependence on digital communication. These attacks manipulate data to compromise the integrity of communication channels and system performance. Recent studies have focused on developing robust control strategies to mitigate the impact of FDI attacks and address actuator faults in MASs [14–16]. These efforts aim to enhance the resilience of MASs against such cyber threats, ensuring reliable and secure operation in networked environments.

This article proposes a NN-based adaptive impulsive control strategies for the leader-following consensus in uncertain MASs. The main contributions are: (1) A novel control framework integrating deception attack detection signals and adaptive mechanisms to address uncertain nonlinear dynamics; (2) Addressing more destructive attacks targeting both continuous and impulsive control channels, with an attack observer that mitigates multi-channel attack impacts and eliminates dependency on attack intensity thresholds. The organization of this paper is outlined as follows: Sect. 2 addresses the model formulation and preliminaries. Section 3 delves into the consensus analysis. The theoretical results are corroborated through simulation studies detailed in Sect. 4.

2 Preliminaries and Problem Statement

This section introduces an attack model characterizing potential threats, dynamic model formulation, and an NN-based adaptive impulsive control algorithm.

2.1 Dynamic Model Formulation

This research examines the consensus problem in nonlinear MASs with a leader-follower structure, comprising N follower agents and a single leader. The dynamics of the ith follower agent are described by

$$\dot{x}_i(t) = f_i(x_i(t)) + u_i(t),\ i \in \{1, 2, \ldots, N\} \tag{1}$$

where the leader's dynamics can be described by

$$\dot{x}_0(t) = f_0(x_0(t)), \tag{2}$$

where $x_i(t)$ and $x_0(t)$ represent the state vectors of the ith follower and the leader, respectively. The function $f_i(x_i(t))$ denotes an unknown smooth nonlinear function, while $f_0(x_0(t))$ is bounded such that $\|f_0(x_0(t))\| \leq \bar{f}_0$.

Definition 1. *In the adaptive control framework, consensus is achieved within a bounded area for the MASs described by (1) and (2) if*

$$\limsup_{t \to +\infty} \|x_i(t) - x_0(t)\| \leq \kappa, \tag{3}$$

where κ denotes a small positive constant upper bound for all $i \in \mathcal{V}$.

Assumption 1. *[17] The leader has at least one path to each follower.*

2.2 Attack Model

The control model, unaffected by attacks, is initially formulated as follows:

$$u_i(t) = u_{i1}(t) + u_{i2}(t), \tag{4}$$

where $u_i(t)$ signifies the hybrid control input, composed of a continuous control term $u_{i1}(t)$ and an impulsive control term $u_{i2}(t)$.

In the context of multiagent systems, a class of actuator attacks is considered. These attacks can inject an unknown false signal $\varrho_i(t) \in \mathbb{R}^n$ into the controller-to-actuator channels, resulting in a compromised control input $u_i^A(t)$. The actuator attacks can be mathematically described as

$$\begin{cases} u_{i1}^A(t) = u_{i1}(t) + \varrho_i(t), \quad t \neq t_k \\ u_{i2}^A(t) = \delta(t-t_k)\big(u_{i2}(t) + \varrho_i(t)\big), \end{cases} \tag{5}$$

where $\delta(\cdot)$ indicates the Dirac delta function and t_k signifies the sequence of impulses.

Assumption 2. *[18] The attack signal $\varrho_i(t)$ is differentiable, and its derivative $\dot{\varrho}_i(t)$ satisfies $\|\dot{\varrho}_i(t)\| \leq \bar{w}_i$ where $\bar{w}_i > 0$ is a constant.*

2.3 NN-Based Adaptive Impulsive Control Algorithm

To tackle the problem of actuator attacks, this paper proposes a hybrid adaptive and impulsive control protocol. Within this framework, an RBFNN-based adaptive algorithm with estimators is employed to handle uncertain nonlinear dynamics.

According to RBFNNs, the uncertain nonlinear function in (1) can be represented as

$$f_i(x_i(t)) = W_i\phi(x_i(t)) + \sigma_i, \tag{6}$$

where W_i denotes the ideal weight matrix, $\phi(x_i(t))$ is a basis function vector, and σ_i is the approximation error.

The continuous controller $u_{i1}(t)$ and the impulsive controller $u_{i2}(t)$ are designed in the following forms:

$$\begin{cases} u_{i1}(t) = -\hat{\varrho}_i(t) - \hat{W}_i(t)\phi(x_i), \quad t \neq t_k \\ u_{i2}(t) = \displaystyle\sum_{k=1}^{\infty} \delta(t-t_k)\left[c_i\left(b_i(x_i(t) - x_0(t)) - \sum_{j=1}^{N} l_{ij}x_j(t)\right) - \hat{\varrho}_i(t)\right], \end{cases} \tag{7}$$

where c_i is the communication strength between agent i and its neighbors, while b_i signifies the connection strength between the leader and agent i. The impulsive interval $\tau_k = t_{k+1} - t_k$ satisfies $0 < h_1 \leq \tau_k \leq h_2 < +\infty$. $\hat{\varrho}_i(t)$ is the estimated attack signal for $\varrho_i(t)$, and $\hat{W}_i(t)$ is the estimate of W_i.

Under the control protocol (4) and (7) with attack signals, (1) can be rewritten as

$$\begin{cases} \dot{x}_i(t) = W_i\phi(x_i) + \sigma_i + u_{i1}^A(t), t \neq t_k \\ x_i(t_k^+) = x_i(t_k) + u_{i2}^A(t_k). \end{cases} \tag{8}$$

According to (8), the ith estimator is designed as

$$\begin{cases} \dot{\hat{x}}_i(t) = \hat{W}_i\phi(x_i) + g_i(x_i(t) - \hat{x}_i(t)) + u_{i1}(t) + \hat{\varrho}_i(t), t \neq t_k \\ \hat{x}_i(t_k^+) = \hat{x}_i(t_k) + u_{i2}(t_k) + \hat{\varrho}_i(t_k), \end{cases} \tag{9}$$

where $\hat{x}_i(t)$ represents the estimated state vector, g_i denotes the distributed gain allocated to each agent $i \in \mathcal{V}$, and $\hat{W}_i(t)$ is updated according to the adaptive law $\dot{\hat{W}}_i(t) = (x_i(t) - \hat{x}_i(t))\phi^{\mathrm{T}}(x_i(t)) - r\hat{W}_i(t)$, with $r > 0$ being a predefined positive constant.

Then, define the auxiliary variable $p_i(t) = \varrho_i(t) - d_i x_i(t)$ where $d_i > 0$ is a designed parameter. Based on (8) and Assumption 2, the estimation of $p_i(t)$ can be designed as

$$\begin{cases} \dot{\hat{p}}_i(t) = -d_i\big(\hat{W}_i\phi(x_i) + \hat{\varrho}_i(t) + u_{i1}(t)\big), t \neq t_k \\ \hat{p}_i(t_k^+) = -d_i\big(\hat{\varrho}_i(t_k) + u_{i2}(t_k)\big) + \hat{p}_i(t_k). \end{cases} \tag{10}$$

From (10), the attack observer is given by

$$\hat{\varrho}_i(t) = \hat{p}_i(t) + d_i x_i(t). \tag{11}$$

3 Main Results

This section examines the boundedness of system estimation and attack observer errors, derives conditions for leader-following consensus, and proposes a resource-efficient impulsive control algorithm that requires state updates only at discrete instants. Stability is verified under analogous conditions.

3.1 Boundedness of Observation and Estimation Errors

Define $\tilde{x}_i(t) = x_i(t) - \hat{x}_i(t)$ and $\tilde{W}_i(t) = W_i - \hat{W}_i(t)$. For simplicity, let $\tilde{W}_i = \tilde{W}_i(t)$, $\hat{W}_i = \hat{W}_i(t)$, and $\phi(x_i) = \phi(x_i(t))$ in the subsequent analysis. According to (8) and (9), we obtain

$$\dot{\tilde{x}}_i(t) = \tilde{W}_i\phi(x_i) + \sigma_i + \tilde{\varrho}_i(t) - g_i\tilde{x}_i(t), \tag{12}$$

where $\tilde{\varrho}_i(t) = \varrho_i(t) - \hat{\varrho}_i(t)$ represents the attack observer estimation error.

Define the estimation errors $\tilde{p}_i(t) = p_i(t) - \hat{p}_i(t)$, which we can derive

$$\begin{cases} \dot{\tilde{p}}_i(t) = \dot{\varrho}_i(t) - d_i(\tilde{W}_i\phi(x_i) + \tilde{\varrho}_i(t) + \sigma_i), \quad t \neq t_k \\ \tilde{p}_i(t_k^+) = (1 - d_i)\tilde{p}_i(t_k). \end{cases} \tag{13}$$

Based on the above analysis, the tracking consensus characteristics of the system (1) can be amenable to further investigation.

Theorem 1. *The control strategy (4) with attack observer (11) can ensure convergence provided that there are positive scalars ν_1, ν_2, γ_1, γ_2, γ_3, ψ, $\hat{\upsilon}$, and that d_i and the distributed gain g_i are properly selected to meet the following conditions:*

$$\frac{\ln \hat{\upsilon}}{h_1} - \psi < 0 \tag{14}$$

where $\psi = \max\{2g_i - \frac{1}{\nu_1} - \frac{1}{\nu_2}, 2d_i - \nu_2 - \frac{1}{\gamma_1} - \frac{1}{\gamma_2} - \frac{1}{\gamma_3}, r - \gamma_2 d_i^2 \phi_s^2\}$ *and* $\|\phi(x_i)\| \leq \phi_s$. *Additionally,* $\hat{\upsilon} = \max\{2 + \upsilon^{-1} + (1-d_i)^2 + \upsilon\}$ *with* $\upsilon > 0$.

Proof: Choose $V_1(t) = \sum_{i=1}^N \tilde{x}_i^T(t)\tilde{x}_i(t) + \sum_{i=1}^N \tilde{p}_i^T(t)\tilde{p}_i(t) + \sum_{i=1}^N \mathrm{Tr}(\tilde{W}_i^T \tilde{W}_i)$. When $t \in (t_{k-1}, t_k]$, according to Assumption 2, there exists

$$\begin{aligned}\dot{V}_1(t) \leq & \sum_{i=1}^N \Bigg((\frac{1}{\nu_1} + \frac{1}{\nu_2} - 2g_i)\tilde{x}_i^T(t)\tilde{x}_i(t) + \nu_1 \sigma_i^T \sigma_i + \nu_2 \tilde{\varrho}_i^T(t)\tilde{\varrho}_i(t) + r\mathrm{Tr}(W_i^T W_i) \\ & - r\mathrm{Tr}(\hat{W}_i^T \hat{W}_i) + (\gamma_2 d_i^2 \phi_s^2 - r)\mathrm{Tr}(\tilde{W}_i^T \tilde{W}_i) + \gamma_1 \bar{w}_i^2 + \gamma_3 d_i^2 \sigma_i^T \sigma_i \\ & + (\frac{1}{\gamma_1} + \frac{1}{\gamma_2} + \frac{1}{\gamma_3} - 2d_i)\tilde{p}_i^T(t)\tilde{p}_i(t) \Bigg) \\ \leq & -\psi V_1(t) + \theta, \end{aligned} \tag{15}$$

where $\theta = \sum_{i=1}^N \left((\nu_1 + \gamma_3 d_i^2)\sigma_i^T \sigma_i + \gamma_1 \bar{w}_i^2 + r\mathrm{Tr}(W_i^T W_i) \right)$.

When it comes to $t = t_k$, we can obtain

$$V_1(t_k^+) \leq \hat{\upsilon} V_1(t_k). \tag{16}$$

Based on (15) and (16), for $t \in (t_0, t_1]$, the following inequality can be established.

$$V_1(t) \leq V(t_0)\exp\left(-\psi(t - t_0)\right) + \theta. \tag{17}$$

Furthermore, at the instant $t = t_1$, the value of $V_1(t_1^+)$ can be bounded by

$$V_1(t_1^+) \leq \hat{\upsilon} V_1(t_1) \leq \hat{\upsilon} V(t_0)\exp\left(-\psi(t - t_0)\right) + \hat{\upsilon}\theta. \tag{18}$$

For $t \in (t_1, t_2]$, the inequality for $V_1(t)$ becomes

$$V_1(t) \leq \hat{\upsilon} V(t_0)\exp\left(-\psi(t - t_0)\right) + \hat{\upsilon}\theta \exp\left(t - t_1\right) + \theta. \tag{19}$$

Similarly, at the instant $t = t_2$, we have

$$V_1(t_2^+) \leq \hat{\upsilon}^2\, V(t_0)\exp\left(-\psi(t_2 - t_0)\right) + \hat{\upsilon}^2\theta \exp\left(-\psi(t_2 - t_1)\right) + \hat{\upsilon}\theta. \tag{20}$$

Finally, the expression for $V_1(t)$ can be generalized as

$$\begin{aligned} V_1(t) \leq & V_1(t_k^+)\exp\left(-\psi(t - t_k)\right) + \theta \\ \leq & \hat{\upsilon}^k V_1(t_0)\exp\left(-\psi(t - t_0)\right) + \hat{\upsilon}\theta \Bigg(\exp\left(-\psi(t - t_k)\right) + \hat{\upsilon}\exp\left(-\psi(t - t_{k-1})\right) \\ & + \ldots + \hat{\upsilon}^{k-1}\exp\left(-\psi(t - t_1)\right) \Bigg). \end{aligned} \tag{21}$$

It is evident that $\hat{\upsilon} > 1$. Within the time interval $t \in (t_k, t_{k+1}]$, the inequality (21) will satisfy

$$\begin{aligned} V_1(t) \leq & \hat{\upsilon}\theta \frac{\hat{\upsilon}^{\frac{t-t_0}{h_1}} \exp(-\psi(t-t_0-h_2))}{1-\hat{\upsilon}^{-1}\exp(\psi h_2)} + \frac{\theta}{\hat{\upsilon}^{-1}\exp(\psi h_2)+1} \\ & + \hat{\upsilon}^{\frac{t-t_0}{h_1}+1} V_1(t_0) \exp(-\psi(t-t_0)) \\ \leq & \Upsilon \exp((\frac{\ln \hat{\upsilon}}{h_1} - \psi)(t-t_0)) + w^*, \end{aligned} \tag{22}$$

where $\Upsilon = \hat{\upsilon}V_1(t_0) + \frac{\theta\hat{\upsilon}}{1-\hat{\upsilon}^{-1}e^{\psi h_2}}$ and $w^* = \hat{\upsilon}\theta\frac{e^{\psi h_2}-\hat{\upsilon}^{-1}}{1-\hat{\upsilon}^{-1}e^{\psi h_2}}$, the inequality (22) converges to $\hat{\upsilon}\theta\frac{e^{\psi h_2}-\hat{\upsilon}^{-1}}{1-\hat{\upsilon}^{-1}e^{\psi h_2}}$ as $t \to +\infty$, assuming that conditions (14) are satisfied. This indicates that the state estimation errors and the weight estimation errors can be bounded above. Additionally, the boundedness of the auxiliary variable $\tilde{p}_i(t)$ allows the attack observer to detect attack variations, albeit with some error.

Remark 1. The framework presented in this study differs from prior works [19] that focus on cyberattacks occurring either during continuous intervals or at discrete impulsive moments. Importantly, additional attack vectors introduced at impulsive instants can significantly affect system stability.

3.2 Analysis of Tracking Error

The system stability under the controller (7) is demonstrated in the subsequent theorem.

Theorem 2. *Suppose there exist positive constants $\alpha < 1$ and ν, along with a compatible positive-definite matrix $P > 0$ that satisfies the following inequality:*

$$(1+\frac{1}{\nu})Q_1^T P Q_1 - \alpha P < 0. \tag{23}$$

Under these conditions, the leader-following consensus of the MASs (1) and (2) can be achieved, with the constant κ bounded by

$$\kappa \leq \frac{\Upsilon^* h_2(w^* + l^*)}{\lambda_{\min}(P)(1-\alpha)}, \tag{24}$$

where $\Upsilon^* = \max(4\Upsilon_1\ell, 4\Upsilon_1, \Upsilon_2)$, with $\Upsilon_1 = (\nu+1)(1+\theta)\lambda_{\max}(P)$ and $\Upsilon_2 = (1+\nu)(1+\theta^{-1})\lambda_{\max}(P)$, and $\ell = \max(\phi_s^2, 1)$. Additionally, $\sigma = (\sigma_1^T, \sigma_2^T, \dots, \sigma_N^T)^T$ and $\tilde{W} = (\tilde{W}_1^T, \tilde{W}_2^T, \dots, \tilde{W}_N^T)^T$. Moreover, $l^* = ||\bar{F}_0||^2 + ||\sigma||^2$, where $F_0(t) = (f_0^T(x_0(t)), f_0^T(x_0(t)), \dots, f_0^T(x_0(t)))^T$.

Proof: Define the consensus error $e_i(t) = x_i(t) - x_0(t)$. Combining (2) and (8), the ith tracking error system is described as

$$\begin{cases} \dot{e}_i(t) = \tilde{f}_i(t) + u_{i1}^A(t), \quad t \neq t_k \\ e_i(t_k^+) = e_i(t_k) + \varrho_i(t_k) - \hat{\varrho}_i(t_k) \\ \qquad + c_i \left(b_i(x_i(t) - x_0(t)) - \sum_{j=1}^{N} l_{ij} x_j(t) \right). \end{cases}$$

By writing the above equation in column form, we can obtain

$$\begin{cases} \dot{e}(t) = \tilde{F}(t) + U_1^A(t), t \neq t_k \\ e(t_k^+) = (Q_1 \otimes I_n) e(t_k) + \tilde{\varrho}(t_k), \end{cases} \tag{25}$$

where $Q_1 = I_N + C(B - L)$, with $C = \text{diag}(c_1, c_2, \dots, c_N)$.

To prove leader-following consensus, consider the following Lyapunov function

$$V_2(t) = e(t)^T (P \otimes I_n) e(t). \tag{26}$$

At $t = t_k$, from the error system (25), it follows

$$e(t_k) = (Q_1 \otimes I_n) e(t_{k-1}) + v_1(t_{k-1}), \tag{27}$$

where $v_1(t_{k-1}) = \int_{t_{k-1}^+}^{t_k} (\tilde{F}(s) + U_1^A(s)) ds + \tilde{\varrho}(t_{k-1})$.

$$\begin{aligned} V_2(t_k) \leq{}& (1 + \frac{1}{\nu}) e^T(t_{k-1})(Q_1^T P Q_1) \otimes I_n) e(t_{k-1}) \\ &+ (1 + \nu) v_1^T(t_{k-1})(P \otimes I_n) v_1(t_{k-1}). \end{aligned} \tag{28}$$

Then, from (28), there is

$$\begin{aligned} &(1 + \nu) v_1^T(t_{k-1})(P \otimes I_n) v_1(t_{k-1}) \\ &\leq \Upsilon_1 \left\| \int_{t_{k-1}^+}^{t_k} (\tilde{W}\phi(x(s)) + \sigma - F_0(s) + \tilde{\varrho}(s)) ds \right\|^2 + \Upsilon_2 \tilde{\varrho}^T(t_{k-1}) \tilde{\varrho}(t_{k-1}) \\ &\leq \Upsilon^* \tau_k (\Upsilon \xi^{k-1} + w^* + l^*). \end{aligned} \tag{29}$$

The existence of a constant ξ satisfying $e^{(\frac{\ln \hat{v}}{h_1} - \psi)\tau_k} < \xi < 1$ ensures the result. Thus, one can have

$$\begin{aligned} V_2(t_k) &\leq \alpha V_2(t_{k-1}) + \Upsilon^* h_2 (\Upsilon \xi^{k-1} + w^* + l^*) \\ &\leq \alpha^{k-1} V_2(t_1) + \Upsilon^* \Upsilon h_2 \sum_{j=1}^{k-1} \alpha^{j-1} \xi^{k-1-j} + \Upsilon^* h_2 \sum_{j=1}^{k-1} \alpha^{j-1} (w^* + l^*) \\ &= \alpha^{k-1} V_2(t_1) + \frac{\Upsilon^* \Upsilon h_2 (\xi^{k-2} - \alpha^{k-1} \xi^{-1})}{1 - \alpha \xi^{-1}} + \frac{\Upsilon^* h_2 (w^* + l^*)(1 - \alpha^{k-1})}{1 - \alpha}. \end{aligned} \tag{30}$$

Apparently, there is

$$V_2(t_1) \leq (1+\frac{1}{\nu})V_2(t_0) + \Upsilon^*\tau_1(\Upsilon + w^* + l^*). \tag{31}$$

By combining (30) and (31), this indicates $\limsup_{k\to+\infty} V_2(t_k) \leq \frac{\Upsilon^* h_2(w^*+l^*)}{1-\alpha}$. In addition, for any $t \in (t_{k-1}, t_k]$

$$e(t) = Q_1 \otimes I_n e(t_{k-1}) + \int_{t_{k-1}^+}^{t} (\tilde{F}(s) + U_1^A(s))ds, \tag{32}$$

which leads to

$$\begin{aligned} V_2(t) \leq & (1+\nu)v_1^T(t_{k-1})(P \otimes I_n)v_1(t_{k-1}) \\ & + (1+\frac{1}{\nu})e^T(t_{k-1})((Q_1^T P Q_1) \otimes I_n)e(t_{k-1}). \end{aligned} \tag{33}$$

By similar deduction with (29) and (30), it is not difficult to determine $\limsup_{t\to+\infty} V_2(t) \leq \frac{\Upsilon^* h_2(w^*+l^*)}{1-\alpha}$ for (33). The error $e(t)$ ultimately converges below a threshold, ensuring the MASs achieves adaptive tracking consensus.

4 Simulation

In this section, numerical simulations are provided to verify the correctness of the aforementioned theoretical results. Consider a communication network comprising a leader and five followers, characterized by the adjacency matrix

$$A = \begin{bmatrix} 0 & 1 & -2 & 0 & 0 \\ 0 & 0 & 1 & 1 & 3 \\ 1 & 0 & 0 & 2 & 1 \\ 1 & 0 & 0 & 0 & 0 \\ 0 & 1 & 1 & 1 & 0 \end{bmatrix}$$

and the matrix $B = \mathrm{diag}(1, 0, 0, 1, 1)$.

Example 1. Consider the following classic Chua's circuit with the nonlinear function

$$f_i(x_i(t)) = \begin{pmatrix} -s_{i1}x_{i1}(t) + s_{i1}x_{i2}(t) - s_{i1}h_i(x_{i1}(t)) \\ x_{i1}(t) + x_{i3}(t) - x_{i2}(t) \\ -s_{i2}x_{i2}(t) \end{pmatrix}. \tag{34}$$

where $f_i(x_i(t))$ characterizes the dynamics of the i-th Chua's circuit, $x_{i1}(t)$, $x_{i2}(t)$, and $x_{i3}(t)$ denote the state variables of the system, $h_i(x_{i1}(t))$ describes the piecewise-linear behavior of the Chua diode, and s_{i1} and s_{i2} are system constants that depend on the specific circuit configuration.

The input space partitioning involves three-dimensional state vectors $x_{ij}(t)$ $(j = 1, 2, 3)$, with some centers $\bar{\mathbf{c}}_i$ set to zero. A shared synchronization gain $r = 0.4$ is used for distributed learning. Additionally, the parameters are set as $\nu_1 = 2.2$, $\nu_2 = 0.2$, $\gamma_1 = 5.9$, $\gamma_2 = 2.1$, and $\gamma_3 = 4.8$. For the control input (7), the values are $c_1 = c_4 = c_5 = 0.2$ and $c_2 = c_3 = 0.4$. Derived from (14), the parameter $h_1 = 0.11$. A positive definite P ($\nu = 9.3$, $\alpha = 0.97$) satisfying Theorem 2's condition is obtained via LMI tools.

Figure 1 depicts the time evolution of observation errors, showing that the observer estimates attacks within a certain error. Figure 2 illustrates the agent state trajectories, indicating that followers can track the leader over time. Therefore, the parameter t_k must be carefully selected within an intermediate range to balance both operational requirements.

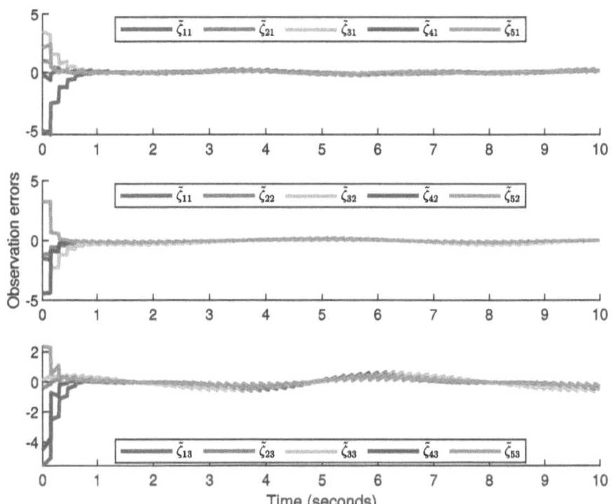

Fig. 1. Trajectories of $\tilde{\varrho}_{ij}(t)$ with $\tau_k = 0.15$.

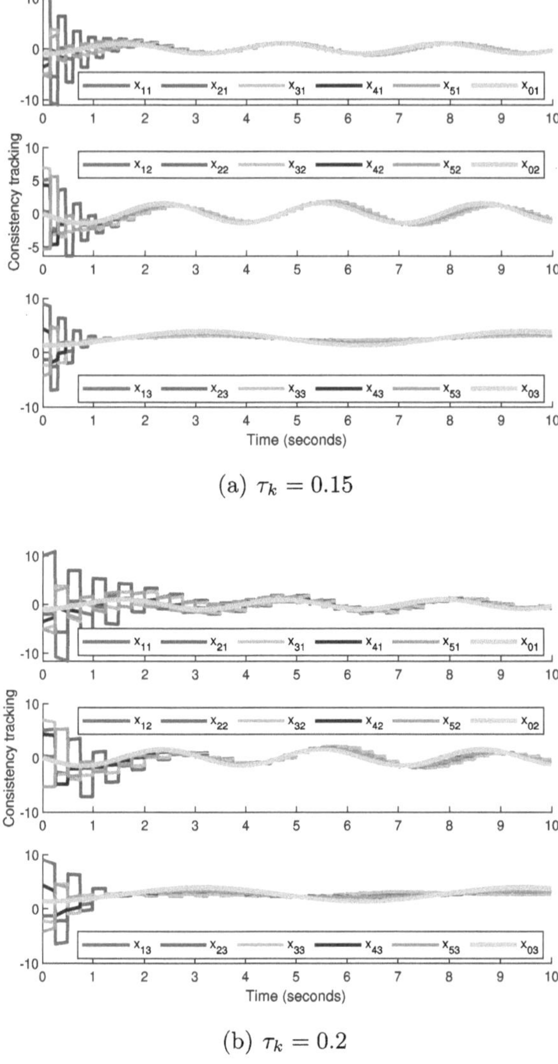

(a) $\tau_k = 0.15$

(b) $\tau_k = 0.2$

Fig. 2. Leader–follower trajectory tracking for two impulsive intervals τ_k.

5 Conclusion

This paper investigates the leader-follower consensus problem of nonlinear MASs under deception attacks using an NN-based adaptive impulsive control framework. A novel control strategy is proposed, which leverages continuous-time state monitoring via an attack observer to effectively mitigate the impacts of uncertain nonlinearities and external attacks, thereby enhancing control precision and system stability. Simulation results demonstrate the effectiveness of the proposed strategy, highlighting the importance of impulsive interval selection for optimal performance. Future work will focus on incorporating more comprehensive attack models and addressing system topological changes caused by malicious attacks.

Acknowledgments. We are particularly grateful to Southwest University for their technical expertise and insightful contributions, which significantly enhanced the quality of our work. Their dedication and support were instrumental in overcoming several challenges encountered during the study. We also acknowledge the School of Electronic and Information Engineering for providing the essential laboratory facilities and resources that were pivotal to the successful completion of this research. The availability of these resources greatly facilitated our experimental and analytical processes.

Disclosure of Interests. The authors declare that they have no relevant competing interests related to this article.

References

1. Liu, Z.-W., Shi, Y.-L., Yan, H., Han, B.-X., Guan, Z.-H.: Secure consensus of multiagent systems via impulsive control subject to deception attacks. IEEE Trans. Circuits Syst. II Express Briefs **70**(1), 166–170 (2023)
2. Ying, W., Chen, M., Li, H., Chadli, M.: Event-triggered-based adaptive nn cooperative control of six-rotorUAVs with finite-time prescribed performance. IEEE Trans. Autom. Sci. Eng. **21**(2), 1867–1877 (2024)
3. Yan, Z., Yan, X.: A multi-agent deep reinforcement learning method for cooperative load frequency control of a multi-area power system. IEEE Trans. Power Syst. **35**(6), 4599–4608 (2020)
4. Zhang, Z., Peng, S., Liu, D., Wang, Y., Chen, T.: Leader-following mean-square consensus of stochastic multiagent systems with rous and rons via distributed event-triggered impulsive control. IEEE Trans. Cybern. **52**(3), 1836–1849 (2020)
5. Han, Y., Zeng, Z.: Impulsive communication with full and partial information for adaptive tracking consensus of uncertain second-order multiagent systems. IEEE Trans. Cybern. **52**(10), 10302–10313 (2021)
6. Zhang, L., Jianquan, L., Jiang, B., Ruan, Q., Lou, J.: Delay-dependent impulsive synchronization control for coupled dynamic networks under hybrid cyberattack. Nonlinear Dyn. **113**(10), 11615–11630 (2025)
7. Taotao, H., Liu, X., He, Z., Zhang, X., Zhong, S.: Hybrid event-triggered and impulsive control strategy for multiagent systems with switching topologies. IEEE Trans. Cybern. **52**(7), 6283–6294 (2020)
8. Zeng-Guang, Hou, L., Cheng, M., Tan: Decentralized robust adaptive control for the multiagent system consensus problem using neural networks. IEEE Trans. Syst. Man Cybern. Part C Appl. Rev. **39**(3), 636–647 (2009)
9. Liang, H., Liu, G., Zhang, H., Huang, T.: Neural-network-based event-triggered adaptive control of nonaffine nonlinear multiagent systems with dynamic uncertainties. IEEE Trans. Neural Netw. Learn. **32**(5), 2239–2250 (2020)
10. Youfeng, S., Huang, J.: Cooperative adaptive output regulation for a class of nonlinear uncertain multi-agent systems with unknown leader. Syst. Control Lett. **62**(6), 461–467 (2013)
11. Qin, J., Zhang, G., Zheng, W.X., Kang, Yu.: Neural network-based adaptive consensus control for a class of nonaffine nonlinear multiagent systems with actuator faults. IEEE Trans. Neural Netw. Learn. **30**(12), 3633–3644 (2019)

12. Ding, D., Wang, Z., Han, Q.-L.: Neural-network-based consensus control for multiagent systems with input constraints: the event-triggered case. IEEE Trans. Cybern. **50**(8), 3719–3730 (2019)
13. Zhao, K., Song, Y., Shen, Z.: Neuroadaptive fault-tolerant control of nonlinear systems under output constraints and actuation faults. IEEE Trans. Neural Netw. Learn. **29**(2), 286–298 (2016)
14. Gao, Y., Zhou, W., Niu, B., Kao, Y., Wang, H., Sun, N.: Distributed prescribed-time consensus tracking for heterogeneous nonlinear multi-agent systems under deception attacks and actuator faults. IEEE Trans. Autom. Sci. Eng. **21**(4), 6920–6929 (2024)
15. Gong, Z., Yang, F., Yuan, Y., Ma, Q., Zheng, W.X.: Secure formation control of multi-agent system against FDI attack using fixed-time convergent reinforcement learning. IEEE Trans. Control Network Syst. (2025)
16. Zuo, Z., Cao, X., Wang, Y.: Security control of multi-agent systems under false data injection attacks. Neurocomputing **404**, 240–246 (2020)
17. Yang, Q., Sam Ge, S., Sun, Y.: Adaptive actuator fault tolerant control for uncertain nonlinear systems with multiple actuators. Automatica **60**, 92–99 (2015)
18. Ren, C.-E., Li, J., Shi, Z., Guan, Y., Philip Chen, C.L.: Adaptive impulsive consensus of nonlinear multiagent systems with limited bandwidth under uncertain deception attacks. IEEE Trans. Syst. Man Cybern. Syst. (2024)
19. Rong, N., Wang, Z.: Event-based impulsive control of IT2 T-S fuzzy interconnected system under deception attacks. IEEE Trans. Fuzzy Syst. **29**(6), 1615–1628 (2021)

Neural Network-Based Adaptive Learning Control of Nonlinear Crane System

Yang Liu[1], Jia-Ke Wang[1], Hui Ma[2(✉)], Yingnan Pan[3], and Mohammed Chadli[4]

[1] School of Automation and Electronics Engineering, Qingdao University of Science and Technology, Qingdao 266061, China

[2] School of Mathematics and Statistics, Guangdong University of Technology, Guangzhou 510006, China
huima2016@163.com

[3] College of Control Science and Engineering, Bohai University, Jinzhou 121013, Liaoning, China

[4] University Paris-Saclay, Univ Evry, IBISC, 91020 Evry, France

Abstract. Crane is an important industrial device, and has been widely applied to many fields such as automobile making and marine equipment. This work proposes a neural network-based adaptive learning control scheme for a class of nonlinear cranes with nonstrict-feedback structure and disturbances. Firstly, a prescribed performance function is introduced to limit the swing angle of the load, thereby removing an angle assumption. Then, a series of modified auxiliary variables are designed to simplify the system model. Further, under the framework of backstepping, a learning control algorithm is presented consisting of NN and adaptive laws to ensure that the error approaches zero and other signals are bounded as the iteration goes to infinity. The simulation experiments demonstrate the effectiveness of the proposed scheme.

Keywords: Nonlinear crane · Neural networks · Adaptive learning control · Backstepping

1 Introduction

Crane has been widely applied to various fields such as ocean engineering [1], metallurgy [2], factories [3] due to its strong transportation capacity. Moreover, the crane transports the load can be viewed as a repeated operation. Noted that the iterative learning control (ILC) is a control method to improve the performance of systems repeatedly performing the same task [4]. ILC has been widely used in industrial applications [5–7], and there have been some results for crane systems [8,9]. However, the achievements in [8,9] are considered for linear crane systems. In practice, the crane is a strongly nonlinear system, which indicates that the linearization may be difficult to achieve accurate control.

Backstepping is an effective tool for nonlinear system control, and has been applied to the nonlinear crane system [10–13]. For example, a backstepping-based

C. Li et al. (Eds.): ICNC 2025, CCIS 2946, pp. 521–536, 2026.
https://doi.org/10.1007/978-981-92-1599-7_44

state feedback controller is proposed to study the positioning and anti-swing problem of a crane with flexible cables [11]. An adaptive backstepping method is proposed in [12,13] to address the tracking problem of the underactuated bridge crane system with unknown parameters and external disturbances. However, the above approaches only are suitable for the crane system with strict-feedback structure. When the crane system has the strong nonlinearity that it is only be modeled as a nonstrict-feedback form, the traditional method will be invalid.

It is well known that neural networks (NNs) is an effective method to deal with the nonstrict-feedback issue [14–17]. This kind of schemes is to employ the property of NNs and inequality technique to transform nonstrict form into a strict one. For instance, an observer is established to tackle the unmeasurable state problem, and NNs are used to solve the adaptive fault-tolerant tracking problem in spite of the system with nonstrict-feedback structure [16]. An nonlinear mapping scheme combines with RBF NN to do with the nonstrict-feedback case in [17].

In terms of the crane control problem, an assumption is necessary in previous methods, that is, the load swing angle δ is constrained as $\delta \in (-\frac{\pi}{2}, \frac{\pi}{2})$. Inspired by the prescribed performance control, a new control algorithm is proposed to limit the range of load swing angle instead of any assumption conditions in [18]. However, a problem is that the designed method is relatively complex since the prescribed performance function is used to restrain the tangent value of the angle rather than angle itself.

Based on the above analysis, an NN-based adaptive tracking control method of a nonlinear crane with nonstrict-feedback structure is proposed under the framework of ILC. Firstly, the load swing angle range is directly constrained by using a prescribed performance function. Then, a series of auxiliary variables are designed to simplify the system model, and RBF NNs are employed to deal with the nonstrict-feedback problem. Further, an adaptive ILC strategy is presented to solve the problem of unknown control direction. Finally, the convergence analysis is shown by using the composite energy function. The main contributions of this paper are given as follows:

(1) Compared with [18], a prescribed performance function is used to directly constraint the swing angle rather than its function. Hence, the designed method has a less conservatism since the assumption condition for the swing angle is removed. Besides, the algorithm complexity is reduced since no additional variables need to be introduced and calculated.
(2) RBF NN is used to approximate the unknown nonlinear function and address the nonstrict-feedback problem. It is noted that the traditional NN-based adaptive control only estimates a constant, however, the proposed adaptive learning control method is able to estimate directly the time-varying weight vector, which broadens the application scope of the algorithm.
(3) A projection operator is introduced to guarantee that the convergence bound of the error does not rely on the iteration k, which is viewed as an expand of [19]. Furthermore, an assumption that the initial value of estimation algorithm at every iteration is set to zero [19] is removed in this paper.

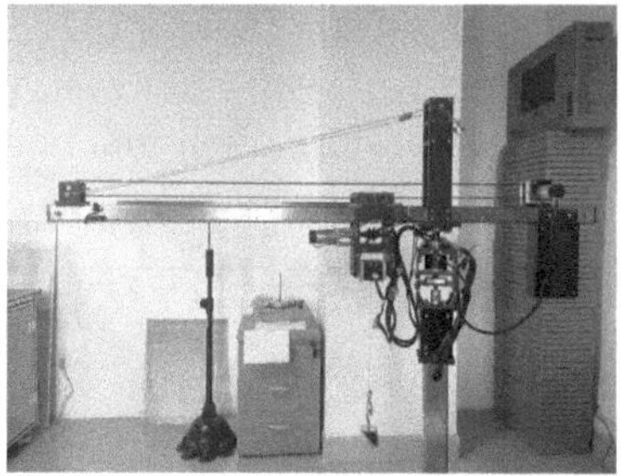

Fig. 1. Overhead crane system.

2 Problem Formulation and Preliminaries

The model of nonlinear crane (see Fig. 1) is established by

$$(m_c + m_l)\ddot{p}_k + m_l l\ddot{\delta}_k \cos\delta_k - m_l l\dot{\delta}_k^2 \sin\delta_k = u_k + f_k + d_k$$
$$m_l l \cos\delta_k \ddot{p}_k + m_l l^2 \ddot{\delta}_k + m_l g l \sin\delta_k = 0 \tag{1}$$

where $k \in \mathcal{N}$ is the index of iterations. p_k and δ_k are position of the trolley and the swing angle of the rope, respectively. u_k is the control input. d_k is the unknown external disturbance. The system runs repeatedly on $[0, T]$. Other parameters can be found in [18], and

$$f_k = -r_1 \tanh\left(\frac{\dot{p}_k}{\nu}\right) + r_2 \dot{p}_k |\dot{p}_k| \tag{2}$$

where r_1, r_2 and ν are friction parameters.

Let $\zeta_{1,k} = p_k$, $\zeta_{2,k} = \dot{p}_k$, $\zeta_{3,k} = \delta_k$ and $\zeta_{4,k} = \dot{\delta}_k$, the original system can be rewritten as

$$\begin{aligned}
\dot{\zeta}_{1,k} &= \zeta_{2,k} \\
\dot{\zeta}_{2,k} &= -g\tan\zeta_{3,k} - \frac{l}{\cos\zeta_{3,k}}\dot{\zeta}_{4,k} \\
\dot{\zeta}_{3,k} &= \zeta_{4,k} \\
\dot{\zeta}_{4,k} &= \varsigma_k u_k + \hbar_k + \upsilon_k
\end{aligned} \tag{3}$$

where
$\varsigma_k = -\frac{\cos\zeta_{3,k}}{(m_c+m_l\sin^2\zeta_{3,k})l}$, $\hbar_k = -\frac{m_l\zeta_{4,k}^2 \sin\zeta_{3,k}\cos\zeta_{3,k}}{m_c+m_l\sin^2\zeta_{3,k}} - \frac{(m_c+m_l)g\sin\zeta_{3,k} - f_k\cos\zeta_{3,k}}{(m_c+m_l\sin^2\zeta_{3,k})l}$, $\upsilon_k = \frac{-\cos\zeta_{3,k}d_k}{(m_c+m_l\sin^2\zeta_{3,k})l}$.

To limit the swing angle $\zeta_{3,k}$ of the load, a prescribed performance function $\eta(t)$ is designed as

$$\eta(t) = (\eta_0 - \eta_\infty)e^{-at} + \eta_\infty \tag{4}$$

where η_0, η_∞ and a are positive constants, and $0 < \eta_\infty < \eta_0$.

Our goal is to achieve $-\eta(t) < \zeta_{3,k} < \eta(t)$, however, the inequality constraint is hard to be satisfied. Hence, a transformation function is introduced, that is

$$H(\epsilon_{3,k}) = \frac{e^{\epsilon_{3,k}} - e^{-\epsilon_{3,k}}}{e^{\epsilon_{3,k}} + e^{-\epsilon_{3,k}}} \tag{5}$$

where $\epsilon_{3,k}$ is a new transition variable.

It follows from (4) and (5) that the inequality constraint can be achieved by

$$\zeta_{3,k} = \eta(t) H(\epsilon_{3,k}) \tag{6}$$

which indicates

$$\epsilon_{3,k} = \frac{1}{2} \ln \left[1 + \frac{\zeta_{3,k}}{\eta(t)}\right] - \frac{1}{2} \ln \left[1 - \frac{\zeta_{3,k}}{\eta(t)}\right] \tag{7}$$

Differentiating $\epsilon_{3,k}$ with respect to time yields

$$\dot{\epsilon}_{3,k} = \frac{\dot{\zeta}_{3,k} - \dot{\eta}(t) \left[H(\epsilon_{3,k})\right]}{\eta(t) \left[\partial H(\epsilon_{3,k}) / \partial \epsilon_{3,k}\right]} \tag{8}$$

By virtue of (3) and (8), we have

$$\begin{aligned}
\dot{\zeta}_{1,k} &= \zeta_{2,k} \\
\dot{\zeta}_{2,k} &= -g \tan\left[\eta(t) H(\epsilon_{3,k})\right] - \frac{l}{\cos\left[\eta(t) H(\epsilon_{3,k})\right]} \dot{\zeta}_{4,k} \\
\dot{\epsilon}_{3,k} &= \frac{\zeta_{4,k} - \dot{\eta}(t) \left[H(\epsilon_{3,k})\right]}{\eta(t) \left[\partial H(\epsilon_{3,k}) / \partial \epsilon_{3,k}\right]} \\
\dot{\zeta}_{4,k} &= \varsigma_k u_k + \hbar_k + \upsilon_k
\end{aligned} \tag{9}$$

In the following, a series of auxiliary signals are defined as

$$\begin{aligned}
x_{1,k} &= \zeta_{1,k} + l \ln\left\{\sec\left[\eta(t) H(\epsilon_{3,k})\right] + \tan\left[\eta(t) H(\epsilon_{3,k})\right]\right\} \\
x_{2,k} &= \dot{x}_{1,k} \\
x_{3,k} &= -g \tan\left[\eta(t) H(\epsilon_{3,k})\right] \\
x_{4,k} &= -\frac{g}{\cos^2\left[\eta(t) H(\epsilon_{3,k})\right]} \zeta_{4,k}
\end{aligned} \tag{10}$$

and computing its derivative has

$$\begin{aligned}
\dot{x}_{1,k} &= x_{2,k} \\
\dot{x}_{2,k} &= x_{3,k} + \varrho_k \\
\dot{x}_{3,k} &= x_{4,k} \\
\dot{x}_{4,k} &= \gamma_k u_k + \rho_k + \psi_k + \vartheta_k
\end{aligned} \tag{11}$$

where $\gamma_k = -\frac{g\varsigma_k}{\cos^2\zeta_{3,k}}$, $\rho_k = \frac{m_l g\zeta_{4,k}^2 \sin\zeta_{3,k}}{(m_c+m_l\sin^2\zeta_{3,k})\cos\zeta_{3,k}} + \frac{(m_c+m_l)g^2\sin\zeta_{3,k}}{(m_c+m_l\sin^2\zeta_{3,k})l\cos^2\zeta_{3,k}} - \frac{r_1\tanh(\zeta_{2,k}/\nu)g+r_2 g\zeta_{2,k}|\zeta_{2,k}|}{(m_c+m_l\sin^2\zeta_{3,k})l\cos\zeta_{3,k}}$, $\psi_k = \frac{-2g\tan\zeta_{3,k}}{\cos^2\zeta_{3,k}}\zeta_{4,k}^2$, $\vartheta_k = \frac{-g\upsilon_k}{\cos^2\zeta_{3,k}}$, $\varrho_k = -\frac{lx_{3,k}x_{4,k}^2}{(g^2+x_{3,k}^2)^{3/2}}$. It follows from the form of γ_k that $0 < \underline{\gamma} \leq \gamma_k \leq \overline{\gamma} < \infty$ with $\underline{\gamma}$ and $\overline{\gamma}$ being the upper and low bound. It is assumed that the initial value of each iteration is equal to the desired trajectory.

Thus far, the objective of this paper is to design an NN-based adaptive learning controller for the crane system such that the system output to completely track the desired trajectory x_d as the iteration k goes to infinity, and effectively suppress the swing of the load.

To obtain this goal, a lemma is required, that is

Lemma 1. *If the projection operator is defined as*

$$\mathcal{P}(\chi) = \begin{cases} \bar{\chi}, \chi > \bar{\chi} \\ \chi, \underline{\chi} \leq \chi \leq \bar{\chi} \\ \underline{\chi}, \chi < \underline{\chi} \end{cases} \tag{12}$$

where χ, $\underline{\chi}$, $\overline{\chi} \in \mathbb{R}$. *Then,* $[\xi - \mathcal{P}(\chi)][\chi - \mathcal{P}(\chi)] \leq 0$ *with* $\xi \in \mathbb{R}$, $\underline{\chi} \leq \xi \leq \overline{\chi}$.

3 Main Results

The controller design and the convergence analysis are given in this section.

3.1 Controller Design

Define the following coordinate transformation

$$\begin{aligned} z_{1,k} &= x_{1,k} - x_d \\ z_{2,k} &= x_{2,k} - \alpha_{1,k} \\ z_{3,k} &= x_{3,k} - \alpha_{2,k} \\ z_{4,k} &= x_{4,k} - \alpha_{3,k} \end{aligned} \tag{13}$$

where $\alpha_{1,k}$, $\alpha_{2,k}$ and $\alpha_{3,k}$ are virtual control signals.

Step 1: Calculating $\dot{z}_{1,k}$ along with (11) and (13) as

$$\dot{z}_{1,k} = z_{2,k} + \alpha_{1,k} - \dot{x}_d \tag{14}$$

Define the following Lyapunov function

$$V_{1,k} = \frac{1}{2}z_{1,k}^2$$

and its derivative is

$$\dot{V}_{1,k} = z_{1,k}\dot{z}_{1,k} = z_{1,k}(z_{2,k} + \alpha_{1,k} - \dot{x}_d) \tag{15}$$

Design the virtual controller $\alpha_{1,k}$ as

$$\alpha_{1,k} = -c_1 z_{1,k} + \dot{x}_d \tag{16}$$

where $c_1 > 0$ is design parameter.

On the basis of (15) and (16), we have

$$\dot{V}_{1,k} = -c_1 z_{1,k}^2 + z_{1,k} z_{2,k} \tag{17}$$

Step 2: Calculating the time derivative of $z_{2,k}$ yields

$$\dot{z}_{2,k} = x_{3,k} + \varrho_k - \dot{\alpha}_{1,k} = z_{3,k} + \alpha_{2,k} + \bar{\hbar}_{2,k} \tag{18}$$

where $\bar{\hbar}_{2,k} = \varrho_k - \dot{\alpha}_{1,k}$.

Based on [17], an RBF NN $W_{2,k}^{\mathrm{T}} \phi_2(\chi_{2,k})$ with $\chi_{2,k} = [\dot{x}_d, \ddot{x}_d, x_{2,k}, x_{3,k}, x_{4,k}]^{\mathrm{T}}$ is used to address unknown function $\bar{\hbar}_{2,k}$, that is

$$\bar{\hbar}_{2,k} = W_{2,k}^{\mathrm{T}} \phi_2(\chi_{2,k}) + \varepsilon_2(\chi_{2,k}) \tag{19}$$

where $W_{2,k} = [w_{21,k}, ..., w_{2\,m,k}]^{\mathrm{T}}$, $\phi_2(\chi_{2,k}) = [\upsilon_1(\chi_{2,k}), ..., \upsilon_m(\chi_{2,k})]^{\mathrm{T}}$ with $m > 1$ and $\upsilon_j(\chi_{2,k})$ are the weigh vector, basis function vector and Gaussian function, respectively. $\varepsilon_2(\chi_{2,k})$ denotes the approximation error.

Furthermore, by Young's inequality, we have

$$\begin{aligned} \bar{\hbar}_{2,k} \leq & \frac{1}{4\sigma} \|W_{2,k}\|^2 \phi_2^{\mathrm{T}}(\chi_{2,k}) \phi_2(\chi_{2,k}) + \varepsilon_2(\chi_{2,k}) + \sigma \\ \leq & \frac{\|W_{2,k}\|^2}{4\sigma \phi_2^{\mathrm{T}}(\bar{\chi}_{2,k}) \phi_2(\bar{\chi}_{2,k})} + \varepsilon_2(\chi_{2,k}) + \sigma \\ \leq & \theta_{2,k} \Gamma_2(\bar{\chi}_{2,k}) \end{aligned} \tag{20}$$

where $\theta_{2,k} = \left\{ \frac{\|W_{2,k}\|^2}{4\sigma}, \|\varepsilon_2(\chi_{2,k}) + \sigma\| \right\}$ is unknown. σ is a positive constant. $\Gamma_2(\bar{\chi}_{2,k}) = 1 + \frac{1}{\phi_2^{\mathrm{T}}(\bar{\chi}_{2,k}) \phi_2(\bar{\chi}_{2,k})}$. $\bar{\chi}_{2,k} = [\dot{x}_d, \ddot{x}_d, x_{1,k}, x_{2,k}]^{\mathrm{T}}$.

Consider the following Lyapunov function

$$V_{2,k} = V_{1,k} + \frac{1}{2} z_{2,k}^2 \tag{21}$$

Differentiating $V_{2,k}$ with respect to time produces

$$\dot{V}_{2,k} = -c_1 z_{1,k}^2 + z_{1,k} z_{2,k} + z_{2,k} \dot{z}_{2,k} \tag{22}$$

By utilizing (18) and (20), (22) is rewritten as

$$\dot{V}_{2,k} \leq -c_1 z_{1,k}^2 + z_{2,k} \left[z_{1,k} + z_{3,k} + \alpha_{2,k} + \mathrm{sign}(z_{2,k}) \theta_{2,k} \Gamma_2(\bar{\chi}_{2,k}) \right] \tag{23}$$

An appropriate virtual controller $\alpha_{2,k}$ is selected as

$$\alpha_{2,k} = -c_2 z_{2,k} - z_{1,k} - \mathrm{sign}(z_{2,k}) \hat{\theta}_{2,k} \Gamma_2(\bar{\chi}_{2,k}) \tag{24}$$

where $c_2 > 0$ is design parameter. $\hat{\theta}_{2,k}$ denotes the estimation of $\theta_{2,k}$, and the estimate error is define as $\tilde{\theta}_{2,k} = \hat{\theta}_{2,k} - \theta_{2,k}$.

According to (24), one has

$$\dot{V}_{2,k} \leq -c_1 z_{1,k}^2 - c_2 z_{2,k}^2 - \tilde{\theta}_{2,k}|z_{2,k}|\Gamma_2(\bar{\chi}_{2,k}) + z_{2,k}z_{3,k} \tag{25}$$

Step 3: From (11) and (13), the time derivative of $z_{3,k}$ is

$$\dot{z}_{3,k} = z_{4,k} + \alpha_{3,k} + \bar{\hbar}_{3,k} \tag{26}$$

where $\bar{\hbar}_{3,k} = -\dot{\alpha}_{2,k}$.

Similar to Step 2, by using the RBF NN to estimate $\bar{\hbar}_{3,k}$, we have

$$\bar{\hbar}_{3,k} = W_{3,k}^{\mathrm{T}}\phi_3(\chi_{3,k}) + \varepsilon_3(\chi_{3,k}) \leq \theta_{3,k}\Gamma_3(\bar{\chi}_{3,k}) \tag{27}$$

where $\theta_{3,k} = \{\frac{1}{4\sigma}\|W_{3,k}\|^2, \|\varepsilon_3(\chi_{3,k}) + \sigma\|\}$ is unkown. $\Gamma_3(\bar{\chi}_{3,k}) = 1 + 1/[\phi_3^{\mathrm{T}}(\bar{\chi}_{3,k})\phi_3(\bar{\chi}_{3,k})]$. $\bar{\chi}_{3,k} = [\dot{x}_d, \ddot{x}_d, x_{1,k}, x_{2,k}, x_{3,k}]^{\mathrm{T}}$.

Select a Lyapunov function as

$$V_{3,k} = V_{2,k} + \frac{1}{2}z_{3,k}^2 \tag{28}$$

Differentiating $V_{3,k}$ with respect to time yields

$$\begin{aligned}\dot{V}_{3,k} \leq & -\sum_{i=1}^{2} c_i z_{i,k}^2 - \tilde{\theta}_{2,k}|z_{2,k}|\Gamma_2(\bar{\chi}_{2,k}) + z_{3,k}\,[z_{2,k} + z_{4,k} \\ & + \alpha_{3,k} + \mathrm{sign}(z_{3,k})\theta_{3,k}\Gamma_3(\bar{\chi}_{3,k})]\end{aligned} \tag{29}$$

The virtual controller $\alpha_{3,k}$ is designed as

$$\alpha_{3,k} = -c_3 z_{3,k} - z_{2,k} - \mathrm{sign}(z_{3,k})\hat{\theta}_{3,k}\Gamma_3(\bar{\chi}_{3,k}) \tag{30}$$

where $c_3 > 0$ is design parameter.

According to (30), (29) can be rewritten by

$$\dot{V}_{3,k} \leq -\sum_{i=1}^{3} c_i z_{i,k}^2 - \sum_{i=2}^{3}\tilde{\theta}_{i,k}|z_{i,k}|\Gamma_i(\bar{\chi}_{i,k}) + z_{3,k}z_{4,k} \tag{31}$$

Step 4: Computing the time derivative of $z_{4,k}$ as

$$\dot{z}_{4,k} = \gamma_k u_k + \psi_k + \bar{\hbar}_{4,k} \tag{32}$$

where $\bar{\hbar}_{4,k} = \rho_k - \dot{\alpha}_{3,k} + \vartheta_k$.

By using the RBF NN to estimate $\bar{\hbar}_{4,k}$, we have

$$\bar{\hbar}_{4,k} = W_{4,k}^{\mathrm{T}}\phi_4(\chi_{4,k}) + \varepsilon_4(\chi_{4,k}) \leq \theta_{4,k}\Gamma_4(\bar{\chi}_{4,k}) \tag{33}$$

where $\theta_{4,k} = \{\frac{1}{4\sigma}\|W_{4,k}\|^2, \|\varepsilon_4(\chi_{4,k}) + \sigma\|\}$ is unknown. $\Gamma_4(\bar{\chi}_{4,k}) = 1 + 1/[\phi_4^{\mathrm{T}}(\bar{\chi}_{4,k})\phi_4(\bar{\chi}_{4,k})]$. $\bar{\chi}_{4,k} = [\dot{x}_d, \ddot{x}_d, x_d^{(3)}, x_{1,k}, x_{2,k}, x_{3,k}, x_{4,k}]^{\mathrm{T}}$.

Design a Lyapunov function as

$$V_{4,k} = V_{3,k} + \frac{1}{2} z_{4,k}^2 \tag{34}$$

The derivative of $V_{4,k}$ is computed as

$$\begin{aligned} V_{4,k} \leq & - \sum_{i=1}^{3} c_i z_{i,k}^2 - \sum_{i=2}^{3} \tilde{\theta}_{i,k} |z_{i,k}| \Gamma_i(\bar{\chi}_{i,k}) + z_{4,k}(\gamma_k u_k + \psi_k + z_{3,k} - \bar{u}_k) \\ & + z_{4,k} \left[\bar{u}_k - \text{sign}(z_{4,k}) \theta_{4,k} \Gamma_4(\bar{\chi}_{4,k}) \right] \end{aligned} \tag{35}$$

The control input $\bar{u}_k$ is designed as

$$\bar{u}_k = -c_4 z_{4,k} - \text{sign}(z_{4,k}) \hat{\theta}_{4,k} \Gamma_4(\bar{\chi}_{4,k}) \tag{36}$$

where $c_4 > 0$ is design parameter.

According to (36), we obtain

$$V_{4,k} \leq - \sum_{i=1}^{4} c_i z_{i,k}^2 - \sum_{i=2}^{4} \tilde{\theta}_{i,k} |z_{i,k}| \Gamma_i(\bar{\chi}_{i,k}) + z_{4,k}(\gamma_k u_k + \psi_k + z_{3,k} - \bar{u}_k) \tag{37}$$

In the following, we will establish the relationship between u_k and $\bar{u}_k$ by the third term of (37). First, it is well known that $z_{4,k}(\psi_k + z_{3,k} - \bar{u}_k - \text{sign}(z_{4,k}) \| \psi_k + z_{3,k} - \bar{u}_k \|) \leq 0$. Hence, we have

$$u_k = -\frac{1}{\gamma_k} \text{sign}(z_{4,k}) \| \psi_k + z_{3,k} - \bar{u}_k \| \tag{38}$$

It is noted that γ_k is unknown, hence, the control input is further designed as

$$u_k = -\hat{\varphi}_k \text{sign}(z_{4,k}) \| \psi_k + z_{3,k} - \bar{u}_k \| \tag{39}$$

where $\hat{\varphi}_k$ is the estimation of φ, and $\hat{\varphi}_k \geq 0$. Define $\tilde{\varphi}_k = \hat{\varphi}_k - \varphi$ as the estimation error. φ is an upper bound on $1/\gamma_k$.

According to (39), the third item in (37) can be transformed into

$$\begin{aligned} & z_{4,k}(\gamma_k u_k + \psi_k + z_{3,k} - \bar{u}_k) \\ & = z_{4,k} [-\text{sign}(z_{4,k}) \gamma_k \hat{\varphi}_k \| \psi_k + z_{3,k} - \bar{u}_k \| + \psi_k + z_{3,k} - \bar{u}_k] \end{aligned} \tag{40}$$

From $\hat{\varphi} \geq 0$, we can get

$$- |z_{4,k}| \hat{\varphi}_k \| \psi_k + z_{3,k} - \bar{u}_k \| \leq 0 \tag{41}$$

As a result, (40) can be transformed into

$$\begin{aligned} & z_{4,k} [-\text{sign}(z_{4,k}) \gamma_k \hat{\varphi}_k \| \psi_k + z_{3,k} - \bar{u}_k \| + \psi_k + z_{3,k} - \bar{u}_k] \\ & \leq z_{4,k} [-\text{sign}(z_{4,k}) \varphi^{-1} (\tilde{\varphi}_k + \varphi) \| \psi_k + z_{3,k} - \bar{u}_k \| + \psi_k + z_{3,k} - \bar{u}_k] \\ & = -|z_{4,k}| \varphi^{-1} \tilde{\varphi}_k \| \psi_k + z_{3,k} - \bar{u}_k \| + z_{4,k} [\psi_k + z_{3,k} \\ & \quad - \bar{u}_k - \text{sign}(z_{4,k}) \| \psi_k + z_{3,k} - \bar{u}_k \|] \end{aligned} \tag{42}$$

According to (42), we obtain

$$V_{4,k} \leq - \sum_{i=1}^{4} c_i z_{i,k}^2 - \sum_{i=2}^{4} \tilde{\theta}_{i,k} |z_{i,k}| \Gamma_i(\bar{\chi}_{i,k}) - |z_{4,k}| \varphi^{-1} \tilde{\varphi}_k \| \psi_k + z_{3,k} - \bar{u}_k \| \tag{43}$$

3.2 Convergence Analysis

The main result of this paper can be summarized as Theorem 1.

Theorem 1. *For the system* (1), *if the virtual control signals, actual control input and adaptive laws are designed as* (16), (24), (30), (39), (48) *and* (50), *respectively. Then, the proposed algorithm can guarantee that (i)* $z_{i,k}$ $(i = 1, 2, 3, 4)$ *approaches to zero as* k *goes to the infinity, and the swing angle* δ_k *satisfies the prescribed performance; (ii) All signals of the closed-loop system are bounded.*

Proof. A composite energy function is constructed as

$$E_k = V_{4,k} + \sum_{i=2}^{4} \frac{1}{2\iota_i} \int_0^t \left(\tilde{\theta}_{i,k}\right)^2 d\tau + \frac{1}{2\lambda\varphi} \int_0^t (\tilde{\varphi}_k)^2 d\tau \tag{44}$$

Define $\Delta E_k = E_k - E_{k-1}$. Based on (44), one has

$$\begin{aligned}\Delta E_k =& V_{4,k} - V_{4,k-1} + \sum_{i=2}^{4} \frac{1}{2\iota_i} \int_0^t \left(\tilde{\theta}_{i,k}\right)^2 - \left(\tilde{\theta}_{i,k-1}\right)^2 d\tau \\ &+ \frac{1}{2\lambda\varphi} \int_0^t (\tilde{\varphi}_k)^2 - (\tilde{\varphi}_{k-1})^2 d\tau\end{aligned} \tag{45}$$

Due to $V_{4,k-1} \geq 0$ and $V_{4,k}(0) = 0$, ΔE_k is expressed as

$$\begin{aligned}\Delta E_k =& \int_0^t \dot{V}_{4,k} d\tau + \sum_{i=2}^{4} \frac{1}{2\iota_i} \int_0^t \left(\tilde{\theta}_{i,k}\right)^2 - \left(\tilde{\theta}_{i,k-1}\right)^2 d\tau \\ &+ \frac{1}{2\lambda\varphi} \int_0^t (\tilde{\varphi}_k)^2 - (\tilde{\varphi}_{k-1})^2 d\tau\end{aligned} \tag{46}$$

The item $\frac{1}{2\iota_i}\left[\left(\tilde{\theta}_{i,k}\right)^2 - \left(\tilde{\theta}_{i,k-1}\right)^2\right]$ can be addressed by

$$\frac{1}{2\iota_i}\left[\left(\tilde{\theta}_{i,k}\right)^2 - \left(\tilde{\theta}_{i,k-1}\right)^2\right] \leq -\frac{1}{\iota_i}\tilde{\theta}_{i,k}\left(\hat{\theta}_{i,k-1} - \hat{\theta}_{i,k}\right)$$

and taking $\tilde{\theta}_{i,k}|z_{i,k}|\Gamma_i(\bar{\chi}_{i,k})$ into consideration gets

$$\begin{aligned}&\frac{1}{2\iota_i}\left[\left(\tilde{\theta}_{i,k}\right)^2 - \left(\tilde{\theta}_{i,k-1}\right)^2\right] \\ &\leq -\frac{1}{\iota_i}\tilde{\theta}_{i,k}\left(\hat{\theta}_{i,k-1} + \iota_i|z_{i,k}|\Gamma_i(\bar{\chi}_{i,k}) - \hat{\theta}_{i,k}\right) + \tilde{\theta}_{i,k}|z_{i,k}|\Gamma_i(\bar{\chi}_{i,k})\end{aligned} \tag{47}$$

Combining with Lemma 1, the learning law is designed as

$$\hat{\theta}_{i,k} = \mathcal{P}\left(\hat{\theta}_{i,k-1} + \iota_i|z_{i,k}|\Gamma_i(\bar{\chi}_{i,k})\right), \hat{\theta}_{i,k}(0) = 0 \tag{48}$$

and

$$\left(\theta_{i,k}-\hat{\theta}_{i,k}\right)\left(\hat{\theta}_{i,k-1}+\iota_i|z_{i,k}|\Gamma_i(\bar{\chi}_{i,k})-\hat{\theta}_{i,k}\right)\leq 0$$

Substituting (48) into (47) has

$$\frac{1}{2\iota_i}\left[\left(\tilde{\theta}_{i,k}\right)^2-\left(\tilde{\theta}_{i,k-1}\right)^2\right]\leq\tilde{\theta}_{i,k}|z_{i,k}|\Gamma_i(\bar{\chi}_{i,k}) \tag{49}$$

which further results in

$$\sum_{i=2}^{4}\frac{1}{2\iota_i}\int_0^t\left(\tilde{\theta}_{i,k}\right)^2-\left(\tilde{\theta}_{i,k-1}\right)^2d\tau\leq\sum_{i=2}^{4}\int_0^t\tilde{\theta}_{i,k}|z_{i,k}|\Gamma_i(\bar{\chi}_{i,k})d\tau$$

Similar to (47)–(49), we obtain

$$\hat{\varphi}_k=\mathcal{P}\left(\hat{\varphi}_{k-1}+\lambda\left|z_{4,k}\right|\left\|\psi_k+z_{3,k}-\bar{u}_k\right\|\right),\hat{\varphi}_k(0)=0 \tag{50}$$

and

$$\frac{1}{2\lambda\varphi}\int_0^t\left(\tilde{\varphi}_k\right)^2-\left(\tilde{\varphi}_{k-1}\right)^2d\tau\leq\int_0^t|z_{4,k}|\varphi^{-1}\tilde{\varphi}_k\|\psi_k+z_{3,k}-\bar{u}_k\|d\tau \tag{51}$$

Based on (48) and (50), it is can be deduced that

$$\Delta E_k\leq-\int_0^t\sum_{i=1}^{4}c_iz_{i,k}^2d\tau\leq 0 \tag{52}$$

which means that E_i^k will not increase along the iteration axis.

Next, the convergence of $z_{1,k}$ will be given. Specifically, when $k=1$, $\dot{E}_1$ satisfies

$$\begin{aligned}\dot{E}_1\leq&-\sum_{i=1}^{4}c_iz_{i,1}^2-\sum_{i=2}^{4}\tilde{\theta}_{i,1}|z_{i,1}|\Gamma_i(\bar{\chi}_{i,1})-|z_{4,1}|\varphi^{-1}\tilde{\varphi}_1\|\psi_1\\&+z_{3,1}-\bar{u}_1\|+\sum_{i=2}^{4}\frac{1}{2\iota_i}\left(\tilde{\theta}_{i,1}\right)^2+\frac{1}{2\lambda\varphi}(\tilde{\varphi}_1)^2\end{aligned} \tag{53}$$

Based on (49) and the following equation

$$\frac{1}{2\iota_i}\left(\tilde{\theta}_{i,1}\right)^2=\frac{1}{2\iota_i}\left[\left(\tilde{\theta}_{i,1}\right)^2-\left(\tilde{\theta}_{i,0}\right)^2\right]+\frac{1}{2\iota_i}(\theta_{i,1})^2 \tag{54}$$

it has

$$-\tilde{\theta}_{i,1}|z_{i,1}|\Gamma_i(\bar{\chi}_{i,1})+\frac{1}{2\iota_i}\left(\tilde{\theta}_{i,1}\right)^2\leq\frac{1}{2\iota_i}(\theta_{i,1})^2 \tag{55}$$

Further, we have

$$- |z_{4,1}|\varphi^{-1}\tilde{\varphi}_1\|\psi_1 + z_{3,1} - \bar{u}_1\| + \frac{1}{2\lambda\varphi}(\tilde{\varphi}_1)^2 \leq \frac{(\varphi_1)^2}{2\lambda\varphi} \tag{56}$$

Then, (53) is changed as

$$\dot{E}_1 \leq \sum_{i=2}^{4} \frac{1}{2\iota_i}(\theta_{i,1})^2 + \frac{1}{2\lambda\varphi}(\varphi_1)^2 \tag{57}$$

In view of the finite operation time and $E_1(0) = 0$, the boundedness of E_1 can be guaranteed. Then, we have

$$E_k = E_1 + \sum_{j=2}^{k} \Delta E_j \leq E_1 - \sum_{j=2}^{k}\sum_{i=1}^{4}\int_0^t \sum_{i=1}^{4} c_i z_{i,j}^2 d\tau \tag{58}$$

which shows that E_k is bounded.

Then, we have

$$\lim_{k\to\infty} \sum_{j=2}^{k}\sum_{i=1}^{4}\int_0^t \sum_{i=1}^{4} c_i z_{i,j}^2 d\tau \leq E_1 - \lim_{k\to\infty} E_k < \infty \tag{59}$$

The convergence of $z_{i,k}$ can be achieved by

$$\lim_{k\to\infty} \sum_{i=1}^{4} \int_0^t c_i z_{i,k}^2 d\tau = 0 \tag{60}$$

i.e., $\lim\limits_{k\to\infty} z_{i,k} = 0$.

From the above analysis, the conclusion (i) of Theorem 1 is gained.

It follows from (16), (24) and (30) that $\alpha_{1,k}$, $\alpha_{2,k}$ and $\alpha_{3,k}$ are bounded, which from (13) further implies that $x_{1,k}$, $x_{2,k}$, $x_{3,k}$ and $x_{4,k}$ are bounded. From (10) we know $\zeta_{1,k}$, $\zeta_{2,k}$, $\epsilon_{3,k}$ and $\zeta_{4,k}$ are bounded. Then, according to (6), we can deduce that $\zeta_{3,k}$ is also bounded. Therefore all signals in the closed-loop system are bounded. Hence, the conclusion (ii) of Theorem 1 is achieved. By the above analysis, the proof of Theorem 1 is completed. □

4 Simulation Results

The model (1), the physical parameters are set as $m_c = 1.852$ kg, $m_l = 0.5$ kg, $l = 0.5$ m, $g = 9.8$ m/s^2, $r_1 = 0.5$, $r_2 = -2$ and $\nu = 0.5$. The running time is $T = 40$ s.

The design parameters are defined as $c_1 = 0.2$, $c_2 = 1$, $c_3 = 2$, $c_4 = 2$, $\iota_2 = 0.1$, $\iota_3 = 2$, $\iota_4 = 0.5$, $\lambda = 1$, $\eta_0 = \frac{\pi}{2}$, $\eta_\infty = 0.05$ and $a = 0.5$. The uncertain disturbance is selected as $d_k(t) = 0.1\sin(k\pi/10)e^{-0.02t}$.

The initial conditions are chosen as: $\zeta_{1,k}(0) = \zeta_{2,k}(0) = \zeta_{3,k}(0) = \zeta_{4,k}(0) = 0$.

The tracking trajectory is set as

$$x_d(t) = \begin{cases} t & ,t \in [0, 5\text{s}) \\ 5 & ,t \in [5\text{s}, +\infty) \end{cases} \tag{61}$$

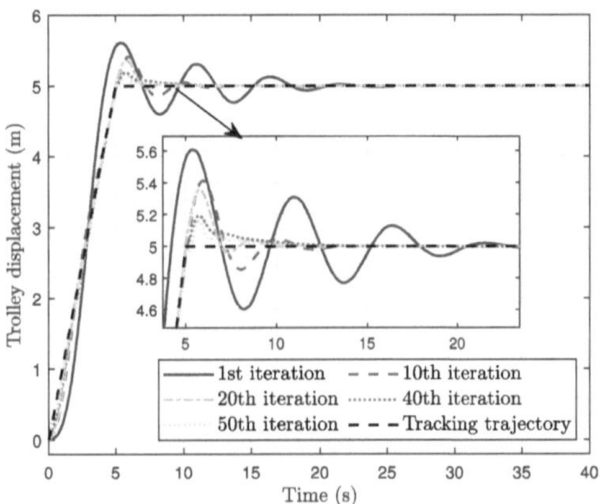

Fig. 2. Curves of the trolley displacement p_k at different iterations.

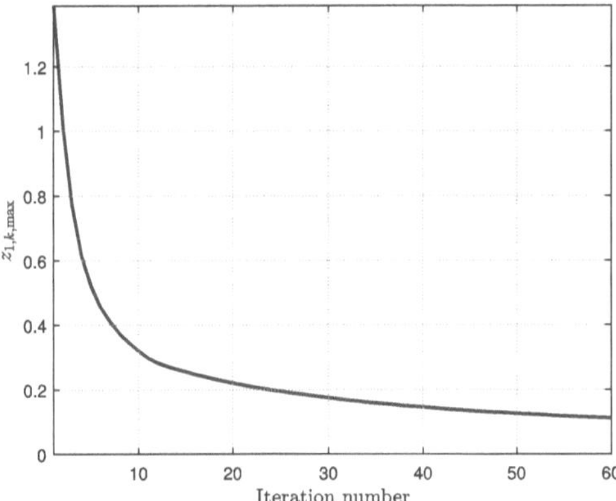

Fig. 3. Curve of $z_{1,k,\max}$ at different iterations.

The simulation results of the developed control scheme are captured in Figs. 2, 3 and 4. Figure 2 illustrates that as the number of iterations increases, the output of the trolley gradually tracks the desired trajectory. Define the maximum absolute error of the trolley displacement as $z_{1,k,\max} = \max_{t\in[0,40]} |z_{1,k}(t)|$. Furthermore, the curve of $z_{1,k,\max}$ is shown in Fig. 3, from which we can see that

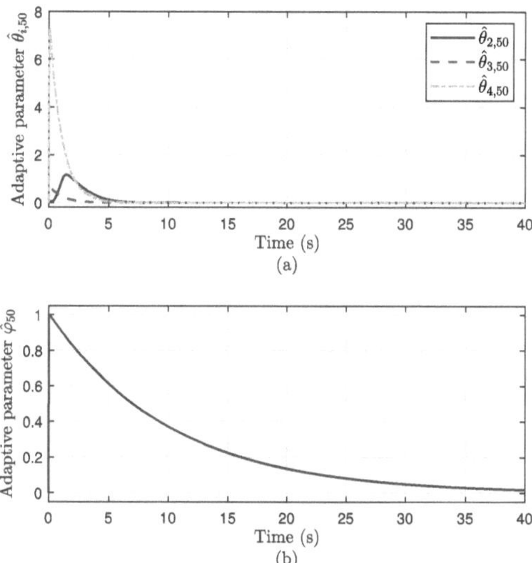

Fig. 4. Curves of $\hat{\theta}_{i,50}$ and $\hat{\varphi}_{50}$ at 50th iteration.

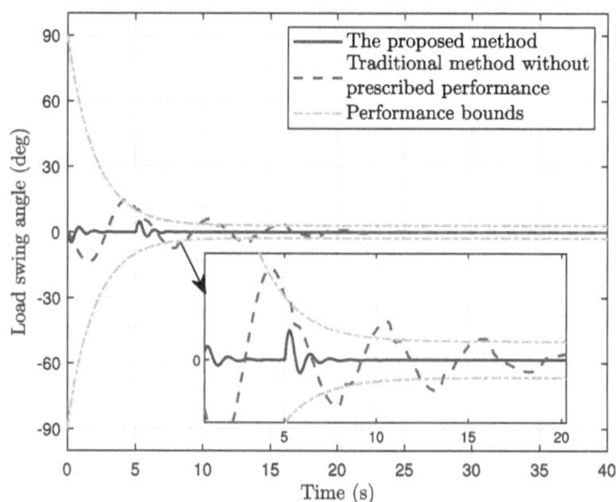

Fig. 5. Curves of the load swing angle δ_k at 50th iteration.

the proposed control method can reduce the tracking error and improve the system performance when the number of iterations increases. The estimates of the unknown parameters $\hat{\theta}_{i,50}(i = 2, 3, 4)$ and $\hat{\varphi}_{50}$ at the 50th iteration are shown in Fig. 4 (a) and Fig. 4 (b), respectively.

To verify the superiority of the designed method, a comparison result with the same simulation condition is shown in Fig. 5 where we can observer that the proposed control strategy not only keeps the angle δ_k always within the desired range, but also makes it have a better steady-state performance by comparing with the traditional method without prescribed performance control.

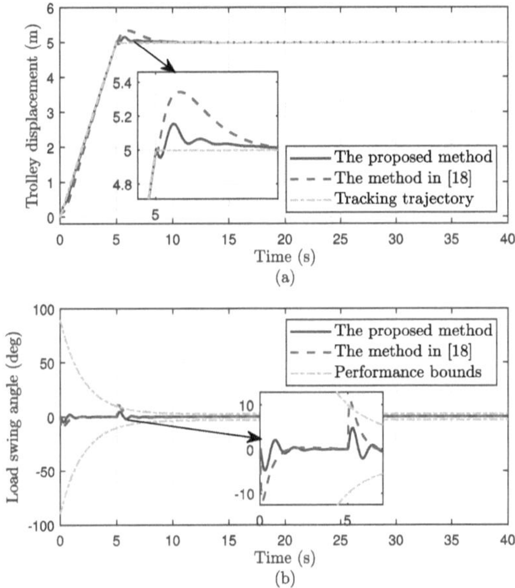

Fig. 6. Simulation comparison results between the proposed method and [18].

Another comparison is made between the designed method and [18], and the result is shown in Fig. 6. From Fig. 6 (a), it can be seen that the designed scheme not only has a faster convergence speed, but also has a smaller overshoot. Moreover, it can be observed from Fig. 6 (b) that the proposed method can better suppress the load swing.

5 Conclusion

An adaptive ILC method is proposed for nonlinear crane system with nonstrict-feedback based on RBF NN and backstepping technique. By using the prescribed performance function, an assumption on the load swing angle existed in the traditional backstepping is eliminated. The auxiliary variables are introduced to simplify the system model, and the RBF NN is utilized to handle the nonstrict-feedback problem. Furthermore, the convergence analysis is given with the help of composite energy function and projection operator. Finally, the effectiveness of the proposed scheme is verified through simulation experiments. Inspired by the cooperative control of multiagent systems, the proposed scheme will be used to address the control problem of double-cranes. The future work mainly focuses on how to realize the cooperative control of multi-cranes.

Acknowledgements. This work was supported in part by the National Natural Science Foundation of China (62373208), the Taishan Scholar program of Shandong Province of China (tsqn202306218), and the National Natural Science Foundation of Shandong Province (ZR2024YQ032).

Disclosure of Interests. The authors have no competing interests to declare that are relevant to the content of this article.

References

1. Masoud, Z., Nayfeh, A., Mook, D.: Cargo pendulation reduction of ship-mounted cranes. Nonlinear Dyn. **35**(3), 299–311 (2004)
2. Feng, Y., Zhang, H., Gu, C.: The prescribed-time sliding mode control for underactuated bridge crane. Electronics **13**(1), 219 (2024)
3. Li, J.-Q., et al.: A hybrid iterated greedy algorithm for a crane transportation flexible job shop problem. IEEE Trans. Autom. Sci. Eng. **19**(3), 2153–2170 (2022)
4. Bristow, D.A., Tharayil, M., Alleyne, A.G.: A survey of iterative learning control. IEEE Control. Syst. **26**(3), 96–114 (2006)
5. Tayebi, A.: Adaptive iterative learning control for robot manipulators. Automatica **40**(7), 1195–1203 (2004)
6. Li, X., Ren, Q., Xu, J.-X.: Precise speed tracking control of a robotic fish via iterative learning control. IEEE Trans. Ind. Electron. **63**(4), 2221–2228 (2016)
7. Jin, X., Xu, J.-X.: A barrier composite energy function approach for robot manipulators under alignment condition with position constraints. Int. J. Robust Nonlinear Control **24**(17), 2840–2851 (2013)
8. Guth, M., Seel, T., Raisch, J.: Iterative learning control with variable pass length applied to trajectory tracking on a crane with output constraints. In: 52nd IEEE Conference on Decision and Control, pp. 6676–6681 (2013)
9. Aschemann, H., Wache, A., Kraegenbring, O.: A discrete-time norm-optimal approach to iterative learning control of a bridge crane. In: 22nd International Conference on Methods and Models in Automation and Robotics, pp. 319–324 (2017)
10. Wen, Y., Lou, X., Wu, W., Cui, B.: Backstepping boundary control for a class of gantry crane systems. IEEE Trans. Cybern. **53**(9), 5802–5814 (2023)
11. Xing, X., Liu, J.: Vibration and position control of overhead crane with three-dimensional variable length cable subject to input amplitude and rate constraints. IEEE Trans. Syst. Man Cybern. Syst. **51**(7), 4127–4138 (2021)
12. Yang, Y., Ye, X., Wen, B., Huang, J., Su, X.: Adaptive control design for uncertain underactuated cranes with nonsmooth input nonlinearities. IEEE Trans. Syst. Man Cybern. Syst. **53**(2), 1074–1083 (2023)
13. Yu, Z., Niu, W.: Flatness-based backstepping antisway control of underactuated crane systems under wind disturbance. Electronics **12**(1), 244 (2023)
14. Wang, L., Meng, Y., Liu, Y., Pan, Y., Mohammed, C.: Event-based adaptive fixed-time containment control of nonlinear multiagent systems with unmodeled dynamics. Nonlinear Dyn. **112**(21), 19095–19109 (2024)
15. Na, J., Wang, S., Liu, Y.-J., Huang, Y., Ren, X.: Finite-time convergence adaptive neural network control for nonlinear servo systems. IEEE Trans. Cybern. **50**(6), 2568–2579 (2020)
16. Wu, C., Liu, J., Xiong, Y., Wu, L.: Observer-based adaptive faulttolerant tracking control of nonlinear nonstrict-feedback systems. IEEE Trans. Neural Netw. Learn. Syst. **29**(7), 3022–3033 (2018)
17. Zhang, J., Niu, B., Wang, D., Wang, H., Duan, P., Zong, G.: Adaptive neural control of nonlinear nonstrict feedback systems with full-state constraints: A novel nonlinear mapping method. IEEE Trans. Neural Netw. Learn. Syst. **34**(2), 999–1007 (2023)

18. Huang, J., Wang, W., Zhou, J.: Adaptive control design for underactuated cranes with guaranteed transient performance: theoretical design and experimental verification. IEEE Trans. Ind. Electron. **69**(3), 2822–2832 (2022)
19. Chen, Q., Shi, H., Sun, M.: Echo state network-based backstepping adaptive iterative learning control for strict-feedback systems: an errortracking approach. IEEE Trans. Cybern. **50**(7), 3009–3022 (2020)

Fixed-Time Bipartite Output Consensus of Heterogeneous Multi-agent Systems via Event-Triggered Intermittent Control

Lifang Guo[1], Zhiyong Yu[1(✉)], and Haijun Jiang[1,2]

[1] College of Mathematics and System Science, Xinjiang University, Urumqi 830017, China
yzygsts@163.com

[2] School of Mathematical Sciences, Xinjiang Normal University, Urumqi 830017, China

Abstract. This paper considers fixed-time (FXT) bipartite output consensus (BOC) for heterogeneous multi-agent systems (MASs) under event-triggered intermittent control. Firstly, a form of FXT bipartite observation system with intermittent control is designed, in which the control intervals are determined according to auxiliary functions and an event-triggered mechanism. Utilizing Lyapunov stability analysis, we prove that the bipartite consensus between all observers and the leader can be reached within FXT. Secondly, considering the measurability of state information, both the state-dependent and output-dependent event-triggered sampling control strategies are proposed, and some sufficient conditions are derived to ensure the achievement of FXT BOC for heterogeneous MASs. In addition, this serves to demonstrate that Zeno behavior is avoidable. Finally, a numerical example is provided to verify the validity of the theoretical findings.

Keywords: Heterogeneous MASs · Fixed-time · Event-triggered intermittent control · Sampled-data control · Bipartite output consensus

1 Introduction

In the recent period, MASs have garnered significant interest due to their efficient collaboration capabilities in complex tasks. However, the traditional MASs usually assume that all agents are isomorphic. Although this assumption simplifies system design and analysis, it also limits its adaptability and expansibility in the actual complex scenarios. Under this background, heterogeneous MASs

This work was supported in part by the National Natural Science Foundation of China (Grant No. 62363033), in part by Tianshan Talent Training Program (Grant No. 2023TSYCCX0102), and in part by the Natural Science Foundation of Xinjiang Uygur Autonomous Region (Grant No. 2023D01C162).

C. Li et al. (Eds.): ICNC 2025, CCIS 2946, pp. 537–552, 2026.
https://doi.org/10.1007/978-981-92-1599-7_45

have gradually become a research hotspot. Heterogeneous MASs allow agents to differ significantly in structure, function, dynamic characteristics, and even goals. Therefore, the diversity of heterogeneous MASs is closer to the complex demands of the real world. In [1], the fully distributed adaptive formation tracking of heterogeneous MASs was studied.

Traditional consensus control achieves gradual convergence of states or outputs across all agents by means of protocol design, but it operates under the fundamental assumption that all agents are fully cooperative. It is difficult to describe the widespread coexistence of competition and cooperation in the real scene. In response to such problems, bipartite consensus emerged and was proposed by Altafini in [2]. In practical, due to sensor limitations or privacy requirements, the system often only needs some key output variables to achieve coordination. Therefore, many scholars have carried out research on the BOC. In [3], the BOC of heterogeneous MASs was explored using output regulation approach.

Earlier studies focused on asymptotic consensus [4] and finite-time consensus [5], but their performance was limited by the dependence of convergence rate and initial state. Consequently, FXT consensus has gained significant attention in recent years, as its convergence time upper bound is independent of initial states. This feature has significant advantages in practical applications, which has aroused the research interest of many scholars. In [6], FXT containment control of nonlinear MASs under external disturbances was studied.

The formulation of control protocols is pivotal to the achievement of consensus. In practical applications, continuous control may lead to excessive consumption of communication resources. Therefore, intermittent control, as a discontinuous control strategy, has gradually become an important direction in the optimization design of MASs because of its significant advantages in reducing communication and computing costs. The application of control strategies in practice typically relies on the system's state. An investigation of hybrid control schemes for nonlinear MASs was presented in [7]. To the best of our knowledge, research into state-based intermittent control remains scarce, which constitutes one of the key motivations for the present study.

As a control method based on discrete signal interaction, sampling control significantly reduces the communication and computing load, and becomes an important research direction of MASs. This mechanism can effectively alleviate the energy consumption pressure caused by continuous communication. As a result, numerous researchers have explored sampled-data control from various perspectives. Periodic sampling control of MASs has been studied in [8]. However, in some cases, periodic sampling control may waste computing resources due to frequent sampling, or the sampling interval may be too large to capture rapid changes in the system. This limitation prompted the development of aperiodic sampling control, while event-triggered synchronization in multi-agent network systems was specifically investigated in [9].

Building on this analysis, we will amalgamate the benefits of intermittent and sampling control, and subsequently introduce a type of control protocol aimed at exploring the FXT BOC of heterogeneous MASs. The main contributions are given as follows:

(1) This study synthesizes the strengths of intermittent and sampling control into a novel event-triggered protocol, thereby effectively decreasing the control duration of the observation system as well as the controller's update frequency.
(2) Compared with the asymptotic consensus in [10], this work addresses FXT BOC for heterogeneous MASs. Several sufficient conditions for attaining FXT BOC are derived, and the estimation of the settling-time is independent of system's initial conditions.
(3) In contrast to [11,12], which merely considered either state feedback control or output feedback control, this research develops both state and output feedback control protocols by incorporating event-trigger mechanisms, thereby enabling heterogeneous MASs to achieve BOC. The adoption of event-triggered control helps conserve communication resources and lower control costs.

Following this introduction, Sect. 2 outlines the necessary preliminaries. Section 3 presents a discussion of the main results. Section 4 demonstrates the theoretical results through a case study. Lastly, Sect. 5 draws the conclusion.

Notations. $\mathcal{R}$, N, $\mathcal{R}^n$, $\mathcal{R}^{n\times n}$ denote the sets of real numbers, natural numbers, n-dimensional real space, $n\times n$ real matrices, respectively. I denotes identity matrix. $|\cdot|$ stands absolute value. $\|\cdot\|$ denotes the 2-norm. For $\mathcal{S}\in\mathcal{R}^{n\times n}$, $\lambda_{\min}(\mathcal{S})$, $\lambda_2(\mathcal{S})$, and $\lambda_{\max}(\mathcal{S})$ respectively represent the minimum, second smallest, and maximum eigenvalues of the matrix $\mathcal{S}$. $Q>0$ refers that the matrix Q is both symmetric and positive definite. $\otimes$ is the kronecker product. The function $sig^{\mu}(\cdot)$ is defined as $sig^{\mu}(x)=sgn(x)|x|^{\mu}$, where $\mu>0$ and $sgn(x)$ is the sign function of x.

2 Preliminary

2.1 Graph Theory

Consider the MAS with one leader and N followers. The network is described by $\overline{\mathcal{G}}=\mathcal{G}\cup\{0\}$, where vertex 0 represents the leader. The follower subgraph is $\mathcal{G}=(\mathcal{W},\mathcal{E},\mathcal{A})$, in which $\mathcal{W}=\{w_1,w_2,\cdots,w_N\}$ represents the node set and $\mathcal{E}=\{(w_i,w_j)|w_i,w_j\in\mathcal{W}\}$ represents the edge set. $(w_i,w_j)\in\mathcal{E}$ indicates a directed edge from w_i to w_j. The adjacency matrix is $\mathcal{A}=[o_{ij}]\in\mathcal{R}^{N\times N}$, in which $o_{ii}=0$, $o_{ij}\neq 0\Leftrightarrow(w_j,w_i)\in\mathcal{E}$ and $o_{ij}=0$, otherwise. The Laplacian matrix L is defined as $l_{ij}=-o_{ij}, i\neq j$ and $l_{ii}=\sum_{j\neq i}|o_{ij}|$. The connectivity between the leader and followers is characterized by a diagonal matrix $R=diag(o_{10},o_{20},\ldots,o_{N0})$, where the diagonal entry $o_{i0}>0$ means that the i-th follower agent can receive information form the leader, otherwise, $o_{i0}=0$. Let $\mathcal{L}=L+R$. The graph $\mathcal{G}$ is undirected when $o_{ij}=o_{ji}$ holds for all $i,j=1,2,\cdots,N$.

2.2 Some Useful Definition and Lemmas

Definition 1. [13] The graph $\mathcal{G} = (\mathcal{W}, \mathcal{E}, \mathcal{A})$ is called structurally balanced if its node set $\mathcal{W}$ can be divided into two mutually exclusive subsets $\mathcal{W}_1$ and $\mathcal{W}_2$, where $\mathcal{W}_1 \cup \mathcal{W}_2 = \mathcal{W}$, $\mathcal{W}_1 \cap \mathcal{W}_2 = \emptyset$, such that all edges have non-negative weights within the same subset, while edges have non-positive weights within the different subset. If not, it is called structurally unbalanced.

Lemma 1. [13] A connected signed directed graph $\mathcal{G}$ is deemed structurally balanced if there exists matrix $D = diag(g_1, g_2, \cdots, g_N)$ with $g_i \in \{1, -1\}$ such that all entries of $D\mathcal{A}D$ are nonnegative.

Lemma 2. [14] Consider the nonlinear system $\dot{m}(t) = f(m(t)), m(0) = m_0$, where $m \in R^n, f(m) : \mathcal{R}^n \to \mathcal{R}^n$ and $f(0) = 0$. Suppose that there exists a radially unbounded positive-definite Lyapunov function $\mathcal{M}(m(t)) : \mathcal{R}^n \to \mathcal{R}$ such that

$$\dot{\mathcal{M}}(m(t)) \leq -\rho_1 \mathcal{M}^{m_1}(m(t)) - \rho_2 \mathcal{M}^{m_2}(m(t)), m(t) \in \mathcal{R}^n \setminus \{0\},$$

where $\rho_1 > 0$, $\rho_2 > 0$, $0 < m_1 < 1$ and $m_2 > 1$. It can be concluded that the origin is FXT stable, and the system settling-time is bounded by

$$\mathcal{T}_{\max} \leq \frac{1}{\rho_1(1 - m_1)} + \frac{1}{\rho_2(m_2 - 1)}.$$

.

Lemma 3. [15] Let $x_1, x_2, \ldots, x_n \geq 0$, then

$$(\sum_{l=1}^{n} x_l)^{\vartheta} \leq \sum_{l=1}^{n} x_l^{\vartheta} \leq n^{1-\vartheta}(\sum_{l=1}^{n} x_l)^{\vartheta}, 0 < \vartheta \leq 1,$$
$$n^{1-\vartheta}(\sum_{l=1}^{n} x_l)^{\vartheta} \leq \sum_{l=1}^{n} x_l^{\vartheta} \leq (\sum_{l=1}^{n} x_l)^{\vartheta}, \vartheta > 1.$$

Lemma 4. [13] For $r = (r_1, r_2, \cdots, r_N)^T \in \mathcal{R}^N$, the following inequalities hold

$$r^T sig^{\alpha}(r) \geq N^{\frac{1-\alpha}{2}} (r^T r)^{\frac{1+\alpha}{2}}, \quad r^T sig^{\beta}(r) \geq (r^T r)^{\frac{1+\beta}{2}},$$

where $\alpha > 1$ and $0 < \beta < 1$.

2.3 Problem Statement

Consider a heterogeneous linear MAS that is associated with the network $\overline{\mathcal{G}}$. The leader is modeled as

$$\dot{x}_0 = Sx_0, y_0 = Fx_0, \tag{1}$$

where $x_0 \in \mathcal{R}^{n_0}$ and $y_0 \in \mathcal{R}^q$ denote the state and the output of the leader. S and F are matrices with compatible dimensions.

The follower dynamics are given by:

$$\begin{aligned} \dot{x}_i &= A_i x_i + B_i u_i, \\ y_i &= C_i x_i, \end{aligned} \tag{2}$$

where $x_i \in \mathcal{R}^{n_i}$ and $y_i \in \mathcal{R}^q$ denote the state and the output of the i-th follower, respectively. $u_i \in \mathcal{R}^{n_i}$ corresponds to the i-th follower's input. A_i, B_i and C_i are matrices with compatible dimensions.

Definition 2. [13] For the MAS (1)-(2), the FXT BOC is said to be achieved if

$$\lim_{t \to T} \|y_i - g_i y_0\| = 0, \tag{3}$$

where T is a bounded constant, which is independent on the initial value.

Assumption 1. The matrix pairs (A_i, B_i) are stable and (A_i, C_i) are observable.

Assumption 2. For MAS (1)–(2), there exist matrix pairs (X_i, U_i) satisfying the following algebraic equations:

$$\begin{aligned} A_i X_i + B_i U_i &= X_i S, \\ C_i X_i - F &= 0. \end{aligned} \tag{4}$$

Assumption 3. At least one follower can obtain information from the leader, and the undirected subgraph $\mathcal{G}$ is connected as well as structurally balanced.

3 Main Results

This section presents solutions to the BOC problem of the heterogeneous MAS (1)–(2), employing state and output feedback control schemes, respectively.

3.1 Design of Bipartite Fixed-Time Observers

We propose the following bipartite FXT observer:

$$\dot{z\eta}_i(t) = \begin{cases} S\zeta_i(t) - \mu_1 \chi_i(t) - \mu_2 sig^{\alpha}(\chi_i(t)) - \mu_3 sig^{\beta}(\chi_i(t)), & t \in [T_k, S_k), \\ S\zeta_i(t), & t \in [S_k, T_{k+1}), \end{cases} \tag{5}$$

where $\zeta_i(t) \in \mathcal{R}^{n_0}$ denotes the ith follower's estimate of the leader's state, $\chi_i(t) = \sum_{j=1}^{N_i} |o_{ij}| \left(\zeta_i(t) - sign(o_{ij})\zeta_j(t)\right) + o_{i0}(\zeta_i(t) - g_i x_0(t))$, $\mu_1 > 0$, $\mu_2 > 0$, $\mu_3 > 0$, $0 < \alpha < 1$, and $\beta > 1$.

The estimation error is defined as $\bar{\zeta}_i(t) = \zeta_i(t) - g_i x_0(t)$. Let $\bar{\zeta}(t) = [\bar{\zeta}_1^T(t), \bar{\zeta}_2^T(t), \ldots, \bar{\zeta}_N^T(t)]^T$ and $\chi(t) = [\chi_1^T(t), \chi_2^T(t), \ldots, \chi_N^T(t)]^T$. Then, we can obtain

$$\dot{\bar{\zeta}}(t) = \begin{cases} (I_N \otimes S)\bar{\zeta}(t) - \mu_1\chi(t) - \mu_2 sig^{\alpha}(\chi(t)) - \mu_3 sig^{\beta}(\chi(t)), & t \in [T_k, S_k), \\ (I_N \otimes S)\bar{\zeta}(t), & t \in [S_k, T_{k+1}), \end{cases} \tag{6}$$

where $\chi(t) = (\mathcal{L} \otimes I_{n_0})\bar{\zeta}(t)$.

For the purpose of determining the control intervals, we formulate two positive-definite and differentiable functions, denoted as $W_1(t)$ and $W_2(t)$, which satisfy

$$\begin{cases} \dot{W}_1(t) = -\hat{\alpha} W_1^{\frac{1+\nu}{2}}(t) - \hat{\beta} W_1^{\frac{1+\mu}{2}}(t), \\ \dot{W}_2(t) = -\tilde{\alpha} W_2^{\frac{1+\nu}{2}}(t) - \tilde{\beta} W_2^{\frac{1+\mu}{2}}(t), \end{cases} \tag{7}$$

where $W_1(0) = \theta_1 \bar{\zeta}^T(0)(\mathcal{L} \otimes I_{n_0})\bar{\zeta}(0)$, $W_2(0) = \theta_2 \bar{\zeta}^T(0)(\mathcal{L} \otimes I_{n_0})\bar{\zeta}(0)$, $0 < \hat{\alpha} < \tilde{\alpha}$, $0 < \hat{\beta} < \tilde{\beta}$ and $0 < \theta_2 < 1 < \theta_1$.

Using Lemma 2, we can get that $W_1(t)$ and $W_2(t)$ are FXT stable, in which the settling-time satisfying $\mathcal{T}(W_1(0)) \leq \frac{2}{\hat{\alpha}(1-\nu)} + \frac{2}{\hat{\beta}(\mu-1)}$ and $\mathcal{T}(W_2(0)) \leq \frac{2}{\tilde{\alpha}(1-\nu)} + \frac{2}{\tilde{\beta}(\mu-1)}$, respectively.

The following strategy determines the controller's active and dormant intervals.

$$\begin{cases} S_k^* = \inf\left\{t \mid t > T_k, W_2(t) \geq \bar{\zeta}^T(t)(\mathcal{L} \otimes I_{n_0})\bar{\zeta}(t)\right\}, \\ S_k = \max\left\{S_k^*, \tau_k\right\}, \\ T_{k+1} = \inf\left\{t \mid t > S_k, W_1(t) \leq \bar{\zeta}^T(t)(\mathcal{L} \otimes I_{n_0})\bar{\zeta}(t)\right\}, \end{cases} \tag{8}$$

where $\tau_k = T_k + \tau$ and $\tau > 0$. At time Tk, the controller is enabled and turned off at S_k.

Theorem 1. Under Assumptions 1–3, if parameter μ_1 satisfies

$$S + S^T - 2\mu_1 \lambda_{\min}(\mathcal{L}) I < 0,$$

then the bipartite observer (6) can achieve estimation of the leader's state in FXT, in which the settling-time is estimated by

$$\mathcal{T}_1 \leq \frac{1}{\mu_2(1-\alpha)\lambda_{\min}^{\frac{1+\alpha}{2}}(\mathcal{L})} + \frac{1}{\mu_3(\beta-1)N^{\frac{1-\beta}{2}} n^{-\beta} \lambda_{\min}^{\frac{1+\beta}{2}}(\mathcal{L})}.$$

Proof. Consider the following Lyapunov candidate function

$$V(t) = \bar{\zeta}^T(t)(\mathcal{L} \otimes I_{n_0})\bar{\zeta}(t). \tag{9}$$

Under Assumption 3, the matrix $\mathcal{L} > 0$. Consequently, exists an orthogonal matrix U satisfying $U\mathcal{L}U^{-1} = diag(\lambda_1, \lambda_2, \ldots, \lambda_N)$. Let $\hat{\zeta}(t) = (U \otimes I_{n_0})\bar{\zeta}(t)$.

For $t \in [T_k, S_k)$, we can derive that

$$\begin{aligned}\dot{V}(t) =& 2\bar{\zeta}^T(t)(\mathcal{L} \otimes I_{n_0})\dot{\bar{\zeta}}(t) \\ =& \bar{\zeta}^T(t)\big[\mathcal{L} \otimes (S + S^T) - 2\mu_1(\mathcal{L}^2 \otimes I_{n_0})\big]\bar{\zeta}(t) \\ &- 2\mu_2\bar{\zeta}^T(t)(\mathcal{L} \otimes I_{n_0})sig^{\alpha}((\mathcal{L} \otimes I_{n_0})\bar{\zeta}(t)) \\ &- 2\mu_3\bar{\zeta}^T(t)(\mathcal{L} \otimes I_{n_0})sig^{\beta}((\mathcal{L} \otimes I_{n_0})\bar{\zeta}(t)) \\ \le& \sum_{i=1}^{N} \hat{\zeta}_i^T(t)\lambda_i\big(S + S^T - 2\mu_1\lambda_{\min}(\mathcal{L})I\big)\hat{\zeta}_i(t) \\ &- 2\mu_2\bar{\zeta}^T(t)(\mathcal{L} \otimes I_{n_0})sig^{\alpha}((\mathcal{L} \otimes I_{n_0})\bar{\zeta}(t)) \\ &- 2\mu_3\bar{\zeta}^T(t)(\mathcal{L} \otimes I_{n_0})sig^{\beta}((\mathcal{L} \otimes I_{n_0})\bar{\zeta}(t)) \\ \le& - 2\mu_2\lambda_{\min}^{\frac{1+\alpha}{2}}(\mathcal{L})V^{\frac{1+\alpha}{2}}(t) - 2\mu_3N^{\frac{1-\beta}{2}}n^{-\beta}\lambda_{\min}^{\frac{1+\beta}{2}}(\mathcal{L})V^{\frac{1+\beta}{2}}(t).\end{aligned} \tag{10}$$

Over the interval $t \in [S_k, T_{k+1})$, the following result is obtained

$$\dot{V}(t) = 2\bar{\zeta}^T(t)(\mathcal{L} \otimes I_{n_0})\dot{\bar{\zeta}}(t) \le 2\left\|S\right\|V(t). \tag{11}$$

For $t \in [T_k, S_k)$, since $V(T_k) = W_1(T_k)$, it follows from (10) and the comparison principle that $0 \le V(t) \le W_1(t)$ throughout this interval. Likewise, during $t \in [S_k, T_{k+1})$, the inequality $0 \le V(t) < W_1(t)$ remains valid. Hence, the inequality $0 \le V(t) \le W_1(t)$ holds for $t > 0$. Since $\lim_{t \to \mathcal{T}_1} W_1(t) = 0$, $\lim_{t \to \mathcal{T}_1} V(t) = 0$ holds. In conclusion, the bipartite observer (6) achieves FXT estimation of the leader's state.

Given that $S_k - T_k \ge \tau > 0$ and the triggering condition implies $W_2(T_k) < V(T_k) = W_1(T_k)$ and $V(t) < W_1(t)$ holds for $t \in [T_k, S_k)$. By the continuity of $V(t)$ and $W_1(t)$, exists $\varpi > 0$ such that $V(S_k) \le W_2(S_k)$ with $S_k = T_k + \varpi$. For $t \in [S_k, T_{k+1})$, the inequality $\dot{V}(t) \le 2\left\|S\right\|V(t)$, leads to the estimate $V(t) \le V(S_k)e^{2\|S\|(t-S_k)}$. Furthermore, since $W_1(T_{k+1}) = V(T_{k+1}) \le V(S_k)e^{2\|S\|(T_{k+1}-S_k)}$ and $V(S_k) \le W_2(S_k)$, we can derive that $T_{k+1} - S_k > 0$. Thus, it is feasible to rule out the Zeno behavior.

Remark 1. This paper improves upon the continuous communication scheme of [16] by introducing resource-efficient intermittent control. Furthermore, our convergence analysis, distinct from [17], is proven via auxiliary functions.

3.2 Fixed-Time Bipartite Output Consensus by State Feedback Controller with Event-Triggered

Utilize the estimated state obtained from the observer, this section presents a state-feedback control strategy to ensure BOC among all follower agents.

The state error is formulated as $\tilde{x}_i(t) = x_i(t) - X_i\zeta_i(t)$. Traditional control schemes typically necessitate continuous control, resulting in inefficient resource utilization. Consequently, an event-triggered control method is presented to relieve the above-mentioned problem.

$$u_i(t) = K_{i1}\tilde{x}_i(t_l^i) + U_i\zeta_i(t_l^i) - c_1K_{i3}sig^{\alpha}(\tilde{x}_i(t_l^i)) - c_2K_{i3}sig^{\beta}(\tilde{x}_i(t_l^i)), \tag{12}$$

where $c_1 > 0$, $c_2 > 0$ and t_l^i represents the l-th triggering time for agent i.

We establish the definition of the measurement error as

$$\bar{e}_i(t) = u_i(t) - K_{i1}\tilde{x}_i(t) - U_i\zeta_i(t) + c_1 K_{i3} sig^{\alpha}(\tilde{x}_i(t)) + c_2 K_{i3} sig^{\beta}(\tilde{x}_i(t)). \tag{13}$$

We formulate the triggering condition as follows

$$t_{l+1}^i = \inf\left\{ t > t_l^i \mid \|\bar{e}_i(t)\| \geq \frac{\omega_1 \|\tilde{x}_i(t)\|}{2\|P_i\|\|B_i\|} \right\}, \tag{14}$$

where $\omega_1 = \mid \lambda_{\max}(P_i(A_i + B_iK_{i1}) + (A_i + B_iK_{i1})^T P_i) \mid$.

When $t > \mathcal{T}_1$, it implies that

$$\begin{aligned} \dot{\tilde{x}}_i(t) =& \dot{x}_i(t) - X_i\dot{\zeta}_i(t) \\ =& (A_i + B_iK_{i1})\tilde{x}_i(t) - c_1 B_i K_{i3} sig^{\alpha}(\tilde{x}_i(t)) \\ & - c_2 B_i K_{i3} sig^{\beta}(\tilde{x}_i(t)) + B_i\bar{e}_i(t). \end{aligned} \tag{15}$$

Theorem 2. Under Assumptions 1–3, if matrices K_{i1} and K_{i3} satisfy

$$P_i(A_i + B_iK_{i1}) + (A_i + B_iK_{i1})^T P_i < 0, \quad B_iK_{i3}P_i = I,$$

where $P_i > 0$, then the MAS (1)–(2) can achieve FXT BOC via the event-triggered control (12), in which the settling-time is evaluated by

$$\mathcal{T}_2 \leq \max_{1\leq i\leq N}\left\{ \frac{\lambda_{\min}(P_i)^{\frac{1+\alpha}{2}}}{c_1 n^{-\alpha} N^{\frac{1-\alpha}{2}}(1-\alpha)} + \frac{\lambda_{\min}(P_i)^{\frac{1+\beta}{2}}}{c_2(\beta - 1)} \right\}.$$

Proof. Consider the Lyapunov function $V_i(t) = \tilde{x}_i^T(t) P_i \tilde{x}_i(t)$, where $P_i > 0$.

Computing the time derivative of $V_i(t)$ leads to

$$\begin{aligned} \dot{V}_i(t) =& 2\tilde{x}_i^T(t)P_i(A_i + B_iK_{i1})\tilde{x}_i(t) - 2c_1\tilde{x}_i^T(t) sig^{\alpha}(\tilde{x}_i(t)) \\ & - 2c_2\tilde{x}_i^T(t) sig^{\beta}(\tilde{x}_i(t)) + 2\tilde{x}_i^T(t)P_iB_i\bar{e}_i(t) \\ \leq& \tilde{x}_i^T(t)\big(P_i(A_i + B_iK_{i1}) + (A_i + B_iK_{i1})^T P_i\big)\tilde{x}_i(t) \\ & - 2c_2(\tilde{x}_i^T(t)\tilde{x}_i(t))^{\frac{1+\beta}{2}} - 2c_1 n^{-\alpha} N^{\frac{1-\alpha}{2}}(\tilde{x}_i^T(t)\tilde{x}_i(t))^{\frac{1+\alpha}{2}} \\ & + 2\|\tilde{x}_i(t)\|\|P_i\|\|B_i\|\|\bar{e}_i(t)\| \\ \leq& - 2c_1 n^{-\alpha} N^{\frac{1-\alpha}{2}} \lambda_{\min}(P_i)^{\frac{-1-\alpha}{2}} V_i^{\frac{1+\alpha}{2}}(t) - 2c_2\lambda_{\min}(P_i)^{\frac{-1-\beta}{2}} V_i^{\frac{1+\beta}{2}}(t). \end{aligned} \tag{16}$$

Using Lemma 2, one can get the FXT stability of $\tilde{x}_i(t)$, ensuring it converges to zero with a settling-time $\mathcal{T}_{i2} \leq \frac{\lambda_{\min}(P_i)^{\frac{1+\alpha}{2}}}{c_1 n^{-\alpha} N^{\frac{1-\alpha}{2}}(1-\alpha)} + \frac{\lambda_{\min}(P_i)^{\frac{1+\beta}{2}}}{c_2(\beta-1)}$. Furthermore, one can conclude that $\tilde{x} = [\tilde{x}_1^T, \tilde{x}_2^T, \cdots, \tilde{x}_N^T]^T$ converges to zero in FXT with a settling-time $\mathcal{T}_2 \leq \max\{\mathcal{T}_{12}, \mathcal{T}_{22}, \cdots, \mathcal{T}_{N2}\}$. Subsequently, it follows that $\lim_{t\to\mathcal{T}_2}(y_i - g_i y_0) = \lim_{t\to\mathcal{T}_2} C_i\tilde{x}_i = 0$.

Subsequently, we shall demonstrate the absence of Zeno behavior.

$$\begin{aligned}\frac{d\left\|\bar{e}_i(t)\right\|}{dt} &\leq \left\|\dot{\bar{e}}_i(t)\right\| \\ &\leq \left\|U_iSx_0(t)\right\| + \left\|c_1\alpha K_{i3}D_{\alpha i}(t)\dot{\tilde{x}}_i(t)\right\| + \left\|c_2\beta K_{i3}D_{\beta i}(t)\dot{\tilde{x}}_i(t)\right\| \\ &\quad + \left\|K_{i1}\dot{\tilde{x}}_i(t)\right\| \\ &\leq (\left\|K_{i1}\right\| + c_1\alpha\left\|K_{i3}\right\|(\frac{V_i(0)}{\lambda_{\min}(P_i)})^{\frac{\alpha-1}{2}} + \left\|U_iSx_0(t)\right\| \\ &\quad + c_2\beta\left\|K_{i3}\right\|(\frac{V_i(0)}{\lambda_{\min}(P_i)})^{\frac{\beta-1}{2}})\left\|\dot{\tilde{x}}_i(t)\right\| \\ &= m\left\|\dot{\tilde{x}}_i(t)\right\| + \iota,\end{aligned} \tag{17}$$

$$\begin{aligned}\left\|\dot{\tilde{x}}_i(t)\right\| &\leq \left\|(A_i + B_iK_{i1})\right\|\left\|\tilde{x}_i(t)\right\| + c_1\left\|P_i^{-1}\right\|\left\|\tilde{x}_i(t)\right\|^{\alpha} + \left\|B_i\right\|\left\|\bar{e}_i(t)\right\| \\ &\quad + c_2\left\|P_i^{-1}\right\|\left\|\tilde{x}_i(t)\right\|^{\beta} \\ &\leq \left\|(A_i + B_iK_{i1})\right\|(\frac{V_i(0)}{\lambda_{\min}(P_i)})^{\frac{1}{2}} + \left\|B_i\right\|\left\|\bar{e}_i(t)\right\|\end{aligned} \tag{18}$$

$$+c_1\left\|P_i^{-1}\right\|(\frac{V_i(0)}{\lambda_{\min}(P_i)})^{\frac{\alpha}{2}} + c_2\left\|P_i^{-1}\right\|(\frac{V_i(0)}{\lambda_{\min}(P_i)})^{\frac{\beta}{2}} = \sigma + \left\|B_i\right\|\left\|\bar{e}_i(t)\right\|,$$

where $D_{\alpha i}(t) = diag(|\tilde{x}_{i1}(t)|^{\alpha-1}, \cdots, |\tilde{x}_{in}(t)|^{\alpha-1})$, $\iota = \|U_iSx_0(t)\|$, $D_{\beta i}(t) = diag(|\tilde{x}_{i1}(t)|^{\beta-1}, \cdots, |\tilde{x}_{in}(t)|^{\beta-1})$, $m = c_1\alpha\left\|K_{i3}\right\|(\frac{V_i(0)}{\lambda_{\min}(P_i)})^{\frac{\alpha-1}{2}} + c_2\beta\left\|K_{i3}\right\|(\frac{V_i(0)}{\lambda_{\min}(P_i)})^{\frac{\beta-1}{2}} + \left\|K_{i1}\right\|$ and $\sigma = \left\|(A_i + B_iK_{i1})\right\|(\frac{V_i(0)}{\lambda_{\min}(P_i)})^{\frac{1}{2}} + c_1\left\|P_i^{-1}\right\|(\frac{V_i(0)}{\lambda_{\min}(P_i)})^{\frac{\alpha}{2}} + c_2\left\|P_i^{-1}\right\|(\frac{V_i(0)}{\lambda_{\min}(P_i)})^{\frac{\beta}{2}}$.

Thus, combine with (17) and (18), we have

$$\frac{d\left\|\bar{e}_i(t)\right\|}{dt} \leq m\left\|B_i\right\|\left\|\bar{e}_i(t)\right\| + m\sigma + \iota. \tag{19}$$

Therefore, $\left\|\bar{e}_i(t)\right\| \leq \frac{m\sigma+\iota}{m\|B_i\|}(e^{m\|B_i\|(t-t_l^i)} - 1)$. According to the trigger condition, $\frac{m\sigma+\iota}{m\|B_i\|}(e^{m\|B_i\|(t_{l+1}^i-t_l^i)} - 1) \geq \frac{|\lambda_{\max}(P_i(A_i+B_iK_{i1})+(A_i+B_iK_{i1})^TP_i)|\left\|\tilde{x}_i(t_{l+1}^i)\right\|}{2\|P_i\|\|B_i\|}$ can be known. Solving the inequality gives $t_{l+1}^i - t_l^i > 0$, this means that the Zeno behavior does not occur.

3.3 Fixed-Time Bipartite Output Consensus by Output Feedback Controller with Event-Triggered

The previous part considers the situation where the state information is measurable. When the state information is unavailable, it cannot be directly used to design control inputs. Therefore, this section considers the design of control strategies based on output information.

Below is the design of state information estimation

$$\begin{aligned}\dot{\check{x}}_i(t) =&A_i\check{x}_i(t) + B_iu_i(t) - R_{i1}(Q_i(y_i(t) - C_i\check{x}_i(t)))\\&+ \zeta_1 M_i^{-1} sig^{\alpha}(Q_i(y_i(t) - C_i\check{x}_i(t)))\\&+ \zeta_2 M_i^{-1} sig^{\beta}(Q_i(y_i(t) - C_i\check{x}_i(t))),\end{aligned} \tag{20}$$

where $\check{x}_i(t)$ is the estimated state of $x_i(t)$, $K_{i1}, K_{i2}, K_{i3}, R_{i1}$ and Q_i are control matrices to be determined. M_i is a positive definite matrix, and $\zeta_1, \zeta_2 > 0$.

The control protocol is formulated in the following manner

$$\begin{aligned}u_i(t) =&K_{i2}\zeta_i(t_l^i) - d_1 K_{i3} sig^{\alpha}(\check{x}_i(t_l^i) - X_i\zeta_i(t_l^i))\\&- d_2 K_{i3} sig^{\beta}(\check{x}_i(t_l^i) - X_i\zeta_i(t_l^i)) + K_{i1}\check{x}_i(t_l^i),\end{aligned} \tag{21}$$

where $d_1 > 0$ and $d_2 > 0$.

The state error $\delta i(t)$ for the i-th agent denotes $\delta_i(t) = \check{x}_i(t) - X_i\zeta_i(t)$.

We define the measurement error as

$$\begin{aligned}\tilde{e}_i(t) =&u_i(t) - K_{i1}\check{x}_i(t) - K_{i2}\zeta_i(t) + d_1 K_{i3} sig^{\alpha}(\check{x}_i(t) - X_i\zeta_i(t))\\&+ d_2 K_{i3} sig^{\beta}(\check{x}_i(t) - X_i\zeta_i(t)).\end{aligned} \tag{22}$$

We formulate the triggering condition as follows

$$t_{l+1}^i = \inf\left\{t > t_l^i \mid \|\tilde{e}_i(t)\| \geq \frac{\omega_2 \|\delta_i(t)\|}{2\|J_i\|\|B_i\|}\right\}, \tag{23}$$

where $\omega_2 =\mid \lambda_{\max}(J_i(A_i + B_iK_{i1}) + (A_i + B_iK_{i1})^T J_i) \mid$.

Theorem 3. Under Assumptions 1–3, if the matrices $A_i+B_iK_{i1}$ and $A_i+R_{i1}C_i$ are Hurwitz stable, and the following equations are satisfied

$$B_iK_{i3} = J_i^{-1}, \quad K_{i2} = U_i - K_{i1}X_i, \quad Q_iC_i = I,$$

where J_i is a positive definite matrix, then FXT BOC of the MAS (1)–(2) is guaranteed via the event-triggered controller (21), and the settling-time can be calculated via

$$\begin{aligned}\mathcal{T}_3 \leq& \max_{1\leq i\leq N}\left\{\frac{2^{\frac{1-\alpha}{2}}\lambda_{\min}(M_i)^{\frac{1+\alpha}{2}} n^{\alpha}}{(1-\alpha)N^{\frac{1-\alpha}{2}}\zeta_1} + \frac{2^{\frac{1-\beta}{2}}\lambda_{\min}(M_i)^{\frac{1+\beta}{2}}}{(\beta-1)\zeta_2}\right\}\\&+ \max_{1\leq i\leq N}\left\{\frac{\lambda_{\min}(J_i)^{\frac{1+\alpha}{2}}}{d_1 n^{-\alpha} N^{\frac{1-\alpha}{2}}(1-\alpha)} + \frac{\lambda_{\min}(J_i)^{\frac{1+\beta}{2}}}{d_2(\beta-1)}\right\}.\end{aligned}$$

Proof. The error is specified as $\varrho_i(t) = x_i(t) - \check{x}_i(t)$ and $\varrho(t) = [\varrho_1^T(t), \varrho_2^T(t), \ldots, \varrho_N^T(t)]^T$, which yields

$$\begin{aligned}\dot{\varrho}_i(t) =&\dot{x}_i(t) - \dot{\check{x}}_i(t)\\=&(A_i + R_{i1}C_i)\varrho_i(t) - \zeta_1 M_i^{-1} sig^{\alpha}(Q_i(y_i(t) - C_i\check{x}_i(t)))\\&- \zeta_2 M_i^{-1} sig^{\beta}(Q_i(y_i(t) - C_i\check{x}_i(t))).\end{aligned} \tag{24}$$

Choose the Lyapunov function as $V_i(t) = \frac{1}{2}\varrho_i^T(t)M_i\varrho_i(t)$. Subsequently, the expression for its time derivative is as follows

$$\begin{aligned}\dot{V}_i(t) \leq &-\zeta_1\varrho_i^T(t)sig^{\alpha}(\varrho_i(t)) - \zeta_2\varrho_i^T(t)sig^{\beta}(\varrho_i(t))\\ \leq &-n^{-\alpha}N^{\frac{1-\alpha}{2}}\zeta_1(\varrho_i^T(t)\varrho_i(t))^{\frac{1+\alpha}{2}} - \zeta_2(\varrho_i^T(t)\varrho_i(t))^{\frac{1+\beta}{2}}\\ = &-n^{-\alpha}N^{\frac{1-\alpha}{2}}\zeta_1(\frac{2}{\lambda_{\min}(M_i)})^{\frac{1+\alpha}{2}}V_i^{\frac{1+\alpha}{2}}(t)\\ &-\zeta_2(\frac{2}{\lambda_{\min}(M_i)})^{\frac{1+\beta}{2}}V_i^{\frac{1+\beta}{2}}(t).\end{aligned} \tag{25}$$

According to Lemma 2, $\varrho_i(t) = 0$ is FXT stable, with the settling-time satisfying $T_{i1} \leq \frac{2^{\frac{1-\alpha}{2}}\lambda_{\min}(M_i)^{\frac{1+\alpha}{2}}n^{\alpha}}{(1-\alpha)N^{\frac{1-\alpha}{2}}\zeta_1} + \frac{2^{\frac{1-\beta}{2}}\lambda_{\min}(M_i)^{\frac{1+\beta}{2}}}{(\beta-1)\zeta_2}$. It can be concluded that $\varrho = [\varrho_1^T, \varrho_2^T, \cdots, \varrho_N^T]^T = 0$ is FXT stable, given that its settling-time meets $T_1 < \max\{T_{11}, T_{21}, \cdots, T_{N1}\}$.

When $t > \max\{T_1, T_1\}$, it implies that

$$\begin{aligned}\dot{\delta}_i(t) =&\dot{x}_i(t) - X_i\dot{\zeta}_i(t)\\ =&(A_i + B_iK_{i1})\delta_i(t) - d_1B_iK_{i3}sig^{\alpha}(\delta_i(t)) + B_i\tilde{e}_i(t)\\ &- d_2B_iK_{i3}sig^{\beta}(\delta_i(t)).\end{aligned} \tag{26}$$

Taking $V_i(t) = \delta_i^T(t)J_i\delta_i(t)$ as the Lyapunov function, its time derivative is then expressed as

$$\begin{aligned}V_i(t) =&2\delta_i^T(t)J_i\dot{\delta}_i(t)\\ \leq&\lambda_{\max}(J_i(A_i + B_iK_{i1}) + (A_i + B_iK_{i1})^TJ_i)\delta_i^T(t)\delta_i(t)\\ &- 2d_2(\delta_i^T(t)\delta_i(t))^{\frac{1+\beta}{2}} - 2d_1n^{-\alpha}N^{\frac{1-\alpha}{2}}(\delta_i^T(t)\delta_i(t))^{\frac{1+\alpha}{2}}\\ &+ 2\left\|\delta_i(t)\right\|\left\|J_i\right\|\left\|B_i\right\|\left\|\tilde{e}_i(t)\right\|\\ \leq&-2d_1n^{-\alpha}N^{\frac{1-\alpha}{2}}\lambda_{\min}(J_i)^{\frac{-1-\alpha}{2}}V_i^{\frac{1+\alpha}{2}}(t) - 2d_2\lambda_{\min}(J_i)^{\frac{-1-\beta}{2}}V_i^{\frac{1+\beta}{2}}(t).\end{aligned} \tag{27}$$

Applying Lemma 2, one can get $\delta_i(t) = 0$ is FXT stable, in which the settling-time satisfying $T_{i2} \leq \frac{\lambda_{\min}(J_i)^{\frac{1+\alpha}{2}}}{d_1n^{-\alpha}N^{\frac{1-\alpha}{2}}(1-\alpha)} + \frac{\lambda_{\min}(J_i)^{\frac{1+\beta}{2}}}{d_2(\beta-1)}$. Therefore, one can conclude that $\delta(t) = 0$ is FXT stable, given that its settling-time meets $T_2 < \max\{T_{12}, T_{22}, \cdots, T_{N2}\}$. Then, it has $\lim_{t\to T_3}(y_i - g_iy_0) = 0$.

Zeno behavior exclusion, analogous to Theorem 2, is not presented here.

Remark 2. The study proposes state and output feedback control strategies for heterogeneous MASs. State feedback utilizes complete internal state data for high accuracy but at the cost of greater resource demands. In contrast, output feedback, which depends only on observable outputs, is more resource-efficient. Event-triggering mechanisms are employed in both strategies to further reduce communication and computation costs.

4 Simulation Example

In this section, a numerical example is given to validate the theoretical findings. **Example 1.** Consider a MAS with one leader and five followers, in which the interaction topology thereof is shown in Fig. 1.

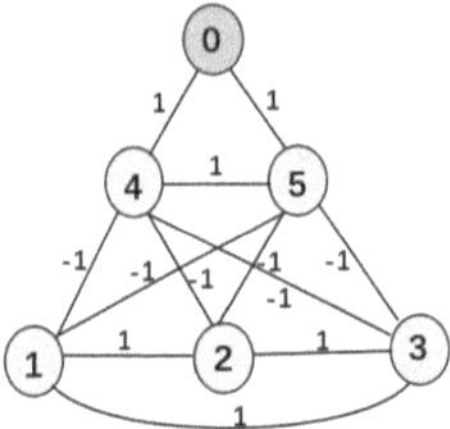

Fig. 1. Communication topology graph.

The system dynamics are characterized by (1)–(2) , in which $A_1 = \begin{bmatrix} -1 & 4 & 5 \\ 0 & -2 & 6 \\ 0 & 0 & -3 \end{bmatrix}$, $A_2 = \begin{bmatrix} -1 & -5 & 7 \\ 0 & -2 & 3 \\ 0 & 0 & -3 \end{bmatrix}$, $A_3 = \begin{bmatrix} 0 & 1 \\ -2 & -2 \end{bmatrix}$, $A_4 = \begin{bmatrix} -2 & 1 & 0 \\ 1 & -2 & 1 \\ 0 & 1 & -2 \end{bmatrix}$, $A_5 = \begin{bmatrix} 0 & 1 \\ -2 & -2 \end{bmatrix}$, $S = \begin{bmatrix} 0 & 1 & 0 \\ 0 & 0 & 1 \\ 0 & -1 & 0 \end{bmatrix}$. It can be found that the topology is structurally balanced, in which $\mathcal{V}_1 = \{4, 5\}$, $\mathcal{V}_2 = \{1, 2, 3\}$, and $D = diag(-1, -1, -1, 1, 1)$. The FXT bipartite observer is shown in (5). In the observer, the control parameters are selected as $\mu_1 = 2, \mu_2 = 3.7, \mu_3 = 4.7$. In auxiliary functions $W_1(t)$ and $W_2(t)$, the parameters are chosen as $\theta_1 = 1.5$ and $\theta_2 = 0.6$. By computation, all the conditions of Theorem 1 are fulfilled, and the upper bound of the settling-time is $\mathcal{T}_1 \leq 28.3383$.

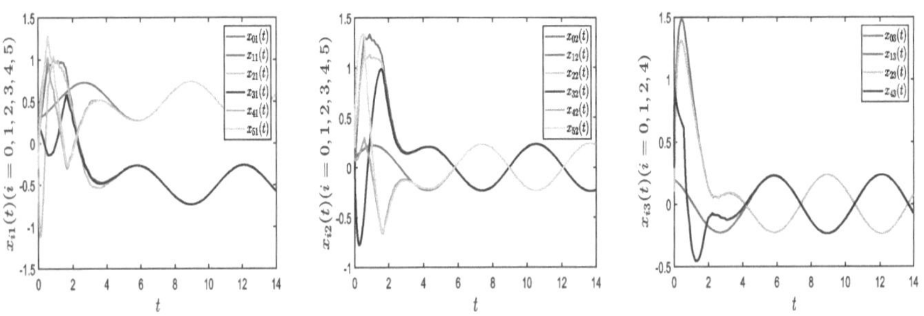

Fig. 2. The state of each agent.

The states of agents $x_i(t)$ for $i = 0, 1, \cdots, 5$ are illustrated in Fig. 2. The trajectories of $V(t)$, $W_1(t)$, $W_2(t)$ are illustrated in Fig. 3. It is clear that as

$W_1(t)$ and $W_2(t)$ approach 0, $V(t)$ also tends towards 0. The activation and rest moments, denoted as T_k and S_k, are presented in Fig. 3. Hence, the bipartite observer allows for the leader's state estimation under FXT, serving to confirm that Theorem 1 is viable.

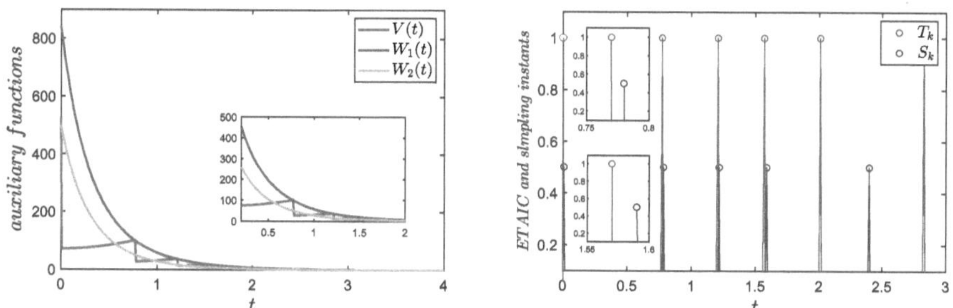

Fig. 3. The $V(t), W_1(t)$ and $W_2(t)$ and intermittent communication moments.

4.1 Bipartite Output Consensus by State-Feedback Controller

We set the parameters and matrices in control protocol (12) as follows: $U_1 = \begin{bmatrix} 1 & -3 & -5 \\ 0 & 2 & -5 \\ 0 & -1 & 3 \end{bmatrix}$, $U_2 = \begin{bmatrix} 1 & 6 & -7 \\ 0 & 2 & -2 \\ 0 & -1 & 3 \end{bmatrix}$, $U_3 = \begin{bmatrix} 0 & 0 & 0 \\ 2 & 2 & 1 \end{bmatrix}$, $U_4 = \begin{bmatrix} 2 & 0 & 0 \\ -1 & 2 & 0 \\ 0 & -2 & 2 \end{bmatrix}$, $U_5 = \begin{bmatrix} 0 & 0 & 0 \\ 2 & 2 & 1 \end{bmatrix}$, $K_{11} = \begin{bmatrix} 0 & -4 & -5 \\ 0 & 1 & -6 \\ 0 & 0 & 2 \end{bmatrix}$, $K_{21} = \begin{bmatrix} -1 & 5 & 7 \\ 0 & 0 & -3 \\ 0 & 0 & 1 \end{bmatrix}$, $K_{31} = \begin{bmatrix} -2 & -1 \\ 2 & 0 \end{bmatrix}$, $K_{41} = \begin{bmatrix} -1 & -1 & 0 \\ -1 & -1 & -1 \\ 0 & -1 & -1 \end{bmatrix}$, $K_{51} = \begin{bmatrix} -2 & -1 \\ 2 & 0 \end{bmatrix}$, $K_{13} = K_{23} = K_{43} = I_3$, $K_{33} = K_{53} = I_2$, $C_i = \begin{bmatrix} 1 & 0 & 0 \end{bmatrix}$ for $i = 1, 2, 4$, $C_i = \begin{bmatrix} 1 & 0 \end{bmatrix}$ for $i = 3, 5$, $F = \begin{bmatrix} 1 & 0 & 0 \end{bmatrix}$, $P_i = I_3$ for $i = 1, 2, 4$, $P_i = I_2$ for $i = 3, 5$, $c_1 = 1.1$ and $c_2 = 1.5$. Through relevant computations, all conditions of Theorem 2 are fulfilled, with the upper bound of the settling-time evaluated to be $\mathcal{T}_2 \leq 8.9889$. Figure 4 depicts the evolution of the output $y_i(t)$. It is evident that each agent's output reaches bipartite consensus within FXT. The event-triggered moments are illustrated in Fig. 4. Therefore, the result of Theorem 2 is valid.

4.2 Bipartite Output Consensus by Output-Feedback Controller

In the state information estimation (20) and control protocol (21), the parameters and matrices are designed as $C_1 = \begin{bmatrix} 0.5 & 0 & -0.5 \\ 1 & 1 & 0 \\ -0.5 & 1 & 0.5 \end{bmatrix}$, $C_2 = \begin{bmatrix} 0.5 & 0 & 0.5 \\ 1 & 0.5 & 0 \\ 0.5 & 1 & 0.5 \end{bmatrix}$,

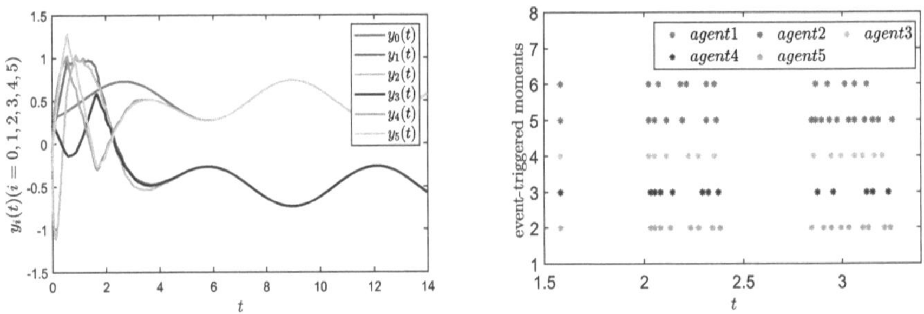

Fig. 4. The output $y_i(t)$ and event-triggered control instants.

$$C_3 = \begin{bmatrix} -1 & 1 \\ 2 & -1 \\ 1 & -1 \end{bmatrix}, C_4 = \begin{bmatrix} 0.5 & 0 & -0.5 \\ -1 & -0.5 & 0 \\ 0.5 & 1 & 0.5 \end{bmatrix}, C_5 = \begin{bmatrix} -1 & 1 \\ 1 & 0 \\ 1 & -1 \end{bmatrix}, R_{11} = \begin{bmatrix} 11 & -5 & 1 \\ 19 & -6 & 7 \\ -6 & 2 & -2 \end{bmatrix},$$

$$R_{21} = \begin{bmatrix} -10 & 2 & 4 \\ 9 & -6 & 3 \\ -3 & 2 & -1 \end{bmatrix}, R_{31} = \begin{bmatrix} -3 & -2 & 0 \\ 2 & 2 & 0 \end{bmatrix}, R_{41} = \begin{bmatrix} -1 & 0 & -1 \\ 2 & 2 & 0 \\ 1 & 0 & -1 \end{bmatrix}, R_{51} = \begin{bmatrix} 2 & 1 & 0 \\ 0 & 1 & 1 \end{bmatrix},$$

$F = \begin{bmatrix} 0 & 1 & 0 \\ 1 & 0 & 1 \\ 0 & -1 & 0 \end{bmatrix}$, $d_1 = 1.4$, $d_2 = 1.2$, $M_i = J_i = I_3$ for $i = 1, 2, 4$, and $M_i = J_i = I_2$ for $i = 3, 5$. It turns out that all the conditions required by Theorem 3 are met, and the estimation of the settling-time is $\mathcal{T}_3 \leq 20.9554$. Figure 5 describes the error trajectories $\varrho_i(t)$ for $i = 1, 2, \cdots, 5$. One can find that the estimation error of the state information converges to 0 within FXT. Under control protocol (21), Fig. 6 shows the output trajectories $y_i(t)$ for $i = 0, 1, \cdots, 5$. From Fig. 6, it is observable that all agents can achieve BOC within FXT. Consequently, Theorem 3 is verified.

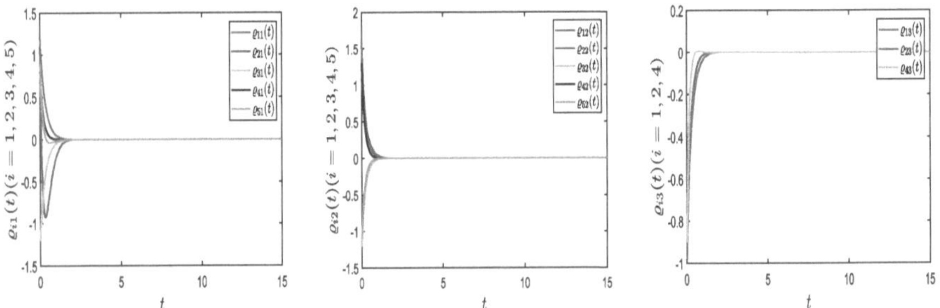

Fig. 5. The trajectories of $\varrho_i(t)$.

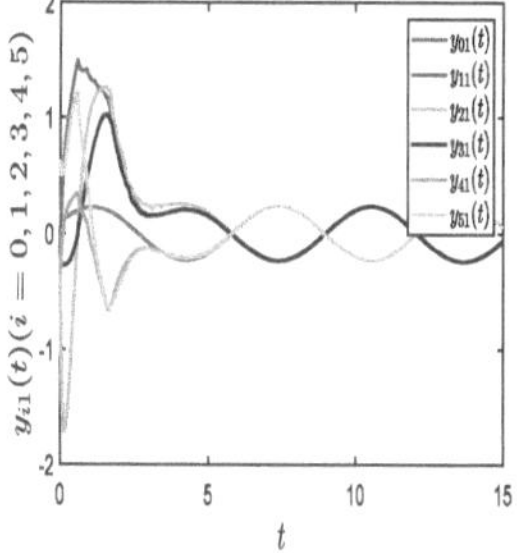

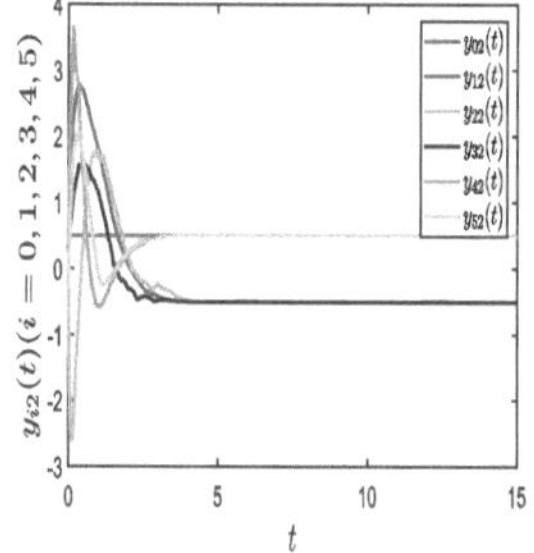

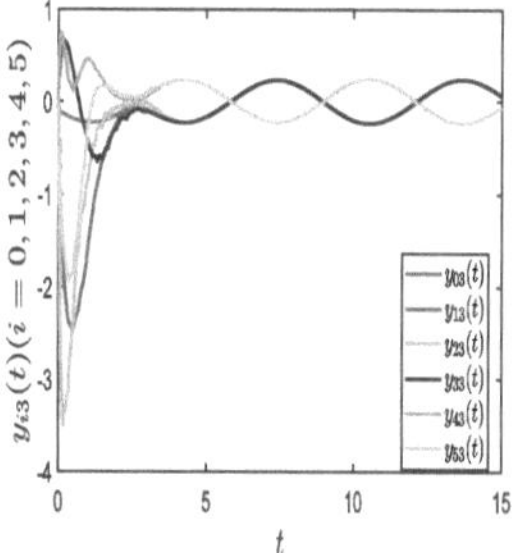

Fig. 6. The output $y_i(t)$.

5 Conclusion

This paper investigated FXT BOC of heterogeneous MASs under event-triggered intermittent control. A bipartite FXT observer with intermittent control was first designed to allow followers to estimate the leader's state. To reduce control costs, an event-triggered state feedback control scheme was proposed to achieve BOC, followed by an output feedback event-triggered controller. Through the adoption of Lyapunov-based stability theory and inequality approaches, sufficient conditions for FXT bipartite consensus were obtained, and Zeno behavior was eliminated. The validity of the proposed control strategy was verified via numerical simulations. Future investigations will explore bipartite consensus in MASs subject to intricate nonlinear dynamics.

References

1. Wang, Q., Dong, X., Lü, J., Ren, Z.: Fully distributed adaptive formation tracking for directed heterogeneous multi-agent systems with unknown input. IEEE Trans. Circ. Syst. II Express Briefs **70**(2), 700–704 (2023)
2. Altafini, C.: Consensus problems on networks with antagonistic interactions. IEEE Trans. Autom. Control **58**(4), 935–946 (2013)
3. Han, T., Zheng, W.: Bipartite output consensus for heterogeneous multi-agent systems via output regulation approach. IEEE Trans. Circ. Syst. II Express Briefs **68**(1), 281–285 (2021)
4. Luo, D., Wang, Y., Li, Z., Song, Y., Lewis, F.: Asymptotic leader-following consensus of heterogeneous multi-agent systems with unknown and time-varying control gains. IEEE Trans. Autom. Sci. Eng. **22**, 2768–2779 (2025)
5. Liu, Z., Zhan, X., Han, T., Yan, H.: Distributed adaptive finite-time bipartite containment control of linear multi-agent systems. IEEE Trans. Circuits Syst. II Express Briefs **69**(11), 4354–4358 (2022)
6. Wang, L., Liu, X., Cao, J., Hu, X.: Fixed-time containment control for nonlinear multi-agent systems with external disturbances. IEEE Trans. Circ. Syst. II Express Briefs **69**(2), 459–463 (2022)
7. Liang, H., Chang, Z., Ahn, C.: Hybrid event-triggered intermittent control for nonlinear multi-agent systems. IEEE Transa. Netw. Sci. Eng. **10**(4), 1975–1984 (2023)

8. Zou, W., Zhou, C., Xiang, Z.: Sampled-data leader-following consensus of nonlinear multi-agent systems subject to impulsive perturbations. Commun. Nonlinear Sci. Numer. Simul. **78**, 104884 (2019)
9. Yi, C., Li, J., Cai, J., You, Z., Jing, G.: Event-triggered synchronization for multi-agent networked systems with time-varying coupling strength and non-differentiable delays. Math. Comput. Simul. **236**, 12–28 (2025)
10. Han, T., Zheng, W.: Bipartite output consensus for heterogeneous multi-agent systems via output regulation approach. IEEE Trans. Circ. Syst. II Express Briefs **68**(1), 281–285 (2021)
11. Qin, J., Fu, W., Zheng, W., Gao, H.: On the bipartite consensus for generic linear multiagent systems with input saturation. IEEE Trans. Cybern. **47**(8), 1948–1958 (2017)
12. Wen, G., Wang, H., Yu, X., Yu, W.: Bipartite tracking consensus of linear multi-agent systems with a dynamic leader. IEEE Trans. Circ. Syst. II Express Briefs **65**(9), 1204–1208 (2018)
13. Zhang, H., Duan, J., Wang, Y., Gao, Z.: Bipartite fixed-time output consensus of heterogeneous linear multiagent systems. IEEE Trans. Cybern. **51**(2), 548–557 (2021)
14. Gong, P., Han, Q.: Fixed-time bipartite consensus tracking of fractional-order multi-agent systems with a dynamic leader. IEEE Trans. Circ. Syst. II Express Briefs **67**(10), 2054–2058 (2020)
15. Duan, S., Yu, Z., Huang, D., Jiang, H.: Distributed fixed-time cluster optimisation for multi-agent systems. Int. J. Syst. Sci. **54**(3), 531–548 (2023)
16. Cai, Y., Zhang, H., Wang, Y., Gao, Z., He, Q.: Adaptive bipartite fixed-time time-varying output formation-containment tracking of heterogeneous linear multiagent systems. IEEE Trans. Neural Netw. Learn. Syst. **33**(9), 4688–4698 (2022)
17. Geng, X., Feng, J., Wang, J., Li, N., Zhao, Y.: Prespecified-time bipartite consensus of multi-agent systems via intermittent control. IEEE Trans. Circ. Syst. I Regul. Pap. **71**(5), 2240–2251 (2024)

Flatness-Based Adaptive Backstepping Control for Underactuated ACVs

Renhai Yu(✉), Wanyu Tang, and Qizheng Zhou

Navigation College, Dalian Maritime University, Dalian 116026, China
{yurenhai,zhouqizheng}@dlmu.edu.cn

Abstract. This paper presents an effective trajectory tracking control strategy for an underactuated air-cushion vehicle (ACV). The differential flatness approach is applied to convert the strongly coupled nonlinear dynamics into a fully actuated equivalent system through appropriate selection of flat outputs, thus reducing the complexity of controller synthesis. On this basis, a backstepping control law is developed with recursive virtual controls and Lyapunov stability analysis to ensure asymptotic error convergence. Numerical simulations demonstrate that the proposed method achieves accurate tracking with fast convergence and robust performance.

Keywords: Underactuated air-cushion vehicle (ACV) · differential flatness · backstepping control · trajectory tracking

1 Introduction

An air-cushion vehicle (ACV) is a unique amphibious craft capable of operating efficiently in various complex environments, such as land, water, mud, and ice [1,2]. It achieves levitation through an air cushion beneath its hull, providing greater operational flexibility and higher speed performance [3]. However, the ACV uses air propellers as its propulsion source, resulting in an underactuated and highly nonlinear dynamic system [4,5]. Specifically, the control inputs are insufficient to directly control all degrees of freedom, while strong coupling and external disturbances further increase the complexity of trajectory tracking control [6,7]. The trajectory tracking control of ACVs has been extensively studied, with various approaches such as robust control [8], sliding mode control [9], and model predictive control [10]. However, these methods often rely on precise dynamic models and do not adequately handle modeling errors and environmental disturbances.

Differential flatness theory provides an effective solution for underactuated systems [11–13]. By introducing flat outputs, the complex nonlinear system can be mapped into a flat space where its states and control inputs are explicitly expressed in terms of these outputs and their derivatives. This characteristic simplifies trajectory planning and controller design while reducing reliance on precise mathematical models. The backstepping method, as a well-established

C. Li et al. (Eds.): ICNC 2025, CCIS 2946, pp. 553–561, 2026.
https://doi.org/10.1007/978-981-92-1599-7_46

nonlinear control approach, addresses control challenges by recursively designing virtual control variables [14–16]. When combined with differential flatness, this method leverages the strengths of both approaches, simplifying system modeling while enhancing control performance.

Motivated by the aforementioned discussion, this work develops an adaptive backstepping control strategy combined with differential flatness for trajectory tracking of an ACV. The main contributions are summarized as follows:

1. By exploiting differential flatness theory, the flat outputs of the ACV model are determined. This approach enables the originally underactuated dynamics to be converted into an equivalent fully actuated representation.
2. An adaptive backstepping control law is constructed, and asymptotic convergence of the tracking errors is rigorously guaranteed via Lyapunov stability theory.

The structure of this paper is arranged as follows. The ACV modeling and differential flatness formulation are presented in Sect. 2. Section 3 describes the backstepping control law design. Simulation results and analysis are given in Sect. 4, while Sect. 5 draws the conclusions.

2 ACV Model and Differential Flatness

2.1 ACV Mathematical Model

According to [17], the ACV dynamics are given by:

$$\begin{cases} \tau = M\dot{\xi} + C(\xi)\xi + D\xi, \\ \dot{\eta} = R(\eta)\xi, \end{cases} \tag{1}$$

where

$$M = \begin{bmatrix} m_1 & 0 & 0 \\ 0 & m_1 & 0 \\ 0 & 0 & m_2 \end{bmatrix}, \quad C(\xi) = \begin{bmatrix} 0 & 0 & -m_1 v \\ 0 & 0 & m_1 u \\ m_1 v & -m_1 u & 0 \end{bmatrix},$$

$$D = \begin{bmatrix} d_1 & 0 & 0 \\ 0 & d_2 & 0 \\ 0 & 0 & d_3 \end{bmatrix}, \quad R(\eta) = \begin{bmatrix} \cos(\psi) & -\sin(\psi) & 0 \\ \sin(\psi) & \cos(\psi) & 0 \\ 0 & 0 & 1 \end{bmatrix},$$

where M represents the inertia matrix, D is the hydrodynamic damping matrix, $C(\xi)$ denotes the Coriolis–centripetal matrix, $R(\eta)$ is the rotation matrix relating the body-fixed and earth-fixed frames, and $\tau = [\tau_1, 0, \tau_3]^{\mathrm{T}}$ denotes the control input vector comprising the surge force and yaw moment generated by the air propeller.

After simplification:

$$\begin{cases} \dot{x} = u\cos(\psi) - v\sin(\psi), \\ \dot{y} = u\sin(\psi) + v\cos(\psi), \\ \dot{\psi} = r, \\ \dot{u} = vr + \tau_u, \\ \dot{v} = -ur - \beta v, \\ \dot{r} = \tau_r, \end{cases} \tag{2}$$

where τ_u, τ_r, and β denote the normalized surge force, yaw moment, and damping ratio, defined as $\tau_u = \frac{\tau_1}{m_1}$, $\tau_r = \frac{\tau_3}{m_3}$, and $\beta = \frac{d_2}{m_2}$, respectively. Note that this formulation represents an underactuated configuration with fewer control inputs than degrees of freedom to be controlled.

2.2 Differential Flatness

The concept of differential flatness is fundamental to the subsequent control design. For a nonlinear system of the form:

$$\begin{cases} \dot{x} = f(x, u), & x \in \mathbb{R}^n,\ u \in \mathbb{R}^m, \\ y = g(x), & y \in \mathbb{R}^n, \end{cases} \tag{3}$$

the system is called differentially flat if there exists an output vector z, termed the flat output, such that:

$$z = \phi\left(x, u, \dot{u}, \cdots, u^{(p)}\right), \quad p \in \mathbb{N}, \tag{4}$$

$$x = \varphi\left(z, \dot{z}, \cdots, z^{(q)}\right), \quad q \in \mathbb{N}, \tag{5}$$

$$u = \theta\left(z, \dot{z}, \cdots, z^{(l)}\right), \quad l \in \mathbb{N}. \tag{6}$$

under these conditions, the underactuated dynamics can be converted into a fully actuated form through appropriate coordinate transformations.

2.3 Selection of Flat Outputs

By choosing the position coordinates x and y as the flat output $A = [x, y]^{\mathrm{T}}$, the remaining state variables and control inputs can be parameterized accordingly. The heading angle is obtained as:

$$\psi = \arctan\left(\frac{\ddot{y} + \beta\dot{y}}{\ddot{x} + \beta\dot{x}}\right). \tag{7}$$

The yaw rate is derived by differentiating ψ:

$$r = \frac{y^{(3)}(\ddot{x} + \beta\dot{x}) - x^{(3)}(\ddot{y} + \beta\dot{y}) - \beta^2(\ddot{x}\dot{y} - \ddot{y}\dot{x})}{(\ddot{x} + \beta\dot{x})^2 + (\ddot{y} + \beta\dot{y})^2}. \tag{8}$$

The surge and sway velocities are expressed as:

$$u = \frac{\dot{x}(\ddot{x} + \beta\dot{x}) + \dot{y}(\ddot{y} + \beta\dot{y})}{\sqrt{(\ddot{x} + \beta\dot{x})^2 + (\ddot{y} + \beta\dot{y})^2}}, \tag{9}$$

$$v = \frac{\dot{y}\ddot{x} - \dot{x}\ddot{y}}{\sqrt{(\ddot{x} + \beta\dot{x})^2 + (\ddot{y} + \beta\dot{y})^2}}. \tag{10}$$

For notational simplicity, we introduce the auxiliary variables:

$$\Omega_x = \ddot{x} + \beta\dot{x}, \quad \Omega_y = \ddot{y} + \beta\dot{y}, \quad \Lambda = \sqrt{\Omega_x^2 + \Omega_y^2}. \tag{11}$$

Then, the control inputs can be written in a more compact form:

$$\tau_u = \frac{\ddot{x}\Omega_x + \ddot{y}\Omega_y}{\Lambda}, \tag{12}$$

$$\begin{aligned} \tau_r = \dot{r} = & \frac{y^{(4)}\Omega_x - x^{(4)}\Omega_y + \beta\left(y^{(3)}\ddot{x} - x^{(3)}\ddot{y}\right) - \beta^2\left(x^{(3)}\dot{y} - y^{(3)}\dot{x}\right)}{\Lambda^2} \\ & - \frac{2\left[y^{(3)}\Omega_x - x^{(3)}\Omega_y - \beta^2\left(\ddot{x}\dot{y} - \ddot{y}\dot{x}\right)\right]}{\Lambda^4} \\ & \cdot \left[\Omega_x\left(x^{(3)} + \beta\ddot{x}\right) + \Omega_y\left(y^{(3)} + \beta\ddot{y}\right)\right]. \end{aligned} \tag{13}$$

The expressions above demonstrate that the system (2) possesses the differential flatness property with $A = [x, y]^{\mathrm{T}}$ serving as the flat output. Leveraging this structural characteristic, a backstepping-based control strategy is developed in the subsequent section.

3 Backstepping Controller Design

3.1 Fully Actuated Model

With flat outputs x and y, differentiating until control inputs appear yields

$$\begin{cases} x^{(4)} = m_1\tau_u, \\ y^{(4)} = m_2\tau_r, \end{cases} \tag{14}$$

where $m_1 = -\tau_r \sin\psi$ and $m_2 = (\beta u + \tau_u)\cos\psi + \beta v \sin\psi$.

Define the state variables

$$\begin{aligned} &x_1 = x,\ x_2 = \dot{x},\ x_3 = \ddot{x},\ x_4 = x^{(3)}, \\ &y_1 = y,\ y_2 = \dot{y},\ y_3 = \ddot{y},\ y_4 = y^{(3)}. \end{aligned} \tag{15}$$

Then the dynamics take the form

$$\begin{cases} \dot{x}_i = x_{i+1},\ i = 1, 2, 3, \\ \dot{x}_4 = m_1\tau_u, \end{cases} \qquad \begin{cases} \dot{y}_i = y_{i+1},\ i = 1, 2, 3, \\ \dot{y}_4 = m_2\tau_r. \end{cases} \tag{16}$$

3.2 Design of τ_u

Define the tracking errors

$$e_{x1} = x_1 - x_{1d}, \qquad e_{xi} = x_i - x_{id},\ i = 2, 3, 4. \tag{17}$$

The backstepping virtual control laws are

$$x_{2d} = \dot{x}_{1d} - k_{x1}e_{x1}, \tag{18}$$
$$x_{3d} = \dot{x}_{2d} - k_{x2}e_{x2} - e_{x1}, \tag{19}$$
$$x_{4d} = \dot{x}_{3d} - k_{x3}e_{x3} - e_{x2}. \tag{20}$$

The control law becomes

$$\tau_u = \frac{1}{\hat{m}_1}\left(\dot{x}_{4d} - k_{x4}e_{x4} - e_{x3} - \eta\,\phi(e_{x4})\right), \tag{21}$$

$$\phi(e_{x4}) = \tanh\left(\frac{e_{x4}}{\delta}\right), \qquad \delta > 0. \tag{22}$$

The adaptive law is

$$\dot{\hat{m}}_1 = \mathrm{Proj}_{\hat{m}_1}(\gamma_1 e_{x4}\tau_u), \tag{23}$$

with the projection operator:

$$\mathrm{Proj}_{\hat{m}_1}(\zeta) = \begin{cases} 0, & \hat{m}_1 \geq m_{1\,\mathrm{max}},\ \zeta > 0, \\ 0, & \hat{m}_1 \leq m_{1\,\mathrm{min}},\ \zeta < 0, \\ \zeta, & \text{otherwise.} \end{cases} \tag{24}$$

3.3 Design of τ_r

Similarly,

$$e_{y1} = y_1 - y_{1d}, \qquad e_{yi} = y_i - y_{id},\ i = 2, 3, 4, \tag{25}$$

and the virtual laws are

$$y_{2d} = \dot{y}_{1d} - k_{y1}e_{y1}, \tag{26}$$
$$y_{3d} = \dot{y}_{2d} - k_{y2}e_{y2} - e_{y1}, \tag{27}$$
$$y_{4d} = \dot{y}_{3d} - k_{y3}e_{y3} - e_{y2}. \tag{28}$$

The control input is

$$\tau_r = \frac{1}{\hat{m}_2}\left(\dot{y}_{4d} - k_{y4}e_{y4} - e_{y3} - \eta\,\phi(e_{y4})\right), \tag{29}$$

$$\phi(e_{y4}) = \tanh\left(\frac{e_{y4}}{\delta}\right), \qquad \delta > 0. \tag{30}$$

The parameter update is

$$\dot{\hat{m}}_2 = \mathrm{Proj}_{\hat{m}_2}(\gamma_1 e_{y4}\tau_r), \tag{31}$$

where the projection operator is given by:

$$\mathrm{Proj}_{\hat{m}_2}(\zeta) = \begin{cases} 0, & \hat{m}_2 \geq m_{2\,\max},\ \zeta > 0, \\ 0, & \hat{m}_2 \leq m_{2\,\min},\ \zeta < 0, \\ \zeta, & \text{otherwise.} \end{cases} \tag{32}$$

3.4 Stability Analysis

Theorem 1. For the closed-loop system under (21) and (29), all tracking errors e_{xi} and e_{yi} $(i = 1, \ldots, 4)$ converge asymptotically to zero.

Proof. Consider the Lyapunov function candidate

$$V = \frac{1}{2}\sum_{i=1}^{4}(e_{xi}^2 + e_{yi}^2) + \frac{1}{2\gamma_1}(\tilde{m}_1^2 + \tilde{m}_2^2), \tag{33}$$

where $\tilde{m}_j = m_j - \hat{m}_j$, $j = 1, 2$.

Since $\tilde{m}_j = m_j - \hat{m}_j$ and m_j is constant, we have

$$\dot{\tilde{m}}_j = -\dot{\hat{m}}_j,$$

and the parameter-error term in $\dot{V}$ becomes

$$\dot{V}_{m_j} = \frac{1}{\gamma_1}\tilde{m}_j\dot{\tilde{m}}_j = -\frac{\tilde{m}_j}{\gamma_1}\dot{\hat{m}}_j.$$

Using the definition of the projection operator and the fact that $m_j \in [m_{j\,\min}, m_{j\,\max}]$, it follows that

$$-\frac{\tilde{m}_j}{\gamma_1}\dot{\hat{m}}_j \leq 0, \qquad j = 1, 2.$$

we obtain

$$\begin{aligned} \dot{V} \leq & -\sum_{i=1}^{4}(k_{xi}e_{xi}^2 + k_{yi}e_{yi}^2) \\ & - \eta\, e_{x4} \tanh\left(\frac{e_{x4}}{\delta}\right) - \eta\, e_{y4} \tanh\left(\frac{e_{y4}}{\delta}\right). \end{aligned} \tag{34}$$

Table 1. Simulation Parameters

Parameter	Value
Initial position (x_0, y_0)	$(1.2, -2)$ m
Initial heading ψ_0	$\pi/3$ rad
Simulation time T	50 s
Time step Δt	0.001 s
Control gains (k_1, k_2, k_3, k_4)	$(0.5, 1.0, 2.0, 1.0)$

Since $z \tanh(z/\delta) > 0$ for all $z \neq 0$, the right-hand side is negative semi-definite and equals zero only when

$$e_{xi} = e_{yi} = 0, \quad i = 1, \ldots, 4.$$

By LaSalle's invariance principle, all tracking errors converge asymptotically to the origin.

4 Simulation Results

Numerical simulations are performed to validate the proposed control scheme. The ACV is required to follow a linear path $y_d = \sqrt{3}(x_d - 1)$. Simulation parameters and control gains are summarized in Table 1.

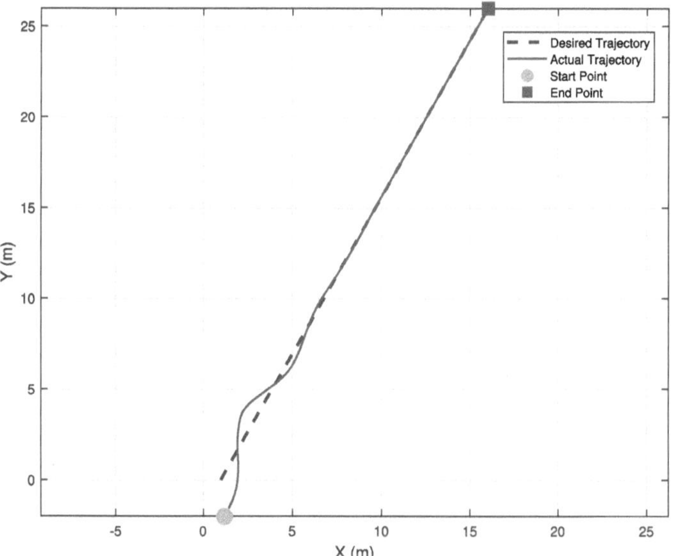

Fig. 1. Trajectory tracking performance.

As shown in the above figures, the ACV successfully tracks the desired trajectory from its initial offset position. The tracking errors converge to near zero within 15 s, and both control inputs remain within saturation limits. These simulation outcomes confirm that the developed adaptive backstepping controller based on differential flatness achieves satisfactory tracking performance (Figs. 1, 2 and 3).

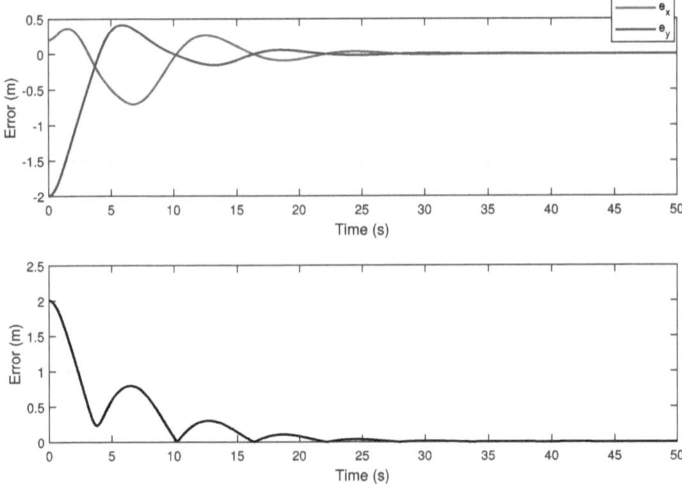

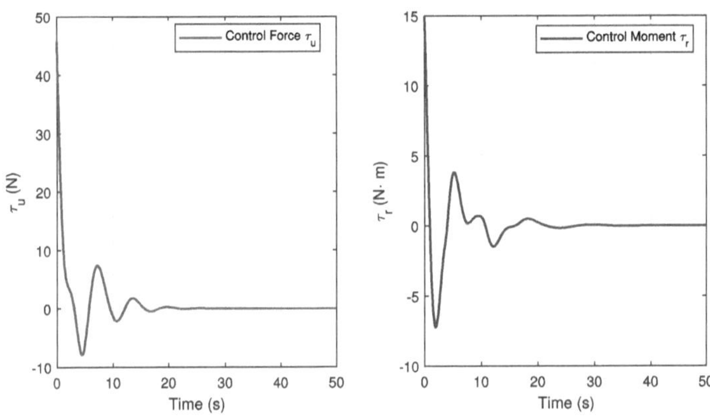

Fig. 2. Tracking errors.

Fig. 3. Control inputs τ_u and τ_r.

5 Conclusions

A trajectory tracking controller for an underactuated ACV has been developed in this work. The differential flatness property has enabled the conversion of the underactuated dynamics into a fully actuated form, thereby simplifying the control design process. An adaptive backstepping approach has been applied to the transformed system, and Lyapunov stability theory has confirmed that all tracking errors tend to zero asymptotically. Through simulation studies, the designed controller has been verified to provide precise trajectory tracking, ensuring accurate trajectory following and adhering closely to the desired spatiotemporal reference.

References

1. Xu, S., Tang, Y., Chen, K.: Numerical investigation on pressure responsiveness properties of the skirt-cushion system of an air cushion vehicle. Int. J. Nav. Archit. Ocean Eng. **12**, 928–942 (2020)
2. Wang, Y., Zhou, H.: Data-driven constrained reinforcement learning algorithm for path tracking control of hovercraft. Ocean Eng. **307**, 118169 (2024)
3. Zuo, Z., Chen, G., Zhou, X.: Analysis of wave load characteristics of hovercraft based on model test. J. Mar. Sci. Eng. **12**, 1537 (2024)
4. Xie, W., Cabecinhas, D., Cunha, R.: Robust motion control of an underactuated hovercraft. IEEE Trans. Control Syst. Technol. **27**, 2195–2208 (2019)
5. Yan, L., Ma, B., Jia, Y.: Adaptive containment control of multiple underactuated hovercrafts subjected to switching and directed topologies. IEEE Syst. J. **17**, 3962–3973 (2023)
6. Yu, R., Tang, W., Li, T.: Speed regulator-based path following control for an underactuated hovercraft considering the sideslip phenomenon. J. Mar. Sci. Eng. **13**(9), 1774 (2025)
7. Su, Y., Teng, F., Li, T., Chen, C.L.P.: Fixed-Time optimal trajectory tracking control for an electric unmanned surface vehicle via reinforcement learning. IEEE/ASME Trans. Mech. **30**(3), 1–12 (2025)
8. Wang, Y., Zhang, H., T.: Lyapunov-based model predictive control for trajectory tracking of hovercraft with actuator constraints and external disturbances. Ocean Eng. **310**, 118631 (2024)
9. Sira-Ramirez, H., T.: Dynamic second-order sliding mode control of the hovercraft vessel. IEEE Trans. Control Syst. Technol. **10**, 860–865 (2002)
10. Zhang, W.; Xu, Y.; Fu, M.; Zhang, G.; Fan, Z., T.: Dynamic event trigger adaptive horizon model free robust predictive control for hovercraft heading tracking using interval predictor. Ocean Eng. **314**, 119602 (2024)
11. Hesser, D.F., Altun, K., Markert, B., T.: Monitoring and tracking of a suspension railway based on data-driven methods applied to inertial measurements. Mech. Syst. Signal Process. **164**, 108298 (2022)
12. Tal, E., Ryou, G., Karaman, S., T.: Aerobatic trajectory generation for a VTOL fixed-wing aircraft using differential flatness. IEEE Trans. Robotics **39**, 4805–4819 (2023)
13. de Vries, E.; Fehn, A.; Rixen, D., T.: Flatness-based model inverse for feed-forward braking control. Vehicle Syst. Dyn.**48**, 353–372 (2010)
14. Su, Y., Shan, Q., Liang, H., Li, T., Zhang, H.: Event-based adaptive optimal fault-tolerant consensus control for uncertain nonlinear multiagent systems with actuator failures. IEEE Trans. Syst. Man Cybern. Syst **55**(10), 7273–7287 (2025)
15. Su, Y., Teng, F., Li, T., Sun, Q.: Adaptive prescribed-time tracking control for an unmanned surface vehicle considering motor-driven propellers. IEEE Trans. Ind. Inf. **21**(2), 1665–1673 (2025)
16. Yu, R., Zhou, Q., Li, T.: Finite-time fault-tolerant tracking control for an air cushion vehicle subject to actuator faults. J. Mar. Sci. Eng. **13**(2), 210 (2025)
17. Kong, X., et al.: Trajectory tracking control for under-actuated hovercraft using differential flatness and reinforcement learning-based active disturbance rejection control. J. Syst. Sci. Complex. **35**(2), 502–521 (2022)

Decoupled Cooperative Encirclement Control for Unmanned Surface Vehicles Using Asymmetric Radial Weighting

Shilong Liu[1], Fei Teng[1(✉)], and Yuanbo Su[2]

[1] School of Marine Electrical Engineering,Dalian Maritime University, Dalian 116026, China
brenda_teng@163.com

[2] School of Navigation, Dalian Maritime University,Dalian 116026, China

Abstract. Cooperative hunting missions involving Unmanned Surface Vehicles (USVs) present a fundamental conflict between maintaining swarm stability and enabling tactically differentiated motion during high-speed maneuvers. To address this, this paper proposes a cooperative control law based on task decoupling and asymmetric radial weighting. The control strategy orthogonally decomposes the global objective into three independent subtasks: collective translation, tangential distribution, and radial shaping. This decomposition effectively eliminates kinematic coupling and enhances system robustness. Specifically, radial shaping is implemented via a weighted pseudo-inverse controller governed by an asymmetric cost function; this mechanism imposes strict penalties on inner USVs while prioritizing the control allocation of outer vessels, resulting in tactically optimized motion. The global asymptotic stability of the system is rigorously guaranteed via Lyapunov analysis. Simulation results demonstrate superior performance in formation establishment, maneuver stability, and asymmetric shaping, validating the method's potential for complex maritime interdiction operations.

Keywords: Cooperative Encirclement · Task Decoupling · Weighted Pseudo-inverse · Asymmetric Cost

1 Introduction

With the advancement of ocean exploration, Unmanned surface vehicles (USVs) cooperative control technology has become a research frontier [8,11]. Among numerous tasks, multi-USV cooperative hunting is a typical challenge [4]. Primarily, this problem demands efficient cooperative strategies for motion coordination. Moreover, the system must contend with complex maritime environments, individual collision avoidance, and the target's intelligent evasion behaviors [7].

Existing strategies, such as leader-follower [3] and the virtual structure method [12], are characterized by structural clarity. However, this clarity is

C. Li et al. (Eds.): ICNC 2025, CCIS 2946, pp. 562–571, 2026.
https://doi.org/10.1007/978-981-92-1599-7_47

derived from rigid formations, which inherently constrain flexibility against dynamic targets. Methods based on the artificial potential field are prone to falling into local optima [9]. In recent years, Jacobian-based self-organizing swarms [1,2] have received attention for their flexibility, but they still have not systematically solved the key challenges under high-speed maneuvers.

Specifically, these approaches are often compromised by inherent coupling interference between different control dimensions. This, in turn, leads to homogenized allocations that lack tactical differentiation, and forces a reliance on passive adjustment for formation uniformity. The proposed cooperative control law, based on task decoupling and asymmetric weighting, is designed to systematically address these challenges. At the kinematic level, this strategy is designed to eliminate coupling interference. This is achieved by orthogonally decomposing the task into three subtasks: collective translation, tangential distribution, and radial shaping. Specifically, a weighted pseudo-inverse controller is introduced for radial shaping. By constructing dynamic weights via asymmetric cost functions, this mechanism achieves differentiated motion allocation aligned with task demands.

The remainder of this paper is organized as follows: Sect. 2 models the system and formulates the control problem. Section 3 details the proposed decoupled control law and asymmetric weighting mechanism. Section 4 provides a stability analysis of the proposed control law. Section 5 validates the effectiveness of the proposed method through simulation experiments, and Sect. 6 concludes the paper.

2 Problem Formulation

To address the inherent inflexibility of rigid formations when responding to dynamic targets, a self-organizing perspective is adopted for the cooperative hunting task [10]. Accordingly, this section first establishes the USV kinematic model. Subsequently, the hunting task is formulated as a macroscopic state tracking problem, laying the theoretical basis for the proposed decoupled control strategy.

Consider a swarm of N homogeneous USVs operating on the horizontal plane. The kinematics of the i-th USV are governed by the standard Three-Degree-of-Freedom (3-DOF) model [5,6]:

$$\dot{\eta}_i = J_R(\psi_i)\nu_i \tag{1}$$

where $\eta_i = [x_i, y_i, \psi_i]^T$ is the pose of the i-th USV in the Earth-fixed frame, comprising the position vector $P_i = [x_i, y_i]^T$ and heading angle ψ_i. The velocity vector $\nu_i = [u_i, v_i, r_i]^T$, defined in the body-fixed frame, represents the surge, sway, and yaw velocities, respectively. $J_R(\psi_i)$ is the velocity transformation matrix from the body-fixed frame to the Earth-fixed frame:

$$J_R(\psi_i) = \begin{bmatrix} \cos(\psi_i) & -\sin(\psi_i) & 0 \\ \sin(\psi_i) & \cos(\psi_i) & 0 \\ 0 & 0 & 1 \end{bmatrix}. \tag{2}$$

Based on this model, the control objective is formulated to ensure the swarm tracks the desired macroscopic state:

$$\lim_{t\to\infty} \|X_{act}(t) - X_d(t)\| = 0 \tag{3}$$

where $X_{act}(t) = [\overline{x}(t), \overline{y}(t), \sigma^2_{act}(t)]^T$ and $X_d(t) = [x_d(t), y_d(t), \sigma^2_d(t)]^T$ represent the actual swarm state and the desired state, respectively. Here, $(\overline{x}, \overline{y})$ represent the position coordinates of the swarm's geometric centroid, and σ^2_{act} denotes the dispersion metric. Correspondingly, (x_d, y_d) and σ^2_d respectively represent the target position $P_{target}(t)$ and the desired mean-squared radius.

Furthermore, the swarm's geometric centroid $X_{act,cen}(t)$ and the dispersion metric $\sigma^2_{act}(t)$ are explicitly derived from the individual USV positions P_i as follows:

$$X_{act,cen}(t) = (\overline{x}(t), \overline{y}(t)) = \frac{1}{N}\sum_{i=1}^{N} P_i(t) \tag{4}$$

$$\sigma^2_{act}(t) = \frac{1}{N}\sum_{i=1}^{N} \|P_i(t) - X_{act,cen}(t)\|^2 \tag{5}$$

The mean-squared form is adopted for this dispersion metric to facilitate Jacobian derivation and enhance numerical stability.

3 Cooperative Hunting Strategy

This section presents a control strategy centered on task decoupling and asymmetric weighting to address the instability of high-speed maneuvers. The proposed strategy systematically resolves the coupling conflict between target tracking and formation maintenance, ensuring both system robustness and tactical differentiation.

3.1 Decoupled Control Law Design

The global convergence objective defined in Eq. (3) necessitates the simultaneous elimination of the centroid error and the dispersion error. Direct coupling of these objectives often leads to kinematic conflicts. Implementation of the decoupled strategy relies on decomposing the control vector into independent kinematic subspaces that map one-to-one to the state errors. As illustrated in Fig. 1, the decoupled control law is synthesized as the following vector sum:

$$\mathbf{u}_i = \mathbf{u}_{trans} + \mathbf{u}_{i,tan} + \mathbf{u}_{i,rad} \tag{6}$$

where $\mathbf{u}_{trans}$, $\mathbf{u}_{i,tan}$, and $\mathbf{u}_{i,rad}$ represent the collective translation velocity, tangential distribution component, and radial shaping velocity, respectively. Specifically, the first two components establish the kinematic foundation for formation stability, while the third realizes tactical differentiation through differentiated

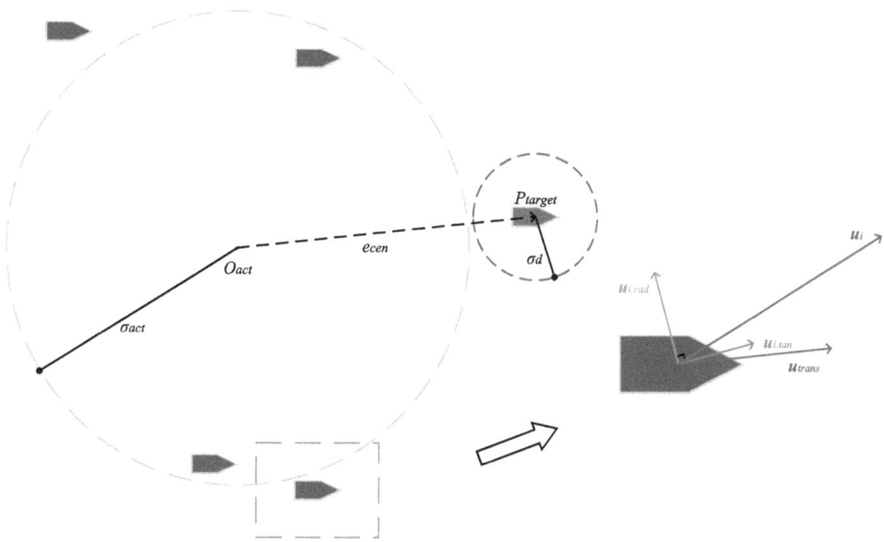

Fig. 1. Schematic diagram of the task decoupling strategy.

radial contraction. Serving as the kinematic reference command, this synthesized vector $\mathbf{u}_i$ provides the desired speed and course.

The translation component $\mathbf{u}_{trans}$ drives the macroscopic target tracking. By defining the centroid tracking error as $e_{cen} = X_{act,cen} - P_{target}$, a unified command is applied to all USVs to eliminate this error while maintaining the swarm's geometric integrity

$$\mathbf{u}_{trans} = \dot{P}_{target} - K_{cen} \cdot e_{cen} \tag{7}$$

where K_{cen} is a positive gain matrix. Since this component acts identically on the entire swarm, it introduces no deformation to the geometry.

The tangential component $\mathbf{u}_{i,tan}$ governs angular spacing regulation. By defining the spacing errors with counter-clockwise and clockwise neighbors as $e_{\theta,i,ccw}$ and $e_{\theta,i,cw}$ respectively, a distributed consensus protocol is established to enforce a uniform distribution $(2\pi/N)$.

$$\mathbf{u}_{i,tan} = K_{unif} \cdot (e_{\theta,i,ccw} - e_{\theta,i,cw}) \cdot \hat{\tau}_i \tag{8}$$

where K_{unif} is a positive control gain and $\hat{\tau}_i$ is the unit tangent vector.

The radial shaping component $\mathbf{u}_{i,rad}$ achieves tactical differentiation. Unlike the explicit expressions of the first two components, this command is synthesized through a constrained optimization process to exploit kinematic redundancy for differentiated allocation.

The formulation begins by establishing the differential relationship between the macroscopic shape evolution $\dot{\sigma}^2_{act}$ and the individual radial velocities $\mathbf{u}_{rad}$:

$$\dot{\sigma}^2_{act} \approx J_{\sigma^2} \mathbf{u}_{rad} \tag{9}$$

where J_{σ^2} represents the Jacobian matrix of the dispersion metric σ_{act}^2 with respect to the swarm configuration. Specifically, by differentiating Eq. (5), this matrix is explicitly derived as:

$$J_{\sigma^2} = \frac{2}{N}\left[(P_1 - X_{act,cen})^T, \ldots, (P_N - X_{act,cen})^T\right] \tag{10}$$

Subsequently, a desired decay rate $\dot{\sigma}_{corr}^2 = -K_{e\sigma}e_{\sigma^2}$, is defined to guarantee rapid convergence of the shape error $e_{\sigma^2} = \sigma_{act}^2 - \sigma_d^2$. Substituting this requirement into Eq. (9) yields the system's linear kinematic constraint:

$$J_{\sigma^2}\mathbf{u}_{rad} = \dot{\sigma}_{corr}^2. \tag{11}$$

Since this system is under-determined $(1 \times 2N)$, addressing the kinematic redundancy requires a specific optimization criterion. The signed radial error $e_{rad,i}$ forms the basis for differentiated motion allocation by quantifying the relative position of each USV:

$$e_{rad,i} = \|P_i - X_{act,cen}\| - \sigma_{act}. \tag{12}$$

Based on this error, an asymmetric cost function C_i is designed to penalize inner USVs while encouraging outer ones.

$$C_i = \begin{cases} 1 + \gamma_{enc} \cdot e_{rad,i} & \text{if } e_{rad,i} > 0 \\ 1 + \gamma_{pun} \cdot (e_{rad,i})^2 & \text{if } e_{rad,i} \leq 0 \end{cases} \tag{13}$$

where γ_{enc} and γ_{pun} are positive constant weighting coefficients designed to regulate the intensity of encouragement and punishment, respectively. By enforcing $\gamma_{pun} \gg \gamma_{enc}$, the weighting matrix $W = \text{diag}(w_1, ..., w_N)$ is constructed using $w_i = 1/C_i$. Consequently, the radial velocity is computed using the weighted pseudo-inverse to satisfy the constraint in Eq. (11) with minimal cost:

$$\mathbf{u}_{rad} = \mathbf{J}_{\sigma^2,W}^{+} \cdot \dot{\sigma}_{corr}^2 \tag{14}$$

where the weighted pseudo-inverse matrix $\mathbf{J}_{\sigma^2,W}^{+}$ is defined as:

$$\mathbf{J}_{\sigma^2,W}^{+} = \mathbf{W}\mathbf{J}_{\sigma^2}^T(\mathbf{J}_{\sigma^2}\mathbf{W}\mathbf{J}_{\sigma^2}^T)^{-1}. \tag{15}$$

This mechanism automatically prioritizes the movement of outer USVs during contraction, realizing the desired tactically differentiated asymmetric shaping.

Finally, considering the physical constraints of USV propulsion systems, the synthesized velocity command $\mathbf{u}_i$ derived in Eq. (6) may exceed the maximum linear speed v_{max}. A saturation constraint is applied to ensure physical feasibility while preserving the direction of the cooperative vector:

$$\mathbf{u}_i^{cmd} = \begin{cases} \mathbf{u}_i, & \text{if } \|\mathbf{u}_i\| \leq v_{max} \\ v_{max}\frac{\mathbf{u}_i}{\|\mathbf{u}_i\|}, & \text{if } \|\mathbf{u}_i\| > v_{max} \end{cases} \tag{16}$$

This synthesized linear velocity vector $\mathbf{u}_i^{cmd}$ provides the final kinematic reference command.

3.2 Stability Analysis

To theoretically guarantee the global asymptotic stability of the proposed decoupled control strategy, the following convergence theorem is established.

Theorem 1. *Consider a swarm of N USVs governed by the kinematic model (1). Assume the communication topology governing the tangential distribution is connected, the control gain matrix* K_{cen} *is positive definite, and the radial gain* $K_{e\sigma}$ *is a positive constant. The proposed decoupled cooperative control law, defined by Eq. (6) and its subtasks (Eqs. 7, 8, 14), guarantees global asymptotic stability of the system. Specifically, the macroscopic tracking errors and angular spacing errors converge asymptotically to zero:*

$$\lim_{t\to\infty} \|X_{act}(t) - X_d(t)\| = 0, \quad \lim_{t\to\infty} \|\mathbf{e}_\theta(t)\| = 0. \tag{17}$$

Proof. A composite Lyapunov function V_{total}, aggregating the energy of the translation, tangential, and radial subsystems, is constructed to verify system stability:

$$V_{total} = \frac{1}{2}e_{cen}^T e_{cen} + \frac{1}{2}\mathbf{e}_\theta^T \mathbf{e}_\theta + \frac{1}{2}e_{\sigma^2}^2 \tag{18}$$

where $\mathbf{e}_\theta$ denotes the angular spacing error vector. The time derivative of V_{total} along the system trajectories driven by the synthesized command $\mathbf{u}_i$ is given by:

$$\dot{V}_{total} = e_{cen}^T \dot{e}_{cen} + \mathbf{e}_\theta^T \dot{\mathbf{e}}_\theta + e_{\sigma^2}\dot{e}_{\sigma^2}. \tag{19}$$

Substituting the kinematic relation of $\mathbf{u}_i = \mathbf{u}_{trans} + \mathbf{u}_{tan} + \mathbf{u}_{rad}$, the cross-coupling terms vanish due to the established kinematic orthogonality. In particular, since the radial and tangential components sum to zero over the swarm, they exert no influence on macroscopic translational dynamics. Consequently, the derivative simplifies to the decoupled form:

$$\dot{V}_{total} = -e_{cen}^T K_{cen} e_{cen} - \mathbf{u}_{tan}^T \mathcal{L} \mathbf{u}_{tan} - K_{e\sigma} e_{\sigma^2}^2 \tag{20}$$

where $\mathcal{L}$ is the Laplacian matrix of the communication graph.

Since K_{cen} is positive definite, $K_{e\sigma}$ is positive, and $\mathcal{L}$ is positive semi-definite for a connected graph, the inequality $\dot{V}_{total} \leq 0$ holds. This confirms the non-increasing nature of the system energy. Furthermore, invoking LaSalle's Invariance Principle, the system trajectories converge to the largest invariant set where $\dot{V}_{total} = 0$. In this set, $e_{cen} \to 0$, $\mathbf{u}_{tan} \to 0$ (hence $\mathbf{e}_\theta \to 0$ due to connectivity), and $e_{\sigma^2} \to 0$ as $t \to \infty$. Thus, the proof is completed. □

Remark 1. The three decoupled subtasks rely on distinct information mechanisms. The collective translation and radial shaping tasks are formulated as macroscopic state tracking problems (Eqs. 7 and 14), which implicitly assume the availability of swarm centroid and target states via global sensing or broadcasting. In contrast, the tangential distribution (Eq. 8) is designed as a distributed consensus protocol involving only local neighbor interactions. Consequently, the connectivity condition in Theorem 1 is specifically imposed on the communication topology governing the tangential layer to ensure the convergence of angular spacing errors.

4 Simulation and Analysis

This section validates the proposed strategy through simulation experiments utilizing a standard 3-DOF USV kinematic model. Low-level controllers are assumed to effectively track the generated kinematic commands. The target acts as a dynamic entity, with specific trajectories defined for each experiment.

4.1 Simulation Parameter Setup

The scenario involves a swarm of four USVs operating cooperatively. The core parameters related to the control law are listed in Table 1.

Table 1. Main parameter settings for the simulation experiments.

Symbol	Value	Symbol	Value
N	4	σ_d	25.0 m
v_{max}	10.0 m/s	$\dot{\psi}_{max}$	50.0°/s
K_{cen}	diag(0.4, 0.4)	K_{unif}	8.0
$K_{e\sigma}$	0.1	Δt	0.05 s
γ_{enc}	0.5	γ_{pun}	5.0

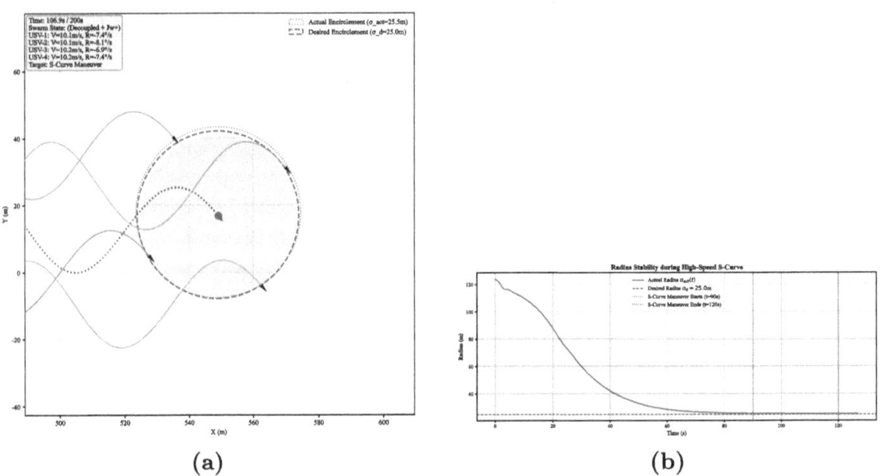

Fig. 2. Formation Robustness under High-Speed Maneuvers. (a) Snapshot of the swarm tracking the S-curve target. (b) Stability of the actual swarm radius $\sigma_{act}(t)$.

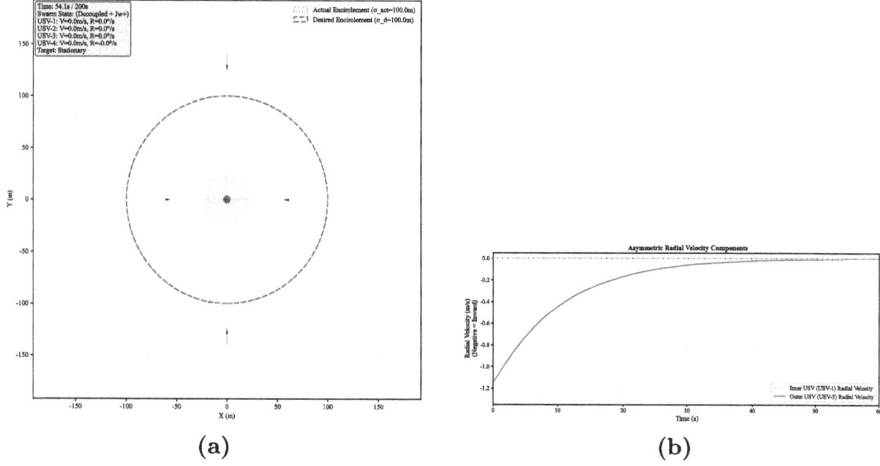

Fig. 3. Asymmetric Radial Shaping. (a) Snapshot of asymmetric contraction. (b) Radial velocity comparison (Inner vs. Outer).

4.2 Formation Robustness Under High-Speed Maneuvers

This experiment verifies the robustness of the decoupled control strategy. The swarm tracks a target executing a continuous high-speed S-turn trajectory. Thus, structural integrity under dynamic stress is rigorously tested, as shown in Fig. 2.

Figure 2(a) visually confirms that the swarm maintains a cohesive and uniform encirclement while closely tracking the high-dynamic S-curve. Crucially, the radius stability near $\sigma_d = 25m$ in Fig. 2(b) verifies the effective isolation of radial shaping from translational interference, proving system robustness.

4.3 Asymmetric-Weighted Radial Shaping

To validate the effectiveness of the Asymmetric-Weighted Control Law, the simulation initializes two USVs inside ($e_{rad} < 0$) and two USVs outside ($e_{rad} > 0$) the target radius of $\sigma_d = 100m$. The results are presented in Fig. 3.

Figure 3(a) visualizes the contraction driven primarily by outer USVs. Figure 3(b) confirms this allocation mechanism: outer USVs are granted higher weights, whereas inner USVs are inhibited by high cost penalties (γ_{pun}), validating the adaptive asymmetric shaping.

5 Conclusion

This paper addresses the stability and rationality challenges in high-speed cooperative hunting by proposing a control strategy based on task decoupling and asymmetric weighting. Coupling interference is effectively eliminated through the orthogonal decomposition of kinematic tasks, and tactically rational motion allocation is ensured via the weighted pseudo-inverse mechanism. The asymptotic stability of these decoupled subsystems is theoretically confirmed using Lyapunov's method. Simulation results validate the strategy's robustness under aggressive maneuvers and its efficacy in achieving intelligent, asymmetric formation control. Future work will focus on integrating dynamic constraints and conducting field validation in maritime environments.

Acknowlegement. This work was supported by the National Natural Science Foundation of China under Grant 52201407.

Disclosure of Interests. The authors have no competing interests to declare that are relevant to the content of this article.

References

1. Liang, X., Qu, X., Wang, N., Li, Y., Zhang, R.: A novel distributed and self-organized swarm control framework for underactuated unmanned marine vehicles. IEEE Access **7**, 112703–112712 (2019). https://doi.org/10.1109/ACCESS.2019.2934190
2. Liang, X., Qu, X., Wang, N., Li, Y., Zhang, R.: Swarm control with collision avoidance for multiple underactuated surface vehicles. Ocean Eng. **191** (2019). https://doi.org/10.1016/j.oceaneng.2019.106516
3. Rezaee, H., Abdollahi, F.: A decentralized cooperative control scheme with obstacle avoidance for a team of mobile robots. IEEE Trans. Industr. Electron. **61**(1), 347–354 (2014). https://doi.org/10.1109/TIE.2013.2245612
4. Sivaneri, V.O., Gross, J.N.: Flight-testing of a cooperative UGV-to-UAV strategy for improved positioning in challenging GNSS environments. Aerosp. Sci. Technol. **82–83**, 575–582 (2018). https://doi.org/10.1016/j.ast.2018.09.035
5. Su, Y., Teng, F., Li, T., Chen, C.L.P.: Fixed-time optimal trajectory tracking control for an electric unmanned surface vehicle via reinforcement learning. In: IEEE/ASME Transactions on Mechatronics, pp. 1–12 (2025). https://doi.org/10.1109/TMECH.2025.3602024
6. Su, Y., Teng, F., Li, T., Sun, Q.: Adaptive prescribed-time tracking control for an unmanned surface vehicle considering motor-driven propellers. IEEE Trans. Industr. Inf. **21**(2), 1665–1673 (2025). https://doi.org/10.1109/TII.2024.3485795
7. Wang, D., Chen, H., Lao, S., Drew, S.: Efficient path planning and dynamic obstacle avoidance in edge for safe navigation of USV. IEEE Internet Things J. **11**(6), 10084–10094 (2024). https://doi.org/10.1109/JIOT.2023.3325234
8. Wang, Y., Liu, W., Liu, J., Sun, C.: Cooperative USV-UAV marine search and rescue with visual navigation and reinforcement learning-based control. ISA Trans. **137**, 222–235 (2023). https://doi.org/10.1016/j.isatra.2023.01.007
9. Wang, Y., Li, J., Zhao, S., Su, P., Fu, H., Niu, H.: Hybrid path planning for USV with kinematic constraints and COLREGS based on APF-RRT and DWA. Ocean Eng. **318** (2025). https://doi.org/10.1016/j.oceaneng.2024.120128
10. Yu, J., et al.: A cooperative hunting method for multi-USVs based on trajectory prediction by OR-LSTM. IEEE Trans. Veh. Technol. **73**(12), 18087–18101 (2024). https://doi.org/10.1109/TVT.2024.3432739

11. Zhang, P., Xue, H., Gao, S., Zhang, J.: Distributed adaptive consensus tracking control for multi-agent system with communication constraints. IEEE Trans. Parallel Distrib. Syst. **32**(6), 1293–1306 (2021). https://doi.org/10.1109/TPDS.2020.3048383
12. Zhang, Q., Lapierre, L., Xiang, X.: Distributed control of coordinated path tracking for networked nonholonomic mobile vehicles. IEEE Trans. Industr. Inf. **9**(1), 472–484 (2013). https://doi.org/10.1109/TII.2012.2219541

Memory-Based Hybrid-Triggered Dissipative DP Control for Switched UMVs

Shixu Guo, Ying Zhao(✉), Shuanghe Yu, Jin Huang, Yan Yan, and Pengyuan Li

Dalian Maritime University, Da Lian 116026, Liao Ning, China
yingz@dlmu.edu.cn

Abstract. The paper addresses the memory-based hybrid event-triggered (MHET) dissipative dynamic positioning (DP) issue for the nonlinear unmanned marine vehicles (UMVs) with mass-switched. Primarily, a switched Takagi-Sugeno (T-S) fuzzy UMV model is established to characterize both the mass-switched behaviors and nonlinear features of the UMVs. Subsequently, a hybrid event-triggered scheme based on fuzzy rules and memory is proposed to optimize the balance between communication resource savings and system performance while reducing the conservativeness of the mechanism. This scheme incorporates dynamic memory elements internally, enabling the effective utilization of historical triggering information. Besides, the DP goal is explained from a quasi-dissipative perspective. Finally, simulation results for actual switched vessels are analyzed to validate the effectiveness of the proposed controller in DP tasks.

Keywords: Unmanned marine vehicles · mass-switched · dynamic positioning · hybrid · memory · event-triggered · quasi-dissipative

1 Introduction

Unmanned marine vehicles (UMVs) are advanced autonomous vessels whose operation relies on dynamic positioning (DP) systems for station-keeping and trajectory tracking [1]. The DP system maintains precise position control under environmental disturbances by adjusting thrust [2], yet faces challenges from nonlinear dynamics, uncertainties, and actuator constraints [3]. While T-S fuzzy methods address UMV nonlinearities, mass-switching behavior during operations introduces additional instability risks [4], which existing studies often overlook. Event-triggered control schemes, including static event-triggered (SET), dynamic event-triggered (DET) [5], and MET approaches [6], have been developed to reduce communication burden. However, memory-based hybrid event-triggered (MHET) schemes remain unexplored for UMVs. Meanwhile, dissipativity theory provides an energy-based stability criterion where system energy growth cannot exceed external energy supply [7]. While progress has been made

C. Li et al. (Eds.): ICNC 2025, CCIS 2946, pp. 572–581, 2026.
https://doi.org/10.1007/978-981-92-1599-7_48

in combining ET schemes with dissipativity analysis, these results remain primarily theoretical without practical application to UMV DP systems. To address these gaps, this paper proposes a memory-based hybrid event-triggered dissipative dynamic positioning (MHETDDP) strategy for switched fuzzy UMVs, introducing an energy-based perspective to the DP problem under MHET scheme.

i) A T-S fuzzy model is constructed for UMVs with nonlinear and mass-switching dynamics, offering improved practicality over models neglecting these features [8]. Event-triggered parameters are also tuned via fuzzy rules to reduce conservatism.

ii) A memory-based hybrid event-triggered (MHET) scheme utilizes historical triggers to ease communication burden. Unlike single-mode schemes [4], MHET combines dynamic MET with static triggering, balancing DP performance and resource savings under bandwidth constraints.

iii) The DP objective is achieved via a multi-Lyapunov approach and quasi-dissipativity analysis, ensuring stability through efficient energy dissipation. Effectiveness is validated with a mass-varying UMV case.

2 Problem Formulation

2.1 Switched Fuzzy Modeling for UMVs

Consider the nonlinear UMV model

$$\dot{\psi}(t) = \mathcal{T}(\kappa(t))\pi(t), \quad R\dot{\pi}(t) = -G\pi(t) + K\phi(t) + \mho(t), \tag{1}$$

where $\psi(t) = [\psi_x(t), \psi_y(t), \kappa(t)]^T$ and $\pi(t) = [\pi_x(t), \pi_y(t), r(t)]^T$ denote the position-yaw and velocity vectors, respectively. $\phi(t)$ and $\mho(t)$ represent propulsion forces and ocean disturbances. R and G are invertible matrices, and $\mathcal{T}(\kappa(t))$ is the rotation matrix. Defining $\ell(t) = [\psi^T(t) \;\; \pi^T(t)]^T$, the system is expressed as

$$\dot{\ell}(t) = M\ell(t) + F\phi(t) + N\mho(t). \tag{2}$$

Parameters R, G, K vary with mass switching. The switching is governed by a piecewise constant function $\eta(t) : [0, \infty) \longrightarrow U = 1, 2, \ldots, u$, leading to the switched nonlinear model:

$$\dot{\ell}(t) = M\eta(t)\ell(t) + F\eta(t)\phi(t) + N_{\eta(t)}\mho(t). \tag{3}$$

To address nonlinearities, a T-S fuzzy approach is adopted. Considering $\kappa(t) \in [-\pi/6, \pi/6]$, premise variables are chosen as $\theta_1(t) \in [-1/2, 1/2]$ and $\theta_2(t) \in [\sqrt{3}/2, 1]$. Four fuzzy rules ($\upsilon \in \mathbb{H} = 1, 2, 3, 4$) are constructed:

Plant Rule υ: IF $\theta_1(t)$ is $\hbar_{\upsilon 1}$ and $\theta_2(t)$ is $\hbar_{\upsilon 2}$, THEN

$$\dot{\ell}(t) = M^{\upsilon}\eta(t)\ell(t) + F^{\upsilon}\eta(t)\phi(t) + N^{\upsilon}\eta(t)\mho(t), \;\; \varOmega(t) \;\; = \mathcal{E}^{\upsilon}\eta(t)\ell(t), \tag{4}$$

where $\Omega(t)$ is the output. Define $\hbar_{\upsilon}(\theta(t)) = \hbar_{\upsilon 1}(\theta_1(t))\hbar_{\upsilon 2}(\theta_2(t))$. The normalized membership function is obtained as $\varphi^{\upsilon}\eta(t)(\theta(t)) = \hbar\upsilon(\theta(t))/\sum_{k=1}^{4}\hbar_k(\theta(t))$.

The overall switched fuzzy UMV (SFUMV) model is:

$$\begin{aligned}\dot{\ell}(t) &= \sum_{\upsilon=1}^{4}\varphi^{\upsilon}\eta(t)(\theta(t))\left[M^{\upsilon}\eta(t)\ell(t) + F^{\upsilon}\eta(t)\phi(t) + N^{\upsilon}\eta(t)\mho(t)\right],\\ \Omega(t) &= \sum_{\upsilon=1}^{4}\varphi^{\upsilon}\eta(t)(\theta(t))\mathcal{E}^{\upsilon}\eta(t)\ell(t).\end{aligned} \tag{5}$$

2.2 Hybrid Memory Event-Triggered Scheme Design

To ensure navigation safety under limited communication resources, a memory-based hybrid event-triggered (MHET) scheme is developed for UMVs.

Case A): If the time-triggered scheme is executed, the triggered instantaneous state can be obtained by $\ell_a(t) = \ell(t)$.

Case B): If the triggering criteria outlined below are met, the execution of the dynamic MET scheme within the hybrid-triggered scheme takes place

$$t_{s+1} = \min\left\{\vartheta \;\middle|\; \sum_{\upsilon=1}^{4}\varphi^{\upsilon}_{\eta(\vartheta)}\mathcal{S}\left(e^T(\vartheta)\Theta^{\upsilon}_{\eta(\vartheta)}\ell_e(\vartheta) - \alpha\ell^T(\vartheta)\Theta^{\upsilon}_{\eta(\vartheta)}\ell(\vartheta)\right) \geq \partial(\vartheta)\right\} \tag{6}$$

where t_s indicates the last ET point, $\upsilon \in \mathbb{H}$, $s \in \mathbb{N}$, $\ell_e(t) = \ell(t_s) - \ell(t)$ refers to the state error. Here, $\mathcal{S} > 0$, $0 < \alpha < 1$, $\partial(t)$ is derived from the dynamics

$$\begin{cases}\dot{\partial}(t) = \sum\limits_{\upsilon=1}^{4}\varphi^{\upsilon}_{\eta(t)}[\alpha\ell^T(t)\Theta^{\upsilon}_{\eta(t)}\ell(t) - \ell_e^T(t)\Theta^{\upsilon}_{\eta(t)}\ell_e(t)\\ \quad +re^{-\epsilon\partial(t-\check{\vartheta})}\partial(t-\check{\vartheta}) - \varepsilon\partial(t)], t \in [0, +\infty),\\ \partial(t) = f(t), t \in [-\check{\vartheta}, 0).\end{cases} \tag{7}$$

Let $\partial(0) = \partial_0 \geq 0$, $r > 0$, $\varepsilon > 0$, $\check{\vartheta} > 0$ (memory span), and $\epsilon > 0$ as specified constants. The matrix $\Theta^{\upsilon}_i > 0$ for $i \in U$ is to be determined. Then, the state output with the added memory functionality is redefined as $\ell_b(t) = \ell(t_s) = \ell(t) + \ell_e(t)$.

Here, to represent the transition probability between the two scheme modes, an adjustable Bernoulli variable $\mathcal{P}(t)$ is introduced. Therefore, the hybridized transmission moment $\bar{\ell}(t)$ is finally defined as

$$\bar{\ell}(t) = \ell(t) + (1 - \mathcal{P}(t))\ell_e(t), \tag{8}$$

where $\mathcal{P}(t)$ follows a Bernoulli distribution with the expectation $\mathbb{E}\{\mathcal{P}(t)\} = \bar{\mathcal{P}}$.

Next, we outline the design process for the MHET controller as follows:

Controller Rule δ: IF $\theta_1(t_s)$ is $\hbar_{\delta 1}$ and $\theta_2(t_s)$ is $\hbar_{\delta 2}$

THEN

$$\phi(t) = \mathcal{L}^{\delta}_{\eta(t)}\bar{\ell}(t_s), \forall t \in [t_s, t_{s+1}),$$

where $\theta_1(t_s)$, $\theta_2(t_s)$ are the premise variables, $\hbar_{\delta 1}$, $\hbar_{\delta 2}$ $(\delta \in \mathbb{H})$ are the fuzzy sets, the controller gains $\mathcal{L}_i^\delta$ for $i \in \eta(t)$ are determined under the switched rule $\eta(t)$ and fuzzy rule δ. The specific form of the MHET controller based on T-S fuzzy approach is as follows

$$\phi(t) = \sum_{\delta=1}^{4} \varphi_{\eta(t)}^{\delta}(\theta(t_s))\mathcal{L}_{\eta(t)}^{m}\bar{\ell}(t_s), \forall t \in [t_s, t_{s+1}). \tag{9}$$

By applying $\ell_e(t) = \ell(t_s) - \ell(t)$, we can derive the closed-loop system

$$\begin{aligned} \dot{\ell}(t) &= \sum_{\upsilon=1}^{4}\sum_{\delta=1}^{4} \varphi_i^{\upsilon}(\theta(t))\varphi_i^{\delta}(\theta(t_s))\Big\{ \left(M_i^{\upsilon} + F_i^{\upsilon}\mathcal{L}_i^{\delta}\right)\ell(t) \\ &\quad + \left([1-\mathcal{P}(t)]F_i^{\upsilon}\mathcal{L}_i^{m}\right)\ell_e(t) + N_i^{\upsilon}\mho(t)\Big\}, \\ \Omega(t) &= \sum_{\upsilon=1}^{4} \varphi_{\eta(t)}^{\upsilon}(\theta(t))\mathcal{E}_i^{\upsilon}\ell(t), \end{aligned} \tag{10}$$

The relevant definitions and lemma regarding the dissipative DP control problem are given here.

Lemma 1. *[6] For the dynamic MET scheme (6), if $X = -2\varepsilon + r + 1$ and the inequality $X + o < 0$ holds for some constant $o > 0$, then, it follows that*

$$0 \le \partial(t) < \partial_m, t \in [-\check{\vartheta}, +\infty), \tag{11}$$

where $\partial_m \ge 0$ is a constant.

Definition 1. *[9] For any two times ϑ_a and ϑ_b such that $\vartheta_b - \vartheta_a > 0$, the mass-switched rule $\eta(t)$ defines $\mathcal{R}_\eta(\vartheta_a, \vartheta_b)$ as the measure of switching within the interval $[\vartheta_a, \vartheta_b]$. If*

$$\mathcal{R}_\eta(\vartheta_a, \vartheta_b) \le \mathcal{R}_0 + \frac{\vartheta_b - \vartheta_a}{\vartheta_d}, \tag{12}$$

holds for $\vartheta_d > 0$, then, ϑ_d is termed the average dwell-time of $\eta(t)$.

Definition 2. *[10] Given two real symmetric matrices Q, R and a real matrix V, where $Q \le 0$. The function $\varpi(\Omega, \mho, t)$ defined as*

$$\varpi(\Omega, \mho, t) = \langle \Omega, Q\Omega \rangle_t + 2\langle \Omega, V\mho \rangle_t + \langle \mho, R\mho \rangle_t, \forall t \ge 0, \tag{13}$$

represents the supply rate of the system (10). Besides, suppose that there exists $\bar{Q} > 0$ such that $-Q = \bar{Q}^T\bar{Q}$ holds.

Definition 3. *[11] If there exist a nonnegative storage function $H(\ell(t))$ and a constant $o > 0$ such that for $\forall t^* \ge 0$,*

$$\mathbb{E}\left\{H\left(\ell\left(t^*\right)\right) - H\left(\ell(0)\right)\right\} \le \mathbb{E}\left\{\int_0^{t^*} \left(\varpi(\Omega, \mho, t) + o\right) dt\right\} \tag{14}$$

holds, then the SFUMV system (10) is termed to exhibit quasi-dissipativity with respect to the supply rate $\varpi(\Omega, \mho, t)$.

2.3 Performance Objective

The control objectives associated with the SFUMV system (10) are summarized as follows

1. **Practical Stability**: When the disturbance $\mho(t)=0$, the practical stability is reflected in the state trajectories of the system (10) across three dimensions.
2. **Quasi-Dissipativity**: When the disturbance $\mho(t)\neq 0$, the system demonstrates quasi-dissipativity, expressed by the inequality:

$$\mathbb{E}\left\{H(\ell(t),\eta(t))-H(\ell(0),\eta(0))\right\}<\mathbb{E}\left\{\int_0^t(\varpi(\Omega,\mho,\xi)+\epsilon_0)d\xi\right\} \tag{15}$$

where $\epsilon_0>0$ is a constant. If such conditions are met, the MHETDDP problem for the SFUMV system (10) is solvable. Thus, the controller defined in (9) is termed the MHET controller.

3 Main Result

Via this section, a solvability sufficient criteria for solving the MHETDDP problem of the SFUMV system (10) is established. Subsequently, the corresponding MHET controller (9) is designed. The absence of Zeno behavior is also verified.

3.1 Stability Analysis for Switched Fuzzy UMVs System

Theorem 1. *Consider the SFUMV system (10). For the given parameters $\gamma_0>0,\rho\geq 1,\varepsilon>0$, $0<\alpha<1$ and matrices $\bar{Q}\geq 0$, $V=[V_1,V_2,V_3]$, $R=R^T=\begin{bmatrix}R_1 & R_2 & R_3\\ * & R_4 & R_5\\ * & * & R_6\end{bmatrix}$, if the $\mathcal{Q}_i>0$, $\Theta_i^{\upsilon}>0$ and matrices $\mathcal{L}_i^{\delta}$ can be determined to satisfy*

$$\begin{bmatrix}\Gamma_{1i}^{\upsilon\delta} & \Gamma_{2i}^{\upsilon\delta} & \sum\limits_{\upsilon=1}^{4}\varphi_i^{\upsilon}\mathcal{Q}_iN_i^{\upsilon} & 0\\ * & -\Theta_i^{\upsilon} & 0 & 0\\ * & * & \Gamma_{3i} & 0\\ * & * & * & \gamma_0-\varepsilon\end{bmatrix}<0, \tag{16}$$

$$\mathcal{Q}_i<\rho\mathcal{Q}_j,\ i,j\in U, \tag{17}$$

where

$$\Gamma_{1i}^{\upsilon\delta}=\sum_{\upsilon=1}^{4}\sum_{\delta=1}^{4}\varphi_i^{\upsilon}\varphi_i^{\delta}\Psi_{1i}^{\upsilon\delta}+\sum_{\upsilon=1}^{4}\varphi_i^{\upsilon}\mathcal{E}_i^{\upsilon^T}\mathcal{E}_i^{\upsilon}+\Psi_{2i}^{\upsilon}+\sum_{\upsilon=1}^{4}\varphi_i^{\upsilon}\alpha\Theta_i^{\upsilon},$$

$$\Gamma_{2i}^{\upsilon\delta}=\sum_{\upsilon=1}^{4}\sum_{\delta=1}^{4}\varphi_i^{\upsilon}\varphi_i^{\delta}(1-\mathcal{P}(t))\mathcal{Q}_iF_i^{\upsilon}\mathcal{L}_i^{\delta},$$

$$\Gamma_{3i}=V^TV-R,\Psi_{2i}^{\upsilon}=\sum_{\upsilon=1}^{4}\varphi_i^{\upsilon}\mathcal{E}_i^{\upsilon^T}\bar{Q}^T\bar{Q}\mathcal{E}_i^{\upsilon},$$

then, under the MHET scheme (8), controller (9) and mass-switched rule $\eta(t)$ *with the constraint*

$$\vartheta_d^* = \frac{\ln \rho}{\gamma} < \vartheta_d, \gamma \in (0, \gamma_0), \tag{18}$$

the MHETDDP issue for the SFUMV system (10) can be resolved.

Proof. When the subsystem is $\eta(t) = i$, we have the Lyapunov function as $\overline{H}_i(\ell(t)) = \ell^T(t)\mathcal{Q}_i\ell(t),\ i \in Q$. For convenience, an auxiliary function is selected

$$H_i(\ell(t)) = \ell^T(t)\mathcal{Q}_i\ell(t) + \eth(t),\ i \in Q. \tag{19}$$

Take the derivative of $H_i(\ell(t))$ and Lemma 1 yields that

$$\begin{aligned}&\mathbb{E}\{\dot{H}_i(\ell(t))\} + \gamma_0\mathbb{E}\{H_i(\ell(t))\} - \Omega^T(t)Q\Omega(t) - 2\Omega^T(t)V\mho(t) - \mho^T(t)R\mho(t)\\ &\le \begin{bmatrix} \ell(t) \\ e(t) \\ \mho(t) \\ \sqrt{\eth}(t) \end{bmatrix}^T \begin{bmatrix} \Lambda_{11i}^{\upsilon\delta} & \mathcal{Q}_i\Lambda_{12i}^{\upsilon\delta} & \mathcal{Q}_iN_i^{\upsilon} & 0 \\ * & -\sum\limits_{\upsilon=1}^{4}\varphi_i^{\upsilon}\Theta_i^{\upsilon} & 0 & 0 \\ * & * & \Lambda_{33} & 0 \\ * & * & * & \Lambda_{44} \end{bmatrix} \begin{bmatrix} \ell(t) \\ \ell_e(t) \\ \mho(t) \\ \sqrt{\eth}(t) \end{bmatrix} + \epsilon_0\end{aligned}$$

with $\Lambda_{11i}^{\upsilon\delta} = \mathcal{Q}_i \sum\limits_{\upsilon=1}^{4}\sum\limits_{\delta=1}^{4}\left[\varphi_i^{\upsilon}\varphi_i^{\delta}\left(M_i^{\upsilon} + F_i^{\upsilon}\mathcal{L}_i^{\delta}\right) + \left(M_i^{\upsilon} + F_i^{\upsilon}\mathcal{L}_i^{\delta}\right)^T\mathcal{Q}_i + \gamma_0\mathcal{Q}_i + \mathcal{E}_i^{\upsilon^T}\mathcal{E}_i^{\upsilon} - \mathcal{E}_i^{\upsilon^T}Q\mathcal{E}_i^{\upsilon}\right] + \sum\limits_{\upsilon=1}^{4}\varphi_i^{\upsilon}\alpha\Theta_i^{\upsilon}$, $\Lambda_{44} = \gamma_0 - \varepsilon$, $\epsilon_0 = r\eth_m$, $\Lambda_{12i}^{\upsilon\delta} = \sum\limits_{\upsilon=1}^{4}\sum\limits_{\delta=1}^{4}\varphi_i^{\upsilon}\varphi_i^{\delta}([1-\bar{\mathcal{P}}]F_i^{\upsilon}\mathcal{L}_i^{\delta})$.

During this process, we apply $-2\Omega^T(t)V\mho(t) = \sum\limits_{\upsilon=1}^{4}\varphi_i^{\upsilon}[-2\ell^T(t)\mathcal{E}_i^{\upsilon^T}V\mho(t)] \le \sum\limits_{\upsilon=1}^{4}\varphi_i^{\upsilon}[\ell^T(t)\mathcal{E}_i^{\upsilon^T}\mathcal{E}_i^{\upsilon}\ell(t) + \mho(t)V^TV\mho(t)]$. Integrating (16) with (17), it generates

$$\mathbb{E}\{\dot{H}_i(\ell(t))\} < -\gamma_0\mathbb{E}\{H_i\}(\ell(t)) + \Gamma(t) + \epsilon_0, \tag{20}$$

where $\Gamma(t) = \Omega^T(t)Q\Omega(t) + 2\Omega^T(t)V\mho(t) + \mho^T(t)R\mho(t)$.

Presently, the proof needs to be discussed in two distinct cases.

Case i): When $\mho(t) = 0$, it is easy to obtain $\ell(t)$ remains confined within the region $\zeta(\ell(t)) = \left\{\ell(t) : \|\ell(t)\| \le \max\limits_{i\in U}\sqrt{\frac{\epsilon_0}{\gamma_0\nu_{\min}(\mathcal{Q}_i)}}\right\}$.

Case ii): For simplicity, $H_{\eta(t_k)}(\ell(t_k))$ is considered as the value of (19) at the mass-switched point t_k, where $k = 0, 1, \ldots, \mathcal{R}_\eta(0, t)$. When $\mho(t) \ne 0$, considering (20), if $t \in [t_k, t_{k+1})$, we can derive

$$\mathbb{E}\{H_{\eta(t)}(\ell(t))\} < e^{-\gamma_0(t-t_k)}\mathbb{E}\{H_{\eta(t_k)}(\ell(t_k))\} + \mathbb{E}\left\{\int_{t_k}^{t} e^{-\gamma_0(t-\xi)}\Phi(\xi)d\xi\right\} \tag{21}$$

with $\Phi(\xi) = \Omega^T(\xi)Qz(\xi) + 2\Omega^T(\xi)V\mho(\xi) + \mho^T(\xi)R\mho(\xi) + \epsilon_0$.
Further, we know that at the switching point t_k,

$$\mathbb{E}\{H_{\eta(t_k)}(\ell(t_k))\} \leq \rho\mathbb{E}\{H_{\eta(t_k^-)}(\ell(t_k^-))\}. \tag{22}$$

Synthesizing (21) and (22) yields that

$$\mathbb{E}\{H_{\eta(t)}(\ell(t))\} < \rho e^{-\gamma_0(t-t_k)}\mathbb{E}\{H_{\eta(t_k^-)}(\ell(t_k^-))\} + \mathbb{E}\left\{\int_{t_k}^{t} e^{-\gamma_0(t-\xi)}\Phi(\xi)\,d\xi\right\}. \tag{23}$$

From now on, n_η is used to represent $\mathcal{R}_\eta(0,t)$. Thus, it yields that

$$\mathbb{E}\{H_{\eta(t)}(\ell(t))\} < e^{-\gamma_0 t + n_\eta \ln \rho}\mathbb{E}\{H_{\eta(0)}(\ell(0))\} + \mathbb{E}\left\{\int_{0}^{t} e^{-\gamma_0(t-\xi)+n_\eta \ln \pi}\Phi(\xi)\,d\xi\right\}. \tag{24}$$

By dividing $e^{n_\eta \ln \rho}$ on both sides of (24), one has

$$\mathbb{E}\{e^{-n_\eta \ln \rho}H_{\eta(t)}(\ell(t))\} - \mathbb{E}\{e^{-\gamma_0 t}H_{\eta(0)}(\ell(0))\} < \mathbb{E}\left\{\int_{0}^{t} \Phi(\xi)\,d\xi\right\}. \tag{25}$$

It yields that

$$\mathbb{E}\{e^{-\gamma_0 t}H_{\eta(t)}(\ell(t))\} \leq \mathbb{E}\{e^{-n_\eta \ln \rho}H_{\eta(t)}(\ell(t))\} \tag{26}$$

Noticing $\mathbb{E}\{e^{-\gamma_0 t}H_{\eta(0)}(\ell(0))\} \leq \mathbb{E}\{H_{\eta(0)}(\ell(0))\}$ for $t > 0$ and utilizing (26), (25) can be further deduced by $\mathbb{E}\{e^{-\gamma_0 t}H_{\eta(t)}(\ell(t)) - H_{\eta(0)}(\ell(0))\} < \mathbb{E}\left\{\int_0^t \Phi(\xi)\,d\xi\right\}$. Letting $\check{H}(t,\ell(t),\eta(t)) = e^{-\gamma_0 t}\{H_{\eta(t)}(\ell(t))\}$, it yields

$$\mathbb{E}\left\{\check{H}(t,\zeta(t),\eta(t)) - \check{H}(0,\zeta(0),\eta(0))\right\}\mathbb{E}\left\{\int_{0}^{t}(\varpi(\Omega,\mho,\xi) + \epsilon_0)d\xi\right\}. \tag{27}$$

Therefore, according to the expression of $\varpi(\Omega,\mho,t)$ in Definition 2, it can be inferred that the SFUMV system (10) is quasi-dissipative.

3.2 DP Control Design Algorithm

In this part, the gains of the MHET controller (9) required to solve the MHET-DDP issue for SFUMV system (10) is computed.

Theorem 2. *For the SFUMV system* (10) *with the given scalars* $\gamma_0 > 0, \rho \geq 1, \varepsilon > 0$, $0 < \alpha < 1$, *and specified matrices* $\bar{Q} \geq 0$, $V = [V_1, V_2, V_3]$, $R = R^T$, *if* $\bar{\mathcal{Q}}_i > 0$, $\hat{\Gamma}_{2i}^{\upsilon} > 0$ *and* $\bar{\mathcal{L}}_i^{\delta}$, $i \in U$, $\delta \in \mathbb{H}$ *can be searched to meet the constraints*

$$\begin{bmatrix} \hat{\Gamma}_{1i}^{\upsilon\delta} & \hat{\Psi}_{2i}^{\upsilon\delta} & \sum\limits_{\upsilon=1}^{4}\varphi_i^{\upsilon}N_i^{\upsilon}\bar{\mathcal{Q}}_i & 0 & \hat{\Psi}_{5i}^{\upsilon} & \hat{\Psi}_{6i}^{\upsilon} \\ * & -\hat{\Gamma}_{2i} & 0 & 0 & 0 & 0 \\ * & * & \hat{\Gamma}_{3i} & 0 & 0 & 0 \\ * & * & * & \Lambda_{44} & 0 & 0 \\ * & * & * & * & -I & 0 \\ * & * & * & * & * & -I \end{bmatrix} < 0, \tag{28}$$

$$\bar{\mathcal{Q}}_i < \rho\bar{\mathcal{Q}}_j,\ i,j \in U, \tag{29}$$

where

$$R = R^T = \begin{bmatrix} R_1 & R_2 & R_3 \\ * & R_4 & R_5 \\ * & * & R_6 \end{bmatrix} \hat{\Gamma}_{1i}^{\upsilon\delta} = \sum_{\upsilon=1}^{4}\sum_{\delta=1}^{4} \varphi_i^{\upsilon}\varphi_i^{\delta}\hat{\Psi}_{1i} + \alpha\hat{\Gamma}_{2i},$$

$$\hat{\Gamma}_{2i}^{\upsilon} = \sum_{\upsilon=1}^{4} \varphi_i^{\upsilon}\bar{\mathcal{Q}}_i\Theta_i^{\upsilon}\bar{\mathcal{Q}}_i, \hat{\Psi}_{5i}^{\upsilon} = \bar{\mathcal{Q}}_i\sum_{\upsilon=1}^{4}\varphi_i^{\upsilon}\mathcal{E}_i^{\upsilon^T}, \hat{\Psi}_{6i}^{\upsilon} = \bar{\mathcal{Q}}_i\sum_{\upsilon=1}^{4}\varphi_i^{\upsilon}\mathcal{E}_i^{\upsilon^T}\bar{\mathcal{Q}}^T,$$

$$\hat{\Gamma}_{3i} = \begin{bmatrix} V_1^2 - R_1 & V_1V_2 - R_2 & V_1V_3 - R_3 \\ * & V_2^2 - R_4 & V_2V_3 - R_5 \\ * & * & V_3^2 - R_6 \end{bmatrix}, \hat{\Psi}_{2i}^{\upsilon\delta} = \sum_{\upsilon=1}^{4}\sum_{\delta=1}^{4}\varphi_i^{\upsilon}\varphi_i^{\delta}[1-\bar{\mathcal{P}}]\mathcal{Q}_iF_i^{\upsilon}\mathcal{L}_i^{m},$$

$$\hat{\Psi}_{1i}^{\upsilon\delta} = \sum_{\upsilon=1}^{4}\sum_{\delta=1}^{4}\varphi_i^{\upsilon}\varphi_i^{\delta}\left[M_i^{\upsilon}\bar{\mathcal{Q}}_i + F_i^{\upsilon}\bar{\mathcal{L}}_i^{\delta} + \bar{\mathcal{Q}}_iM_i^{\upsilon^T} + \bar{\mathcal{L}}_i^{\delta^T}F_i^{\upsilon^T} + \bar{\mathcal{Q}}_i\gamma_0\right],$$

then, by combining the MHET scheme (8), mass-switched rule $\eta(t)$ ensuring (12) and MHET controller (9), SFUMV system (10) attains the MHETDDP control objective.

4 Simulation Examples

In marine engineering, mass-switched UMVs serve as key tools for seafloor shape detection. The vessel exhibits mass-switching characteristics during detector deployment and recovery, necessitating DP to maintain station-keeping.

Let $\mathcal{E}_1^{\upsilon} = \mathcal{E}_2^{\upsilon} = [0, 0, -0.2, 0, 0, 0.01], \upsilon \in \mathbb{H}$. Using the parameters $R_{\eta(t)}$, $G_{\eta(t)}$, $F_{\eta(t)}$, $N_{\eta(t)}$, the matrices for the SFUMV system (10) are obtained.

The adjustable parameters of the MHET scheme (8) are set as $\varepsilon = 3$, $\mathcal{S} = 0.8$, $\epsilon = 1$, $r = 2$, $\breve{\vartheta} = 0.5$, $\alpha = 0.95$. Initial conditions are $\psi(0) = [2\text{m}, 1\text{m}, 1.2°]^T$ and $\pi(0) = [1.5\text{m/s}, 0\text{m/s}, 0°/\text{s}]^T$.

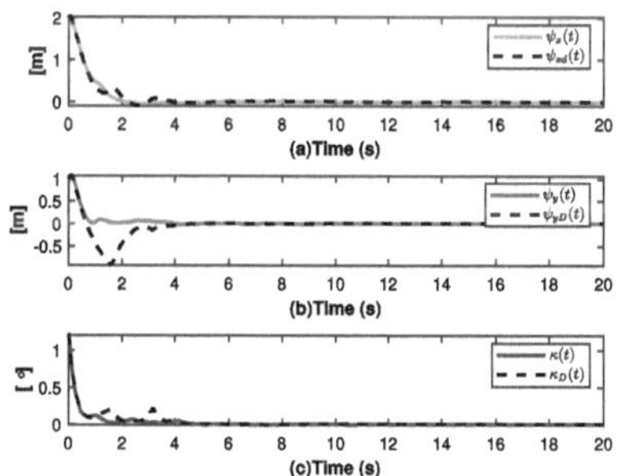

Fig. 1. Position-yaw information.

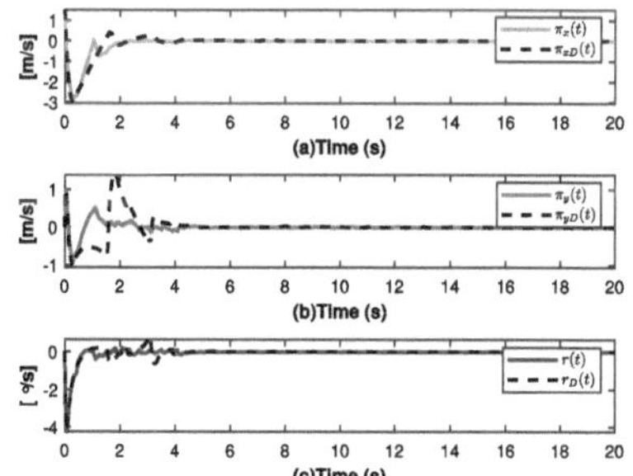

Fig. 2. Velocity and rate information.

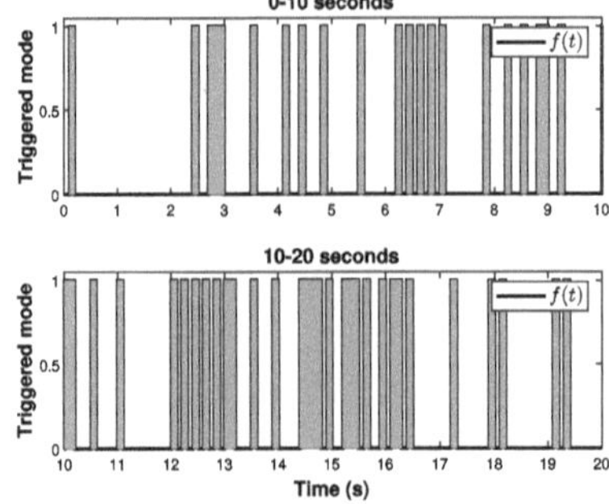

Fig. 3. Trigger mode switching.

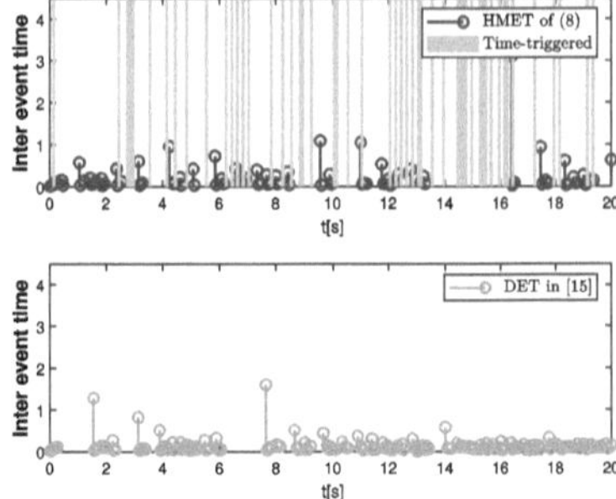

Fig. 4. Comparison of two schemes.

The operational outcomes of SFUMV system are depicted in Figs. 1-4. Figures 1 and 2 compare the performance of the SFUMV under different ET schemes. The small peak-to-trough variation and reduced volatility underscore the superiority of the proposed mechanism in this study. Figure 3 demonstrates the switching pattern of the proposed mechanism, while Fig. 4 clearly shows a significant reduction in the usage of communication resources. The gray shaded regions indicate the active periods of time-triggered operation, during which the event-triggered mechanism remains inactive. Note that the proposed MHET scheme in this work seeks a compromise between system performance levels and the economy of communication resources, thereby significantly enhancing DP performance by reducing triggering times.

5 Conclusion

This paper proposes a mixed/hybrid event-triggered dynamic positioning control strategy for the switched fuzzy UMV system. A switched T-S fuzzy model is established to characterize the UMV with mass switching and nonlinearities. To reduce conservatism under switching rules, a fuzzy rule-based MHET scheme is developed. Its dynamic memory component effectively utilizes historical triggering data. By integrating time-triggered and memory-based event-triggered mechanisms under limited bandwidth, the scheme balances system performance and communication efficiency. From a quasi-dissipativity perspective, the DP objective of the SFUMV is achieved, ensuring stable energy conversion and positioning. Finally, a case study on a vessel with multiple mass blocks confirms the strategy's reliability.

Acknowledgments. This work was supported in part by the National Natural Science Foundation of China under Grant 62373072 and 62003070. Fundamental Research Funds for Central Universities of China under Grand 3132024109. China Postdoctoral Science Foundation with grant number 2024M764265.

Disclosure of Interests. The authors have no competing interests to declare that are relevant to the content of this article.

References

1. Guo, G., Zhao, Z., Zhang, R.: Distributed trajectory optimization and fixed-time tracking control of a group of connected vehicles. IEEE Trans. Veh. Technol. **72**(2), 1478–1487 (2023)
2. Zheng, M., Yang, S., Li, L.: Network-based sampled-data control for unmanned marine vehicles with dynamic positioning system. Int. J. Vehicle. Des. **84**(1–4), 118–136 (2021)
3. Wang, Y., Jiang, B., Wu, Z.-G., Xie, S., Peng, Y.: Adaptive sliding mode fault-tolerant fuzzy tracking control with application to unmanned marine vehicles. IEEE Trans. Syst. Man Cybern. Syst. **51**(11), 6691–6700 (2021)
4. Song, W., Tong, S.: Event-triggered fuzzy finite-time reliable control for dynamic positioning of nonlinear unmanned marine vehicles. Ocean Eng. **266**, 113139 (2022)
5. Fei, Z., Shi, S., Ahn, C.K., Basin, M.V.: Finite-time control for switched T-S fuzzy systems via a dynamic event-triggered mechanism. IEEE Trans. Fuzzy Syst. **29**(12), 3899–3909 (2021)
6. Hou, Q., Dong, J.: Finite-time membership function-dependent H-infinity control for T-S fuzzy systems via a dynamic memory event-triggered mechanism. IEEE Trans. Fuzzy Syst. (2023). https://doi.org/10.1109/TFUZZ.2023.3273080
7. Shi, P., Su, X., Li, F.: Dissipativity-based filtering for fuzzy switched systems with stochastic perturbation. IEEE Trans. Automat. Contr. **61**(6), 1694–1699 (2015)
8. Hu, X., Zhu, G., Ma, Y., Li, Z., Malekian, R., Sotelo, M.: Dynamic event-triggered adaptive formation with disturbance rejection for marine vehicles under unknown model dynamics. IEEE Trans. Veh. Technol. **72**(5), 5664–5676 (2023)
9. Zhao, Y., Gao, Y., Sang, H., Fu, J., Li, Y.: Event-triggered adaptive antidisturbance switching control for switched systems with dynamic neural network disturbance modeling. IEEE Trans. Neural Netw. Learn. Syst. (2023). https://doi.org/10.1109/TNNLS.2023.3307389
10. Gassara, H., El Hajjaji, A., Kchaou, M., Chaabane, M.: Observer based (Q, V, R)-alpha-dissipative control for T-S fuzzy descriptor systems with time delay. J. Franklin Inst. **351**(1), 187–206 (2014)
11. Polushin, I.G., Marquez, H.J.: Boundedness properties of nonlinear quasi-dissipative systems. IEEE Trans. Automat. Contr. **49**(12), 2257–2261 (2004)

Dynamic Positioning Control of Unmanned Surface Vehicles Based on an Improved Long Short-Term Memory Network

Xin Yang[1], Li-Ying Hao[2(✉)], Tieshan Li[3], and Yang Xiao[4]

[1] College of Navigation, Dalian Maritime University, Dalian 116026, China
[2] College of Marine Electrical Engineering, Dalian Maritime University, Dalian 116026, China
haoliying_0305@163.com
[3] College of Automation Engineering, University of Electronic Science and Technology of China, Chengdu 611731, China
[4] Department of Computer Science, The University of Alabama, Tuscaloosa, AL 35487, USA
yangxiao@ieee.org

Abstract. This article addresses dynamic positioning control for an unmanned surface vehicle subject to state time delays and ocean disturbances, and develops a long short-term memory (LSTM) learning-based approach. A Lyapunov matrix-based Lyapunov–Krasovskii functional is constructed to incorporate delay information, on which a delay-compensation strategy and a dynamic positioning controller are designed. External disturbances are estimated by an enhanced LSTM model with a selective state-update mechanism and an adaptive mixed-gradient rule, improving the learning of rapidly varying disturbance components while reducing computational burden. The resulting scheme ensures closed-loop stability and satisfactory positioning performance, and the proposed LSTM-based estimator achieves more accurate reconstruction of nonlinear time-varying ocean disturbances than existing methods. Finally, simulations demonstrate the effectiveness and advantages of the control strategy.

Keywords: Unmanned surface vehicle · long short-term memory · Lyapunov matrix

1 Introduction

As a new generation of intelligent marine platforms, unmanned surface vehicles (USVs) offer small size, low operating cost, and fully autonomous operation, making them attractive for tasks such as ocean surveying, environmental sensing, and maritime search and rescue [1–3]. To accomplish these missions, USVs often rely on dynamic positioning (DP) systems, which maintain the vessel near

C. Li et al. (Eds.): ICNC 2025, CCIS 2946, pp. 582–591, 2026.
https://doi.org/10.1007/978-981-92-1599-7_49

a desired position and heading and are therefore central to their high-precision station-keeping capability. Nevertheless, practical DP implementations are frequently subject to communication and feedback delays, which can deteriorate transient performance and even jeopardize closed-loop stability if not properly accounted for [4]. In networked USV, sensor and control signals are transmitted over communication links, so delays inevitably enter the loop and may threaten stability. This motivates delay-compensation methods that exploit detailed delay information rather than coarse bounds. Recent work shows that complete-type Lyapunov–Krasovskii functionals based on Lyapunov matrices can encode richer delay information, enabling less conservative analysis and controller design for USVs and other nonlinear systems with state delays [5,6].

The highly variable marine environment inevitably introduces model uncertainties and external disturbances, which complicate motion control for USVs. Although recurrent neural networks have been adopted to approximate these unknown dynamics, their limited memory depth makes it difficult to capture rapidly varying temporal patterns, leading to degradation of long-horizon disturbance representation. To alleviate this, the long short-term memory-enhanced forget-gate (LSTM-EFG) architecture refines the forget gate and employs peephole connections, thereby strengthening the exploitation of past information under fluctuating conditions [7,8]. These developments suggest that carefully designed LSTM variants provide a promising tool for accurate disturbance learning in USV control.

This paper studies the DP control of USVs subject to unknown disturbances and state time delays. An LSTM neural network with a selective state-update mechanism and an adaptive mixed-gradient optimizer is employed to estimate the disturbances while reducing computational cost. Furthermore, a Lyapunov matrix-based complete Lyapunov-Krasovskii functional with double integral terms is constructed, from which a delay-dependent compensation strategy and a DP controller are derived, ensuring that the closed-loop error system is uniformly ultimately bounded under both unknown disturbances and state delays. The main contributions are: (1) A Lyapunov matrix-based complete Lyapunov–Krasovskii functional that exploits richer delay and parameter information to reduce the conservatism of time-delay compensation; (2) A selective state-update LSTM combined with an adaptive mixed-gradient learning rule, which improves disturbance-learning capability while lowering computational complexity.

2 USV Movement Model

In this paper, the DP system of the USV, under the influence of state time delays and ocean disturbances, is modeled as

$$\begin{aligned}
\dot{\eta}(t) &= R\big(\theta(t)\big)\, v(t) + \eta(t-h) \\
\dot{v}(t) &= -M^{-1}C\, v(t) - M^{-1}D\, \eta(t) + M^{-1}B\, u(t) + M^{-1}d(t) + \eta(t-h) + v(t-h)
\end{aligned} \tag{1}$$

where $\upsilon(t) = [\upsilon_1(t)\ \upsilon_2(t)\ \upsilon_3(t)]^{\mathrm{T}} \in \mathbb{R}^3$ denotes the velocity vector of the USV in the body-fixed frame, where $\upsilon_1(t)$, $\upsilon_2(t)$, and $\upsilon_3(t)$ represent the surge velocity, sway velocity, and yaw rate, respectively. The vector $\eta(t) = [x_p(t)\ y_p(t)\ \theta(t)]^{\mathrm{T}} \in \mathbb{R}^3$ denotes the position and heading of the USV in the North–East coordinate frame, where $x_p(t)$ and $y_p(t)$ are the position coordinates and $\theta(t)$ is the heading angle. The matrix $R(\theta(t)) \in \mathbb{R}^{3\times3}$ is the rotation matrix from the body-fixed frame to the North–East frame. $M \in \mathbb{R}^{3\times3}$ is a symmetric positive definite inertia matrix, $C \in \mathbb{R}^{3\times3}$ is the hydrodynamic damping matrix, $D \in \mathbb{R}^{3\times3}$ is the mooring stiffness matrix, and $B \in \mathbb{R}^{3\times m}$ is the thruster configuration matrix. The vector $u(t) = [u_1(t)\ \cdots\ u_6(t)]^{\mathrm{T}} \in \mathbb{R}^m$ denotes the thruster control input, and $h > 0$ is a constant time delay. The terms $\eta(t-h)$ and $\upsilon(t-h)$ represent the delayed position–heading vector and velocity vector, respectively. The disturbance $d(t) \in \mathbb{R}^3$ represents the ocean environmental forces induced by wind, waves, and currents.

Let $x(t) = [\eta^{\mathrm{T}}(t)\ \upsilon^{\mathrm{T}}(t)]^{\mathrm{T}}$. Then, (1) can be further rewritten as

$$\dot{x}(t) = Ax(t) + A_1x(t-h) + Bu(t) + B_1d(t), \tag{2}$$

where $A = \begin{bmatrix} 0 & I \\ -M^{-1}D & -M^{-1}C \end{bmatrix}^{\mathrm{T}}$, $A_1 = \begin{bmatrix} I & 0 \\ 0 & I \end{bmatrix}$, $B = \begin{bmatrix} 0 \\ M^{-1}B \end{bmatrix}$, $B_1 = \begin{bmatrix} 0 \\ M^{-1} \end{bmatrix}$. Let the position–attitude reference signal and the velocity reference signal be denoted by $x_{\mathrm{ref}} = \left[\eta_{\mathrm{ref}}^{\mathrm{T}}(t)\ \upsilon_{\mathrm{ref}}^{\mathrm{T}}(t)\right]^{\mathrm{T}}$, and define the state error vector as $e(t) = x(t) - x_{\mathrm{ref}} = \begin{bmatrix} \eta(t) - \eta_{\mathrm{ref}}(t) \\ \upsilon(t) - \upsilon_{\mathrm{ref}}(t) \end{bmatrix}$. The error system can be written as

$$\begin{aligned} \dot{e}(t) &= Ae(t) + A_1e(t-h) + B\big(u(t) + \chi_r(t, u(t))\big) + B_2\omega(t), \\ e(s) &= \kappa(s), \quad s \in [-h, 0], \end{aligned} \tag{3}$$

where $B_2\omega(t) = (A + A_1)x_{\mathrm{ref}} + B_1d(t)$, $\omega(t) = \begin{bmatrix} x_{\mathrm{ref}} \\ d(t) \end{bmatrix}$, $B_2 = \begin{bmatrix} A + A_1 & B_1 \end{bmatrix}$.

Assumption 1. [6]: For system (3), it is assumed that there exists a constant matrix B_0 such that

$$B_2 = BB_0. \tag{4}$$

3 LSTM Network Based on Selective State Update

In this section, a selective state-update LSTM network combined with an adaptive mixed-gradient optimizer is designed for estimation of ocean disturbances. As shown in Fig. 1, the network input is a length-10 time series $X^k = [z_{k-9}, \ldots, z_k] \in \mathbb{R}^{9\times10}$, where $z_k = [x_k^{\mathrm{T}}\ \omega_k^{\mathrm{T}}]^{\mathrm{T}}$, $x_k \in \mathbb{R}^6$ is the system state and $\omega_k \in \mathbb{R}^3$ is the disturbance. The data are organized via a sliding window and normalization into a unified 9×10 tensor, and a fully connected layer extracts features from the six state and three disturbance components at each time step. The temporal features are then transformed by two dense layers with dimensions 64 → 72 → 36, forming an "expand–compress" structure that enhances and compacts the sequence representation. Finally, the 36-dimensional feature vector is mapped to a three-dimensional output to generate the disturbance estimates $\hat{\omega}_1, \hat{\omega}_2, \hat{\omega}_3$, and the network is trained in a one-step regression manner using the mean absolute error between predictions and true disturbance components as the loss.

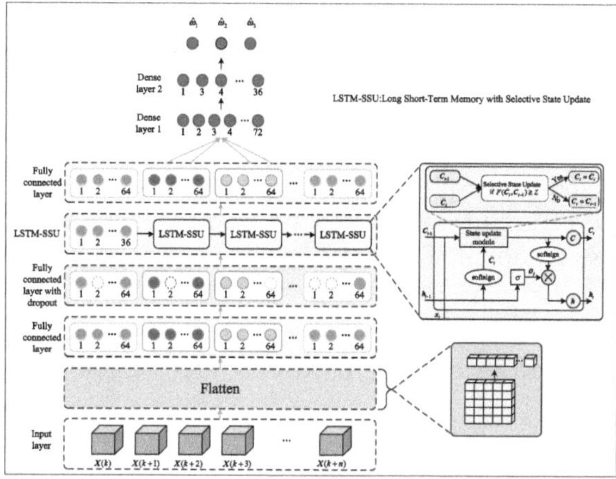

Fig. 1. Structure of the LSTM network based on a selective state update mechanism.

3.1 Cellular Units with Selective Update Strategies

In the proposed LSTM cell, a candidate cell state is first computed from the current input and the previous hidden state as $\tilde{C}_t = \sigma\big(W_c[x_t, h_{t-1}] + b_c\big)$, and is then compared with the previous cell state C_{t-1} through a discrepancy measure $F(\tilde{C}_t, C_{t-1})$. The actual cell state is updated only when this discrepancy exceeds a prescribed threshold $Z > 0$, $C_t = \begin{cases} \tilde{C}_t, & F(\tilde{C}_t, C_{t-1}) > Z, \\ C_{t-1}, & \text{otherwise,} \end{cases}$ so that small fluctuations do not trigger redundant updates. Afterwards, the output gate $o_t = \sigma\big(W_o[h_{t-1}, x_t] + b_o\big)$ and the softsign activation $h_t = o_t \odot \text{softsign}(C_t)$, $\quad \text{softsign}(x) = \frac{x}{1+|x|}$ are used to generate the hidden state, which helps smooth the activation and alleviate gradient explosion while allowing the network to focus on meaningful temporal variations. Here, $x_t \in \mathbb{R}^{n_x}$ denotes the input vector at time t, $h_{t-1} \in \mathbb{R}^{n_h}$ and h_t are the previous and current hidden states, C_{t-1} and C_t are the previous and current cell states, W_c and W_o are the corresponding weight matrices, b_c and b_o are the bias vectors, $\sigma(\cdot)$ is the logistic sigmoid function, and $F(\tilde{C}_t, C_{t-1})$ is a user-defined discrepancy measure used to decide whether the state update should be performed.

3.2 Adaptive Mixed Gradient Optimization Method

To improve the training of the disturbance-prediction network, an adaptive mixed-gradient optimization strategy is employed. Instead of relying only on the instantaneous steepest-descent direction, the optimizer blends the current gradient with a filtered history and uses a globally decaying learning rate η_k. This design allows relatively large steps and fast decrease of the loss at the early stage of training, while providing more cautious updates and stable convergence afterwards. By preserving temporal correlations in the gradient flow rather than destroying them via simple clipping, the proposed optimizer is better suited for learning long-term dependencies associated with nonlinear ocean disturbances.

4 DP Controller Design

In this section, a DP controller for the USV is designed using the time-delay and LSTM-based disturbance estimates, based on a Lyapunov-matrix complete Lyapunov–Krasovskii functional that exploits delay information for stability analysis.

4.1 Improved Complete-Type Lyapunov-Krasovskii Functional

We choose the following complete-type Lyapunov–Krasovskii functional $V_3(e(t))$:

$$V_3(e(t)) = V_1(e(t)) + V_2(e(t)), \qquad e \in \mathcal{C}_p([-h, 0], \mathbb{R}^n), \tag{5}$$

where

$$V_1(e(t)) = e^{\mathrm{T}}(t)U(0)e(t) + 2e^{\mathrm{T}}(t)\,\Gamma_1(e(t)) + \int_{-h}^{0}\int_{-h}^{0} e^{\mathrm{T}}(t+\theta_1)\,A_1^{\mathrm{T}}U(\theta_1-\theta_2)A_1\,e(t+\theta_2)\,\mathrm{d}\theta_1\mathrm{d}\theta_2, \tag{6}$$

$$V_2(e(t)) = \int_{-h}^{0}\int_{\theta}^{0} e^{\mathrm{T}}(t+s)\,A_1^{\mathrm{T}}U^{\mathrm{T}}(-h-\theta)HU(-h-\theta)A_1\,e(t+s)\,\mathrm{d}s\mathrm{d}\theta + \int_{-h}^{0} e^{\mathrm{T}}(t+\theta)\,W_1\,e(t+\theta)\,\mathrm{d}\theta, \tag{7}$$

and

$$\Gamma_1(e(t)) = \int_{-h}^{0} U(-h-\theta)\,A_1 e(t+\theta)\,\mathrm{d}\theta, \tag{8}$$

where H and W_1 are positive definite matrices satisfying $H^{\mathrm{T}} = H$ and $W_1^{\mathrm{T}} = W_1$, U is the Lyapunov matrix satisfying the following definition.

According to [10], for a given positive definite matrix W, the Lyapunov matrix $U(\cdot) : [-h, h] \to \mathbb{R}^{n\times n}$ associated with W is defined as the solution of

$$\dot{U}(s) = U(s)A_0 + U(s-h)A_1, \qquad s \geq 0, \tag{9}$$

$$U(-s) = U^{\mathrm{T}}(s), \qquad s \geq 0, \tag{10}$$

$$-W = U(0)A_0 + U(-h)A_1 + A_0^{\mathrm{T}}U(0) + A_1^{\mathrm{T}}U(h). \tag{11}$$

4.2 Lyapunov Matrix-Based DP Controller

The control input is decomposed into three components

$$u(t) = u_1(t) + u_2(t) + u_3(t), \tag{12}$$

where

$$u_1(t) = K_1 P^{-1} e(t), u_2(t) = \tfrac{1}{2} K_2 B^{\mathrm{T}} \left[U(0)e(t) + \Gamma_1(e(t)) \right] + \tfrac{1}{2} K_3 e(t-h),$$
$$u_3(t) = -\frac{\|\hat{\omega}\|^2 B^{\mathrm{T}} \left[U(0)e(t) + \Gamma_1(e(t)) \right] B_0}{\left\| B^{\mathrm{T}} \left[U(0)e(t) + \Gamma_1(e(t)) \right] \right\| \|\hat{\omega}\| + \sigma_1}.$$

Here, $u_1(t)$ is introduced to ensure the existence of the Lyapunov matrix $U(\cdot)$. The term $u_2(t)$ is designed to counteract the delay-dependent and nonlinear terms appearing in the error system (3), whereas $u_3(t)$ compensates the effect of ocean disturbances in the same error dynamics. The matrices K_1, K_2, and K_3 are design gain matrices.

4.3 Stability Analysis

In this subsection, we investigate the stability of the USV DP error system in the presence of state time delays and disturbances.

Theorem 1. *Consider the error system* (3) *of the USV. Suppose that Assumptions 1 hold. For given positive definite matrix $W \in \mathbb{R}^{n \times n}$. If there exist positive definite matrices $H, W_1 \in \mathbb{R}^{n \times n}$ and gain matrices $K_2 \in \mathbb{R}^{p \times p}$ and $K_3 \in \mathbb{R}^{p \times n}$ such that $W - W_1 - P_1 > 0$ and, the matrix inequality*

$$E := \begin{bmatrix} -W + W_1 + P_1 - E_1 & E_2 & E_3 \\ E_2^T & -W_1 & K_1^T B \\ E_3^T & BK_2 & E_4 \end{bmatrix} < 0 \tag{13}$$

holds, where

$$E_1 = -\frac{1}{2} U(0) B \left(K_2 + K_2^T \right) B^T U(0), E_2 = \frac{1}{2} U(0) B K_3,$$
$$E_3 = U(0) B K_2 B^T, E_4 = -\frac{H}{h} + B K_2 B^T,$$

then the state $e(t)$ of the closed-loop system (3) *is uniformly ultimately bounded.*

Proof. Along the solution of system (3), the time derivative of the function V_3 with respect to t can be expressed as

$$\begin{aligned} \dot{V}(e) = &-w_0(e(t)) + 2\left[U(0)e(t) + \Gamma_1(e(t)) \right]^{\mathrm{T}} B_2 \omega(t) \\ &+ 2\left[U(0)e(t) + \Gamma_1(e(t)) \right]^{\mathrm{T}} B\left((u_2(t) + u_3(t)) \right) \end{aligned} \tag{14}$$

$$w_0\big(e(t)\big) = e^{\mathrm{T}}(t)(W - W_1 - P_1)e(t) + e^{\mathrm{T}}(t-h)W_1 e(t-h) + \int_{-h}^{0} e^{\mathrm{T}}(t+\theta)\, A_1^{\mathrm{T}} U^{\mathrm{T}}(-h-\theta)\, HU(-h-\theta) A_1 e(t+\theta)\, \mathrm{d}\theta \tag{15}$$

$$\int_{-h}^{0} e^{\mathrm{T}}(t+\theta)\, A_1^{\mathrm{T}} U^{\mathrm{T}}(-h-\theta)\, HU(-h-\theta) A_1 e(t+\theta)\, \mathrm{d}\theta \geq \frac{1}{h} \varGamma_1^{\mathrm{T}}\big(e(t)\big)\, H\, \varGamma_1\big(e(t)\big) \tag{16}$$

Substituting inequalities (12), (15), and (16) into (14) yields

$$\dot{V}(e) \leq \varGamma^{\mathrm{T}}(t)E\varGamma(t) + 2\big[U(0)e(t) + \varGamma_1\big(e(t)\big)\big]^{\mathrm{T}} B_2 \omega(t) - \frac{2\big\|B^{\mathrm{T}}[U(0)e(t) + \varGamma_1(e(t))]\big\|^2 \|\hat{\omega}\|^2 \, \|B_0\|}{\big\|B^{\mathrm{T}}[U(0)e(t) + \varGamma_1(e(t))]\big\| \|\hat{\omega}\| + \sigma_1}. \tag{17}$$

where

$$-\frac{2\big\|B^{\mathrm{T}}[U(0)e(t) + \varGamma_1(e(t))]\big\|^2 \|\hat{\omega}\|^2 \, \|B_0\|}{\big\|B^{\mathrm{T}}[U(0)e(t) + \varGamma_1(e(t))]\big\| \|\hat{\omega}\| + \sigma_1} + 2[U(0)e(t) + \varGamma_1(e(t))]^{\mathrm{T}} B_2 \omega(t) \leq 2\|B_0\|\sigma_1.$$

Thus, we have

$$\dot{V}(e) \leq \varGamma^{\mathrm{T}}(t)E\varGamma(t) + 2\|B_0\|\sigma_1. \tag{18}$$

The state $e(t)$ of the closed-loop error system (3) is uniformly ultimately bounded.

5 Illustrative Example

In this section, the proposed DP control strategy is evaluated by simulations using Fossen's ship model [9]. Two cases are considered: Case 1 compares the LSTM network with selective state updates against several representative neural networks to assess the disturbance estimation performance, while Case 2 compares the designed DP controller with existing methods to verify the overall control effectiveness. The simulation parameters are given below.

The reference signal is set to $x_{\mathrm{ref}} = [0.5\ 0\ 0.17\ 0\ 0\ 0]^{\mathrm{T}}$, The time delay is $h = 1s$, The ocean disturbance acting on the USV in a complex marine environment is specified as

$$\omega(t) = [\omega_1(t)\ \omega_2(t)\ \omega_3(t)]^{\mathrm{T}},$$

where $\omega_1(t) = 1.1{+}2\sin(0.4t){+}0.5\sin\!\big(1.2t{+}\frac{\pi}{3}\big){+}\chi_3(t)$, $\omega_2(t) = 0.5{+}2\cos(0.6t){+}1.5\cos(1.5t) + \chi_4(t)$, $\omega_3(t) = 2\sin t - 1.2\sin(0.2t) + \chi_5(t)$. where $\omega_1(t)$ and $\omega_2(t)$

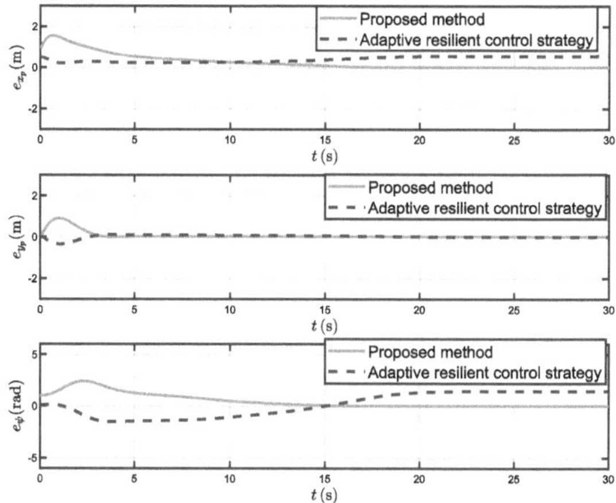

Fig. 2. Response curves comparing USV position and yaw angle errors with the method proposed in [5].

are measured in newtons (N) and $\omega_3(t)$ is measured in newton-meters (N·m). The disturbance terms $\chi_i(t)$ $(i = 3, 4, 5)$ are zero-mean Gaussian white noises with $\chi_i(t) \sim \mathcal{N}(0, \sigma_1^2)$, where $\sigma_1 = 0.02$, i.e., $\chi_i(t) = 0.02\,\nu_i(t)$ with $\nu_i(t) \sim \mathcal{N}(0, 1)$.

Case 1: Performance validation of the proposed LSTM neural network

From Table 1, the proposed LSTM-SSU attains the best metrics, with $R^2 =$ 0.978162, RMSE = 0.103296 (about 4.5% lower than the standard LSTM), and MAE = 0.087883, indicating more accurate disturbance prediction than the other compared networks.

Table 1. Comparison of evaluation metrics of different neural networks

Algorithm	RMSE	MAEe	R^2
LSTM [11]	0.108233	0.088309	0.977869
MultiLayer_LSTM [12]	0.108391	0.110635	0.977804
BiGRU [13]	0.108934	0.088981	0.977581
GRU [14]	0.109680	0.089388	0.977273
BiLSTM [15]	0.110262	0.089855	0.977031
LSTM-SSU	**0.103296**	**0.087883**	**0.978162**

Case 2: Comparison of DP strategies.

To evaluate the effectiveness of the proposed DP controller, it is benchmarked against the adaptive DP controller reported in [5]. In Fig. 2 and Fig. 3, the blue dashed curves correspond to the simulation results of the controller in

[5]. As observed from Fig. 2, that controller fails to keep the USV tightly around the desired position and a noticeable steady-state offset remains. In contrast, under the proposed controller the positioning errors exhibit only a small transient increase and then decay rapidly, converging within a short time to a small neighborhood of the origin. This confirms that the proposed DP scheme achieves markedly improved positioning performance.

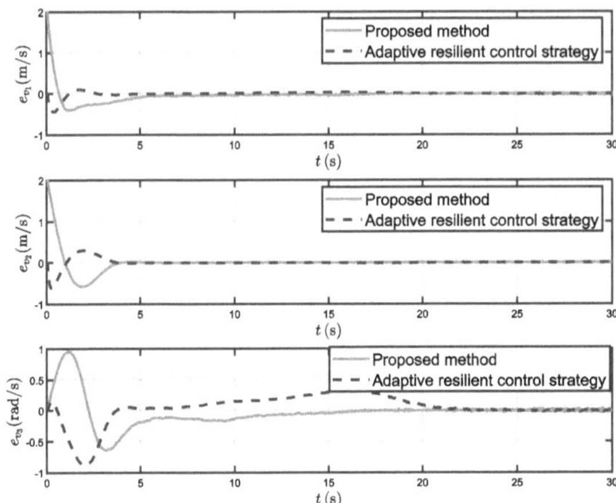

Fig. 3. Response curves comparing USV velocity errors with the method proposed in [5].

6 Conclusion

This paper has presented an LSTM-based DP scheme for the USV operating under state time delays and ocean disturbances. A Lyapunov matrix-based Lyapunov–Krasovskii functional was constructed to incorporate delay information and to guide the design of a delay-compensation strategy and DP controller. An enhanced LSTM with selective state updates and an adaptive mixed-gradient rule was developed to estimate rapidly varying disturbances while reducing computational burden. The resulting closed-loop system achieves stable and accurate station-keeping performance in the presence of delays and nonlinear time-varying ocean disturbances. Simulation results demonstrate that the proposed learning-based DP strategy outperforms existing methods in both disturbance estimation and positioning accuracy.

Acknowledgment. This work was supported in part by the National Natural Science Foundation of China under Grant 52471376, Grant 51939001, Grant 52171292, and Grant 61976033; by the Fundamental Research Funds for the Central Universities under Grant ZYGX2024Z018; and in part by the Outstanding Young Talent Program under Grant 2022RJ05.

Disclosure of Interests. It is now necessary to declare any competing interests or to specifically state that the authors have no competing interests.

References

1. Liang, Z.X., Wen, H., Yao, B., Mao, Z.H., Lian, L.: Event-triggered adaptive fault-tolerant control for marine vehicles with multiple faults and environmental disturbance. Ocean Eng. **313**, 119473 (2024)
2. Wang, Y.-L., Han, Q.-L.: Network-based modelling and dynamic output feedback control for unmanned marine vehicles in network environments. Automatica **91**, 43–53 (2018)
3. Hu, X., Zhu, G.-B., Ma, Y., Li, Z.-X., Malekian, R., Sotelo, M.: Event-triggered adaptive fuzzy setpoint regulation of surface vessels with unmeasured velocities under thruster saturation constraints. IEEE Trans. Intell. Transp. Syst. **23**(8), 13463–13472 (2021)
4. Wang, X., Park, J.H., Liu, H., Zhang, X.: Cooperative output-feedback secure control of distributed linear cyber-physical systems resist intermittent DoS attacks. IEEE Trans. Cybern. **51**(10), 4924–4933 (2021)
5. Yang, X., Hao, L.-Y., Xiao, Y., Li, T.S.: Lyapunov matrix-based adaptive resilient control for unmanned marine vehicles with sensor and thruster attacks. ISA Trans. **153**, 70–77 (2024)
6. Yang, X., Wang, Y.-T., Zhang, X.: Lyapunov matrix-based method to guaranteed cost control for a class of delayed continuous-time nonlinear systems. IEEE Trans. Syst. Man Cybern: Syst. **52**(1), 554–560 (2020)
7. Yu, R.G., Gao, J., Yu, M., Lu, W.H., et al.: LSTM-EFG for wind power forecasting based on sequential correlation features. Future Gener. Comput. Syst. **93**, 33–42 (2019)
8. Zhang, B., Zou, G.J., Qin, D.M., Lu, Y.J., et al.: A novel encoder-decoder model based on read-first LSTM for air pollutant prediction. Sci. Total Environ. **765**, 144507 (2021)
9. Fossen, T.I.: Guidance and Control of Ocean Vehicles. Wiley, New York, NY, USA (1994)
10. Kharitonov, V.L.: Time-Delay Systems: Lyapunov Functionals and Matrices. Birkhäuser, New York, Heidelberg, Dordrecht, London (2013)
11. Pascual-Valdunciel, A., Lopo-Martínez, V., Sendra-Arranz, R., et al.: Prediction of pathological tremor signals using long short-term memory neural networks. IEEE J. Biomed. Health Inform. **26**(12), 5930–5941 (2022)
12. Çelebi, S.B., Karaman, Ö.A.: Multilayer LSTM model for wind power estimation in the SCADA system. Eur. J. Tech. (EJT) **13**(2), 116–122 (2023)
13. Hu, X., Huo, Y., Dong, X., et al.: Channel prediction using adaptive bidirectional GRU for underwater MIMO communications. IEEE Internet Things J. **11**(2), 3250–3263 (2023)
14. Jia, C., Ma, J., Kouw, W.M.: Multiple variational Kalman-GRU for ship trajectory prediction with uncertainty. IEEE Trans. Aerosp. Electron. Syst. **61**(2), 3654–3667 (2024)
15. Dai, T., Zhu, L., Wang, Y., et al.: Attentive stacked denoising autoencoder with Bi-LSTM for personalized context-aware citation recommendation. IEEE/ACM Trans. Audio, Speech, Lang. Process. **28**, 553–568 (2019)

Adaptive Neural Network Optimal Fault-Tolerant Trajectory Tracking Control for Underactuated Autonomous Underwater Vehicles

Huibin Gong and Li-Ying Hao(✉)

College of Marine Electrical Engineering, Dalian Maritime University, 1 Linghai Road, Dalian 116026, Liaoning, China
haoliying_0305@163.com

Abstract. For the control of underactuated autonomous underwater vehicles, it is necessary to consider not only fault-tolerance but also optimal performance. Thus, an adaptive neural network-based optimal fault-tolerant trajectory tracking control method is proposed. First, the system output is redefined using a coordinate transformation method to address the underactuation issue. Subsequently, a control law integrating an adaptive disturbance observer and an improved backstepping approach is designed, achieving online estimation and compensation of composite disturbances caused by both actuator faults and external disturbances. On this basis, a radial basis function neural network is utilized to construct an optimal control law, which effectively suppresses the impact of actuator faults on the closed-loop system stability while accomplishing trajectory tracking control. Theoretical analysis demonstrates that all signals of the closed-loop system are uniformly ultimately bounded, and numerical simulations verify the superiority and feasibility of the proposed method.

Keywords: Underactuated autonomous underwater vehicles · Fault-tolerant control · Neural networks · Adaptive dynamic programming

1 Introduction

Autonomous underwater vehicles (AUVs) play a vital role in ocean exploration and monitoring [1,2]. Underactuated AUVs, which have fewer thrusters than degrees of freedom, offer efficiency but present control challenges [3]. They must operate reliably in the presence of ocean disturbances and possible actuator faults. Thus, designing control strategies that deliver optimal performance in terms of precise trajectory tracking, fault-tolerance, and energy efficiency remains a key challenge in underwater robotics.

C. Li et al. (Eds.): ICNC 2025, CCIS 2946, pp. 592–602, 2026.
https://doi.org/10.1007/978-981-92-1599-7_50

In recent years, a range of control strategies have been proposed to mitigate the effects of environmental disturbances on 3D trajectory tracking for underactuated AUVs. Examples include PID-based sliding mode control [4] and a fixed-time backstepping scheme with a fixed-time disturbance observer [5]. Leveraging the approximation capabilities of intelligent systems, neural networks (NNs) and fuzzy logic systems have also been employed for disturbance compensation—e.g., fuzzy dynamic surface control [6], adaptive NN approaches [7,8], a finite-time method using self-organizing NNs [9]. However, none of these works consider actuator faults—a critical issue in real-world AUV operations. Although a fault-tolerant adaptive control scheme was later proposed in [10] to compensate for actuator faults via adaptive laws, none of the aforementioned studies address the optimality of the control system. Thus, achieving optimal, high-performance trajectory tracking under disturbances and actuator faults remains an open challenge.

An iterative ADP control scheme based on an actor-critic architecture was introduced in [11] to achieve optimal 3D trajectory tracking for underactuated AUVs. Subsequent studies [12] and [13] further addressed actuator faults. However, all these approaches neglect the critical issue of environmental disturbances, which are unavoidable in real-world marine operations. Consequently, achieving optimal trajectory tracking of underactuated AUVs in the simultaneous presence of actuator faults and external disturbances remains an open problem.

Building upon the aforementioned work, this paper addresses the challenges posed by actuator faults and environmental disturbances in 3D trajectory tracking control of underactuated AUVs within an ADP-based optimal control framework. The key contributions of this study are outlined as follows:

1) The proposed controller simultaneously handles actuator faults and disturbances while optimizing energy consumption and tracking performance.

2) Actuator faults and disturbances are uniformly modeled as a composite disturbance and compensated through an adaptive observer within the backstepping control framework.

3) A parameter update law without requiring initial tuning is introduced, relaxing the requirement for initially admissible control and improving usability.

2 Preliminaries and Problem Formulation

2.1 Notations

$\lambda_{\min}(\cdot)$ and $\lambda_{\max}(\cdot)$ denote the minimum eigenvalue and the maximum eigenvalue of a matrix, respectively, $\mathrm{diag}(o_1, o_2, \ldots, o_n)$ denote a diagonal matrix formed by elements $o_1, o_2, \ldots, o_n$, and $\boldsymbol{I}_n$ denote the $n \times n$ identity matrix.

2.2 Problem Formulation

Incorporating the effects of actuator faults and environmental disturbances, the kinematic and dynamic models of an underactuated AUV can be formulated as follows:

$$\dot{\boldsymbol{\eta}} = \boldsymbol{J}(\boldsymbol{\eta})\boldsymbol{\nu}, \tag{1}$$

$$\boldsymbol{M}\dot{\boldsymbol{\nu}} = -\boldsymbol{C}(\boldsymbol{\nu})\boldsymbol{\nu} - \boldsymbol{D}(\boldsymbol{\nu})\boldsymbol{\nu} - \boldsymbol{g}(\boldsymbol{\eta}) + \boldsymbol{\tau}_e + \boldsymbol{\tau}, \tag{2}$$

where $\boldsymbol{\eta} = [x, y, z, \theta, \psi]^{\mathrm{T}}$ is a vector composed of the AUV's positions x, y and z and the AUV's attitude angles θ and ψ; $\boldsymbol{J}(\boldsymbol{\eta})$ denotes the transformation matrix between the body-fixed coordinate frame and the fixed coordinate frame; $\boldsymbol{\nu} = [u, v, w, q, r]^{\mathrm{T}}$ represents a vector composed of the AUV's linear velocities u, v and w, and the AUV's angular velocities q and r; $\boldsymbol{M} = \mathrm{diag}(m_{11}, m_{22}, m_{33}, m_{55}, m_{66})$ represents the matrix combining the rigid-body inertia and hydrodynamic added inertia, with the parameters m_{11}, m_{22}, m_{33}, m_{55} and m_{66}; $\boldsymbol{C}(\boldsymbol{\nu})$ is the Coriolis and centripetal matrix; $\boldsymbol{D}(\boldsymbol{\nu}) = \mathrm{diag}\,(d_{11}, d_{22}, d_{33}, d_{55}, d_{66})$ denotes the hydrodynamic damping matrix; $\boldsymbol{g}(\boldsymbol{\eta}) = [0, 0, 0, \rho g \nabla \overline{GM}_L \sin(\theta), 0]^{\mathrm{T}}$ is the restoring force and moment vector, where ρ, g, ∇, and $\overline{GM}_L$ represent the water density, gravitational acceleration, displaced water volume, and longitudinal metacentric height, respectively; $\boldsymbol{\tau}_e = [\tau_{eu}, \tau_{ev}, \tau_{ew}, \tau_{eq}, \tau_{er}]^{\mathrm{T}}$ represents a vector composed of the environmental disturbance forces τ_{eu}, τ_{ev} and τ_{ew}, and moments τ_{eq} and τ_{er}; $\boldsymbol{\tau} = (\boldsymbol{I}_5 - \boldsymbol{\beta})\boldsymbol{\tau}_c + \boldsymbol{\tau}_\delta$ denotes the actual control input vector that incorporates both partial loss of actuator effectiveness and bias faults, and $\boldsymbol{\beta} = \mathrm{diag}(\beta_u, 0, 0, \beta_q, \beta_r)$ $(0 \leq \beta_i < 1, i = u, q, r)$ represents the efficiency factor matrix, $\boldsymbol{\tau}_\delta = [\tau_{\delta u}, 0, 0, \tau_{\delta q}, \tau_{\delta r}]^{\mathrm{T}}$ denotes the bias fault vector, and $\boldsymbol{\tau}_c = [\tau_{cu}, 0, 0, \tau_{cq}, \tau_{cr}]^{\mathrm{T}}$ is the control input vector comprising the surge force τ_{cu} along with the pitch and yaw moments τ_{cq} and τ_{cr}.

Assumption 1. The 3-D desired trajectory $\boldsymbol{\eta}_{1d}$ is sufficiently smooth and its first-order derivatives are bounded.

Assumption 2. The disturbances $\boldsymbol{\tau}_e$ are time-varying and bounded.

Control Objective: Considering an underactuated autonomous underwater vehicle subject to actuator faults and environmental disturbances, and under Assumptions 1-2, this paper aims to design an adaptive neural network optimal fault-tolerant trajectory tracking control scheme. The scheme should enable the vehicle to track a 3-D desired trajectory while simultaneously optimizing tracking error and energy consumption.

3 Main Results

3.1 Coordinate Transformation

To address the underactuation issue of AUVs, the state vector $\boldsymbol{\eta}$ is reformulated via the following method:

$$\boldsymbol{\eta}_1 = [\bar{x}, \bar{y}, \bar{z}]^{\mathrm{T}} = [x + l_\nu \cos(\psi)\cos(\theta), y + l_\nu \sin(\psi)\cos(\theta), z - l_\nu \sin(\theta)]^{\mathrm{T}}, \tag{3}$$

where $l_\nu > 0$ is a design parameter.

By combining (1), (2), and (3), the system can be expressed in the following form:

$$\dot{\boldsymbol{\eta}}_1 = \boldsymbol{J}_1(\boldsymbol{\eta}_2)\boldsymbol{\nu}_1 + \boldsymbol{J}_2(\boldsymbol{\eta}_2, \boldsymbol{\nu}_2), \tag{4}$$

$$\boldsymbol{M}_1\dot{\boldsymbol{\nu}}_1 = -\boldsymbol{C}_1(\boldsymbol{\nu}_2)\boldsymbol{\nu}_1 - \boldsymbol{D}_1(\boldsymbol{\nu}_1)\boldsymbol{\nu}_1 - \boldsymbol{g}_1(\theta) + \boldsymbol{\tau}_{e1} + \boldsymbol{\tau}_1, \tag{5}$$

where $\boldsymbol{\nu}_1 = [u, q, r]^{\mathrm{T}}$, $\boldsymbol{\eta}_2 = [\theta, \psi]^{\mathrm{T}}$, $\boldsymbol{\nu}_2 = [v, w]^{\mathrm{T}}$, $\boldsymbol{J}_1(\boldsymbol{\eta}_2)$, $\boldsymbol{J}_2(\boldsymbol{\eta}_2, \boldsymbol{\nu}_2)$, and $\boldsymbol{C}_1(\boldsymbol{\nu}_2)$ are specifically represented in [14]. The remaining terms are specified as $\boldsymbol{M}_1 = \mathrm{diag}(m_{11}, m_{55}, m_{66})$, $\boldsymbol{D}_1(\boldsymbol{\nu}_1) = \mathrm{diag}(d_{11}, d_{55}, d_{66})$, $\boldsymbol{g}_1(\theta) = [0, \rho g \nabla \overline{GM}_L \sin(\theta), 0]^{\mathrm{T}}$, $\boldsymbol{\tau}_{e1} = [\tau_{eu}, \tau_{eq}, \tau_{er}]^{\mathrm{T}}$, $\boldsymbol{\tau}_1 = (\boldsymbol{I}_3 - \boldsymbol{\beta}_1)\boldsymbol{\tau}_{c1} + \boldsymbol{\tau}_{\delta 1}$ with $\boldsymbol{\tau}_1 = [\tau_u, \tau_q, \tau_r]^{\mathrm{T}}$, $\boldsymbol{\beta}_1 = \mathrm{diag}(\beta_u, \beta_q, \beta_r)$, $\boldsymbol{\tau}_{\delta 1} = [\tau_{\delta u}, \tau_{\delta q}, \tau_{\delta r}]^{\mathrm{T}}$ and $\boldsymbol{\tau}_{c1} = [\tau_{cu}, \tau_{cq}, \tau_{cr}]^{\mathrm{T}}$.

Next, (5) can be reformulated as

$$\dot{\boldsymbol{\nu}}_1 = \boldsymbol{f}(\boldsymbol{\nu}_1, \boldsymbol{\nu}_2, \theta) + \boldsymbol{b}(\boldsymbol{f}_c + \boldsymbol{\tau}_{c1}) \tag{6}$$

where $\boldsymbol{f}(\boldsymbol{\nu}_1, \boldsymbol{\nu}_2, \theta) = \boldsymbol{M}_1^{-1}(-\boldsymbol{C}_1(\boldsymbol{\nu}_2)\boldsymbol{\nu}_1 - \boldsymbol{D}_1(\boldsymbol{\nu}_1)\boldsymbol{\nu}_1 - \boldsymbol{g}_1(\theta))$, $\boldsymbol{b} = \boldsymbol{M}_1^{-1}$, $\boldsymbol{f}_c = -\boldsymbol{\beta}_1\boldsymbol{\tau}_{c1} + \boldsymbol{\tau}_{\delta 1} + \boldsymbol{\tau}_{e1}$ is the composite disturbance term consisting of actuator faults and disturbances.

3.2 Design of Adaptive Disturbance Observer and Backstepping Controller

An adaptive disturbance observer is formulated as

$$\dot{\hat{\boldsymbol{\nu}}}_1 = \boldsymbol{f}(\hat{\boldsymbol{\nu}}_1, \boldsymbol{\nu}_2, \theta) + \boldsymbol{b}(\hat{\boldsymbol{f}}_c + \boldsymbol{\tau}_{c1}) + \boldsymbol{K}_{o1}\boldsymbol{E}_o, \tag{7}$$

where $\hat{\boldsymbol{\nu}}_1$ denotes the estimate of $\boldsymbol{\nu}_1$, $\boldsymbol{K}_{o1} \in \mathbb{R}^3$ is the diagonal observer gain, and $\boldsymbol{E}_o = \boldsymbol{\nu}_1 - \hat{\boldsymbol{\nu}}_1$ denotes the observation error.

Define the estimation error of the composite disturbances as $\tilde{\boldsymbol{f}}_c = \boldsymbol{f}_c - \hat{\boldsymbol{f}}_c$. Combining (6) with (7), the resulting observer error dynamics can be expressed as

$$\dot{\boldsymbol{E}}_o = \boldsymbol{f}(\boldsymbol{\nu}_1, \boldsymbol{\nu}_2, \theta) - \boldsymbol{f}(\hat{\boldsymbol{\nu}}_1, \boldsymbol{\nu}_2, \theta) + \boldsymbol{b}\tilde{\boldsymbol{f}}_c - \boldsymbol{K}_{o1}\boldsymbol{E}_o. \tag{8}$$

The position error vector $\boldsymbol{E}_\eta$ and velocity error vector $\boldsymbol{E}_\nu$ are defined as $\boldsymbol{E}_\eta = \boldsymbol{\eta}_1 - \boldsymbol{\eta}_{1d}$ and $\boldsymbol{E}_\nu = \boldsymbol{\nu}_1 - \boldsymbol{\alpha}_1$, where $\boldsymbol{\eta}_{1d} = [x_d, y_d, z_d]^{\mathrm{T}}$ is the desired 3D trajectory vector, $\boldsymbol{\alpha}_1 = \boldsymbol{\alpha}_1^b + \boldsymbol{\alpha}_1^*$ represents the composite virtual control input, consisting of the backstepping-based component $\boldsymbol{\alpha}_1^b$ and the optimal component $\boldsymbol{\alpha}_1^*$.

The following virtual control law based on the backstepping approach can be formulated:

$$\boldsymbol{\alpha}_1^b = \boldsymbol{J}_1^{-1}(\boldsymbol{\eta}_2)(-\boldsymbol{K}_\eta\boldsymbol{E}_\eta - \boldsymbol{J}_2(\boldsymbol{\eta}_2, \boldsymbol{\nu}_2) + \dot{\boldsymbol{\eta}}_{1d}) \tag{9}$$

with $\boldsymbol{K}_\eta \in \mathbb{R}^{3\times 3}$ representing a design matrix.

The following presents the design of the backstepping control law along with the update law for composite disturbances:

$$\boldsymbol{\tau}_{c1}^b = \boldsymbol{b}^{-1}(-\boldsymbol{K}_\nu\boldsymbol{E}_\nu - \boldsymbol{J}_1(\boldsymbol{\eta}_2)\boldsymbol{E}_\eta + \dot{\boldsymbol{\alpha}}_1 + \boldsymbol{h}(\boldsymbol{E}_\nu, \boldsymbol{\nu}_2) - \boldsymbol{f}(\boldsymbol{\nu}_1, \boldsymbol{\nu}_2, \theta)) - \hat{\boldsymbol{f}}_c, \tag{10}$$

$$\dot{\hat{\boldsymbol{f}}}_c = \boldsymbol{K}_{o2}\boldsymbol{b}^{\mathrm{T}}(\boldsymbol{E}_o + \boldsymbol{K}_{o3}\boldsymbol{E}_\nu), \tag{11}$$

where $\boldsymbol{h}(\boldsymbol{E}_\nu, \boldsymbol{\nu}_2) = \boldsymbol{f}(\boldsymbol{\nu}_1, \boldsymbol{\nu}_2, \theta) - \boldsymbol{f}(\boldsymbol{\alpha}_1, \boldsymbol{\nu}_2, \theta)$, $\boldsymbol{K}_\nu \in \mathbb{R}^{3\times 3}$ is a positive-definite design matrix.

By substituting (9), (10), and (11) into the time derivative of the Lyapunov function candidate $L_{V2} = \boldsymbol{E}_\eta^{\mathrm{T}}\boldsymbol{E}_\eta/2 + \boldsymbol{E}_o^{\mathrm{T}}\boldsymbol{K}_{o3}^{-1}\boldsymbol{E}_o/2 + \tilde{\boldsymbol{f}}_c^{\mathrm{T}}\boldsymbol{K}_{o2}^{-1}\boldsymbol{K}_{o3}^{-1}\tilde{\boldsymbol{f}}_c\big/2 + \boldsymbol{E}_\nu^{\mathrm{T}}\boldsymbol{E}_\nu/2$, and leveraging the fact that the nonlinear function $\boldsymbol{f}(\boldsymbol{\nu}_1, \boldsymbol{\nu}_2, \theta)$ satisfies the Lipschitz condition—implying the existence of a positive constant ζ_f such that $\|\boldsymbol{f}(\boldsymbol{\nu}_1, \boldsymbol{\nu}_2, \theta) - \boldsymbol{f}(\hat{\boldsymbol{\nu}}_1, \boldsymbol{\nu}_2, \theta)\| \leq \zeta_f \|\boldsymbol{E}_o\|$ holds, together with the boundedness of the matrix $\boldsymbol{J}_1(\boldsymbol{\eta}_2)$, i.e., there exists $\zeta_J^M > 0$ satisfying $\|\boldsymbol{J}_1(\boldsymbol{\eta}_2)\| \leq \zeta_J^M$. the derivative $\dot{L}_{V2}$ can be further simplified using the Cauchy-Schwarz inequality as

$$\begin{aligned} \dot{L}_{V2} \leq & -(\lambda_{\min}(\boldsymbol{K}_\eta) - \zeta_J^M)\|\boldsymbol{E}_\eta\|^2 - (\lambda_{\min}(\boldsymbol{K}_\nu) - \zeta_J^M)\|\boldsymbol{E}_\nu\|^2 \\ & - (\lambda_{\min}(\boldsymbol{K}_{o3}^{-1}\boldsymbol{K}_{o1}) - \lambda_{\max}(\boldsymbol{K}_{o3}^{-1})\zeta_f)\|\boldsymbol{E}_o\|^2 \\ & + \boldsymbol{E}^{\mathrm{T}}\left(\begin{bmatrix} \mathbf{0}_{3\times 1} \\ \boldsymbol{h}(\boldsymbol{E}_\nu, \boldsymbol{\nu}_2) \end{bmatrix} + \begin{bmatrix} \boldsymbol{J}_1(\boldsymbol{\eta}_2) & \mathbf{0}_{3\times 3} \\ \mathbf{0}_{3\times 3} & \boldsymbol{b} \end{bmatrix}\boldsymbol{U}\right), \end{aligned} \tag{12}$$

where $\boldsymbol{E} = [\boldsymbol{E}_\eta^{\mathrm{T}}, \boldsymbol{E}_\nu^{\mathrm{T}}]^{\mathrm{T}}$ and $\boldsymbol{U} = [\boldsymbol{\alpha}_1^{*\mathrm{T}}, \boldsymbol{\tau}_{c1}^{*\mathrm{T}}]^{\mathrm{T}}$.

3.3 NN Optimal Controller Design

According to (12), the following equation can be obtained:

$$\dot{\boldsymbol{E}} = \boldsymbol{F} + \boldsymbol{G}\boldsymbol{U}, \tag{13}$$

where $\boldsymbol{F} = \begin{bmatrix} \mathbf{0}_{3\times 1} \\ \boldsymbol{h}(\boldsymbol{E}_\nu, \boldsymbol{\nu}_2) \end{bmatrix}$ and $\boldsymbol{G} = \begin{bmatrix} \boldsymbol{J}_1(\boldsymbol{\eta}_2) & \mathbf{0}_{3\times 3} \\ \mathbf{0}_{3\times 3} & \boldsymbol{b} \end{bmatrix}$.

Consider the following infinite-horizon cost function:

$$V(\boldsymbol{E}(t)) = \int_t^\infty e^{-\gamma(s-t)} N(\boldsymbol{E}(s), \boldsymbol{U}(s))\mathrm{d}s, \tag{14}$$

where $\gamma > 0$ represents design parameter, $N(\boldsymbol{E}, \boldsymbol{U}) = \boldsymbol{E}^{\mathrm{T}}\boldsymbol{Q}\boldsymbol{E} + \boldsymbol{U}^{\mathrm{T}}\boldsymbol{R}\boldsymbol{U}$ with the positive-definite matrices $\boldsymbol{Q} \in \mathbb{R}^{6\times 6}$ and $\boldsymbol{R} \in \mathbb{R}^{6\times 6}$.

To derive the optimal control law, define the Hamiltonian function as follows:

$$H(\boldsymbol{E}, \boldsymbol{U}) = N(\boldsymbol{E}, \boldsymbol{U}) + (\nabla V(\boldsymbol{E}))^{\mathrm{T}}(\boldsymbol{F} + \boldsymbol{G}\boldsymbol{U}) - \gamma V(\boldsymbol{E}) \tag{15}$$

where $\nabla V(\boldsymbol{E}) = \partial V(\boldsymbol{E})/\partial \boldsymbol{E}$ denotes the gradient of $V(\boldsymbol{E})$ with respect to $\boldsymbol{E}$.

The optimal control law can then be obtained as

$$\boldsymbol{U}^* = -\frac{1}{2}\boldsymbol{R}^{-1}\boldsymbol{G}^{\mathrm{T}}\nabla V^*(\boldsymbol{E}) \tag{16}$$

where $\nabla V^*(\boldsymbol{E}) = \partial V^*(\boldsymbol{E})/\partial \boldsymbol{E}$ stands for the gradient of $V^*(\boldsymbol{E})$ with respect to $\boldsymbol{E}$.

By inserting (16) into (15), the resulting Hamilton-Jacobi-Bellman equation is obtained as follows:

$$\boldsymbol{E}^{\mathrm{T}}\boldsymbol{Q}\boldsymbol{E} + (\nabla V^*(\boldsymbol{E}))^{\mathrm{T}}\boldsymbol{F} - \frac{1}{4}(\nabla V^*(\boldsymbol{E}))^{\mathrm{T}}\boldsymbol{R}_1\nabla V^*(\boldsymbol{E}) - \gamma V^*(\boldsymbol{E}) = 0, \tag{17}$$

where $\boldsymbol{R}_1 = \boldsymbol{G}\boldsymbol{R}^{-1}\boldsymbol{G}^{\mathrm{T}}$.

A critic NN is designed using a radial basis function NN, i.e.,

$$V^*(\boldsymbol{E}) = \boldsymbol{W}^{*\mathrm{T}}\boldsymbol{\sigma}(\boldsymbol{E}) + \varepsilon(\boldsymbol{E}), \tag{18}$$

where $\boldsymbol{W}^* \in \mathbb{R}^{l_F}$ denotes the ideal weight vector, $\boldsymbol{\sigma}(\boldsymbol{E}) \in \mathbb{R}^{l_F}$ is the activation function vector, l_F stands for the number of activation functions and $\varepsilon(\boldsymbol{E})$ represents the approximation error.

The gradient of the optimal value function $V^*(\boldsymbol{E})$ with respect to $\boldsymbol{E}$ is then given by:

$$\nabla V^*(\boldsymbol{E}) = (\nabla\boldsymbol{\sigma}(\boldsymbol{E}))^{\mathrm{T}}\boldsymbol{W}^* + \nabla\varepsilon(\boldsymbol{E}), \tag{19}$$

with $\nabla\boldsymbol{\sigma}(\boldsymbol{E}) = \partial\boldsymbol{\sigma}(\boldsymbol{E})/\partial\boldsymbol{E}$ and $\nabla\varepsilon(\boldsymbol{E}) = \partial\varepsilon(\boldsymbol{E})/\partial\boldsymbol{E}$ representing the gradients of $\boldsymbol{\sigma}(\boldsymbol{E})$ and $\varepsilon(\boldsymbol{E})$ with respect to $\boldsymbol{E}$, respectively.

Combining (19) and (16), the resulting optimal control law is derived as follows:

$$\boldsymbol{U}^* = -\frac{1}{2}\boldsymbol{R}^{-1}\boldsymbol{G}^{\mathrm{T}}\left[(\nabla\boldsymbol{\sigma}(\boldsymbol{E}))^{\mathrm{T}}\boldsymbol{W}^* + \nabla\varepsilon(\boldsymbol{E})\right]. \tag{20}$$

By substituting (18), (19), and (20) into the Hamiltonian function (15), we obtain:

$$H^*(\boldsymbol{E},\boldsymbol{W}^*) = \boldsymbol{E}^{\mathrm{T}}\boldsymbol{Q}\boldsymbol{E} + \boldsymbol{W}^{*\mathrm{T}}\nabla\boldsymbol{\sigma}(\boldsymbol{E})\boldsymbol{F} - \frac{1}{4}\boldsymbol{W}^{*\mathrm{T}}\boldsymbol{R}_2\boldsymbol{W}^* - \gamma\boldsymbol{W}^{*\mathrm{T}}\boldsymbol{\sigma}(\boldsymbol{E}) + e_{\varepsilon} = 0 \tag{21}$$

where $\boldsymbol{R}_2 = \nabla\boldsymbol{\sigma}(\boldsymbol{E})\boldsymbol{R}_1(\nabla\boldsymbol{\sigma}(\boldsymbol{E}))^{\mathrm{T}}$, $e_{\varepsilon} = (\nabla\varepsilon(\boldsymbol{E}))^{\mathrm{T}}\dot{\boldsymbol{E}}^* + (1/4)(\nabla\varepsilon(\boldsymbol{E}))^{\mathrm{T}}\boldsymbol{R}_1\nabla\varepsilon(\boldsymbol{E}) - \gamma\varepsilon(\boldsymbol{E})$ denotes the residual error with $\dot{\boldsymbol{E}}^* = \boldsymbol{F} + \boldsymbol{G}\boldsymbol{U}^*$ representing the ideal error equation.

Let $\hat{\boldsymbol{W}}$ denote the estimate of the ideal weight vector $\boldsymbol{W}^*$. An approximation of the optimal value function can then be expressed as

$$\hat{V}(\boldsymbol{E}) = \hat{\boldsymbol{W}}^{\mathrm{T}}\boldsymbol{\sigma}(\boldsymbol{E}). \tag{22}$$

Computing the gradient of $\hat{V}(\boldsymbol{E})$ with respect to $\boldsymbol{E}$ results in

$$\nabla\hat{V}(\boldsymbol{E}) = (\nabla\boldsymbol{\sigma}(\boldsymbol{E}))^{\mathrm{T}}\hat{\boldsymbol{W}}. \tag{23}$$

Based on (23) and (20), an approximate optimal control law can be derived as follows:

$$\hat{\boldsymbol{U}}^* = -\frac{1}{2}\boldsymbol{R}^{-1}\boldsymbol{G}^{\mathrm{T}}(\nabla\boldsymbol{\sigma}(\boldsymbol{E}))^{\mathrm{T}}\hat{\boldsymbol{W}} \tag{24}$$

with $\hat{\boldsymbol{U}}^* = [\hat{\boldsymbol{\alpha}}_1^{*\mathrm{T}}, \hat{\boldsymbol{\tau}}_{c1}^{*\mathrm{T}}]^{\mathrm{T}}$.

By inserting (22), (23), and (24) into the Hamiltonian function (15), the resulting approximate Hamiltonian is expressed as

$$\hat{H}(\boldsymbol{E},\hat{\boldsymbol{W}}) = \boldsymbol{E}^{\mathrm{T}}\boldsymbol{Q}\boldsymbol{E} + \hat{\boldsymbol{W}}^{\mathrm{T}}\nabla\boldsymbol{\sigma}(\boldsymbol{E})\boldsymbol{F} - \frac{1}{4}\hat{\boldsymbol{W}}^{\mathrm{T}}\boldsymbol{R}_2\hat{\boldsymbol{W}} - \gamma\hat{\boldsymbol{W}}^{\mathrm{T}}\boldsymbol{\sigma}(\boldsymbol{E}). \tag{25}$$

To obtain the update law of the critic NN weights $\hat{\boldsymbol{W}}_c$, the objective function is chosen as $E_c = \frac{1}{2}e_c^2$, where $e_c = \hat{H}(\boldsymbol{E},\hat{\boldsymbol{W}}) - H^*(\boldsymbol{Z},\boldsymbol{W}^*) = \hat{H}(\boldsymbol{E},\hat{\boldsymbol{W}})$.

Subsequently, the adaptation law for $\hat{\boldsymbol{W}}$ is formulated using the normalized gradient descent approach, namely:

$$\dot{\hat{\boldsymbol{W}}} = -\kappa_c \frac{\boldsymbol{m}_{cw}}{(\boldsymbol{m}_{cw}^{\mathrm{T}} \boldsymbol{m}_{cw} + 1)^2} e_c, \tag{26}$$

where $\kappa_c > 0$ represents a parameter and $\boldsymbol{m}_{cw} = \nabla \boldsymbol{\sigma}(\boldsymbol{Z}) \boldsymbol{F} - (1/2) \boldsymbol{G}_2 \hat{\boldsymbol{W}} - \gamma \boldsymbol{\sigma}(\boldsymbol{Z})$.

Define the weight estimation error as $\tilde{\boldsymbol{W}} = \boldsymbol{W}^* - \hat{\boldsymbol{W}}$, and its time derivative satisfies $\dot{\tilde{\boldsymbol{W}}} = -\dot{\hat{\boldsymbol{W}}}$. Using the (21) and (25), we have

$$e_H = -\tilde{\boldsymbol{W}}^{\mathrm{T}} \nabla \boldsymbol{\sigma}(\boldsymbol{E}) \boldsymbol{F} - \frac{1}{4} \tilde{\boldsymbol{W}}^{\mathrm{T}} \boldsymbol{R}_2 \tilde{\boldsymbol{W}} + \frac{1}{2} \tilde{\boldsymbol{W}}^{\mathrm{T}} \boldsymbol{R}_2 \boldsymbol{W}^* + \gamma \tilde{\boldsymbol{W}}^{\mathrm{T}} \boldsymbol{\sigma}(\boldsymbol{E}) - e_\varepsilon. \tag{27}$$

According to (26), the weight estimation error dynamics is expressed as

$$\begin{aligned} \dot{\tilde{\boldsymbol{W}}} =& k_{w1} \zeta_m \left(\nabla \boldsymbol{\sigma}(\boldsymbol{E}) \boldsymbol{F} - \frac{1}{2} \boldsymbol{R}_2 \hat{\boldsymbol{W}} - \gamma \boldsymbol{\sigma}(\boldsymbol{E}) \right) \\ & \times \left(-\tilde{\boldsymbol{W}}^{\mathrm{T}} \nabla \boldsymbol{\sigma}(\boldsymbol{E}) \boldsymbol{F} - \frac{1}{4} \tilde{\boldsymbol{W}}^{\mathrm{T}} \boldsymbol{R}_2 \tilde{\boldsymbol{W}} + \frac{1}{2} \tilde{\boldsymbol{W}}^{\mathrm{T}} \boldsymbol{R}_2 \boldsymbol{W}^* + \gamma \tilde{\boldsymbol{W}}^{\mathrm{T}} \boldsymbol{\sigma}(\boldsymbol{E}) - e_\varepsilon \right). \end{aligned} \tag{28}$$

4 Stability Analysis

Theorem 1. *For an underactuated AUV described by* (1) *and* (2)*, and under Assumptions 1 and 2, implementing* (3)*, the observer* (7) *with* (11)*, the backstepping controller* (9) *and* (10)*, and the approximate optimal control law* (24) *with* (26) *ensures all signals in the closed-loop system are uniformly ultimately bounded.*

Proof. The Lyapunov function candidate is chosen as follows:

$$L_V = L_{V2} + L_{V3} + L_{V4}, \tag{29}$$

where $L_{V3} = V^*(\boldsymbol{E})$, $L_{V4} = (1/2) \tilde{\boldsymbol{W}}^{\mathrm{T}} \tilde{\boldsymbol{W}}$.

Since the boundedness of $\boldsymbol{W}^*$, $V^*(\boldsymbol{E})$, $\nabla \varepsilon(\boldsymbol{E})$, $\boldsymbol{R}_1$ and $\boldsymbol{R}_2$, there exist positive constants ζ_W^M, ζ_V^M, $\zeta_{\nabla\varepsilon}^M$, ζ_{R1}^M, ζ_{R2}^m and ζ_{R2}^M such that the relations $\|\boldsymbol{W}^*\| \leq \zeta_W^M$, $\|V^*(\boldsymbol{E})\| \leq \zeta_V^M$, $\|\nabla \varepsilon(\boldsymbol{E})\| \leq \zeta_{\nabla\varepsilon}^M$, $\|\boldsymbol{R}_1\| \leq \zeta_{R1}^M$ and $\zeta_{R2}^m \leq \|\boldsymbol{R}_2\| \leq \zeta_{R2}^M$ hold. In view of (17), (19), a (24), the Cauchy-Schwarz and Young's inequalities, $\dot{L}_{V3}$ can be reduced to the expression

$$\begin{aligned} \dot{L}_{V3} \leq & -\lambda_{\min}(\boldsymbol{Q}) \|\boldsymbol{E}\|^2 + \frac{1}{2} \left\| \tilde{\boldsymbol{W}} \right\|^2 + \frac{1}{4} (\zeta_{R1}^M \zeta_{\nabla\varepsilon}^M)^2 \\ & + \frac{1}{4} (\zeta_{R2}^M \zeta_W^M)^2 + \frac{1}{4} (\zeta_{R1}^M \zeta_{\nabla\varepsilon}^M)^2 + \gamma \zeta_V^M. \end{aligned} \tag{30}$$

According to $\nabla \boldsymbol{\sigma}(\boldsymbol{E}) \dot{\boldsymbol{E}}^*$, $\nabla \boldsymbol{\sigma}(\boldsymbol{E}) \boldsymbol{R}_1 \nabla \varepsilon(\boldsymbol{E})$, $\boldsymbol{\sigma}(\boldsymbol{E})$ and e_ε are bounded, it is known that there exist positive constants $\zeta_{\sigma E}^M$, $\zeta_{\sigma\varepsilon}^M$, ζ_σ^M and ζ_e^M such that $\|\nabla \boldsymbol{\sigma}(\boldsymbol{E}) \dot{\boldsymbol{E}}^*\| \leq \zeta_{\sigma E}^M$, $\|\nabla \boldsymbol{\sigma}(\boldsymbol{E}) \boldsymbol{R}_1 \nabla \varepsilon(\boldsymbol{E})\| \leq \zeta_{\sigma\varepsilon}^M$, $\|\boldsymbol{\sigma}(\boldsymbol{E}) t\| \leq \zeta_\sigma^M$ and $|e_\varepsilon| \leq \zeta_e^M$

hold, respectively. Next, combining the Cauchy-Schwarz inequality and Young's inequality, $\dot{L}_{V4}$ is simplified as

$$\dot{L}_{V4} \leq -\zeta_1 \left\| \tilde{\boldsymbol{W}} \right\|^4 + \zeta_2 \left\| \tilde{\boldsymbol{W}} \right\|^2 + \zeta_3, \tag{31}$$

where ζ_1, ζ_2 and ζ_3 are theoretically guaranteed positive constants.

Substituting (12), (30) and (31) into $\dot{L}_V = \dot{L}_{V2} + \dot{L}_{V3} + \dot{L}_{V4}$, and the inequality $\|\boldsymbol{F} + \boldsymbol{G}\boldsymbol{U}\| \leq \zeta_F \sqrt{\|\boldsymbol{E}\|}$ with a constant $\zeta_F > 0$, we have

$$\dot{L}_V \leq -\boldsymbol{Z}^{\mathrm{T}} \boldsymbol{\kappa}_z \boldsymbol{Z} + \zeta_{Lz} \leq -\lambda_{\min}(\boldsymbol{\kappa}_z) \|\boldsymbol{Z}\|^2 + \zeta_{Lz}. \tag{32}$$

where $\boldsymbol{Z} = [\boldsymbol{E}^{\mathrm{T}}, \boldsymbol{E}_o^{\mathrm{T}}, \|\tilde{\boldsymbol{W}}\|^2 - \zeta_4/2\zeta_1]^{\mathrm{T}}$ with $\zeta_4 = \zeta_2 + 1/2$, $\boldsymbol{\kappa}_z = \mathrm{diag}(\kappa_1 \boldsymbol{I}_6, \kappa_2 \boldsymbol{I}_3, \kappa_3)$ with $\kappa_1 = \lambda_{\min}(\boldsymbol{Q}) + k - (3/4)\zeta_F$, $k = \min\{\lambda_{\min}(\boldsymbol{K}_\eta) - \zeta_J^M, \lambda_{\min}(\boldsymbol{K}_\nu) - \zeta_J^M\}$, $\kappa_2 = \lambda_{\min}(\boldsymbol{K}_{o3}^{-1} \boldsymbol{K}_{o1}) - \lambda_{\max}(\boldsymbol{K}_{o3}^{-1})\zeta_f$ and $\kappa_3 = \zeta_1$, $\zeta_{L1} = (1/4)(\zeta_{R1}^M \zeta_{\nabla\varepsilon}^M)^2 + (1/4)(\zeta_{R1}^M \zeta_{\nabla\varepsilon}^M)^2 + (1/4)(\zeta_{R2}^M \zeta_W^M)^2 + (1/4)\zeta_F + {\zeta_4}^2/(4\zeta_1) + \gamma \zeta_V^M + \zeta_3$.

Thus, when $\lambda_{\min}(\boldsymbol{\kappa}_z) > 0$ and $\|\boldsymbol{Z}\| \geq \sqrt{\zeta_{Lz}/\lambda_{\min}(\boldsymbol{\kappa}_z)}$, $\dot{L}_V \leq 0$ holds. Thus, the proposed control method guarantees that all signals in the closed-loop system are uniformly ultimately bounded.

4.1 Simulation Studies

To validate the proposed method, comparative MATLAB simulations are conducted with the AFTDC approach from [10], using the underactuated AUV model presented in [15].

In the proposed control method, the relevant design parameters are selected as follows: $l_\nu = 0.25(\mathrm{m})$; the adaptive disturbance observer parameters are $\boldsymbol{K}_{o1} = 80\boldsymbol{I}_3$, $\boldsymbol{K}_{o2} = 3 \times 10^4 \mathrm{diag}(50, 10, 1)$ and $\boldsymbol{K}_{o3} = 0.015\boldsymbol{I}_6$; the backstepping controller parameters are $\boldsymbol{K}_\eta = 0.3\boldsymbol{I}_3$ and $\boldsymbol{K}_\nu = \boldsymbol{I}_3$; the optimal controller parameters are $\boldsymbol{Q} = 1 \times 10^4 \boldsymbol{I}_3$, $\boldsymbol{R} = \boldsymbol{I}_6$, $k_{w1} = 2 \times 10^{-4}$, $k_{w2} = 0.1$, the critic NN structure is 6-15-1, the centers of the Gaussian functions are evenly spaced in ± 30, the widths are 50, and the initial weight vector is $\hat{\boldsymbol{W}}(0) = [1, 1, ..., 1]^{\mathrm{T}} \in \mathbb{R}^{15\times 1}$. Environmental disturbances are denoted as $\boldsymbol{\tau}_e = [2\sin(0.1\pi t)(\mathrm{N}), 0.5\sin(0.1\pi t)(\mathrm{N}), 0.5\cos(0.1\pi t)(\mathrm{N}), 2\cos(0.1\pi t)(\mathrm{N} \cdot \mathrm{m}), \sin(0.05\pi\ \mathrm{t}) + \cos(0.05\pi t)(\mathrm{N} \cdot \mathrm{m})]^{\mathrm{T}}$. The simulation time is set to 300 seconds, and actuator faults are described as $\beta_i = \begin{cases} 0.3, t \geq 100(\mathrm{s}) \\ 0, 0 \leq t < 100(\mathrm{s}) \end{cases}$, $\tau_{\delta u} = \begin{cases} 4, t \geq 100(\mathrm{s}) \\ 0, 0 \leq t < 100(\mathrm{s}) \end{cases}$ and $\tau_{\delta q}, \tau_{\delta r} = \begin{cases} 1.2, t \geq 100(\mathrm{s}) \\ 0, 0 \leq t < 100(\mathrm{s}) \end{cases}$. The desired trajectory is $\boldsymbol{\eta}_{1d} = [5\sin\ (0.05t)(\mathrm{m}), 5\sin(0.05t)(\mathrm{m}), -0.05t(\mathrm{m})]^{\mathrm{T}}$. The initial states are $\boldsymbol{\eta}(0) = [3(\mathrm{m}), 0.5(\mathrm{m}), -0.2(\mathrm{m}), 0(\mathrm{rad}), 0.1(\mathrm{rad})]^{\mathrm{T}}$ and $\boldsymbol{\nu}(0) = [0(\mathrm{m/s}), 0(\mathrm{m/s}), 0(\mathrm{m/s}), 0(\mathrm{rad/s}), 0(\mathrm{rad/s})]^{\mathrm{T}}$.

Simulation results are given in Figs. 1 and 2. Figure 1 illustrates the capability of the proposed method to enable the underactuated AUV to track the desired trajectory with high accuracy. In comparison with the AFTDC method, the present approach exhibits superior tracking performance. As shown in Fig. 2, the proposed method yields smoother control inputs with smaller initial values and less chattering than the AFTDC method, thus reducing actuator wear and improving energy efficiency.

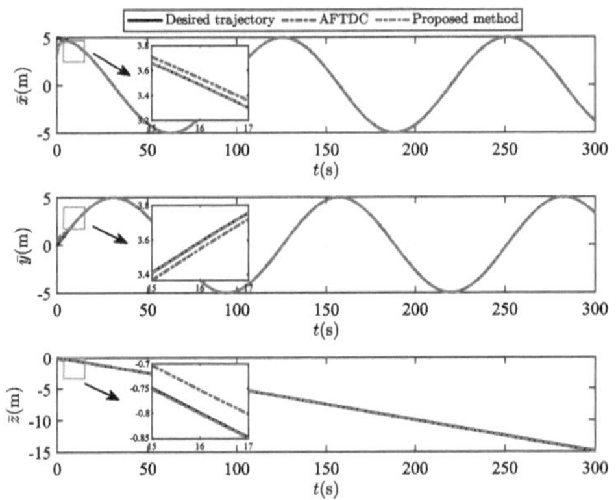

Fig. 1. Comparison of the positions $\bar{x}$, $\bar{y}$ and $\bar{z}$.

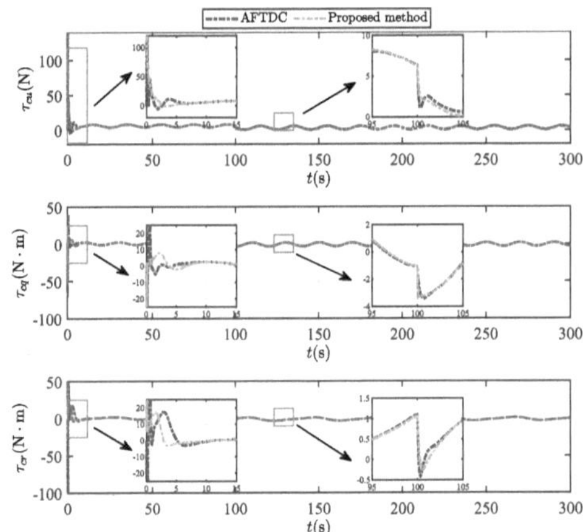

Fig. 2. Comparison of control inputs τ_{cu}, τ_{cq}, and τ_{cr}.

5 Conclusions

In this paper, an adaptive neural network-based optimal fault-tolerant control method is developed for 3D trajectory tracking of underactuated AUVs.

The approach effectively handles actuator faults and environmental disturbances, ensuring stable and accurate tracking performance, as confirmed by theoretical analysis and simulations.

Acknowledgment. This work was supported in part by the National Natural Science Foundation of China under Grant 52171292 and in part by Dalian Outstanding Young Talents Program under Grant 2022RJ05.

Disclosure of Interests. It is now necessary to declare any competing interests or to specifically state that the authors have no competing interests.

References

1. Yang, Y., Xiao, Y., Li, T.: A survey of autonomous underwater vehicle formation: performance, formation control, and communication capability. IEEE Commun. Surveys Tuts. **23**(2), 815–841 (2021)
2. Hao, L.-Y., Wang, R.-Z., Shen, C., Shi, Y.: Trajectory tracking control of autonomous underwater vehicles using improved tube-based model predictive control approach. IEEE Trans. Ind. Informat. **20**(4), 5647–5657 (2023)
3. Er, M.J., Gong, H., Liu, Y., Liu, T.: Intelligent trajectory tracking and formation control of underactuated autonomous underwater vehicles: a critical review. IEEE Trans. Syst., Man Cybern. Syst. **54**(1), 543–555 (2024)
4. Yu, H., Guo, C., Yan, Z.: Globally finite-time stable three-dimensional trajectory-tracking control of underactuated UUVs. Ocean Eng. **189** 106329 (2019)
5. An, S., Wang, L., He, Y.: Robust fixed-time tracking control for underactuated AUVs based on fixed-time disturbance observer. Ocean Eng. **266** 112567 (2022)
6. Liang, X., Qu, X., Wang, N., Zhang, R., Li, Y.: Three-dimensional trajectory tracking of an underactuated AUV based on fuzzy dynamic surface control. IET Intell. Transp. Syst. **14**(5), 364–370 (2020)
7. Wang, J., Wang, C., Wei, Y., Zhang, C.: Command filter based adaptive neural trajectory tracking control of an underactuated underwater vehicle in three-dimensional space. Ocean Eng. **180**, 175–186 (2019)
8. Xu, R., Tang, G., Xie, D., Han, L., Huang, H.: Neural network for 3D trajectory tracking control of a CMG-actuated underwater vehicle with input saturation. ISA Trans. **123**, 152–167 (2022)
9. Liu, H., Zhuo, J., Tian, X., Mai, Q.: Finite-time self-structuring neural network trajectory tracking control of underactuated autonomous underwater vehicles. Ocean Eng. **268** 113450 (2023)
10. Tabatabaee-Nasab, F.S., Moosavian, S.A.A., Khalaji, A.K.: Adaptive fault-tolerant control for an autonomous underwater vehicle. Robotica **40**(11), 4076–4089 (2022)
11. Che, G., Liu, L., Yu, Z.: Nonlinear trajectory-tracking control for autonomous underwater vehicle based on iterative adaptive dynamic programming. J. Intell. Fuzzy Syst. **37**(3), 4205–4215 (2019)
12. Che, G., Yu, Z.: ADP based output-feedback fault-tolerant tracking control for underactuated AUV with actuators faults. J. Intell. Fuzzy Syst. **45**(4), 5871–5883 (2023)
13. Che, G.: Single critic network based fault-tolerant tracking control for underactuated AUV with actuator fault. Ocean Eng. **254**, 111380 (2022)

14. Shojaei, K., Arefi, M.M.: On the neuro-adaptive feedback linearising control of underactuated autonomous underwater vehicles in three-dimensional space. IET Control Theory Appl. **9**(8), 1264–1273 (2015)
15. Gong, H., Er, M.J., Liu, Y.: Fuzzy optimal fault-tolerant trajectory tracking control of underactuated AUVs with prescribed performance in 3-D space. IEEE Trans. Syst. Man Cybern. Syst. **55**(1), 170–182 (2025)

An Agentic Framework for 3D Semantic Gaussian Maps

Lei Xia, Shuangjie Yuan, Lubo Li, Huifu Chen, Haoyu Liu, Tieshan Li, and Lu Yang(✉)

University of Electronic Science and Technology of China, No. 2006, Xiyuan Ave, West Hi-Tech Zone, Chengdu, China
yanglu@uestc.edu.cn

Abstract. Egocentric 3D object localization driven by natural language is a pivotal capability for next-generation Embodied AI and Augmented Reality (AR) assistants. However, existing approaches often treat linguistic reasoning, 2D visual grounding, and 3D scene reconstruction as isolated tasks, hindering the ability to intuitively locate objects in physical space through speech alone. To bridge this gap, we present a novel unified agentic framework that seamlessly integrates Large Language Model (LLM) agents with video understanding and semantic 3D reconstruction. Our system captures egocentric video and user speech, transcribing audio commands into text via Automatic Speech Recognition (ASR). We employ LangManus as the central intelligent agent to interpret user intent and orchestrate task planning. Upon receiving a localization query, the system utilizes Sa2VA (Marrying SAM2 with LLaVA) to perform dense grounded understanding within the video stream, while concurrently deploying SegAnyGaussian (SAGA) to reconstruct the scene and lift 2D segmentation features into a semantically segmented 3D Gaussian Splatting field. By synergizing the spatiotemporal reasoning of Sa2VA with the interactive 3D semantic capabilities of SAGA, our framework effectively translates spoken instructions into precise 3D object coordinates. This end-to-end pipeline significantly enhances human-environment interaction, enabling accurate and intuitive object retrieval in complex 3D scenarios.

Keywords: 3D Gaussian Splatting · 3D Semantic Segmentation · Object Localization · Multi-Agent Collaboration

1 Introduction

In the realm of Embodied AI and Augmented Reality (AR), the ability to perceive surroundings from a first-person (egocentric) perspective is crucial [1,2]. Modern head-mounted devices are increasingly capable of capturing high-fidelity video streams [3] and interacting with users. However, a significant gap remains between recording a video and truly "understanding" the 3D physical space. Users envision an intelligent assistant that can not only recognize objects in a

C. Li et al. (Eds.): ICNC 2025, CCIS 2946, pp. 603–614, 2026.
https://doi.org/10.1007/978-981-92-1599-7_51

video but also precisely locate them in the 3D world based on spoken instructions (e.g.,"Where is my bottle?"). Achieving this requires a system that seamlessly integrates natural language processing, visual grounding in videos, and semantically rich 3D scene reconstruction.

Current approaches typically address these challenges in isolation. Large Language Model (LLM) agents [4–6] excel at task planning but lack direct visual perception. Vision-Language Models (VLMs) [7,8] have made strides in 2D video understanding—outputting bounding boxes or text descriptions—but they operate in the image plane, devoid of spatial depth. Conversely, novel 3D reconstruction techniques like 3D Gaussian Splatting (3DGS) [9] provide photorealistic rendering but traditionally lack semantic understandability. While recent works like Segment Any 3D Gaussians (SAGA) [10] have introduced the capability to lift 2D segmentation masks into 3D semantic features, enabling interactive segmentation, there is currently no unified framework that connects this 3D semantic capability with an intelligent agent and video understanding modules to automate speech-driven 3D object localization.

To bridge these gaps, we propose a novel agentic framework that orchestrates three specialized components to realize speech-driven 3D localization. We utilize LangManus as the central brain of our system. It processes the user's voice input (transcribed via ASR) [11] to interpret intent. When a localization request is identified, LangManus acts as a planner, decomposing the user's natural language instruction into executable tasks. To understand the visual context, LangManus dispatches the query to Sa2VA [12]. Leveraging its architecture that marries SAM2 with LLaVA, Sa2VA performs dense grounded understanding on the egocentric video, confirming the object's presence and providing a textual and 2D spatial description of the target within the video frames. Parallel to video analysis, our system leverages SegAnyGaussian for 3D spatial processing which reconstructs the scene using 3D Gaussian Splatting and, crucially, employs its inherent capability to map 2D segmentation features onto 3D Gaussians. This allows SAGA to perform 3D semantic segmentation directly within the reconstructed model.

By integrating these modules, the system uses the target object information derived from the agent to query SAGA's semantically segmented 3D scene, thereby outputting the precise 3D position of the object for user interaction.

Overall, this work represents a significant step towards fully autonomous embodied perception. By effectively bridging the semantic gap between natural language instructions, temporal video understanding, and static 3D scene reconstruction, we provide a robust solution for real-world object localization. Unlike previous isolated approaches, our system creates a synergistic loop where the agent's intent drives the spatial analysis, transforming the wearable camera from a passive recorder into an active, intelligent assistant. The specific contributions of this research are outlined as follows:

- System Integration: We present a cohesive pipeline that integrates ASR, an LLM-based agent (LangManus), Video-LMMs (Sa2VA), and semantic 3D reconstruction (SAGA) into a unified wearable system.

- Automated 3D Localization: We demonstrate how to leverage SAGA's internal 2D-to-3D feature lifting capability, driven by agentic commands, to achieve accurate 3D object localization without manual intervention.
- Semantic Interactivity: Our approach enables users to interact with their physical environment through natural language, effectively translating spoken queries into precise 3D spatial coordinates within a reconstructed 3D scene.

2 Related Work

2.1 3D Scene Reconstruction

The field of 3D scene representation has undergone a paradigm shift from explicit mesh-based methods to implicit neural representations. Neural Radiance Fields (NeRF) [13] demonstrated exceptional capability in photorealistic novel view synthesis but suffered from slow training and rendering speeds, limiting their utility in real-time wearable applications. Recently, 3D Gaussian Splatting has emerged as a superior alternative, representing scenes as a set of anisotropic 3D Gaussians. Unlike implicit fields, 3DGS utilizes an explicit representation that supports real-time rasterization while maintaining high visual fidelity. This explicit nature is particularly advantageous for downstream tasks, as it allows for the direct attachment of semantic attributes to individual primitives. In our work, we leverage 3DGS as the underlying spatial representation, exploiting its efficiency to construct 3D scene of the environment from egocentric video streams.

2.2 From 2D Masks to 3D Segmentation

While 3DGS provides excellent geometric and photometric reconstruction, raw Gaussians lack semantic meaning. A growing body of research focuses on "lifting" 2D semantic features into 3D space to enable 3D semantic segmentation. Early approaches distilled feature fields from pre-trained 2D encoders into NeRFs or 3DGS [14–16]. However, these methods often struggle with boundary precision.

More recent works, such as Segment Any 3D Gaussians, distilling the powerful segmentation capabilities of the 2D Segment Anything Model (SAM) [17] into 3D Gaussian features. SAGA employs a contrastive training strategy to learn multi-granular semantic features, allowing for interactive 3D segmentation where 2D prompts can be accurately mapped to 3D structures. Our framework adopts SAGA to endow the reconstructed scene with semantic interactability, enabling the system to query and retrieve the 3D spatial occupancy of specific objects based on user instructions.

2.3 Dense Grounded Understanding in Video

Accurate object localization in dynamic video streams—often termed visual grounding—is a prerequisite for mapping objects into 3D space. Traditional

Vision-Language Models typically operate on static images, which is insufficient for continuous egocentric video inputs. The evolution of Large Multimodal Models has led to architectures that can process temporal sequences.

Notably, Sa2VA signifies a paradigm shift in this domain by synergizing the robust segmentation capabilities of SAM2 [18] with the multimodal reasoning faculties of LLaVA [19]. This architecture unifies referring segmentation, visual question answering, and conversational interaction into a monolithic framework, facilitating dense grounded understanding within video streams. Diverging from conventional trackers prone to semantic drift, Sa2VA employs instruction tuning to ensure robust temporal consistency. In our proposed pipeline, we leverage the spatiotemporal grounding capabilities of Sa2VA to precisely identify and localize target objects within the 2D egocentric video stream based on natural language queries, ensuring accurate semantic verification in the visual domain.

2.4 LLM-Based Agents and Multi-Agent Collaboration

The ability to interpret vague or complex human instructions requires reasoning capabilities beyond simple pattern matching. LLMs have demonstrated remarkable potential as autonomous agents capable of task planning and tool usage [20,21]. Frameworks like LangManus illustrate how LLMs can function as a central brain to orchestrate various modules for diverse tasks. LangManus excels in parsing natural language, decomposing abstract intents into structured, executable tasks, and dispatching them to specialized sub-modules. In our proposed system, LangManus serves as the high-level planner that coordinates the workflow: it interprets the user's speech via ASR, assigns the localization task to the vision expert (Sa2VA), and subsequently triggers the 3D query mechanism in SAGA, effectively bridging the gap between human interaction and spatial perception.

3 Methodology

3.1 Overview

Figure 1 illustrates the proposed end-to-end framework for egocentric 3D object localization. We propose a unified agentic framework that bridges the gap between natural language voice commands and precise 3D spatial localization in egocentric environments. The pipeline comprises three interconnected modules: (1) An Input Processing and Task Dispatcher driven by LangManus; (2) A Visual Verification module utilizing Sa2VA; and (3) A 3D Spatial Localization module leveraging SAGA.

3.2 Agentic Planning and Intent Recognition

The workflow initiates with the user's voice command captured by a head-mounted device. An Automatic Speech Recognition (ASR) model transcribes the

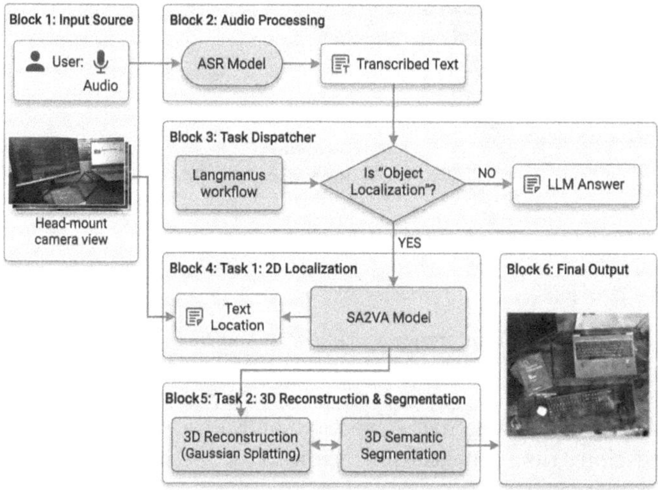

Fig. 1. The architecture of the proposed agentic framework.

audio stream into a text sequence T_{input}. This text is processed by LangManus, a Large Language Model (LLM)-based agent acting as the central planner.

LangManus analyzes the semantic content of T_{input} to classify the user's intent. The decision logic is twofold:

- General Interaction: If the input is conversational, the system bypasses the vision pipeline and generates a direct LLM response.
- Object Localization: If the input implies a search task, the agent extracts the target semantic entity, denoted as E_{target}, and triggers the downstream visual perception modules.

3.3 Spatiotemporal Visual Verification via Sa2VA

To ensure the target object exists within the current field of view before attempting expensive 3D processing, we employ Sa2VA for dense 2D localization. As illustrated in Fig. 2, our implementation focuses specifically on the multimodal reasoning path of Sa2VA. Unlike the original architecture which performs simultaneous segmentation, we explicitly deactivate the segmentation branch in this stage. This allows the system to focus solely on verifying object presence and generating spatial descriptions through the text generation pipeline.

Visual-Language Encoding and Fusion. The data processing begins with feature encoding. The input egocentric video frames, denoted as tensors of shape $T \times H \times W$, are encoded into a sequence of visual tokens $\mathbf{V}_{tokens}$ with dimensions $N_{vis} \times D$ using a Hiera-based image encoder. Simultaneously, the query text derived from LangManus is tokenized into text tokens $\mathbf{T}_{tokens}$ of shape $N_{txt} \times D$.

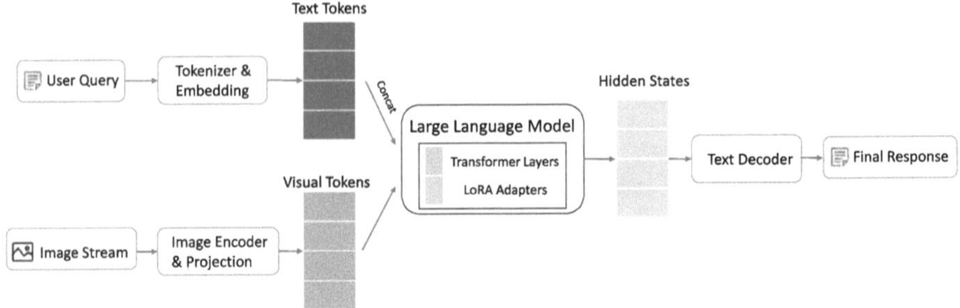

Fig. 2. The inference workflow of the Sa2VA module.

As depicted in the "Token Strips" section of Fig. 2, these two modalities are concatenated along the sequence dimension to form a unified multimodal input:

$$\mathbf{X}_{input} = [\mathbf{V}_{tokens}; \mathbf{T}_{tokens}] \tag{1}$$

This concatenation effectively maps the visual data into the language embedding space.

Generative Reasoning. This fused sequence is subsequently processed by the LLM backbone equipped with LoRA adapters [22]. The LLM performs deep reasoning on the multimodal context to generate high-dimensional hidden states. Instead of decoding these states into segmentation masks, we utilize the text-generation capability of the model. A text decoder transforms the hidden states into the final textual output L_{2D}. This step acts as a semantic filter, ensuring that only visually verified objects and their 2D spatial contexts are passed to the subsequent 3D reconstruction module.

3.4 3D Semantic Reconstruction and Localization via SAGA

The core of our spatial awareness lies in the SAGA module. Figure 3 depicts the concrete data flow of SAGA, highlighting how 2D semantic features are lifted into the 3D Gaussian field to enable interactive querying.

Scale-Aware Feature Gating and Distillation. A key challenge in 3D segmentation is multi-granularity. SAGA addresses this via a Scale-Gated mechanism [10]. As the foundational step, each 3D Gaussian in the scene is assigned a learnable base affinity feature vector (forming a tensor of shape $N \times D$). To adapt to different object scales, a Multi-Layer Perceptron (MLP) generates a scale-dependent gate vector $\mathcal{S}(s) \in [0,1]^D$ based on a physical scale input s. This gate vector performs an element-wise product with the base features to produce multi-granular gated features $\mathbf{f}_s$:

$$\mathbf{f}_s = \mathcal{S}(s) \odot \mathbf{f}_g \tag{2}$$

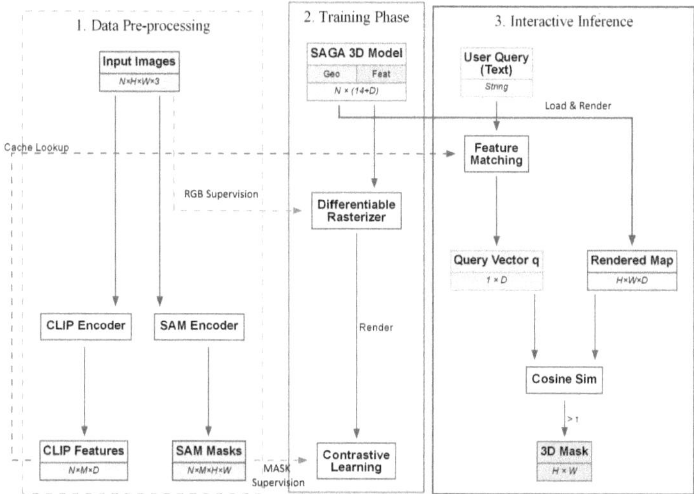

Fig. 3. The concrete data flow of the SAGA module.

This gating mechanism allows the system to dynamically suppress or activate specific feature dimensions. To ensure these features carry semantic meaning, they are distilled from a 2D foundation model (e.g., SAM). During training, we minimize the difference between the rendered features $\mathbf{F}(p)$ and the reference features $\mathbf{F}_{SAM}$ extracted from the image encoder:

$$\mathcal{L}_{feat} = \sum_{p \in \mathcal{R}} \|\mathbf{F}(p) - \mathbf{F}_{SAM}(p)\|_1 \tag{3}$$

where $\mathcal{R}$ represents the set of sampled pixels/rays. This supervision aligns the 3D Gaussian attributes with robust 2D semantic priors.

Differentiable Feature Rendering. The processed 3D features are subsequently rendered into a 2D feature map via differentiable rasterization [9]. As depicted in the "Rasterization" block of Fig. 3, the high-dimensional affinity features are projected onto the image plane. For a pixel p, the rendered feature vector $\mathbf{F}(p)$ is computed by blending the overlapping ordered Gaussians:

$$\mathbf{F}(p) = \sum_{i \in \mathcal{N}_p} \mathbf{f}_{s,i} \cdot \alpha_i \prod_{j=1}^{i-1} (1 - \alpha_j) \tag{4}$$

This process results in a dense feature map where each pixel contains a semantic descriptor derived directly from the 3D scene structure.

Semantic Query and Coordinate Extraction. This rendered map supports interactive querying by matching with the user's text-derived query vector. In the inference stage, the target text entity E_{target} is converted into a query vector $\mathbf{q}$ using a CLIP text encoder [23]. The system computes a similarity map M_{sim} by normalizing the dot product between the rendered features and the query:

$$M_{sim}(p) = \frac{\mathbf{F}(p) \cdot \mathbf{q}}{\|\mathbf{F}(p)\| \|\mathbf{q}\|} \tag{5}$$

Pixels exceeding a confidence threshold τ form a segmentation mask $\mathcal{M} = \{p \mid M_{sim}(p) > \tau\}$. To achieve physical localization, we identify the set of 3D Gaussians $\mathcal{G}_{target}$ corresponding to these mask regions. The final 3D coordinate P_{xyz} is computed as the centroid of these activated Gaussians:

$$P_{xyz} = \frac{1}{|\mathcal{G}_{target}|} \sum_{k \in \mathcal{G}_{target}} \mu_k \tag{6}$$

where μ_k represents the mean position of the k-th Gaussian. This step effectively translates the semantic segmentation into precise physical coordinates for agent navigation.

4 Experiments

4.1 Experimental Setup

To validate the feasibility of our proposed framework, we deployed the system on Drift X3. The camera can capture egocentric video at 1920 × 1080 resolution (30 FPS). To ensure efficient processing, the computationally intensive inference tasks for Sa2VA and SAGA are offloaded to a local server equipped with an NVIDIA RTX 4060 GPU.

Our experiments were conducted using the -NeRF 360 dataset [24] alongside a custom real-world dataset collected via our wearable device to evaluate system performance in diverse scenarios.

4.2 Comparative Evaluation

We evaluated the visual understanding capabilities of various VLMs using the uniform query:"Where is the piano?" in a same scene shown in Fig. 4. As detailed in Table. 1, larger parameter sizes did not guarantee success. Models such as Gemma3:4B and Llava:7B failed to detect the target, with Llava exhibiting severe hallucinations by describing non-existent elements instead of the piano.

Fig. 4. The scene used for visual understanding evaluation.

Table 1. Visual Understanding capability of Different Models

Model	Answer	Object Location	Time(s)
Qwen3-vl:2b	There is no piano visible in the provided image. The scene depicts a living room with furniture such as a couch, a chair, a TV, a coffee table, and other household items, but no piano.	×	7.38
Qwen2.5-vl:3b	The piano is located in the living room.	✓	5.70
Gemma3:4b	There is no piano visible. There is a television and sound system on a dark cabinet to the right of the room.	×	14.79
Llava:7b	The image appears to be a screenshot from a video, likely taken by someone using their smartphone or another device, as indicated by the presence of a status bar at the top showing signal strength and battery life. In the photo, there isn't an obvious piano visible in the living room. However, it seems like there is a small white dog on a rug near a couch in the background, but no sign of a piano or musical instrument. The focus of the image is on the interior of a cozy living room with a couch, a coffee table, and various pieces of furniture.	×	3.93
Sa2VA:1b	The piano is located in the corner of the living room.	✓	3.57

Table 2. Comparison on the SPIn-NeRF dataset.

Method	MVSeg	SA3D	OmniSeg3D	GauGroup	SA-GS	SA3D-GS	SAGA
mIoU (%)	90.9	92.4	**94.3**	86.5	89.9	93.2	93.4
mAcc (%)	98.9	98.9	**99.3**	98.9	98.7	99.1	99.2

In contrast, Sa2VA-1B successfully identified and located the "piano," matching the accuracy of Qwen2.5-VL:3B but with significantly higher efficiency. As the lightest model, Sa2VA achieved the lowest inference latency 3.57 s, outperforming other VLMs. This result confirms that Sa2VA-1B provides the optimal balance of accuracy and speed required for real-time wearable deployments.

We utilize SAGA as the core framework for 3D semantic reconstruction. As demonstrated in Table 2, SAGA delivers superior segmentation accuracy among 3D Gaussian Splatting-based approaches, achieving an mIoU of 93.4% and an mAcc of 99.2% on the SPIn-NeRF dataset. This high geometric fidelity justifies its selection as the spatial grounding backbone for our proposed pipeline (Table 3).

Table 3. Results of the proposed system in different scenarios.

Original Scene	Query	Answer from Sa2VA
	1. Where is the blue book? 2. Where is the pair of keys?	1. The blue book is placed on the desk. It is next to a purple box. 2. The pair of keys is next to the book on the desk.
3D-GS Scene	Different Semantic Segmentaion in 3D scene	
	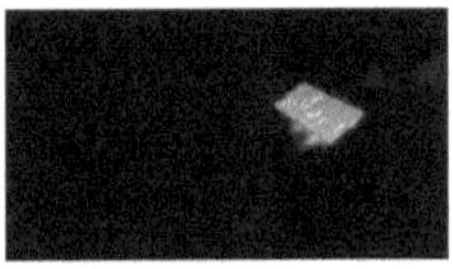	
Original Scene	Query	Answer from Sa2VA
	1. Where is the table? 2. Where is the flower vase?	1. The table is located on the patio, next to a wall. 2. The flower vase is placed on top of the table.
3D-GS Scene	Different Semantic Segmentaion in 3D scene	
		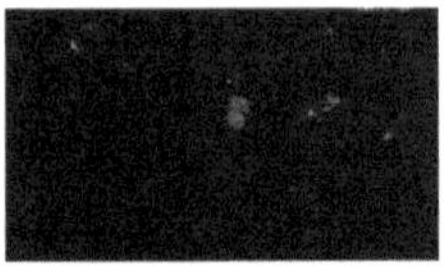
Original Scene	Query	Answer from Sa2VA
	1. Where are the shoes? 2. Where is the piano?	1. The shoes are next to the couch. 2. The piano is located in the corner of the living room.
3D-GS Scene	Different Semantic Segmentaion in 3D scene	
		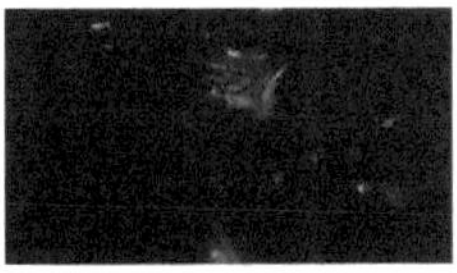

4.3 Evaluation of the Whole System

Our system effectively grounds natural language queries into 3D space. In scenarios featuring objects with clear boundaries, such as the "blue book" and the "table", the pipeline exhibits exceptional precision, successfully filtering out unrelated background elements to generate accurate 3D masks.

A closer inspection of the "shoes" case, however, reveals a notable limitation regarding object-surface separation. Although the system correctly recognized the shoes in response to the user's query, the 3D reconstruction struggled to define the precise lower boundary, erroneously merging the underlying white carpet into the segmented volume. This occurrence indicates that while the system excels at semantic localization, strictly separating small objects from texture-similar contact surfaces remains a challenging frontier for 3D Gaussian segmentation.

5 Conclusion

In this paper, we presented an agentic framework for speech-driven 3D object localization in egocentric scenarios. By orchestrating LangManus for intent understanding, Sa2VA for dense video grounding, and SAGA for semantic 3D reconstruction, we established a cohesive pipeline that translates natural language directly into physical 3D coordinates.

Our system effectively processes both explicit commands and abstract functional queries. It successfully verifies objects in video streams and localizes them within the reconstructed 3D scene.

This work stands as a foundational proof-of-concept for endowing 3D Gaussian Splatting with agentic intelligence. Future work will focus on improving fine-grained geometric segmentation, conducting quantitative benchmarking on large-scale datasets, and optimizing the pipeline for dynamic, real-time wearable applications.

References

1. Pei, B., et al.: EgoThinker: unveiling egocentric reasoning with spatio-temporal cot. ArXiv abs/2510.23569 (2025). https://api.semanticscholar.org/CorpusID:282389525
2. Chen, H., et al.: Symbridge: a human-in-the-loop cyber-physical interactive system for adaptive human-robot symbiosis (2025). https://api.semanticscholar.org/CorpusID:276258639
3. Zou, Y., Yuan, S., Liu, H., Cheng, X., Zhu, T., Yang, L.: Vision language model based panel digit recognition for medical screen data acquisition. Displays **92**, 103282 (2026). https://doi.org/10.1016/j.displa.2025.103282, https://www.sciencedirect.com/science/article/pii/S0141938225003191
4. Yang, A., et al.: Qwen3 technical report. arXiv preprint arXiv:2505.09388 (2025)
5. Liu, A., et al.: DeepSeek-V3 technical report (2024). arXiv:2412.19437 arXiv preprint
6. Leon, M.: GPT-5 and open-weight large language models: advances in reasoning, transparency, and control. Inf. Syst., 102620 (2025)
7. Zhu, J., et al.: InternVL3: exploring advanced training and test-time recipes for open-source multimodal models. ArXiv abs/2504.10479 (2025). https://api.semanticscholar.org/CorpusID:277780955
8. Bai, S., et al.: Qwen3-VL technical report (2025). https://api.semanticscholar.org/CorpusID:283262018

9. Kerbl, B., Kopanas, G., Leimkuehler, T., Drettakis, G.: 3D gaussian splatting for real-time radiance field rendering. ACM Trans. Graph. (TOG) **42**, 1–14 (2023). https://api.semanticscholar.org/CorpusID:259267917
10. Cen, J., Fang, J., Yang, C., Xie, L., Zhang, X., Shen, W., Tian, Q.: Segment any 3D gaussians. ArXiv abs/2312.00860 (2023). https://api.semanticscholar.org/CorpusID:265609986
11. Gao, Z., et al.: FUNASR: a fundamental end-to-end speech recognition toolkit. ArXiv abs/2305.11013 (2023). https://api.semanticscholar.org/CorpusID:258762651
12. Yuan, H., et al.: Sa2va: marrying SAM2 with LLAVA for dense grounded understanding of images and videos (2025). https://api.semanticscholar.org/CorpusID:276317403
13. NERF: Commun. ACM **65**, 99–106 (2020). https://api.semanticscholar.org/CorpusID:213175590
14. Zhang, Y., Luo, H., Lei, Y.: Towards clip-driven language-free 3d visual grounding via 2D-3D relational enhancement and consistency. In: 2024 IEEE/CVF Conference on Computer Vision and Pattern Recognition (CVPR), pp. 13063–13072 (2024). https://api.semanticscholar.org/CorpusID:272696083
15. Ye, J., Wang, N., Wang, X.: FeatureNERF: learning generalizable nerfs by distilling foundation models. In: 2023 IEEE/CVF International Conference on Computer Vision (ICCV), pp. 8928–8939 (2023). https://api.semanticscholar.org/CorpusID:257663574
16. Peng, Y., Wang, H., Liu, Y., Wen, C., Dong, Z., Yang, B.: Gags: granularity-aware feature distillation for language gaussian splatting. ArXiv abs/2412.13654 (2024). https://api.semanticscholar.org/CorpusID:274822820
17. Kirillov, A., et al.: Segment anything. In: 2023 IEEE/CVF International Conference on Computer Vision (ICCV), pp. 3992–4003 (2023). https://api.semanticscholar.org/CorpusID:257952310
18. Ravi, N., et al.: Sam 2: segment anything in images and videos. ArXiv abs/2408.00714 (2024). https://api.semanticscholar.org/CorpusID:271601113
19. Liu, H., Li, C., Wu, Q., Lee, Y.J.: Visual instruction tuning. ArXiv abs/2304.08485 (2023). https://api.semanticscholar.org/CorpusID:258179774
20. Yao, S., et al.: React: synergizing reasoning and acting in language models. ArXiv abs/2210.03629 (2022). https://api.semanticscholar.org/CorpusID:252762395
21. Yang, H., Yue, S., He, Y.: Auto-GPT for online decision making: benchmarks and additional opinions. ArXiv abs/2306.02224 (2023). https://api.semanticscholar.org/CorpusID:259075577
22. Hu, J.E., et al.: LORA: low-rank adaptation of large language models. ArXiv abs/2106.09685 (2021). https://api.semanticscholar.org/CorpusID:235458009
23. Radford, A., et al.: Learning transferable visual models from natural language supervision. In: International Conference on Machine Learning (2021). https://api.semanticscholar.org/CorpusID:231591445
24. Barron, J.T., Mildenhall, B., Verbin, D., Srinivasan, P.P., Hedman, P.: MIP-NERF 360: unbounded anti-aliased neural radiance fields. In: 2022 IEEE/CVF Conference on Computer Vision and Pattern Recognition (CVPR), pp. 5460–5469 (2021). https://api.semanticscholar.org/CorpusID:244488448

Distributed Economic Dispatch for Port Energy Systems with Offshore Wind

Linxue Zhang, Qihe Shan(✉), Yuxin Zhang, and Fei Teng

School of Navigation, Dalian Maritime University, Dalian 116026, China
{zhanglinxue1011,shanqihe}@dlmu.edu.cn

Abstract. Offshore wind energy is a widely applied clean energy source and an important pathway to achieve port decarbonization. It ensures the sustainable development of the marine economy. This paper proposes a distributed economic dispatch strategy for port integrated energy systems considering offshore wind power to reduce port carbon emissions and promote sustainable development. Firstly, considering the distributed characteristics of heterogeneous energy supply equipment in ports, a Port Integrated Energy system is established. Secondly, the transmission losses exist in the power lines of offshore wind farms, an optimal schedule model of PIES considering offshore wind farms is developed. Then, based on distributed mixed-integer linear programming, a distributed algorithm is proposed. Finally, the effectiveness of the proposed method is verified through simulation.

Keywords: Offshore wind power · Port integrated energy system · Transmission line loss · Distributed energy management

1 Introduction

As global climate change intensifies, reducing greenhouse gas emissions and transitioning to clean energy have become international consensus [1]. Offshore wind energy, particularly in deep-sea regions, offers abundant clean renewable resources is an ideal alternative to fossil fuels. To advance renewable energy development and enable multi-energy utilization, Port Integrated Energy Systems (PIES) have emerged as a critical solution. PIES is a multi-energy coupled architecture integrating distributed energy resources, energy storage systems, and flexible loads through advanced economic dispatch strategies [2]. Unlike traditional port power systems, PIES features electricity-heat coupling capabilities to meet diverse port operational needs. Therefore, research on its economic dispatch is essential to improve energy utilization efficiency.

PIES economic dispatch aims to minimize operational costs while satisfying load demands and safety constraints. Existing research has explored various aspects: integrated schemes for coordinated thermal and renewable energy [3], low-carbon operation through carbon benefits and demand response [4], distributed neural dynamics accommodating intermittent renewable energy [5],

C. Li et al. (Eds.): ICNC 2025, CCIS 2946, pp. 615–624, 2026.
https://doi.org/10.1007/978-981-92-1599-7_52

multi-stage dispatch under frequency security constraints [6], and decentralized optimization for island systems with pumped hydro storage [7]. Given the substantial potential of offshore wind energy, transmission line losses must be considered in model development.

Economic dispatch can be solved through centralized and distributed methods. Traditional centralized method approaches [8,9] face limitations including single-point failure risks [8] and reduced privacy protection [9]. With the increse of renewable energy utilization, the PIES performs a distributed and flat structure. Hence, the centralized method are unsuitable, distributed algorithms have gained significant attention. Existing research include consensus-based strategies for flexible loads [10], fast and economical port dispatch algorithms [11], and game-theory-based approaches for distributed energy resource coalition formation [12].

However the mentioned algorithms ignore the 0–1 integer variables related to the start-stop issues of equipment in PIES. Therefore, how to design a distributed mixed-integer linear programming for the PIES is crucial to ensuring its safe operation.

2 System Framework of PIES

The PIES is an innovative energy supply architecture designed to enhance energy efficiency and reduce carbon emissions in port operations. The PIES, based on advanced distributed optimization algorithms, coordinates multiple energy flows including electricity, heat and gas to meet the diverse energy demands of port facilities, berthed vessels and port equipment under varying operational conditions. The basic structure of the PIES is illustrated in Fig. 1.

The PIES includes various devices, such as offshore wind farm (OWF), gas boiler (GB), gas turbine (GT), electric boiler (EB), solar photovoltaic (PV) and energy storage device (ESD). Their operating characteristics are as follows:

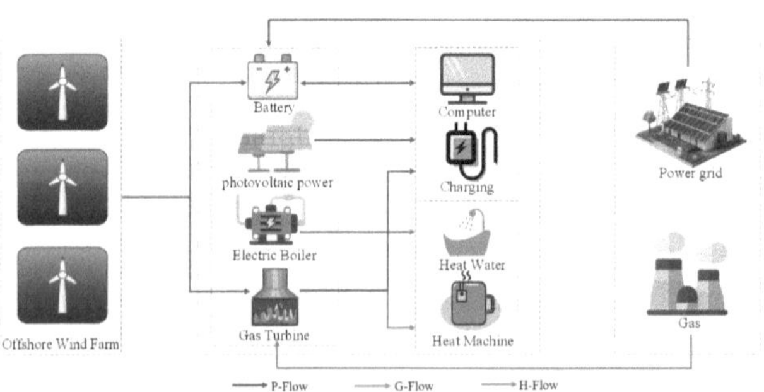

Fig. 1. System framework of PIES.

2.1 OWF

OWF transmits electricity to ports via submarine cables, and the line loss cost of the ith OWF during time t is represented as $C^t_{OWF,i}$. Referring to the loss cost model in [13], the line loss cost formula is as follows

$$C^t_{OWF,i} = a_{OWF,i} P^t_{OWF,i} + b_{OWF,i} \tag{1}$$

where $a_{OWF,i}$ and $b_{OWF,i}$ are line transmission loss cost coefficients for OWF.

$$P^{\min}_{OWF,i} \leq P^t_{OWF,i} \leq P^{\max}_{OWF,i} \tag{2}$$

where $P^{\max}_{OWF,i}$ and $P^{\min}_{OWF,i}$ represent the maximum and minimum electricity production of ith OWF in the t-period.

2.2 GB

GB generates heat by consuming natural gas. The natural gas consumption cost of the ith GB during time t is represented as $C^t_{GB,i}$, and is calculated using the following formula

$$C^t_{GB,i} = C_g Q^t_{GB,i} \tag{3}$$

$$H^t_{GB,i} = \mu_{GB} P^t_{GB,i} \tag{4}$$

$$P^{\min}_{GB,i} \leq P^t_{GB,i} \leq P^{\max}_{GB,i} \tag{5}$$

$$H^{\min}_{GB,i} \leq H^t_{GB,i} \leq H^{\max}_{GB,i} \tag{6}$$

where $Q^t_{GB,i}$ and $P^t_{GB,i}$denotes the natural gas consumption and electricity consumption of the ith GB during time t. where $P^{\max}_{GB,i}$, $P^{\min}_{GB,i}$, $H^{\max}_{GB,i}$ and $H^{\min}_{GB,i}$ represent the maximum and minimum electricity and heat production of ith GT in the t-period.

2.3 GT

The GT generates electricity and heat by consuming natural gas. The cost function of the GT is as follows

$$C^t_{GT,i} = C^t_{GT-E,i} + C^t_{GT-H,i} \tag{7}$$

where $C^t_{GT-E,i}$ represents the electricity production cost of the ith GT during time t, $C^t_{GT-H,i}$ represents the heat production cost of the ith GT during time t. And they further expressed as

$$C^t_{GT-E,i} = C_g Q^t_{GT-E,i} \tag{8}$$

$$C^t_{GT-H,i} = C_g Q^t_{GT-H,i} \tag{9}$$

where $Q^{n}_{GT-E}(t)$ and $Q^{n}_{GT-H}(t)$ represent the consumption of natural gas for electricity and heat production of the ith GT during time t. The energy transition relationship of the GT is described as

$$P^{t}_{GT,i} = \mu_{GT-E} Q^{t}_{GT-E,i} \tag{10}$$

$$H^{t}_{GT,i} = \mu_{GT-H} Q^{t}_{GT-H,i} \tag{11}$$

where $P^{t}_{GT,i}$ and $H^{t}_{GT,i}$ represent the electricity and heat output of the ith GT at time step t, μ_{GT-E} and μ_{GT-H} are the electrical and heat energy conversion efficiencies.

To ensure safe operation, the GT must comply with the following restrictions:

$$P^{\min}_{GT,i} \le P^{t}_{GT,i} \le P^{\max}_{GT,i} \tag{12}$$

$$H^{\min}_{GT,i} \le H^{t}_{GT,i} \le H^{\max}_{GT,i} \tag{13}$$

where $P^{\max}_{GT,i}$, $P^{\min}_{GT,i}$, $H^{\max}_{GT,i}$ and $H^{\min}_{GT,i}$ represent the maximum and minimum electricity and heat production of ith GT in the t-period.

2.4 EB

The EB is a device that converts electricity into thermal energy. The cost function of EB is expressed as follows:

$$C^{t}_{EB,i} = C_e P^{t}_{EB,i} \tag{14}$$

where $C^{t}_{EB,i}$ represents the operational cost of the ith EB at time step t, C_e represents the electricity price coefficient, $P^{t}_{EB,i}$ represents the electrical power consumption of the ith EB at time t.

$$H^{t}_{EB,i} = \mu_{EB} P^{t}_{EB,i} \tag{15}$$

where μ_{EB} represents the thermal conversion efficiency, and $H^{t}_{EB,i}$ represents the heat output of the ith EB at time t.

To ensure safe operation, the EB must comply with the following restrictions:

$$P^{\min}_{EB,i} \le P^{t}_{EB,i} \le P^{\max}_{EB,i} \tag{16}$$

$$H^{\min}_{EB,i} \le H^{t}_{EB,i} \le H^{\max}_{EB,i} \tag{17}$$

where $P^{\max}_{EB,i}$, $P^{\min}_{EB,i}$, $H^{\max}_{EB,i}$ and $H^{\min}_{EB,i}$ represent the maximum and minimum electricity consumption and heat production of the ith EB at time t.

2.5 PV

Photovoltaic power generation equipment, which utilizes solar energy to generate electricity, is a key device for reducing environmental pollution. It should comply with the following restrictions:

$$P^{\min}_{PV,i} \le P^{t}_{PV,i} \le P^{\max}_{PV,i} \tag{18}$$

where $P^{\max}_{PV,i}$, $P^{\min}_{PV,i}$ and $P^{t}_{PV,i}$ represent the maximum and minimum limits of power generation and electrical energy output ability of the ith PV in t-period.

2.6 ESD

The ESD serves as a power source in PIES. The cost function of the ESD is as follows:

$$C_{ESD,i}^{t} = P_{ESD,i}^{t}(C_a \alpha_i^t - C_b \beta_i^t) \tag{19}$$

where C_a and C_b are the charging and discharging costs, $P_{ESD,i}^{t}$, α_i^t and β_i^t represent the charge and discharge power for every time period and the corresponding charging and discharging states.

Its operating limitations can be depicted as:

$$e_i^0 = E_{init,i} \tag{20}$$

$$e_i^{t+1} = e_i^t + P_{ESD,i}^{t}\left[\theta_{A,i}\alpha_i^t - \theta_{B,i}\beta_i^t\right] \tag{21}$$

$$E_i^T \geq E_{ref,i} \tag{22}$$

$$E_i^{\min} \leq e_i^t \leq E_i^{\max} \tag{23}$$

$$\alpha_i^t + \beta_i^t \leq 1 \tag{24}$$

where the initial energy storage is donated by $E_{init,i}$, and the energy storage at time $t+1$ is calculated based on the energy inputs and outputs at time t, while the energy capacity of the ith ESD at time T, must exceed or match the baseline energy $E_{ref,i}$. $E_i^{\min}$, $E_i^{\max}$ represent the maximum and minimum limits of energy storage at any time. These restrictions ensure that the ESD operates within its designated energy spectrum and fulfills the necessary energy demand. The parameters $E_{init,i}$, $E_{ref,i}$, $E_i^{\min}$, and $E_i^{\max}$ are all quantified in kWh.

3 Energy Dispatching Method for Port Integrated Energy System

To minimize the operating cost, an economic dispatch strategy of PIES is established in this section.

3.1 Model

The optimal schedule model of PIES can be constructed as follows

Objective Function

$$\begin{aligned} \min C_{\text{total}} = \sum_{t=1}^{T} \Bigg(& \sum_{i=1}^{n_1} C_{OWF,i}^{t} + \sum_{i=1}^{n_2} C_{GB,i}^{t} + \sum_{i=1}^{n_3} C_{GT,i}^{t} \\ & + \sum_{i=1}^{n_4} C_{EB,i}^{t} + \sum_{i=1}^{n_5} C_{ESD,i}^{t} \Bigg) \end{aligned} \tag{25}$$

where n_1 represents the number of OWF, n_2 represents the number of GB, n_3 represents the number of GT, n_4 represents the number of EB, n_5 represents the number of ESD and n_6 represents the number of PV. Especially, $n_1 + n_2 + n_3 + n_4 + n_5 + n_6 = N$.

Constraints

(1) Secure Operation Constraints: Since excessive use of natural gas exacerbates the greenhouse effect, it is necessary to impose constraints on carbon dioxide emissions. And to prevent power accidents and other safety incidents, energy equipment must possess robust safety performance.

$$\sum_{i=1}^{n_2} Q_{GB,i}^t + \sum_{i=1}^{n_3} (Q_{GT-E,i}^t + Q_{GT-H,i}^t) \leq Q_{\max} \tag{26}$$

$$P_{ESD,i}^{\min,t} \leq \sum_{i=1}^{n_5} P_{ESD,i}^t(\alpha_i^t - \beta_i^t) \leq P_{ESD,i}^{\max,t} \tag{27}$$

where $Q_{\max}$ denotes the maximum threshold for gas acquisition across all equipment within the PIES infrastructure, while $P_{ESD,t}^{\min,i}$ and $P_{ESD,t}^{\max,i}$ represent the minimum and maximum limits of ESD capacity, respectively, for the power generation from the ith ESD in the PIES during time interval t. The measurement unit for $Q_{\max}$, $P_{ESD,t}^{\min,i}$, and $P_{ESD,t}^{\max,i}$ is expressed in KW.

(2) Energy Supply-Demand Equilibrium Constraints: To satisfy the power requirements of various consumers within a PIES, achieving equilibrium between energy provision and consumption is essential.

$$\sum_{i=1}^{n_1} P_{OWF,i}^t + \sum_{i=1}^{n_3} P_{GT,i}^t + \sum_{i=1}^{n_6} P_{PV,i}^t = P_{\text{load},i}^t \tag{28}$$

where $P_{\text{load},i}^t$ is the electrical load of each Power generation equipment at time t, and the unit is KW.

$$\sum_{i=1}^{n2} H_{GB,i}^t + \sum_{i=1}^{n3} H_{GT,i}^t + \sum_{i=1}^{n4} H_{EB,i}^t = H_{\text{load}}^t \tag{29}$$

where $H_{\text{load},i}^t$ is the heat load of each heat generation equipment at time t, and the unit is KW.

3.2 Main Algorithm

According to (1)–(29), A distributed mixed-integer linear programming economic dispatch model for the PIES is established, which is further simplified into (30) and (31).

$$\min_{\{y_i \in \xi_i\}_{i=1}^L} \sum_{i \in L} c_i^T y_i \tag{30}$$

$$\sum_{i=1}^{L} A_i y_i \leq b \tag{31}$$

where L was the amount of nodes, and $c_i^T y_i$ was the cost function corresponding to node i, where y_i represented the output of node i.

Furthermore, the Lagrange function and the dyadic problem can be expressed as follows:

$$L(y, \lambda) = \sum_{i=1}^{L} L_i(y_i, \lambda) = \sum_{i=1}^{L} \left(c_i^T y_i + \lambda^T (A_i y_i - b) \right) \tag{32}$$

$$d(\lambda) = \min_{y} L(y, \lambda) \tag{33}$$

$$d(\lambda) = \sum_{i=1}^{L} d_i(\lambda) = \sum_{i=1}^{L} \min_{y_i \in \xi_i} L_i(y_i, \lambda) \tag{34}$$

$$D : \max_{\lambda \geq 0} \sum_{i=1}^{L} d_i(\lambda) \tag{35}$$

where λ is the Lagrange multiplier.

Algorithm 1.

Initialization:
$k = 0$
Consider $\hat{y}_i(0) \in Y_i$, for all $i = 1, \ldots, I$
Consider $\lambda_i(0) \in \mathbb{R}^{d_i}$, for all $i = 1, \ldots, I$
while not converged **do**
    for $i = 1, \ldots, m$ **do**
        $\zeta_i = \sum_{j=1} a_j^i \lambda_j$
        $y_i[k+1] \in \arg\min_y \left(c_i^T y_i + \zeta_i^T A_i y_i - \zeta_i^T b \right)$
        $\lambda_i[k+1] = \left[0, \zeta_i + c_k A_i y_i[k+1] - c_k \frac{b}{\zeta} \right]^+$
        $\hat{y}_i[k+1] = \frac{\sum_{r=0}^{k-1} c_r y_{ir} + c_k y_i[k+1]}{\sum_{r=0}^{k} c_r + c_k}$
    end for
    $k \leftarrow k+1$
end while

4 Simulation Analysis

To verify the effectiveness of the proposed PIES method, a simulation case is presented. And to improve energy efficiency and reduce operational costs, two different methods are adopted to solve the economic dispatch problem of the PIES constructed by (1)–(25), which are solved using a centralized algorithm and a distributed algorithm, respectively. The centralized algorithm employed is the intlinprog algorithm. Considering the distinct characteristics of the output of distributed devices, the classification problem is reformulated as (29) and further transformed into (30) (Table 1).

Table 1. Device parameters and numerical values

Device Parameter	Numerical Value	Device Parameter	Numerical Value
η_{GT-E}	0.40	η_{GT-H}	0.40
η_{GB}	0.95	C_g	0.385 (CNY/m^3)
$P_{GT,i}^{\min}$	250 (kWh)	$H_{GB,i}^{\min}$	250 (kWh)
$P_{GB,i}^{\max}$	100 (kWh)	$E_i^{\max}$	[8; 16] (kWh)
$E_i^{\min}$	1 (kWh)	E_i^{init}	[0.2; 0.5]$E_n^{\max}$ (kWh)
E_i^{ref}	[0.55; 0.8]$E_n^{\max}$ (kWh)	ΔT	60 (min)
$P_{ESD}^{\max}$	50 kW	$P_{ESD}^{\min}$	−50 (kW)
C_a	[0.6; 1.1] (CNY/kWh)	C_b	1.1C_a (CNY/kWh)
ζ_i	[0.015; 0.075]	$P_{OWF,i}^{\max}$	100 (kW)

4.1 The Performance of Centralized Algorithm

For the 24-h optimal dispatch problem of the PIES, a traditional centralized strategy is adopted for solution and implemented in the intlinprog tool of MATLAB. Figures 2a–2c present the dispatch schemes for different energy flows. Fig. 2a and Fig. 2b illustrate the electrical and thermal energy dispatch plan under the centralized algorithm. Figure 2c discribes the state of ESD charging and discharging.

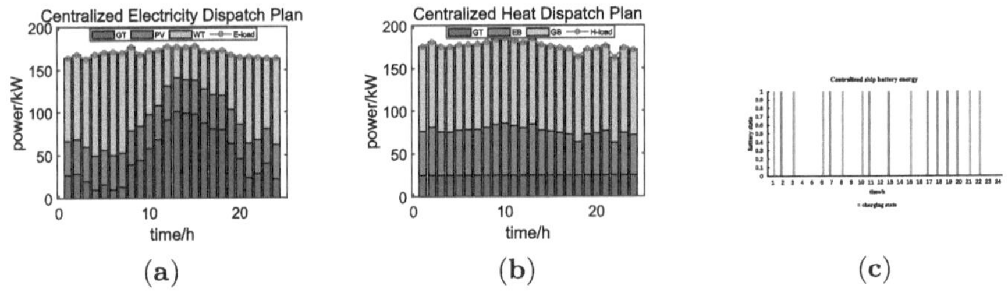

Fig. 2. Centralized Algorithm: (**a**) Electricity. (**b**) Heat. (**c**) State of ESD.

In the economic dispatch process under the centralized algorithm, OWF serves as the primary energy source (50–120 kW), while PV maintains stable output at 30–50 kW. For thermal energy dispatching, GT operates at 25–50kW, GB serves as the main heat source (80–100 kW), and EB maintains 30–50 kW. The ESD charges at 2:00, 7:00, 11:00, 13:00, 17:00–20:00, and discharges at 1:00, 3:00, 6:00, 8:00, 10:00, 15:00, 21:00–22:00, with no operation during other periods.

4.2 The Performance of Distributed Algorithm

In the economic dispatch process under the distributed algorithm, the various energy transfer pathways are illustrated in Figs. 3a–3c .

Fig. 3a and Fig. 3b illustrate the electrical and thermal energy dispatch plan under the distributed algorithm. Figure 3c describes the charging and discharging states of ESD.

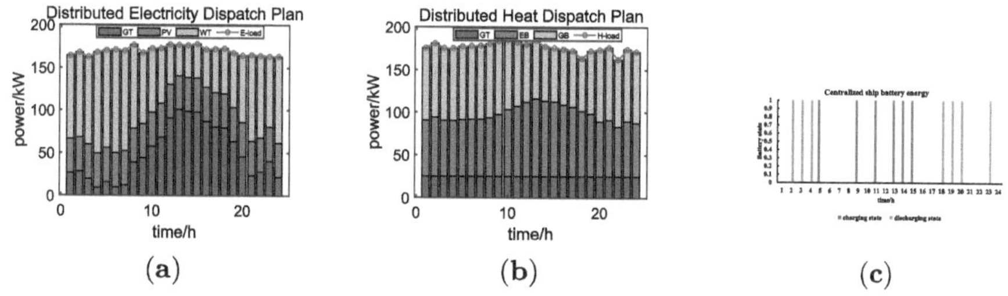

(a) (b) (c)

Fig. 3. Distributed Algorithm: (**a**) Electricity. (**b**) Heat. (**c**) State of ESD.

The results under the distributed algorithm show that, OWF serves as the main source, while PV remains stable at around 50 kW. When peaks occur, GT output reaches its maximum, and as the peaks pass, GT output decreases significantly. EB is the primary heat production equipment. GT output remains stable at around 25 kW, while GB output decreases as peaks appear. During the 1–24 h period, the battery was charged at 5:00, 9:00, 11:00, 13:00, 14:00, 15:00, and discharged occurred at 2:00, 3:00, 4:00, 18:00, 19:00, 20:00, 23:00. At the rest of the time, there were no charging or discharging operation. The results obtained by the distributed algorithm are close to those of the centralized one, maintaining high accuracy while enabling a distributed communication architecture.

5 Conclusion

In this study, an optimization scheduling approach was designed for PIES utilizing a distributed mixed-integer linear programming framework. The objective is to guarantee system reliability while improving energy utilization efficiency. By analyzing the distributed operational characteristics of PIES, an economic dispatch model for PIES was constructed. Then based on distributed mixed-integer linear programming, a distributed algorithm was proposed to deal with the optimal scheduling problem of PIES. Finally, comparative simulation cases verified that the results obtained by the distributed algorithm are close to those of the centralized approach, thereby validating the effectiveness of the proposed algorithm.

Acknowledgement. This work was funded by the Liaoning Provincial Science and Technology Project–Provincial Doctoral Research Startup Funding Project under Grant 2025-BS-0220.

References

1. Acciaro, M., Ghiara, H., Cusano, M.I.: Energy management in seaports: a new role for port authorities. Energy Policy **71**, 4–12 (2014)
2. Mianaei, P.K., Aliahmadi, M., Faghri, S., Ensaf, M., Ghasemi, A., Abdoos, A.A.: Chance-constrained programming for optimal scheduling of combined cooling, heating, and power-based microgrid coupled with flexible technologies. Sustain. Cities Soc. **77**, 103502 (2022)
3. Roldán, A., Manteca, P., Ozkat, R., Siano, P.: Integration of cold ironing and renewable energy sources for maritime vehicles in smart ports. IEEE Trans. Ind. Appl. **55**(6), 7198–7206 (2019)
4. Liu, Z., Yang, M., Jia, W., Ding, T.: Low-carbon economic dispatch of integrated electricity–heat–hydrogen systems considering integrated demand response. In: 2022 IEEE/IAS Industrial and Commercial Power System Asia (I&CPS Asia), pp. 1173–1177 (2022)
5. Yi, Z., Xu, Y., Hu, J., Zhou, M., Sun, H.: Neural-dynamics-based distributed economic dispatch method for integrated energy systems. IEEE Trans. Ind. Informat. **16**(4), 2245–2257 (2020)
6. Kou, Y., Zhao, Z., Li, Y., Liu, J., Deng, H., Qi, X.: Multi-stage optimal scheduling for offshore wind power grid-connected systems considering frequency security constraints. In: Proceedings of the CSU-EPSA, pp. 1–10. https://doi.org/10.19635/j.cnki.csu-epsa.001623
7. Liu, Y., Deng, C., Wu, H., Yang, Q., Ma, Q.: A decentralized optimal dispatch method for island interconnected grid with seawater pumping storage, pp. 1–11. Engineering Journal of Wuhan University (2023)
8. Zhao, B., et al.: PriMPSO: a privacy-preserving multiagent particle swarm optimization algorithm. IEEE Trans. Cybern. **53**(11), 7136–7149 (2023)
9. Pourbabak, H., Luo, J., Chen, T., Su, W.: A novel consensus-based distributed economic dispatch algorithm based on local power mismatch estimation. IEEE Trans. Smart Grid **9**(6), 5930–5942 (2018)
10. Chen, K., Xie, J., Wang, L., Yue, D., Yong, T., Li, Y.: A fully distributed economic dispatch strategy for power systems considering flexible loads. In: 34th Chinese Control Conference (CCC), pp. 6602–6607 (2015)
11. Chen, G., Lei, Y.: A distributed solution for economic dispatch problem in finite time. In: 35th Chinese Control Conference (CCC), pp. 2907–2912 (2016)
12. Moafi, M., et al.: Optimal coalition formation and maximum profit allocation for distributed energy resources in smart grids based on cooperative game theory. Int. J. Electr. Power Energy Syst. **144**, 108492 (2023)
13. Sørensen, J.D., Sørensen, J.N.: Wind Energy Systems: Optimising Design and Construction for Safe and Reliable Operation. Woodhead Publishing Series in Energy, No. 10. Woodhead Publishing, Oxford, UK (2010). ISBN 978-1-84569-580-4

Author Index

C. Li et al. (Eds.): ICNC 2025, CCIS 2946, pp. 625–627, 2026.
https://doi.org/10.1007/978-981-92-1599-7

Z

Zeitfracht Medien GmbH
Ferdinand-Jühlke-Straße 7
99095 Erfurt, Deutschland
produktsicherheit@kolibri360.de